Introduction to INSTRUMENTATION and MEASUREMENTS

Robert B. Northrop
Electrical and Systems Engineering
University of Connecticut
Storrs, Connecticut

CRC Press
Boca Raton New York

Library of Congress Cataloging-in-Publication Data

Northrop, Robert B.
Introduction to instrumentation and measurements / Robert B. Northrop
p. cm. -- (Electronic engineering systems series)
Includes bibliographical references and index.
ISBN 0-8493-7898-2 (alk. paper)
1. Electronic instruments. 2. Electronic measurements.
I. Title. II. Series.
TK7870.N66 1997
621.3815'48--dc21

97-2462
CIP

Direct all inquiries to CRC Press LLC, 2000 Corporate Blvd., N.W., Boca Raton, Florida 33431.

International Standard Book Number 0-8493-7898-2
Library of Congress Card Number 97-2462
Printed in the United States of America 1 2 3 4 5 6 7 8 9 0
Printed on acid-free paper

PREFACE

Purpose: This text is intended to be used in a classroom course for engineers that covers the theory and art of modern instrumentation and measurements (I&M). There is more than enough material to support two semesters work. Thus the instructor has the option of choosing those topics and the depth of coverage that suit his or her interests and curriculum. Because of its breadth, *Introduction to Instrumentation and Measurements* will also be useful as a reference for the practicing engineer and scientist interested in I&M.

Why have a classroom course in I&M? Over the past decade or so, in the US, many electrical engineering (EE) departments have discontinued classroom courses on the theory and practice of instrumentation and measurements. In the past decade we have also seen the swift development of new and exciting means of measurement using new technologies, the adoption of new standards, and concurrently, the lack of development of a coherent educational base to support their understanding and use. Using an instrument in the laboratory is not the same as understanding the physical and electronic principles underlying its design and its functional limitations. Clearly, there is now more than ever a need for classroom experience in the new I&M that will give students the necessary technical background to use and design sensors, signal conditioning systems, and I&M systems. We feel that this text suppports that need.

This text was written based on the author's 33 years of experience in teaching a classroom course (EE 230), Electrical Instrumentation, to juniors and seniors in the Electrical and Systems Engineering Department at the University of Connecticut, Storrs. Our EE 230 is a required course for our seniors; it supports their capstone senior design laboratory experience.

Obviously, in 33 years, we have seen the field of instrumentation and measurements evolve with the rest of electrical engineering technology. Because of the rapid pace of technical development, it generally has been difficult to find an up-to-date text for our Electrical Instrumentation course. After years of frustration in trying to match a text to course content, I decided to write one that would not only encompass the "traditional" aspects of I&M, but also include material on modern integrated circuits (IC) and photonic sensors, microsensors, signal conditioning, noise, data interfaces, digital signal processing (DSP), etc.

Reader Background: Readers are assumed to have taken core EE curriculum courses, or their equivalents. The reader should be skilled in basic, linear circuit theory; i.e., he or she has mastered Thevenin's and Norton's theorems, Kirchoff's laws, superposition, dependent sources, ideal op amps, and should know how to describe DC and AC steady-state circuits in terms of loop and node equations. An introductory systems course should have given him/her familiarity with both time- and frequency-domain methods of descibing linear dynamic systems characterized by ordinary, linear, differential, or difference equations, including state space, Fourier and Laplace transforms, transfer functions, steady-state frequency response of systems, as well as Bode plots. From physics or an EE course in electromagnetics, the reader should have a basic knowledge of electric and magnetic fields, inductance, capacitance, reluctance, etc. There should also be familiarity with electromagnetic waves, Maxwell's equations, transmission lines, and polarization. From a first course in electronics, there should be basic knowledge of bipolar junction transistor amplifiers (BJTs), junction field-effect transistors (JFETs), diodes, and photodiodes and their simpler circuit models.

Scope of the Text: A major feature of *Introduction to Instrumentation and Measurements* is its breadth of coverage. Throughout the text, a high level of mathematical analytical detail is maintained. It is not a "picture book"; we assume that readers have already had contact with basic electrical instruments, including oscilloscopes and meters, in their introductory EE laboratories and in physics labs.

In the following paragraphs, we give an overview of the contents. Chapter 1, *Measurement Systems*, is introductory in nature. In it, we illustrate measurement system architecture and describe sensor dynamics, signal conditioning, and data display and storage. Errors in measurements are discussed, including the meaning of accuracy and precision, limiting error, etc. The recent (1990) quantum standards adopted for the volt and the ohm are described, as well as other electrical and physical standards.

In Chapter 2, *Analog Signal Conditioning,* we describe, largely at the systems level, means of conditioning the analog outputs of various sensors. Op amps and differential, instrumentation, autozero, and isolation amplifiers are covered. Applications of op amps as active filters, differential instrumentation amps, charge amplifiers, phase-sensitive rectifiers, etc. are shown. We also give practical considerations of errors caused by offset voltage, bias currents, input impedance, slew rate and gain*bandwidth, etc. There is also a section on nonlinear signal processing with op amps.

Noise and Coherent Interference in Measurements are treated in depth in Chapter 3. An heuristic yet rigorous approach is used in which we define and use one-sided, noise voltage, and current power density spectra to describe the effect of noise in instruments and measurement systems. Noise factor and figure are covered, and output signal-to-noise ratios are used to evaluate system noise performance. Examples are given of calculations of the noise-limited resolution of the quantity under measurement (QUM). Techniques are shown for the minimization of coherent interference.

The traditional topics of *DC Null Methods of Measurement,* are presented in Chapter 4, and *AC Null Measurements* in Chapter 5. Wheatstone and Kelvin bridges, and potentiometers are described in Chapter 4, and the major AC bridges used to measure inductance, Q, capacitance and D are treated in Chapter 5.

Sensors are treated in two robust chapters in this text: Chapter 6, *Survey of Sensor Mechanisms,* and Chapter 7, *Applications of Sensors to Physical Measurements.* These are large and substantive chapters, covering a broad range of sensor mechanisms and types. Of special note is the introduction of certain fiber optic and electro-optic sensors, as well as selected chemical and ionizing radiation sensors. There is a detailed analysis of mechanical gyroscopes, clinometers, and accelerometers. The Sagnac effect is introduced, and the basic fiber optic gyro is described. The Doppler effect in ultrasonic velocimetry and laser Doppler velocimetry is covered.

In Chapter 8, *Basic Electrical Measurements,* the classic means of measuring electrical quantities are presented, as well as newer methods such as Faraday magneto-optic ammeters and Hall effect gaussmeters and wattmeters. Electronic means of measuring stored charge and static electric fields are described.

Digital Interfaces in Measurement Systems are the topic of Chapter 9. This chapter begins with a description of the sampling theorem and its significance in the process of periodically sampling and digitizing data. Quantization noise and its significance to the resolution and accuracy of digitized data are discussed. The design of digital-to-analog and analog-to-digital converters are described, along with the trade-offs between speed and accuracy in data conversion. The important serial and parallel data bus structures are presented. Finally, the role of wire and fiber optic transmission lines in limiting data transfer rates is given.

Because digitized, measured data is processed and stored on computers in modern instrumentation practice, Chapter 10, *Introduction to Digital Signal Conditioning,* was written to acquaint the reader with this specialized field. The z-transform and its use in describing filtering operations on discrete, digitized data in the frequency domain is introduced. Examples of finite impulse response (FIR) and infinite impulse response (IIR) digital filters are given, including numerical integration and differentiation routines, viewed both in the time- and frequency domains. The discrete and fast Fourier transforms are covered, and the effect of data windows on spectral resolution is discussed. Finally, the use of splines in interpolating discrete data sequences and in estimating missing data points is described.

In Chapter 11, *Examples of the Design of Measurement Systems,* four examples of complex measurement systems developed by the author and his students are given to illustrate design philosophy.

References and Bibliography: The references cited encompass a wide time span; from the 1950s through the mid-1990s. There are many recent entries of review articles and specialized texts that should lead the reader interested in pursuing a specialized area of I&M further into that particular field.

Acknowledgments: I thank the reviewers who have helped guide the development of my manuscript: Charles E. Nunnally, Virginia Polytechnic Institute; Derald O. Cummings, Pennsylvania State University; Richard S. Marleau, University of Wisconsin; and Burt Robinson, Purdue University. The help and guidance from my editors (Navin Sullivan, Felicia Shapiro) at CRC are appreciated. I am greatly indebted to my wife, Adelaide, who has given me moral support and encouragement.

Dedication: I dedicate this text to Dr. Gregory S. Timoshenko, Professor Emeritus and former EE Department Head at UCONN.

INTRODUCTION TO INSTRUMENTATION AND MEASUREMENTS

CONTENTS

1

Measurement Systems

1.0 INTRODUCTION

In this introductory chapter we examine the architecture of "typical" measurement systems, and discuss how noise, calibration errors, sensor dynamic response, and nonlinearity can affect the accuracy, precision, and resolution of measurements. We also discuss the modern, physical, and electrical standards used by NIST (National Institute of Standards and Technology, formerly the National Bureau of Standards) and discuss how these standards are used to create secondary standards used for practical calibration of measurement systems.

Measurement systems are traditionally used to measure physical and electrical quantities, such as temperature, pressure, capacitance, and voltage. However, they also can be designed to locate things or events, such as the epicenter of an earthquake, employees in a building, partial discharges in a high-voltage power cable, or a land mine. Often a measurement system is called upon to discriminate and count objects, such as red blood cells, or fish of a certain size swimming past a checkpoint. A measurement system is often made part of a control system. The old saying that "if you can't measure it, you can't control it" is certainly a valid axiom for the control engineer as well as for the instrumentation engineer.

The reader should realize that the fields of instrumentation and measurements are rapidly changing, and new standards, sensors, and measurement systems are continually being devised and described in the journal literature. The *IEEE Transactions on Instrumentation and Measurement*, the *Review of Scientific Instruments*, the *IEEE Transactions on Biomedical Engineering*, and the *Journal of Scientific Instruments* are four of the more important periodicals dealing with the design of new measurement systems and instruments.

1.1 MEASUREMENT SYSTEM ARCHITECTURE

Figure 1.1 illustrates the block diagram of a typical measurement system. The quantity under measurement (QUM) is converted to a useful form, such as a voltage, current, or physical displacement by an input transducer or sensor. The output of the sensor, here assumed to be a voltage, is amplified and filtered by a signal conditioning subsystem. The purposes of the signal conditioning subsystem are to amplify, give a low or matched output impedance, and to improve the signal-to-noise ratio of the analog signal proportional to the QUM. The conditioned signal, V_1, can be distributed to various display and recording devices. V_1 can be displayed on an analog or digital oscilloscope, or a strip-chart recorder. It can also be recorded on a magnetic tape recorder, or it can be low-pass filtered to prevent aliasing and then periodically converted to digital words by an analog-to-digital converter (ADC). The ADC output, D_N, is generally made an input to a digital computer through an appropriate interface. Once in the computer environment, the sampled and digitized QUM can be further (digitally) filtered and processed, and stored on a magnetic hard and/or floppy disk, or on an optical disk in digital form. Often the sensing instrument contains an ADC and is connected to a computer by a special instrumentation bus called the GPIB or IEEE-488 bus. (Hewlett-Packard calls the GPIB the HPIB.) The GPIB is organized so that many GPIB-compatible sensing instruments can be managed by one master computer. More will be said about the GPIB in Chapter 10 of this text.

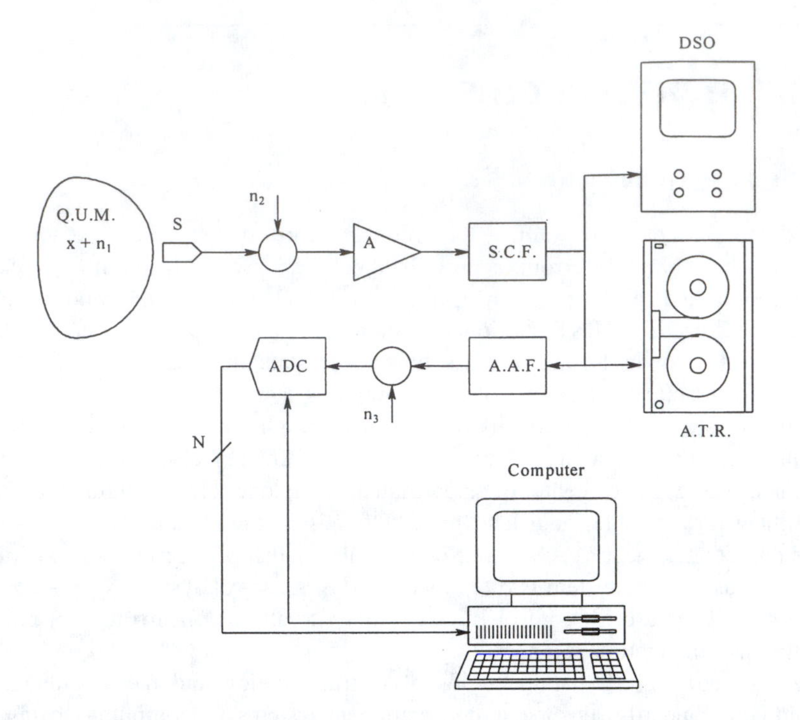

FIGURE 1.1 A generalized measurement system. Abbreviations: Q.U.M., quantity under measurement; S, sensor; A, amplifier; S.C.F., signal conditioning filter; DSO, digital storage oscilloscope; A.T.R., analog tape recorder; A.A.F., anti-aliasing (lowpass) filter, n_1, noise accompanying the QUM; n_2, noise from electronics, n_3, equivalent quantization noise; ADC, analog-to-digital converter.

Note that there are three major sources of noise in the measurement system of Figure 1.1; noise sensed along with the QUM (environmental noise), noise associated with the electronic signal conditioning system (referred to its input), and equivalent noise generated in the analog-to-digital conversion process (quantization noise). These noise sources generally provide a limit to measurement system resolution and to its accuracy.

Many sensors provide the rate-limiting element in a measurement system. That is, the response speed of the sensor is much slower than any of the other elements in the system. As a result of sensor dynamics, a measurement system will have a settling time before a stable measurement can be made when there is a step input of the QUM. That is, a settling time will be required for the sensor to reach a steady-state output.

1.1.1 Sensor Dynamics

Sensor dynamics generally fall into one of three categories: first-order low-pass, second-order low-pass, and band-pass. A general example of a first-order, low-pass sensor is shown in Figure 1.2A. The QUM is x(t), and the sensor output is $v_x(t)$. The ODE relating v_x and x is

$$\dot{v}_x(t) = -av_x(t) + Kx(t) \qquad 1.1A$$

or

$$\dot{v}_x(t)\tau = -v_x(t) + K\tau x(t) \qquad 1.1B$$

Here a is the natural frequency of the sensor in r/s, and $\tau = 1/a$ is its time constant in seconds. Solution of the ordinary differential equation (ODE) above for a step input, $x(t) = X_o U(t)$, is

$$v_x(t) = \frac{X_o K}{a}\left(1 - e^{-at}\right) = X_o K\tau\left(1 - \exp[-t/\tau]\right) \tag{1.2}$$

The sensor step response is plotted in Figure 1.2B. Note that the response is within 5% of the steady-state value for t = 3τ, and within 2% of steady state for t = 4τ. The transfer function for the first-order, low-pass sensor is found by Laplace transforming the ODE, Equation 1.1A,

$$\frac{V_X(s)}{X(s)} = \frac{K}{s + a} \tag{1.3}$$

and its frequency response is found by letting s = jω in Equation 1.3.

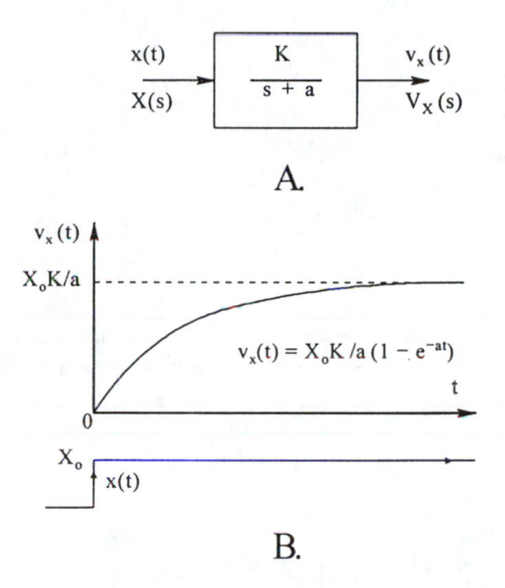

FIGURE 1.2 (A) Transfer function of sensor with first-order dynamics in LaPlace format. x(t) is the QUM, X(s) is its LaPlace transform, $v_X(t)$ is the sensor's voltage output, $V_X(s)$ is its LaPlace transform, a is the natural frequency of the first-order sensor. (B) Step response of the first-order sensor.

Second-order sensor dynamics fall into one of three categories, depending on the location of the roots of the characteristic equation of the ODE. These categories are underdamped (with complex-conjugate roots), critically damped (with two equal real roots), and overdamped (with unequal real roots). In all cases, the real parts of the ODE's roots are negative, as the sensor is assumed stable. The ODEs for the three damping conditions are:

$$\ddot{v}_x = -\dot{v}_x(2\zeta\omega_n) - v_x(\omega_n^2) + Kx(t) \tag{1.4A}$$

$$\ddot{v}_x = -\dot{v}_x(2a) - v_x a^2 + Kx(t) \tag{1.4B}$$

$$\ddot{v}_x = -\dot{v}_x(a + b) - v_x ab + Kx(t) \tag{1.4C}$$

In the underdamped case, Equation 1.4A above, ζ is the damping factor and ω_n is the natural frequency in r/s. For an underdamped sensor, 0 ≤ ζ ≤ 1, and the roots of the ODE are located at $s = \pm\omega_n \pm j\omega_n\sqrt{1-\zeta^2}$ The critically damped sensor has two roots at $s = -\omega_n$, and the overdamped sensor's ODE has roots at s = –a and s = –b. These conditions are shown in Figure 1.3. Using

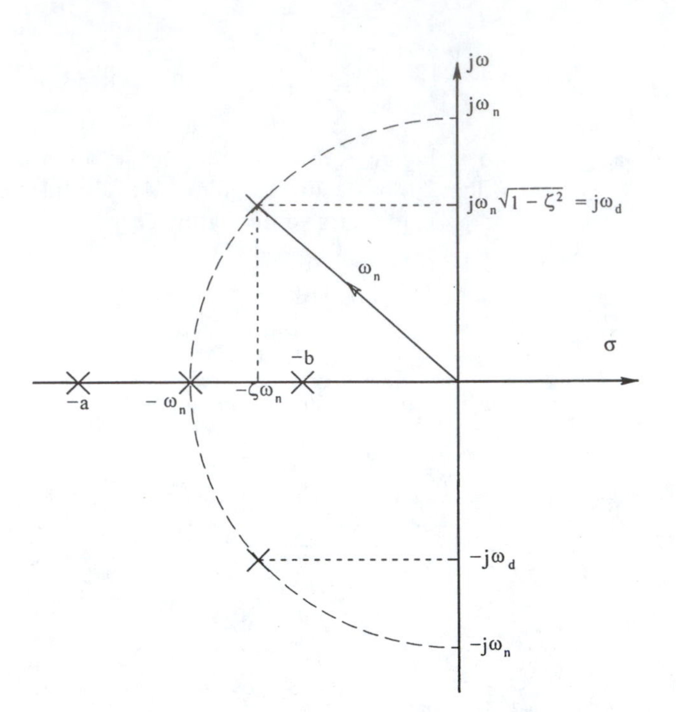

FIGURE 1.3 Pole locations of a second-order sensor's transfer function for the cases of overdamping (poles at s = –a, –b), critical damping (two poles at $s = -\omega_n$), and underdamping (poles at $s = -\zeta\omega_n \pm j\omega_n\sqrt{1-\zeta^2}$).

Laplace transforms, we can find the step response of the second-order sensor for the three conditions above. The step response for the underdamped case is

$$v_x(t) = \frac{KX_o}{\omega_n^2}\left\{1 - \frac{1}{\sqrt{1-\zeta^2}}\exp(-\zeta\omega_n t)\sin\left[\omega_n\sqrt{1-\zeta^2}\,t + \phi\right]\right\} \quad 1.5$$

where

$$\phi = \tan^{-1}\left(\frac{\sqrt{1-\zeta^2}}{\zeta}\right) \quad 1.6$$

The underdamped, second-order system step response is illustrated in Figure 1.4A. The step response of the critically damped sensor ($\zeta = 1$) is shown in Figure 1.4B, and is

$$v_x(t) = \frac{KX_o}{a^2}\left[1 - e^{-at} - ate^{-at}\right] \quad 1.7$$

Finally, the step response of an overdamped sensor having two real poles is shown in Figure 1.4C, and can be written as

$$v_x(t) = \frac{KX_o}{ab}\left[1 - \frac{1}{b-a}\left(be^{-at} - ae^{-bt}\right)\right], \quad b > a \quad 1.8$$

While the four types of low-pass sensor dynamics described above are most commonly encountered, sensors exist which do not respond to a constant (DC) QUM. The QUM must in fact be time-

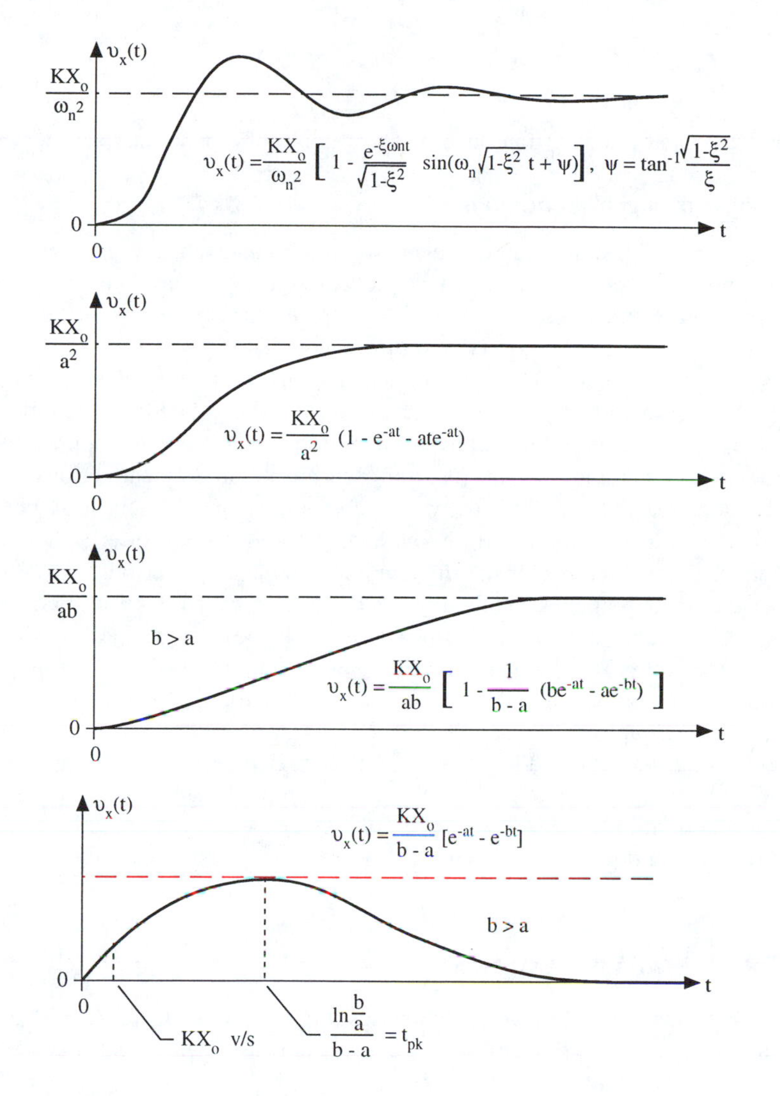

FIGURE 1.4 (A) Step response of an underdamped, second-order sensor. (B) Step response of a critically damped, second-order sensor. (C) Step response of an overdamped, second-order sensor. (D) Step response of an overdamped, band-pass, second-order sensor.

varying to produce a sensor output. Such sensors are said to have a band-pass frequency response characteristic, and can be described by the ODE

$$\ddot{v}_x = -\dot{v}_x(a+b) - v_x ab + K\dot{x}, \quad b > a \tag{1.9}$$

The corresponding transfer function for a band-pass sensor with two real poles is

$$\frac{V_x(s)}{X(s)} = \frac{Ks}{(s+b)(s+a)} \tag{1.10}$$

The output voltage step response of a band-pass sensor rises from zero to a peak, then goes to zero in the steady state; it is illustrated in Figure 1.4D, and is given by

$$v_x(t) = \frac{KX_o}{b-a}\left(e^{-at} - e^{-bt}\right) \tag{1.11}$$

The piezoelectric crystal pressure transducer is an example of a band-pass sensor.

1.1.2 Overview of Signal Conditioning

The voltages or currents obtained directly at the output of a sensor are generally low-level and contain additive noise and coherent interference picked up from the environment of the QUM, and from the sensor itself. Sometimes the measurement process or the sensor introduces a nonlinear distortion of the QUM (as in the case of hot wire anemometers) which must be linearized. The analog signal conditioning module following the sensor thus must amplify the sensor output voltage, as well as perform linear filtering on it in order to improve the signal-to-noise ratio. Such amplification and filtering is usually performed by a low-noise instrumentation amplifier followed by op amp active filters. Compensation for inherent nonlinearities, including corrections for temperature changes, etc., is most easily done digitally, with a computer. Thus the sensor output, after analog conditioning, is converted to digital form by an analog-to-digital converter (ADC) and read into the computer as a binary number. If the digitization is done on a periodic basis, it is common practice to use the analog signal conditioning system to low-pass filter the input to the ADC so that the ADC input voltage contains no significant energy having frequencies in excess of one half of the sampling frequency. This low-pass filtering before sampling is called anti-aliasing filtering, and is necessary for accurate digital signal processing. Sampling and anti-aliasing filters are discussed in detail in Chapter 10.

Digital signal conditioning can also be used to effectively remove coherent interference accompanying the QUM, and to calculate functions of the sampled signal such as its rms value, its autocorrelation function, or its root power spectrum. It is now common practice to store digitized measurement records on floppy or hard magnetic disks, on optical disks, or on video magnetic tape in digital form.

1.2 ERRORS IN MEASUREMENTS

Errors in measurements can arise from many causes; there are remedies for some types of errors, but others haunt us as intrinsic properties of the measurement system under use, and often can be mitigated by system redesign.

Gross errors in a measurement can arise from such human mistakes as:

1. Reading the instrument before it has reached its steady state. Such a premature reading produces a dynamic error.
2. Not eliminating parallax when reading an analog meter scale, and incorrect interpolation between analog meter scale markings.
3. Mistakes in recording measured data, and in calculating a derived measurand.
4. Misuse of the instrument. A simple example of this source of error is when a 10 volt full-scale analog voltmeter of typical sensitivity (20,000 ohms per volt) is connected to a voltage source having a high Thevenin resistance (100,000 ohms). Significant voltage is dropped across the internal (Thevenin) resistor.

System errors can arise from such factors as:

1. The instrument is not calibrated, and has an offset; i.e., its sensitivity is off, and it needs zeroing. Loss of calibration and zero can occur because of long-term component value changes with aging, or changes associated with temperature rise.
2. Reading uncertainty from the presence of random noise. This noise can accompany the measurand, and can arise from the signal conditioning electronics in the system.

   a. Noise from without is called environmental noise, and often can be reduced by appropriate electric and magnetic shielding, and proper grounding and guarding practices. Environmental noise is often coherent in nature, and can come from poorly shielded radio frequency sources, such as computers or radio stations. Powerline frequency electric and magnetic fields can also be troublesome, as can ignition noise.
   b. The significant, internally generated random noise may be shown to arise in the first stage of an instrument's signal conditioning amplifier. Some of this noise comes from resistors (Johnson or thermal noise), some comes from the active elements (transistors) in the headstage, and some from the quantization or round-off inherent in the operation of analog-to-digital converters in modern digital instruments.
3. Slow or long-term drift in the system can destroy the certainty of static measurements, i.e., measurements of measurands which are not varying in time. Drifts can cause slow changes of system sensitivity and/or zero. Drift may arise as the result of a slow temperature change as a system "warms up." Many parameter values can change with temperature, such as capacitance, resistance, inductance, the electromotive force (EMF) of standard cells, the avalanche voltage of pn junctions, etc. Humidity changes can also affect circuit parameter values, including resistance and capacitance. Good system design involves temperature regulation of the system, and the use of low- or zero-tempco, sealed components. In integrators such as charge amplifiers, output drift can also be caused by the integration of the op amp's DC bias current. Drift or system offset can also arise from DC static charges affecting analog electronic circuits. In some cases, the face of an analog meter can become charged with static electricity which attracts or repels the meter pointer, causing an error.

Below, we discuss the concepts of *accuracy, precision, resolution, limiting error*, and various error statistics. To begin our quantitative discussion of errors in measurements, we define the *error* in the nth measurement as:

$$\varepsilon_n \equiv X_n - Y_n \tag{1.12}$$

and

$$\%\varepsilon \equiv \left|\varepsilon_n / Y_n\right| 100 \tag{1.13}$$

Here Y_n is the actual, true, defined, or calculated value of the QUM, and X_n is the nth measured value of the QUM. Philosophically, the use of Y_n in the definitions above may present an ontological problem because one can argue that the *true* value can never be known, as it is the result of a non-ideal measurement process, in which case, we might take Y_n to be defined by a high-resolution, primary standard. Nevertheless, error as given in Equation 1.12 is generally an accepted definition. In some cases, the absolute value signs in the $\%\varepsilon$ expression may be omitted, and the percent error can be negative.

The *accuracy*, A_n, of the nth measurement is defined as:

$$A_n \equiv 1 - \left|\frac{Y_n - X_n}{Y_n}\right| \tag{1.14}$$

The percent accuracy is simply

$$\%A_n = 100 - \%\varepsilon = A \times 100 \tag{1.15}$$

The *precision* of the nth measurement is defined as:

$$P_n \equiv 1 - \left| \frac{X_n - \overline{X}}{\overline{X}} \right| \tag{1.16}$$

$\overline{X}$ is defined as the *sample mean* of N measurements:

$$\overline{X} \equiv \frac{1}{N} \sum_{n=1}^{N} X_n \tag{1.17}$$

Low noise accompanying the QUM and in the signal conditioning system is required as a necessary (but not sufficient) condition for *precision measurements.* Precision measurements also require a measurement system capable of resolving very small changes in the QUM, say one part in 10^7. Consequently, many precision, high-resolution measurement systems rely on null methods, such as used with AC and DC bridges, potentiometers, etc. The high resolution inherent in null measurements comes from the fact that the bridge components themselves are known to high precision, and the null sensor is capable of resolving very small voltage or current differences around null. The null condition equation(s) allow calculation of the QUM to high precision, given high accuracy components (parameters).

Accurate measurements require the use of a precision measurement system which is calibrated against a certified, accurate standard.

Other statistics used to describe the quality of a series of measurements are the *deviation*, d_n, the *average deviation*, D_N, and the *standard deviation*, S_N, defined as:

$$d_n \equiv X_n - \overline{X} \tag{1.18}$$

$$D_N \equiv \frac{1}{N} \sum_{n=1}^{N} d_n \tag{1.19}$$

$$S_N \equiv \sqrt{\frac{1}{N} \sum_{n=1}^{N} d_n^2} = \sigma_x \tag{1.20}$$

The variance of X is simply S_N^2, and can be shown to be equal to:

$$S_N^2 \equiv \frac{1}{N} \sum_{n=1}^{N} X_n^2 - (\overline{X})^2 = \sigma_x^2 \tag{1.21}$$

If we assume that samples of the QUM, X_n, have a normal or Gaussian probability density function, then the probable error, pe, in any one observation is defined such that there is a 50% probability that X_n lies between $\overline{X}$ – pe and $\overline{X}$ + pe. From the normal probability density function, it is easy to show that pe = 0.6745 σ_x to satisfy the condition above.

A noisy set of measurements will have a small $\overline{X}/\sigma_x$, which may be considered to be a signal-to-noise ratio. If we increase the total number of measurements, N, the signal-to-noise ratio can be shown to improve by a factor of $\sqrt{N}$.

The *resolution* in measuring the QUM is related to the precision of the measurement, and is basically the smallest unit of the QUM that can reliably be detected.

Analog indicating instruments with scales covering about 90° of arc provide a fundamental limitation to measurement system accuracy, even when null measurements are performed. Large, mirror-scale analog meters with knife-edge pointers can usually be read to no better than 0.2%, or 2 parts/thousand. Oscilloscopes generally offer slightly poorer resolution. As a result of this physical limitation to resolution, direct-reading instruments with digital outputs are used when accuracies better than 0.2% are needed. Null instruments have taps and verniers to permit high resolution.

Limiting error (LE) is an important parameter used in specifying instrument accuracy. The limiting error, or guarantee error, is given by manufacturers to define the outer bounds or worst case expected error. For example, a certain voltmeter may be specified as having an accuracy of 2% of its full-scale reading. If on the 100 V scale, the meter reads 75 V, the LE in this reading is $(2/75) \times 100 = 2.67\%$.

In many cases, such as in the determination of a resistor value by Wheatstone bridge, the QUM must be calculated from a formula in which various system parameters, each having a specified accuracy, appears. Thus we must derive a formula for the determination of LE in the value of the calculated QUM. Let the QUM be a function of N variables, i.e.,

$$Q = f(X_1, X_2, \ldots X_N) \tag{1.22}$$

Let us assume each variable, X_j, is in error by $\pm\Delta X_j$. Hence the calculated QUM will be "noisy," and will be given by

$$\hat{Q} = f(X_1 \pm \Delta X_1, X_2 \pm \Delta X_2, \ldots X_N \pm \Delta X_N) \tag{1.23}$$

$\hat{Q}$ can be expanded into a Taylor's series for N variables. For one variable,

$$f(X \pm \Delta X) = f(X) + \frac{df}{dX}\frac{\Delta X}{1!} + \frac{d^2 f(\Delta X)^2}{dX^2 2!} + \ldots + \frac{d^{n-1}f}{dX^{n-1}}\frac{(\Delta X)^{n-1}}{(n-1)!} + R_n \tag{1.24}$$

Hence for the N variable case:

$$\begin{aligned} \hat{Q} = f(X_1, X_2, \ldots X_N) &+ \left\{ \frac{\partial f}{\partial X_1}\Delta X_1 + \frac{\partial f}{\partial X_2}\Delta X_2 + \ldots + \frac{\partial f}{\partial X_N}\Delta X_N \right\} \\ &+ \frac{1}{2!}\left\{ \frac{\partial^2 f}{\partial X_1^2}(\Delta X_1)^2 + \frac{\partial^2 f}{\partial X_2^2}(\Delta X_2)^2 + \ldots + \frac{\partial^2 f}{\partial X_N^2}(\Delta X_N)^2 \right\} + \ldots \\ &+ \frac{1}{3!}\left\{ \frac{\partial^3 f}{\partial X_1^3}(\Delta X_1)^3 + \ldots \right\} + \ldots \end{aligned} \tag{1.25}$$

The second and higher derivative terms are assumed to be negligable numerically. Thus the maximum or worst-case uncertainty in Q can be finally approximated by:

$$\Delta Q_{MAX} = \left| Q - \hat{Q} \right| = \sum_{j=1}^{N} \left| \frac{\partial f}{\partial X_j} - \Delta X_j \right| \tag{1.26}$$

As an example of the use of Equation 1.26 for worst-case uncertainty, let us examine the limiting error in the calculation of the DC power in a resistor:

$$P = I^2R \tag{1.27}$$

so

$$\Delta P_{MAX} = 2IR\Delta I + I^2\Delta R \tag{1.28}$$

and

$$\frac{\Delta P}{P}MAX = 2\left|\frac{\Delta I}{I}\right| + \left|\frac{\Delta R}{R}\right| \tag{1.29}$$

Thus if the LE in R is 0.1%, the 0 – 10 A ammeter has 1% of full-scale accuracy, the resistor value is 100 ohms and the ammeter reads 8 A, and the nominal power dissipated in the resistor is 6,400 W, then LE in the power measurement is

$$\frac{\Delta P}{P}MAX = 2\frac{0.1}{8} + 0.001 = 0.026 \tag{1.30}$$

or 2.6%.

As a second example, consider finding the limiting error in calculating the Potassium Nernst potential, given by Equation 1.31 below:

$$E_K = \frac{RT}{\Im}\ln(C_o/C_i) \tag{1.31}$$

Here R is the gas constant, T is the Kelvin temperature, $\Im$ is the Faraday number, C_o is the potassium ion concentration outside a semipermeable membrane, and C_i is the potassium concentration inside a compartment bounded by the membrane. To find the LE in calculating E_K, we differentiate Equation 1.31 according to Equation 1.26 above:

$$\Delta E_K = \left|\frac{R}{\Im}\ln(C_o/C_i)\Delta T\right| + \left|\frac{RT}{\Im}\frac{\Delta C_o}{C_o}\right| + \left|\frac{RT}{\Im}\frac{\Delta C_i}{C_i}\right| \tag{1.32}$$

Now if we divide Equation 1.32 by Equation 1.31, we find:

$$\frac{\Delta E_K}{E_K}MAX = \left|\frac{\Delta T}{T}\right| + \left|\frac{\Delta C_o}{C_o}\frac{1}{\ln(C_o/C_i)}\right| + \left|\frac{\Delta C_i}{C_i}\frac{1}{\ln(C_o/C_i)}\right| \tag{1.33}$$

Let the temperature at 300 K be known to ±0.1°, and C_o be 100 m*M*, C_i be 10 m*M*, and the concentrations be known to ±2%. The LE in E_K is thus

$$\frac{\Delta E_K}{E_K}MAX = 0.1/300 + 0.02/2.3 + 0.02/2.3 = 1.77 \times 10^{-2} \tag{1.34}$$

or 1.77%.

In conclusion, we have seen that the Taylor's series approach to the calculation of LE for a derived (calculated) measurand provides us with a systematic means of finding the worst-case error

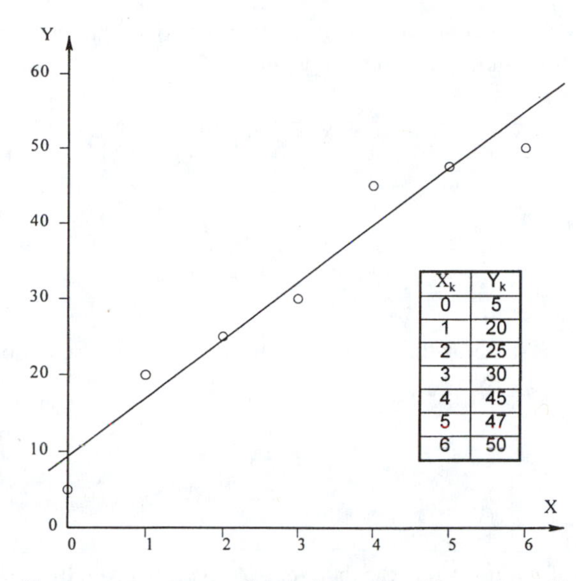

FIGURE 1.5 Illustration of least mean square error fitting of a number of data points with a linear regression line.

in that quantity. In actuality, the uncertainties in the parameters may fortuitously algebraically combine to give a low degree of error. The details of this combination of errors is unknown to us, however, so we generally take the LE as the outer bound in evaluating the accuracy of a measurement.

Least mean square linear graph fitting, or linear regression, is a technique used to obtain an optimum linear fit of the form y = mx + b to a noisy set of N measurement points, $\{Y_k, X_k\}$, k = 1, … N. (Such a fit is illustrated for a typical case in Figure 1.5.) For example, we may wish to measure the optical rotation of polarized light, Y_k, as it passes through sugar solutions of various concentrations, X_k. In the optimum, least mean square (LMS) error fitting of y = mx + b to the data points, we wish to choose m and b to minimize the function σ_y^2:

$$\sigma_y^2 = \frac{1}{N}\sum_{k=1}^{N}\left\{\left[mX_k + b\right] - Y_k\right\}^2 \qquad 1.35$$

Equation 1.35 can be expanded to give:

$$\sigma_y^2 = \frac{1}{N}\sum_{k=1}^{N}\left\{m^2X_k^2 + b^2 + Y_k^2 + 2mX_x b - 2Y_k\left[mX_k + b\right]\right\} \qquad 1.36$$

In order to minimize σ_y^2, we set the derivatives equal to zero:

$$\frac{\partial\sigma_y^2}{\partial b} = 0 = 2Nb + 2m\sum_{k=1}^{N}X_k - 2\sum_{k=1}^{N}Y_k \qquad 1.37A$$

$$\frac{\partial\sigma_y^2}{\partial m} = 0 = 2m\sum_{k=1}^{N}X_k^2 + 2b\sum_{k=1}^{N}X_k - 2\sum_{k=1}^{N}X_kY_k \qquad 1.37B$$

The two Equations, 1.37A and B above can be written as two simultaneous linear equations in m and b, and solved by conventional methods to obtain:

$$b = \frac{\overline{Y}\overline{X^2} - \overline{X}R_{xy}(0)}{\sigma_x^2} \qquad 1.38A$$

$$m = \frac{R_{xy}(0) - \overline{X}\,\overline{Y}}{\sigma_x^2} \qquad 1.38B$$

for the y-intercept and the slope of the LMS linear fit, respectively. $R_{xy}(0)$ is the cross-correlation function evaluated at $\tau = 0$, given by:

$$R_{xy}(0) = \frac{1}{N}\sum_{k=1}^{N} X_k Y_k \qquad 1.39$$

The goodness of the fit of $y = mx + b$ to the data set, $\{Y_k, X_k\}$, is given by the correlation coefficient for the LMS fit, r, defined as:

$$r \equiv \left[R_{xy}(0) - \overline{X}\,\overline{Y}\right] / \sigma_x \sigma_y, \quad 0 \le r \le 1 \qquad 1.40$$

r^2 is called the *coefficient of determination* of the linear regression fit. Obviously, $r = 1$ indicates a perfect fit, i.e., all $Y_k = mX_k + b$.

It should be pointed out that not all measurands are linear functions of a single independent variable, and the analysis above represents one of the simpler cases in statistical data analysis.

1.3 STANDARDS USED IN MEASUREMENTS

One concern of everyone who has occasion to make a measurement is, is the instrument calibrated? As we have seen above, calibration is necessary, along with precision, to enable accurate measurements to be made. Calibration implies observing the instrument's performance when measuring a standard of some sort. Major changes have taken place in the late 1980s in the definitions of standards, and on January 1, 1990, several of these new standards were adopted by the international community. In the U.S., the responsibility for maintaining primary and secondary standards lies with the National Institute for Standards and Technology (NIST), formerly the National Bureau of Standards (NBS). NIST also actively seeks to establish new and more accurate standards, and means of transferring their accuracy when calibrating instruments.

1.3.1 Electrical Standards

A standard is a physical representation of the QUM whose true value is known with great accuracy. Standards can be classified as:

1. International standards
2. Primary standards
3. Secondary standards
4. Working standards

International standards are defined by international agreement, and are kept at the International Bureau of Weights and Measures. An example of an international standard is the kilogram mass.

(More will be said about the mass standard below.) International standards are normally not available on a daily basis for calibration or comparison.

Primary standards are maintained in national standards laboratories in countries around the world. Primary standards, representing some of the fundamental physical and electrical units, as well as some derived quantities, are independently measured and calibrated at the various national laboratories, and compared against each other. This process leads to grand or world average figures for the standards. Primary standards are used continually, but generally do not leave the national standards labs. Secondary standards are reference standards which are initially calibrated from primary standards, and then used in industry and research labs on a daily basis to calibrate their working standards, which are used on a daily basis to check and calibrate working laboratory instruments.

1.3.1.1 The Volt

The defined international standard for the volt which was in effect from 1908 to 1990 is based on the EMF of the saturated Weston Standard Cell, first developed in 1892 by Edward Weston. The Weston cell is an electrochemical battery which consists of two half-cell electrodes in a sealed, glass H-tube (see Figure 1.6). In the "Normal" Weston standard cell, the electrolyte is a saturated aqueous solution of cadmium sulfate ($CdSO_4$). The saturated $CdSO_4$ electrolyte gives the EMF of the Normal cell a relatively high temperature coefficient, about –40 μV/°C. This form of Weston cell shows higher, long-term, output EMF stability than do cells with unsaturated electrolyte. The EMF of the saturated Weston cell drifts about 1 μV/year, and the useful life of a well-treated normal cell is 10 to 20 years (Helfrick and Cooper, 1990).

Under nearly open circuit conditions, the EMF of the Normal cell in absolute volts is given by

$$E(\Delta T) = E_{20} - 4.6 \times 10^{-5}(\Delta T) - 9.5 \times 10^{-7}(\Delta T)^2 - 1 \times 10^{-8}(\Delta T)^3 \qquad 1.41$$

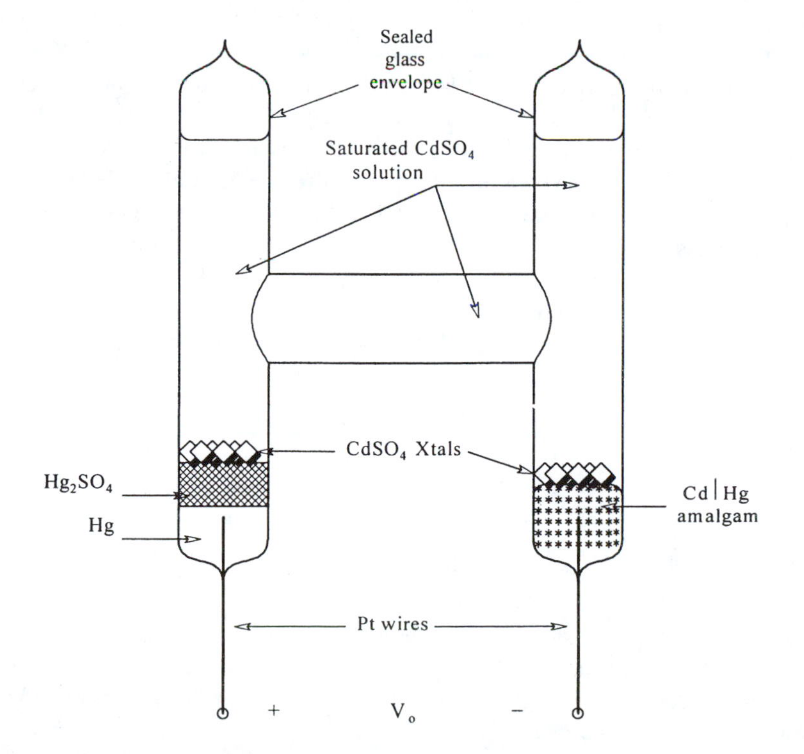

FIGURE 1.6 Diagram of a saturated Weston standard cell battery.

where E_{20} is the normal cell EMF at exactly 20°C ($E_{20} \equiv 1.01858$ Absolute volts), and $\Delta T = (T°C - 20)$. One volt was defined as 1/1.01858 times the EMF of a normal Weston cell at exactly 20°C. Because of their high temperature sensitivity, saturated Weston cells are kept in temperature-regulated baths in which the temperature is kept constant to within ±0.000 010°.

The present international standard for the volt was adopted January 1, 1990, and is based on a quantum-effect phenomenon which takes place at liquid helium temperatures, the Josephson effect. One form of a Josephson junction (JJ) consists of two thin films of superconducting lead separated by a 1 nm thick layer of lead oxide (Taylor, 1990). Another JJ design uses superconducting niobium separated by a thin film of Al_2O_3. Such JJs can be fabricated into series arrays of JJs using integrated circuit technology (Pöpel et al., 1991). Pöpel et al. reported on the performance of integrated circuit (IC) arrays with 2,000 and 20,160 JJs which were useful as standards in the range of 1 V DC and 10 V DC, respectively.

A single, superconducting JJ has the unique property such that when it is irradiated with microwave energy in the frequency range from 9 to 100 GHz, and biased with a DC current, a precisely known DC voltage appears across the dc portion of the JJ having a stepwise volt-ampere curve, as shown in Figure 1.7. The voltage of each stable step, E_J, is given by

$$E_J = nf/(2e/h) \text{ volts} \tag{1.42}$$

where n is the step number, f is the microwave frequency in GHz, e is the electron charge, and h is Planck's constant. The quantity (2e/h) is known as K_J, the Josephson constant. The Comité International des Poids et Mesures (CIPM) established a universal standard value for the Josephson constant, $K_{J\text{-}90} = 483\ 597.9$ GHz/V on January 1, 1990. For a typical microwave frequency of 94 GHz, derived from a Gunn oscillator, the step voltage is seen to be 194.4 μV. Exactly one volt can be defined as the EMF of a Josephson junction when it is irradiated with electromagnetic radiation at a frequency of 483 597.9 GHz. Of course, such a frequency is a mathematical convenience; the Josephson junction would not work when irradiated with 620 nm light, which has that frequency.

When using a JJ array as a primary standard, the microwave frequency is seen to be quite critical. Thus, the Gunn oscillator is phase-locked to an ultra-stable frequency source, such as a cesium beam clock. The frequency source which has been used for several designs of JJ array voltage standards has been a 50- to 60-mW, 93- to 95-GHz Gunn oscillator.

Figure 1.8 gives a simple schematic of a JJ array (after Endo et al. 1983) which was used to calibrate other standard DC voltage sources, and DC potentiometers. One of the problems in operating a JJ array is knowing which quantum step, n, the JJs are all on. One possible way of solving this problem is to obtain a stable array output voltage, then vary the dc bias current until the array jumps to the next step level, and then adjust the microwave frequency down to f_2 until the same output voltage is obtained. Using this paradigm, we can solve for n:

$$nf_1 = (n+1)f_2 \tag{1.43}$$

so

$$n = \text{INT}\left[(f_1 - f_2)/f_2\right] \tag{1.44}$$

One can also simply count the steps as the bias current is slowly increased under conditions of constant microwave frequency and power.

The output of an array of N JJs in series all having the same DC bias current and subject to the same microwave frequency and energy is

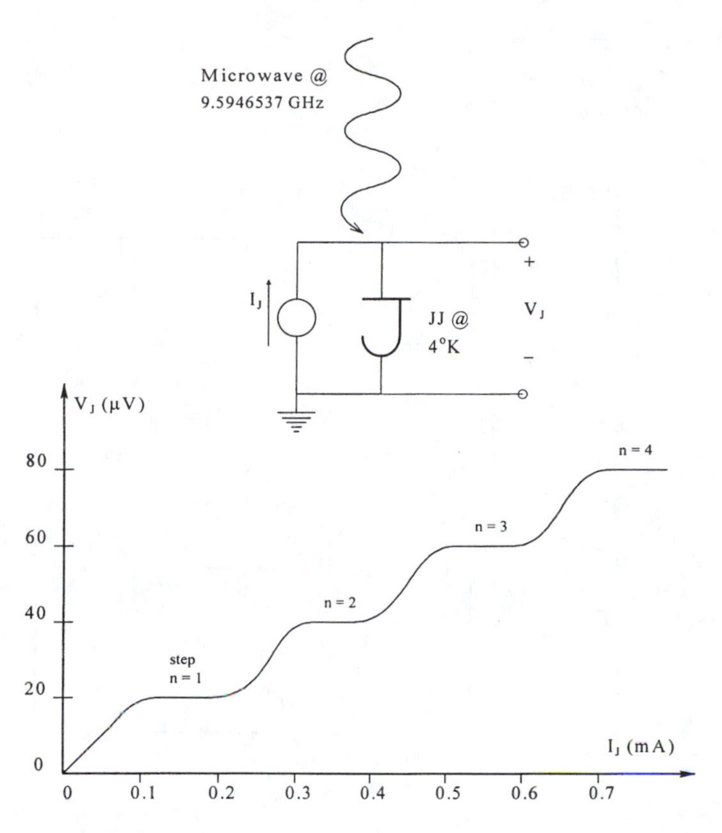

FIGURE 1.7 Volt-ampere curve of a typical Josephson junction voltage standard. (Drawn from data given by Ibuka et al., 1983.) The microwave frequency was 9.5946537 GHz. V_J is approximately 1.9 mV at the 97th step, with I_J = 24 mA, at 4.2 K.

$$V_O = NE_J(n, f) \tag{1.45}$$

Although the microwave frequency is the critical input parameter to a JJ array voltage standard, the microwave voltage and DC current must be kept within certain bounds for stable operation of the JJ array as a quantized voltage source. Hamilton et al. (1991) have found relations for critical values of the DC current density, J_C, the length, l, and width, W, of a JJ for maximum stability of the voltage steps (to prevent noise-induced jumps between steps). They also give an empirical relation for the current range for the nth step, under optimum conditions. Hamilton et al. (1991) have designed JJ arrays having stable operation at 24 GHz, and producing up to 1.2 V.

Most of the Josephson array voltage standards which have been described in the literature (see, for example, a number of papers on JJ standards in the April 1991 *IEEE Transactions on Instrumentation and Measurement*, Vol. 40, No. 2) have uncertainties around 2×10^{-8}. Reymann (1991) describes a JJ array system which is used to measure Weston normal cell EMFs to better than 1 nV accuracy. A problem with the operation of JJ array voltage standards is that the noise-induced instability of the voltage steps leads to jumps to unknown or uncertain step numbers. It is apparent that the Josephson junction array provides an accurate, flexible, DC voltage standard which can be easily adapted for the calibration of secondary standards and DC voltmeters.

1.3.1.2 Resistance

In 1884, an international congress in Paris adopted a reproducible standard for the "legal ohm" consisting of a column of ultra-pure mercury 106 cm in length, and 1 mm^2 in cross-sectional area,

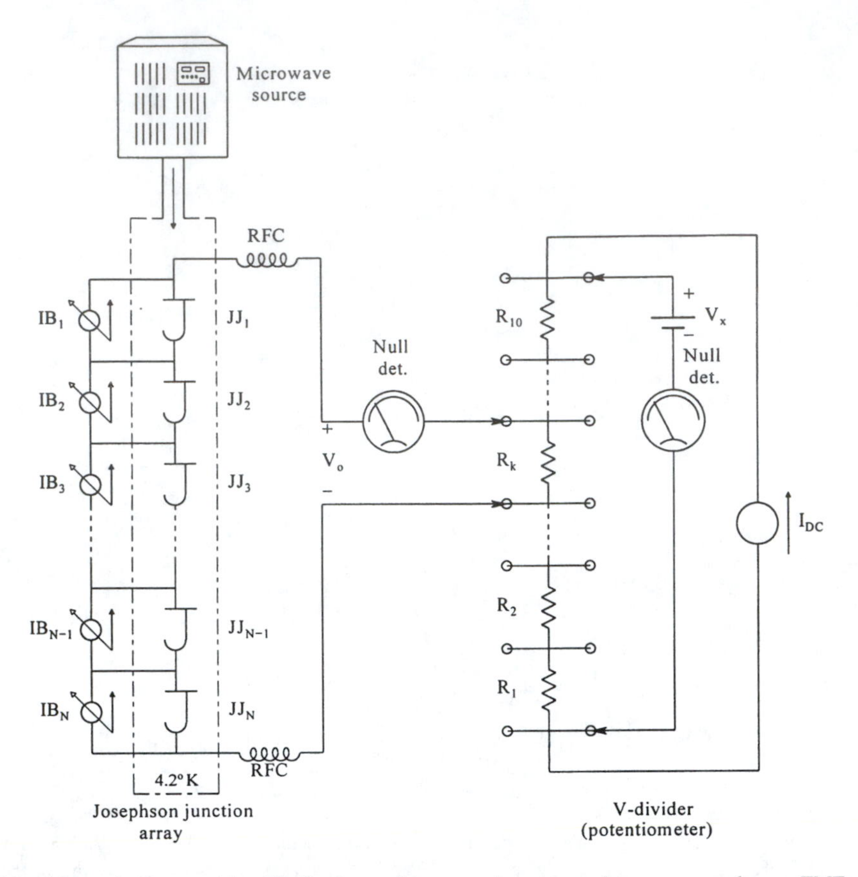

FIGURE 1.8 Schematic diagram (simplified) of a precision potentiometer used to measure unknown EMFs. The potentiometer is calibrated with the standard voltage from the Josephson junction array by adjusting I_{DC}.

measured at 0°C. In 1893, the Chicago Congress specified the length of mercury column of the "international ohm" to be 106.3 cm, and specified the weight of the mercury to be 14.4521 g. The mercury ohm proved to be a poor standard. Mercury has a high temperature coefficient, requiring close thermoregulation of the standard in an ice bath. Slight impurities in the mercury also contribute to errors in certainty, and variations in the design of the end terminals caused errors in the resistance. According to Stout (1950), a major cause of error in the design of the international ohm was nonuniformity in the diameter of the glass tubing. It was found that a 30 ppm change in the measured resistance could result from an undulation in the axis of the glass tube too small to be detected by eye.

In the 1920s, work was done on defining an "absolute" ohm as a derived quantity based on the fundamental units of length, mass, and time. The absolute ohm was determined through the use of a rotating commutator or conductor whose speed must be known, giving the time dependency. The determination also used either a self- or mutual inductance, whose value was determined by calculations involving the dimensions. Not surprisingly, the value of the absolute ohm was found to differ significantly from that of the international ohm. One absolute ohm was equal to 0.999 505 2 international ohms. The absolute ohm and other absolute electrical units were formally adopted January 1, 1948.

After 1892, working standard, wirewound, one ohm resistors made of the alloy manganin were developed, and in 1897 were calibrated in terms of the mercury ohm as defined in 1893 by the Chicago Congress. The U.S. National Bureau of Standards (NBS) adopted the Rosa design, one ohm, manganin, wirewound resistor in 1910 as the international ohm. In 1931, an improved, standard resistor design by J.L. Thomas was adopted at NBS. The cross section of a Thomas-type,

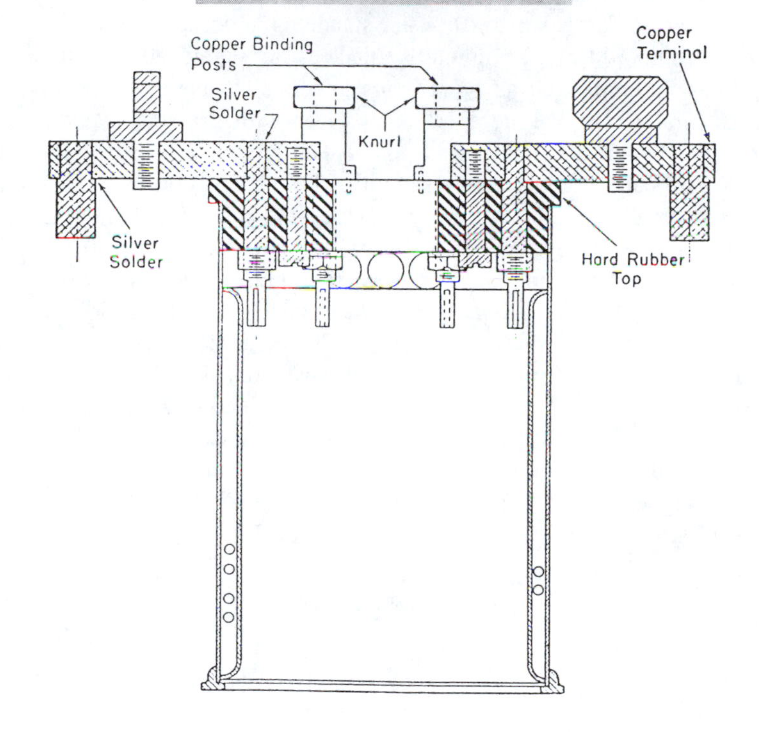

FIGURE 1.9 Top shows double-walled Thomas-type standard resistor. Below is cross-sectional view. Figure courtesy of the National Bureau of Standards.

one ohm standard resistor is shown in Figure 1.9. This resistor uses a double-walled construction, with the resistance wire wound on the inner cylinder, and the space between the cylinders is filled with dry air and sealed. The center cylinder is filled with oil for temperature stabilization and control. In the Rosa-type resistor, the resistance wire windings are in direct contact with dry oil in a sealed can. Rosa resistors are available in decade sizes, typically 1, 10, 100, 1 k, and 10 k ohms.

In 1966, Hewlett-Packard (HP) developed an improved standard resistor design, the 11100 series, which achieved superior temperature stability and precision through the use of new materials for insulation, resistance coil support, and the resistance wire. HP used the alloy Evanohm for the coil (75% nickel, 20% chromium, 2.5% aluminum, 2.5% copper). According to HP, Evanohm has an α tempco of 0 to +2 ppm/°C, a high resistivity, a relatively flat resistance vs. temperature curve, and insensitivity to moisture. The Evanohm wire coil of the HP 11100 resistors is suspended in sealed, dry oil, and supported by an inner and an outer polyester coil form, rather than a metal form such as used in the Rosa or Thomas resistors. HP claims the following specifications for its 11100 series resistors: Limit of error at 25°C and 100 mW power dissipation is ±6 ppm (0.0006%)

with reference to the legal ohm (1966) maintained by NBS. Rated tempco is ±4 ppm. Drift per year is <20 ppm (100 kΩ). Power rating is 100 mW. Internal trim resistance permits adjustment of ±25 ppm. Connections are NBS type, 4-terminal, oxygen-free copper, nickel-rhodium plated. Note that the HP 11100 standard resistors can be set to within ±0.15 ppm of their nominal (true) resistance, but the 1966 NBS primary standard has an uncertainty of ±6 ppm.

At the present time (1991), the Fluke 742 series, 4-terminal, working standard resistors are available in values from 1 Ω to 19 MΩ. The Fluke 742 series resistors are used at ambient room temperatures, from 18 to 28°C, with resistance changes ranging from ±1.5 (10 k) to ±4 (19 M) ppm, showing excellent temperature stability – no oil baths are required. Calibration uncertainty at 23°C ranges from ±1.0 ppm for 1 Ω to 100 Ω sizes, to ±10.0 ppm (10 MΩ) and ±20.0 ppm (19 MΩ). Stability over 1 year ranges from ±8.0 ppm (1 to 10 Ω) to ±10.0 ppm (19 MΩ).

Electro Scientific Industries (ESI) also makes transfer standard resistors. The ESI model SR1060 resistance transfer system consists of six transfer standards in decades from 1 Ω to 100 kΩ. Each decade standard is composed of 12 nominally equal resistors matched to within ±10 ppm. Each decade standard can be configured as 10 resistors in series, 10 parallel resistors, or 9 of the 10 resistors in series/parallel. By making a 1:1 comparison with the 10th resistor, it is possible to resolve a series/parallel value to better than ±1 ppm. All of the standards except the 100 kΩ standard are immersed in an oil bath for temperature stability. The ESI SR1060 resistance transfer standard system is initially calibrated to ±10 ppm traceable to the NIST primary standard. Long-term stability is ±35 ppm for 1 year, and tempcos are ±15 ppm/°C for 1 Ω, and ±5 ppm for 100 Ω to 100 kΩ.

On January 1, 1990, the international intrumentation and measurement community adopted a standard for the ohm based on the *quantum Hall effect* (QHE). This definition of the ohm, like the Josephson junction definition for the volt, is based on fundamental physical constants, rather than an artifact (such as a column of mercury). The QHE was first described by the noted physicist, Klaus von Klitzing in 1980. The operation of a basic Hall effect sensor at room temperature is described in Section 6.3.5.2 of this text. The ordinary Hall sensor is a four-terminal device; one pair of terminals is used to inject current into the thin, doped semiconductor bar which is ordinarily perpendicular to a magnetic field, B. The second pair of terminals, as shown in Figure 6.22, is used to pick off the Hall EMF. It can be shown that the Hall EMF is given by

$$E_H = R_H B_y I_x / h \qquad 1.46$$

where B_y is the B field component in the y direction, orthogonal to the current density vector; J_x, I_x is the injected current; h is the thickness of the doped semicon bar in the y direction; and R_H is the Hall coefficient. $R_H = -1/qn$ for n-doped semicon, and 1/qp for p-doped semicon. q is the magnitude of the electron charge in coulombs, n is the electron donor doping density, and p is the hole donor doping density. The Hall sensor was seen to be useful in measuring magnetic fields, and indirectly, electric power. As will be illustrated below, the properties of the Hall sensor change markedly as its temperature approaches 0 K, and it is subject to strong DC magnetic fields.

Figure 1.10 illustrates an isometric view of a quantum Hall resistor (QHR). Modern QHRs are generally fabricated from GaAs/AlGaAs heterostructures. The channel width may range from 150 to 250 μm, and the distance between lateral voltage contacts on the same side may range from 300 to 600 μm (Piquemal et al., 1991). The thickness of the aluminum-doped conducting channel may range from 10 to 100 nm (von Klitzing, 1986). The QHR has six or eight terminals; two for current, two for voltage, and two or four for measuring body resistivity or voltage in the x direction. Under normal operation, the QHR is operated at a constant, DC current bias, I_x. I_x values ranging from 10 to 50 μA are typically used. A very strong DC magnetic field, By, on the order of 4 to 15 Tesla, is generated by a superconducting coil in close proximity to the QHR. A QHR is generally operated at temperatures ranging from 0.3 to 2 K, with 1.20 K being commonly used. As can be seen from Figure 1.11A, as the magnetic field is varied under conditions of constant DC bias current, the Hall

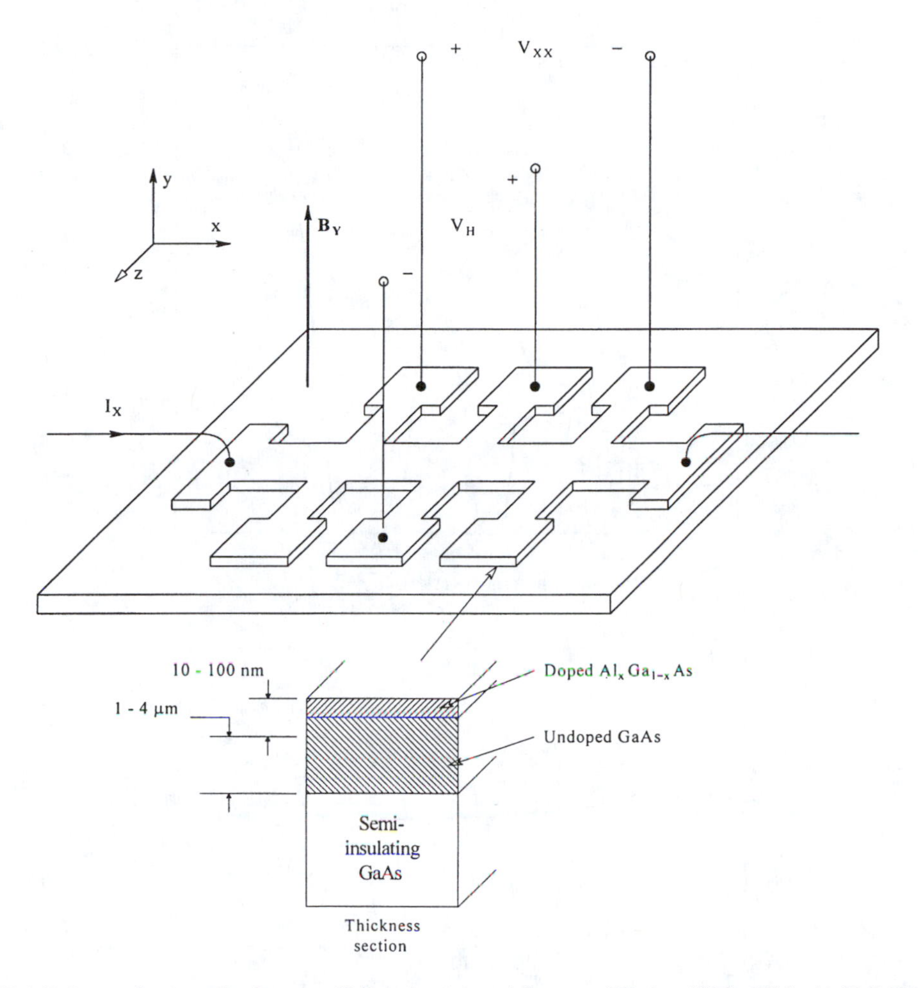

FIGURE 1.10 Isometric view of a Quantum Hall device (adapted from von Klitzing, 1986). EH Is the Hall EMF, V_X is the voltage drop due to the DC current, I_X, B_y is the strong, DC magnetic field, 2DEG = two-dimensional electron gas. A gallium-aluminum-arsenide heterostructure is used. Operating temperature is typically 0.3 K.

resistance assumes a series of steps of precisely known values, regardless of the exact B_y, I_x, or temperature and material of the QHR semiconductor. The quantum Hall resistance, R_H, has been shown by von Klitzing (1986) to be given by:

$$R_H(j) = E_H/I_x = R_{K\text{-}90}/j = h/(q^2 j) \text{ ohms} \quad 1.47$$

where h is Planck's constant, q is the electron charge, j is the step number (integer), $R_{K\text{-}90}$ is the von Klitzing constant (universally adopted as 25 812.807 ohms January 1, 1990), and E_H is the measured DC Hall voltage. For all QHR devices, the quantum Hall resistances are

$$R_H(1) = 25\,812.807 \text{ ohms}, \qquad R_H(2) = 12\,906.400 \text{ ohms},$$

$$R_H(3) = 8\,604.269\,0 \text{ ohms}, \qquad R_H(4) = 6\,453.201\,8 \text{ ohms},$$

$$R_H(5) = 5\,162.561\,4 \text{ ohms}, \qquad R_H(6) = 4\,302.134\,5 \text{ ohms}$$

The value of $R_H(1)$ is defined as one von Klitzing.

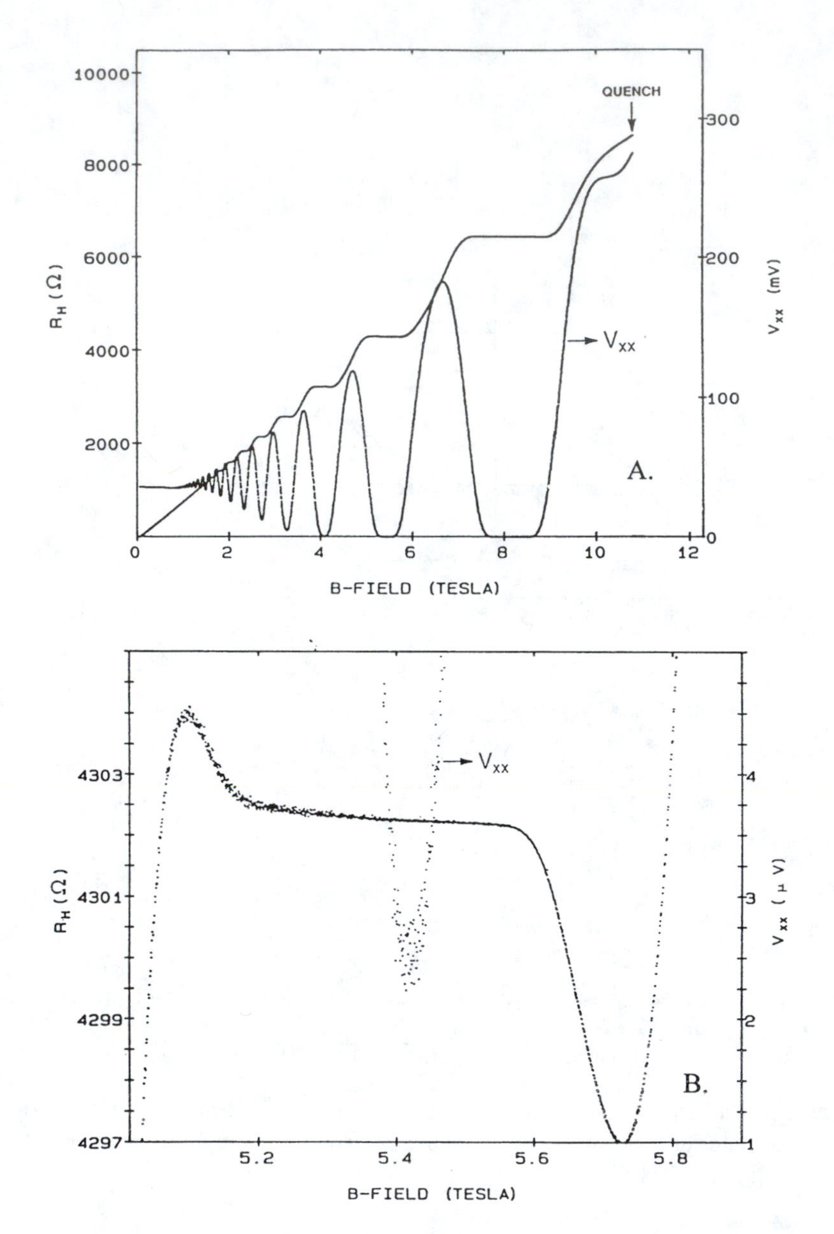

FIGURE 1.11 (A) Plot of the quantum Hall resistance, R_H, and V_{XX} vs. B_Y. Note that there are plateaus of $R_H = E_H/I_X$ over a range of B_Y on which R_H is precisely defined. In this case, T = 0.3 K, and I_X = 50 μA. (B) Detail of the i = 6 quantum Hall step. Note that there is a distinct minimum of V_{XX} at the flattest part of the R_H curve. Same conditions as in (A). (From Jaeger, K.B. et al, *IEEE Trans. Instrum. Meas.*, 40(2), 256, 1991. With permission.)

Also plotted in Figure 1.11A is the voltage drop in the x direction. Note that V_{xx} has minima which occur at the centers of the E_H (B) or R_H(j, B) steps. This is shown more clearly in Figure 1.11B in which the value of V_{xx} is shown as B_y is varied over the range producing the R_H(4) step (Jaeger et al. 1991). Figure 1.12 illustrates the schematic of a simple series circuit which is used to compare the QHR with a secondary standard, using Ohm's law and precision DC nanovoltmeters and DC picoammeters.

Comparisons between quantum Hall resistance measurements made in various standards laboratories around the world suggest that the QHR method of defining the ohm has an uncertainty of less than one part in 10^8.

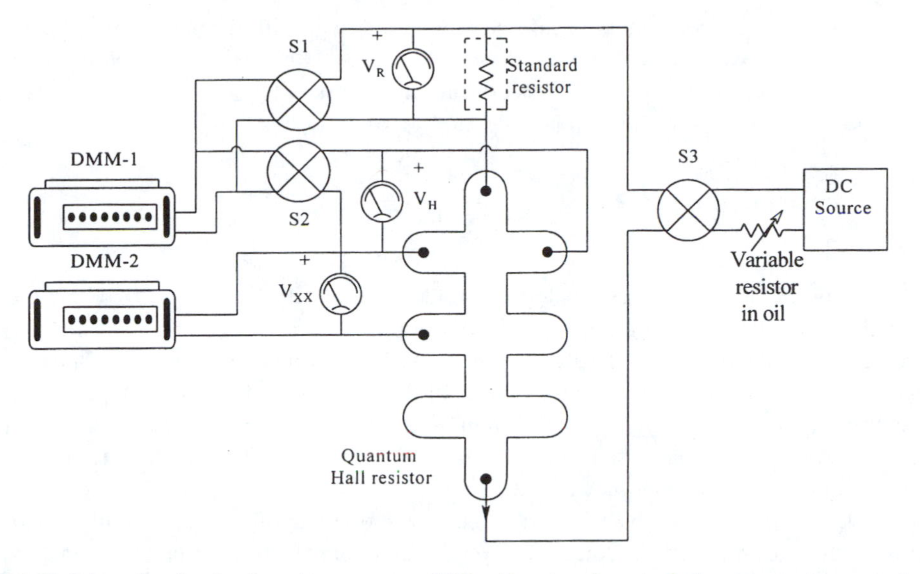

FIGURE 1.12 Schematic of a circuit used to compare a QHR with a secondary standard resistor by Jaeger et al., 1991.

1.3.1.3 Current and Charge

In a metallic conductor, such as a copper wire, the moving charge is electrons, whose average velocity is in the opposite direction to defined (in the U.S.) current flow. In semiconductors, two kinds of mobile charges are involved: holes and electrons. In p-semiconductor material, the intrinsic semiconductor is doped with electron-accepting atoms, and the majority charge carrier is holes. In n-material, electron donor atoms provide electrons as the majority carrier. In wet (chemical and biochemical) materials, the carriers are generally ions, although electrons are involved at metal/metal salt electrode interfaces.

In the SI system of measurements, the ampere is defined to be the basic unit of current, and current is considered to be a fundamental unit of measurement. Before the SI protocols were adopted, current was considered to be a derived quantity, namely one coulomb of charge per second flowing past a perpendicular plane through a conductor, and charge was considered to be a fundamental quantity. In the SI system, charge is derived, and has the units of current times time. One coulomb is the charge transferred when one ampere flows for exactly one second. Charge can also be measured by using the relationship that the voltage on a capacitor is equal to the charge divided by the capacitance, or that the change in voltage on a capacitor is equal to the change in charge divided by the capacitance.

Early definitions of the ampere made use of the electrolytic reduction of silver ions from a standard silver nitrate solution at a platinum cathode. Metallic silver was electroplated out on the cathode at a rate proportional to the electric (DC) current flowing in the external circuit. After a known time, the cathode was removed from the electroplating cell, dried, and weighed. Since the gram molecular weight of silver is known, as is Avogadro's number, the accumulated weight of silver on the cathode was proportional to the total charge transferred in the circuit over the known time. At the London Conference of 1908, the International Ampere was defined as that (DC) current which deposits silver at the rate of 0.001 118 00 g/s from a standard $AgNO_3$ solution. This definition was based on a knowledge of the gram molecular weight of silver (107.868), and Avogadro's number, N_A, which is the number of molecules (hence Ag^+ ions) in one gram molecular weight of a substance. The value is presently known to be $6.022\ 137 \pm 0.000\ 007 \times 10^{23}$ mol^{-1}.

Because of the lack of precision and repeatability of the silver electroplating method, the International Ampere was superceded in 1948 by the Absolute Ampere. The determination of the Absolute Ampere is made using a current balance which weighs the force between two current-carrying coils. By definition, the SI Absolute Ampere is the constant current which, if maintained in two, straight, parallel conductors of infinite length and negligible circular cross section placed 1 m apart *in vacuo*, will produce between these conductors a force of 2×10^{-7} N/m length. If the same current is travelling in the same direction in the two, parallel conductors, it may be shown that the force exerted between the conductors is attractive, and is given by

$$F = \frac{\mu_o I^2}{2\pi d} \text{ newtons} \qquad 1.48$$

where d is the separation of the conductors in meters, and μ_o is defined as $4\pi \times 10^{-7}$ webers/amp meter. Such a theoretical definition is unrealizable in practice.

Working standards to measure the absolute ampere generally involve a current balance, in which the force between a movable coil or coils and a set of fixed coils or a permanent magnet is measured by a weighing balance. This type of determination of the ampere requires accurate measurement of force. The value of the mass used and g, the Earth's gravitational acceleration at the location of the balance, must be accurately known.

Kibble et al. (1983) describe the design and development of a moving-coil apparatus for determination of the ampere at the National Physics Laboratory (Britain). A large permanent magnet with maximum flux density of about 0.7 T is used. Two series-connected, rectangular coils, each of 3,362 turns, are mounted one above the other on an armature which moves vertically in the magnet's air gap. The armature is attached by a vertical rod to the balance arm, and passes coaxially through a force-producing solenoid coil at the bottom of the balance case. The current in the solenoid is controlled by a servo system so it can generate axial forces to cause the measurement coils to move up and down through the magnetic field at a constant velocity, u = 0.002 m/s. This motion generates an EMF in the series measurement coils whose magnitude is given by $V = K_v u$ volts. The value of K_v depends on B in the magnet's air gap, the number of turns, and other geometrical factors. In the second phase of measuring the ampere, a DC current, I, is passed through the series measurement coils which are positioned at the center of the B field, where the determination of K_v was made. A Lorentz force is generated, given by $F = KF_I$. F(I) is measured with the balance. The value of I is determined from the fact that in the MKS (SI) system, K_v is equal to K_F. If the relation $V = K_v u$ is divided by $F = K_F I$, we can obtain a relation equating electrical and mechanical power:

$$VI = Fu = Mgu \qquad 1.49$$

Now we assume that the generated electrical power is dissipated in a resistor R such that V/R = I. Thus

$$VI = I^2R = Mgu \qquad 1.50$$

Solving for I, we find that

$$I = \sqrt{Mgu/R} \text{ SI amperes} \qquad 1.51$$

Although under development, Kibble et al. (1983) reported that their moving-coil current system for determination of the ampere had ±1 ppm reproducibility, and that they eventually expected to obtain 0.1 ppm accuracy.

A system utilizing the magnetic levitation of a superconducting mass to determine the SI ampere was proposed by Kibble (1983). This system is shown schematically in Figure 1.13. A current source, I, causes current to flow in a superconducting coil surrounding a superconducting mass. Induced current in the mass produces a magnetic field which reacts with the field of the coil to

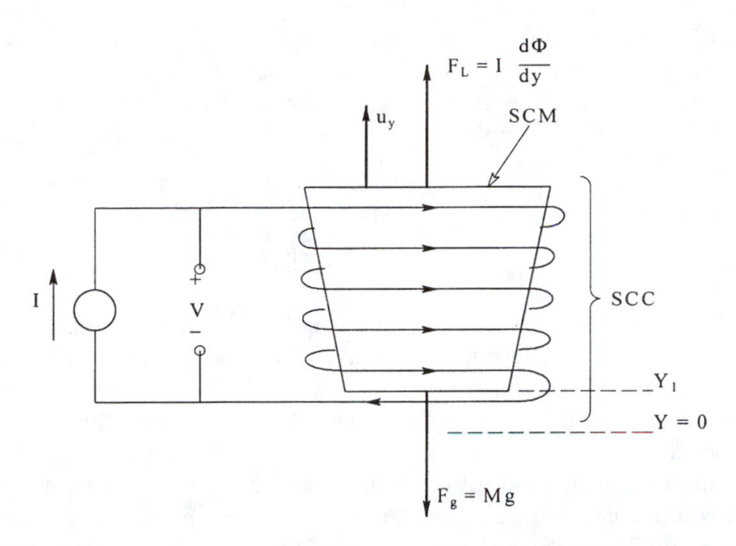

FIGURE 1.13 Schematic of a superconducting, mass-balance means of measuring current proposed by Kibble, 1983. SCC, superconducting current-carrying coil; SCM, superconducting mass, M. Means of inducing constant velocity, u_y, not shown. The induced voltage, V, is measured potentiometrically, and is related to the current by Equation 1.53.

produce an upward force until the mass reaches a position Y1 where the downward gravitational force, Mg, equals the upward levitation force, $F_L = I\ (d\Phi/dy)$. Now if the mass is given a vertical velocity, $u_y = dy/dt$, around Y1, a voltage $V = d\Phi/dt$ will be induced in the superconducting coil, across the current source. We also note that $d\Phi/dt = (d\Phi/dy)(dy/dt)$. Now from these relations it is easy to see that $Mg/I = d\Phi/dy = V/uy$. Solving this relation for I gives:

$$I = \left(Mgu_y\right)/V \qquad 1.52$$

Kibble's proposed method of measuring DC current thus requires accurate knowledge of the mass, M, and the local acceleration of gravity, g. A constant, linear velocity, of the mass, u_y, around its equilibrium displacement, Y1, must be generated, either by mechanical means or by electrostatic forces. A "no-touch" means of measuring u_y could be designed using laser interferometry. Lastly, measurement of V can be done potentiometrically by comparison with a Josephson voltage standard.

The reader should be aware that the bases for electrical and physical standards are constantly under development, and are changing with the advances in quantum physics, laser technology, etc. We expect to see a trend in metrology away from complex, expensive, ultra-precision measurement systems toward simpler, more elegant solutions for the determination of standard quantities. Perhaps we should consider making the quantum Hall ohm and the Josephson volt basic SI quantities, rather than the ampere. Is charge a more easily measured quantity than current? Should we return to the earlier MKS system where charge was a basic quantity, rather than current which is now used in the present SI definitions? There are no simple answers to these questions, and the points of debate change constantly with evolving technology.

1.3.1.4 Capacitance

There are several ways of defining the basic unit of capacitance, the Farad. A capacitor is said to have a capacitance of one Farad if the potential across its plates is exactly one volt when it has been charged by one coulomb of electric charge. This definition of the Farad requires an accurate voltage measurement in which a negligible charge is exchanged with the capacitor, and accurate measurement of the total charge put into the capacitor. The latter measurement is more difficult, since it involves precision integration of the capacitor's charging current.

Another way of measuring capacitance is through the use of a commutated-capacitor bridge (cf. Section 5.4.1.6 and Figures 5.9 and 5.10). This technique requires a precision DC null detector, accurate resistances, and a switching clock with an accurately known period, T. Under the assumption that $R_4 C_X << T/2 << R_2 C_X$, we show in Section 5.4.1.6 that at null (set by R1),

$$C_X = \frac{T(R_3 + R_1)}{4R_2(R_3 - R_1)} \qquad 1.53$$

Here we assume that the switch has zero closed resistance and infinite open resistance, the waveforms at V_2 are exactly rectangular and triangular, and the clock duty cycle is exactly 50%. In practice, the duty cycle can be adjusted to be very close to 50%, taking into consideration the switching waveform rise and fall times, and the time delays in operating the metal oxide semiconductor (MOS) switch.

Probably the most accurate means of measuring capacitance is by comparison of the capacitor to be measured with a *calculable capacitor*. Calculable capacitors are based on the electrostatics theorem developed by Thompson and Lampard (1956). A calculable capacitor is typically made from four parallel, hollow metal cylinders arranged in a square, with a movable shield device which effectively regulates the length of the active (exposed) cylinders. Capacitance is measured between opposite cylinders, *in vacuo*. Calculable capacitors have been made with capacitances ranging from about 0.1 to 1.0 pF.

The Thompson and Lampard theorem states:

> Let the closed curve S be the cross-section of a conducting cylindrical shell, which cross-section has one axis of symmetry AC but is otherwise arbitrary. Further, let this shell be divided into four parts by two planes at right angles, the line of intersection of the planes being parallel to the generators of the cylinder, and one of the planes containing the symmetry axis AC. Then the direct capacitance, per unit length of the cylinder, between opposing parts of the shell (for example, $\alpha\beta$ to $\gamma\delta$) due to the field inside (or outside) the shell, is a constant:

$$C_o = \frac{\ln(2)}{4\pi^2} \text{ e.s.u.} = 0.0175576 \text{ e.s.u./m} \qquad 1.54$$

Note that 1 pF of capacitance in SI units is equal to 0.898 797 e.s.u. capacitance units. Thus it is easy to show that C_o = 1.953 549 pF/m (or attofarad/μm). A consequence of this theorem is that the absolute cross-sectional area of the closed cylindrical shell is not important. What is important is that the sides be perfectly parallel and that the structure have perfect symmetry. Figure 1.14A illustrates the geometry discussed in the theorem stated above, and Figure 1.14B illustrates the shielded, four-tube, cross-sectional geometry of a practical calculable capacitor (Igarishi et al., 1968). The length of the calculable capacitor at the U.S. NBS (now NIST) is measured interferometrically using a stabilized HeNe laser. Overall uncertainty in the measurement of the NBS Farad, based on the NBS calculable capacitor, is given as 0.014 ppm (Shields et al., 1989).

Calculable capacitor geometry has not been limited to four-tube devices. Delahaye et al. (1987) have used five tubes in a pentagonal cluster, each 75.5 mm in diameter and 450 mm long. A cylindrical central screen is displaced axially by a stepping motor, and its position is measured with a laser inteferometer. The calculable capacitor of Delahaye et al. was used to establish a standard for the farad, and for the quantum Hall resistance R_H (2). The total one standard deviation uncertainty in the measurement of $R_H(2)$ was 2.2×10^{-7}.

Calculable capacitors have been used as standards by others to calibrate the quantum Hall ohm, inductors and transfer standard capacitors through the use of precision AC bridges (Shida et al., 1989; Dahake et al., 1983).

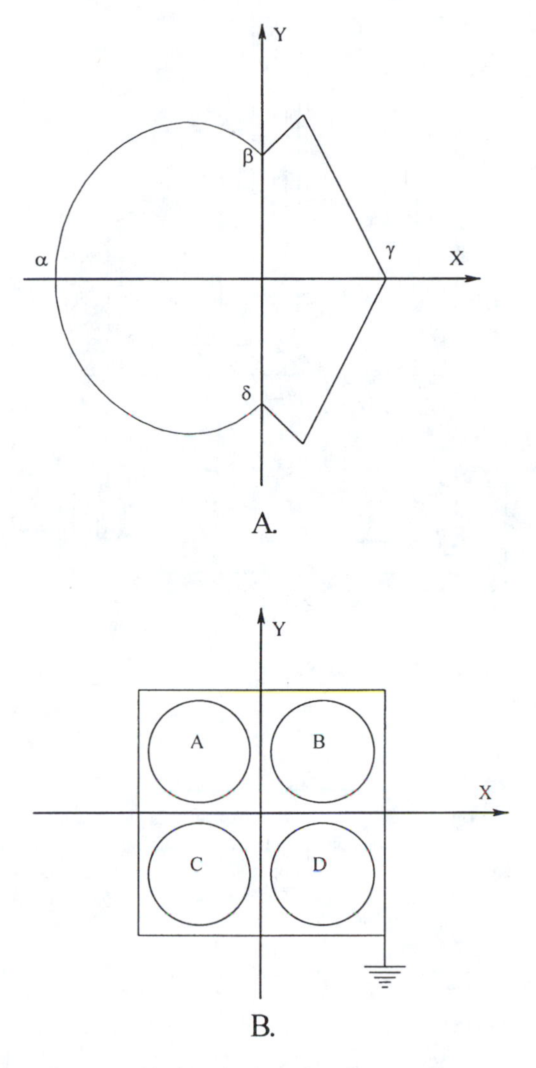

FIGURE 1.14 (A) Cross section of a symmetrical conducting tube relevant to the Thompson-Lampard theorem. The T-L theorem defines the capacitance between sections αβ and γδ on the tube as equal to $\ln(2)/(4\ \pi^{2})$ e.s.u./m, regardless of the absolute cross-sectional area of the tube. (B) Cross section of a practical, four-tube, shielded calculable capacitor, such as built by Igarishi et al., 1968. Tube diameters were about 1 cm.

Secondary and transfer standard capacitors are generally on the order of 1 to 10 pF, and are made with air, quartz (silica), or mica dielectrics. Such capacitors are generally made as three-terminal devices, so the capacitance from either terminal to the grounded case can be compensated for. Air dielectric capacitors of the guard well geometry (see Figure 1.15) have been constructed with values down to 1 fF. A 0.1 pF capacitor of this design has an accuracy of 0.1% (Stout, 1960). For secondary standards with larger values, hermetically sealed, silvered mica capacitors offer long-term stability and low tempcos. For example, a General Radio type 1409, 1 μF standard silver mica capacitor has its value guaranteed to ±0.05%, and its tempco is 35 ± 10 ppm/°C. Silver mica capacitors have very low D values at 1 kHz. Most bridges used to transfer calibration from calculable capacitors to working and transfer standard capacitors are of the ratio transformer type, operated at frequencies of 2,500 and 5,000 r/s (Delahaye et al., 1987) and at 10,000 r/s (Igarishi et al., 1968; Dahake et al., 1983).

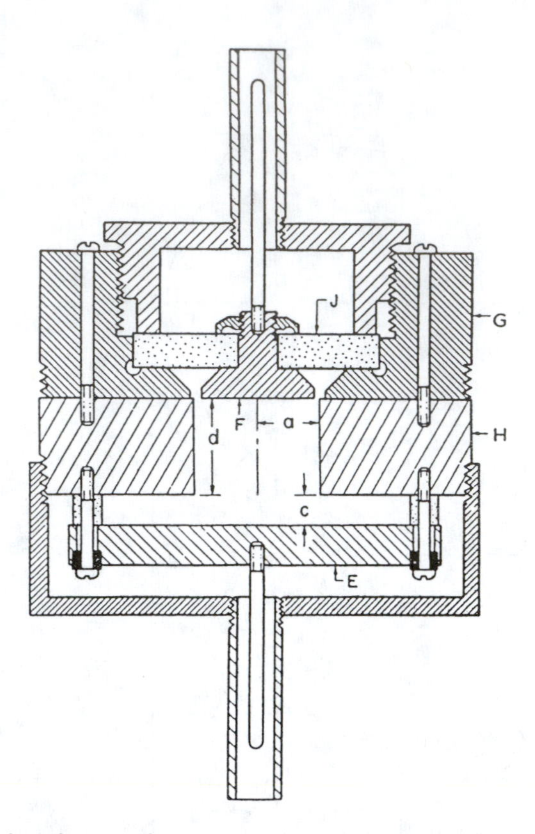

FIGURE 1.15 Cross section through a secondary standard, air dielectric, guard-well capacitor. Key: F and E, insulated electrodes; J, borosilicate glass dielectric; G, H, K, and L, grounded shield conductor pieces. Note that this capacitor is a three-terminal device; both top and bottom electrodes have capacitances to ground, as well as to each other. (From Stout, M.B., *Basic Electrical Measurements*, 2nd ed., 1960. Adapted by permission of Prentice-Hall, Inc., Upper Saddle River, NJ.)

1.3.1.5 Inductance

The primary inductance standard is a derived quantity, being measured using an AC bridge circuit using standard resistors and capacitors. We have seen how the quantum Hall ohm can be used to calibrate working standard resistors, and the calculable capacitor may be used to establish working standard capacitors. By using a bridge, such as the Maxwell (cf. Section 5.4.2.1), having three accurate resistors and an accurate capacitor, unknown inductances can be measured to a fair degree of accuracy.

Fixed-value, standard inductors are generally wound in solenoidal form on glass or ceramic forms having spiral grooves cut into their surfaces to precisely space the windings. Low-value (μH) inductors have single windings; however, millihenry and larger inductors may have groups of series-connected windings set into evenly spaced slots in a ceramic core. Inductors with solenoidal windings are subject to the external influences of magnetic materials and time-varying magnetic fields, and thus care must be taken when using them not to place them on or near steel tables or cabinets, and to keep them well away from other current-carrying solenoidal coils. Modern secondary standard inductors, such as the General Radio model 1490 series of switchable decade inductors, are wound on toroidal ferrite cores, from which there is little flux leakage, and small influence from nearby magnetic materials or time-varying magnetic fields. Another design which has been used for secondary standard inductors with fixed values in the millihenry range and larger, is the use of two series-connected coils wound on semicircular, "D" shaped forms, mounted so that one coil can be rotated with respect to the other so that the area of overlap of the enclosed areas of each coil can be adjusted. This rotation allows variation of the mutual inductance of the coils, and provides a means of adjusting the inductance of the series coils in accordance with the relation

$$L = L_1 + L_2 + 2M \tag{1.55}$$

assuming that the mutual inductance is aiding. The same principle of varying the mutual inductance of two series coils has also been used to make continuously variable inductors (Brooks inductometer). The Brooks inductometer's calibration is ±0.3% of its maximum value (Stout, 1960).

A problem in the construction of all standard inductors is that their inductance depends on their dimensions, so any small change in dimensions caused by temperature changes can lead to errors. Also, as we demonstrate in Section 5.1, inductors, however made, are fraught with stray capacitance between their windings, and between layers and groups of windings. The wire from which they are wound also has a finite resistance. As the operating frequency of an inductor increases, the capacitive effects become more dominant, and can cause significant errors in the apparent inductance. Thus all standard inductors are generally calibrated at a convenient frequency such as 1 kHz, and are designed to be used at a constant temperature, such as 25°C. Inductors wound on laminated iron cores or sintered ferrite cores also have the additional limitations of core losses due to hysteresis and eddy currents, and core saturation at magnetizing force (ampere turns).

It may certainly be said that of all the electrical quantities, natural inductance is the least pure, being always associated, at a given frequency, with a series resistance and distributed interwinding capacitance. Inductor equivalent circuits are described in Section 5.1 of this text. At superconducting temperatures, for aircore coils of certain metals, the coil resistance disappears, but the interwinding capacitance remains.

Another way of obtaining a very pure inductance, ranging in values from microhenries to kilohenries in the audio frequency range, is by means of the generalized impedance converter (GIC) active circuit element (see Section 2.3.1.3). One end of the GIC circuit must be grounded. It is shown that the GIC's input impedance is given by:

$$\mathbf{Z}_{11} = \frac{\mathbf{Z}_1\mathbf{Z}_3\mathbf{Z}_5}{\mathbf{Z}_2\mathbf{Z}_4} \tag{1.56}$$

Thus if the circuit element for $\mathbf{Z}_2$ is made a precisely known capacitor of high quality, such as a silver mica secondary standard, the remaining GIC impedance elements are resistors calibrated from the quantum Hall ohm, and the op amps are high quality ICs having large gain-bandwidth products, high input impedances, and low noise, we will see a virtually lossless inductance with value given by:

$$L_{eq} = \frac{R_1C_2R_3R_5}{R_4} \text{ henrys} \tag{1.57}$$

It should be pointed out that the GIC circuit can also be used to rescale the size of capacitors. In this case $Z_{11} = 1/\omega C_{eq}$. It is easy to see that if $\mathbf{Z}_1$ is from a capacitor, then:

$$C_{eq} = \frac{C_1R_2R_4}{R_3R_5} \text{ farads} \tag{1.58}$$

In using the GIC circuit we must be careful to avoid op amp voltage or current saturation, and to avoid operating the circuit at high frequencies near the f_T of the op amps.

1.3.2 Time and Frequency Standards

The basic unit of time is the second. The basic unit of frequency is the hertz, or cycle per second of a periodic waveform. Obviously, time and frequency are related, in a measurement sense. The earliest definitions of the second were based on the Earth's rotational velocity, which is now known to vary over the period of a year due to seasonal changes in the Earth's moment of inertia due to the build-up of polar ice caps, etc. The second was first defined as 1/86,400 of the length of a mean solar day. The mean solar day is the average time of 365 consecutive days. Refinements in the measurement of the period of the Earth's rotation led to the definitions for universal time (UT). In a search for a more precise definition for the second, astronomers developed the units of ephemeris time (ET). The ephemeris second was defined as 1/31,556,925.9747 of the tropical year (Helfrick and Cooper, 1990). The ET standard was impractical as it took several years to determine the length of the tropical year, and it required precise sightings of the positions of the sun and the moon.

The primary time/frequency standard in current use is the cesium 133 atomic beam clock. A diagram of the cesium clock is shown in Figure 8.61. This standard is discussed in detail in Chapter 2. To summarize the cesium clock's properties, it oscillates at 9.192 631 770 GHz, having an effective Q of 2×10^8. Coordinated universal time (UTC) whose basic unit is the second, is defined as 9,192,631,770 periods of the cesium 133 beam oscillator. This international standard was adopted in October 1967, and is more accurate than any clock calibrated by astronomical measurements.

The Hewlett-Packard HP5061B(Opt 004) Cesium Beam Frequency Standard has a long-term stability of $\pm 2 \times 10^{-12}$ over the life of the cesium beam tube. Accuracy is also $\pm 2 \times 10^{-12}$. The HP5051B cesium clock has sinusoidal outputs at 10, 5, 1, and 0.1 MHz. HP cesium clocks are used to calibrate and synchonize the SATNAV, Omega, and LORAN-C radio navigation systems for boats and aircraft. Cesium beam clocks are ordinarily used to adjust secondary rubidium and quartz oscillators used as secondary standards for frequency or period determination.

Rubidium frequency standards are second in the hierarchy of accuracy. Similar to the operation of a cesium beam clock, the atomic resonance of a rubidium vapor cell is used to synchronize a quartz crystal oscillator in a frequency-lock loop. The long-term stability of the rubidium vapor oscillator is $\pm 1 \times 10^{-11}$/month. It, too, has outputs at 5, 1, and 0.1 MHz.

1.3.3 Physical Standards

The physical standards discussed below include mass, length, volume, and temperature. We have seen above that the present trend is to redefine standard quantities in terms of the (presumably) universal constants of quantum physics, rather than to use physical artifacts (e.g., the present use of the Josephson junction to define the volt instead of the Weston normal cell).

1.3.3.1 Mass

The present (SI) unit of mass is the kilogram. It is presently defined in terms of the standard kilogram artifact which has been kept at the Bureau International des Poids et Mesures (BIPM) at Sèvres, France, for over 100 years. Physically, the standard kilogram is a cylinder of platinum-10% iridium. Comparison of a working standard kilogram mass with the primary standard must be made by means of a precision balance. The current "gold standard" balance is the NBS-2, single arm, two knife balance which has a resolution of 1 μg or one part in 10^9 (Quinn, 1991). What has been discovered over the years is that the international prototype kilogram and its copies systematically loose mass when they are cleaned and washed in preparation for weighing. On the other hand, the masses tend to increase with time following cleaning and washing. These changes are on the order of tens of micrograms (Quinn, 1991). The long-term stability of the standard kilogram appears to be on the order of 5 μg per year. This figure is based on the assumption that the international prototype and its copies are not drifting in mass by more than ten times the rate at which they are drifting apart from each other, at about 0.5 μg per year.

A number of workers are considering ways to replace the present artifact standard kilogram at BIPM with a mass standard based on the fundamental physical constants, which include the electron charge (q), Planck's constant (h), the speed of light (c), and the permeability of vacuum (μ_o). Other constants which can be used are the von Klitzing constant ($R_{K\text{-}90}$), the Josephson constant ($K_{J\text{-}90}$), the fine structure constant (α), the Faraday number ($\Im$), and the Avogadro constant (N_A).

Taylor (1991) argued that the Avogadro constant, which is the number of molecules in a gram molecular weight, or mole, of some pure element (X) can be one determining factor, along with the molar mass of the element (M_X) in kg/mol. For example, M(^{12}C) is equal to 0.012 kg, by definition (^{12}C is the common isotope of the element carbon). Thus

$$n_x = N_A/M_X \qquad 1.59$$

where n_X is the number of free Xs at rest required to make the mass of an international kilogram prototype. There are a number of interesting ways to determine the Avogadro number. Taylor (1991) proposed that N_A can be found from the X-ray crystal density of a pure, grown silicon crystal from the relation:

$$N_A = M(Si)\Big/\left[\delta(Si)\left(d_{220}\sqrt{8}\right)^3/8\right] \qquad 1.60$$

where M(Si) is the mean molar mass of silicon in kg/mol, δ(Si) is the crystal density in kg/m^3, and d_{220} is the 2,2,0 silicon lattice spacing in meters. M(Si), δ(Si), and d_{220} must be measured. In a practical determination of NA, the ratio of the three naturally occurring isotopes (^{28}Si, ^{29}Si, ^{30}Si) in a crystal must be measured. Another way to find N_A indirectly is to use the relation

$$N_A = K_{J\text{-}90} R_{K\text{-}90}\, \Im/2 \qquad 1.61$$

where $K_{J\text{-}90}$ = (2q/h), the Josephson constant; $R_{K\text{-}90}$ = (h/q), the von Klitzing constant; and $\Im$ = 96,486.7, the Faraday number (the coulomb charge of an Avogadro number of electrons).

Perhaps in the near future, progress will be made in replacing the standard kilogram artifact with a standard kilogram defined by fundamental constants. Uncertainty in determining such a new standard may be on the order of 1×10^{-8}.

1.3.3.2 Length

The SI standard unit of length is the meter. Formerly, the meter was defined as one ten millionth of the arc distance of a meridian passing from the north pole through Paris, to the equator. It was represented by an artifact consisting of a platinum-iridium bar with lines scribed on it one meter apart, kept at constant temperature at the BIPM near Paris. In 1960, the meter was redefined in terms of the wavelength of monochromatic light. One meter was exactly 1,650,763.73 wavelengths in vacuum of the orange radiation corresponding to the transition between the levels $2p_{10}$ and $5d_5$ of the krypton-86 atom. However, this definition of the meter by wavelength assumes the permanence and invariance of the atomic energy levels, Planck's constant, and the speed of light. These quantities are basic to the relation for wavelength:

$$\lambda = c/\nu = ch/E = ch/(E_2 - E_1) \qquad 1.62$$

To eliminate the requirement for h and ΔE, the invariant speed of light, c, was used in 1983 to redefine the standard meter as the distance light travels in free space in 1/299 729 458 seconds

(3.336 342 0 ns). Frequency-stabilized laser sources are used to realize this definition (Sirohi and Kothiyal, 1991).

Gauge blocks are used as secondary standards for length. These are generally platinum alloy or stainless steel cubes or blocks whose dimensions are established interferometrically to within a fraction of a wavelength of light, and whose faces are polished to optical smoothness. Gauge blocks are used to calibrate micrometers and to set up precision mechanical and optical systems.

1.3.3.3 Temperature

The SI unit of temperature is the Kelvin. The standard reference temperature is defined by the triple point of water, at which pressure and temperature are adjusted so that ice, water, and water vapor simultaneously exist in a closed vessel. The triple point of pure water occurs at +0.0098°C (273.16 K) and 4.58 mmHg pressure. The Kelvin was defined in 1967 at the thirteenth CGPM to be the unit of thermodynamic temperature, equal to 1/273.16 of the thermodynamic temperature of the triple point of water. Other primary, fixed temperature points are also used for temperature calibration: The boiling point of O_2 (–182.97°C), the boiling point of sulfur (444.6°C), the freezing point of silver (960.8°C), and the freezing point of gold (1,063°C), all at atmospheric pressure. Absolute zero is at 0 K, or –273.15°C. Obviously, Celcius and Kelvin degrees are the same "size".

Because so many physical and chemical phenomena are strong functions of temperature, there are many ways to measure temperature, as will be seen in Chapters 6 and 7 of this text.

1.4 CHAPTER SUMMARY

In this introductory chapter, we have introduced the architecture of typical measurement systems, and have shown that they not only contain sensors, signal conditioning operations, and data display, storage, and retrieval options, but also sources of noise. A measurement error is shown to occur if sensor dynamics are not considered, and the sensor is read before it reaches a steady-state output.

The types of errors in measurements are considered. Elementary statistics are discussed and the important concept of limiting error in a measurement is introduced and derived; examples are given of calculating limiting error.

Standards in electrical and physical measurements are described. The important SI standards for voltage (the Josephson junction), resistance (the Quantum Hall resistor), and capacitance (the Thompson-Lampard theorem) are discussed in detail. The reader is advised that primary standards are constantly evolving with science and technology, and accuracies of one part in 10^7 are routine; many systems exceed this accuracy.

2

Analog Signal Conditioning

2.0 INTRODUCTION

Practically all instrumentation systems require some type of analog signal conditioning between the input transducer and the data processing, display, and storage systems. In its simplest form, analog signal conditioning can be voltage amplification with a change in impedance level between the conditioning amplifier's input and output. Analog signal conditioning may also involve linear filtering in the frequency domain, such as band-pass filtering to improve signal-to-noise ratio at the amplifier's output. In other cases, the analog input to the signal conditioning system may be processed nonlinearly. For example, depending on system requirements, the output of the analog signal conditioner may be proportional to the square root of the input, to the RMS value of the input, to the logarithm of the input, to the cosine of the input, etc.

Analog signal conditioning is often accomplished by use of the ubiquitous operational amplifier, as well as special instrumentation amplifiers, isolation amplifiers, analog multipliers, and/or dedicated nonlinear processing integrated circuits (ICs).

In the following sections of this chapter, we examine the properties of the integrated circuit systems used in analog signal conditioning in instrumentation systems, beginning with the differential amplifier which has wide use in the headstages of nearly all types of op amps, instrumentation amplifiers, isolation amplifiers, analog multipliers, etc.

2.1 DIFFERENTIAL AMPLIFIERS

Differential amplifiers (DAs) are widely used as input stages in a variety of amplifier types, including op amps, analog comparators, analog multipliers, instrumentaion amplifiers, oscilloscope vertical and horizontal amplifiers, and other specialized ICs. There are two good reasons for their wide use in signal conditioning system design. The first and foremost is the ability of the DA to respond to the difference in the input signals ($V_1 - V_1'$) and to discriminate against a signal common to both inputs. Such a commonmode (CM) input voltage is often a DC level, hum, or other coherent interference which is desired to be eliminated. Another reason for using a DA headstage is that it inherently discriminates against changes in the amplifier's DC power supply voltages.

2.1.1 Analysis of Differential Amplifiers

Figure 2.1 illustrates the most general form of differential amplifier. Note that including ground, it is a four-port circuit. Most practical DAs have only a single-ended output, V_o, and are thus three-port circuits. The deflection amplifier circuits of most analog oscilloscopes are an exception to this rule, preserving the differential circuit architecture from input to the output to the CRT deflection plates.

In the analysis of DAs it is expedient to define difference-mode and common-mode input and output signals.

$$V_{id} = (V_i - V_i')/2 \qquad 2.1$$

$$V_{ic} = (V_i + V_i')/2 \qquad 2.2$$

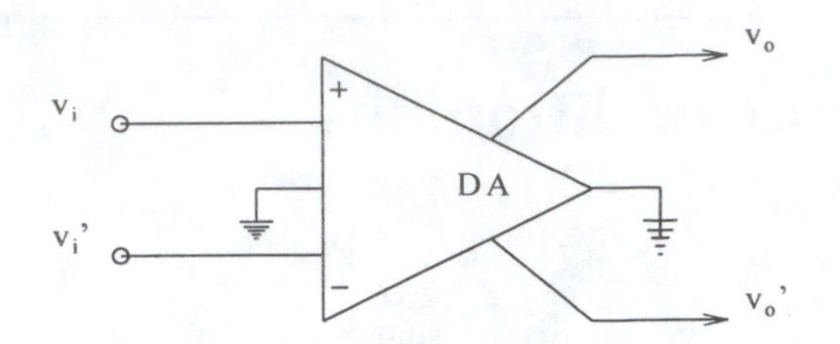

FIGURE 2.1 Generalized differential amplifier with differential output.

$$V_{od} = (V_o - V'_o)/2 \qquad 2.3$$

$$V_{oc} = (V_O - V'_O)/2 \qquad 2.4$$

The input/output relationships for a general DA can be expressed in terms of Middlebrook equations:

$$V_{od} = A_{dd}V_{id} + A_{dc}V_{ic} \qquad 2.5$$

$$V_{oc} = A_{cd}V_{id} + A_{cc}V_{ic} \qquad 2.6$$

The four gains in Equations 2.5 and 2.6 are written in vector format to stress that they are functions of frequency.

It is easy to show that the single-ended outputs are given by:

$$V_o = (A_{dd} + A_{cd})V_{id} + (A_{dc} + A_{cc})V_{ic} \qquad 2.7$$

$$V'_o = (A_{cd} - A_{dd})V_{id} + (A_{cc} - A_{dc})V_{ic} \qquad 2.8$$

The single-ended output, V_o, can also be written in terms of the actual DA's inputs:

$$V_o = \frac{A_{dd}A_{cd} + A_{dc}A_{cc}}{2}V_i + \frac{A_{cc}A_{dc} - A_{dd}A_{cd}}{2}V'_i \qquad 2.9$$

Often it is expedient to write:

$$V_o = A_D V_{id} + A_C V_{ic} \qquad 2.10$$

where A_o and A_c are given by the corresponding terms in Equation 2.7. In an ideal DA, $A_{cc} = A_{dc} = A_{cd} = 0$, so $A_C = 0$ and $A_D = A_{dd}$ in Equation 2.10 above.

2.1.2 Common-Mode Rejection Ratio

The common-mode rejection ratio (CMRR) is a figure of merit for DAs, usually given in decibels. It describes how well the behavior of a real differential amplifier approaches that of an ideal DA. The scalar CMRR, measured at some frequency, is defined as:

$$\text{CMRR} \equiv (V_{ic} \text{ to give a certain output magnitude})/(V_{id} \text{ to give the same output}) \qquad 2.11$$

Using the definition for CMRR and Equation 2.7, we find

$$CMRR = \left(A_{dd} + A_{cd}\right)/\left(A_{cc} + A_{dc}\right) \qquad 2.12$$

If the DA circuit is perfectly symmetrical, then $A_{dc} = A_{cd} = 0$, and the CMRR magnitude at a given frequency reduces to:

$$CMRR = A_{dd}/A_{cc} = A_D/A_C \qquad 2.13$$

State-of-the-art instrumentation and isolation amplifiers may have CMRRs of 120 dB or more ($>10^6$), under ideal conditions. Generally, CMRR decreases with frequency, and, as we will see, an inbalance in Thevenin source resistances at the input of a DA can lead to either a loss or increase in CMRR over the manufacturer's specified value.

2.1.3 Measurement of CMRR, A_D, and A_C

Often, to save money, an engineer puts together a differential instrumentation amplifier from op-amp "building blocks." (The design of such amplifiers is covered in Section 2.2.2.) In order to measure the DA's CMRR, we make use of Equation 2.11. The two inputs are shorted together and a sinusoidal signal of frequency f and amplitude V_{s1} is applied to them, generating a CM input signal, V_{s1c}, which is adjusted to give an output, V_o. Next, the negative input is grounded, so $V_i' = 0$. The sinusoidal signal source of frequency f and amplitude V_{s2} is now connected to the + input terminal of the DA. This procedure is necessitated because of the difficulty in generating a perfectly DM input signal. The single-ended input generates both DM and CM signals: $V_{id} = V_{s2}/2$, and $V_{1c} = V_{s2}/2$. V_{s2} is adjusted to make $V_{o2} = V_o$ of the CM case. Thus from Equation 2.10,

$$A_C = V_o/V_{s1} \qquad 2.14$$

and also, considering the single-ended input, V_{s2}, we have

$$V_o = A_D V_{s2}/2 + A_C V_{s2}/2 \qquad 2.15$$

If Equation 2.14 for AC is substituted into Equation 2.15, we may solve for the DM gain, A_D:

$$A_D = V_O\left(2/V_{s2} - 1/V_{s1}\right) \qquad 2.16$$

Hence the experimentally found CMRR can be expressed as Equation 2.17 using relations 2.14, 2.16, and 2.13:

$$CMRR = A_D/A_C = \left(2V_{s1}/V_{s2} - 1\right) \qquad 2.17$$

If the amplifier's A_D is very large, a precision attenuator may be needed to reduce the output of the signal source to an appropriately small V_{s2}.

2.1.4 Effect of Source Resistance Asymmetry on CMRR

Not only does CMRR decrease with input signal frequency (largely due to an increase of A_D with frequency), but it is also severely affected by unbalance in the Thevenin resistance (or impedance) of the sources connected to the DA's input leads. Figure 2.2 illustrates the DC input circuit of a

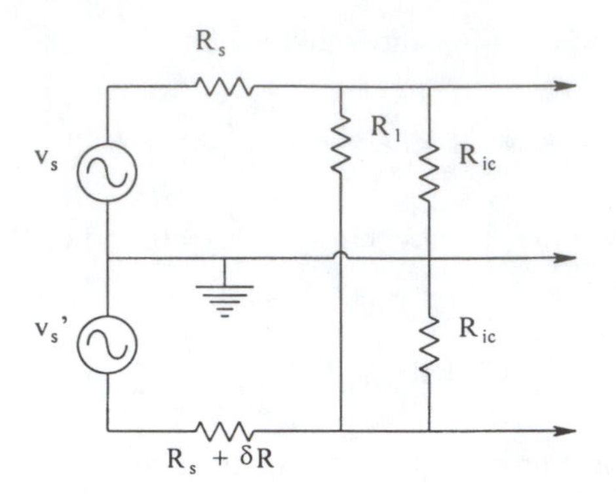

FIGURE 2.2 Differential amplifier input circuit at DC showing source resistance imbalance.

typical DA. Manufacturers typically specify a common-mode input resistance, R_{ic}, measured from one input lead to ground under CM excitation, and a difference-mode input resistance, R_{id}, measured under DM excitation from either input tc ground. By Ohm's law, the DM current into the noninverting input node is just:

$$I_d = 2V_{id}/R_1 + V_{id}/R_{ic} \quad 2.18$$

from which we can write

$$I_d/V_{id} = 1/R_{id} = 2/R_1 + 1/R_{ic} \quad 2.19$$

Solving for the shunting resistance in Equation 2.19, we get

$$R_1 = 2R_{id}R_{ic}/(R_{ic} - R_{id}) \quad 2.20$$

In many differential amplifiers, $R_{ic} > R_{id}$. If $R_{ic} = R_{id}$, then from Equation 2.20, $R_1 = \infty$.

Let us assume that $R_{ic} = R_{id}$. Thus R_1 may be eliminated from Figure 2.2, which illustrates two Thevenin sources driving the DA through unequal source resistances, R_s and $R_s + \Delta R$. Using superposition and the definitions in Equations 2.1 and 2.2, it is possible to show that a purely CM excitation, V_{sc}, produces an unwanted difference-mode component at the DA's input terminals

$$V_{id}/V_{sc} = R_{ic}\Delta R/2(R_{ic} + R_s)^2 \quad 2.21$$

Also, V_{sc} produces a large CM component; the ΔR term is numerically negligible.

$$V_{ic}/V_{sc} = R_{ic}/(R_{ic} + R_s) \quad 2.22$$

For purely DM excitation in V_s, we can also show that

$$V_{id}/V_{sd} = R_{ic}/(R_{ic} + R_s) \quad 2.23$$

and

$$V_{ic}/V_{sd} = R_{ic}\Delta R/2(R_{ic} + R_s)^2 \quad 2.24$$

In order to find the CMRR of the circuit of Figure 2.2, we will use Equation 2.10 for V_o, and the definition for CMRR, relation 2.11. Thus

$$V_o = A_D V_{sc} R_{ic} \Delta R / 2(R_{ic} + R_s)^2 + A_C V_{sc} R_{ic} / (R_{ic} + R_s) \quad 2.25$$

and

$$V_O = A_D V_{sd} R_{ic} / (R_{ic} + R_s) + A_C V_{sd} R_{ic} \Delta R / 2(R_{ic} + R_s)^2 \quad 2.26$$

After some algebra, we find that the circuit's CMRR, CMRRsys, is given by:

$$\text{CMRRsys} = [A_D + A_C \Delta R / 2(R_{ic} + R_s)] / [A_D \Delta R / 2(R_{ic} + R_s) + A_C] \quad 2.27$$

Equation 2.27 may be reduced to the hyperbolic relation:

$$\text{CMRRsys} = [(A_D + A_C) + \Delta R / 2(R_{ic} + R_s)] / [(A_D / A_C) \Delta R / 2(R_{ic} + R_s) + 1] \quad 2.28$$

which can be approximated by Equation 2.29:

$$\text{CMRRsys} = \text{CMRR}_A / [\text{CMRR}_A \Delta R / 2(R_{ic} + R_s) + 1] \quad 2.29$$

in which the manufacturer-specified CMRR = $\text{CMRR}_A = A_D/A_C$, and $\text{CMRR}_A >> \Delta R/2(R_{ic} + R_s)$.

A plot of CMRRsys vs. $\Delta R/Rs$ is shown in Figure 2.3. Note that when the Thevenin source resistances are matched, CMRRsys = CMRR_A. Also, when

$$\Delta R / R_s = -2(R_{ic} / R_s + 1) / \text{CMRR}_A \quad 2.30$$

CMRRsys $\to \infty$. This implies that a judicious addition of an external resistance in series with one input lead or the other to introduce a ΔR may be used to increase the effective CMRR of the system. For example, if R_{ic} = 100 MΩ, R_s = 10 kΩ, and CMRR_A = 100 dB, then $\Delta R/Rs = -0.2$ to give ∞ CMRRsys. Since it is generally not possible to reduce R_s', it is easier to externally add a similar δR to R_s.

Again, we stress that an amplifier's CMRR_A is a decreasing function of frequency because of the frequency dependence of the gains, A_D and A_C. Also, the AC equivalent input circuit of a DA contains capacitances in parallel with R_1, R_{ic}, and R_{ic}', and the source impedances often contain a reactive, frequency-dependent component. Thus in practice, in a given range of frequencies, CMRRsys can often be maximized by the ΔR method, but seldom can be drastically increased because of reactive imbalances in the input circuit.

2.2 OPERATIONAL AMPLIFIERS

In this section, we review the systems properties of integrated circuit operational amplifiers of various types, and show the many ways engineers use them for signal conditioning in the design of instrumentation systems.

A typical op amp is a differential amplifier with a single output. It has a very high DC gain, K_{VO}, and high CMRR. Because op amps are generally used with large amounts of negative feedback, the open-loop transfer function of most op amps is designed to be of the form

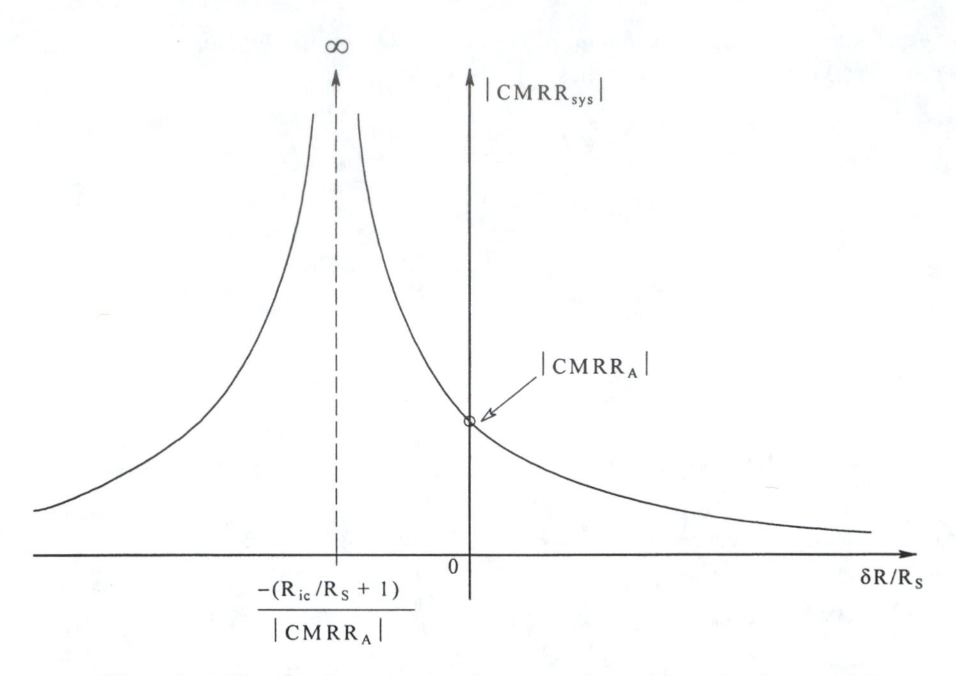

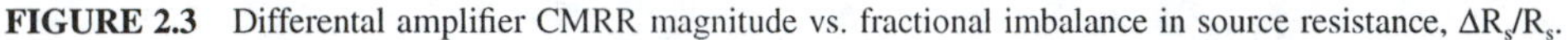

FIGURE 2.3 Differental amplifier CMRR magnitude vs. fractional imbalance in source resistance, $\Delta R_s/R_s$.

$$A_D = V_o/(V_i - V_i') = K_{VO}/[(\tau_1 s + 1)(\tau_2 s + 1)] \qquad 2.31$$

The second break frequency, $f_2 = 1/2\pi\tau_2$, generally is made to occur so that $|\mathbf{A_D}(jf_2)| < 1$. This ensures closed-loop stability over a wide range of feedback conditions. The lower break frequency, $f_1 = 1/2\pi\tau_1$, occurs at a relatively low value. However, a critical parameter governing the closed-loop, high-frequency response of an op amp circuit is the small-signal gain*bandwidth product, which for most op amps can be approximated by

$$\text{GBWP} \cong K_{VO}/2\pi\tau_1 \qquad 2.32$$

The symbol, f_T, is the unity gain or 0 dB frequency of the open-loop op amp. f_T is approximately equal to the gain-bandwidth product (GBWP). The larger the GBWP, the higher the –3 dB cutoff frequency can be made in a broad-band amplifier made with an op amp.

Another manufacturer-specified parameter which determines the large-signal dynamic behavior of an op amp is its slew rate, η. η is defined as the largest possible attainable rate of change of the output voltage, and its units are generally given in volts/microsecond. Op amp slew rates range from less than 1 v/μs in older models such as the venerable LM741, to 3000 V/μs for the Comlinear CLC300A which is intended to condition video frequency signals. As with the case of GBWP or f_T, the cost of op amps increases with increasing slew rate. If an op amp signal conditioner circuit is driven by a high frequency *and* high amplitude sinusoidal source such that $|dV_o/dt|$ exceeds the amplifier's slew rate, then the output will appear to be a triangle wave with rounded (sinusoidal) peaks. When it occurs, slew rate limiting generates harmonic distortion in V_o. A prudent designer will always check to see that the maximum expected dV_o/dt in an op amp circuit does not exceed η.

While the parameters K_{VO}, f_T, η, f_1, and f_2 govern an op amp's dynamic behavior, manufacturers also give specifications on an op amp's output resistance (R_o), CM and DM input resistances, DC bias current (I_B), DC offset voltage (V_{OS}), temperature coefficients for I_B and V_{OS}, short-circuit voltage noise root power spectrum ($e_{na}(f)$), and equivalent input noise current root power spectrum ($i_{na}(f)$), to mention the most important ones.

2.2.1 Types of Op Amps

Op amps may be categorized by characteristics and applications. As you will see in Chapter 3, low-noise op amps are best used in the headstages of analog signal conditioning systems, because their noise characteristics determine the over-all noise performance of the system. When great DC stability and freedom from thermally caused DC drift are required, then designers often specify *chopper-stabilized* op amps, providing other criteria for dynamic response and noise are met. *Electrometer* op amps are used when input bias currents of 10^{-13} A or less and input resistances of greater than 10^{13} Ω are required, other conditions being met. *Fast* op amps are defined here as those having slew rates in excess of 25 V/μs, and small-signal GBWPs of 75 MHz or more. Most op amps are designed to operate with ±15 or ±12 V DC supplies, and can source about ±10 mA to a load. Because of the rapid growth of battery-operated (portable) telecommunications equipment, camcorders, and notebook computers, engineers now have available a number of new op amp designs that operate on ±5 V, or lower, and which consume little power.

We define *power* and *high voltage* op amps as those which can source in excess of 10 mA to a load, or which can operate with high voltage supplies of greater than ±15 V (typically ±40 to ±150 V). Some power op amps can deliver as much as 30 A at 68 V (Apex PA03). In instrumentation systems, power op amps can be used to drive strip-chart recorder pen motors, or other electromagnetic transducers. It is possible for an op amp to fall in two or more of the categories described above; for example, the Apex PA84 has a GBWP of 75 Mhz, a slew rate of 200 V/μs, a peak output voltage of 143 V, and a peak output current of 40 mA. Thus it falls in the high voltage and high power categories, as well as the fast category.

A final type of op amp which is widely encountered but, unfortunately, is not commercially available, is the *ideal* op amp. The ideal op amp assumption is used in pencil-and-paper design calculations. The ideal op amp has the following properties: infinite differential gain, CMRR, input resistance, slew rate, and GBWP; zero noise, bias currents, offset voltage, and output resistance. Probably the most important of these assumptions is that $A_D = \infty$. In order to realize a finite output voltage under this condition, $(V_i - V_i') = 0$, or V_i must equal V_i'. In summary, an ideal op amp is a differential voltage-controlled voltage source (VCVS) with infinite gain and frequency response.

The designer of an analog signal conditioning system for an instrumentation system faces a bewildering array of op amp types and specifications. In the following sections we try to give some rationale for choosing one op amp over another. In making such choices, we are generally balancing meeting all minimum system specifications with dollar cost.

2.2.2 Basic Broad-Band Amplifier Design Using Op Amps

In many instances, the output signal of a sensor is too small to effectively record, observe, or store. The signal is assumed to be a voltage or current. Thus the designer has the task of amplifying the weak signal, while introducing as little noise or distortion as possible. Often impedance levels are a concern. For example, the transducer output may be represented by a Thevenin equivalent circuit in which the Thevenin (series) impedance may not be negligible compared to a signal conditioning amplifier's input impedance. Thus the design problem to be resolved includes the selection of an op-amp type and circuit which will minimize loading of the sensor's output.

Another problem in designing high-gain, broad-band signal conditioners with op amps is the gain*bandwidth trade-off. With simple op amp amplifiers, the GBWP of the system is approximately equal to the op amp's GBWP. This implies that to realize a signal conditioning gain of 1000 with a bandwidth of 100 kHz using low-noise op amps with f_T = 10 MHz, the strategy would be to cascade two amplifier circuits with their gains set to 31.62 (overall gain of 1000). The bandwidth (–3 dB point) of each stage is just 107/31.62 = 316.2 kHz. It can be shown that when N identical stages with transfer function

$$A_V(s) = K_{VO}/(\tau s + 1) \qquad 2.33$$

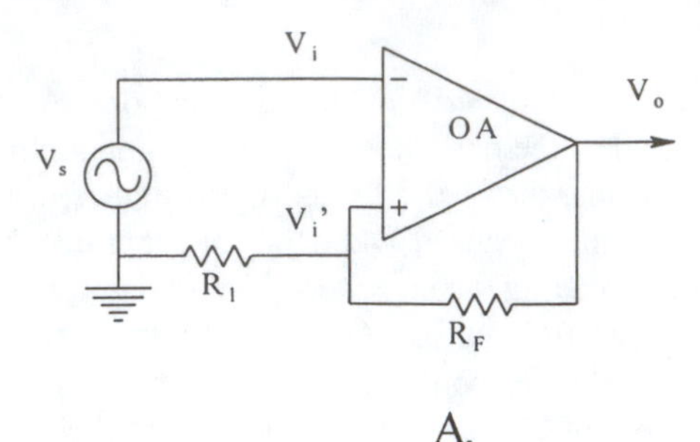

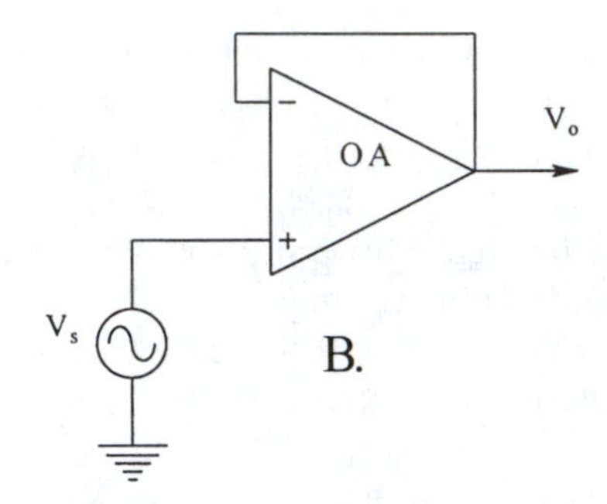

FIGURE 2.4 (A) Non-inverting differential amplifier circuit, with gain set by R_1 and R_F. (B) Unity gain buffer op amp circuit.

are cascaded, then the –3 dB frequency of the cascade, f_c, is given by relation 2.34.

$$f_c = (1/2\pi\tau)\left(2^{1/N} - 1\right)^{1/2} \qquad 2.34$$

Thus in the example above, f_c = 203.5 kHz, and the frequency response design criterion is well met. If a single op amp circuit were used with gain of 1000, then f_c would obviously be 10^4 Hz.

2.2.2.1 Non-Inverting Amplifier

Figure 2.4A illustrates the common, single-ended, non-inverting amplifier configuration. Because of the op amp's high K_{VO}, $V_i \cong V_i'$ and the input impedance seen by the source circuit is R_{ic}, the op amp's CM input resistance (cf. Figure 2.2). In a modern, FET input op amp, such as the Texas Instruments TL071, R_{ic} can be on the order of 10^{12} ohms. Thus, in this case, unless the Thevenin source resistance approaches 10^8 ohms, loading effects and offset due to DC bias current (about 30 pA) will be negligible.

To simplify calculations, we will treat R_{ic} at the V_i' node as infinite, and consider the op amp's output resistance to be negligible compared to R_F. The amplifier's closed-loop transfer function can be found by noting that

$$V_o = \left(V_s - V_i'\right)K_{VO}/(\tau s + 1) \qquad 2.35$$

and

$$V_i' = V_o R_1/\left(R_1 + R_F\right) = \beta V_o \qquad 2.36$$

If Equation 2.36 is substituted into Equation 2.35, we can solve for V_o/V_s:

$$V_o/V_s = \frac{K_{VO}/(\tau s+1)}{1+\beta K_{VO}/(\tau s+1)} \quad 2.37$$

which reduces to

$$V_o/V_s = \frac{1/\beta}{s(\tau/\beta K_{VO})+1} \quad 2.38$$

if we assume that βK_{VO} is >> 1, which is generally valid.

The gain*bandwidth product for this amplifier is found from Equation 2.37 to be

$$GBWP = \left[K_{VO}/(1+\beta K_{VO})\right]\left[1+\beta K_{VO}/2\pi\tau\right] = K_{VO}/2\pi\tau \text{ Hz} \quad 2.39$$

which is the same as for the op amp by itself.

If R_1 is made infinite, and R_F is made a short circuit, the non-inverting follower circuit of Figure 2.4B becomes a unity gain voltage follower. Unity gain followers are often used as output buffers to achieve impedance isolation.

2.2.2.2 The Inverting Amplifier and Summer

Figure 2.5 illustrates the common inverting amplifier configuration. Here we have assumed two inputs, V_{s1} and V_{s2}. The amplifier output can be found by superposition. The output voltage can be written:

$$V_o = (0 - V_i')K_{VO}/(\tau s+1) \quad 2.40$$

V_i' is found by solving the node equation:

$$(V_i' - V_{s1})G_1 + (V_i' - V_{s2})G_1 + (V_i' - V_o)G_F = 0 \quad 2.41$$

When V_i' is substituted into Equation 2.40, we may solve for V_o/V_{s1}, letting $V_{s2} = 0$:

$$V_o = \frac{-V_{s1}(G_1 K_{VO}/\Sigma G)/(1+K_{VO}G_F/\Sigma G)}{s\left[\tau/(1+K_{VO}G_F/\Sigma G)\right]+1} \quad 2.42$$

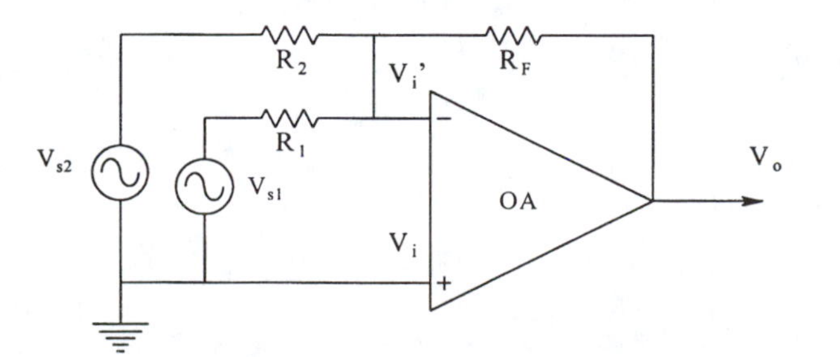

FIGURE 2.5 Summing inverter op amp configuration.

where $\Sigma G \equiv G_1 + G_2 + G_F$. When we assume that $K_{VO}G_F/\Sigma G >> 1$, we can write:

$$V_o/V_{s1} = \frac{-(G_1/G_F)}{s(\tau\,\Sigma G/K_{VO}G_F)+1} \quad 2.43$$

In general, the gain for the kth inverting input is, by superposition:

$$V_o/V_{sk} = \frac{-(G_k/G_F)}{s(\tau\,\Sigma G/K_{VO}G_F)+1} \quad 2.44$$

The GBWP for the kth inverting input is easily seen to be:

$$GBWP_k = (K_{VO}/2\pi\tau)(G_k/\Sigma G) \text{ Hz} \quad 2.45$$

Note that in the inverting amplifier case, the amplifier GBWP, given by Equation 2.45, is not equal to the op amp's GBWP, except for very large G_k.

2.2.3 Current Feedback Op Amps

Current feedback op amps (CFOAs) constitute a relatively new and growing class of op amp that offers certain design advantages in terms of independent control of gain and bandwidth, and very high slew rates. The internal systems architecture of a CFOA connected as a simple non-inverting amplifier is shown in Figure 2.6. A nearly ideal, unity gain VCVS replaces the high impedance input circuit of a conventional op amp. Thus the non-inverting (V_i) terminal has a very high input impedance, but the V_i' node is the output of the VCVS and therefore presents a very low input (Thevenin) impedance, in the order of 50 ohms. The output current of the VCVS, In, is the input for a CCVS, Ωs), which determines the CFOS output voltage, V_o. The CCVS is designed so that:

$$\Omega(s) = \Omega_o/(\tau s+1) \text{ Ohms} \quad 2.46$$

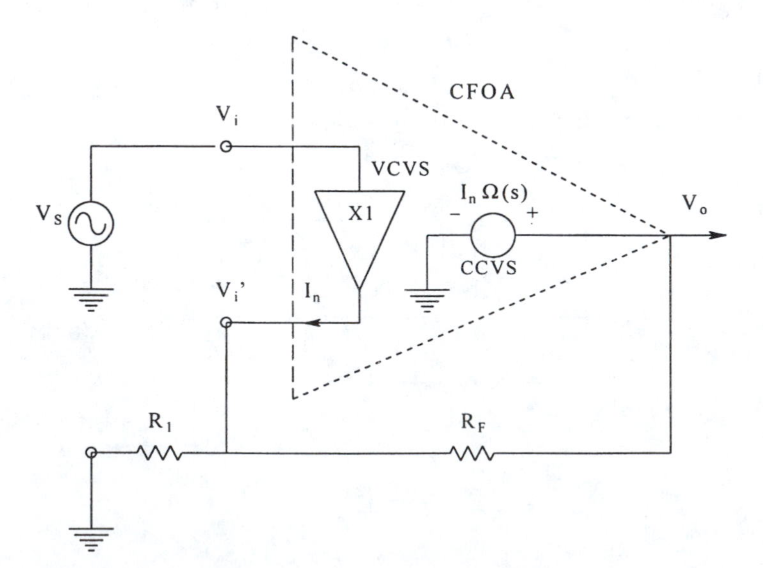

FIGURE 2.6 Current feedback op amp non-inverting amplifier circuit.

Hence

$$V_o = I_n[\Omega_o/(\tau s + 1)] \text{ Volts} \tag{2.47}$$

The control current, I_n, may be found from Kirchoff's current law applied to the V_i' node (here we neglect the CCVS output resistance).

$$I_n = V_s G_1 + (V_s - V_o)G_F \tag{2.48}$$

If Equation 2.48 for In is substituted into Equation 2.47, we may solve for the transfer function for the non-inverting CFOA amplifier.

$$V_o/V_s = \frac{(G_1 + G_F)\Omega_o/(1 + G_F\Omega_o)}{s[\tau/(1 + G_F\Omega_o)] + 1} \tag{2.49}$$

If we assume that $G_F\Omega_o >> 1$, then the transfer function reduces to

$$V_o/V_s = \frac{1 + R_F/R_1}{s(\tau R_F/\Omega_o) + 1} \tag{2.50}$$

The DC gain of the circuit, A_{VO}, is just

$$A_{VO} = (R_1 + R_F)/R_1 \tag{2.51}$$

The GBWP of the circuit is

$$\text{GBWP} = \frac{\Omega_o(R_F + R_1)}{2\pi\tau R_F R_1} = \Omega_o A_{VO}/2\pi\tau R_F \text{ Hz} \tag{2.52}$$

The –3 dB frequency of the circuit is given by Equation 2.53:

$$f_b = (\Omega_o + R_F)/2\pi\tau R_F \approx \Omega_o/2\pi\tau R_F \text{ Hz} \tag{2.53}$$

Note that the break frequency, f_b, is dependent on R_F alone. Thus the DC gain can be set with R_1, holding the break frequency constant with R_F. Note that with a conventional op amp non-inverting amplifier, the break frequency varies inversely with the DC gain of the circuit.

Although CFOAs offer advantages in terms of high slew rate and f_T, and the absence of the customary trade-off between gain and bandwidth found with conventional op amps, they are not as flexible in terms of stability, and the ability to synthesize transfer functions. First, let us examine the transfer function of a CFOA used in the CFOA inverting configuration, shown in Figure 2.7. Here the positive voltage input (V_i) node is tied to ground, causing the output of the unity gain buffer to be zero. Neglecting R_o, $V_i' = 0$, and I_n is given simply by:

$$I_n = -(V_s G_1 + V_o G_F) \tag{2.54}$$

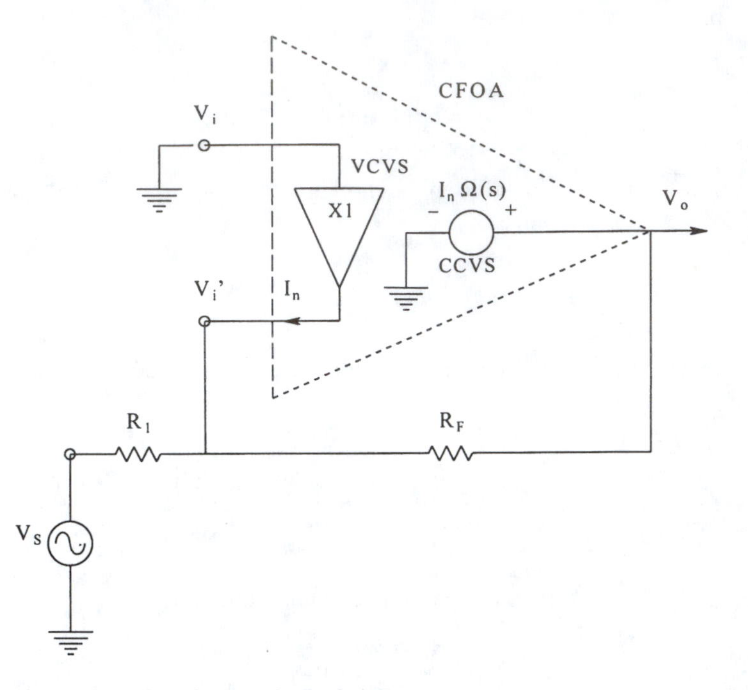

FIGURE 2.7 Inverting configuration of current feedback op amp.

If this expression for I_n is substituted into Equation 2.47 for V_o, we can solve for the inverting amplifier's transfer function, assuming that $\Omega_0 G_F >> 1$.

$$V_o/V_s = \frac{-R_F/R_1}{s(\tau R_F/\Omega_o)+1} \qquad 2.55$$

Note that for the inverter, the DC gain is simply $-R_F/R_1$, the break frequency, f_b, is $\Omega_o/2\pi\tau R_F$ Hz, and the GBWP is $\Omega_o/2\pi\tau R_1$ Hz. In using a conventional op amp, it is a simple extension of the gain relation to replace R_1 and/or R_F with capacitors or series or parallel R-C circuits; stability is generally preserved. In the case of CFOAs, however, the designer must avoid applying direct capacitive feedback between the V_o and V_i' nodes; some manufacturers provide an internal RF to avoid this problem.

To examine what happens when a capacitor is connected in place of R_F in the inverting CFOA amplifier in Figure 2.7, we note that the current I_n can be written:

$$I_n = -(V_s G_1 + V_o sC) \qquad 2.56$$

When Equation 2.56 for I_n is substituted into Equation 2.47 for V_o, we can derive the transfer function 2.57:

$$V_o/V_s = \frac{-\Omega_0 G_1}{s(\tau + C\Omega_o)+1} \qquad 2.57$$

If we set R_1 = 1 Megohm, C = 1 μF, and $\Omega_o = 10^8$ ohms, the transfer function 2.57 becomes:

$$V_o/V_s = -100/(s100+1) \qquad 2.58$$

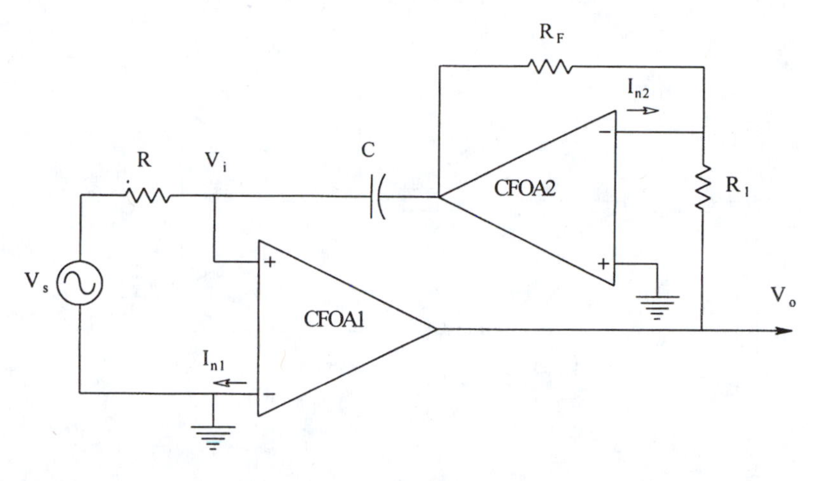

FIGURE 2.8 Non-inverting integrator using two current feedback op amps.

This is certainly not an effective integrator; it is best characterized as an inverting low-pass filter with a break frequency of 10^{-2} r/s.

An integrator which can be made from CFOAs is shown in Figure 2.8. This is a noninverting integrator. If we let the CFOA's $\tau s \rightarrow 0$, the integrator's gain may be shown to be approximately

$$V_o/V_s \cong \frac{\Omega_{o1}/R_{o1}}{sRC(R_F\Omega_{o1}/R_{o1}R_1)+1} \quad 2.59$$

Note that a Thevenin output resistance, R_{o1}, must be assumed for CFOA1's unity gain buffer to permit analysis in this case. At frequencies above the break frequency, f_b, the integrator's gain becomes:

$$V_o/V_s = R_1/sRCR_F \quad 2.60$$

The break frequency for this integrator is quite low, about 5×10^{-7} r/s for $\Omega_o = 10^8$ ohms, C = 1 μF, R = 1 Megohm, $R_F = R_1 = 10^3$ ohms and $R_{o1} = 50$ ohms.

As a means to synthesize simple active filters using CFOAs, consider the general circuit shown in Figure 2.9. Here we have replaced the integrator's R and C in Figure 2.8 with admittances Y_1 and Y_2, respectively. Using the same assumptions we made above in the analysis of the integrator, we find

$$V_o/V_s = \frac{Y_1\Omega_1 G_{o1}}{Y_1 + Y_2(1+\Omega_1\Omega_2 G_{o1} G_A/G_F\Omega_2)} \quad 2.61$$

If we let $\Omega_1 = \Omega_2$, $R_F > R_A$, and assume $\Omega_1 G_{o1} >> 1$, the gain expression Equation 2.61 reduces to:

$$V_o/V_s \cong Z_2 R_A/Z_1 R_F \quad 2.62$$

(A more detailed analysis would include the real poles of the CFOAs in Equation 2.61.) Note that using this circuit architecture, it is possible to synthesize a simple, two-pole, one-zero band-pass filter if we let $Z_1 = R_1 + 1/sC_1$ and $Y_2 = G_2 + sC_2$.

In summary, we see that in simple inverting and non-inverting configurations, CFOAs offer high frequency response and slew rates. However, care must be exercised in using them in circuits having

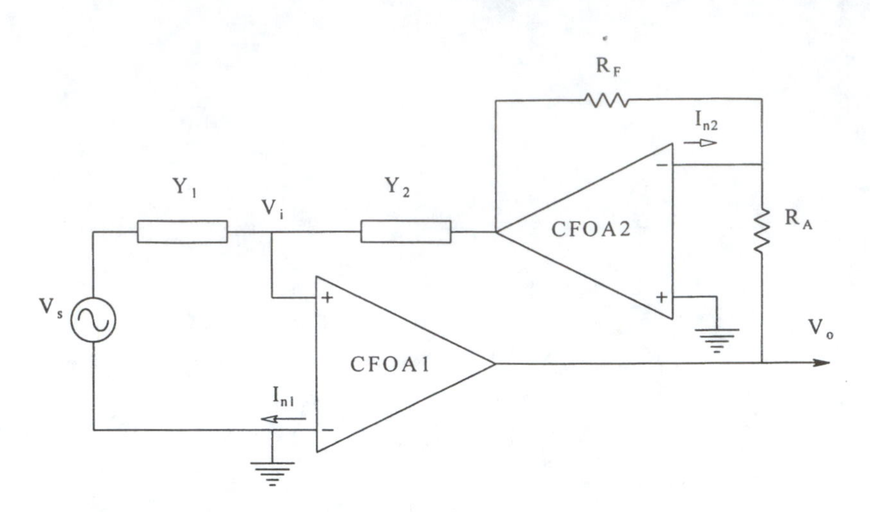

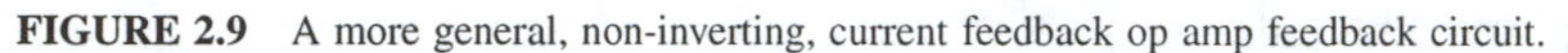
FIGURE 2.9 A more general, non-inverting, current feedback op amp feedback circuit.

reactive elements. The basic inverting gain relationship of conventional op amps, $V_o/V_s = -Z_F/Z_{IN}$, does not apply to CFOAs. Special circuit architecture must be used in many cases to realize even simple transfer functions, such as integration or band-pass filtering.

2.3 ANALOG ACTIVE FILTER APPLICATIONS USING CONVENTIONAL OP AMPS

2.3.1 Introduction

In the previous sections of this chapter we have stressed the use of op amps to make single-ended, broad-band amplifiers. There are many instances in instrumentation systems where it is also necessary to filter a time-varying signal by passing it through a linear, frequency-dependent gain device. Such filtering is often used to improve signal-to-noise ratio by attenuating those frequencies containing noise but which do not contain significant signal information. Coherent interference may also be reduced or eliminated by filtering. Finally, before analog-to-digital conversion (periodic sampling), analog signals must be low-pass filtered to prevent errors in subsequent digital processing caused by aliasing. (Sampling and aliasing are discussed in detail in Chapter 9.) Low-pass filtering is also used to smooth or average the outputs of nonlinear circuits, such as rectifiers, phase sensitive rectifiers, true RMS converters, etc.

Analog active filters (AFs) fall into several broad categories dependent on the shape of the filter pass band. Figure 2.10 illustrates the major categories we will consider in the following sections. These include high-pass, broad band-pass, narrow band-pass (high Q), low-pass, notch (band-reject), and all-pass (used for phase compensation, the transfer function magnitude remains constant).

Our treatment of analog AFs is by no means complete; it is provided as a review of AF circuits which find common use for analog signal conditioning in instrumentation systems. We will not go into the design details of high-order AFs in this text (pole and zero placement). The reader should appreciate that the analysis and design of analog AFs is an extensive subject and books and chapters of books have been written on the subject. Many references on analog active filter design are included in the bibliography.

2.3.2 Analog Active Filter Architectures

The reader should be aware that there are many active filter circuit architectures which can be used to realize a given transfer function. Some circuits offer the advantage of few components or minimum sensitivity to op amp gain changes, while others offer design flexibility in terms of being

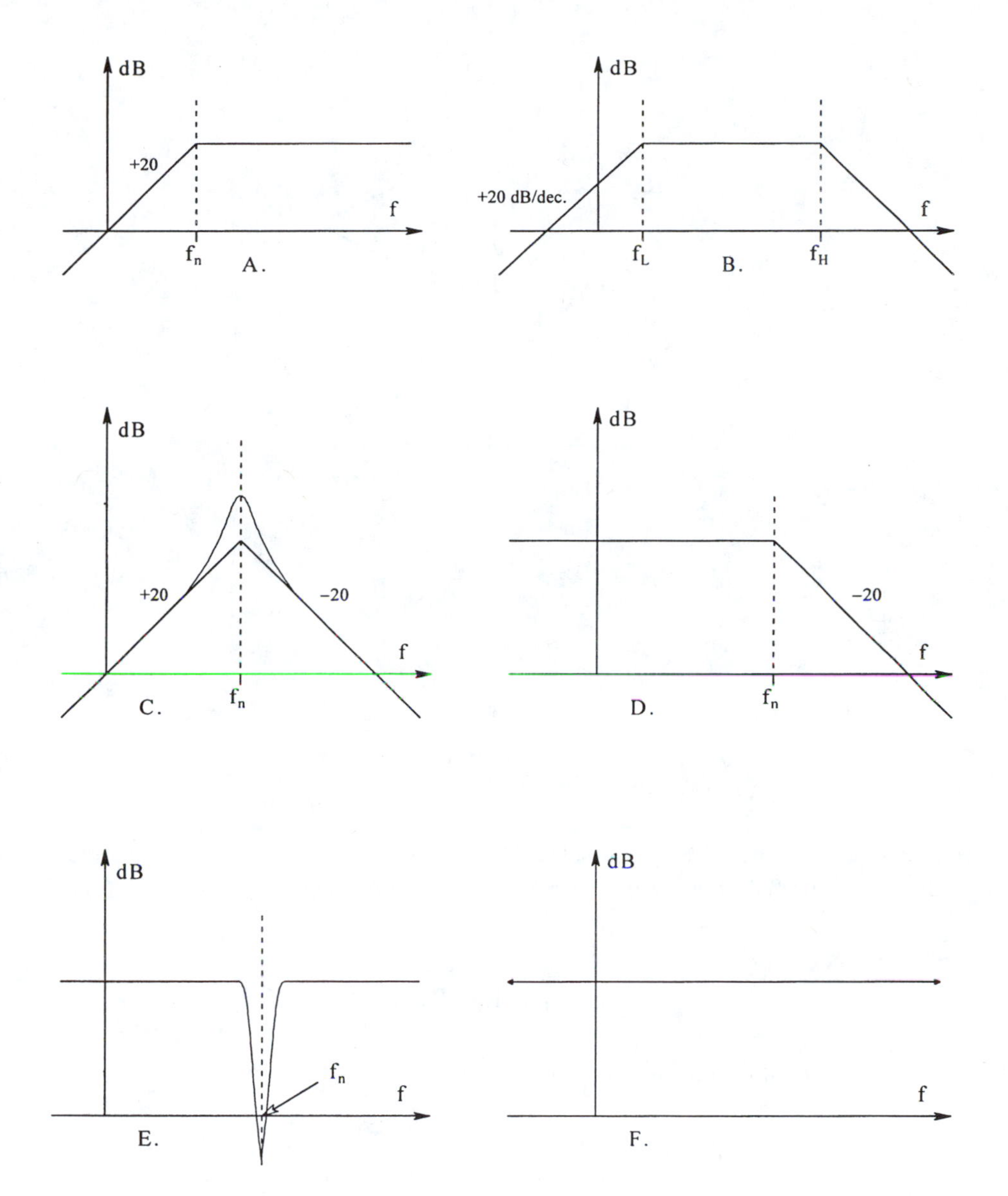

FIGURE 2.10 Bode asymptotes of common types of filters: (A) high-pass, (B) broad band-pass, (C) narrow (tuned) band-pass, (D) low-pass, (E) band reject or notch, (F) all-pass.

able to set one or more of the filter parameters (mid-band gain, break frequency, damping factor or Q) by adjusting one unique R or C for each parameter. In this section, we stress certain active filter architectures which allow easy design. To expedite analysis, we assume all of the op amps are of conventional design, and are ideal.

2.3.2.1 Controlled Source Active Filters

This type of filter can be used to realize band-pass, low-pass, and high-pass quadratic transfer functions. A single op-amp is used as a low or unity-gain VCVS with four or five resistive or capacitive feedback elements. Figure 2.11 illustrates the well-known *Sallen and Key low-pass filter.* To derive the transfer function of this filter, we observe that the op amp behaves as a VCVS with gain $K_v = 1 + R_F/R_A$. Thus:

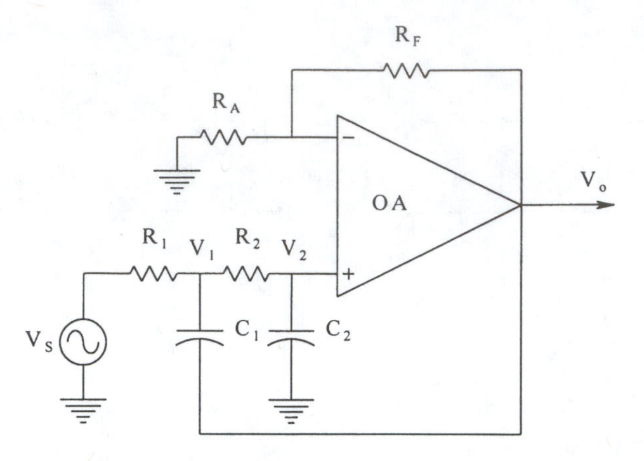

FIGURE 2.11 Sallen and Key quadratic low-pass filter.

$$V_2 = \frac{V_1(1/sC_2)}{R_2 + 1/sC_2} = \frac{V_1}{1 + sR_2C_2} \qquad 2.63$$

We write a node equation on the V_1 node:

$$V_1\left(G_1 + \frac{1}{R_2 + 1/sC_2} + sC_1\right) - sC_1V_o = V_sG_1 \qquad 2.64$$

Equation 2.63 can be solved for V_1 and substituted into Equation 2.64, allowing us to solve for V_o in terms of V_s. The Sallen and Key low-pass filter transfer function can thus be written:

$$V_o/V_s = \frac{K_v}{s^2C_1C_2R_1R_2 + s\left[C_2(R_1 + R_2) + R_1C_1(1 - K_v)\right] + 1} \qquad 2.65$$

If K_v is made unity, then the transfer function becomes

$$V_o/V_s = \frac{1}{s^2C_1C_2R_1R_2 + sC_2(R_1 + R_2) + 1} \qquad 2.66$$

The transfer function denominator has the "standard" quadratic form:

$$s^2/\omega_n^2 + s2\zeta/\omega_n + 1$$

The undamped natural frequency, ω_n, for the Sallen and Key low-pass filter is:

$$\omega_n = \frac{1}{C_1C_2R_1R_2} \text{ r/s} \qquad 2.67$$

The damping factor, ζ, is

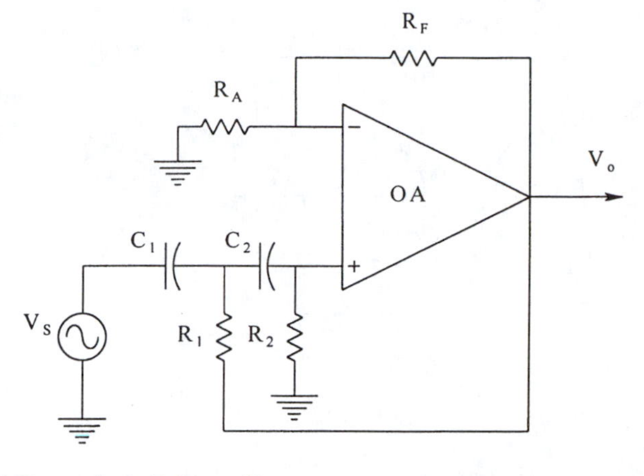

FIGURE 2.12 Sallen and Key quadratic high-pass filter.

$$\zeta = \frac{C_2(R_1 + R_2)}{2C_1C_2R_1R_2} \tag{2.68}$$

If R_1 is made equal to R_2, then the damping factor reduces to the simple expression

$$\zeta = \sqrt{\frac{C_2}{C_1}} \tag{2.69}$$

Conveniently, the damping factor of the low-pass filter is set by the square root of the ratio of the capacitors. Simultaneous adjustment of $R_1 = R_2$ with a ganged variable resistor can set ω_n independently from the damping factor.

Figure 2.12 illustrates a *Sallen and Key high-pass filter*. Analysis of this circuit follows that for the low-pass filter above. If K_v is set to unity, and $C_1 = C_2 = C$, then it is easy to show that the transfer function is:

$$V_o/V_s = \frac{s^2C_1C_2R_1R_2}{s^2C_1C_2R_1R_2 + sR_1 2C + 1} \tag{2.70}$$

The undamped natural frequency, ω_n, of the high-pass filter is also given by Equation 2.67. The damping factor is given by:

$$\zeta = \frac{R_1 2C}{2}\sqrt{\frac{1}{C^2R_1R_2}} = \sqrt{\frac{R_1}{R_2}} \tag{2.71}$$

It is also possible to make a *controlled-source band-pass filter*, as shown in Figure 2.13. Although a general analysis is interesting, it is more useful to realize this circuit with $K_v = 1$ and $C_1 = C_2 = C$. The transfer function of the band-pass filter is:

$$V_o/V_s = \frac{s[CG_1/(G_2G_1 + G_2G_3)]}{s^2[C^2/(G_2G_1 + G_2G_3)] + s[C(G_2 + 2G_1 + G_3)/(G_2G_1 + G_2G_3)] + 1} \tag{2.72a}$$

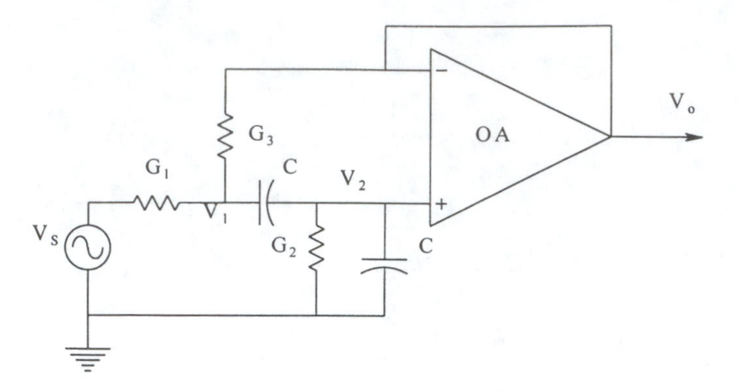

FIGURE 2.13 Controlled source quadratic bandpass filter.

Here

$$\omega_n = \frac{\sqrt{G_2(G_1 + G_3)}}{C} \text{ r/s} \tag{2.72b}$$

The gain at resonance is:

$$A_{VPK} = \frac{G_1}{G_2 + 2G_1 + G_3} \tag{2.73}$$

And the Q of the tuned circuit is:

$$Q = \frac{\sqrt{G_2(G_1 + G_3)}}{G_2 + 2G_1 + G_3} \tag{2.74}$$

Note that the peak response, A_v, and Q are independent of C; $C_1 = C_2 = C$ can be used to set ω_n at constant Q and A_v.

2.3.2.2 Biquad Active Filters

Another active filter achitecture with which it is extremely easy to meet design criteria is the biquad AF. There are several versions of the biquad, but Figure 2.14 shows the most versatile, the two-loop biquad. This circuit can act as a high-pass, sharply tuned band-pass or low-pass filter, depending on which output is selected. It also allows independent adjustment of the filter's peak gain, break frequency, and damping factor or Q. All-pass and band-reject (notch) filters can also be synthesized from the basic biquad filter with the addition of another op amp.

First, we will examine the low-pass filter transfer function of the biquad AF in Figure 2.14. Note that the subunits of this filter are two inverting summers and two integrators. The low-pass filter transfer function can be found from an application of Mason's rule:

$$V_5/V_s = \frac{(-R/R_1)(-R/R_2)(-1/sCR_3)(-1/CR_3)}{1 + [(-1)(-R/R_2)(-1/CR_3) + (-1)(-1/sCR_3)(-1/sCR_3)]} \tag{2.75}$$

This transfer function can be simplified to standard quadratic (time-constant) form:

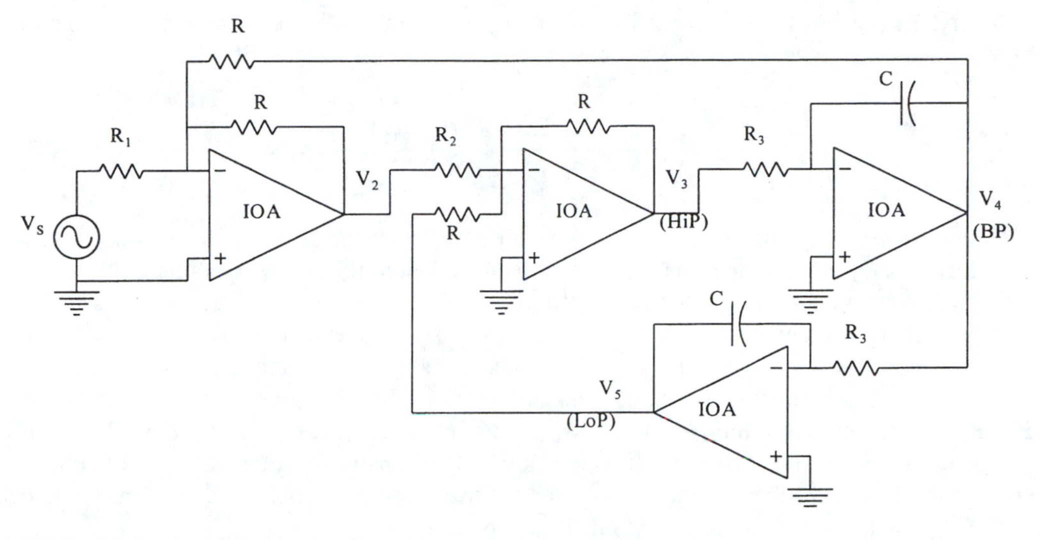

FIGURE 2.14 Two-loop, biquad active filter.

$$V_o/V_s = \frac{R^2/(R_1R_2)}{s^2R_3^2C^2 + sR_3CR/R_2 + 1} \quad 2.76$$

Inspection of Equation 2.76 shows that the low frequency gain of this low-pass filter is

$$A_{VLOW} = R^2/(R_1R_2) \quad 2.77$$

The natural frequency is

$$\omega_n = 1/CR_3 \text{ radians/second} \quad 2.78$$

and the damping factor is

$$\zeta = R/2R_2 \quad 2.79$$

The same biquad filter can be used as a tuned band-pass filter. In this case, the output is taken from the V_4 node. The transfer function again can be found by application of Mason's rule:

$$V_4/V_s = \frac{-sR_3CR^2/(R_1R_2)}{s^2R_3^2C^2 + sR_3CR/R_2 + 1} \quad 2.80$$

The band-pass filter has a peak response at the natural frequency given by Equation 2.78. The peak gain is

$$A_{vPK} = -R/R_1 \quad 2.81$$

The band-pass filter's Q, a measure of the sharpness of its tuning, is given by

$$Q = R_2/R \quad 2.82$$

Finally, the biquad AF of Figure 2.14 can produce a quadratic high-pass characteristic. Here the output is taken from the V_3 node. The transfer function can be shown to be:

$$V_3/V_s = \frac{s^2R_3^2C^2R^2/(R_1R_2)}{s^2R_3^2C^2 + sR_3CR/R_2 + 1} \qquad 2.83$$

The high-pass filter's natural frequency and damping factor are the same as for the low-pass filter. Its high frequency gain is the same as the low-frequency gain of the low-pass filter.

It is also possible through the use of a fourth op amp inverting summer to generate a notch filter which will sharply attenuate a selected frequency and pass all other frequencies. Such notch filters often find application in eliminating unwanted coherent interference, such as power line frequency hum. The basic operation performed by a notch filter is illustrated in Figure 2.15. Simply stated, the output of a band-pass filter is subtracted from its input. The peak gain of the band-pass filter is made unity. As a result of this operation, the notch filter transfer function has a pair of conjugate zeros on the jω axis in the S-plane at $\pm j\omega_n$, and the usual complex-conjugate, quadratic pole pair at $s = \pm j\omega_n$. Its transfer function is given by Equation 2.84.

$$V_6/V_s = \frac{s^2/\omega_n^2 + 1}{s^2/\omega_n^2 + s(1/Q\omega_n) + 1} \qquad 2.84$$

A biquad realization of a notch filter is shown in Figure 2.16. It is left as an exercise to derive its transfer function (note that R_1 should equal R to obtain the form of Equation 2.84).

As a final example of the versatility of the biquad AF architecture, we examine the all-pass filter design shown in Figure 2.17. All-pass filters have a flat frequency magnitude response; only the phase varies with frequency. All-pass filters are used to modify the phase response of electronic signal conditioning systems without introducing frequency-dependent gain changes.Here the band-

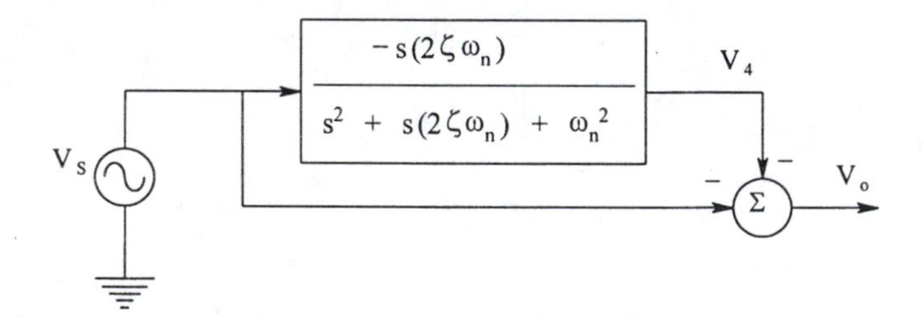

FIGURE 2.15 Block diagram for biquad notch filter. V_4 is the inverting band-pass output of the biquad filter of Figure 2.14.

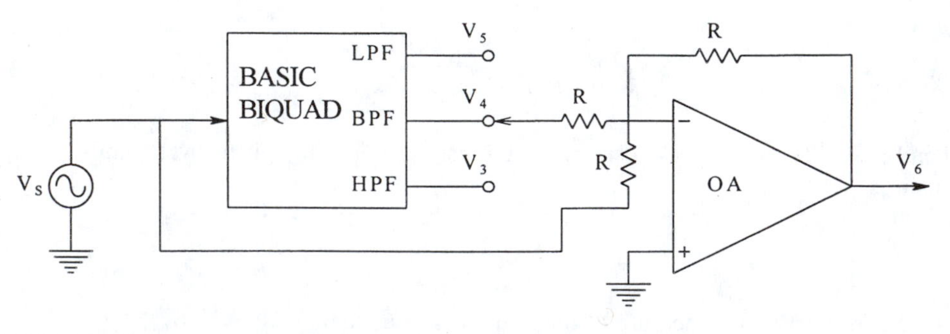

FIGURE 2.16 Realization of biquad notch filter of Figure 2.15.

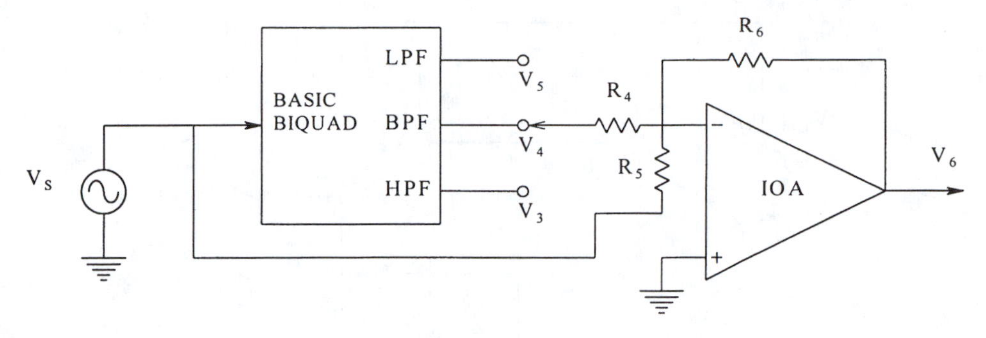

FIGURE 2.17 Realization of biquad all-pass filter. See text for special relations for R_4, R_5, and R_6.

pass output, V_4, of the biquad of Figure 2.14 is summed with the filter input, V_s. The all-pass filter output, V_6, is easily seen to be

$$V_6 = -R_6\left(V_s/R_5 + V_4/R_4\right) \quad 2.85$$

When the relation for V_4/V_s, Equation 2.80, is substituted into Equation 2.85, we can show that

$$V_6/V_s = \frac{\left(R_5/R_6\right)\left(s^2R_3^2C^2 + sR_3CR/R_2 + 1\right) - \left(R_6/R_4\right)sR_3CR^2/R_1R_2}{s^2R_3^2C^2 + sR_3CR/R_2 + 1} \quad 2.86$$

This relation may be reduced to the standard all-pass filter format given by Equation 2.87 if we let $R_1 = R$, $R_5 = 2R_4$ and $R_6 = R_5$:

$$V_6/V_s = -\frac{s^2R_3^2C^2 - sR_3CR/R_2 + 1}{s^2R_3^2C^2 + sR_3CR/R_2 + 1} \quad 2.87$$

In summary, note that the frequency response magnitude function for the all-pass filter is unity over all frequencies; only the phase varies. When $s = j\omega$ in Equation 2.87, the phase of the frequency response function is given by:

$$\phi = -180° - 2\tan^{-1}\left(\frac{1 - \omega^2R_3^2C^2}{\omega CR_3R/R_2}\right) \quad 2.88$$

2.3.2.3 Generalized Impedance Converter Active Filters

Generalized impedance converter (GIC) circuits, using two op amps and five two-terminal elements (either Rs or Cs), can be used to synthesize impedances to ground of the form, Z(s) = sX, or Z(s) = $1/s^2D$. These impedances in turn can be combined with other Rs, Cs, and op amps to generate various quadratic transfer functions, such as discussed for the biquad AF architecture above. Figure 2.18 illustrates the basic GIC architecture. Analysis of the GIC driving point impedance is made easier if we assume that the op amps are ideal. Under this condition, we can write that the input impedance of the GIC to ground is:

$$Z_{11} = V_1/I_1 = V_1/\left[\left(V_1 - V_2\right)/Z_1\right] \quad 2.89$$

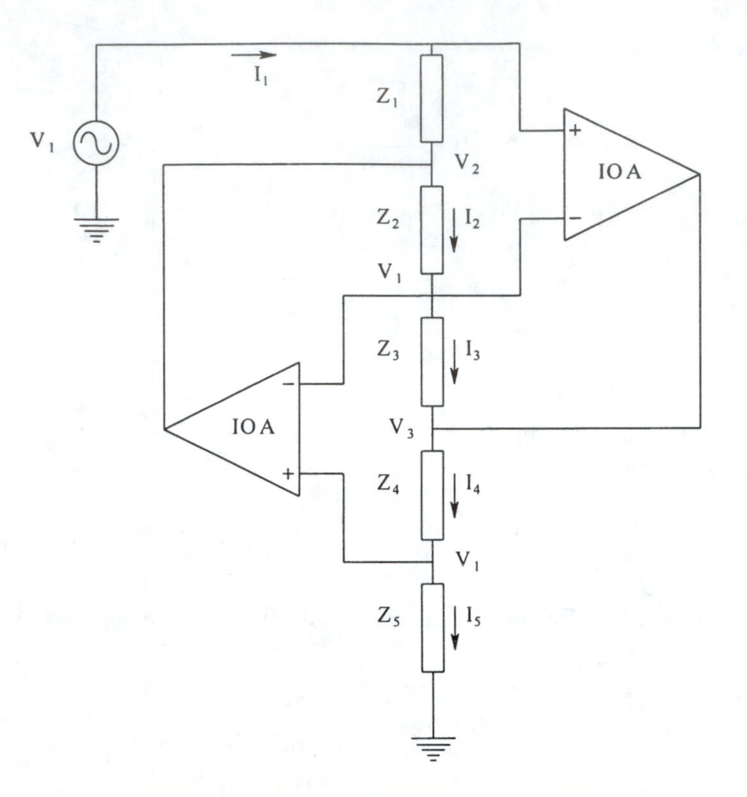

FIGURE 2.18 Basic generalized impedance converter (GIC) architecture.

Also, from the ideal op amp assumption and Ohm's law we have:

$$I_2 = (V_2 - V_1)/Z_2 = I_3 = (V_1 - V_3)/Z_3 \qquad 2.90$$

$$I_4 = (V_3 - V_1)/Z_4 = I_5 = V_1/I_5 \qquad 2.91$$

From the three equations above we can show

$$V_3 = V_1(1 + Z_4/Z_5), \text{ and} \qquad 2.92$$

$$V_2 = V_1\left(\frac{Z_3Z_5 - Z_2Z_4}{Z_3Z_5}\right) \qquad 2.93$$

Finally, the GIC driving point impedance can be shown to be:

$$\mathbf{Z_{11} = V_1/I_1 = Z_1Z_3Z_5/Z_2Z_4} \qquad 2.94$$

If we let $\mathbf{Z_2} = 1/j\omega C_2$ (a capacitor), and the other Zs be resistors, then $\mathbf{Z_{11}}$ is in the form of a high-Q inductive reactance

$$\mathbf{Z_{11}} = j\omega[C_2R_1R_3R_5/R_4] \qquad 2.95$$

where the equivalent inductance is given by

$$L_{eq} = \left[C_2 R_1 R_3 R_5 / R_4\right] \text{ Henrys} \tag{2.96}$$

If both Z_1 and Z_5 are made from capacitors, then the GIC input impedance becomes a *frequency-dependent negative resistance*, or FDNR, element:

$$\mathbf{Z}_{11} = -R_3/s^2 C_1 C_5 R_2 R_4 = -1/s^2 D \tag{2.97}$$

Thus the GIC "D element" is

$$D = C_1 C_5 R_2 R_4 / R_3 \tag{2.98}$$

A number of AF designs based on the inductive and "D" Z_{11} forms have been devised. Band-pass, all-pass, notch, low-pass, and high-pass designs are possible. Examples of these designs may be found in Chapter 4 of the text by Franco (1988). We illustrate a two-pole band-pass filter in Figure 2.19, and an FDNR based, two-pole low-pass filter in Figure 2.20.

The transfer function for the band-pass filter can be shown to be:

$$V_o/V_s = \frac{sL_{eq}/R}{s^2 CL_{eq} + sL_{eq}/R + 1} \tag{2.99}$$

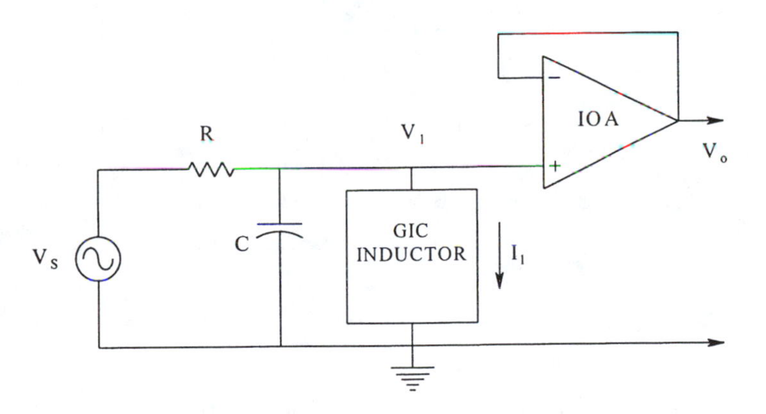

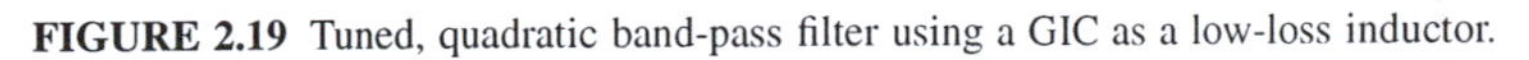

FIGURE 2.19 Tuned, quadratic band-pass filter using a GIC as a low-loss inductor.

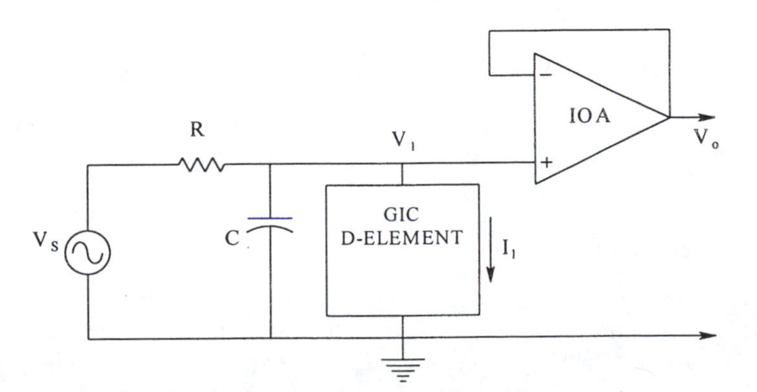

FIGURE 2.20 Quadratic low-pass filter realized with a GIC frequency-dependent negative resistance element.

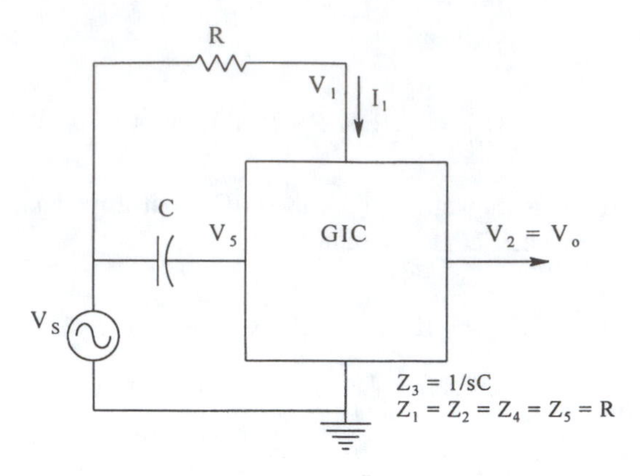

FIGURE 2.21 GIC-derived all-pass filter.

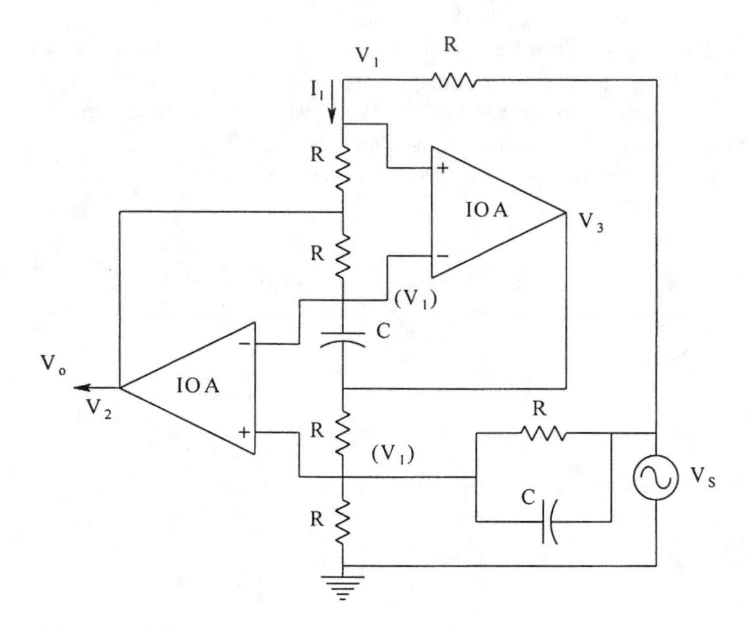

FIGURE 2.22 GIC-derived notch filter.

where L_{eq} is given by Equation 2.96.

The low-pass filter transfer function is given by Equation 2.100.

$$V_o/V_s = \frac{1}{s^2RD + sRC + 1} \quad 2.100$$

where D is given by Equation 2.98 above. Figure 2.21 shows a GIC-derived all-pass filter and Figure 2.22 a GIC-derived notch filter. Derivations of their transfer functions are left as exercises for the reader.

2.3.3 High-Order Active Filters

Often the simple two-pole (quadratic) active filter designs which we have described above do not meet the signal conditioning requirements of an instrumentation system. In these cases, quadratic fiters may be cascaded to create high-order filters having four or more poles. Such high-order filters are described according to how they perform in the time and frequency domains. Below we list the more commonly used types of high-order active filters.

Butterworth filters: The frequency response magnitude for this type of high-order AF has no ripple in the pass-band. Its phase vs. frequency response is nonlinear, so that the signal propagation time through the filter is not constant over frequency. There is considerable output overshoot in the time domain response of Butterworth filters of order greater than n = 4 to transient inputs. The slope of the stop-band attenuation is –20 n dB/decade.

Chebychev filters: In the frequency domain, the Chebychev filter has no ripple in its stopband, but has a specified pass-band ripple. Generally, a Chebychev low-pass filter having the same attenuation slope in the stop-band as a comparable Butterworth low-pass filter requires a lower order (fewer poles).

Elliptic or Cauer filters: In order to attain a higher attenuation (for a given order) in the transition band of frequencies (between pass-band and stop-band), the elliptic filter allows ripple in the frequency response magnitude in both the pass- and stop-bands. High-order elliptic filters have oscillatory transient responses.

Bessel or Thompson filters: In order to achieve a transient response which is free from overshoot and ringing, Bessel filters trade off steep attenuation slope in the transition band for linear phase response. For any order, n, Bessel filters make good low-pass filters where faithful time domain signal reproduction is required.

2.3.4 Operational Amplifier Integrators and Differentiators

Often in conditioning analog measurement records, we wish to either integrate or differentiate these signals. Simple *analog integration* of an input signal can present two practical problems: one is that if the signal contains a DC or average level, the integrator will eventually saturate. That is, the output voltage of the op amp is bounded at levels less than its power supply voltage. The second problem with a practical analog integrator is that even with its input grounded, it will integrate its own DC bias current and offset voltage. Using Figure 2.23, and assuming an otherwise ideal op amp, it is easy to show that the integrator's output due to its bias current and offset voltage is given by:

$$v_o(t) = -t\,I_B/C_F + V_{OS} + t\,V_{OS}/RC_F \qquad 2.101$$

Here we assume zero initial conditions until t = 0. Clearly, $v_o(t)$ will linearly approach the + or – saturation voltage of the op amp. Use of electrometer or autozero op amps can minimize analog integrator drift; however, as we will see in Chapter 10, digital integration can eliminate most of the bias current and drift problems.

Analog differentiation also presents practical problems. In *op amp differentiators,* bias current and drift present little problem. However, a real op amp's finite bandwidth can give high frequency transient problems. Refer to Figure 2.24. At the summing junction, we can write the node equation:

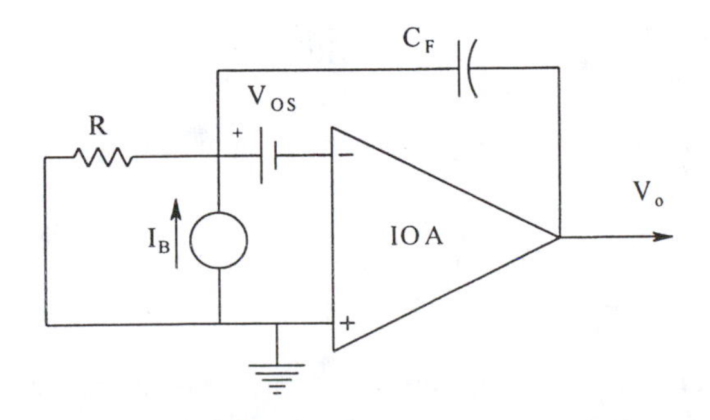

FIGURE 2.23 Op amp integrator circuit showing sources of DC drift error. The signs and magnitudes of I_B and V_{OS} depend on the op amp design and temperature.

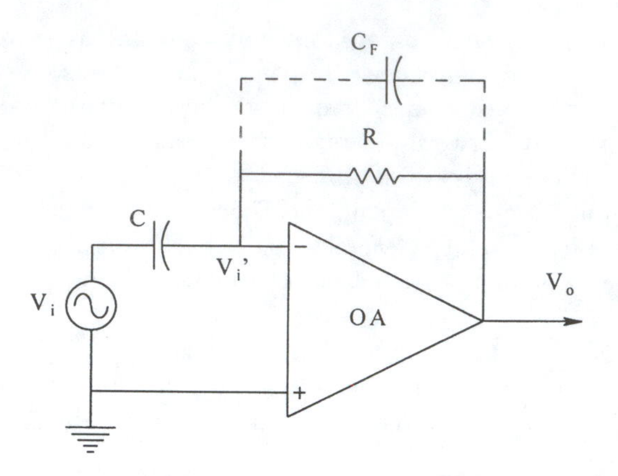

FIGURE 2.24 Op amp differentiator. C_F is used to damp the high-frequency response.

$$V_i'(G + sC) - V_oG = V_isC \tag{2.102}$$

The op amp output is given by:

$$\frac{V_o}{V_i'} = \frac{-K_V}{\tau_A s + 1} \tag{2.103}$$

If we combine Equations 2.101 and 2.102, we can show:

$$\frac{V_o}{V_i} = \frac{-[K_V/(K_V + 1)]sRC}{s^2\tau_A RC/(K_V + 1) + s(\tau_A + RC)/(K_V + 1) + 1} \tag{2.104}$$

At low input frequencies,

$$\frac{V_o}{V_i} \cong -sRC \tag{2.105}$$

At high input frequencies, the quadratic denominator of the transfer function, Equation 2.104, is significant. The break frequency of the transfer function is:

$$\omega_n = \sqrt{\frac{K_V + 1}{\tau_A RC}} \text{ r/s} \tag{2.106}$$

The damping factor of the analog differentiator can be shown to be:

$$\zeta = \frac{(\tau_A + RC)}{2(K_V + 1)}\sqrt{\frac{K_V + 1}{\tau_A RC}} \tag{2.107}$$

To see where the problem arises, let us substitute typical numerical values for the circuit: R = 1 MΩ, C = 1 μF, $\tau_A = 0.001$ s, $K_V = 10^6$. Now we calculate $\omega_n = 3.16 \times 10^4$ r/s, and the damping factor is $\zeta = 1.597 \times 10^{-2}$, which is quite low. Thus any transients in V_i will excite a poorly damped, sinusoidal transient at the output of the differentiator. This behavior can present a problem if the differentiator input is not bandwidth limited. If the differentiator input is not bandwidth limited, we must limit the bandwidth of the circuit by placing a small capacitor in parallel with R. It is left as an exercise to compute the transfer function of the practical differentiator with a parallel RC feedback circuit.

2.3.5 Summary

In the preceding sections we have reviewed the architectures of three commonly used, quadratic, active filter "building blocks" used in the design of analog signal conditioning systems. There are many other AF architectures which filter designers also use. However, lack of space prohibits their description here. The interested reader should consult the bibliography.

2.4 INSTRUMENTATION AMPLIFIERS

As defined by current practice, instrumentation amplifiers (IAs) are generally described as direct-coupled, low-noise, differential amplifiers (DAs) having very high input impedances, high common-mode rejection ratio, user settable gain from 0 to 60 dB, and bandwidth from DC to the hundreds of kHz, depending on design and operating gain. Many manufacturers offer IAs as dual in-line package (DIP) integrated circuits. IAs are often used at the output of a transducer, such as a bridge circuit, as the first stage in an analog signal conditioning system. As headstages, IAs must have low-noise characteristics in noise-critical applications. We illustrate below two IA circuits which can be made from low-noise op amps in order to obtain custom low noise IA designs.

Table 2.1 summarizes the systems characteristics of some commercially available instrumentation amplifiers. Note that IA bandwidths are not generally very broad, generally covering the audio spectrum, with some exceptions. Their slew rates are also modest. IAs are generally characterized by high CMRRs and high input impedances, however.

2.4.1 Instrumentation Amplifiers that Can Be Made from Op Amps

Figure 2.25 illustrates the well-known, three op amp IA circuit. The differential gain of this circuit can be set from 1 to 1000 by adjusting the value of R_1. Note that op amp 3 is set up as a DA using resistors R_3, R_3', R_4, and R_4'. Assuming OA3 is ideal, it is easy to show that its output is given by

$$V_o = (V_4 - V_3)(R_4/R_3) \tag{2.108}$$

This also assumes that the resistors R_4 and R_4' and R_3 and R_3' are perfectly matched. The DA circuit of OA3, although having an apparent high CMRR and differential gain, is unsuitable as an instrumentation amplifier because of its unequal and relatively low input resistances. V_3 sees an input impedance of R_3 to the summing junction of OA3, which is at virtual ground, while V_4 sees an input impedance of $R_3 + R_4$. Op amps 1 and 2 form a symmetrical circuit which can be analyzed for the cases of pure common-mode (CM) input signals and pure difference-mode input signals. (Note that any combination of V_s and V_s' can be broken down into CM and DM input components to OA1 and OA2, and the responses to these components can be summed at the output by superposition.)

For purely CM inputs, $V_s = V_s'$, and symmetry considerations lead to the conclusions that the summing junction potentials of OA1 and OA2 are equal, hence there is no current flowing in R_1. (Here we assume that OA1 and OA2 are ideal op amps, and that $R_2 = R_2'$.) Because no current flows in R_1 under CM input excitation, we may remove it from the circuit (i.e., replace it with an open circuit). Thus, OA1 and OA2 are seen to be unity-gain followers under CM excitation. Thus, their outputs V_3 and V_4 are seen to be equal and equal to the CM input, V_{sc}. The following DA circuit of OA3 subtracts V_3 from V_4, producing, ideally, a $V_o = 0$ for CM inputs. Because the op

TABLE 2.1 Specification Summary of Certain Instrumentation Amplifiers

IA Mfgr/ Model	CMRR,dB /@ gain	$R_{in}\|\|C_{in}$	Small-signal BW(kHz)/ @ gain	Slew rate V/μs	e_{na} @ 1 kHz/ @ gain nV/√Hz	i_{na} pA/√Hz	V_{OS} Tempco	I_{BIAS}	I_{BIAS} Tempco
PMI/ AMP-01	100/1 120/10 130/100 130/10^3	20 GΩ CM 1 GΩ, DM	570/1 100/10 82/100 26/10^3	4.5 @ G = 10	540/1 59/10 10/100 5/10^3	0.15 @ G = 10^3	0.3 μV/°C	3 nA	40 pA/°C
PMI/ AMP-02	90/1 110/10 120/100 120/10^3	10 GΩ DM 16.5 GΩ CM	1200/1 300/10 200/100 200/10^3	6 @ G = 10	120/1 18/10 10/100 9/10^3	0.4 pA @ G = 10^3	1 μV/°C	2 nA	150 pA/°C
PMI/ AMP-05	98/1 110/10 115/≥100	10^{12} || 8 pF (CM & DM)	3000/1 120/≥10	7.5 @ G = 10	350/1 38/10 16/≥100	10 fA/√Hz	5 μV/°C	20 pA	—
BB/ INA 102-CG	90/1 100/≥10	10^{10} || 2 pF (CM & DM)	30/1 3/10 0.3/100 0.03/10^3	0.2 @ G = 100	25/10^3	0.15 pA/√Hz	0.2 ± 5/G μV/°C	6 nA	0.1 nA/°C
BB/ INA 110-BG	100/1 112/10 116/100- 500	5E12 || 6 pF (CM) 2E12 || 1 pF (CM)	2500/1-10 470/100 240/200 100/500	17 @ G = 1-200	10 nV/√Hz @ 10 kHz	1.8 fA/√Hz	1+ 10/G μV/°C	10 pA	—
BB/ INA 116	85/1 90/10 94/100-10^3	>10^{15} || .2 CM >10^{15} || 7 DM	800/1 500/10 70/100 7/1000	0.8 @ G = 10-200	12 nV/√Hz @ 1 kHz	0.1 fA/√Hz @ 1 kHz	±20 μV/°C	±2 nA	3 fA/°C @ 25°C
AD/ AD625-C	90/1 115/10 125/100 140/10^3	1E9 || 4 pF CM & DM	650/1 400/10 150/100 25/10^3	5	4	60 pA PPK, 0.1-10 Hz	0.1 μV/°C	±10 nA	±50 pA/°C
AD/ AD621	110/10 130/100	10^{10} || 2	200/10 800/100	1.2	13 nV/√Hz @ G = 10 9 nV/√Hz @ G = 100	100 fA/√Hz @ G = 10, 100	<1 μV/°C	0.5 nA	3 pA/°C

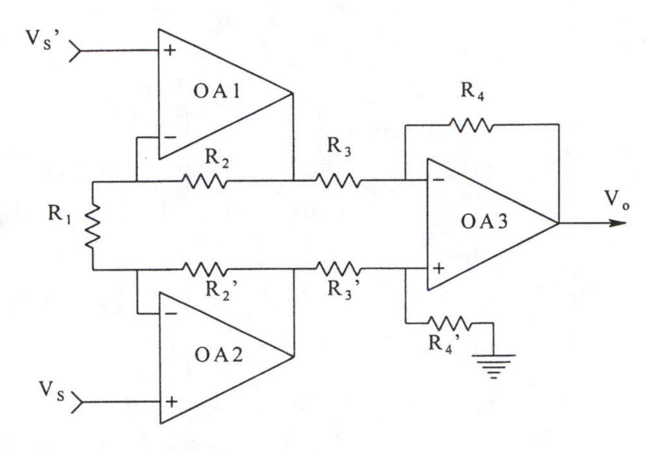

FIGURE 2.25 a three op amp instrumentation amplifier circuit.

amps are not ideal and have finite CMRRs, gains, bandwidths, and input and output impedances, and because resistor matching may not be perfect, there will be a small V_o for a pure CM input.

In the case of a purely DM input, $V_s' = V_s$, and the summing junctions of OA1 and OA2 are at $-V_s$ and V_s, respectively. Hence, a current of $2V_s/R_1$ amps flows in R_1. Furthermore, consideration of Kirchoff's voltage law leads to the conclusion that the center of R_1 (i.e., $R_1/2$ ohms from either end) is at 0 V potential with respect to ground, and thus may be tied to ground without disturbing the circuit. Now the DM OA1 and OA2 circuits are seen to consist of noninverting amplifiers with feedback resistor R_2 and resistance $R_1/2$ to ground from the summing junction. Hence both the DM OA1 and OA2 circuits can easily be shown to have gains

$$V_3/V_s' = V_4/V_s = (1 + 2R_2/R_1) \tag{2.109}$$

Thus, the output for DM inputs is just

$$V_o = (1 + 2R_2/R_1)(R_4/R_3)(V_s - V_s') \tag{2.110}$$

The three op amp IA configuration is used in some commercial IA designs. It provides the designer with some flexibility if special circuit behavior is desired, such as low noise or wide bandwidth. Resistor matching is critical for obtaining a high CMRR. Often a resistor, such as R_4, can be made variable to tweak the CMRR to a maximum.

Figure 2.26 illustrates a two op amp IA design. Again, two matched op amps must be used, and resistors, R, must be closely matched to obtain a high CMRR. This circuit may be analyzed using superposition and the ideal op amp assumption. First we assume that $V_s' > 0$ and $V_s \to 0$. Thus $V_3 = V_s'$ and $V_5 = 0$. Now we write node equations on the V_3 and the V_5 nodes:

$$V_s'(2G + G_F) - V_4 G - V_5 G_F = 0 \tag{2.111}$$

$$V_5(2G + G_F) - V_3 G_F - V_O G - V_4 G = 0 \tag{2.112}$$

Simultaneous solution of Equations 2.111 and 2.112 give us:

$$V_O = -V_s' 2(1 + R/R_F) \tag{2.113}$$

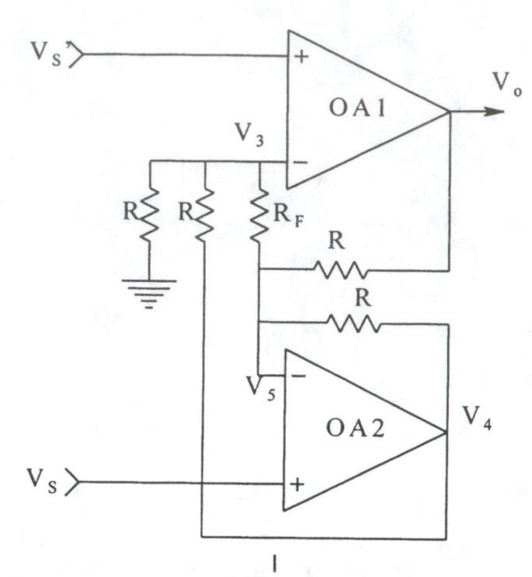

FIGURE 2.26 A two op amp instrumentation amplifier circuit.

In a similar manner, we set $V_s > 0$ and $V_s' \to 0$, and note that $V_5 = V_s$ and $V_3 = 0$. Node equations are written on the V_3 and V_5 nodes:

$$V_3(2G + G_F) - V_4G - V_5G_F = 0 \tag{2.114}$$

$$V_5(2G + G_F) - V_3G_F - V_oG - V_4G = 0 \tag{2.115}$$

Solution of these equations gives us:

$$V_o = V_s2(1 + R/R_F) \tag{2.116}$$

Hence, by superposition, the two op amp IA output can be written:

$$V_o = (V_s - V_s')2(1 + R/R_F) \tag{2.117}$$

Note that for both IA designs above, the inputs are made directly to the noninverting inputs of op amps. Hence the input impedance of these IAs will essentially be that of the op amp. If field effect transistor (FET) input op amps are used, then input resistances in excess of 10^{12} ohms can be attained with very low bias currents.

2.4.2 Isolation Amplifiers

Isolation amplifiers are a special class of instrumentation amplifier which find extensive application in biomedical instrumentation; specifically, the recording of low-level bioelectric signals from human and animal subjects. Their use is generally dictated by electrical safety considerations. Their use is also called for when the input signal is found in the presence of a high common-mode voltage, and also in the situation where ground loop currents can introduce errors in the signals under measurement. Isolation amplifiers have an extremely low conductance between the reference ground of the output signal and the ground and input terminals of the differential headstage. The DC conductive isolation between the headstage ground and the output stage ground is generally accomplished by a magnetic transformer coupling or by opto-coupling techniques.

In the Analog Devices (Norwood, MA) family of isolation amplifiers, the output of the input DA is coupled to the output stage of the isolation amplifier by an isolation transformer. The operating

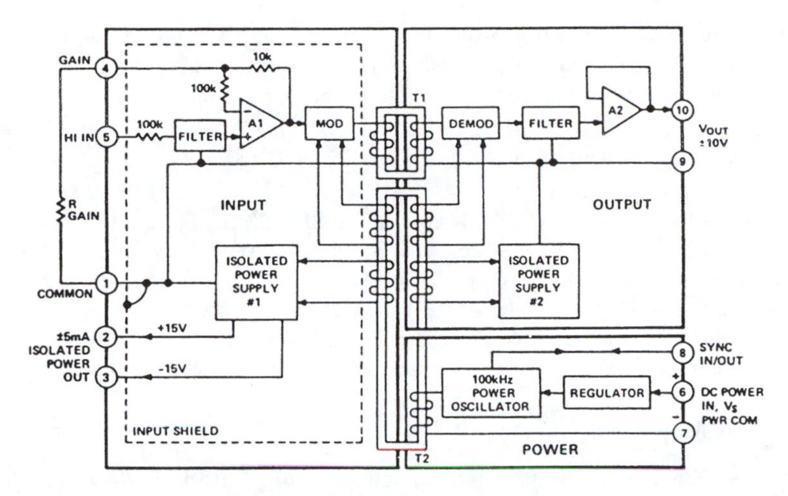

FIGURE 2.27 Block diagram of an Analog Devices AD289 precision, wide bandwidth isolation amplifier. (Figure used with permission of Analog Devices, Norwood, MA.)

power for the input DA is also transformer-coupled from a high-frequency power converter oscillator which derives its power from the non-isolated or normal supply for the isolation amplifier. The block diagram of an Analog Devices AD289 precision hybrid IA is shown in Figure 2.27. Note that in this design, both the input DA and the output circuit have isolated (internal) power supplies.

To illustrate the differences between ordinary instrumentation amplifiers and isolation amplifiers, we review some of the key specifications for the Analog Devices Model AD294A Medical Isolation Amplifier:

The overall gain is the product of G_{IN} and G_{OUT}. $G_{IN} = (1 + 10^5/R_G)$, $R_G \geq 1$ kΩ, $G_{INmax} = 100$, $G_{OUT} = (1 + R_A/R_B)$, $1 < G_{OUT} < 10$. Hence the overall gain may be 1000. The CMRR of the AD294A is 100 dB, not exceptionally high for an IA. The maximum continuous common-mode voltage (CMV) is ±3500 Vpk. The amplifier is rated to stand a ±8000 Vpk, 10 ms pulse at the rate of 1 pulse every 10 seconds. The input impedance is 10^8 Ω || 150 pF. The input (DC) bias current is 2 nA. The small-signal bandwidth is 2.5 kHz (gain 1 – 100 V/V). The slew rate is 9.1 V/μs. The maximum leakage current between headstage and output stage is 2 μArms, given 240 Vrms 60 Hz applied to the input.

The AD 294A is typically used to record electrocardiographic or electromyographic signals using body surface electrodes.

Another medical isolation amplifier suitable not only for ECG recording, but also EEG signal conditioning, is the Intronics Model IA296. This isolation amplifier also uses the two-transformer architecture to couple power into the differential headstage and signals out. Its specifications are summarized below:

> The IA296's gain is fixed at 10. The CMRR is 160 dB with a 5 kΩ source resistance imbalance. The maximum safe DC common-mode input voltage is ±5000 V. The CM input impedance is 10^{11} Ω || 10 pF; the DM input impedance is 3×10^8 Ω || 2.2 pF. The input bias current is ±200 pA; the I_B tempco is ±5 pA/°C. The –3 dB bandwidth is 1 kHz. Input noise with a single-ended input with $R_s = 5$ kΩ is 0.3 μV in a 10 to 1 kHz bandwidth (about 9.5 nV/$\sqrt{\text{Hz}}$); the current noise is about 0.13 pA/$\sqrt{\text{Hz}}$ in a 0.05 Hz to 1 kHz bandwidth. Maximum leakage current due to component failure is 10 μA.

Because of its low noise, the Intronics IA296 is better suited for measurements of low-level biological signals such as fetal ECGs, EEGs, and evoked brain potentials.

Finally, as an example of a photo-optically coupled isolation amplifier, we examine the Burr-Brown Model 3652JG (Tucson, AZ). Because this amplifier does not use a built-in transformer to couple power to the input DA, an external, isolating DC/DC converter must be used, such as the Burr-Brown Model 722. The BB 722 supply uses a 900 kHz oscillator to drive a transformer which is coupled to two independent rectifiers and filters. It can supply ±5 VDC to ±16 VDC at about 16 mA to each of the four outputs. It has isolation test ratings of ±8000 Vpk for 5 s, and ±3500 Vpk continuous between inputs and outputs. The leakage current maximum for this power supply is 1 μA for 240 Vrms at 60 Hz, ensuring medical safety. The Model 3652 isolation amplifier has the following specifications when used with the BB 722 DC/DC power supply:

> Gain settable from 1 to 1000. The CMRR is 80 dB. The BB3652 can block a CM voltage of ±2000 Vdc, and a pulse of ±5000 V for 10 ms. The input resistance is given as 10^{11} ohms, CM and DM. The isolation impedance between input and output is 10^{12} Ω || 1.8 pF. The input bias current is 10 pA. The offset voltage tempco is 25 μV/°C. The –3 dB bandwidth is 15 kHz at all gains. The short-circuit voltage noise is 5 μVrms measured over a 10 Hz to 10 kHz bandwidth. No figure is given for the input current noise. Maximum leakage current is 0.35 μA.

2.4.3 Autozero Amplifiers

Yet another type of IA merits our attention. The commutating autozero (CAZ) amplifier employs a rather unique design to obtain an almost negligible DC input offset voltage drift. The Intersil ICL7605/7606 CAZ amplifiers make use of two novel, switched capacitor networks to convert a true differential input to a single-ended signal which is the actual input to the CAZ amplifier. The second pair of switched capacitors is used in conjunction with two matched op amps to cancel out each op amp's offset voltage. This process is illustrated in Figure 2.28A and B. During the "A" clock cycle, C_2 is forced to charge up to $-V_{os2}$, while C_1 charged to $-V_{os1}$ is switched in series with the input, V_1, and V_{os1} to cancel V_{os1}. In the "B" position, C_1 is charged up to $-V_{os1}$, and C_2 now charged to $-V_{os2}$ is put in series with V_{os2} and V_1 to cancel V_{os2}, etc.

The maximum CMRR is 104 dB, and the amplifier bandwidth is 10 Hz. The offset voltage tempco is an amazing 10 nV/°C. The commutation clock frequency of the Intersil CAZ amplifier can range from 160 to 2560 Hz. Overall gain can be set between 1 and 1000. The ICL7605/7606 CAZ amplifier is intended for use with DC measurement applications such as strain gauge bridges used in electronic scales.

National Semiconductor (Santa Clara, CA) offers the LMC669 autozero module which can be used with any op amp configured as a single-ended inverting summer or as a noninverting amplifier. Figure 2.29A and B illustrates two applications of the LMC669 autozero. Note that the autozero module does not limit the dynamic performance of the op amp circuit to which it is attached; the amplifier retains the same DC gain, small-signal bandwidth, and slew rate. The effective offset voltage drift of any op amp using the autozero is 100 nV/°C. The maximum offset voltage using the autozero is ±5 μV. The autozero bias currents are 5 pA (these must be added to the op amp's IB); its clock frequency can be set from 100 Hz to 100 kHz. The NS LMC669 autozero is very useful to compensate for DC drift in circuits using special-purpose op amps, such as electrometer op amps, which normally have large offset voltages (200 μV) and large V_{os} tempcos (5 μV/°C). The autozero does contribute to an increase in the circuit's equivalent input voltage noise. However, choice of sampling rate and step size, and the use of low-pass filtering in the feedback path, can minimize this effect.

2.4.4 Summary

In this section we have examined some examples of instrumentation amplifiers. IAs are generally defined as differential amplifiers with high input impedances, high CMRR, and programmable gain, usually set by choice of one or two resistors. Their bandwidths generally depend on the gain setting, and roughly follow the constant gain*bandwidth rule. Typical unity-gain bandwidths are in the hundeds of kHz, although some are now available in the 3 MHz range. Special isolation IAs and autozeroing amplifiers are also available.

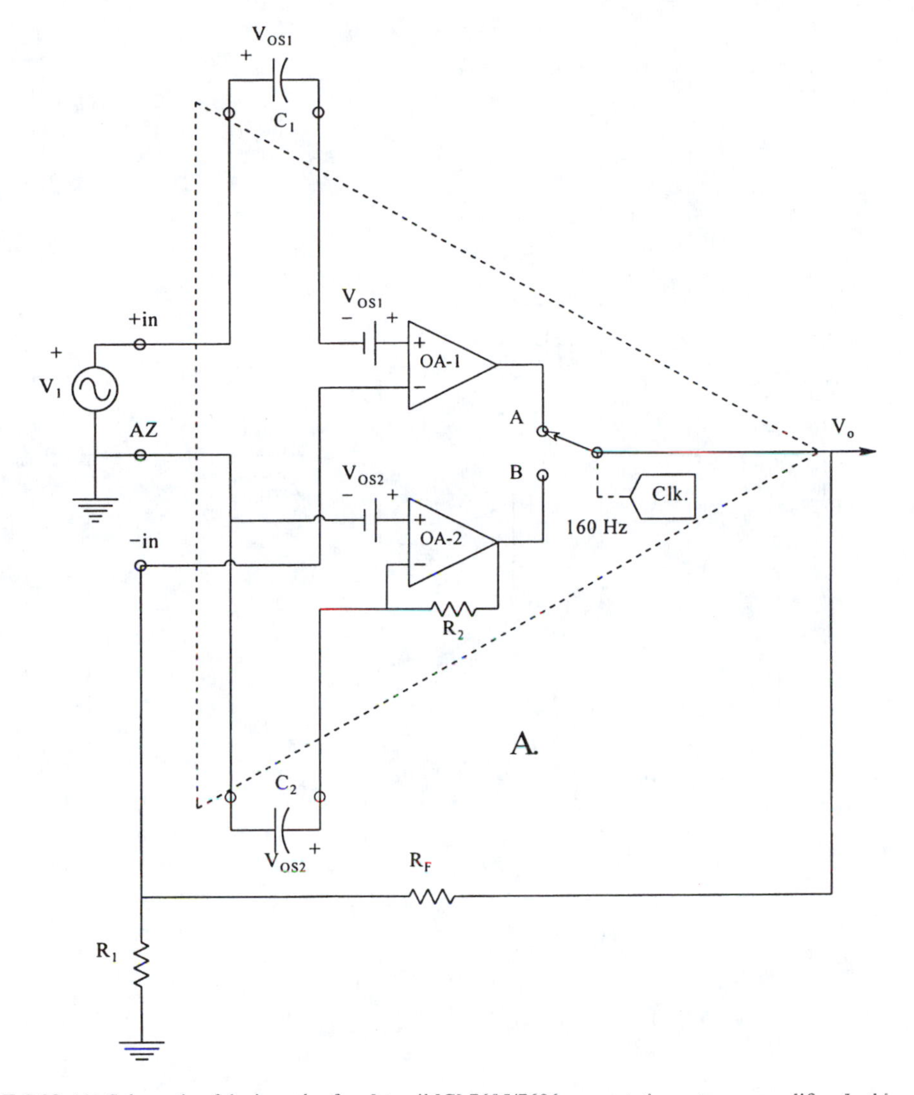

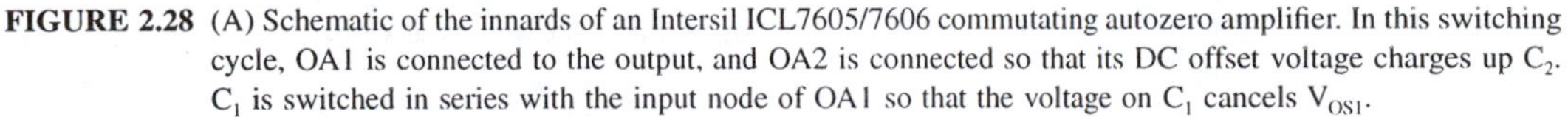

FIGURE 2.28 (A) Schematic of the innards of an Intersil ICL7605/7606 commutating autozero amplifier. In this switching cycle, OA1 is connected to the output, and OA2 is connected so that its DC offset voltage charges up C_2. C_1 is switched in series with the input node of OA1 so that the voltage on C_1 cancels V_{OS1}.

2.5 NONLINEAR ANALOG SIGNAL PROCESSING BY OP AMPS AND BY SPECIAL FUNCTION MODULES

2.5.1 Introduction

Perhaps the best way to define what we mean by nonlinear signal processing is to first describe the properties of a linear system. Linear system (LS) outputs can be found, in general by the operation of *real convolution*. Real convolution can be derived from the general property of linear systems, *superposition:* the operation of convolution can be expressed by the well-known definite integral, sic:

$$y_1(t) = x_1 \otimes h = \int_{-\infty}^{\infty} x_1(\sigma)h(t-\sigma)\,d\sigma \qquad 2.118$$

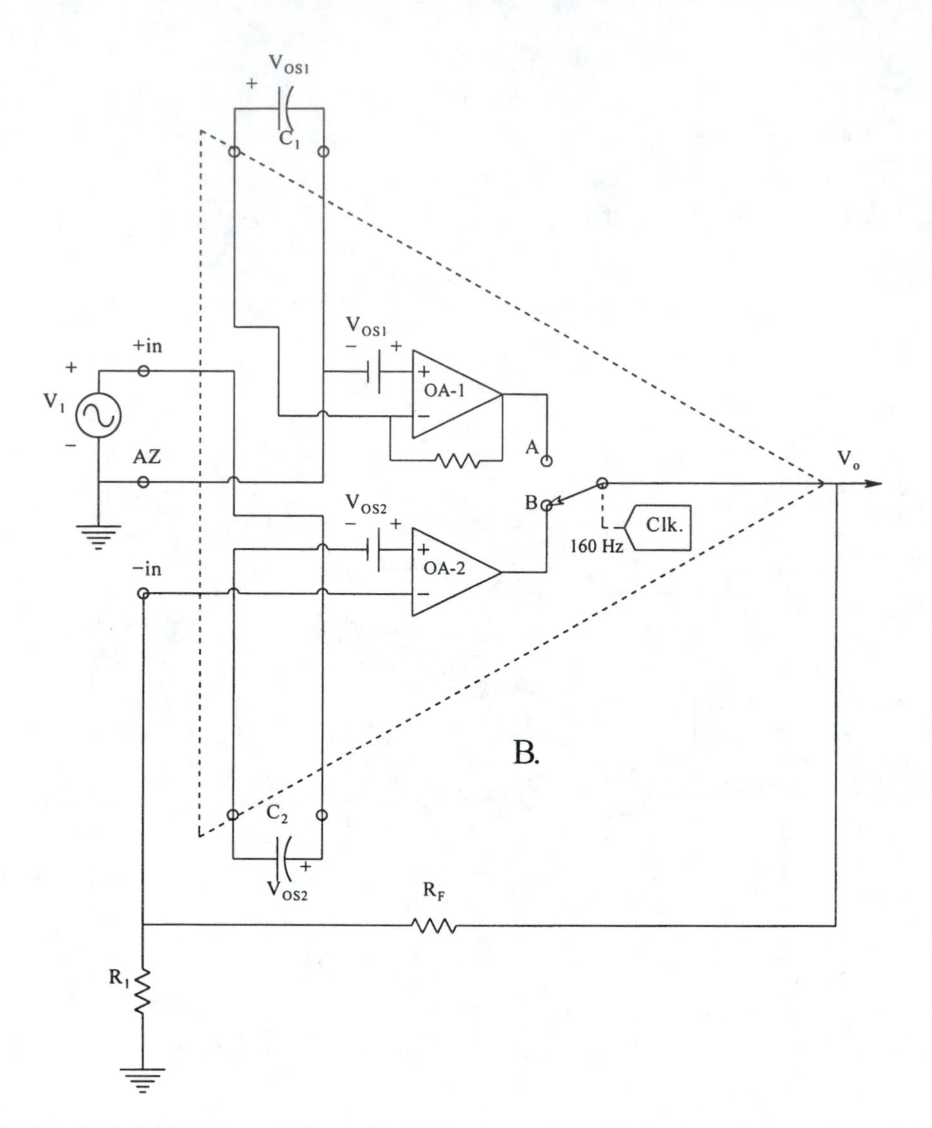

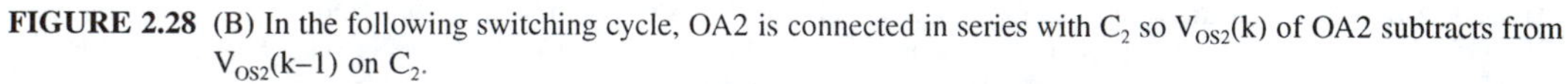

FIGURE 2.28 (B) In the following switching cycle, OA2 is connected in series with C_2 so $V_{OS2}(k)$ of OA2 subtracts from $V_{OS2}(k-1)$ on C_2.

where $y_1(t)$ is the system output, given an input, $x_1(t)$ to an LS with an impulse response, h(t). $\otimes$ denotes the operation of real convolution.

The property of *superposition* is illustrated by Equation 2.119. We note that:

$$y_3 = (a_1 y_1 + a_2 y_2) \qquad 2.119$$

given input $x_3 = (a_1\ x_1 + a_2\ x_2)$.

Linear system outputs can be *scaled,* i.e., if $x_2 = a\ x_1$, then:

$$y_2 = a(x_1 \otimes h) \qquad 2.120$$

Finally, linear systems obey the property of *shift invariance.* If the input x is delayed t_o seconds, then the output is also shifted:

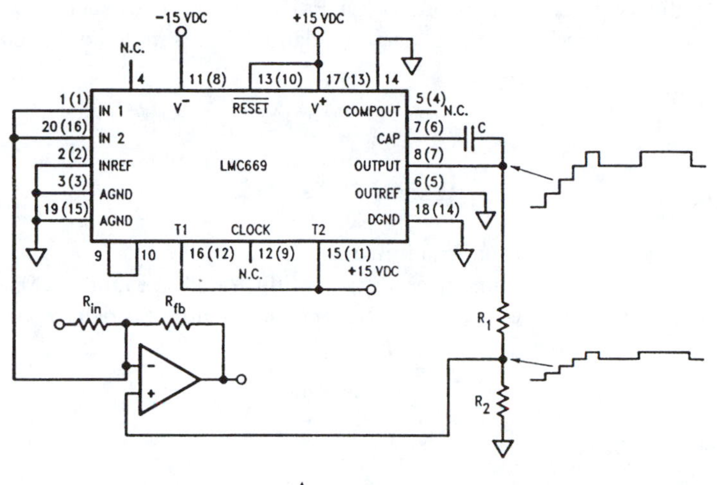

A.

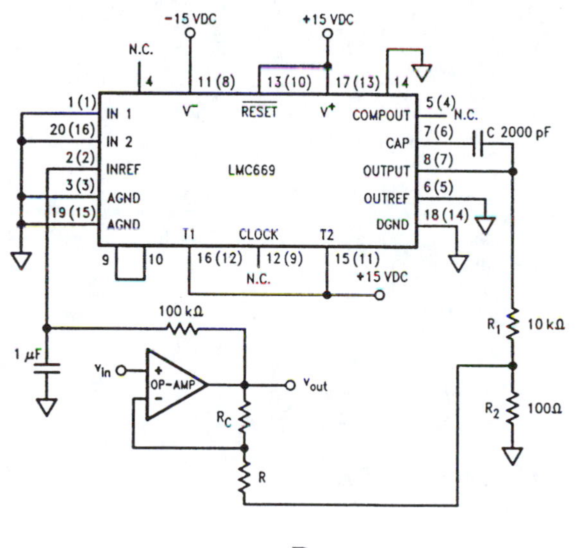

B.

FIGURE 2.29 (A) The LM669 autozero integrated circuit used to cancel out offset drift in an inverting op amp application. The LM669 continuously samples the DC voltage at the op amp's summing junction, and then generates a compensatory DC voltage at the non-inverting input. This action nulls the op amp's V_{OS} to effectively ±5 μV. This offset voltage correction compensates for the drift of VOS with temperature and power supply voltage changes. (B) The LM669 acts as a DC-servo integrating feedback loop around an op amp used as a non-inverting amplifier for ac signals. The DC output error of the op amp is reduced to about 5 μV, the V_{OS} of the LM669. The LM669 replaces the ground reference for the resistor, "R". (Figures used with permission of National Semiconductor, Santa Clara, CA.)

$$y(t - t_o) = x(t - t_o) \otimes h \qquad 2.121$$

Nonlinear systems do not obey the properties of linear systems discussed above. Nonlinear systems can be dynamic, static, or a combination of both. A static nonlinear system is one in which the output, y, is some nonlinear function of the input, x. For example, y = sin(x), y = sgn(x), $y = a + bx + cx^2$, and y = abs(x) describe nonlinear, static relations between input x and output y. y = mx + b is a linear static function of x. A dynamic nonlinearity is one described by a set of first-order, nonlinear ordinary differential equations (ODEs). An example of such a system is the well-known Volterra equations describing predator-prey relations in a simple ecosystem. These are:

$$\dot{x}_1 = k_1 x_1 - k_3 x_1 x_2 \quad 2.122$$

$$\dot{x}_2 = k_2 x_2 - k_4 x_1 x_2 \quad 2.123$$

The Volterra system responds to initial conditions on x_1 and x_2.

Another example of a dynamic nonlinear system is illustrated in Figure 2.30. A linear parallel LC circuit is connected to a negative resistance element whose current is determined by the relation:

$$i_{nl} = -av + bv^3 \quad 2.124$$

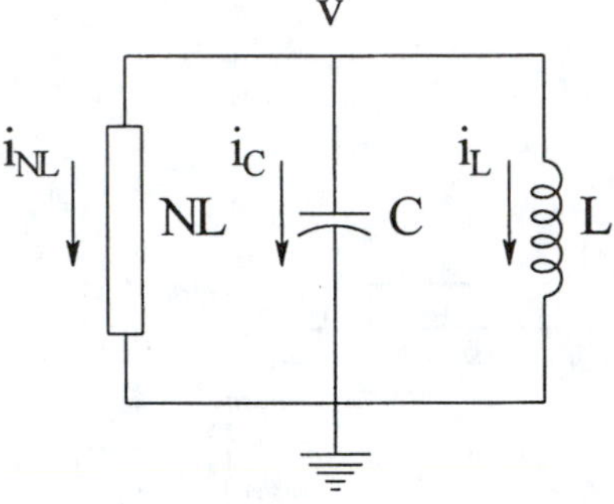

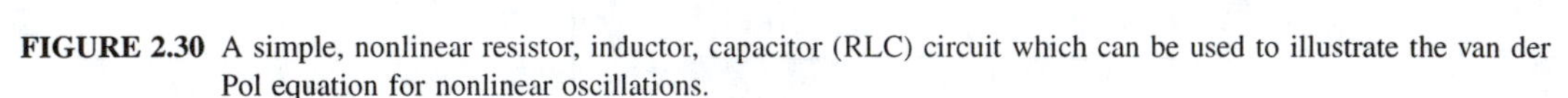

FIGURE 2.30 A simple, nonlinear resistor, inductor, capacitor (RLC) circuit which can be used to illustrate the van der Pol equation for nonlinear oscillations.

Hence, by Kirchoff's current law, we can write:

$$0 = \ddot{v} - \dot{v}\left(1 - 3bv^2/a\right)(a/C) + v/LC \quad 2.125$$

This relation can be written as two first-order ODEs:

$$\dot{x}_1 = -x_2/LC + \left(1 - 3bx_2^2/a\right)(a/C) \quad 2.126$$

$$\dot{x}_2 = x_1 \quad 2.127$$

where obviously, x_2 = v. Equation 2.125 is the well-known van der Pol equation.

Often nonlinear systems can be modeled by separating the linear dynamics from a no-memory nonlinearity, such as y = tanh(x), which can either precede or follow the linear block, depending on circumstances.

There are several important nonlinear analog signal processing operations which are used in instrumentation. These include, but are not limited to: precision full-wave rectification or absolute value circuits, True RMS-to-Dc conversion circuits, peak detection circuits, sample and hold or track and hold circuits, square root circuits, special function modules which give an analog output proportional to $x(y/z)^m$, log ratio modules, and trigonometric function modules.

We will review the applications of each of the nonlinear signal processing circuits before discussing its architecture and design details. Certain nonlinear operations may easily be realized with op amp circuits; for others, it is more expedient to use off-the-shelf, special-purpose microcircuits. We cite examples of these microcircuits, but for practical reasons, it is not possible to list all of a given type available.

2.5.2 Precision Absolute Value (Absval) Circuits

As the name implies, this type of nonlinear circuit takes the absolute value of the input waveform. Thus, the output approximates:

$$y(t) = k\ \text{abs}[x(t)] = k|x(t)| \tag{2.128}$$

where k is a gain or scaling constant. Precision absval circuits can be used for transfer standards. That is, they perform nearly ideal, full-wave rectification of symmetrical sinusoidal signals, which after low-pass filtering, appear as a DC voltage proportional to the peak or RMS value of the input sinewave. This DC voltage can be measured generally with greater accuracy than the corresponding AC voltage, especially when the peak value is less than 1 V. Precision absval circuits have been used in adaptive analog active filters, such as the Dolby B(TM) audio noise suppression system. Here the absval/low-pass filter system output is proportional to the signal power in a certain range of high frequencies, and is used to adjust the high frequency response of the filter. Precision absval circuits can also be used to precede square-law circuits used to square time-variable signals. When this is done, only half of a parabolic nonlinearity needs to be used, because:

$$y = [\text{abs}(x)]^2 = x^2 \tag{2.129}$$

Many precision absval circuits using op amps exist (see, for example, Sections 5.2 through 5.4 in Graeme, 1974); we will examine several of the more useful designs here. Figure 2.31 illustrates the circuit of a commonly used absval operator which uses two op amps. Its input resistance is just R/2, and its output is ideally given by:

$$V_o = +(R_2/R_1)|V_1| \tag{2.130}$$

However, at low levels of V_1, some tweaking of the R/2 resistor may be required to obtain equal and opposite slopes of the absval curve.

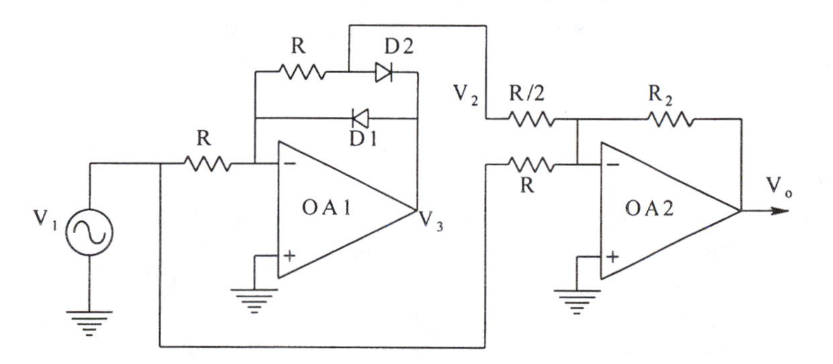

FIGURE 2.31 A precision, operational, full-wave rectifier or absval circuit.

Another absval circuit having high input resistance is shown in Figure 2.32. In this circuit, correct resistor values are important. The parameter n is the rectifier gain, and must be >1. A small capacitor, C, is used to improve high frequency response of the circuit. R_1 and R_3 can be made variable in order to adjust the absval operation to be symmetrical (i.e., have slopes of + or –n).

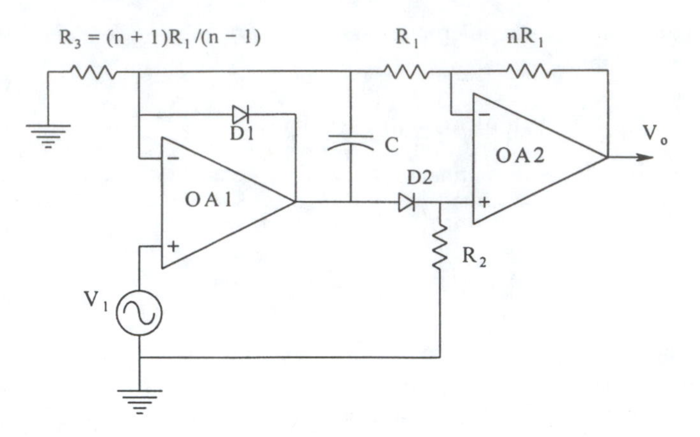

FIGURE 2.32 A precision absval circuit having high input impedance.

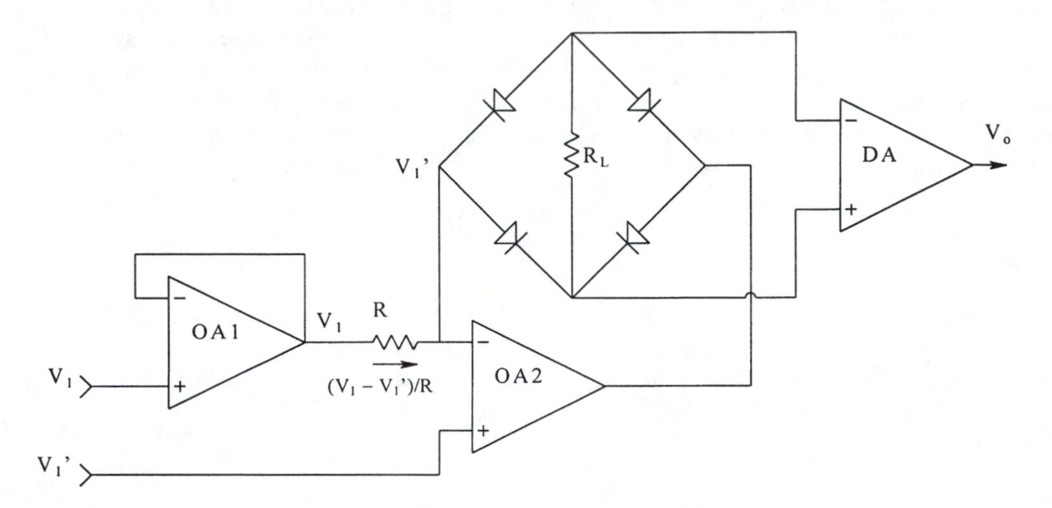

FIGURE 2.33 A precision absval circuit having a high impedance, differential input.

Finally, Figure 2.33 illustrates an absval circuit having a differential input. An instrumentation amplifier with high CMRR and differential gain K_D is used to condition the floating absval voltage across R_L. OA1 is used as a follower, OA2 forces a unidirectional current through R_L. The output of this circuit can easily be shown to be:

$$V_o = K_D(R_L/R)|V_1 - V_1'| \tag{2.131}$$

In describing the absval circuits above, we have neglected second-order errors caused by non-zero offset voltages and bias currents. Also neglected are the effects of finite op amp dynamic response and diode capacitance. If an absval circuit is to be used at high audio frequencies or above, then the designer must choose the op amps with care, and use high frequency switching diodes in the design. The designer should also simulate the high frequency behavior of the absval circuit in the time domain using an electronic circuit analysis program, such as one of the versions of SPICE or Micro-Cap™ in order to verify that it will meet specifications.

2.5.3 Multifunction Converters

Multifunction converters are large-scale integration (LSI) analog integrated circuits which perform the operation

$$V_o = KV_Y\left(V_Z/V_X\right)^m \qquad 2.132$$

The input voltages can range from 0 to 10.0 V, and from DC to 400 or 500 kHz. The exponent, m, can range from 0.2 to 5.0. Commercial examples of multifunction converters are the Burr-Brown (Tucson, AZ) Model 4302 and the Analog Devices (Norwood, MA) AD538 Analog Computational Unit.

The multifunction converter is a useful IC which can be used as a one-quadrant multiplier, squarer, square-rooter, or exponentiator, depending on the value of m and the accessory op amp circuits used. The multifunction converter can also be used to produce an output voltage proportional to the sine, cosine, or arctangent of an input voltage using an external op amp. In addition, using implicit analog computational techniques involving feedback, the true RMS of an input signal may be computed (discussed in the following section), as well as real-time vector magnitude, as shown in Figure 2.34, where:

$$V_o = \sqrt{V_1^2 + V_2^2} \qquad 2.133$$

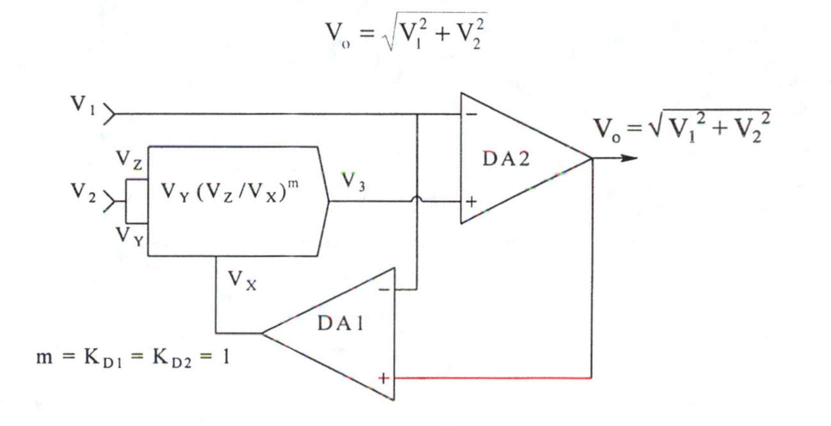

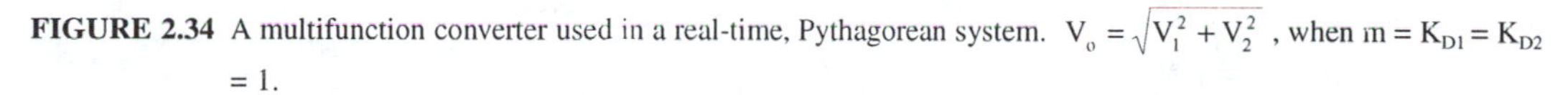
FIGURE 2.34 A multifunction converter used in a real-time, Pythagorean system. $V_o = \sqrt{V_1^2 + V_2^2}$, when $m = K_{D1} = K_{D2} = 1$.

2.5.4 True RMS to DC Converters

True RMS converters are widely used in the quantitative measurement of non-sinusoidal signals and noise. Electrical noise may arise from resistors, devices (e.g., transistors), capacitive or magnetic pickup from environmental noise sources such as sparking motor commutators, or slip-ring contacts. Acoustical noise is also important to measure; e.g., noise on a factory floor, mechanical vibrations of bearings, etc. Often it is useful to measure the RMS value of noise-like bioelectric waveforms, such as muscle action potentials (EMGs) in order to diagnose problems with a patient's neuromuscular system. EMGs can also be used to actuate prosthetic limbs and robotic manipulators. Still another use for true RMS converters is in the design of analog adaptive fiters, where the RMS voltage measured in some frequency band is used to adjust some feature of the over-all frequency response of a filter.

The operations on a signal done by an RMS converter are, in order of operation, squaring, smoothing, or averaging by low-pass filtering, and then square-rooting. The direct method of computing the analog true RMS of a time-varying signal is shown in Figure 2.35. Another way to find the RMS value of a time-varying signal is to use the implicit method illustrated in Figure 2.36. Here a multifunction module is used with feedback of the DC output voltage to force square rooting to occur; m = 1 in this application. Yet another implicit method of computing the true RMS value of a signal is the feedback thermocouple method shown in Figure 2.37. The thermocouple method

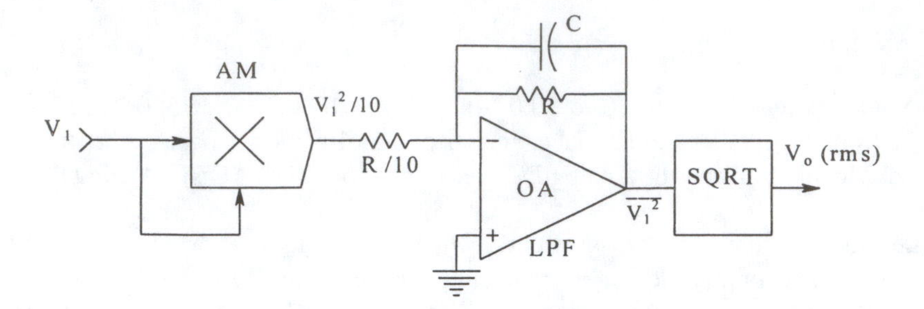

FIGURE 2.35 Circuit for the direct generation of the RMS value of an AC signal. Circuit output is a DC voltage.

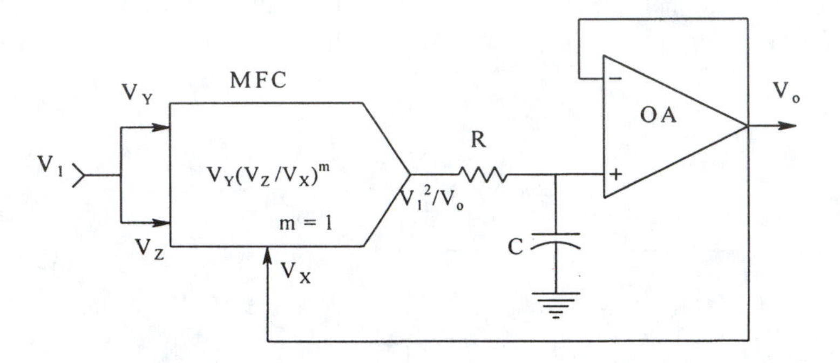

FIGURE 2.36 R and C form a low-pass filter which performs the "mean" operation in this feedback, true RMS converter which uses a multifunction converter. m ≡ 1.

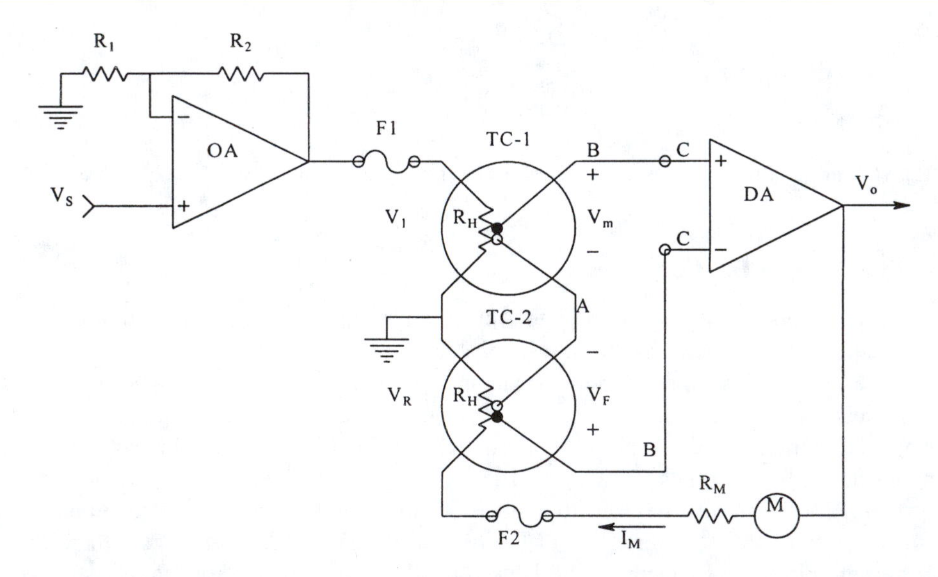

FIGURE 2.37 Basic schematic diagram of a true RMS voltmeter using two matched vacuum thermocouples in a feedback system. See text for analysis.

is also used in the design of true RMS voltmeters. In the feedback thermocouple circuit, we may write:

$$V_{m(dc)} = K_T\left(\overline{V_1^2}\right) \tag{2.134A}$$

$$V_{F(dc)} = K_T\left(\overline{V_R^2}\right) = K_T I_m^2 R_H^2 \qquad 2.134B$$

Now the DC output current of the DA is just

$$I_m = K_D\left(V_m - V_F\right)/R = K_D K_T\left(V_I^2 - I_m^2 R_H^2\right)/R \qquad 2.135$$

where R is the sum of the thermocouple heater resistance R_H, the microammeter resistance R_m, and the DA's output resistance R_o. Solution of the quadratic equation in I_m above yields:

$$I_m = -R/\left(2K_D K_T R_H^2\right) \pm (1/2)\sqrt{\frac{R^2}{K_D^2 K_T^2 R_H^4} + \frac{4\left(\overline{V_I^2}\right)}{R_H^2}} \qquad 2.136$$

The exact solution for I_m reduces to

$$I_m = V_{I(RMS)}/R_H \qquad 2.137$$

for the condition

$$\overline{V_I^2} \gg \frac{R_2}{4K_D^2 K_T^2 R_H^4} \qquad 2.138$$

which is easily met. If an analog microammeter is not used to read the system output, then it is easy to see that the DA's output voltage is given by:

$$V_o = V_{I(RMS)} R/R_H \qquad 2.139$$

Several specialized true RMS-to-dc converter ICs are available. The Burr-Brown Model 4341 (Tucson, AZ) operates by first taking the absval of the input signal, then forming a voltage proportional to twice the log of the absval. An antilog transistor is used in a feedback circuit. The DC output voltage, V_o, of the converter is fed back to the base of the antilogging transistor, so that its collector current is given by

$$i_c = k\log^{-1}\left(\log V_I^2 - \log V_o\right) = k\left(V_I^2/V_o\right) \qquad 2.140$$

$i_c(t)$ is averaged by an op amp current-to-voltage low-pass filter to form V_o. It is easy to see that V_o is proportional to the true RMS value of V_I. The BB 4341 true RMS converter of course works with DC inputs, and is down 1% in frequency response at 80 kHz, and –3 dB at 450 kHz. To be useful at low frequencies, the output must not contain ripple. Hence the capacitor in the low-pass (averaging) filter must be made very large. For example, for negligible output ripple given a 1 V, 10 Hz input sinewave, the filter capacitor should be over 100 μF.

The Analog Devices also offers several IC, true RMS-to-DC converters having basically the same internal circuit architecture as the BB 4341. For example, the frequency response of the high-accuracy AD637 extends to 8 MHz. Its maximum output is 7 VRMS, and its accuracy is 5 mV ± 0.2% of reading, with 0.02% nonlinearity for 0 to 2 VRMS input.

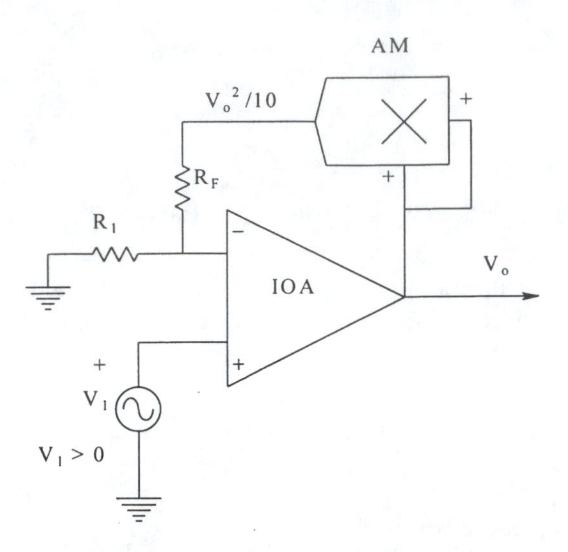

FIGURE 2.38 Feedback square rooting circuit; V_1 must be non-negative.

2.5.5 Square Root Circuits and Dividers

The square root of a voltage signal may be taken in several ways. Obviously, a multi-function converter with m = 0.5 can be used. Another implicit method of square rooting is shown in Figure 2.38. Here an analog multiplier is used to square the output and feed it back to the summing junction. The output voltage can be found from the node equation below, assuming the summing junction is at $V_1 \geq 0$:

$$V_1/R_1 = V_o^2/10R_F \tag{2.141}$$

This leads to the desired result:

$$V_o = \sqrt{V_1(10R_F/R_1)} \tag{2.142}$$

Analog division may also be accomplished by an implicit feedback circuit, one of which is shown in Figure 2.39. Here, V_1 and V_2 must be >0. The node equation at the summing junction is, assuming an ideal op amp:

$$V_1/R_1 = V_2V_o/10R_F \tag{2.143}$$

Hence:

$$V_o = (V_1/V_2)(10R_F/R_1) \tag{2.144}$$

Of course, a multifunction module with m = 1 will also give the same result.

2.5.6 Peak Detectors and Track and Hold (T&H) Circuits

These circuits are closely related, and will be treated together. Peak detectors follow an irregular waveform until it reaches a maximum (or minimum, if a negative peak detector is used), and then holds that value at its output until an even greater maximum comes along, then holds that value,

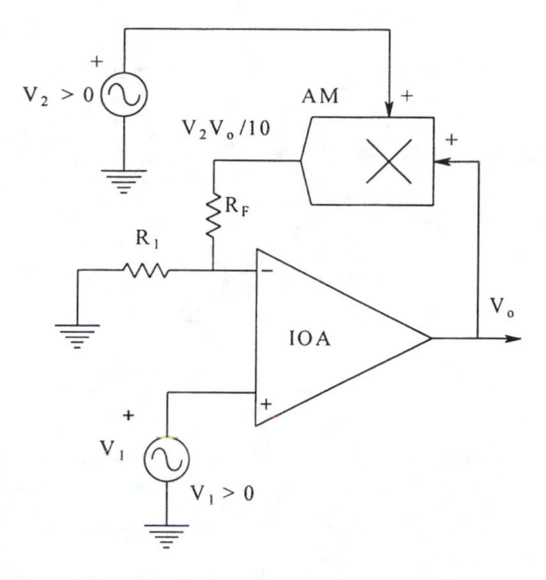

FIGURE 2.39 The circuit of Figure 2.38 is modified to make an analog divider. Here V_1 must be non-negative, and V_2 must also be positive and greater than $V_{2(MIN)} = (V_{1(MIN)}/V_{o(SAT)})(10R_F/R_1)$ to prevent op amp output saturation, and less than the supply voltage of the op amp (e.g., +15 V) to avoid damaging the op amp's input transistors.

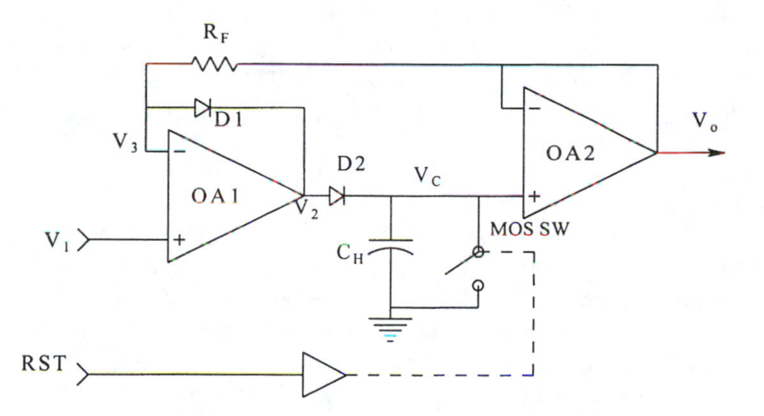

FIGURE 2.40 A MOSFET switch, SW1, is used to reset the two op amp peak detector.

etc. Peak detectors are used in peak-reading voltmeters, and in applications where measurement of the peak value of a waveform has significance (as opposed to the RMS value or rectified average value); an example of a waveform where the maximum and minimum values are important is arterial blood pressure recorded during surgery. The peak value is the systolic pressure, the minimum value is the diastolic pressure. Figure 2.40 illustrates a two-op amp peak detector of conventional design. Note that OA2 is used as a voltage follower. It should be an FET input type with a high input resistance and low bias current. Before a peak is reached in V_1, D_1 is reverse-biased by the forward drop on D2 and $V_3 = V_C = V_o = V_1$. After the peak in V_1 is reached, $V_1(t)$ decreases, D2 is reverse-biased, and D1 is forward-biased so that OA1 does not saturate. The output voltage remains at V_{1pk}, assuming zero capacitor leakage current, diode D2 reverse current, and op amp bias current. In practice, these currents will cause V_C to slowly drift. The reset switch, SW1, is generally a MOSFET switch which is used to reset V_o to zero when required.

The addition of a third diode and a resistor R_2 can improve the performance of the peak detector by eliminating the effective leakage current through D2. Reference to Figure 2.41 shows that D3 is held at zero bias by R_2 which pulls the D3 anode to $V_o = V_C$ after the peak is reached. The overall rate of change in the peak detector output voltage is thus practically

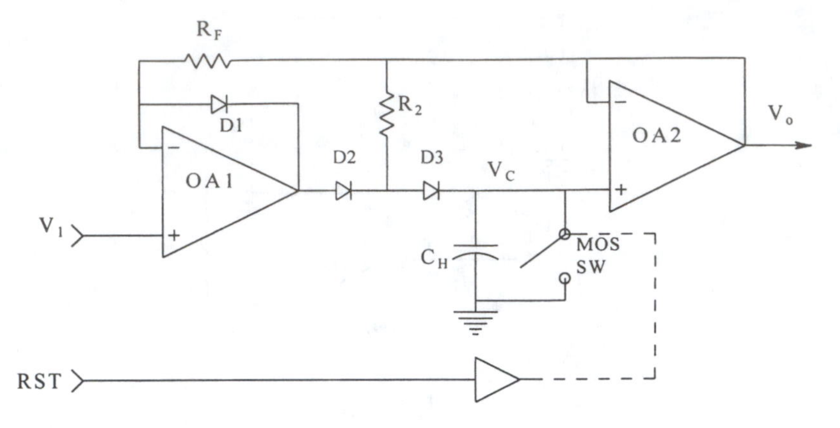

FIGURE 2.41 An improved peak detector design.

$$dV_o/dt = I_B/C_H \quad V/s \tag{2.145}$$

There is a tradeoff in determining the proper size of C_H. A large, low-leakage C_H will minimize drift, but is expensive. Also, too large a C_H will not allow V_C to follow V_1 below the peak due to current saturation in OA1. Typical C_H values range from 1 nF to to 0.1 μF. A good section on the design of peak detectors can be found in Franco (1988).

Track and hold (T&H) (also called sample and hold) circuits are important components of many analog-to-digital conversion systems. In operation, the output of a T&H circuit follows the conditioned analog input signal until the hold command is given. After a short delay, the T&H output assumes a constant level, during which the A/D conversion takes place. Constancy is required because certain ADC designs compare the analog input signal to an analog signal from a digital-to-analog converter in the ADC whose output is determined from the output digital "word". If the ADC's analog input changes during the conversion process, errors will result. Ideally, the digital output of an ADC is the result of an impulse modulation where the analog input signal is multiplied by a periodic train of unit impulses. The output of the impulse modulator is a periodic sequence of numbers whose values represent the values of the analog input signal only at sampling instants. A T&H circuit allows a closer approximation of true impulse modulation to take place.

Figure 2.42 illustrates a basic T&H circuit. When the FET sampling switch is closed, the system is in the tracking mode. OA2 acts as a unity-gain buffer so $V_o = V_C = V_1$ as long as op amp slew rates are not exceeded, and the input signal frequencies lie well below the gain*bandwidth products of the op amps. C_H provides a capacitive load for OA1, and thus must be able to provide enough output current to satisfy $[C_H(dV_1/dt)MAX]$. When the FET switch opens, the charge on C_H is trapped there, so V_o is held at V_C. However, things are not that simple; V_C changes slowly due to charge leaking off C_H through its leakage resistance, and charge transfer to or from due to the bias current of OA2, I_B. This slow drift in V_C is called droop, and its slope can be either positive or negative, depending on the sign of I_B. In addition to the droop, there is also a feedthrough error caused by the fact that the FET switch does not have an infinite off resistance. When the FET switch is switched off, there is also an unwanted transfer of charge from the FET gate to C_H through the FET gate-to-drain capacitance. This charge can produce an unwanted pedistal voltage which is given approximately by

$$\delta V_o = \left(C_{gd}/C_H\right)\left(V_{OB} - V_o\right) \tag{2.146}$$

where V_{OB} is the off level bias voltage on the FET gate, and V_o is of course the value of V_1 at the time the hold command is given. Thus the pedistal voltage depends on the input voltage, and as

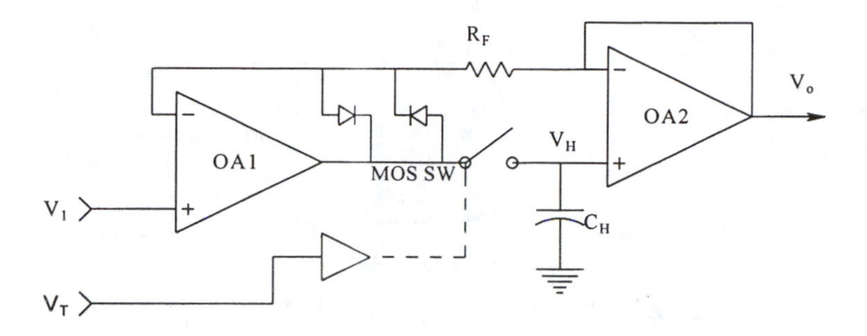

FIGURE 2.42 A basic, unity-gain, track and hold circuit.

numerical calculations reveal, can prove to be quite an objectionable source of T&H circuit error (see, for example, Franco, 1988, Section 7.11). A number of schemes have been devised to minimize pedistal error. One is to inject an equal and oposite switching charge into C_H by means of an auxillary active circuit. Another is illustrated in the circuit achitecture of the Burr-Brown SHC803BM T&H (Tucson, AZ), shown in Figure 2.43. In this T&H system, a capacitor $C'_H = C_H$ is used to compensate for pedistal and droop at the same time, using the differential gain of OA2. Capacitor leakage, OA2 DC bias currents and FET switch charge injection on switching cause equal voltage errors on both capacitors, which are subtracted out. C_H is operated in the integrator mode in the Burr-Brown design. In the hold mode, both SW1 and SW2 are off, hence the charge on C_H is trapped, and V_o remains constant. In the tracking mode, we assume the on resistance of SW1 is 0 ohms. Thus V_o follows V_1 with the transfer function:

$$V_o/V_1 = -1/(1000C_H s + 1) \tag{2.147}$$

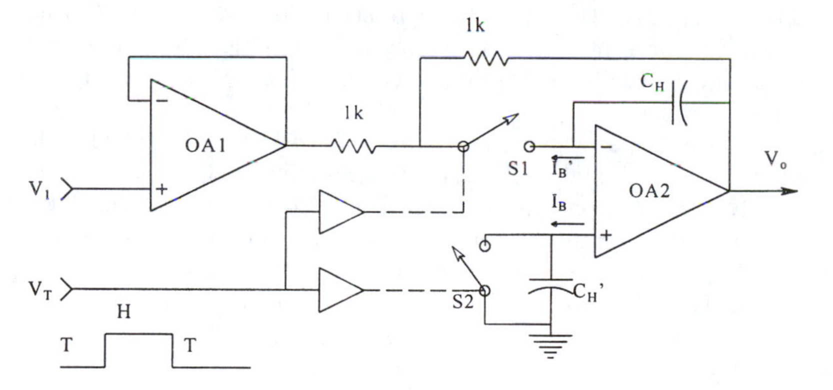

FIGURE 2.43 Basic architecture of the Burr-Brown SHC803BM (Tucson, AZ) track and hold circuit. S2 and capacitor C'_H are used for pedistal voltage cancellation.

If C_H is 100 pF, then the –3 dB frequency is 1.6 MHz. In general, because C_H appears in the feedback loop of OA2, the tracking-mode frequency response of the integrator type of T&H is lower than that attainable from the grounded C_H configuration shown in Figure 2.42. Figure 2.44 illustrates the T&H output errors discussed above. Illustrated are droop, signal feedthrough due to finite FET switch off resistance, pedistal, and several important operating times specified by manufacturers (see caption).

2.5.7 Log Ratio and Trigonometric ICs

Another class of nonlinear analog signal processing circuit which should be considered is the log ratio IC. The analog output voltage of this IC is given by

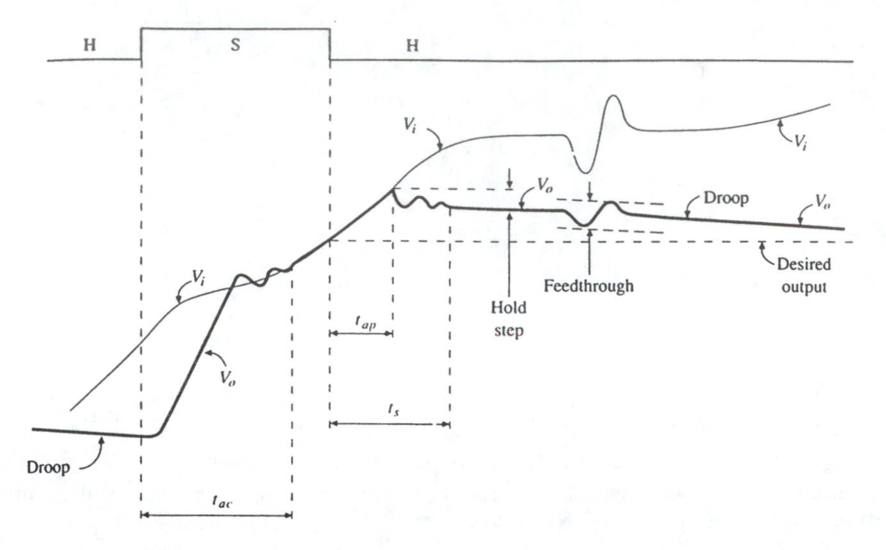

FIGURE 2.44 Waveforms seen in the operation of a typical track and hold circuit. (From Franco, S., *Design with Operational Amplifiers and Analog Integrated Circuits*, McGraw-Hill, New York, 1988. with permission.)

$$V_o = K \log_{10}(I_1/I_2) \qquad 2.148$$

Typically, K is set to some convenient integer value such as 1, 2, 3, or 5. Note that the inputs are currents which may vary over a six decade range (e.g., 1 nA to 1 mA).

This nonlinear circuit module finds applications in spectrophotometry (analytical chemistry) where parameters such as absorbance or optical density need to be simply calculated and displayed. Also, if I_1 and I_2 are made to be DC currents proportional to the RMS value of the sinusoidal output and input, respectively, of a linear system, the log ratio converter output is proportional to the frequency response magnitude (Bode log magnitude) in decibels. Of course if I_2 is fixed, then we have a simple log converter.

One commercially available log ratio converter is the Burr-Brown LOG100 (Tucson, AZ); Figure 2.45 shows the internal circuit of this IC. Analog Devices (Norwood, MA) offer a similar unit, the AD757, which will accept either current or voltage inputs. Note that log ratio modules are normally used with DC inputs. Their frequency responses are a function of the input signal amplitude, ranging from about 100 Hz for nanoamp peak input currents to over 40 kHz for milliamp peak input currents.

Often, the nonlinear signal processing requirements of the system demand a broadband logarithmic nonlinearity which can operate on video-frequency signals. The Analog Devices AD640 is an example of an IC which can produce an output of the form

$$V_o = V_y \log(V_1/V_{sc}) \qquad 2.149$$

(where V_{sc} is the intercept scaling voltage and V_y is the slope voltage, both constants) over a 46 dB conversion range and an 80 MHz bandwidth. Two cascaded AD640s are claimed to have a 95 dB dynamic range in a 20 Hz to 100 kHz bandwidth; a 70 dB range is claimed in a 50 MHz to 150 MHz bandpass. The IC is direct-coupled, however, and will work from DC through the audio spectrum.

The Analog Devices AD639 trigonometric converter is another specialized, nonlinear analog processing module which has applications such as generation of a sine wave from a triangle wave at frequencies up to 1.5 MHz, coordinate conversion and vector resolution, design of quadrature and variable phase oscillators, and imaging and scanning linearization circuits. The nonlinear transfer function of the AD639 is as follows:

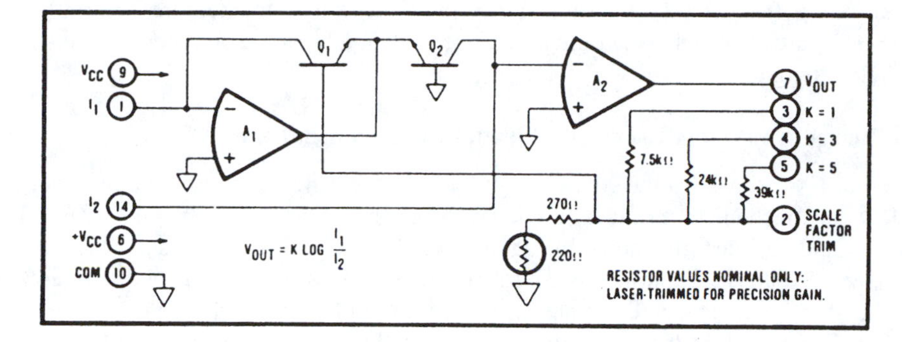

FIGURE 2.45 Schematic of the Burr-Brown LOG100 log ratio converter. (Figure used with permission of Burr-Brown, Tucson, AZ.)

$$V_o = U \frac{\sin(V_{x1} - V_{x2})}{\sin(Y_{y1} - Y_{y2})} \qquad 2.150$$

The scaling factor for the differential voltage inputs is 50°/volt. The IC has on board a 1.8 VDc bandgap voltage reference which can be used to preset either angle argument to 90°. The output of the AD639 is scaled by the constant, U, which can be set between zero and +15 V by either external or internal means. A wide variety of trig functions can be modeled: sin(x), cos(x), tan(x), cosec(x), sec(x), and cotan(x), as well as certain useful inverse trig functions. For the arctangent:

$$V_o = K\theta = k \tan^{-1} \frac{(V_{z2} - V_{z1})}{(V_{u1} - V_{u2})} \qquad 2.151$$

where K = 20 mV/degree. With minor changes in the external circuit architecture, the AD639 will also calculate the arcsine and arctangent functions with the same ratio of voltage differences argument as the arctangent function above. With the exception of sinewave generation, it would appear that this versatile IC is best used with inputs which are DC or at best in the low audio frequency range.

2.5.8 Conclusion

We began this section by discussing some of the properties of linear and nonlinear systems. There are many uses for nonlinear analog signal processing ICs. The present trend is to sample and digitize the input signal and then use a computer to calculate the nonlinear operation, then realize the result in analog form by digital-to-analog conversion. However, there are many instances where the instrumentation system cost will be less expensive if an all-analog approach is taken, and use is made of a specialized analog function module. Computers, while versatile, can be expensive. Obviously, there is a trade-off between accuracy, cost, and convenience in choosing between an all-analog vs. an A/D to computer to D/A type of system for nonlinear signal processing in an instrumentation system. The instrumentation system designer should be aware that all-analog signal processing ICs are available which may compete with the digital approach in terms of function, accuracy, and cost.

2.6 CHARGE AMPLIFIERS

Charge amplifiers make use of FET-input (electrometer) op amps which have ultralow DC bias currents at their inputs, and ultra-high input resistances. They find primary application in isolating and conditioning the outputs of piezoelectric transducers used to measure pressure and force transients. They are also used to measure the charge in coulombs stored on capacitors, and indirectly, electric field strength.

2.6.1 Charge Amplifiers Used with Piezoelectric Transducers

In Chapter 6 we show that the equivalent circuit of a piezoelectric pressure transducer operated well below its mechanical resonant frequency is given by the simple parallel current source, conductance and capacitance, shown in Figure 2.46. Note that d is a constant which depends on the piezoelectric material and how it is cut relative to the crystal's axes to form the transducer, and that in general, current is the rate of transfer of charge through a plane through which it is passing. Current is conventionally taken as positive in the direction of motion of positive charges, such as holes in semiconductors, or positive ions in solutions or plasma discharges. Hence in a conductor, such as copper, where charge is carried by electrons, the current direction is opposite to the direction of electron motion.

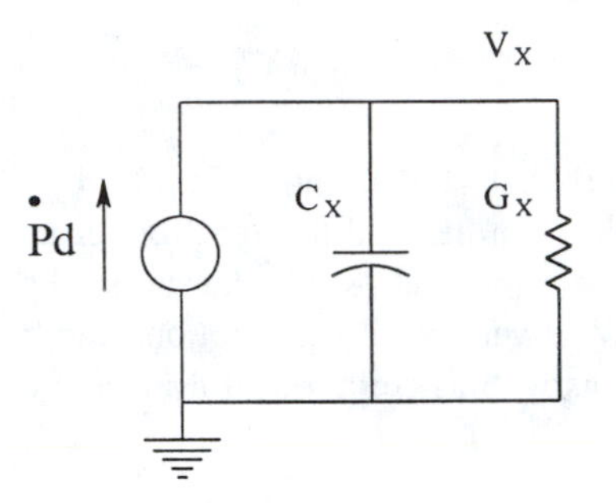

FIGURE 2.46 Equivalent circuit of a piezoelectric crystal responding to pressure on its active surface at frequencies well below its mechanical resonance frequency.

Figure 2.47 illustrates an ideal op amp used as a charge amplifier. The op amp is assumed to have infinite gain, frequency response and input resistance, and zero bias current, offset voltage, noise, and output resistance. Under the ideal op amp assumptions, it is easy to see that because the summing junction is at zero volts, all of the current, $\dot{P}d$, flows into the feedback capacitor, C_F. Hence V_o is proportional to P(t), and is given by:

$$V_o(t) = -\int_0^t \left[\dot{P}(t)\,d/C_F\right] dt \qquad 2.152$$

This result of the ideal case is seen to become less simple when we face reality and note that the electrometer op amp has a non-infinite gain given by:

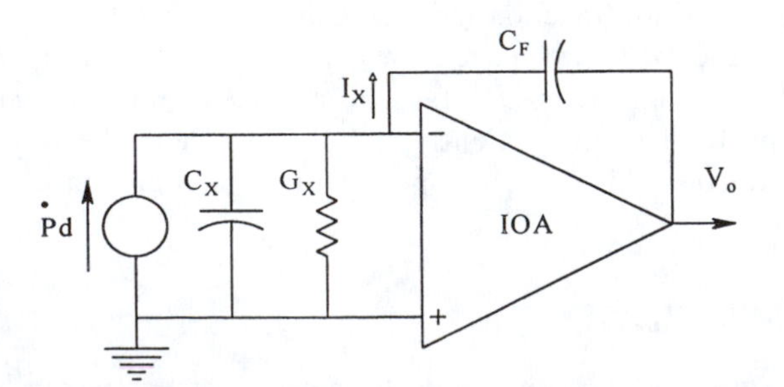

FIGURE 2.47 A charge amplifier circuit used to condition the output of a piezoelectric crystal sensor.

$$K_V = \frac{K_{Vo}}{(\tau s + 1)}(V_i - V_i') \tag{2.153}$$

Also, we assume that there is a small but finite conductance, G_F, in parallel with C_F. G_F is in the order of 10^{-10} Siemens, or smaller. Now if we write the node equation at the summing junction, and substitute the gain expression above to eliminate V_i', we can show that the charge amplifier's transfer function is given by (approximately):

$$V_o/P = \frac{-s(d/G_F)}{s^2\left[\tau(C_T + C_F)/K_{Vo}G_F\right] + s\left[\tau(G_T + G_F) + K_{Vo}C_F\right]/K_{vo}G_F + 1} \tag{2.154}$$

This is the transfer function of a band-pass system. The low-frequency pole is at about:

$$f_{low} \approx 1/2\pi C_F R_F \text{ Hz} \tag{2.155}$$

The mid-band gain is:

$$A_{Vmid} = -K_{vo}d/\left[\tau(G_T + G_F) + K_{vo}C_F\right] = -d/C_F \tag{2.156}$$

The high-frequency pole is at:

$$f_{hi} = K_{vo}C_F/\left[2\pi\tau(C_T + C_F)\right] \text{ Hz} \tag{2.157}$$

Note that $K_{vo}/2\pi\tau$ is simply the small-signal gain*bandwidth product of the electrometer op amp. Figure 2.48 shows the circuit of the non-ideal charge amplifier. Here C_T is the total capacitance to ground which includes the piezoelectric crystal capacitance, the capacitance of the cable connecting the transducer to the charge amp, and the charge amp's input capacitance. Typically C_T might be about 400 pF. G_T is the total shunt conductance to ground at the op amp summing junction. G_T includes the crystal leakage conductance, the cable leakage conductance, and the input conductance of the op amp. Typically, G_T might be about 10^{-13} S. If, in addition, we take C_F to be 2×10^{-9} F, G_F to be 10^{-10} S, $K_{vo} = 5 \times 10^4$, $d = 2 \times 10^{-12}$ coulomb/psi for quartz, and the op amp's gain*bandwidth product = 350 kHz, the system ends up having a mid-band gain of 1 mV/psi, a high-frequency pole at 1.5×10^5 Hz, and a low-frequency pole at 8×10^{-3} Hz. Note that the low-frequency pole is set by C_F/G_F, and is independent of the input circuit. Thus, the piezoelectric transducer charge amplifier is not a system which can respond to static (DC) pressures; it must be used to measure transient pressure changes. In addition to the low-frequency limitations imposed on the transfer function by non-ideal electrometer op amp behavior, there are also DC errors at the output due to the op amp's bias current and DC offset voltage. Figure 2.49 shows the DC equivalent circuit of the charge amplifier. Note that the capacitors drop out from the circuit at DC because they carry no current. It is easy to see that the DC error in the output is:

$$V_{oDC} = I_B R_F + V_{OS}(1 + G_T/G_F) \tag{2.158}$$

Note that I_B and V_{OS} can have either sign. Typical values of V_{OS} are about ±200 μV, and I_B values are around 75 fA (75×10^{-15}A).

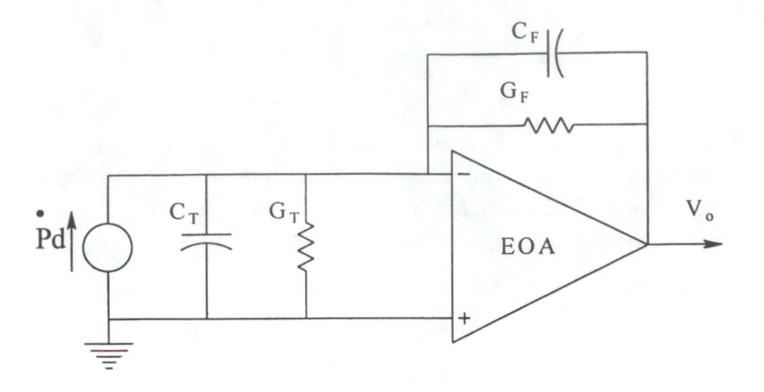

FIGURE 2.48 Circuit of a practical charge amplifier and crystal sensor.

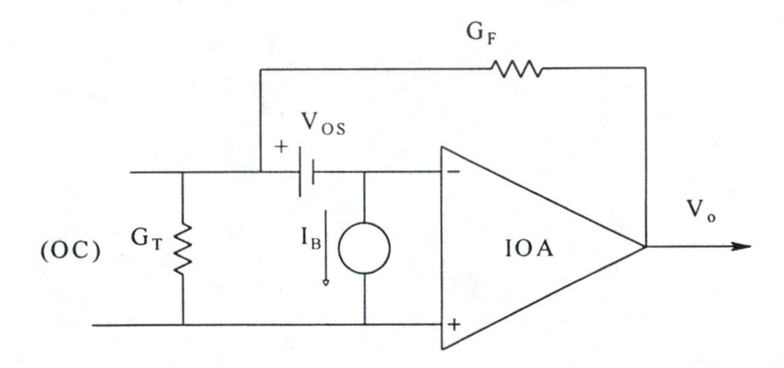

FIGURE 2.49 The DC equivalent circuit of a practical charge amplifier.

2.6.2 The Charge Amplifier as an Integrating Coulombmeter

A charge of Q_I coulombs is trapped on a capacitor, C_I. The capacitor can be a real capacitor, or the capacitance of a charged object such as the fuselage of a helicopter or a human body. We wish to measure Q_I. The voltage across C_I is given by the well-known relation, $V_I = Q_I/C_I$. In Figure 2.50, C_I is connected to the summing junction of an electrometer op amp at the same time the short is removed from the feedback capacitor, C_F, allowing it to charge from the op amp's output to a value which maintains the summing junction at 0 V. Assuming the op amp to be ideal, the output voltage of the op amp, after three or four R_IC_I time constants, is easily shown to be

$$V_{o(SS)} = -Q_I/C_F \text{ DC volts} \tag{2.159}$$

Hence, the system can measure the charge on a small capacitor, providing the peak current $i_{IPK} = Q_I/R_IC_I$ does not exceed the current sourcing or sinking capability of the op amp.

In practice, this circuit is subject to DC errors caused by a practical op amp's bias current and offset voltage. The output error caused by I_B and V_{OS} can be found by superposition, and is given by

$$\delta V_o = T(I_B/C_F) + V_{OS}(1 + C_I/C_F) \text{ volts} \tag{2.160}$$

where T is the time elapsed between closing switch 1 and opening switch 2, and reading the steady-state V_o. In practice, δV_o can be made quite small by judicious choice of circuit parameters.

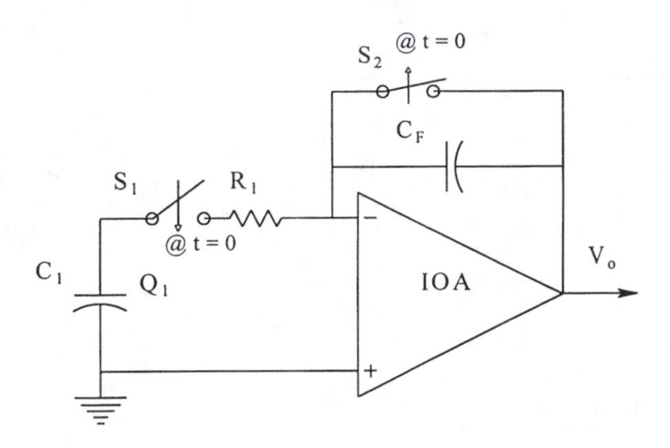

FIGURE 2.50 An integrating coulombmeter circuit. An electrometer op amp is generally used.

2.6.3 Summary

In this section, we have presented two examples of applications of the op amp charge amplifier configuration. The basic charge amplifier consists of an ideal op amp with an ideal feedback capacitor to integrate charge transferred to the summing junction. In practice, we use electrometer amplifiers which have ultra-high input resistances and finite gains, bandwidths, offset voltages, and bias currents. Because a pure C_F will integrate the bias current of the summing junction, a large resistance, R_F, must be put in parallel with C_F to give a fixed DC error proportional to I_B. The parallel combination of R_F and C_F sets the low-frequency pole of a band-pass frequency response for the piezoelectric transducer. In the mid-band range of frequencies, the output voltage is proportional to the pressure on the transducer, rather than its derivative.

2.7 PHASE-SENSITIVE RECTIFIERS

Phase-sensitive rectifiers (PSRs), also called phase-sensitive demodulators or synchronous rectifiers, are an important class of analog circuit widely used in instrumentation, control, and communications systems. The PSR is used to convert the output of certain AC, carrier-operated sensors to DC proportional to the quantity under measurement. One example of the use of PSRs is to condition the output of a resistive Wheatstone bridge using strain gauges to measure mechanical strain, pressure, or force. The strain gauge bridge excitation is generally an audio-frequency sinewave. The bridge output is a sinewave of the same frequency with amplitude proportional to the strain, etc. The phase of the bridge output is 0° with respect to the excitation phase if $\Delta X > 0$, and is –180° with respect to the excitation if $\Delta X < 0$. The PSR uses the bridge excitation voltage to sense this phase change, and to produce a rectified signal which has amplitude proportional to ΔX, and the sign of ΔX.

2.7.1 Double-Sideband, Suppressed Carrier Modulation

The outputs of many sensor systems have information encoded on them in what is called double-sideband, suppressed-carrier modulation (DSBSCM). A DSBSCM signal is formed by multiplying the AC carrier by the modulating signal. To illustrate this process, we examine a Wheatstone bridge with one variable element and AC excitation, as shown in Figure 2.51. It is easy to show (see Chapter 4) that the bridge output is given by:

$$V_o'(t) = K_D\left[V_b \sin(\omega_c t)\right]\left[\frac{X}{X+P} - \frac{N}{M+N}\right] \qquad 2.161$$

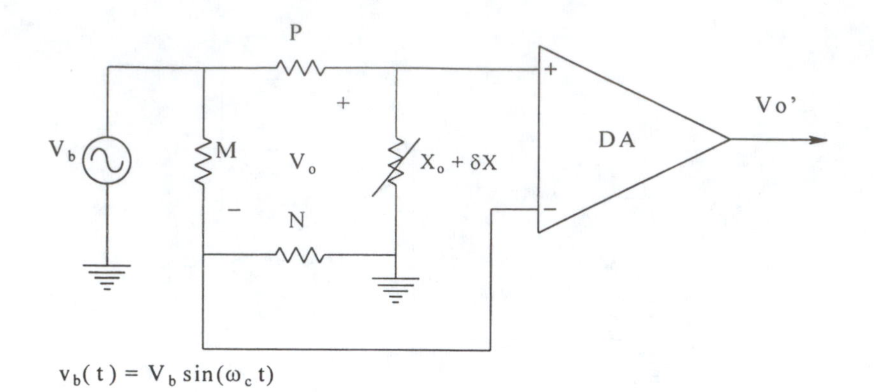

FIGURE 2.51 A Wheatstone bridge circuit with AC excitation generates a double sideband, suppressed carrier-modulated output.

Now we let X vary in time so:

$$X(t) = X_o + \Delta X(t) \quad 2.162$$

Furthermore, we balance the bridge and make it maximally sensitive. This requires that

$$M = N = P = X_o \quad 2.163$$

Thus around null,

$$V_o'(t) = \Delta X(t) K_D V_b (1/4X_o) \sin(\omega_c t) \quad 2.164$$

Note that we have the product of $\Delta X(t)$ and the excitation sinusoid at the bridge output. If $\Delta X(t)$ is sinusoidal with $\omega_m << \omega_c$, we can write, using the well-known trig identity, $[\sin(\alpha)\sin(\beta)] = [\cos(\alpha - \beta) - \cos(\alpha + \beta)]/2$:

$$V_o'(t) = K_D V_b [\Delta X/8X_o][\cos((\omega_c - \omega_m)t) - \cos((\omega_c + \omega_m)t)] \quad 2.165$$

Thus from Equation 2.165, we see that the conditioned bridge output, $V_o'(t)$, consists of two sidebands and no carrier frequency term, hence the designation DSBSCM.

Another example of a transducer which produces a DSBSCM output is the linear variable differential transformer (LVDT), a device which is used to sense small, linear mechanical displacements. Some LVDTs come with a built-in PSR and provide a DC output directly proportional to core displacement; others must be used with an external PSR. (LVDTs are discussed in detail in Chapter 6.)

2.7.2 Demodulation of DSBSCM Signals by Analog Multiplier

Figure 2.52 illustrates one of the simplest ways to demodulate a DSBSCM signal and recover $\Delta X(t)$. An analog multiplier IC is used to multiply a reference signal which is in-phase with the transducer's AC excitation by the transducer's DSBSCM output. This product can be written:

$$V_z(t) = \frac{\Delta X K_D V_b V_r}{(10)8X_o}[\sin(\omega_c t)\cos((\omega_c - \omega_m)t) - \sin(\omega_c t)\cos((\omega_c + \omega_m)t)] \quad 2.166$$

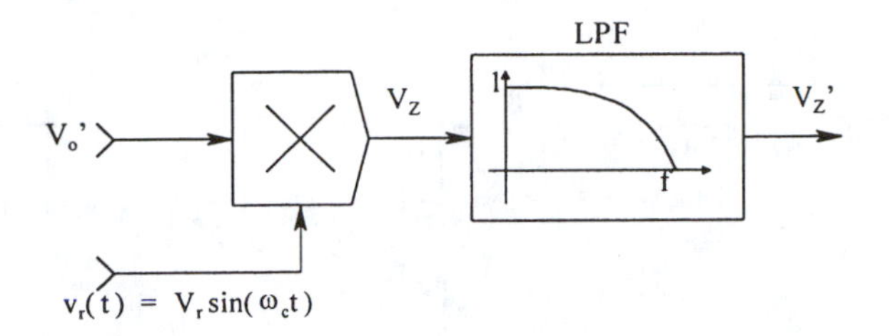

FIGURE 2.52 An analog multiplier and low-pass filter used to demodulate a DSBSC signal.

The bracketed term can again be expanded by trig identity:

$$V_z(t) = \frac{\Delta X K_D V_b V_r}{(10)16 X_o}\left[\sin((2\omega_c - \omega_m)t) + \sin(\omega_m t) - \sin((2\omega_c + \omega_m)t) - \sin(-\omega_m t)\right] \quad 2.167$$

$V_z(t)$ is now passed through a unity-gain low-pass filter which removes the two high-frequency terms. The output of the LPF is the desired, demodulated signal, multiplied by a scale factor.

$$V_z'(t) = \frac{K_D V_b V_r}{80\, X_o}\left[\Delta X \sin(\omega_m t)\right] \quad 2.168$$

2.7.3 Other PSR Designs

Figure 2.53A illustrates a simple PSR which can be made from two op amps and a bidirectional metal oxide semiconductor (MOS) switch operated by a logical square wave derived from the sinusoidal excitation source by first conditioning it with a comparator used as a zero-crossing detector, and then passing the transistor-transistor logic (TTL) waveform through two one-shot multivibrators (e.g., a 74LS123) to adjust the reference signal's phase. Because the MOS switch closes at the same time in the reference signal cycle, when the DSBSCM signal changes phase 180°, the rectifier output changes sign, producing a negative output. A low-pass filter is used to smooth the full-wave rectified output of the switching circuit. The relevant waveforms are shown in Figure 2.53B.

Phase-sensitive rectifier ICs are available from a variety of manufacturers, and include the Analog Devices AD630 balanced modulator/demodulator, The Precision Monolithics GAP-01 general-purpose analog signal processing subsystem, the National Semiconductor LM1596 balanced modulator-demodulator and the Signetics NE/SE5520 LVDT signal conditioner, to mention a few.

2.7.4 The Lock-in Amplifier

The lock-in amplifier (LIA) is used to condition and demodulate DSBSCM signals, and recover the modulating signal under conditions of abominable signal-to-noise ratios, where the rms noise voltage can exceed that of the modulated carrier by about 60 dB. Noise is reduced in the operation of a LIA by three mechanisms: band-pass filtering of the input signal around the carrier frequency, rejection of coherent interference at frequencies other than that of the carrier, and low-pass filtering at the PSR output. Carrier frequencies are best chosen to lie between 10 and 40 Hz, and 200 to 10^5 Hz. Carrier frequencies lying too close to the 60 Hz power line frequency and its first two harmonics are avoided because they may "leak" through the LPF following the PSR. Very low carrier frequencies are also avoided as they may place LIA operation in the range of 1/f noise from the analog electronic circuits.

A block diagram of an LIA is shown in Figure 2.54, along with some pertinent wave-forms. We see that the basic LIA performs four basic operations on the DSB modulated carrier with additive, broad-band noise: The incoming DSBSCM signal plus noise is generally conditioned by

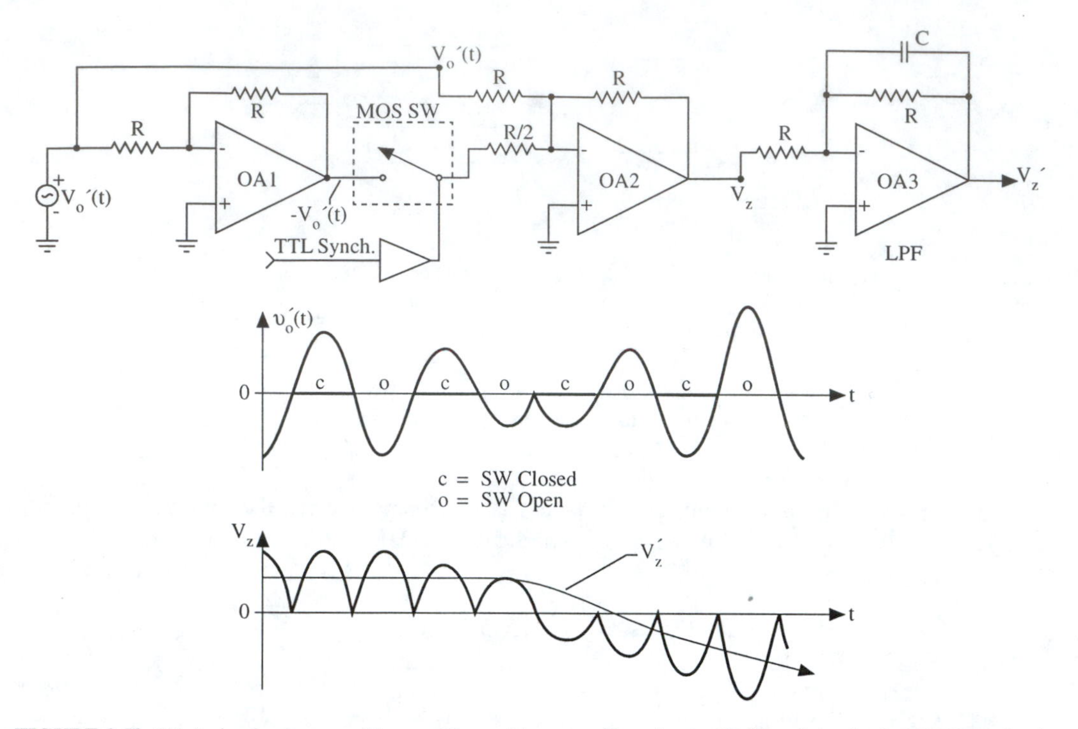

FIGURE 2.53 (A) A simple phase-sensitive rectifier and low-pass filter circuit. (B) Waveforms in the PSR/LPF circuit.

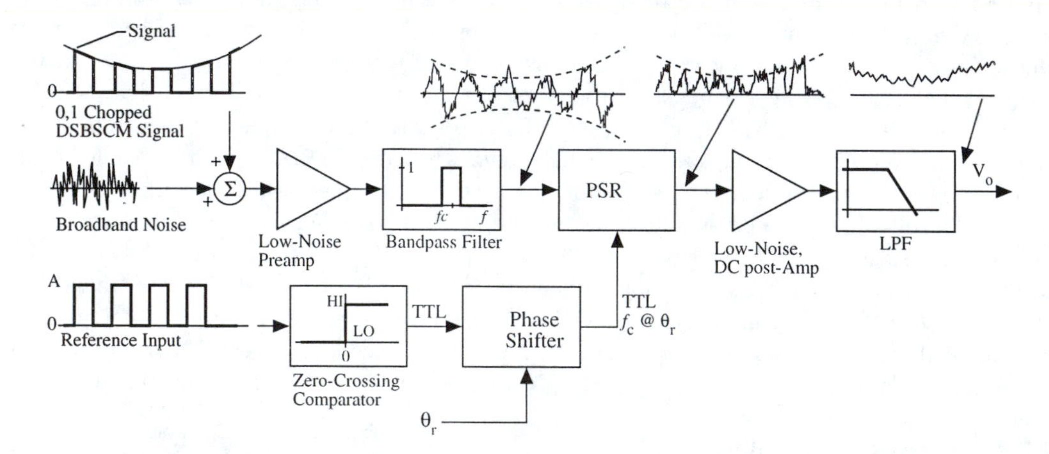

FIGURE 2.54 Block diagram of a lock-in amplifier. Input is a 0,1 chopped, noisy signal, such as the output of a photosensor. Amplification and band-pass filtering at the chopper frequency converts the input signal to a noisy AC signal which is then rectified by the PSR. The low-frequency modulating signal is seen at the LIA output. The low-pass filter bandwidth is adjusted to pass the desired modulating signal with the least amount of noise.

a band-pass amplifier whose center frequency is the same as that of the carrier. The band-pass signal conditioning improves the signal-to-noise ratio of the modulated carrier, which next passes through a PSR, usually of the switching type shown in Figure 2.53A. The PSR output is next conditioned by a DC pass low-pass filter, which generally has an adjustable cutoff frequency and attenuation slope. Some DC post-amplification by a low-noise, low-drift amplifier may precede the low-pass filter, if required. A reference signal derived directly from the carrier is used to operate the PSR.

For optimum performance of the PSR, the phase of the reference signal must be adjusted to be exactly in-phase with the carrier in the DSBSCM input. To illustrate this principle of LIA operation in the noise-free case, we let the modulating signal be a DC voltage:

$$v_m(t) = V_m \quad 2.169$$

The carrier and reference signal are sinusoids, respectively:

$$v_c(t) = V_c \sin(\omega_c t + \theta_c) \quad 2.170$$

$$v_r(t) = V_r \sin(\omega_c t + \theta_r) \quad 2.171$$

The DSBCSM signal is:

$$v_{mc}(t) = V_m V_c \sin(\omega_c t + \theta_c) \quad 2.172$$

Demodulation, as we have shown above, requires the multiplication of the modulated carrier with the reference signal. A scale factor K_D is introduced. Thus the demodulated signal before low-pass filtering can be written:

$$v_d(t) = K_D V_m V_c V_r \sin(\omega_c t + \theta_c)\sin(\omega_c t + \theta_r) \quad 2.173$$

After the sin(α)sin(β) trig. identity is applied, we have:

$$v_d(t) = (K_D V_m V_c V_r/2)\left[\cos(\theta_c - \theta_r) - \cos(2\omega_c t + \theta_c + \theta_r)\right] \quad 2.174$$

After low-pass filtering, the DC output of the LIA can be written:

$$V_o = V_m\left[K_D K_F V_c V_r/2\right]\cos(\theta_c - \theta_r) \quad 2.175$$

where K_F is the DC gain of the LPF. Note that the output is maximum when the phase of the reference signal and the carrier are equal, i.e., there is zero phase shift between the carrier and reference signal. Also, for the LIA to remain in calibration, K_D, K_F, V_c, and V_r must remain constant and known. Most modern LIAs have adjustable phase shifters in the reference channel to permit output sensitivity maximization, others perform this task automatically using phase-lock circuitry (McDonald and Northrop, 1993).

Many applications of LIAs lie in the area of electo-optical or photonics instrumentation. In these systems, the DSBSCM signal is derived by optically chopping a light beam. The chopper wheel generates a square-wave, 0,1 carrier, rather than a sinusoidal carrier. An LED and photodiode on the chopper can generate the reference signal, also a squarewave. The LIA's PSR demodulates squarewave DSBSCM signals by the same means that is used with a sinusoidal carrier.

Commercially available LIAs are multipurpose instruments, and as such, are expensive. For a dedicated LIA application, one can assemble the basic LIA system components from low-noise op amps and a commercial PSR IC, such as the AD630 or LM1596. The reference signal is converted to a TTL wave by a zero-crossing comparator. The TTL reference phase can be adjusted in a carrier frequency independent manner by using a phase-lock loop as a phase shifter (Northrop, 1990). Such a dedicated LIA can be built for less than U.S. $150, exclusive of power supplies.

2.7.5 Summary

The phase-sensitive rectifier is a circuit widely used in instrumentation systems to demodulate double-sideband, suppressed-carrier signals. We have presented a description of the DSBSCM signal and how it is created, and have shown two simple circuits that can be used to demodulate it. We have also mentioned a few of the commercially available ICs suitable for PSR use.

Probably the most important instrument based on the PSR is the lock-in amplifier, which can be used to recover the signals from DSBSCM carriers severely contaminated with noise, even noise in the signal passband. We algebraically illustrate the LIA demodulation process, and show the subsystems of an LIA.

2.8 CHAPTER SUMMARY

Chapter 2 has dealt with the analysis and design of analog signal conditioning circuits. Our treatment has been subdivided to include both linear and nonlinear signal processing systems. Integrated circuit designs have been stressed, and examples of commercially available ICs have been given. The chapter was opened with a basic treatment of differential amplifiers which are widely used at the front ends of op amps and instrumentation amplifiers to reduce common-mode interference.

Section 2.2 dealt with the ubiquitous op amp which finds many applications in linear and nonlinear signal processing. Basic design types of op amps were reviewed, and examples of their applications in linear signal processing were discussed. In Section 2.3, we reviewed basic quadratic analog active filter designs using conventional op amps. Although many, many active filter designs exist, we have focused on three of the more popular and easy to use designs: controlled-source active filters (including the well-known Sallen and Key designs), biquad active filters, and GIC-based active filter designs.

The design, characteristics, and uses of instrumentation amplifiers, including isolation amplifiers and autozero amplifiers, were reviewed.

Section 2.5 deals with nonlinear analog signal processing. In it we cite examples of commercially available function modules, and nonlinear circuits which can be built with op amps are presented. Considered are precision absval circuits, multifunction converters, true RMS converters, square root and divider circuits (other than multifunction converters), peak detectors, track and hold circuits, log ratio circuits, and a generalized trigonometric system.

Two examples of the use of the op amp charge amplifier configuration are given in Section 2.6: conditioning the output of piezoelectric crystal pressure transducers, and measuring charge on a capacitance by integration. The demodulation of double-sideband, suppressed-carrier signals generated by such systems as Wheatstone bridges with AC excitation and LVDTs is considered in Section 2.7 on phase-sensitive rectifiers.

3

Noise and Coherent Interference in Measurements

3.0 INTRODUCTION

Because of the differences in the sources and means of reduction of noise vs. coherent interference, we will treat them separately in sections of this chapter. Both noise and interference provide a major limitation to the precision of measurements, and the detectability of the quantity under measurement (QUM). Thus, a knowledge of how to design low-noise instrumentation systems is an important skill that a practicing instrumentation engineer should learn.

First, let us examine the distinction between noise and coherent interference. Noise is considered to arise in a circuit or measurement system from completely random phenomena. Any physical quantity can be "noisy," but we will generally restrict ourselves to the consideration of noise voltages and currents in this chapter. Such noise voltages and currents will be considered to have zero means, meaning that they have no additive DC components. The unwanted DC components are best considered to be drift or offset, and their reductions are more appropriately treated in a text on DC amplifier design or op amp applications.

Coherent interference, as the name suggests, generally has its origins in periodic, man-made phenomena, such as power line frequency electric and magnetic fields, radio frequency sources such as radio and television station broadcast antennas, certain poorly shielded computer equipment, spark discharge phenomena such as automotive ignitions and motor brushes and commutators, and inductive switching transients such as silicon controlled rectifier (SCR) motor speed controls, etc. As you will see, minimization of coherent interference is often "arty," and may involve changing the grounding circuits for a system, shielding with magnetic and/or electric conducting materials, filtering, the use of isolation transformers, etc.

Minimizing the impact of incoherent noise in a measurement system often involves a prudent choice of low-noise components, certain basic electronic design principles, and filtering. The incoherent noise that we will consider is random noise from within the measurement system; coherent noise usually enters a system from without.

3.1 DESCRIPTIONS OF RANDOM NOISE IN CIRCUITS

In the following discussions, we assume that the noise is *stationary*. That is, the physical phenomena giving rise to the noise are assumed not to change with time. When stationarity is assumed for a noise, averages over time are equivalent to ensemble averages. An example of a nonstationary noise source is a resistor which at time $t = 0$, begins to dissipate power so that its temperature slowly rises, as does its resistance.

Several statistical methods for describing random noise exist. These include, but are not limited to: the probability density function, the cross- and autocorrelation functions and their Fourier transforms, the cross- and auto-power density spectra, and the root power density spectrum, used to characterize amplifier equivalent input noise.

3.1.1 The Probability Density Function

The probability density function (PDF) considers only the amplitude statistics of a noise waveform, n(t), and not how n(t) varies in time. The PDF of n(t) is defined as:

$$p(x) = \frac{\text{probability that } x < n \le (x + dx)}{dx} \quad 3.1$$

where x is a specific value of n taken at some time t, and dx is a differential increment in x. The PDF is the mathematical basis for many formal derivations and proofs in statistics. The PDF has the properties:

$$\int_{-\infty}^{V} p(x)\,dx = \text{Prob}[x \le v] \quad 3.2$$

$$\int_{V_2}^{V_1} p(x)\,dx = \text{Prob}[x_1 < n \le x_2] \quad 3.3$$

$$\int_{-\infty}^{\infty} p(x)\,dx = 1 = \text{Prob}[x \le \infty] \quad 3.4$$

Several PDFs are widely used to describe the amplitude characteristics of electrical and electronic circuit noise. These include:

$$p(x) = \left(\frac{1}{\sigma_x \sqrt{2\pi}}\right) \exp\left[-\frac{(x - \bar{x})^2}{2\sigma_x^2}\right] \quad \text{Gaussian or normal PDF} \quad 3.5$$

$$\left.\begin{array}{l} p(x) = 1/2a, \text{ for } -a < x < a, \text{ and} \\ p(x) = 0, \text{ for } |x| > a \end{array}\right\} \quad \text{Rectangular PDF} \quad 3.6$$

$$p(x) = \left(\frac{x}{\alpha^2}\right) \exp\left[\frac{-x^2}{2\alpha^2}\right] \quad \text{Rayleigh PDF} \quad 3.7$$

Under most conditions, we assume that the random noise arising in electronic circuits has a *Gaussian* PDF. Many mathematical benefits follow this approximation; for example, the output of a linear system is Gaussian with variance σ_y^2 given the input to be Gaussian with variance σ_x^2. If Gaussian noise passes through a nonlinear system, the output PDF in general, will not be Gaussian.

Our concerns with noise in instrumentation systems will be focussed in the frequency domain. Here we will be concerned with the power density spectrum and the root power spectrum, as treated below. We will generally assume the noise PDFs are Gaussian.

3.1.2 The Power Density Spectrum

We will use a heuristic approach to illustrate the meaning of the power density spectrum (PDS) of noise. Formally, the two-sided PDS is the Fourier transform of the *autocorrelation function*, defined below by:

$$R_{nn}(\tau) = \lim_{T\to\infty} \frac{1}{2T} \int_{-T}^{T} n(t)n(t+\tau)\,dt \qquad 3.8$$

The two-sided PDS can be written as:

$$\Phi_{nn}(\omega) = \frac{1}{2\pi} \int_{-\infty}^{\infty} R_{nn}(\tau) e^{-j\omega\tau}\,d\tau \qquad 3.9$$

Because $R_{nn}(\tau)$ is an even function, its Fourier transform, $\Phi_{nn}(\omega)$ is also an even function, which stated mathematically, means:

$$\Phi_{nn}(\omega) = \Phi_{nn}(-\omega) \qquad 3.10$$

In the following discussion, we will consider the *one-sided PDS*, $S_n(f)$, which is related to the two-sided PDS by:

$$S_n(f) = 2\Phi_{nn}(2\pi f) \quad \text{for } f \geq 0, \text{ and} \qquad 3.11$$

$$S_n(f) = 0 \quad \text{for } f < 0 \qquad 3.12$$

To see how we may experimentally find the one-sided PDS, examine the system shown in Figure 3.1. Here, a noise voltage is the input to an ideal low-pass filter with adjustable cut-off frequency, f_c. The output of the ideal low-pass filter is measured with a broad-band, true RMS, AC voltmeter. We begin with $f_c = 0$, and systematically increase f_c, each time recording the square of the RMS meter reading, or mean squared output voltage, $\overline{v_{on}^2}$, of the filter. As f_c is increased, $\overline{v_{on}^2}$ increases monotonically, as shown in Figure 3.2. Because of the finite bandwidth of the noise source, $\overline{v_{on}^2}$ eventually reaches an upper limit which is the total mean-squared noise voltage of the noise source. The plot of $\overline{v_{on}^2}(f_c)$ vs. f_c is called the *cumulative mean-squared noise characteristic* of the noise source. Its units are mean squared volts, in this example. A simple interpretation of the one-sided, noise power density spectrum is that it is the derivative, or slope, of the mean squared noise characteristic curve. Mathematically stated, this is:

$$S_n(f) = \frac{d\overline{v_{on}^2}(f)}{df} \quad \text{mean squared volts/Hz} \qquad 3.13$$

A plot of a typical noise PDS is shown in Figure 3.3. Note that a practical PDS generally drops off to zero at high frequencies.

A question which is sometimes asked by those first encountering the PDS concept is, why is it called a power density spectrum? The power concept has its origin in the consideration of noise in communication systems, and has little meaning in the consideration of noise in instrumentation systems. One way to rationalize the power term is to consider a noise voltage source with a 1 ohm

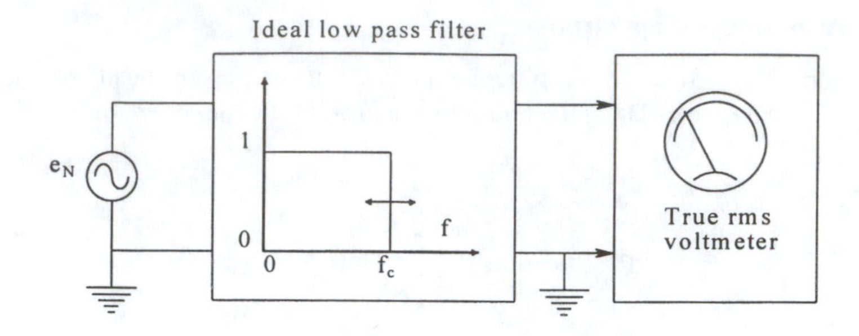

FIGURE 3.1 System relevant to the derivation of the one-sided power density spectrum.

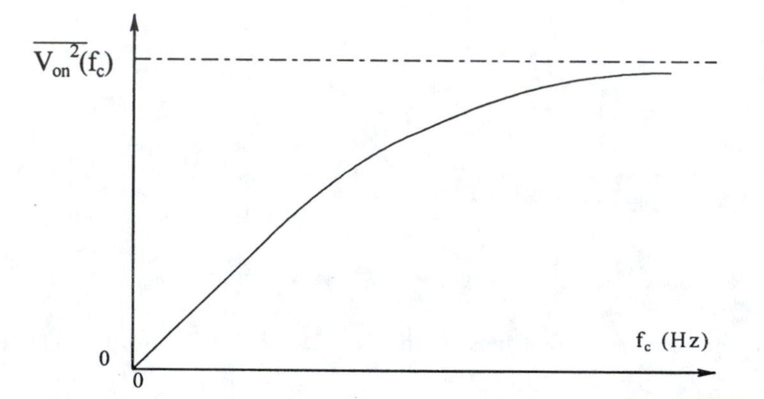

FIGURE 3.2 Mean squared output noise of the system shown in Figure 3.1 as a function of ideal low-pass filter cut-off frequency.

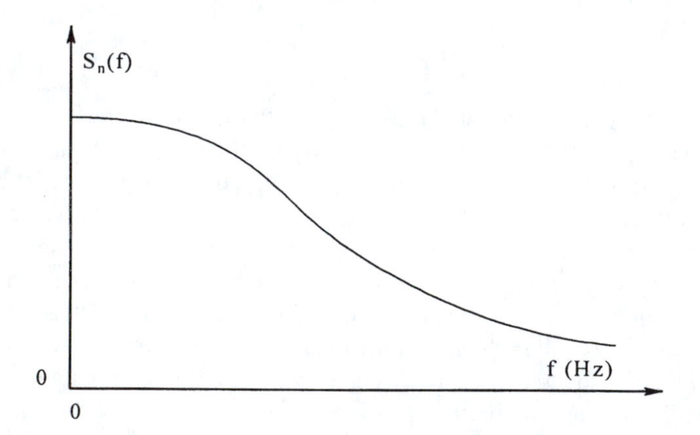

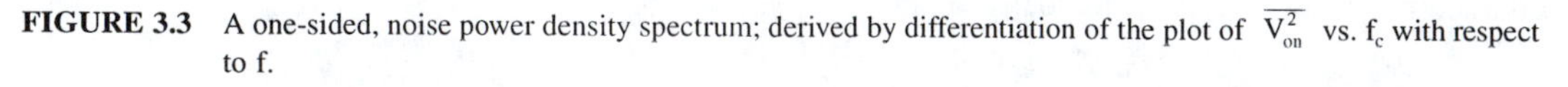

FIGURE 3.3 A one-sided, noise power density spectrum; derived by differentiation of the plot of $\overline{V_{on}^2}$ vs. f_c with respect to f.

load. In this case, the average power dissipated in the resistor is simply the total mean squared noise voltage.

From our heuristic definition of the PDS above, we see that the total mean squared voltage in the noise voltage source can be written as:

$$\overline{v_{on}^2} = \int_0^{\infty} S_n(f)\, df \tag{3.14}$$

The mean squared voltage in the frequency interval, (f_1, f_2), is found by:

$$\overline{v_{on}^2}(f_1, f_2) = \int_{f_1}^{f_2} S_n(f)\, df \tag{3.15}$$

Often noise is specified or described using *root power density spectra*, which are simply plots of the square root of $S_n(f)$ vs. f, and which have the units of RMS volts (or other units) per root Hz.

Special (ideal) PDSs are used to model or approximate portions of real PDSs. These include the white noise PDS, and the one-over-f PDS. A white noise PDS is shown in Figure 3.4. Note that this PDS is flat; this implies that

$$\int_{0}^{\infty} S_{nw}(f)\, df = \infty \tag{3.16}$$

which is clearly not realistic. A 1/f PDS is shown in Figure 3.5. The 1/f spectrum is often used to approximate the low-frequency behavior of real PDSs. Physical processes which generate 1/f-like noise include surface phenomena associated with electrochemical electrodes, carbon composition resistors carrying direct current (metallic resistors are substantially free of 1/f noise), and surface imperfections affecting emission and diffusion phenomena in semiconductor devices. The presence of 1/f noise can present problems in the electronic signal conditioning sytems used for low-level, low-frequency, and DC measurements.

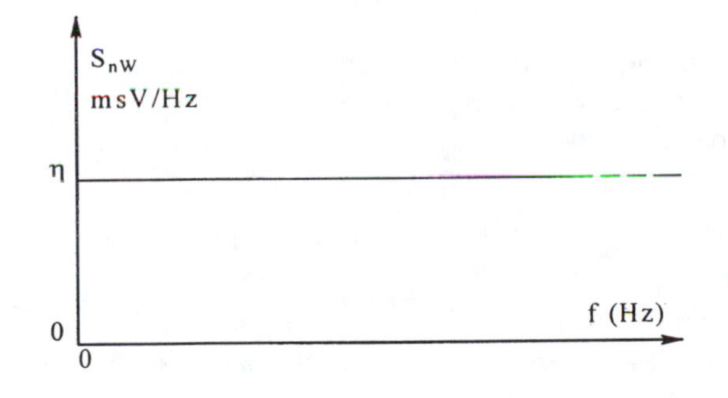

FIGURE 3.4 A white noise power density spectrum.

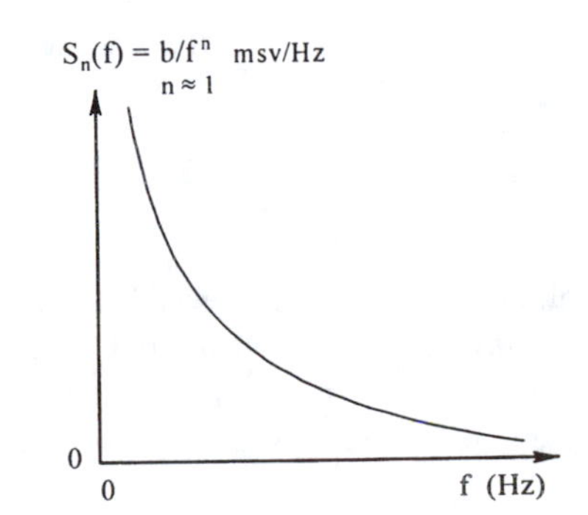

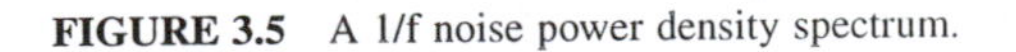

FIGURE 3.5 A 1/f noise power density spectrum.

3.1.3 Sources of Noise in Signal Conditioning Systems

Sources of noise in signal conditioning systems can be separated into two major categories: noise from passive resistors, and noise from active circuit elements such as bipolar junction transistors, field effect transistors, and vacuum tubes. In most cases, the Gaussian assumption for noise amplitude PDFs is valid, and the noise generated can generally be assumed to be white over a major portion of spectrums.

3.1.3.1 Noise from Resistors

From statistical mechanics, it can be shown that any pure resistance at some temperature T, K will have a zero mean noise voltage associated with it. This noise voltage appears in series with the (noiseless) resistor as a Thevenin equivalent circuit. From DC to radio frequencies where the resistor's capacitance to ground and its lead inductance can no longer be neglected, the resistor's noise is well-modeled by a Gaussian white noise source. Noise from resistors is called *thermal* or *Johnson noise*. Its one-sided PDS is given by:

$$S_n(f) = 4kTR \quad \text{mean squared volts/Hz} \tag{3.17}$$

where k is Boltzmann's constant (1.38×10^{-23}), T is in degrees Kelvin, and R is in ohms. In a given bandwidth, the mean squared noise from a resistor can be written:

$$v_{on}^2(B) = \int_{f_1}^{f_2} S_n(f)\, df = 4kTR(f_2 - f_1) = 4kTRB \quad \text{mean squared volts} \tag{3.18}$$

B is the equivalent Hz noise bandwidth.

A Norton equivalent of the Johnson noise source from a resistor can be formed by assuming a white noise current source with PDS:

$$S_{ni}(f) = 4kTG \quad \text{MS Amps/Hz} \tag{3.19}$$

in parallel with a noiseless conductance, G = 1/R.

The Johnson noise from several resistors connected in a network may be combined into a single noise voltage source in series with a single, noiseless, equivalent resistor. Figure 3.6 illustrates some of these reductions for two-terminal circuits.

It has been observed that when DC (or average) current is passed through a resistor, the basic Johnson noise PDS is modified by the addition of a 1/f spectrum.

$$S_n(f) = 4kTR + AI^2/f \quad \text{MSV/Hz} \tag{3.20}$$

where I is the average or DC component of current through the resistor, and A is a constant that depends on the material from which the resistor is constructed (e.g., carbon composition, resistance wire, metal film, etc.). An important parameter for resistors carrying average current is the crossover frequency, f_c, where the 1/f PDS equals the PDS of the Johnson noise. This is:

$$f_c = AI^2/4kTR \quad \text{Hz} \tag{3.21}$$

It is possible to show that the f_c of a noisy resistor can be reduced by using a resistor of the same type, but having a higher wattage or power dissipation rating. As an example of this principle,

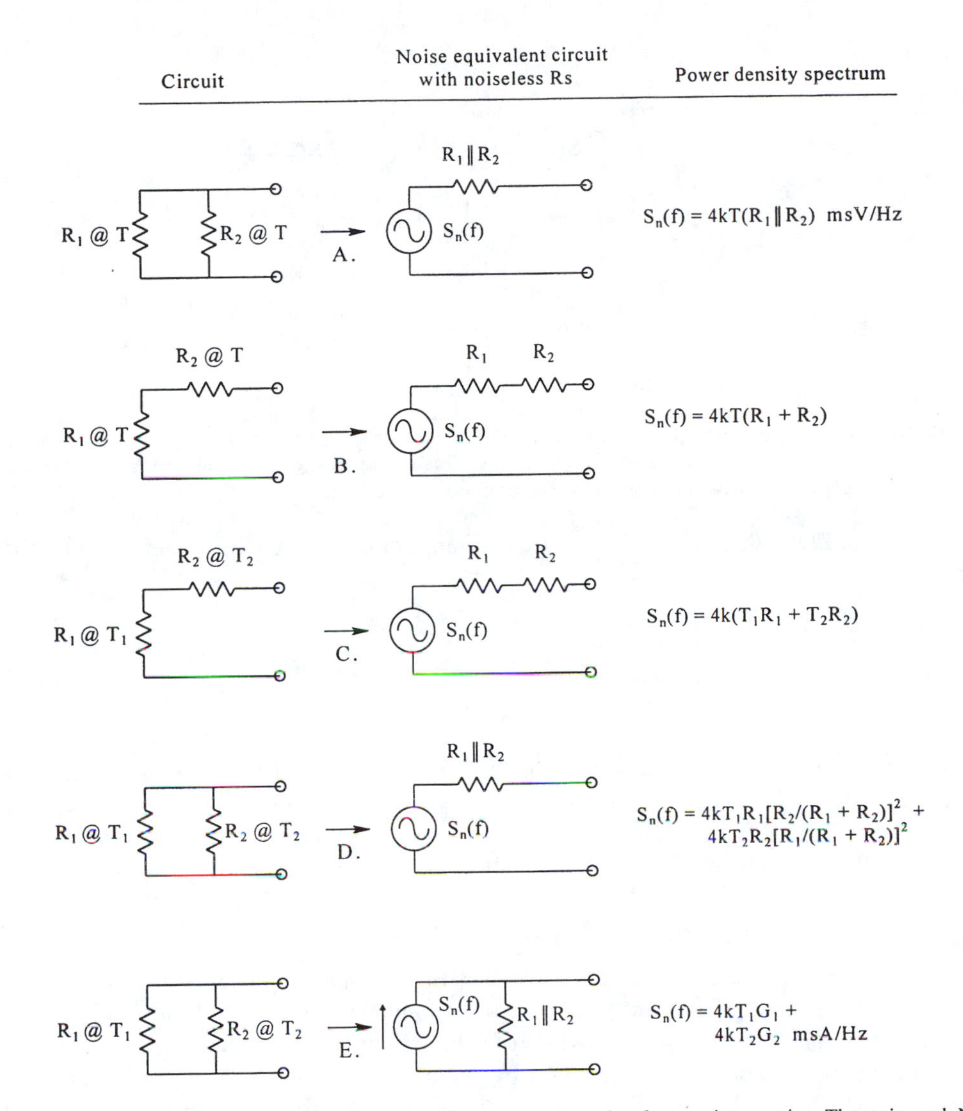

FIGURE 3.6 Means of representing the thermal or Johnson white noise from resistors using Thevenin and Norton equivalent circuits. Noises are always added as mean squared quantities.

consider the circuit of Figure 3.7, in which a single resistor of R ohms, carrying a DC current I is replaced by *nine* resistors of resistance R connected in series-parallel to give a net resistance R which dissipates nine times the power of a single resistor R. The noise PDS in any one of the nine resistors is seen to be:

$$S_n'(f) = 4kTR + A(I/3)^2/f \quad \text{MSV/Hz} \tag{3.22}$$

Each of the nine PDSs given by Equation 3.22 above contributes to the net PDS seen at the terminals of the composite 9 W resistor. Each resistor's equivalent noise voltage source "sees" a voltage divider formed by the other eight resistors in the composite resistor. The attenuation of each of the nine voltage dividers is given by:

$$\frac{3R/2}{3R/2 + 3R} = 1/3 \tag{3.23}$$

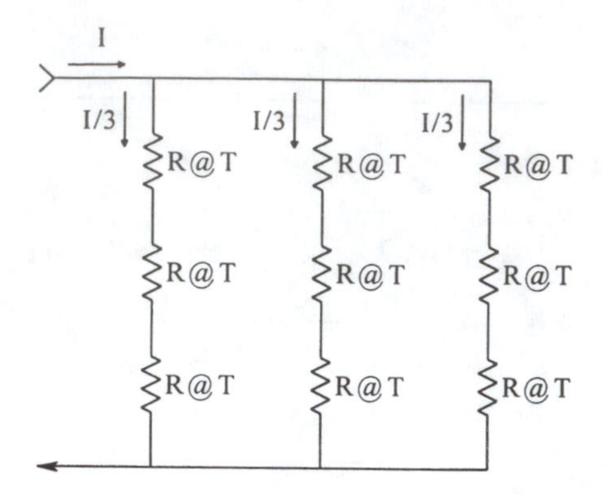

FIGURE 3.7 Equivalent circuit for a ninefold increase in resistor wattage rating, the resistance value remaining the same. Note that an N-fold increase in wattage does not affect the thermal noise from the resistor, but does give an N-fold reduction in the 1/f noise spectrum from the resistor.

The net voltage PDS at the composite resistor's terminals may be found by superposition of PDSs. It is:

$$S_{n(9)}(f) = \sum_{j=1}^{9} \left[4kTR + A(I/3)^2/f\right](1/3)^2 = 4kTR + AI^2/9f \quad \text{MSV/Hz} \qquad 3.24$$

Thus the composite 9 W resistor enjoys a ninefold reduction in the 1/f spectral power because the DC current density through each element is reduced by 1/3. The Johnson noise PDS remains the same, however. It is safe to generalize that the use of high wattage resistors of a given type and resistance will result in reduced 1/f noise generation when the resistor carries DC (average) current.

3.1.3.2 The Two-Source Noise Model for Active Devices

Noise arising in junction field effect transistors (JFETs), bipolar junction transistor amplifiers (BJTs), and other complex integrated circuit (IC) amplifiers is generally described by the *two-source input model.* The total noise observed at the output of an amplifier, given that its input terminals are short-circuited, is accounted for by defining an *equivalent short-circuited input noise voltage*, e_{na}, which replaces all internal noise sources affecting the amplifier output under short-circuited input conditions. The amplifier, shown in Figure 3.8, is now considered noiseless. e_{na} is specified by manufacturers for many low-noise, discrete transistors and IC amplifiers. e_{na} is a root PDS, i.e., it is a function of frequency, and has the units of RMS volts per root Hz. Figure 3.9 illustrates a plot of a typical $e_{na}(f)$ vs. f for a low-noise JFET. Also shown in Figure 3.9 is a plot of $i_{na}(f)$ vs. f for the same device.

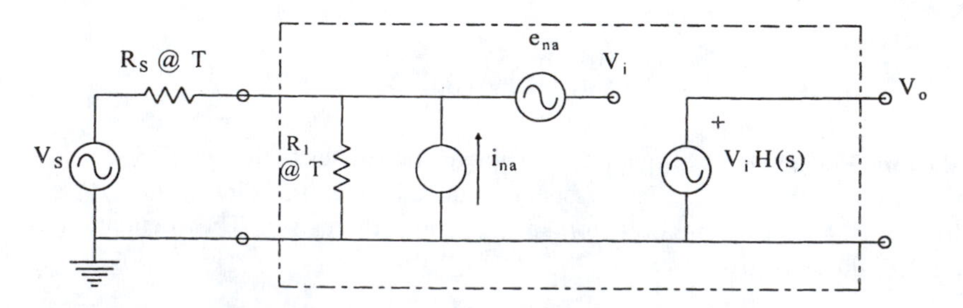

FIGURE 3.8 Equivalent two-source model for a noisy amplifier. See text for discussion.

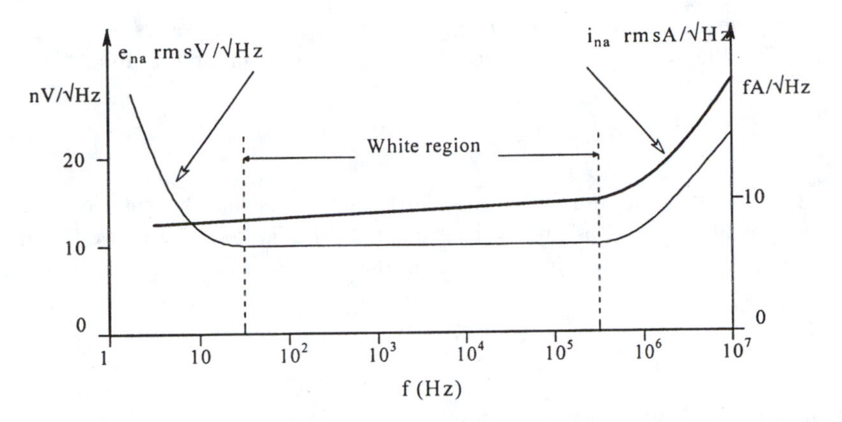

FIGURE 3.9 Plots of typical input noise root power spectrums for an FET input amplifier.

In addition to the equivalent short-circuited input noise voltage, the modeling of the net noise characteristics of amplifiers requires the inclusion of a current noise source, i_{na}, as shown in Figure 3.8. $i_{na}(f)$ is the root PDS of the *input equivalent noise current*; its units are RMS Amps per root Hz. Note that both $e_{na}(f)$ and $i_{na}(f)$ have flat, mid-frequency portions which invite approximation by white noise sources. At high frequencies, both equivalent noise root PDSs slope upward. $e_{na}(f)$ for discrete JFETs and BJTs, and IC amplifiers shows a $1/\sqrt{f}$ region at low frequencies.

3.1.3.3 Noise in JFETs

Certain selected, discrete JFETs are used in the design of low-noise headstages for signal conditioning in measurement systems. Some JFETs give exceptional low-noise performance in the audio and sub-audio frequency regions of the spectrum, while others excel in the video and radio frequency end of the spectrum giving them applications for RF oscillators, mixers, and tuned amplifiers.

van der Ziel (1974) has shown that the theoretical thermal noise generated in the conducting channel of a JFET can be approximated by an equivalent short-circuited input noise with PDS given by:

$$e_{na}^2 = 4kT/g_m = 4kT \Big/ g_{m0}\sqrt{\frac{I_{DSS}}{I_{DQ}}} \quad \text{MSV/Hz} \qquad 3.25$$

where g_{m0} is the FET's small-signal transconductance measured when $V_{GS} = 0$, $I_D = I_{DSS}$, I_{DSS} = the DC drain current measured for $V_{GS} = 0$ and $V_{DS} > V_P$, and I_{DQ} = the quiescent DC drain current at the FET's operating point where $V_{GS} = V_{GSQ}$. In reality, due to the presence of 1/f noise, the theoretical short-circuited input voltage PDS expression can be modified:

$$e_{na}^2(f) = (4kT/g_m)(1 + f_c/f^n) \quad \text{MSV/Hz} \qquad 3.26$$

The exponent n has the range $1 < n < 1.5$, and is device and lot determined. n is usually set equal to 1 for algebraic simplicity. The origins of the $1/f^n$ effect in JFETs are poorly understood. Note that e_{na} given by Equation 3.26 is temperature-dependent. Heat sinking or actively cooling the JFET will reduce e_{na}. The parameter f_c used in Equation 3.26 is the corner frequency of the 1/f noise spectum. Depending on the device, it can range from below 10 Hz to above 1 kHz. f_c is generally quite high in RF and video frequency JFETs because in this type of transistor, $e_{na}(f)$ dips to around 2 nV/$\sqrt{\text{Hz}}$ in the 10^5 to 10^7 Hz region, which is quite good.

Reverse-biased JFET gates have a DC leakage current, I_{GL}, which produces shot noise which is superimposed on the leakage current. This noise component in I_{GL} is primarily due to the random occurrence of charge carriers that have enough energy to cross the reverse-biased gate-channel diode junction. The PDF of the gate current shot noise is Gaussian, and its PDS is given by:

$$i_{na}^2 = 2qI_{GL} \quad \text{MSA/Hz} \qquad 3.27$$

where $q = 1.602 \times 10^{-19}$ coulomb (electron charge) and I_{GL} is the DC gate leakage current in amperes. I_{GL} is typically about 2 pA, hence i_{na} is about 1.8 fA RMS/$\sqrt{\text{Hz}}$ in the flat mid-range of $i_{na}(f)$. $i_{na}(f)$, like $e_{na}(f)$, shows a 1/f characteristic at low frequencies. This can be modeled by:

$$i_{na}^2(f) = 2qI_{GL}\left(1 + f/f_{ic}\right) \quad \text{MSA/Hz} \qquad 3.28$$

where f_{ic} is the current noise corner frequency.

For some transistors, the measured $e_{na}(f)$ and $i_{na}(f)$ have been found to be greater than the predicted theoretical values; in other cases, they have been found to be less. No doubt the causes for these discrepancies lie in the oversimplifications used in their derivations.

3.1.3.4 Noise in BJTs

The values of e_{na} and i_{na} associated with bipolar junction transistor amplifiers (BJTs) depend strongly on the device operating (Q) point. This is because there are shot noise components superimposed on the quiescent base and collector currents. The small-signal model of a grounded-emitter BJT amplifier is shown in Figure 3.10B. In this circuit, we assume negligible noise from R_L, the voltage-controlled current source, g_m v_{be}, and the small-signal base input resistance, r_π. The shot noise sources are

$$i_{nb}^2 = 2qI_{BQ} \quad \text{MSA/Hz} \qquad 3.29$$

$$i_{nc}^2 = 2q\beta I_{BQ} \quad \text{MSA/Hz} \qquad 3.30$$

In this example, it is algebraically simpler to not find the equivalent input e_{na} and i_{na}. Rather, we work directly with the two white shot noise sources in the mid-frequency, hybrid pi, small-signal model. It can be shown (Northrop, 1990) that the total output noise voltage PDS is given by:

$$S_{NO} = 4kTR_L + 2q\left(I_{BQ}\beta\right)\left(\beta R_L\right)^2 + \frac{4kTR_s'\left(\beta R_L\right)^2 + 2qI_{BQ}R_s'^2\left(\beta R_L\right)^2}{\left(V_T/I_{BQ} + R_s'\right)^2} \quad \text{MSV/Hz} \qquad 3.31$$

where it is clear that r_π is approximated by V_T/I_{BQ}, $R_s' = r_x + R_s$, and we will neglect the Johnson noise from R_L because it is numerically small compared to the other terms. It is easy to show that the mean squared output signal can be written as

$$\overline{v_{os}^2} = v_s^2\left(\beta R_L\right)^2 / \left(V_T/I_{BQ} + R_s'\right)^2 \quad \text{MSV} \qquad 3.32$$

Thus, the mean squared signal-to-noise ratio at the amplifier output can be written:

$$SNR_O = \frac{\overline{v_s^2}/B}{4kTR_s' + 2qI_{BQ}R_s'^2 + 2q\left(I_{BQ}\beta\right)\left(V_T/I_{BQ} + R_s'\right)^2} \qquad 3.33$$

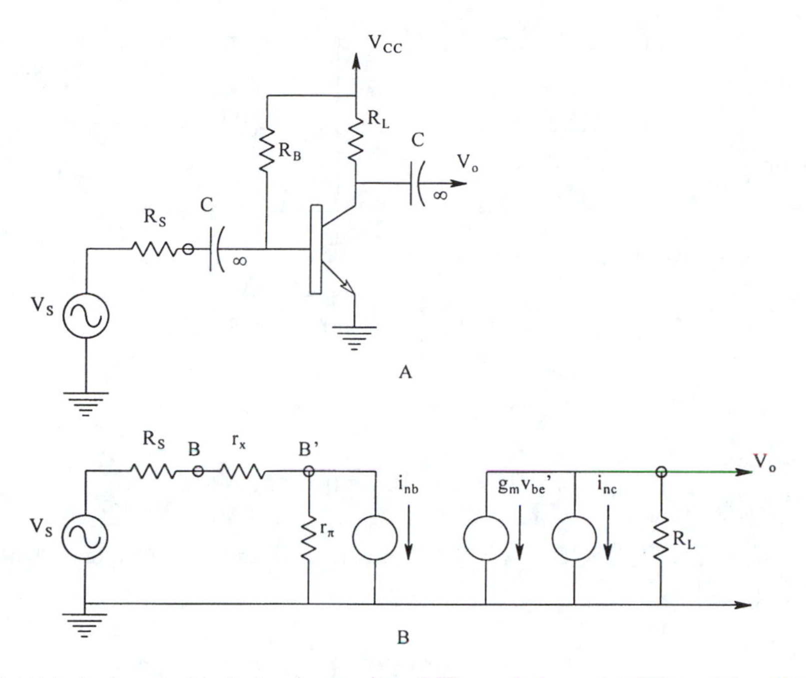

FIGURE 3.10 (A) A simple, noisy bipolar junction transistor (BJT), reactively coupled (RC) amplifier. (B) Mid-frequency equivalent circuit of the BJT amplifier. We assume that $R_B >> r_\pi$. r_π and R_L are noiseless, and $R_S' = r_x + R_S$ makes thermal noise. i_{nb} and i_{nc} are white shot noise current sources.

where B is the specified, equivalent (Hz) noise bandwidth for the system.
The SNR_O given by Equation 3.33 has a maximum for some non-negative I_{BQMAX}. The I_{BQMAX} which will give this maximum can be found by differentiating the denominator of Equation 3.33 and setting the derivative equal to zero. This gives:

$$I_{BQMAX} = \frac{V_T}{R_s'}\sqrt{\frac{\beta}{\beta+1}} \text{ Amperes} \tag{3.34}$$

What we should remember from the above exercise is that the best noise performance for BJT amplifiers is a function of quiescent biasing conditions (Q-point). Often these conditions have to be found by trial and error when working at high frequencies. While individual BJT amplifiers may best be modeled for noise analysis with the two shot noise current sources, as we did above, it is more customary when describing complex, BJT IC amplifier noise performance to use the more general and more easily used e_{na} and i_{na}, two-source model.

3.2 PROPAGATION OF GAUSSIAN NOISE THROUGH LINEAR FILTERS

In a formal, rigorous treatment of noise in linear systems, it is possible to show that the PDF of the output of a linear system is Gaussian, given a Gaussian input. In addition, the PDS of the system's output noise is given by:

$$S_y(f) = S_x(f)\left|\mathbf{H}(2\pi f j)\right|^2 \tag{3.35}$$

This is the scalar product of the positive real input PDS and the magnitude squared of the system's transfer function. This relation can be extended to include two or more cascaded systems, as shown in Figure 3.11.

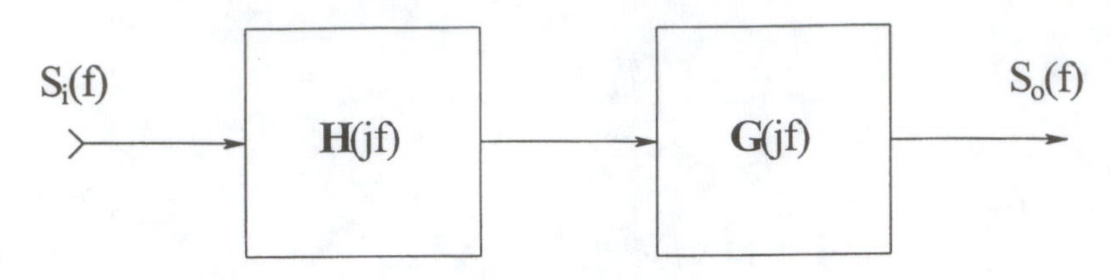

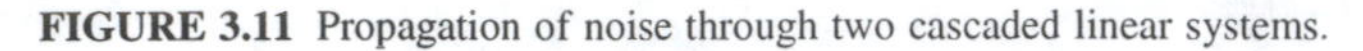

FIGURE 3.11 Propagation of noise through two cascaded linear systems.

$$S_y(f) = S_x(f)\left|\mathbf{H}(2\pi fj)\right|^2 \left|\mathbf{G}(2\pi fj)\right|^2 \quad \text{MSV/Hz} \tag{3.36}$$

or

$$S_y(f) = S_x(f)\left|\mathbf{H}(2\pi fj)\,\mathbf{G}(2\pi fj)\right|^2 \quad \text{MSV/Hz} \tag{3.37}$$

If white noise with a PDS, $S_x(f) = A$ MSV/Hz is the input to a linear system, then the output PDS is simply

$$S_y(f) = A\left|\mathbf{H}(2\pi fj)\right|^2 \quad \text{MSV/Hz} \tag{3.38}$$

The total mean squared output noise of the system is

$$v_{on}^2 = \int_0^\infty S_y(f)\,df = A\int_0^\infty \left|\mathbf{H}(2\pi fj)\right|^2 df \quad \text{MSV} \tag{3.39}$$

The right-hand integral of Equation 3.39 may be shown to be, for transfer functions with one more finite pole than zeros, the product of the transfer function's low-frequency or mid-band gain squared times the filter's equivalent Hz noise bandwidth. The filter's *gain squared-bandwidth* product is thus:

$$\text{GAIN}^2\text{BW} = \int_0^\infty \left|H(2\pi fj)\right|^2 df \tag{3.40}$$

Gain^2-bandwidth integrals have been evaluated using complex variable theory for a number of transfer functions (see James, Nichols, and Philips, 1947). In Table 3.1, we give the gain^2-bandwidth integrals for five common transfer functions. Note that the equivalent noise bandwidths (in brackets in each case) are in hertz, not radians/second. Also note the absence of 2π factors in these expressions. Gain^2-bandwidth integrals are used to estimate the total mean squared output noise from amplifiers having (approximate) white noise input sources, and are thus useful in calculating output signal-to-noise ratios.

3.3 BROADBAND NOISE FACTOR AND NOISE FIGURE OF AMPLIFIERS

An amplifier's *noise factor* is defined as the ratio of the signal-to-noise ratio at the amplifier's input to the signal-to-noise ratio at the amplifier's output. Since a real amplifier is noisy and adds noise to the signal as well as amplifying it, the output signal-to-noise ratio (SNR_O) is always less than the input signal-to-noise ratio (SNR_I), hence the *noise factor* is always greater than one for a noisy amplifier.

TABLE 3.1 Noise Gain² Bandwidth Products for Some Common Transfer Functions

	Transfer Function, H(jω)	Gain² Bandwidth (Hz)
1.	$\frac{K_V}{j\omega\tau + 1}$	$K_V^2[1/4\tau]$
2.	$\frac{K_V}{(j\omega\tau_1 + 1)(j\omega\tau_2 + 1)}$	$K_V^2[1/(4\{\tau_1 + \tau_2\})]$
3.	$\frac{K_V}{(1 - \omega^2/\omega_n^2) + j\omega(2\zeta/\omega_n)}$	$K_V^2[\omega_n/8\zeta]$
4.	$\frac{j\omega(2\zeta/\omega_n)K_V}{(1 - \omega^2/\omega_n^2) + j\omega(2\zeta/\omega_n)}$	$K_V^2[\omega_n\zeta] = K_V^2[\omega_n/2Q]$
5.	$\frac{j\omega\tau_1 K_V}{(j\omega\tau_1 + 1)(j\omega\tau_2 + 1)}$	$K_V^2[1/4\tau_2(1 + \tau_2/\tau_1)]$

$$F \equiv \frac{SNR_I}{SNR_O} \tag{3.41}$$

The *noise figure* is defined as

$$NF = 10\log_{10}(F) \tag{3.42}$$

when the SNRs are in terms of mean squared quantities.

Figure 3.12 illustrates a minimum model for a noisy amplifier. Here we assume the spectra of e_{na} and i_{na} are white, and that $R_1 >> R_s$. The mean squared input signal is $S_i = v_s^2$, the mean squared output signal is $S_O = K_V^2 v_s^2$, where K_V^2 is the amplifier's mid-band gain. The mean squared input noise is simply that associated with v_s. (here zero) plus the Johnson noise from the source resistance, R_s, in a specified Hz noise bandwidth, B. It is:

$$N_i = 4kTR_sB \quad \text{MSV} \tag{3.43}$$

The mean squared noise at the amplifier's output, N_O, is composed of three components: one from the R_s Johnson noise and two from the equivalent noise sources. N_O can be written as:

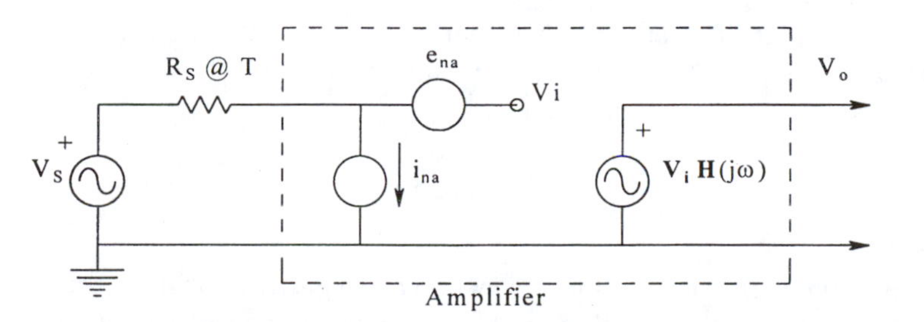

FIGURE 3.12 Minimum noisy amplifier model relevant to the calculation of amplifier noise figure.

$$N_O = \left(4kTR_s + e_{na}^2 + i_{na}^2 R_s^2\right)\int_0^\infty \left|\mathbf{H}(2\pi fj)\right|^2 df$$
$$= \left(4kTR_s + e_{na}^2 + i_{na}^2 R_s^2\right)K_V^2 B \quad \text{MSV} \qquad 3.44$$

Using the definition for F, we find that the noise factor for the simple noisy amplifier model can be written as:

$$F = 1 + \frac{e_{na}^2 + i_{na}^2 R_s^2}{4kTR_s} \qquad 3.45$$

Note that this expression for F contains no bandwidth terms; they cancel out. When NF is given for an amplifier, R_s must be specified, as well as the Hz bandwidth B over which the noise is measured. The temperature should also be specified, although common practice usually sets T at 300 K.

For practical amplifiers, NF and F are functions of frequency because e_{na} and i_{na} are functions of frequency (see Figure 3.9). For a given R_s, F tends to rise at low frequencies due to the 1/f components in the equivalent input noise sources. F also increases at high frequencies; again, due to the high-frequency increases in e_{na} and i_{na}. Often, we are interested in the noise performance of an amplifier in either low or high frequencies where NF and F are not minimum. To examine the detailed noise performance of an amplifier at low and high frequencies, we use the *spot noise figure,* discussed below.

3.4 SPOT NOISE FACTOR AND FIGURE

Spot noise measurements are made through a narrow band-pass filter (BPF) in order to evaluate an amplifier's noise performance in a certain frequency range, particularly where $e_{na}(f)$ and $i_{na}(f)$ are not constant, such as in the 1/f range. Figure 3.13A illustrates a set of spot noise figure contours for a commercial, low-noise preamplifier having applications at audio frequencies.

A system for determining an amplifier's spot noise figure is shown in Figure 3.14. An adjustable white noise source is used at the amplifier's input. Note that the output resistance of the white noise source plus some external resistance must add up to R_s, the specified Thevenin equivalent input resistance. The system is used as follows: first, the BPF is set to the desired center frequency and the white noise generator is set to $e_N = 0$. We assume that the total mean squared noise at the system output under these conditions can be written as:

$$N_o(f_c) = \left[4kTR_s + e_{na}^2(f_c) + i_{na}^2(f_c)R_s^2\right]K_V^2 B_F \quad \text{MSV} \qquad 3.46$$

where $e_{na}(f_c)$ is the value of e_{na} at the center frequency, f_c, B_F is the equivalent noise bandwidth of the BPF, and K_V is the combined gain of the amplifier under measurement at f_c, the BPF at f_c, and the post-amplifier. K_V can be written:

$$K_V = \left|H(2\pi f_c j)\right| K_F K_A \qquad 3.47$$

Second, the white noise source is made non-zero and adjusted so that the true RMS meter reads $\sqrt{2}$ higher than in the first case with $e_N = 0$. The MS output voltage can now be written:

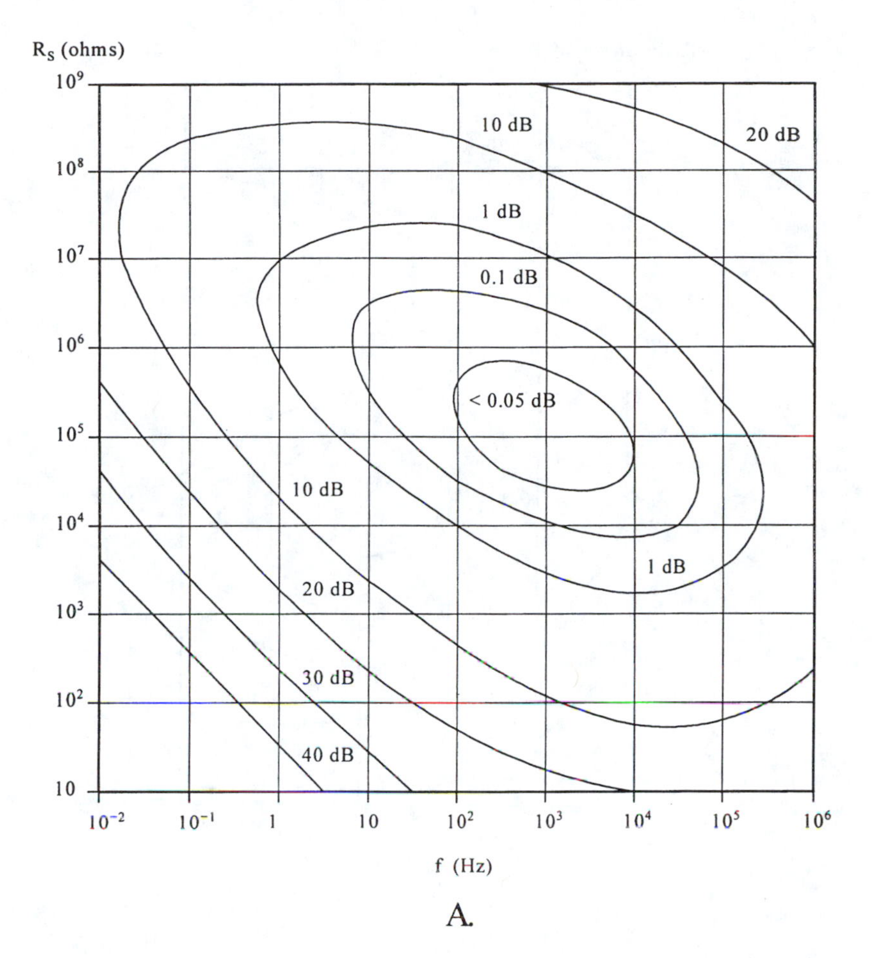

FIGURE 3.13 (A) Spot noise figure contours for a typical amplifier with a direct input having a Thevenin source resistance R_S at temperature T K.

$$\begin{aligned} N_o'(f_c) &= 2N_o(f_c) = 2\left[4kTR_s + e_{na}^2(f_c) + i_{na}^2(f_c)R_s^2\right]K_V^2B_F \\ &= \left[e_N^2 + 4kTR_s + e_{na}^2(f_c) + i_{na}^2(f_c)R_s^2\right]K_V^2B_F \end{aligned} \tag{3.48}$$

Under this condition, it is evident that

$$e_N^2 = 4kTR_s + e_{na}^2(f_c) + i_{na}^2(f_c)R_s^2 \tag{3.49}$$

Hence:

$$\left[e_{na}^2(f_c) + i_{na}^2(f_c)R_s^2\right] = \left[e_N^2 + 4kTR_s\right] \tag{3.50}$$

If the left-hand side of Equation 3.50 is substituted into Equation 3.45 for the noise factor F, we find:

$$F_{spot} = e_N^2/4kTR_s \tag{3.51}$$

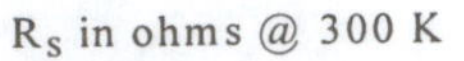

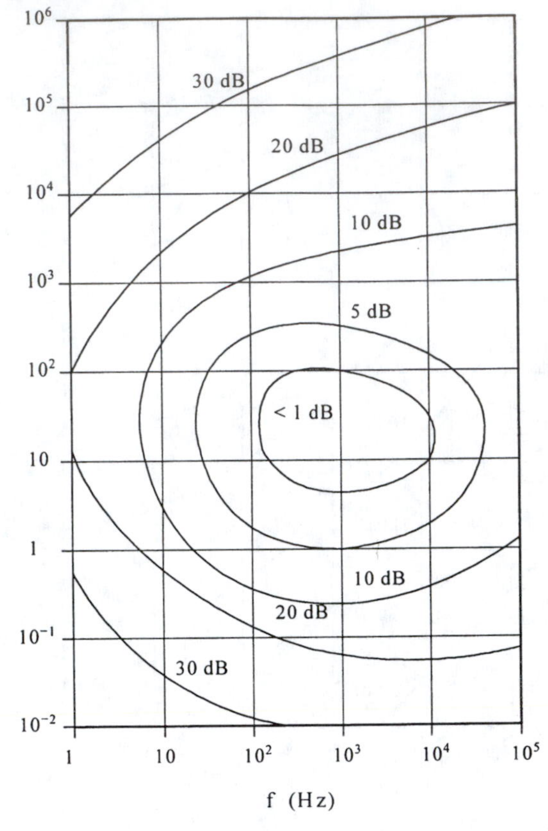

B.

FIGURE 3.13 (B) Noise figure contours for the same amplifier using a 1:100 turns ratio input transformer. Note the shift of the minimum contour to a lower range of R_S.

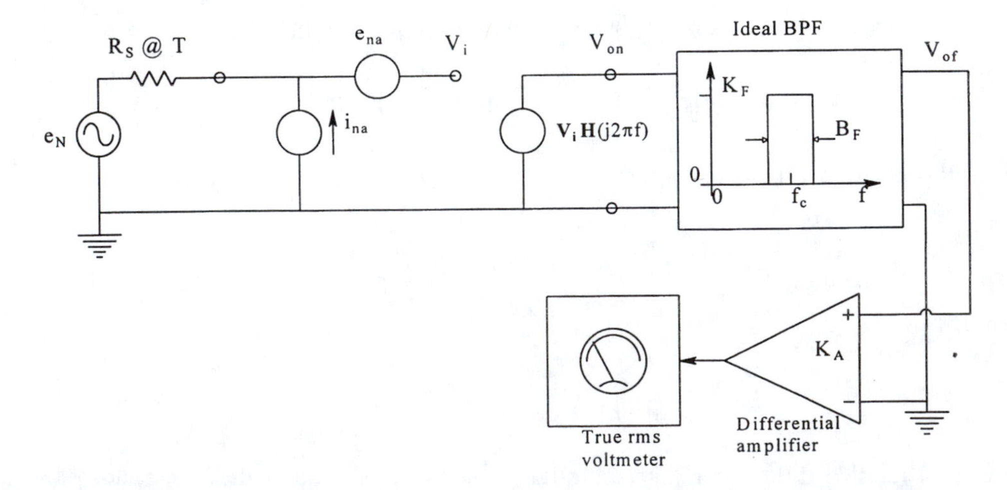

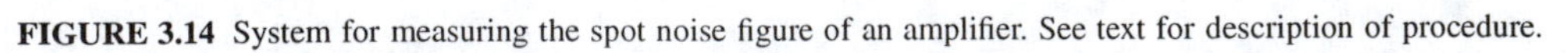
FIGURE 3.14 System for measuring the spot noise figure of an amplifier. See text for description of procedure.

Note that this expression for the spot noise factor does not contain specific terms for the measurement center frequency, f_c, or the BPF Hz noise bandwidth, B_F, yet these parameters should be

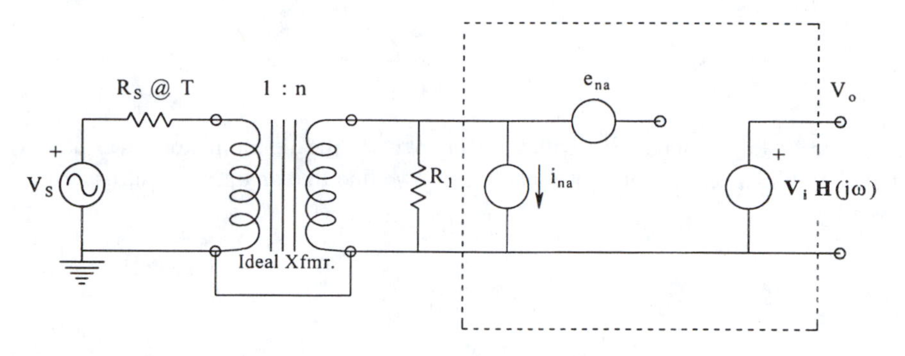

FIGURE 3.15 Use of an input transformer to improve an amplifier's output SNR when R_S is small but finite.

specified when giving F_{spot} for an amplifier. F_{spot} is actually calculated by setting f_c and R_s, then determining the e_N^2 value that doubles the mean squared output noise. This value of e_N^2 is then divided by the calculated white noise spectrum from the resistor R_s to give a value for F_{spot} at a given frequency and R_s.

It is also possible to measure F_{spot} using a sinusoidal source of frequency f_c instead of the calibrated white noise source, e_N.

3.5 TRANSFORMER OPTIMIZATION OF AMPLIFIER F_{SPOT} AND OUTPUT SNR

Figure 3.13A illustrates the fact that for a given set of internal biasing conditions, a given amplifier will have an optimum operating region where NF_{spot} is a minimum in R_s, f_c-space. In some instances, the input sensor to which the amplifier is connected has an R_s that is far smaller than the R_s giving the lowest NF_{spot} on the amplifier's spot NF contours. Consequently, the system (i.e., sensor and amplifier) is not operating to give either the lowest NF or the highest output SNR. One way of improving the output SNR is to couple the input sensor to the amplifier through a low-noise, low-loss transformer, as shown in Figure 3.15. Such coupling, of course, presumes that the signal coming from the transducer is AC and not DC, for obvious reasons. A practical transformer is a band-pass device, it loses efficiency at low and high frequencies, limiting the range of frequencies over which output SNR can be maximized.

The output mean squared SNR can be calculated for the circuit of Figure 3.15 as follows: The mean squared input signal is simply $\overline{v_s^2}$. In the case of a sinusoidal input, it is well known that $\overline{v_s^2} = V_s^2/2$ MSV. The mean squared signal at the output is:

$$S_o = \overline{v_s^2} n^2 K_V^2 \qquad 3.52$$

where n is the transformer's secondary-to-primary turns ratio, and K_V is the amplifier's mid-band gain. The transformer is assumed to be ideal (and noiseless). In practice, transformer windings have finite resistance, hence they make Johnson noise, and their magnetic cores contribute Barkhausen noise to their noisiness. The ideal transformer, besides having infinite frequency response, is lossless. From this latter assumption, it is easy to show that the amplifier "sees" a transformed Thevenin equivalent circuit of the input transducer having an open-circuit voltage of n $v_s(t)$, and a Thevenin resistance of n^2 Rs. Thus, the mean squared output noise can be written:

$$N_o = \left[n^2 kTR_s + e_{na}^2 + i_{na}^2 \left(n^2 R_s\right)^2\right] K_V^2 B \quad \text{MSV} \qquad 3.53$$

and the output SNR is:

$$SNR_O = \frac{\overline{v_s^2}/B}{4kTR_s + e_{na}^2/n^2 + i_{na}^2 n^2 R_s^2} \qquad 3.54$$

The SNR_O clearly has a maximum with respect to n. If the denominator of Equation 3.54 is differentiated with respect to n^2 and set equal to zero, we find that an optimum turns ratio, n_o, exists and is given by:

$$n_o = \sqrt{e_{na}/(i_{na}R_s)} \qquad 3.55$$

If the noiseless (ideal) transformer is given the turns ratio of n_o, then it is easy to show that the maximum output SNR is given by:

$$SNR_O = \frac{\overline{v_s^2}/B}{4kTR_s + 2e_{na}i_{na}R_s} \qquad 3.56$$

The general effect of transformer SNR maximization on the system's noise figure contours is to shift the locus of minimum NFspot to a lower range of R_s; there is no obvious shift along the f_c axis. Also, the minimum NFspot is higher with the transformer. This is because a practical transformer is noisy, as discussed above.

As a rule of thumb, use of a transformer to improve output SNR and reduce NF_{spot} is justified if ($e_{na}^2 + i_{na}^2 R_s^2$) > 20 $e_{na}i_{na}R_s$ in the range of frequencies of interest.

3.6 CASCADED NOISY AMPLIFIERS

In this section, we examine a rule for the design of low-noise signal conditioning systems. Figure 3.16 illustrates three, cascaded noisy amplifiers. The i_{na} terms are assumed to be negligible because $i_{na}^2 R_s^2 \ll e_{na}^2$ in each amplifier. We may write the mean squared output signal as:

$$S_{out} = \overline{v_s^2} K_{V(1)}^2 K_{V(2)}^2 K_{V(3)}^2 \quad \text{MSV} \qquad 3.57$$

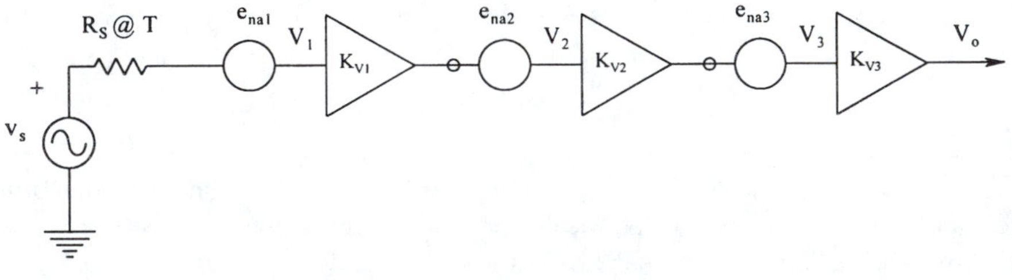

FIGURE 3.16 Three cascaded stages of noisy voltage amplifiers.

The mean squared output noise can be written as:

$$N_{out} = \left(4kTR_s + e_{na1}^2\right) K_{V(1)}^2 K_{V(2)}^2 K_{V(3)}^2 B + e_{na2}^2 K_{V(2)}^2 K_{V(3)}^2 B + e_{na3}^2 K_{V(3)}^2 B \qquad 3.58$$

The three-amplifier system's mean squared output SNR is thus:

$$SNR_{out} = \frac{\overline{v_s^2}/B}{4kTR_s + e_{na1}^2 + e_{na2}^2/K_{V(1)}^2 + e_{na3}^2/K_{V(1)}^2 K_{V(2)}^2} \tag{3.59}$$

Note that the output SNR is set essentially by the noise characteristics of the input amplifier (headstage), as long as $K_{V(1)} > 5$, i.e., the two right-hand terms in the denominator of the SNR_{out} expression are not numerically significant.

The point to remember from this section is that when it is necessary to design a low-noise, signal conditioning system, use the lowest noise amplifier in the headstage position, and give it a gain of at least five.

3.7 EXAMPLES OF CALCULATIONS OF THE NOISE-LIMITED RESOLUTION OF CERTAIN SIGNAL CONDITIONING SYSTEMS

In designing instrumentation systems, it is often necessary to be able to predict the noise-limited, threshold resolvable measurand, or alternately, what input measurand level will produce a given system output SNR. In this section, and in the chapter home problems, we present some examples of situations in which a noise-limited, minimum input signal level is found.

3.7.1 Calculation of the Minimum Resolvable AC Input Voltage to a Noisy, Inverting Op Amp Amplifier

Figure 3.17 illustrates a simple inverting op amp circuit. The noisy op amp is modeled as a frequency-compensated op amp with a white, equivalent, short-circuit input noise, e_{na}. i_{na} is not included in the model because $i_{na} R_F << e_{na}$. We assume the input signal to be a sinusoidal voltage with peak value V_s and frequency f_s. The mean squared output signal voltage is

$$S_{out} = (V_s^2/2)(R_F/R_1)^2 \quad \text{MSV} \tag{3.60}$$

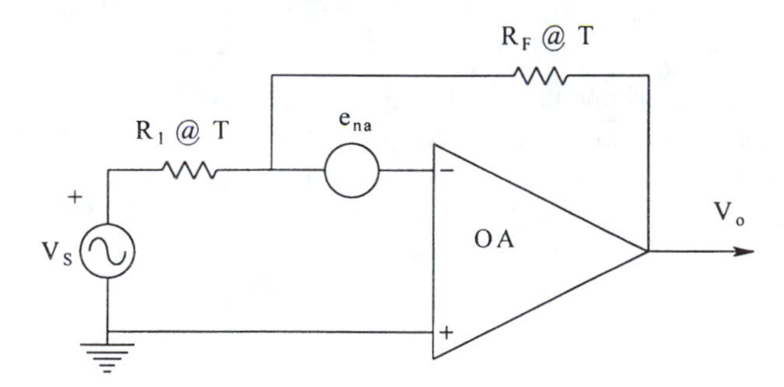

FIGURE 3.17 A noisy inverting op amp amplifier.

The total mean squared output noise is

$$N_{out} = 4kTR_1(R_F/R_1)^2 B + 4kTR_F B + e_{na}^2(1 + R_F/R_1)^2 B \quad \text{MSV} \tag{3.61}$$

The equivalent Hz noise bandwidth, B, may be calculated as follows: a frequency-compensated op amp operated as a simple inverting amplifier has been shown (Northrop, 1990) to have a frequency response given by

$$\frac{V_o}{V_s} = \mathbf{A}_V(j\omega) = \frac{-A_{VO}R_F/\left[R_F + R_1(1+A_{VO})\right]}{j\omega\dfrac{\tau_a(R_F+R_1)}{R_F+R_1(1+A_{VO})}+1} \quad 3.62$$

In this expression, A_{VO} is the DC gain of the open-loop op amp. and τ_a is its open-loop time constant. τ_a may be expressed in terms of the op amp's small-signal, unity gain frequency, f_T.

$$\tau_a = A_{VO}/(2\pi f_T) \text{ seconds} \quad 3.63$$

Hence the closed-loop, inverting amplifier has a time constant which can be expressed approximately by

$$\tau_{CL} = \frac{(1+R_F/R_1)}{2\pi f_T} \text{ seconds} \quad 3.64$$

Thus, the equivalent Hz noise bandwidth for the amplifier of Figure 3.17 may be found using Table 3.1. It is

$$B = 1/(4\tau_{CL}) = \frac{\pi f_T}{2(1+R_F/R_1)} \text{ Hz} \quad 3.65$$

The output mean squared SNR can be written:

$$SNR_{out} = \frac{(V_s^2/2)(R_F/R_1)^2 2(1+R_F/R_1)}{\left[4kTR_1(R_F/R_1)^2 + 4kTR_F e_{na}^2(1+R_F/R_1)^2\right]\pi f_T} \quad 3.66$$

In this SNR calculation, we assume that the frequency of the input signal, f_s, is much less than the closed-loop amplifier's break frequency, $1/2\pi\tau_{CL}$ Hz, so that there is no attenuation of the output signal due to the low-pass filter characteristic of the amplifier.

Using Equation 3.66 above, we now calculate the peak value of V_s. that will give a mean squared SNR of unity, given the following typical system parameters: $R_F = 10^5$, $R_1 = 10^3$, $f_T = 12$ MHz, $4kT = 1.66 \times 10^{-20}$, and $e_{na} = 10$ nVRMS/$\sqrt{\text{Hz}}$. Calculations yield $V_s = 6.65$ μV, peak. This input value is relatively high because of the broad-band nature of the amplifier's output noise ($B = 5.94 \times 10^4$ Hz).

3.7.2 Calculation of the Minimum Resolvable DC Current in White and 1/f Noise

In this example, a very small DC current is to be measured, using an electrometer op amp transresistor circuit, shown in Figure 3.18. The DC current source is represented by a Norton equivalent circuit, having current source, I_s, and conductance, G_s. The thermal noise from G_s and R_F, and i_{na} are assumed to have white spectrums. e_{na}, however, is assumed to have a 1/f component:

$$e_{na}^2(f) = e_{na0}^2 + b/f \quad \text{MSV/Hz} \quad 3.67$$

The PDS of the amplifier's output noise can be written as:

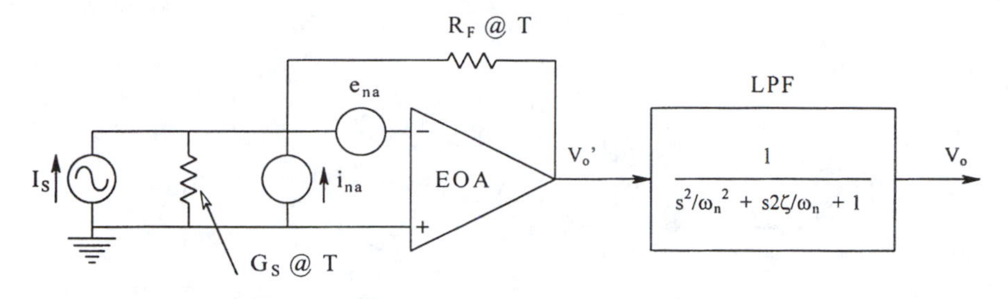

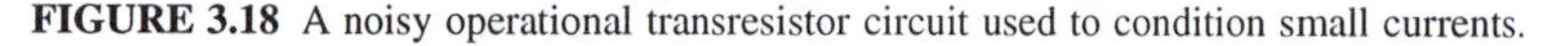
FIGURE 3.18 A noisy operational transresistor circuit used to condition small currents.

$$S_{no}(f) = \left(e_{na0}^2 + b/f\right)\left(1 + R_F G_s\right)^2 + \left[4kT\left(G_F + G_s\right) + i_{na}^2\right]R_F^2 \quad 3.68$$

This noise, is of course, broadband, and is conditioned by a noiseless, unity-gain, quadratic low-pass filter with transfer function:

$$F(s) = \frac{1}{s^2/\omega_n^2 + s(2\zeta)/\omega_n + 1} \quad 3.69$$

Reference to No. 3 in Table 3.1 shows that the quadratic low-pass filter has an equivalent noise bandwidth of $B = \omega_n/(8\zeta)$ Hz. The total mean squared output noise voltage can thus be found by integrating the white noise terms of $S_{no}(f)$ from 0 to $\omega_n/(8\zeta)$. Integration of the b/f term between the same limits leads to the embarassing result of infinity. To avoid this problem, we must impose a practical low-frequency limit for the b/f integral. One arbitrary but effective solution to this problem is to take the low-frequency limit to be the reciprocal of the time it takes to confidently read the output voltage from the low-pass filter with an analog or digital voltmeter. Let us assume this takes 4 s, so the low-frequency limit is 0.25 Hz. Thus:

$$N_o = \left\{\left(e_{na0}^2\right)\left(1 + R_F G_s\right)^2 + \left[4kT\left(G_F + G_s\right) + i_{na}^2\right]R_F^2\right\}\left[\omega_n/(8\zeta)\right] + b\ln\left(\frac{4\omega_n}{8\zeta}\right)\left(1 + R_F G_s\right)^2 \text{ MSV} \quad 3.70$$

The mean squared, DC output signal is $S_o = I_s^2 R_F^2$ MSV.

Now let us find the I_s to give an output mean squared SNR of 4. We assume the following parameters for an AD549 electrometer op amp (Analog Devices, Norwood, MA): $4\text{ kT} = 1.66 \times 10^{-20}$ at 300 K, $\omega_n/(8\zeta) = 2$ Hz, $R_F = 10^{10}$ ohms, $G_s = 10^{-7}$ S, $e_{na0} = 35$ nV RMS/$\sqrt{\text{Hz}}$, $i_{na} = 0.2$ fA RMS/$\sqrt{\text{Hz}}$, $b = 1.36 \times 10^{-17}$ MSV. Evaluating the noise terms in Equation 3.70, we find that the dominant term comes from the Johnson noise in R_F and R_s; this is 3.3233×10^{-7} MSV. The mean squared output noise from the b/f term in e_{na}^2 is 2.8336×10^{-11} MSV, and the total mean squared output noise from e_{na0}^2 and i_{na}^2 is 8.2455×10^{-12} MSV. When we set the mean squared output SNR equal to four, we find that the minimum $I_s = 0.1153$ μA DC, and the DC voltage at the system output is 1.153 mV.

3.7.3 Calculation of the Minimum Resolvable AC Input Signal to Obtain a Specified Output SNR in a Transformer-Coupled, Tuned Amplifier

Figure 3.19 illustrates a signal conditioning system used to amplify the 10 kHz, sinusoidal voltage output of a low impedance sensor. A transformer (here assumed to be ideal and noiseless) is used

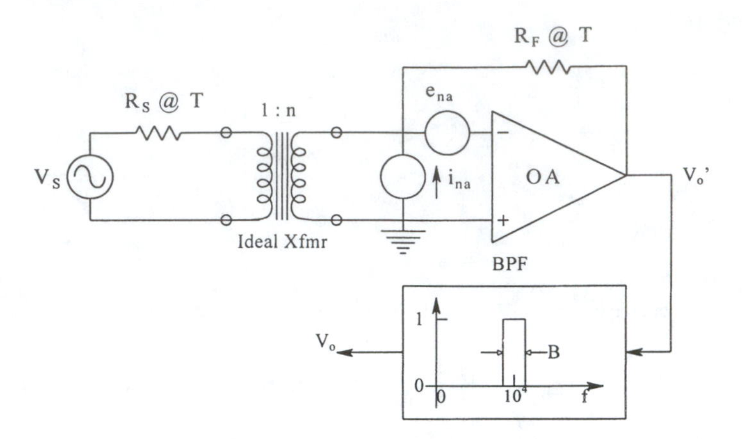

FIGURE 3.19 Use of a transformer to improve the output SNR when conditioning a low-level sinusoidal voltage from a low-resistance source. See text for details.

to maximize the output SNR. A noiseless, quadratic band-pass filter, tuned to 10 kHz, follows the low-noise op amp input stage. The peak gain of the tuned filter is unity. Looking toward the source through the ideal transformer, the op amp summing junction sees a Thevenin equivalent circuit consisting of an open-circuit voltage, $v_{oc}(t) = nv_s(t)$, and an equivalent resistance of $R_{TH} = n^2 R_s$. Thus, the mean squared output voltage is thus

$$S_{out} = (V_s^2/2)\left[R_F^2/(nR_s)^2\right] \text{ MSV} \tag{3.71}$$

All sources of noise in the circuit are assumed to be white. The noise bandwidth is determined by the BPF, and is found in Table 3.1 to be $B = \omega_n\zeta = \omega_n/2Q$ Hz. The total mean squared output noise can thus be written:

$$N_{out} = \left[4kTR_sR_F^2/(nR_s)^2 + e_{na}^2\left(1 + R_F/(n^2R_s)\right)^2 + i_{na}^2R_F^2 + 4kTR_F\right](\omega_n/2Q) \text{ MSV} \tag{3.72}$$

Therefore, the mean squared output SNR is:

$$SNR_{out} = \frac{(V_s^2/2)/B}{4kTR_s + e_{na}^2(nR_s/R_F + 1/n)^2 + i_{na}^2R_s^2n^2 + 4kTG_FR_s^2n^2} \tag{3.73}$$

The denominator of Equation 3.73 has a minimum for some non-negative n_o, hence the SNR has a maximum for that n_o. To find n_o, we differentiate the denominator with respect to n^2, set the derivative equal to zero, and then solve for n_o. This yields:

$$n_o = \frac{\sqrt{e_{na}}}{\sqrt{R_s}\left[e_{na}^2/R_F^2 + 4kTG_F + i_{na}^2\right]^{1/4}} \tag{3.74}$$

We now examine a numerical solution for n_o and V_1 to give an SNR_{out} equal to unity. We use the following parameters: e_{na} = 3 nV RMS/$\sqrt{\text{Hz}}$, i_{na} = 0.4 pA RMS/$\sqrt{\text{Hz}}$, $R_F = 10^4$ ohms, R_s = 10 ohms, the BPF Q = 100, and 4 kT = 1.66×10^{-20}. n_o is found to be 14.73. When n_o is substituted for n in Equation 3.73, and a numerical solution is obtained, V_s to get unity SNR is 13 peak nV. If the transformer is not used (n = 1), then V_s to give SNR = 1 is 76.0 peak nV.

3.7.4 Calculation of the Smallest ΔR/R in a Wheatstone Bridge to Give a Specified SNR_{out}

In this example, we calculate the smallest ΔR required to produce a mean squared output SNR of 10 in the AC-powered, Wheatstone bridge circuit shown in Figure 3.20. Two arms of the bridge are variable, one having increasing resistance while the resistance of the other decreases a corresponding amount. Such a configuration is found in certain unbonded strain gauges used to measure forces. The output of the bridge is amplified by a PMI AMP-01 low-noise, differential instrumentation amplifier, which is followed by a noiseless, quadratic BPF. The purpose of the BPF is to pass the amplified AC output of the bridge while restricting the bandwidth of the noise, thus achieving an output SNR which is higher than that at the output of the differential amplifier.

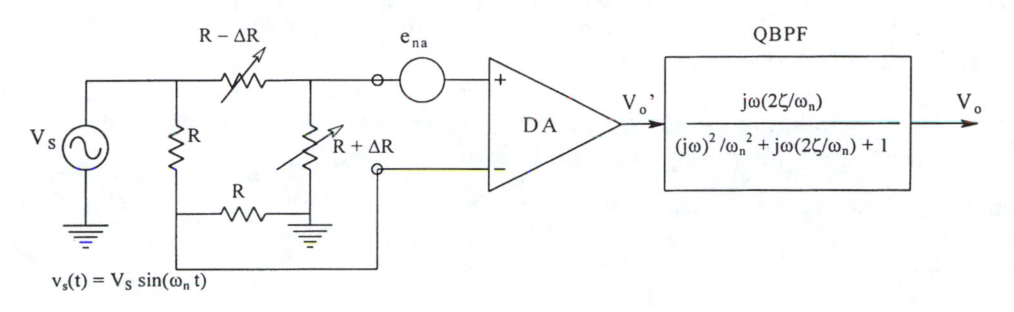

FIGURE 3.20 Use of a quadratic band-pass filter to improve the output SNR and resolution of a Wheatstone bridge with AC excitation.

The signal output of the system is a sinewave whose amplitude and phase is determined by v_s. and ΔR/R. It may be found as follows: let the bridge excitation be $v_s(t) = V_s \sin(2\pi 400t)$. The voltage at the left-hand corner of the bridge, v_1', is easily seen to be $v_s(t)/2$. The voltage at the right-hand corner of the bridge, v_1, can be written

$$v_1(t) = [v_s(t)/2](1 + \Delta R/R) \qquad 3.75$$

Hence, the output of the differential amplifier is

$$v_o'(t) = K_d(\Delta R/2R)V_s \sin(2\pi\, 400t) \qquad 3.76$$

The center frequency of the BPF is 400 Hz, so $v_o(t) = v_o'(t)$. Thus the mean squared signal output is:

$$S_{out} = (V_s^2/2)(K_d^2/4)(\Delta R/R)^2 \quad \text{MSV} \qquad 3.77$$

The mean squared noise at the system output is assumed to be due to the white thermal noise in the bridge resistors, and the equivalent input short-circuit voltage noise, e_{na}, which is assumed to be white. The total mean squared output noise is found by adding up the mean squared noise contributions from all the noise sources. The equivalent Hz noise bandwidth is set by the BPF, and from Table 3.1, is $2\pi f_n/2Q$ Hz. Thus, the mean squared output noise voltage can be written:

$$N_{out} = (4kTR/2 + 4kTR/2 + e_{na}^2)K_d^2(2\pi f_n/2Q) \quad \text{MSV} \qquad 3.78$$

The mean squared SNR at the output of the system is thus:

$$\text{SNR}_{out} = \frac{(V_s^2/2)(K_d^2/4)(\Delta R/R)^2}{(4kTR + e_{na}^2)K_d^2(2\pi f_n/2Q)} \quad 3.79$$

We now calculate the $\Delta R/R$ required to give a mean squared output SNR of 10. We let $f_n = 400$ Hz, $Q = 5$, $R = 1000$ ohms, $K_d = 1000$, $4\,kT = 1.66 \times 10^{-20}$, $e_{na} = 5$ nV RMS/$\sqrt{\text{Hz}}$, $V_s = 4$ V peak. Using Equation 3.79 above, we find $\Delta R/R = 7.23 \times 10^{-8}$, or $\Delta R = 7.23 \times 10^{-5}$ ohms. Using Equation 3.76, we calculate that this ΔR will produce a 400 Hz sinusoidal output with peak value $V_o' = 1.446 \times 10^{-4}$ V.

As can be seen, AC operation of a Wheatstone bridge with a tuned output filter can result in considerable sensitivity. Normally, the slow changes in $\Delta R(t)$ would be extracted with a phase-sensitive rectifier following the BPF.

3.7.5 Determination of the Conditions for Maximum Output SNR Given a Simple Inverting Op Amp Amplifier with Known e_{na} and i_{na}

Figure 3.21 shows a simple op amp inverter circuit having a signal source that lies well within the Hz noise bandwidth, B, for the circuit. Both R_1 and R_F are assumed to make white thermal noise. The mean squared signal at the output can be written:

$$S_{out} = \overline{v_s^2}(R_F/R_1)^2 \quad \text{MSV} \quad 3.80$$

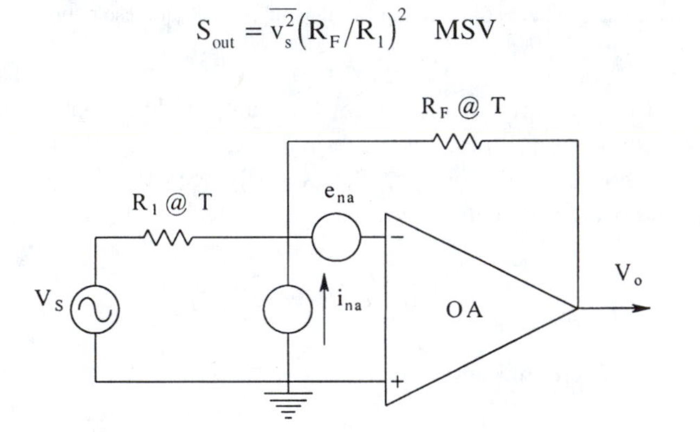

FIGURE 3.21 A simple, noisy, inverting op amp amplifier.

The total mean squared noise at the output is:

$$N_{out} = \left[4kTR_1(R_F/R_1)^2 + (i_{na}^2 + 4kTG_F)R_F^2 + e_{na}^2(1 + R_F/R_1)^2\right]B \quad \text{MSV} \quad 3.81$$

After some algebra, we can write the mean squared output SNR as:

$$\text{SNR}_{out} = \frac{\overline{v_s^2}/B}{4kTR_1(1 + R_1/R_F) + e_{na}^2(1 + R_1/R_F)^2 + i_{na}^2 R_1^2} \quad 3.82$$

Clearly, from Equation 3.82, we see that the SNR is larger for high ratios of R_F/R_1. That is, for high signal gains. Also, the SNR_{out} is larger for small R_1. Small R_1, means, of course, that V_s sees a low effective input resistance (i.e., R_1). A practical limit on the gain is set by the necessary closed-loop bandwidth, which will be approximately $f_T(R_1/R_F)$ Hz. Good low-noise design often involves compromise of other design parameters.

3.8 MODERN, LOW-NOISE AMPLIFIERS FOR USE IN INSTRUMENTATION SIGNAL CONDITIONING SYSTEMS

In the past 10 years, as the result of advances which have been made in transistor fabrication technology and in electronic circuit design, we have seen a rise in the number of available, low-noise, IC amplifiers suitable for use in signal conditioning in measurement systems. "Low-noise", as used in this section, shall mean amplifiers with e_{na}s less than 10 nV RMS/ $\sqrt{Hz}$. It is always risky to attempt to list and categorize low-noise amplifiers in a text because: (1) of the danger of omitting one or more of the more obscure models, and (2) by the time the list is read, new, better low-noise amplifiers will have been put on the market. Facing these risks, we do list below certain low-noise operational amplifiers and instrumentation amplifiers which may have application in the design of low-noise signal conditioning systems.

In specifying the noise performance of IC amplifiers, manufacturers typically give the following information: (1) e_{na} measured in the *flat* portion of its spectrum; (2) the peak-to-peak or RMS output noise measured under short-circuited input conditions, over a specified, low-frequency portion (1/f region) of the $e_{na}^2(f)$ spectrum; (3) The value of i_{na} measured in its flat region. For some amplifiers, the manufacturers also give a plot of the spot noise figure contours vs. R_s and f.

The noise performance of an IC amplifier is largely set by the designer's choice of headstage transistors. Consequently, low-noise amplifiers may be broadly characterized by whether they use low-noise BJTs or field effect transistors (FETs) in their input circuits. Those using FETs generally have high input resistances (over 10^{12} ohms) and low i_{na}s (less than 1 fA RMS/ $\sqrt{Hz}$). Amplifiers with BJT headstages typically have lower input resistances and higher i_{na}s, but in some cases, exhibit e_{na}s an order of magnitude lower than the e_{na}s from junction field effect transistor (JFET) amplifiers. Table 3.2 below is a partial listing of the more widely used, low-noise, IC amplifiers; it is subdivided into instrumentation amps and op amps.

3.9 COHERENT INTERFERENCE AND ITS MINIMIZATION

In this section, we consider another common form of unwanted input which degrades the resolution and accuracy of measurement systems, i.e., periodic or coherent interference. Coherent interference, unlike random noise, has narrow-band power density spectrums, often with harmonic peaks at integral multiples of the fundamental frequency. A serious problem is created when the coherent interference spectrum has power in the signal frequency band. In the following sections, we will discuss common sources of coherent interference, and various means of minimizing its input to a measurement system. The techniques of minimizing coherent interference often appear arcane to the inexperienced, but ultimately, all such techniques follow the laws of physics and circuit theory.

3.9.1 Sources of Coherent Interference

There are, broadly speaking, four major sources of coherent interference: (1) power line frequency interference, often coupled to the measurement system by an electric field and/or a magnetic field. Obvious sources of powerline interference are unshielded power transformers, poorly filtered AC-to-DC power supplies, and electromagnetic field pickup from unshielded, high-voltage 60 Hz power lines. (2) Radio frequency electromagnetic interference (RFI) coupled into the measurement system from local radio transmitters, poorly shielded computers, and other digital instruments having MHz clock frequencies. RFI may be picked up through the AC powerline, by magnetic induction in internal wiring loops, or by magnetic induction in the cables connecting sensors to signal conditioning systems. (3) Of a more transient nature is the periodic noise coupled into a measurement system from sparking motor brushes, gasoline engine ignition systems, or high-frequency power line transients caused by SCR and triac switching in motor speed controls and ovens. Although hardly periodic, the broad-band electromagnetic pulses produced by lightning in thunderstorms are

Table 3.2

Amp. Model	Type	e_{na}	i_{na}	Low Freq V_n	R_{in}	f_T	A_{VO}
PMI 0P-27	OP/BJT	3	0.4 pA	0.08 μVppk	6 M	5	10^6
PMI OP-61A	OP/BJT	3.4	0.8 pA	—	—	200	4×10^5
HA 5147A	OP/BJT	3	0.4 pA	0.08 μVppk	6 M	140	10^6
BB OPA111BM	OP/FET	6	0.4 fA	1.6 μVppk	10^{13} (DM) 10^{14} (CM)	2	125 dB
LT1028	OP/BJT	0.85	1 pA	35 nVppk	20 M (DM) 300 M (CM)	75	7×10^6
AM-427	OP/BJT	3	0.4 pA	0.18 μVppk	1.5 M (DM)	5	120 dB
NE5532	OP/BJT	5	0.7 pA	—	300 K	10	10^5
OEI AH0014	OP/FET	3	—	—	10^{11} (DM&CM)	10^3	10^5
PMI AMP-011	IA/BJT	5	0.15 pA	0.12 μVppk	10 G (DM) 20 G (CM)	26	$1 - 10^3$
PMI AMP-021	IA/FET	9	0.4 fA	0.4 μVppk	10 G (DM) 16.5 G (CM)	5	$1 - 10^3$
AD6251	IA/BJ	4	0.3 pA	0.2 μV	1 G (DM&CM)	25	$1 - 10^3$
BB INA1102	IA/FET	10	1.8 fA	1 μVppk	2×10^{12} (DM) 5×10^{12} (CM)	12	1 – 500
PMI AMP-052	IA/FET	16	10 fA	4 μVpp	10^{12}(DM&CM)	3	$1 - 10^3$
ZN459CP	IA/BJT	0.8	1 pA	—	7k (Single-ended)	15(–3)	60 dB
ZN424	IA/BJT	5.5	0.3 pA	—	200k	4	2×10^4

Note: Partial listing of commercially available, low-noise op amps and instrumentation amplifiers having low noise characteristics. Type refers to whether the amplifier is an op amp (OP) or an instrumentation amplifier (IA), and whether it has a BJT or FET headstage. e_{na} is given in nV RMS/ $\sqrt{Hz}$ measured at 1 kHz, i_{na} is given either in picoamps (pA) or femtoamps (fA) RMS/ $\sqrt{Hz}$ measured at 1 kHz. Low Freq V_n is the equivalent, peak-to-peak, short-circuit input noise measured over a standard 0.1 to 10 Hz bandwidth. R_{in} is the input resistance, f_T is the unity gain band-width in MHz, unless followed by (–3), in which case it is the high frequency –3 dB frequency. A_{VO} for op amps is their open-loop, DC gain; the useful gain range is given for IAs.

often picked up by electromagnetic interference-sensitive electronic signal conditioning systems. (4) In a complex electronic measurement system, coherent interference often may arise from within the system. Such interference may be the result of poor grounding and/or shielding practices allowing the digital portion of the system to "talk" to the low-level analog front end of the system. Internal coherent interference may also arise through poor decoupling of a common power source which supplies a low-level analog front end and a high-frequency digital circuit.

3.9.1.1 Direct Electrostatic Coupling of Coherent Interference

To illustrate this phenomenon, we assume two conductors lie in close proximity. One conductor is at a sinusoidal potential, $v_1(t) = V_1 \sin(\omega t)$. There is a capacitance C_{12} between the two conductors, and conductor 2, which picks up the interference from conductor 1, has a resistance R_2 in parallel with a capacitance C_2 to ground, as shown in Figure 3.22A. The discrete component circuit describing this simple capacitive coupling situation can be redrawn as a simple voltage divider, shown in Figure 3.22B. We may write a transfer function for the coherent interference induced on conductor 2:

$$\frac{V_2}{V_1}(j\omega) = \frac{j\omega C_{12} R_2}{j\omega(C_{12} + C_2)R_2 + 1} \quad 3.83$$

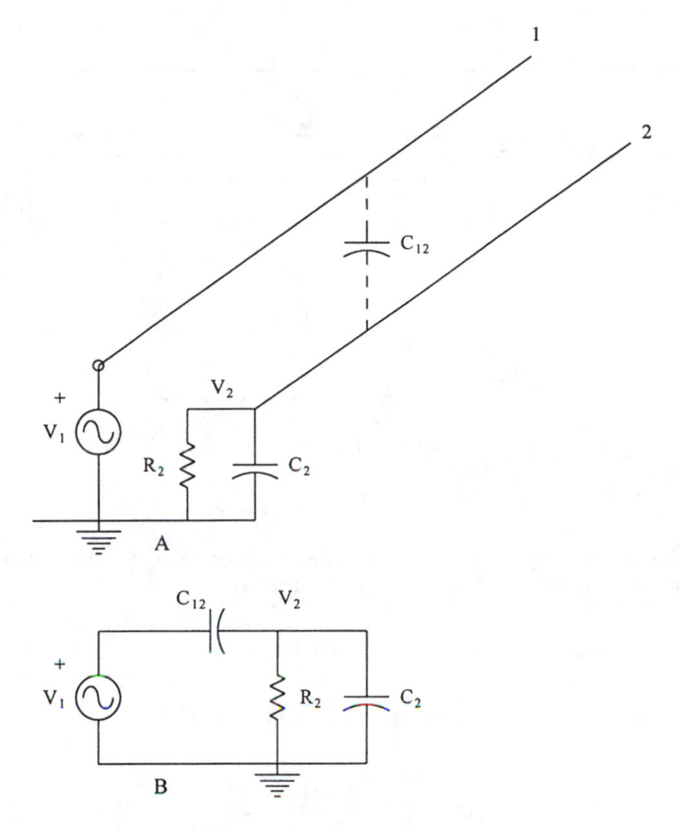

FIGURE 3.22 (A) A circuit describing the electrostatic coupling between two adjacent conductors. A sinusoidal voltage source is assumed. (B) Lumped parameter equivalent circuit of (A).

By inspection, we see that this is a simple, high-pass transfer function. C_{12} is typically on the order of single picofarads, C_2 may be in the tens of pF, and R_2 may vary widely, depending on the application. For a specific example, let C_{12} = 2.5 pF, C_2 = 27.5 pF, R_2 = 50 ohms. The break frequency is found to be 6.67×10^8 r/s. Hence at frequencies below the break frequency,

$$V_2 = j\omega C_{12} R_2 V_1 \tag{3.84}$$

From the equations above, we gather the following lesson: to minimize V_2/V_1, C_{12} and R_2 should be made small. C_{12} is minimized by arranging the wiring geometry so that high-level or output signal wires or printed circuit lines are far from sensitive, high impedance, low-level circuit wiring. Wire crossings, if necessary, should be at right angles. Also, sensitive wires should be shielded, if possible. Resistance R_2 to ground should be as low as possible. Often the sensor output resistance is several kilohms, so the best protection for the headstage is to shield it and the wire connecting it to the sensor. As you will see below, what would appear to be a simple solution to the capacitive pickup of coherent interference by coaxial shielding is not that simple; details of shield grounding determine the effectiveness of the shield.

3.9.1.2 Direct Magnetic Induction of Coherent Interference

Figure 3.23A illustrates the basic mechanism whereby a time-varying magnetic field caused by current in conductor 1 induces an EMF in a second conductor. To model this happening with a lumped-parameter circuit, we assume that a mutual inductance, M, exists between the two conductors, and that the conductors can be modeled by a transformer, as shown in Figure 3.23B. The EMF induced in wire 2 is simply

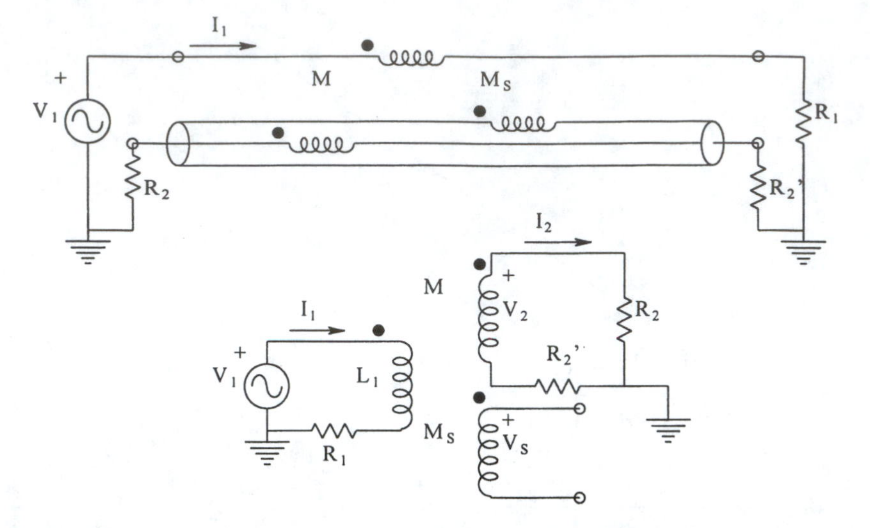

FIGURE 3.23 (A) A circuit describing magnetic induction of coherent interference. The shield of the coaxial cable is not grounded. (B) Lumped parameter equivalent circuit of (A).

$$V_2 = j\omega M I_1 \tag{3.85}$$

Thus the current induced in loop 2 is given by Ohm's law to be

$$\frac{I_2}{I_1}(j\omega) = \frac{j\omega M/(R_2 + R_2')}{[1 + j\omega L_2/(R_2 + R_2')]} \tag{3.86}$$

and the voltage observed across R_2 is simply $I_2\ R_2$. L_2 is the very small self-inductance of wire 2. The break frequency of the pole in Equation 3.86 is generally quite large, so for most cases,

$$I_2 = I_1[j\omega M/(R_2 + R_2')] \tag{3.87}$$

In some cases of induced magnetic interference, a time-varying B-field exists in space in the vicinity of a loop of area A, as shown in Figure 3.24. The B field can arise from an unshielded power transformer, or at high frequencies, can be the result of a strong, radio frequency electromagnetic field. By Faraday's law, the EMF induced in the loop can be written as:

$$E = -\frac{d}{dt}\int \mathbf{B}\cdot\mathbf{ds} \tag{3.88}$$

For the case of a B field which varies sinusoidally in time, and which intersects the plane of the loop at an angle θ, we may write:

$$\mathbf{E} = j\omega BA\cos(\theta) \tag{3.89}$$

and the interference current induced in the loop is simply

$$\mathbf{I}_2 = \frac{j\omega BA\cos(\theta)}{(R_2 + R_2')} \tag{3.90}$$

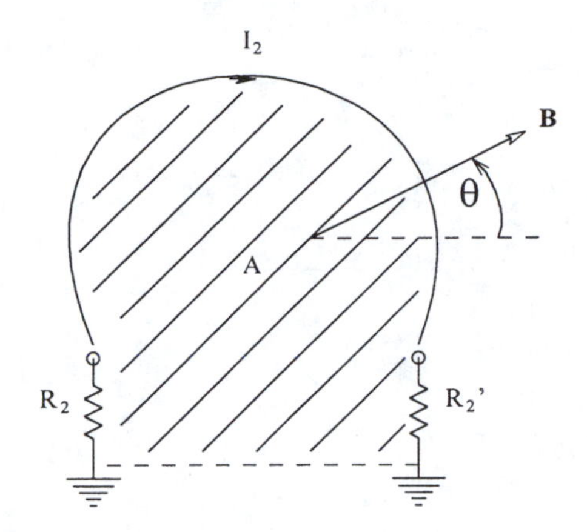

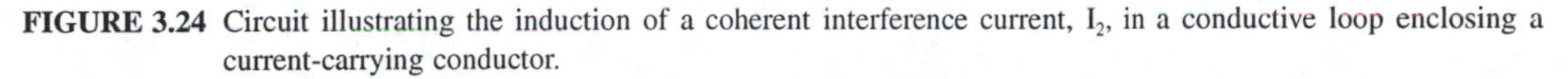

FIGURE 3.24 Circuit illustrating the induction of a coherent interference current, I_2, in a conductive loop enclosing a current-carrying conductor.

(Here we have neglected the inductance in the loop.) Clearly, steps taken to reduce B and A will reduce the pickup of magnetic coherent interference.

3.9.1.3 Ground Loops

Whenever a ground conduction path is forced to carry a power supply return current which has a significant high-frequency component added to the DC, and this ground path is also used in common for a low-level signal ground, it is possible for the AC power component of ground current to develop a small voltage across the small but finite impedance of the ground path. This situation is illustrated in Figure 3.25. There are, of course, causes other than power supply currents which can make the ground potentials of two subsystems not be isopotential.

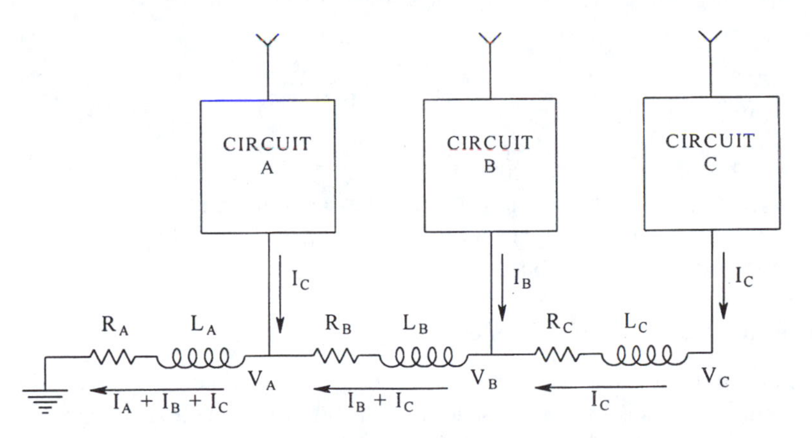

FIGURE 3.25 Circuit illustrating how one circuit's ground current can affect the ground potential of other circuits sharing the ground line. This condition is called a ground loop.

The fact that two ground points are not at exactly zero potential means that an unwanted ground loop current can flow in a coaxial cable shield which is grounded at each end to a different chassis or enclosure. The ground loop current in the shield can induce coherent interference on the center conductor. Figure 3.26 illustrates how a difference in ground potential can add a coherent noise voltage directly to the desired signal. Here we assume that $R_{IN} >> R_S >> R_{C1}, R_{C2}, R_{SH}$. The coherent interference voltage from $V_{GL(OC)}$ appears at the amplifier's input as V_N, and may be shown to be equal to:

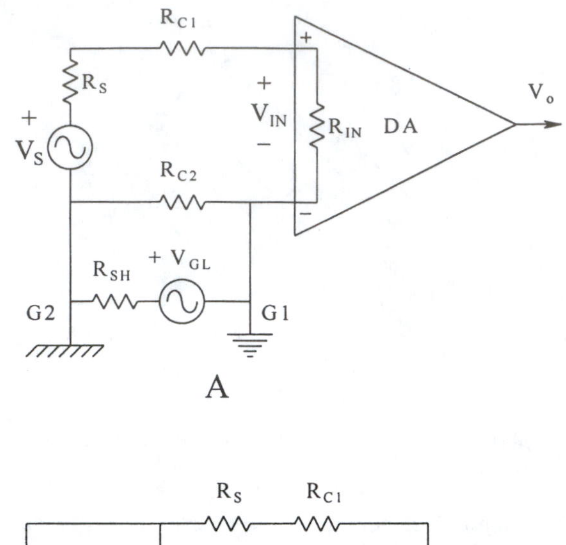

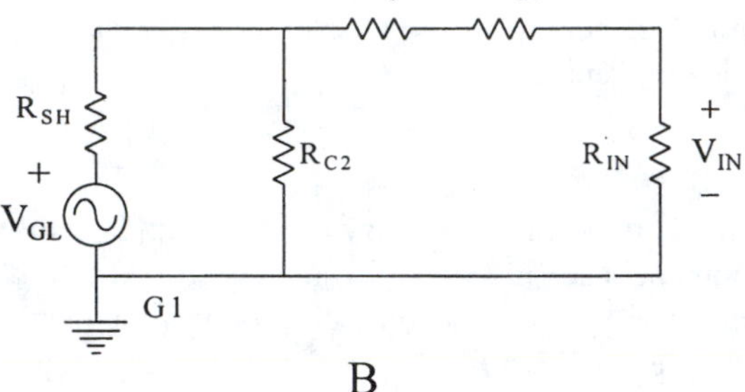

FIGURE 3.26 (A) Circuit showing how a ground loop voltage can affect the input to an amplifier. (B) Equivalent circuit illustrating the two voltage dividers acting on the ground loop voltage.

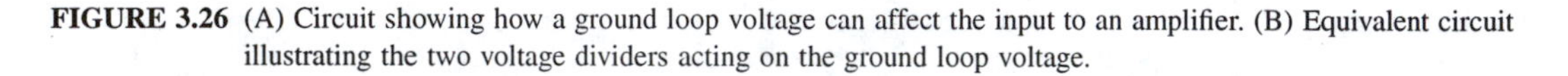

$$V_N = V_{GL(OC)} \frac{R_{C2}}{R_{C2} + R} \frac{R_{IN}}{R_{IN} + R_S} \qquad 3.91$$

The numerical value of the product of the two voltage divider terms depends mostly on the values of R_{C2} and R_{SH}, and is typically about 1/5. Thus we see that an unwanted voltage, 0.2 $V_{GL(OC)}$, appears in series with the desired signal, V_S, at the amplifier's input as the result of a ground loop and grounding the coax's shield at both ends.

3.9.2 Cures for Coherent Interference

As described above, there are many sources of coherent interference; the means to reduce a given type of interference depends on the physical mechanism(s) by which it is coupled into the measurement system. Comprehensive treatments of the causes of, and means for reducing coherent interference may be found in texts by Ott (1976) and by Barnes (1987).

3.9.2.1 Power Line Low-Pass Filters

In many cases, the cure for coherent interference lies in preventing it from escaping from a coherent noise source such as a computer or digital system, an electronic system with a switching power supply, or a radio transmitter. Obviously, in the latter case there must be electromagnetic energy radiated from an antenna; however, great care must be taken in all the systems mentioned above to keep the coherent interference from entering the power mains where it can affect other line powered instruments. Power line filters are generally multistage, LC low-pass filters which attenuate frequencies from about 10 kHz up to about 40 MHz. Maximum attenuation at about 30 MHz ranges from –40 to –65 dB, depending on the number of LC stages in the filter. A typical line filter schematic is shown in Figure 3.27. Power line filters are effective in keeping high-frequency coherent noise from entering a system on the power line, as well as keeping noise generated internally from escaping from the system.

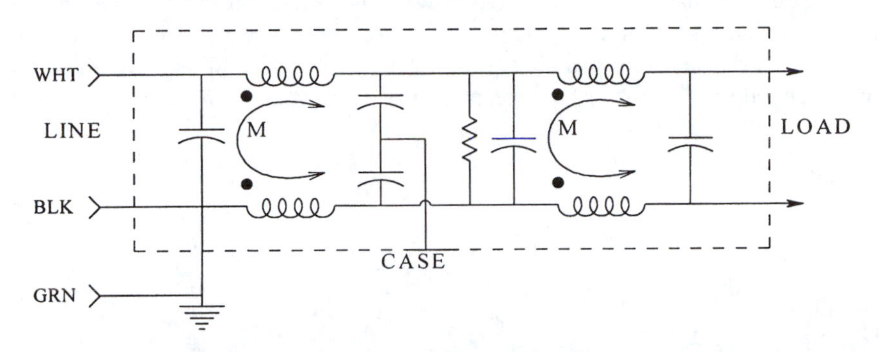

FIGURE 3.27 Schematic of a typical power line low-pass filter, designed to keep high-frequency coherent interference on the power line out of a system, and to prevent high-frequency interference arising within the system from escaping on to the power line.

High-frequency interference on power lines can be described in two forms: commonmode (CM) and difference-mode (DM) interference. In CM interference, the unwanted voltage has the same value on both the black and white wires with respect to ground (green wire). The CM interference voltage can be written as

$$V_{i(CM)} = \left(V_{iB} + V_{iW}\right)/2 \qquad 3.92$$

The DM interference voltage is given by

$$V_{i(DM)} = \left(V_{iB} - V_{iW}\right)/2 \qquad 3.93$$

High-frequency power line filters are designed to attenuate both $V_{i(CM)}$ and $V_{i(DM)}$. Power line filters must be used in conjunction with a robust, grounded metal instrument case; the metal case of the power line filter must have a solid, multipoint electrical connection with the grounded instrument case for best results.

One design trade-off in the application of power line low-pass filters is that the capacitors used to attenuate CM noise inject a current into the green wire power line ground. Generally this current must be kept less than 3.5 mA in nonmedical applications. Hence, the size of the line-to-ground capacitors, and thus the amount of CM interference attenuation, is limited by this ground current limit.

3.9.2.2 Transient Voltage Suppressors

Transient voltage suppressors are used to prevent high voltage, spike-like transients occurring on the power main input to an instrument system from causing physical damage to system power supply components and system circuits, or causing damage and anomalous results in associated computer equipment. There are many sources of power line spikes. Some spikes are periodic, occurring once or more every power frequency cycle. These spikes are generally of a biphasic nature and their voltages may be as much as 30 to 50 V above the instantaneous line voltage. They are typically several microseconds in duration, and arise from SCR or TRIAC switching of inductive loads attached to the power line, such as in motor speed controls. Such "SCR spikes" generally pose no safety problem, but do constitute a major source of coherent interference if allowed to get into the signal ground path or the power supply outputs. SCR spikes are of course attenuated by line filters, but in some cases may be large enough after filtering to pose a problem. Other power line spikes occur randomly as the result of lightning strikes on power lines, or from infrequent switching of inductive loads attached to the power line. Figure 3.28 illustrates the statistical occurrence of singular spikes on a 120 V, 60 Hz power line in the U.S. (Hey and Kram, 1978). The large, singular, transient spikes are best reduced by the use of nonlinear circuit elements such as varistors or zener diodes. Varistors have steady-state current-voltage curves given by the relations:

$$I = G_{OFF}V \quad 0 < V < V_i \tag{3.94A}$$

$$I = KV^{\alpha} \quad V > V_i \tag{3.94B}$$

$$I = G_{ON}V \quad I > I_m \tag{3.94C}$$

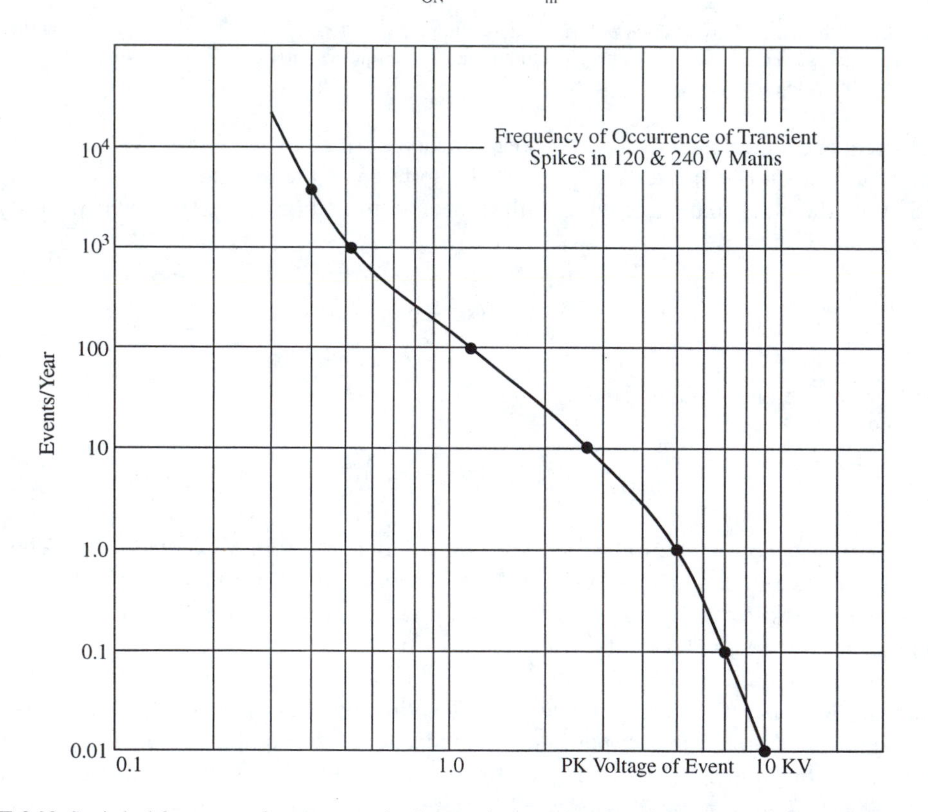

FIGURE 3.28 Statistical frequency of occurrence of spike transients on the 120V, 60 Hz mains vs. the spike peak voltage. (Adapted from Figure 1.7 in the *GE Transient Voltage Suppressor Manual,* 2nd ed., Fairfield, CT, 1978.)

A typical varistor volt-ampere curve is shown in Figure 3.29A. An equivalent, lumped-parameter circuit for a GE MOV is shown in Figure 3.29B. Note that in a linear (Ohm's law) resistor, $\alpha = 1$. In varistors, α can range from 2 to 30, depending on the application and construction of the device. For efficient transient suppression, we generally want α to be as large as possible. Metal oxide varistors were originally developed by Matsushita Electric Co., and are sintered, polycrystalline semiconductor devices. General Electric Corp. (Fairfield, CT) manufactures and sells a family of zinc oxide, metal oxide varistors (MOVTM) under license from Matsushita. The α of GE MOVs ranges from 15 to 30 measured over several decades of surge current. GE MOVs are designed to operate at AC voltages from 10 to 1000 V, and can handle current surges of up to

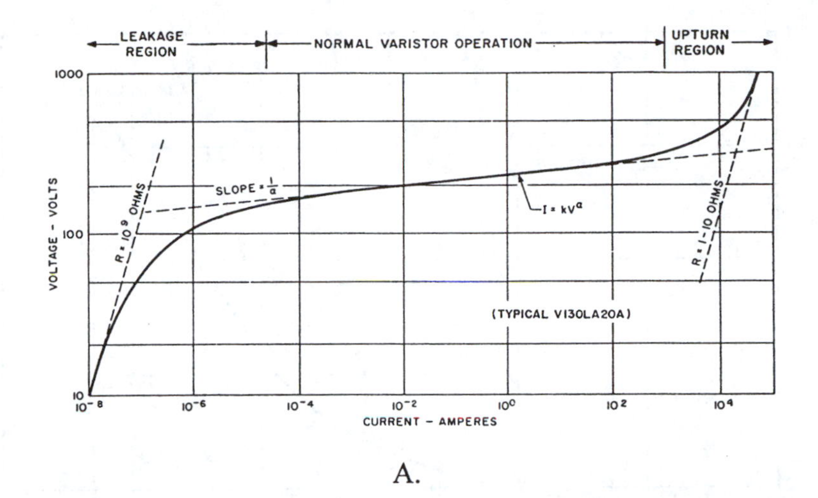

A.

L (LEAD INDUCTANCE)
(TYPICAL V130LA20A)
C (0.002 μF)
V
R_X (0 TO ∞)
R_{OFF} (1000 MΩ)
R_{ON} (1 Ω)

B.

FIGURE 3.29 (A) Volt-ampere curve for a GE V130LA20A metal oxide varistor (MOV) spike transient clipper. Device is symmetrical, only first quadrant behavior is shown here. (B) Equivalent lumped-parameter circuit for the GE MOV device of (A). (From the *GE Transient Voltage Suppressor Manual*, 2nd ed., Fairfield, CT, 1978. with permission.)

25 kA and energy capability of over 600 J in the larger units. Of primary consideration in the design of MOV transient suppression systems is the ability of the MOV device to conduct the transient current that the source can supply under conditions of overvoltage. MOVs are best used in applications where the current rise times (from the power line) are longer than 0.5 μsec.

General Semiconductor Industries, Inc. (Tempe, AZ) manufactures and sell the TransZorb™ transient voltage suppressor. This silicon, PN junction device acts similar to the MOV, with the exception that its alpha is higher (35), and its response time is faster than MOV devices; the manufacturer claims 60 ps, vs. 1 to 50 ns for varistors. The TransZorb has a relatively flat volt-ampere curve vs. varistors. See Figure 3.30. TransZorb AC operating voltages range from 5 to 340 peak volts; 5 V devices clamp at 9.6 V, 340 V devices clamp at 548 V.

Figure 3.31 shows a schematic for an extremely simple six-outlet, "power bar" or "line monitor power conditioner" typically used with personal computers and related equipment to protect them from line spike transients. One MOV-type varistor is used to protect each pair of line outlets. The power bar also has a switch, a neon pilot light, and an inexpensive circuit breaker for 15 A. MOVs or TransZorb-type devices are also used inside sensitive equipment when the presence of a protective power bar cannot be guaranteed.

3.9.2.3 Coherent Interference Induced in Coaxial Cables by Magnetic Coupling

Shielded coaxial cables and twinaxial cables are widely used in instrumentation systems to couple sensors to preamplifiers and signal conditioning subsystems together. Unfortunately, coaxial cables can act as magnetic loop antennas under certain grounding conditions. Referring to Figure 3.32A,

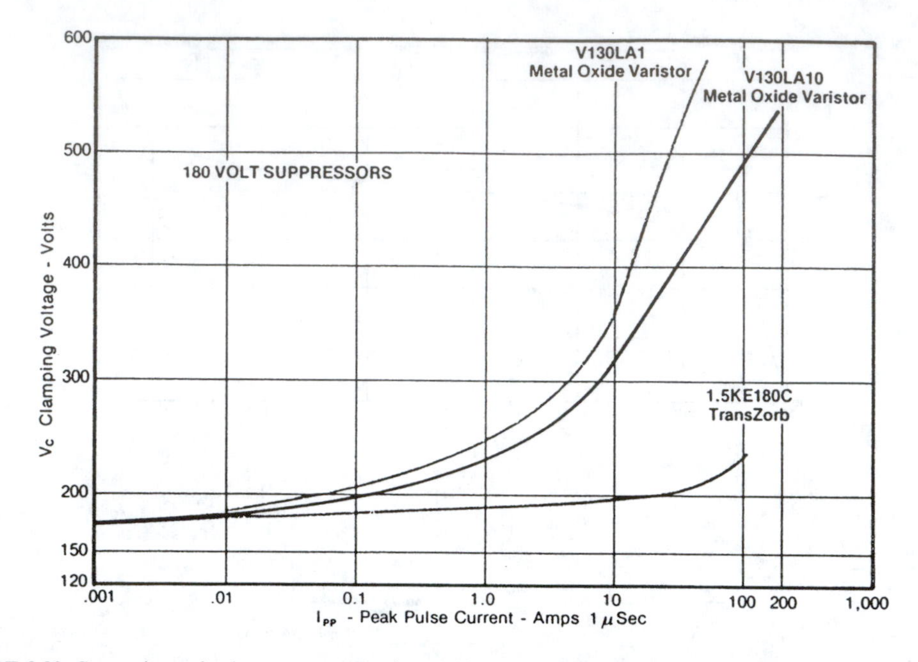

FIGURE 3.30 Comparison of volt-ampere curves of two MOVs and a TransZorb+ pn junction transient suppressor made by General Semiconductor Industries, Inc. (Figure used with permission of General Semiconductor Industries, Inc., Tempe, AZ.)

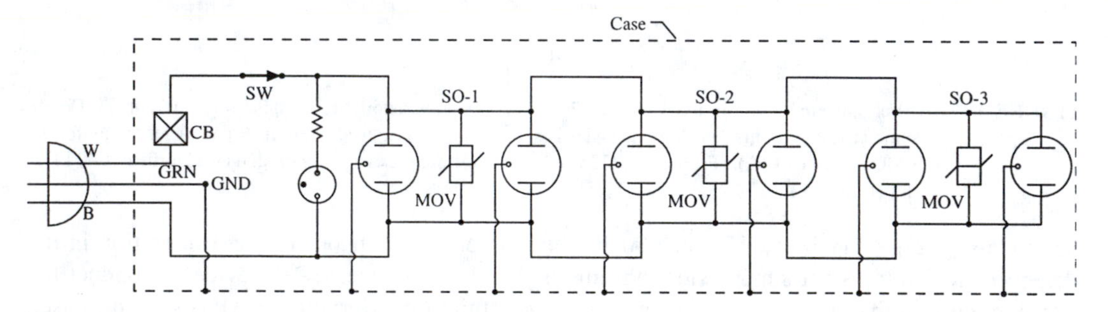

FIGURE 3.31 Schematic of a transient-protected, 6-outlet "power bar", such as used with personal computer systems.

we see that an AC current I1 in an adjacent wire creates flux lines which link the coaxial cable and its shield, creating a mutual inductance between conductor 1 and the coaxial cable. Both ends of the shield are assumed to be grounded in this case. This system can be represented by the lumped parameter circuit in Figure 3.32B. The open-circuit coherent interference voltage induced in the center conductor is simply written as

$$V_{2(OC)} = j\omega M_{12} I_1 - j\omega M_{S2} I_S \tag{3.95}$$

Because of the symmetrical coaxial geometry, it may be argued (Ott, 1976) that the mutual inductance between the center conductor and the shield, M_{S2}, is simply the self-inductance of the shield, L_S. Also, we assume that $M_{1S} = M_{12}$. Now the current in the shield loop is given by

$$I_s = \frac{j\omega M_{12} I_1}{j\omega L_S + R_S} = \frac{(j\omega M_{12}/R_S) I_1}{j\omega L_S/R_S + 1} \tag{3.96}$$

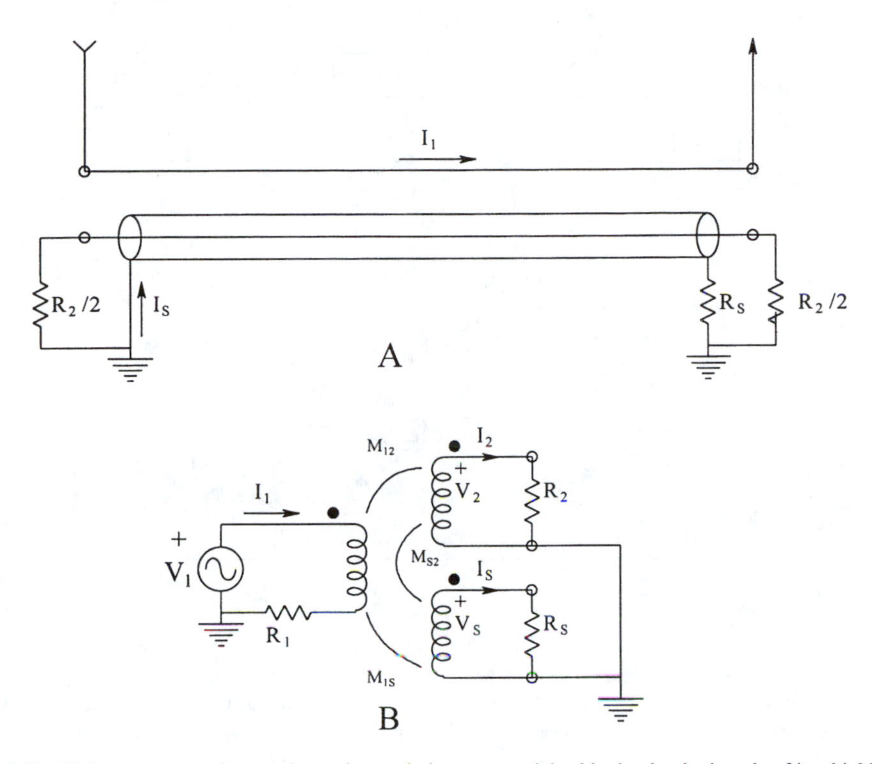

FIGURE 3.32 (A) A current-carrying conductor in proximity to a coaxial cable that has both ends of its shield grounded. This architecture is shown to minimize induced interference in the center conductor, and to prevent magnetic interference from the cable. (B) Lumped-parameter equivalent circuit of the coaxial cable and wire shown in (A).

The shield cutoff frequency, $R_S/2\pi L_S$, is generally on the order of 2 kHz for common RG-58C coaxial cable, and is lower for other types of coax. If we assume that the radian frequency of the interference lies above the shield's break frequency, we find that $I_S \approx I_1$. Thus, from Equation 3.95, we see that $V_{2(OC)} \approx 0$. The lesson here is that to prevent magnetically induced coherent interference on a coaxially shielded line, both ends of the shield must be securely grounded. As you will see below, grounding both ends of the coax shield can, in some cases, cause interference from a ground loop.

Just as we seek to minimize pickup of magnetically induced coherent interference, it is also important to prevent radiation of magnetic interference from a current-carrying coaxial conductor. This situation is illustrated in Figure 3.33. At frequencies above the shield cut-off frequency, we have:

$$I_S = \frac{j\omega L_S I_1}{j\omega L_S + R_S} = \frac{(j\omega L_S/R_S)I_1}{j\omega L_S/R_S + 1} \approx I_1 \qquad 3.97$$

Thus, the induced shield current is equal to the center conductor current and flows in the opposite direction. Hence, outside the coax, the B fields from I_S and I_1 cancel vectorially, and magnetic interference is not generated.

3.9.2.4 Single Grounding of Coax Shields to Prevent Ground Loop Interference

In the circuit describing ground loop interference, Figure 3.34A, we can easily show, using Thevenin's theorem, that the ground loop EMF, V_{GL}, produces a substantial interference voltage at the amplifier's input, given by

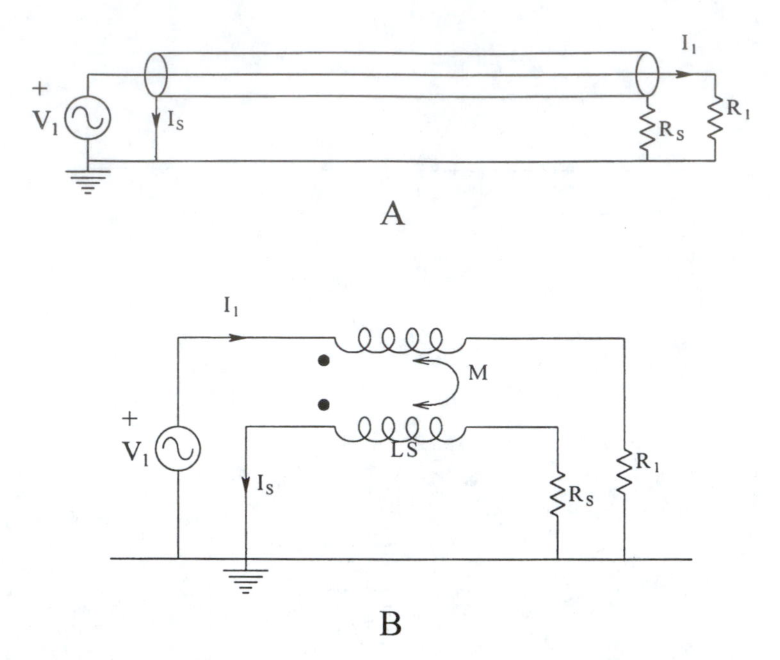

FIGURE 3.33 (A) A coaxial cable carrying current I_1 on its center conductor. (B) Lumped-parameter equivalent circuit of the coaxial cable showing the current induced in the shield loop, I_S. At distances greater than several cable diameters from the cable, the magnetic fields from I_1 and I_S sum effectively to zero.

$$V_{CI} = V_{GL}\left[R_{C2}/\left(R_G + R_{C2}\right)\right] \quad 3.98$$

We can easily examine the case where the input source is floating with respect to ground. This case is illustrated in Figure 3.34B. Because there is a very low conductance pathway between $V_{GL(OC)}$ and V_S, there is very little interference from the ground loop EMF. In this case, the coherent interference voltage at the amplifier input is given by:

$$V'_{CI} \approx V_{GL}\left[R_{C2}/Z_{SG}\right] \quad 3.99$$

Here it is obvious that $V_{CI} >> V'_{CI}$, and we have assumed that $R_{IN} >> R_S >> (R_{C1}, R_{C2}, \text{and } R_G)$. In this case, a differential or instrumentation amplifier should be used to condition V_S, and a shielded, twisted-pair ("twinax") cable is used to minimize electrostatic and magnetically induced interference.

3.9.2.5 Use of a Longitudinal Choke or Neutralizing Transformer to Attenuate Common-Mode Coherent Interference

Figure 3.35A illustrates the use of a toroidal ferrite magnetic core to form a *longitudinal choke.* A longitudinal choke is used to reduce coherent interference caused by high-frequency, common-mode interference picked up by wires connecting sensors to signal conditioning modules, or connecting subsystems in an instrumentation system. The longitudinal choke may be used with pairs of current-carrying wires, twisted pairs of wires, or coaxial cables. More than one pair of wires or coaxial cable may be wound on the same core without the circuits cross-talking. In the simplest form, the wires are passed though the center of a cylindrical ferrite core to form a one-turn longitudinal choke. This geometry is often seen on accessory cables used with PCs. To examine the operation of the longitudinal choke, refer to Figure 3.35B. First, we use superposition to find

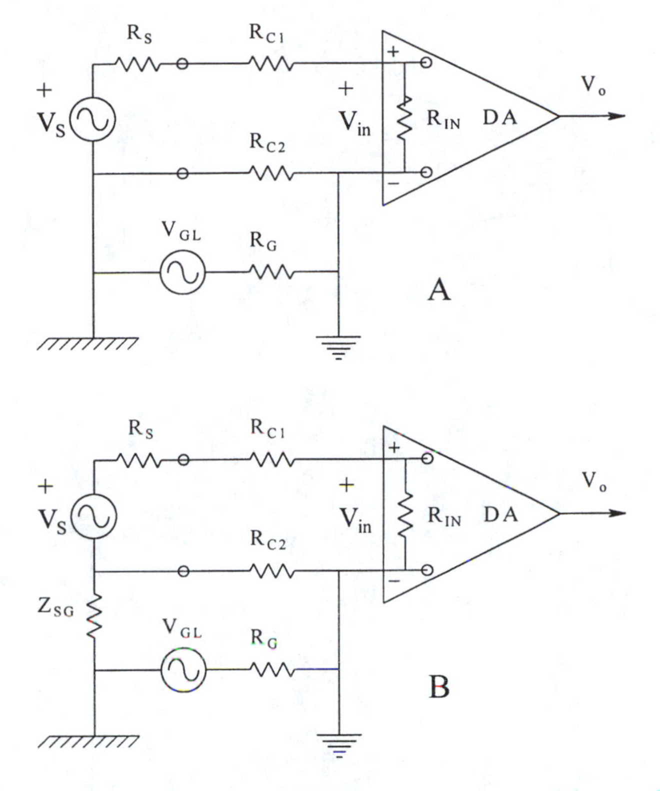

FIGURE 3.34 (A) A differential amplifier is used to condition the voltage of a grounded source. See text for description of circuit. (B) In this case, the DA conditions the voltage of a floating source. The effect of V_{GL} is minimal.

the signal voltage across R_{IN} of the signal conditioning amplifier. Setting $V_{GL} = 0$, we write two loop equations: Assume that $L_1 = L_2 = M_{12} = M_{21} = L$.

$$V_S = I_1[R_S + R_{IN} + R_{C1} + R_{C2} + j\omega 2L] + I_2[R_{C2} + j\omega L] - j\omega 2MI_1 - j\omega MI_2 \qquad 3.100A$$

$$0 = I_1[R_{C2} + j\omega L] + I_2[R_G + R_{C2} + j\omega L_2] - j\omega MI_1 \qquad 3.100B$$

Cancelling terms and using Cramer's rule to solve the loop equations, we find

$$\Delta = (R_S + R_{IN})(R_G + R_{C2} + j\omega L_2) \qquad 3.101$$

and V_{IN} due to V_S is given by:

$$\frac{V_{IN}}{V_S} = \frac{I_1 R_{IN}}{V_S} = \frac{R_{IN}(R_G + R_{C2} + j\omega L_2)}{(R_S + R_{IN})(R_G + R_{C2} + j\omega L_2)} \approx 1 \qquad 3.102$$

Here we have assumed as before that $R_{IN} >> R_S >> R_{C1}, R_{C2}, R_G$. Of great significance is that the signal component at the amplifier input undergoes no high-frequency attenuation due to the longitudinal choke.

Next, we set $V_S = 0$, and consider the amplifier input caused by the ground loop voltage, V_{GL}. Again, writing loop equations:

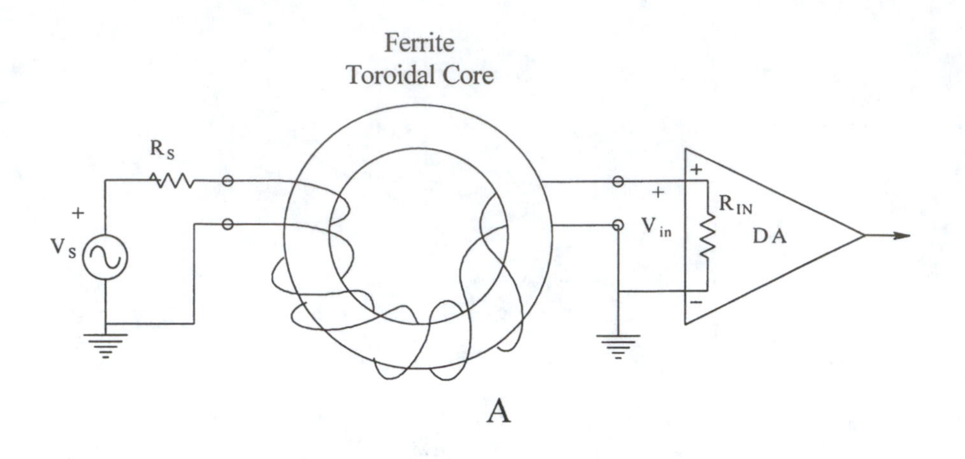

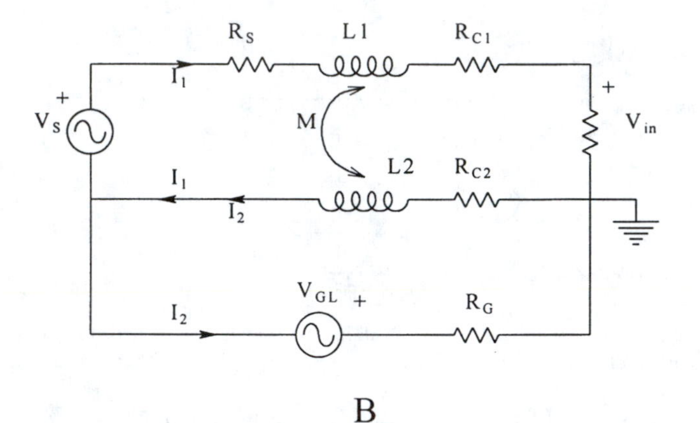

FIGURE 3.35 (A) A toroidal, longitudinal choke. (B) Equivalent circuit of the longitudinal choke. See text for analysis.

$$0 = I_1[R_{IN} + R_S + R_{C1} + R_{C2} + j\omega 2L] + I_2[R_{C2} + j\omega L_2] - j\omega 2MI_1 - j\omega MI_2 \qquad 3.103A$$

$$V_{GL} = I_1[R_{C2} + j\omega L_2] + I_2[R_G + R_{C2} + j\omega L_2] - j\omega MI_1 \qquad 3.103B$$

Cancelling terms, we again solve for $V_{IN} = I_1R_{IN}$ using Cramer's rule:

$$\frac{V_{IN}}{V_{GL}} = (j\omega) = \frac{R_{C2}}{(R_{C2} + R_G)} \frac{1}{[1 + j\omega L_2/(R_G + R_{C2})]} \qquad 3.104$$

From Equation 3.104, we see that the interference due to V_{GL} is attenuated and low-pass filtered by the action of the longitudinal choke.

3.9.2.6 Experimental Verification of Cabling and Grounding Schemes to Achieve Minimum Noise Pickup

Ott (1976) described the effectiveness of 12 different cabling schemes in minimizing the pickup of a 50 kHz magnetic field. The unwanted coherent interference was measured across a 1 megohm resistor. The shielded cables were wound helically (3 turns, 7" diameter) inside the helical field-generating coil (10 turns of 20 ga. wire, 9" diameter, carrying 0.6 A at 50 kHz). Ten volts peak-to-peak were impressed across the field generating coil. In his tests, Ott reports that 0 dB was 0.8 V. Figure 3.36 summarizes his results. We have arranged the cables in the order of increasing magnetic (and electric) shielding effectiveness. Note that the simple coaxial cable with ungrounded shield performed the poorest. The better performing cables had circuits in which the 100 ohm Thevenin source reistance was not tied to ground. Although Ott's evaluation was focused on evaluating the shielding from a magnetic field, his system also produced an electric field. Caution should be used in over-generalizing his results. Many subtle factors contribute to the pickup of coherent interference, including circuit geometry, the frequency and the spatial distributions of E and B fields.

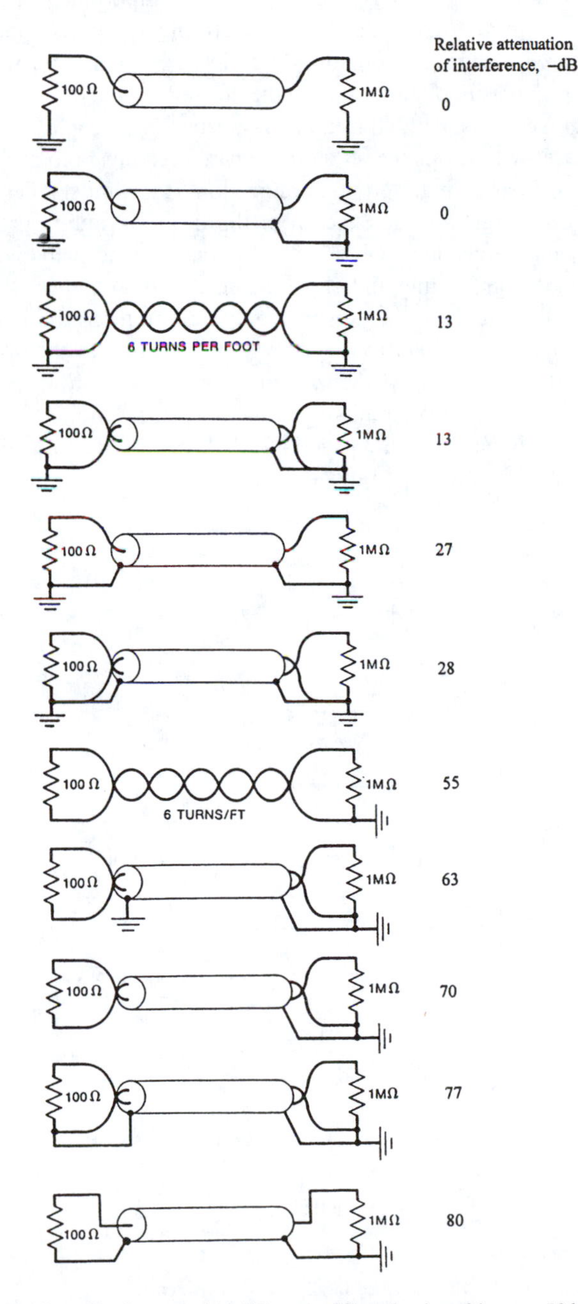

FIGURE 3.36 Relative shielding effectiveness of 12 different cabling circuits subject to a 50 kHz magnetic field determined by Ott, 1976. See text for description.

Ott's results show that the use of shielded twinax cables with single grounds on the shield gave the best results (lowest pickup of coherent interference from a predominantly magnetic source). He attributed the –80 dB attenuation shown by the simple coaxial cable to be anomalous, and due to the fact that this circuit had a smaller net loop area, A, than the twinax cables. This anomalous result underscores what often appears to be the arcane nature of coherent noise reduction techniques, which can appear to the inexperienced as being "arty." However, the laws of physics apply to coherent noise reduction practices, as well as other branches of electrical engineering and instrumentation.

3.9.2.6.1 Circuit Grounding. Probably the single most important principle of reducing the pickup of coherent noise from internal sources in an electronic system is good grounding practice. Appreciate that typical ground currents contain a large DC component on which rides a coherent (AC) component which is the result of conditioning the desired signal, or is from the high-frequency switching of digital (logic) circuits. At high frequencies, ground wires appear as series R-L circuits, whose impedance increases with frequency. If a circuit card contains both digital circuitry and low-level analog amplifiers, and has a single ground connection, it is almost a certainty that the analog system will pick up the coherent interference from the digital circuits' ground current flowing through the common ground impedance. Obviously, separate ground paths will help this situation. Figure 3.37 illustrates a good grounding and electrostatic shielding architecture. Note that there is a common tie point for all grounds at the power supply common terminal. This terminal may or may not be tied to the metal instrument case which is always tied to the green wire (power line ground). It is important to note that separate grounds are used for low-level analog, high-level analog, digital, and inductive (arcing) circuits such as motors, solenoids, and relays. The ground wire itself should be of heavy gauge; if the + and –15 V DC supplies to a circuit card use No. 18 wire, then the ground should be No. 14 stranded wire.

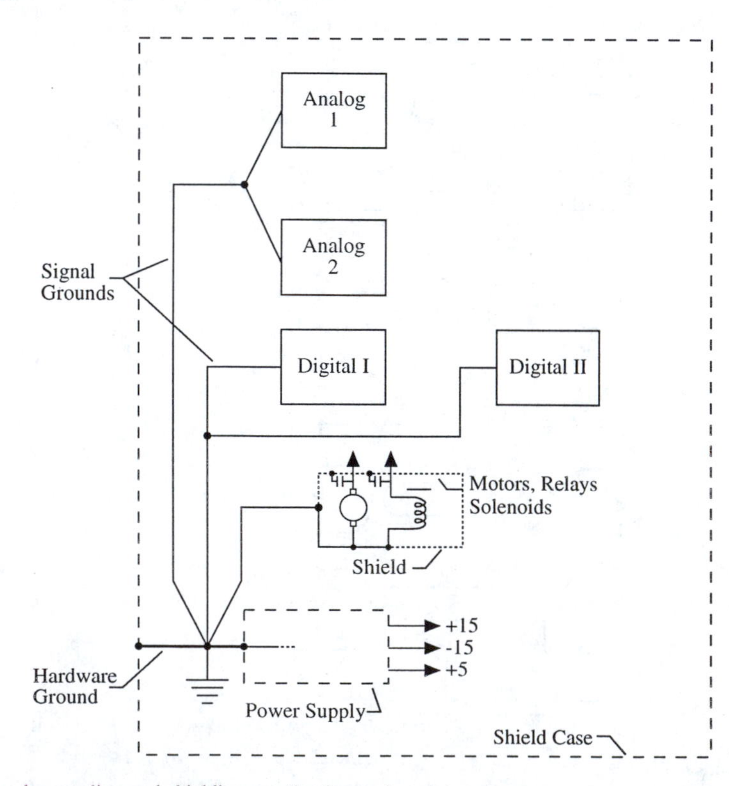

FIGURE 3.37 Good grounding and shielding practice in an electronic system.

3.9.2.6.2 Ferrite Beads and Feed-Through Capacitors. An effective way of attenuating unwanted coherent interference on signal and power lines at frequencies above 1 MHz is through the use of one or more (seldom more than three) ferrimagnetic beads strung on the wire carrying the offending interference. Ferrite beads act as small chokes, making the wire upon which they are strung appear to be a series R-L circuit. Because of their internal losses, some ferrite beads cause their wire to have a nearly constant (resistive) impedance of about 30 to 75 ohms from <1 to 100 MHz. Other beads are made to appear more inductive, and their Z increases linearly with frequency, going from 20 ohms at 5 Mhz to 110 ohms at 100 Mhz (Ott, 1976). The "type 64" material bead made by Amidon Associates, North Hollywood, CA, is designed to operate above 200 MHz. Ferrite beads make effective, well-damped, second-order low-pass filters when combined with appropriate bypass capacitors to ground. One widely used bypass capacitor is the feed-through capacitor, used to decouple RF noisy sources from the power supplies. Typical feed-through capacitors have cylindrical, threaded metal bodies which are mounted in the metal case or shield wall, and terminals at both ends to permit the DC current to pass through. Capacitance from the center through-conductor to the case can range from 0.5 to 5 nF, depending on the application.

In some cases, the use of ferrite beads will produce more harm than good; misuse of beads can cause unwanted, high-frequency resonances and oscillations. Properly applied, ferrite beads and feed-through capacitors can provide a simple and low-cost means of reducing coherent interference and parasitic oscillations.

3.9.2.6.3 Interruption of Ground Loops by the Use of Isolation Transformers and Photo-optic Couplers. Isolation transformers are a simple, add-on means which can often improve measurement system signal-to-noise ratio when used between analog signal conditioning sub-systems. A typical isolation transformer used with 50 ohm coaxial cables has terminating and source impedances of 50 ohms, and a frequency response of 25 Hz to 20 MHz. There can be several thousand megohms impedance between the primary and secondary coax shields, hence the common-mode EMF has little effect (see Equation 3.99) on the signal.

Photo-optic couplers are nonlinear devices, primarily used in digital interfaces where it is desired to have complete ground and signal isolation between digital circuits. A typical optocoupler consists of an LED on a chip in optical, but not electrically conductive intimacy with an output circuit which consists of a phototransistor or photodiode which is caused to conduct when illuminated by the input LED. The primary use of optocouplers is isolated, high-speed, serial digital data transmission. A simple transistor-transistor logic (TTL) logic circuit coupled with a photo-optic coupler is shown in Figure 3.38A.

Photo-optic couplers can be modified to give optically isolated, analog signal transmission by DC biasing the photodiode and phototransistor to operate in the middle of their dynamic regions, and using negative feedback to reduce nonlinearity. Such linear photo-optic coupler circuits can have small-signal bandwidths from DC to over 1 MHz. A feedback analog photo-optic coupler circuit, adapted from a circuit in the Hewlett-Packard (San Jose, CA) Optoelectronics Designer's Catalog, is shown in Figure 3.38B. In this circuit, an HCPL-2530 dual optocoupler is used. Feedback forces the AC signal current through D2 to track the input signal current in D1, effectively cancelling out much of the nonlinearity inherent in the photodiodes and phototransistors.

3.9.2.6.4 The Use of Guarding and Shielding to Reduce Common-Mode, Coherent Interference. When extremely low-level signals are being measured, or when very high CM coherent noise is present, as in the case of certain biomedical applications, strain gauge bridge applications, or thermocouple measurements, the use of a guard shield can effectively reduce the amplifier's input capacitance to ground, hence coherent interference. Figure 3.39 illustrates a conventional differential instrumentation amplifier (DA) connected to a Thevenin source (V_S, R_S), by a shielded, twinaxial cable. The cable shield is tied to the source ground, and not the power supply ground. A large, coherent, CM voltage, V_{GL}, exists between the power ground and the source ground. If no currents flow in the twisted pair wires, then V_{GL} appears as the CM voltage at the amplifier's input, and is rejected by the amplifier's CMRR; there is no differential-mode component

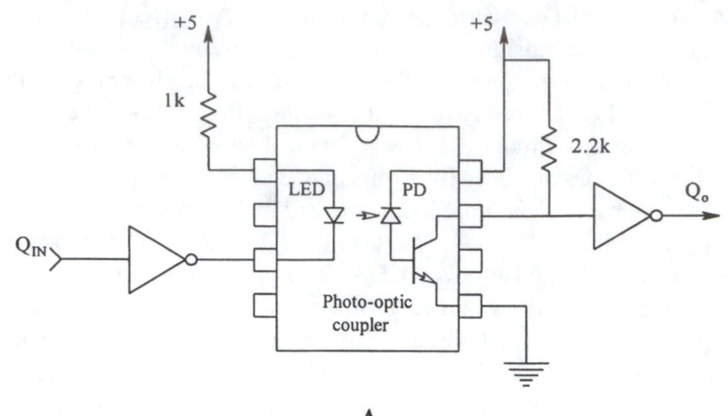

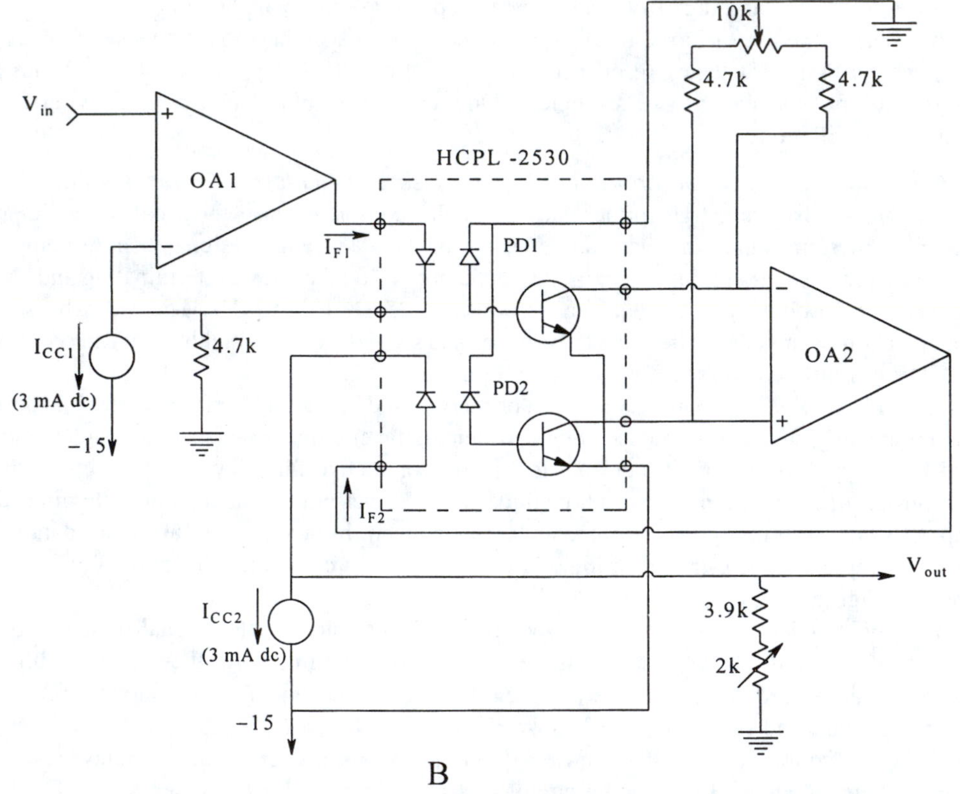

FIGURE 3.38 (A) A photo-optic coupler used to isolate a TTL digital signal line. (B) Use of a dual optocoupler in a feedback circuit in order to realize isolated, analog signal transmission.

of V_{GL}. However, the input capacitances of the DA allow input currents to flow. Note that there is a capacitance C_{1S} between the HI side of the source and the cable shield and conductor 2. Because of asymmetry in the circuit, a DM component of V_{GL} appears across the amplifier's input. This DM interference component may be calculated: by inspection of the circuit of Figure 3.39, $V_- = V_{GL}$. V_+ is given by the voltage divider

$$V_+ = V_{GL} \frac{1/j\omega C_{1G}}{\left[1/j\omega C_{1G} + 1/\left(j\omega C_{1S} + G_S\right)\right]} = V_{GL} \frac{j\omega C_{1S} R_S + 1}{j\omega\left(C_{1S} + C_{1G}\right)R_S + 1} \qquad 3.105$$

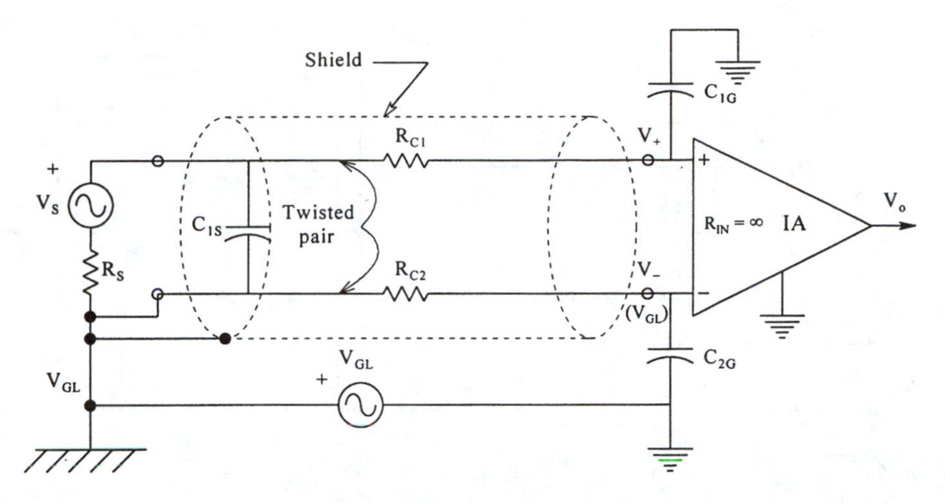

FIGURE 3.39 An instrumentation amplifier (IA) connected to a Thevenin source by twinax shielded cable.

The differential interference voltage at the DA's input is simply:

$$V_d = (V_+ - V_-)/2 \tag{3.106}$$

Substituting from the relations above, we find:

$$V_d = -V_{GL}(j\omega C_{1G} R_S) \quad \text{for } \omega \ll \omega_b = 1/[R_S(C_{1G} + C_{1S})] \text{ r/s} \tag{3.107A}$$

$$= -(V_{GL}/2)\frac{C_{1G}}{C_{1G} + C_{1S}} \quad \text{for } \omega \gg \omega_b \tag{3.107B}$$

For a numerical example, let R_S = 1 k, C_{1G} = 20 pF, C_{1S} = 280 pF, and ω = 377 (60 Hz ground loop voltage). Assume $R_{IN} = \infty$, and $R_{C1} = R_{C2} = R_{GL} = 0$. f_b is found to be 531 kHz, and from Equation 3.107A, the magnitude of the DM interference component is

$$V_d = V_{GL}(3.77 \times 10^{-6}) \tag{3.108}$$

This DM interference can seriously degrade output signal-to-noise ratio when V_S is in the single microvolt range, and V_{GL} is in the hundreds of millivolts. To reduce the interference V_d, we take steps to reduce the input capacitances to ground. This can be accomplished by using an isolation amplifier (see Section 2.4.2) in conjunction with a guard shield and external shield, as shown in Figure 3.40. Here the guard shield is tied to the amplifier's ground, and is made to float with respect to "true ground" as a consequense of the isolation amplifier's RF power supply. The current in the ground loop, and hence the magnitude of the DM interference voltage, is reduced at least 20-fold because the capacitance from either amplifier input to true ground (C_{1G}, C_{2G}) is now reduced by the external shield, and is on the order of 1 pF. The same voltage divider is used to calculate V_d.

Figure 3.41 illustrates the input circuit of a guarded meter. Note that the following analysis of the guarded meter circuit is also valid for a DA circuit in which the common-mode point, CP, is taken to the guard terminal and to the ground through impedances Z_1 and Z_2, as shown in Figure 3.41. In this circuit, four meter terminals are of interest: HI, LOW, GUARD, and GROUND. Here, as in the preceding analysis, an unwanted, coherent, common-mode, ground loop voltage, V_{GL}, is

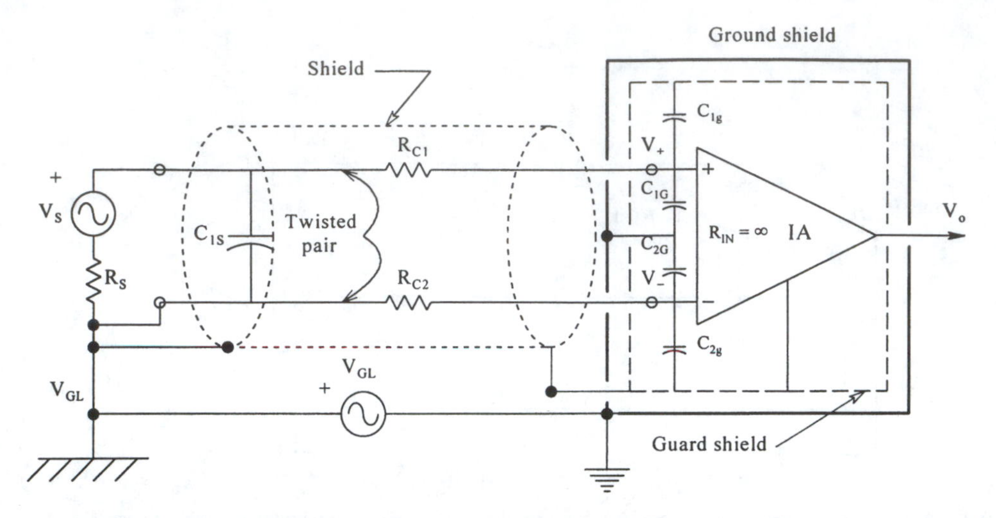

FIGURE 3.40 An isolation amplifier (ISOA) is used with a guard shield and a ground shield to reduce ground loop interference.

present between true ground and the remote source ground, as well as some other, undesired, common-mode voltage associated with the source, V_N.

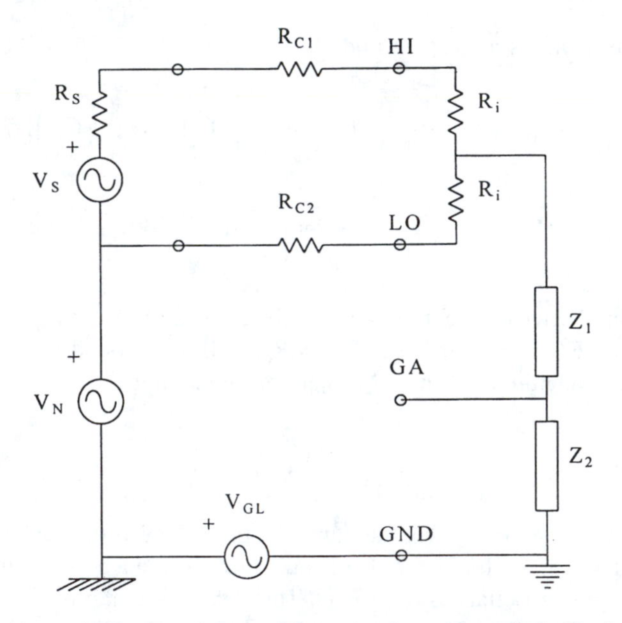

FIGURE 3.41 Equivalent input circuit of a voltmeter with guard and ground terminals. V_{GL} is the ground-loop voltage and V_N is another common-mode noise source.

Five ways of connecting the guard of the guarded meter will be considered. The analysis of each case is simplified by calculating the input voltage at the meter, $(V_{HI} - V_{LO})$, as a function of circuit parameters.

In the first case, we will leave the guard terminal open-circuited. We will assume that the input resistances of the meter or DA, R_i, are very large — at least 10^8 ohms. The resistances of the twisted pair wires, $R_{C1} = R_{C2} = R_C$ are about 0.1 ohm. The source resistance is 1000 ohms, and Z_1 and Z_2 are typically capacitive reactances of about 10^5 ohms. As can be seen from Figure 3.42, the differential input voltage to the meter, V_d, can be considered to be the unbalance voltage of a

Wheatstone bridge whose four arms are R_C, $(R_S + R_C)$, R_i, and R_i. The voltage across the bridge, V_B, is given approximately by:

$$V_B = (V_{GL} + V_N)\frac{R_i/2}{R_i/2 + Z_1 + Z_2} = (V_{GL} + V_N) \qquad 3.109$$

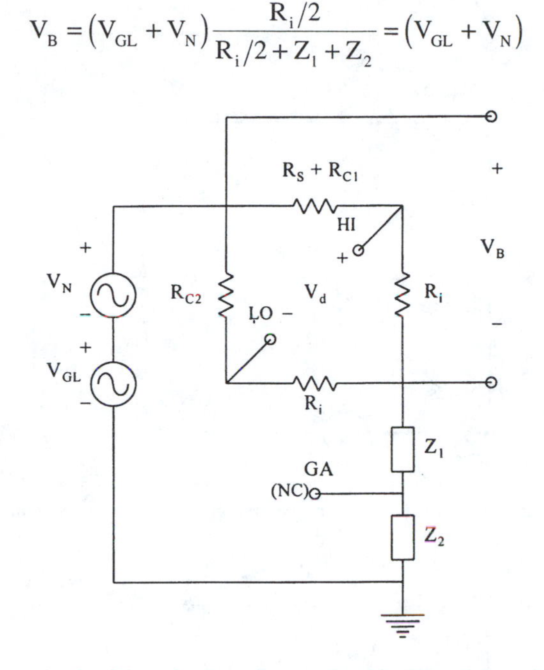

FIGURE 3.42 Equivalent input circuit of the voltmeter when the guard is not used. $V_S = 0$. Note the bridge architecture.

Again, taking into account the sizes of the resistors in the circuit, it is easy to show that V_d is given by:

$$V_d = (V_{GL} + V_N)(-R_S/R_i) = (V_{GL} + V_N)(-10^{-5}) \qquad 3.110$$

In the second case, we assume that the guard is incorrectly connected to the source's ground. This connection effectively shorts out V_{GL}; however, V_N still produces a DM interference voltage at the meter input. The equivalent circuit for this case is shown in Figure 3.43. Again, a bridge circuit is formed with an unbalance voltage, V_d, of:

$$V_d = V_N(-10^{-5}) \qquad 3.111$$

In the third case, shown in Figure 3.44, the guard is incorrectly connected to the meter's LOW terminal. Here it is easy to see that:

$$V_d = (V_{GL} + V_N)[1 - Z_2/(Z_2 + R_C)] = (V_{GL} + V_N)[R_C/Z_2] = (V_{GL} + V_N)[10^{-6}] \qquad 3.112$$

Figure 3.45 illustrates the *fourth case* where the guard is incorrectly connected to the meter ground, shorting out Z_2. The differential interference voltage is found from the bridge circuit to be approximately

$$V_d = (V_{GL} + V_N)(-R_S/R_i) = (V_{GL} + V_N)(-10^{-5}) \qquad 3.113$$

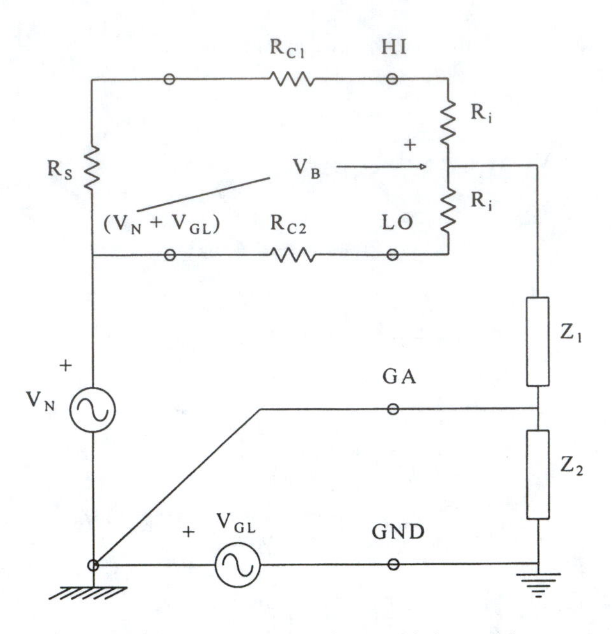

FIGURE 3.43 Equivalent input circuit of the voltmeter when the guard is incorrectly connected to the source ground.

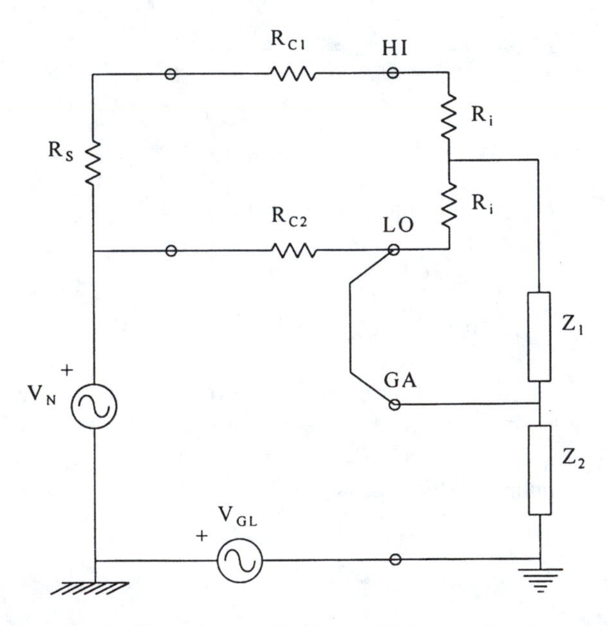

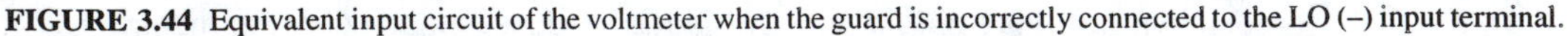

FIGURE 3.44 Equivalent input circuit of the voltmeter when the guard is incorrectly connected to the LO (–) input terminal.

In the fifth example, the guard terminal is correctly connected to the source low (i.e., the node between R_{C2}, R_S, and V_N). This situation is shown in Figure 3.46. Clearly in this case, $V_B = V_d = 0$, and there is no DM component at the meter input from the coherent noise voltages.

3.9.3 Summary

The practice of the art and science of coherent noise reduction should begin with the elimination of local noise sources. Obviously, these are noise sources which have been identified, and which we have authority over. These sources might include digital and computer equipment operating in the immediate environment, or interference caused by sparking motor commutators, etc. The offending noise sources must be enclosed in appropriate, grounded, electrostatic shields. Their

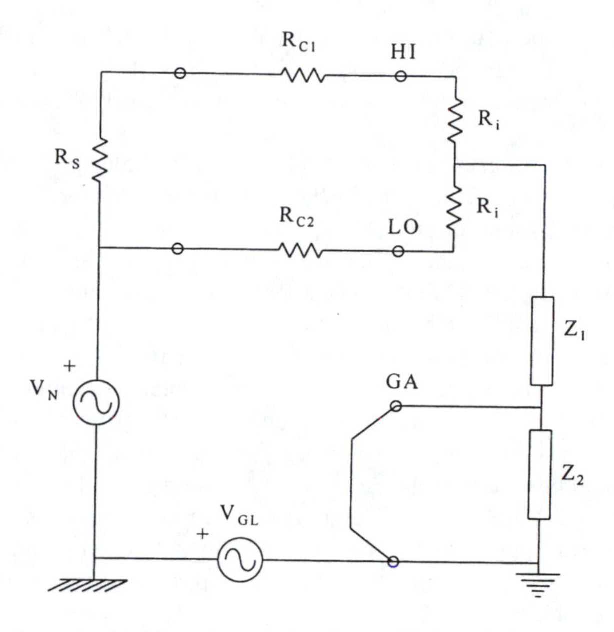

FIGURE 3.45 Equivalent input circuit of the voltmeter when the guard is incorrectly connected to the meter ground.

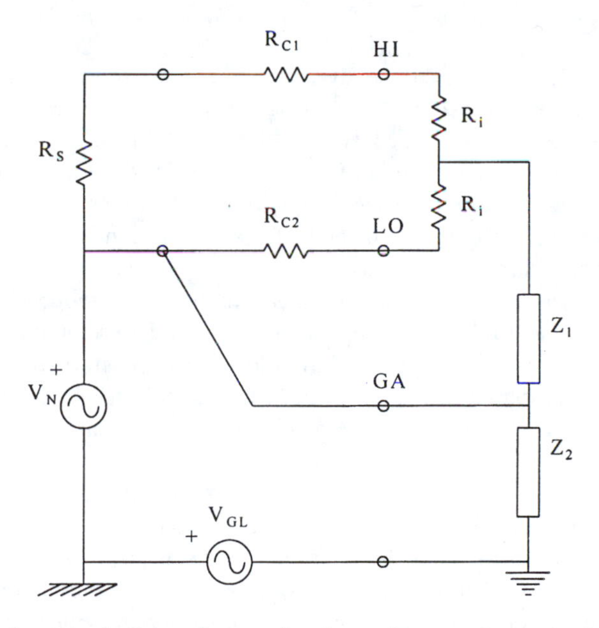

FIGURE 3.46 Equivalent input circuit of the voltmeter when the guard is correctly connected to the source low. There is no difference-mode input component from V_N and V_{GL}.

power leads must be low-pass filtered or bypassed as they leave the shield; the use of ferrite beads and feed-through capacitors may be indicated. In addition, power leads should be twisted so there is minimum magnetic field created. Inductive spikes induced on power lines can be minimized by using varistors or diodes across switched relay and solenoid coils. Both ends of coaxial cables carrying RF should be grounded to minimize the radiation of magnetic field interference.

As we have seen above, there are many means of reducing the pickup of coherent interference in low-level analog signal conditioning systems. Lead dress is important; separate low-level leads from wires carrying high currents and voltages. Low-level leads should be twisted and routed near the grounded case. If twinaxial (twisted and shielded) cable is used, it should be grounded at one end only, and insulated. Alternate grounds and signal leads on ribbon cables and in connectors. High-impedance, low-level leads should be kept as short as possible; this is especially true in wiring op amps. The summing junction should never have an exposed lead of more than 1/2" connecting components to it.

Often, the pickup of coherent interference can be traced to the magnetic component of a low-frequency electromagnetic field. Such pickup may occur in spite of rigorous electrostatic shielding practice. In this case, a mu metal or other lossy, non-saturating magnetic shield may be effective, as will be reducing the area and changing the spatial orientation of the pickup loop.

We have seen above that ground loops can be a serious source of interference in instrumentation systems. The effect of ground loop EMFs often can be minimized by the use of photo-optical couplers (for digital circuits), isolation transformers (for signals in the 20 Hz to 200 kHz range), longitudinal chokes (to attenuate high-frequency CM interference), differential isolation amplifiers, and guarded inputs. Good grounding practice can significantly reduce ground loops. We have seen that a single ground point is the best practice (except for RF circuits). Individual, insulated grounds from subsystems should be brought separately to the common ground point.

Just as separate grounds are important for interference reduction, so is the use of decoupled, low impedance, DC power supplies with proper power ratings. Dedicated, low-noise, analog power supplies for signal conditioning system headstages can help reduce hum and the cross-coupling of interference from other stages.

3.10 CHAPTER SUMMARY

In this chapter, we have presented a heuristic, yet mathematically sound, treatment of random noise in signal conditioning systems based on the one-sided power density spectrum. Sources of random noise in active and passive circuits were considered: thermal (Johnson) noise in resistors, shot noise in bipolar junction transistors, and noise in FETs. The dependence of transistor noise performance on DC operating point was stressed. The standard, two-generator noise source model for amplifiers was introduced, as well as the concepts of noise factor, noise figure, and output signal-to-noise ratio as figures of merit with respect to noise.

The propagation of Gaussian noise through linear filters was introduced, and the concept of a filter's gain^2*bandwidth was applied to finding the mean squared noise output of signal conditioning systems, given white input noise. We showed how a low-noise transformer can be used to increase output SNR when a signal source with a low source resistance drives the input of a noisy amplifier.

Five practical, numerical examples of noise calculations in measurement system design are given.

We presented, in tabular form, a list of commonly used, low-noise, IC amplifiers. Here we note that the lowest system e_{na} values are generally found in amplifiers having BJT input stages, hence relatively low input resistances, high bias currents, thus large i_{na} values. FET input amplifiers have slightly higher e_{na} values, but offer much higher input resistances, lower DC bias currents, hence lower i_{na} values than do BJTs. The wise designer of low-noise signal conditioning equipment thus must make compromises between e_{na}, i_{na}, input impedance, bandwidth, and other performance factors when choosing an IC amplifier for headstage use.

In Section 3.9, we considered coherent interference, its sources, and some of the many ways to minimize its effect in instrumentation systems. Coherent interference can appear at the output of a signal conditioning system as the result of electrostatic or magnetic coupling to input leads. We also show that it can enter an amplifier through the power supply, or through improper grounding. We discuss some of the more common means of mitigating coherent interference, and analyze equivalent circuits relevant to those means.

4

DC Null Methods of Measurement

4.0 INTRODUCTION

DC null techniques are used with Wheatstone bridges, Kelvin bridges, and potentiometers to obtain increased measurement accuracy based on the fact that the human observer can estimate the occurrence of a DC null on a null voltmeter with greater precision than he or she can read the meter scale directly in volts. The accuracy of a DC null measurement system is derived from the known, calibrated accuracy of the resistors making up the bridges or potentiometer, and has little dependence on the null meter. The human eye is quite sensitive in detecting the microscopic, transient deflections of the null meter's pointer under non-null conditions.

Wheatstone bridges are traditionally used to make precise, accurate resistance measurements (generally in the range of 1 to 10^6 ohms), or to measure some physical quantity, such as temperature, light intensity, or strain which causes a known change in resistance. Figure 4.1 illustrates the complete circuit of a DC Wheatstone bridge.

In earlier days, the bridge null was sensed by a sensitive galvanometer, a current-measuring, electromechanical transducer in which the bridge's unbalance current is passed through a coil which is delicately suspended in a strong permanent magnetic field. The magnetic flux caused by the small coil current interacts with the permanent magnet's flux and causes a deflection torque proportional to the coil current. The deflection torque acts on the torsion spring of the suspension, and produces an angular deflection which is usually sensed optically by reflecting a collimated light beam off a small mirror attached to the galvanometer's coil. Needless to say, although sensitive, a mirror galvanometer is not a mechanically robust instrument, and consequently is seldom used in modern DC null measurement systems. Present-day DC null sensing devices make use of electronic amplification and display devices; they generally have differential, high-impedance ($>10^{12}$ Ω) input stages so the unbalance voltage is sensed as an open-circuit voltage.

Kelvin bridges are null devices used to measure low values of resistance between 10 microohms and 10 ohms. The null condition for the Kelvin bridge is described below.

Potentiometers, described in more detail below, are used to make very precise DC voltage measurements in the range of 1 V or higher. They are basically accurate voltage dividers that allow differential comparison of their output with an unknown DC EMF. A simplified potentiometer circuit is shown in Figure 4.2. Traditionally, potentiometers were calibrated with the saturated Weston cell, an electrochemical battery of known open-circuit voltage or EMF. The same null detection means is used to compare the calibrated potentiometer's open-circuit voltage output with the unknown EMF as is used with Wheatstone and Kelvin bridges.

4.1 WHEATSTONE BRIDGE ANALYSIS

If we assume that the null detector in Figure 4.1 is a high input impedance voltmeter, then in the null condition, $0 = V_o = V_2 - V_1$. The voltages at the corners of the bridge are given by the voltage divider relations:

$$V_1 = V_b N/(M + N) \qquad 4.1$$

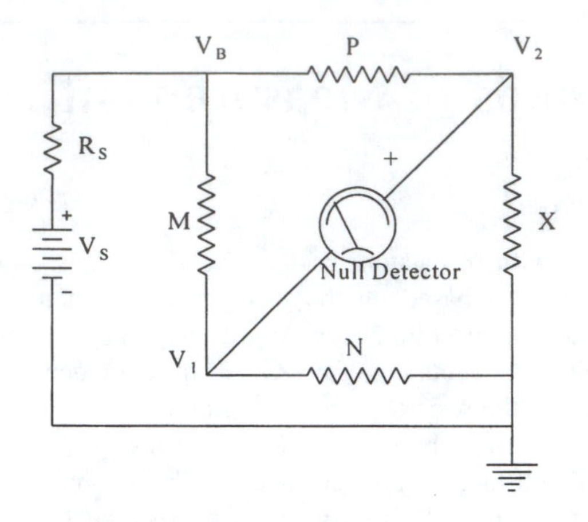

FIGURE 4.1 A basic Wheatstone bridge with DC excitation. The null is detected with a sensitive, center-scale, DC voltmeter which draws negligible current.

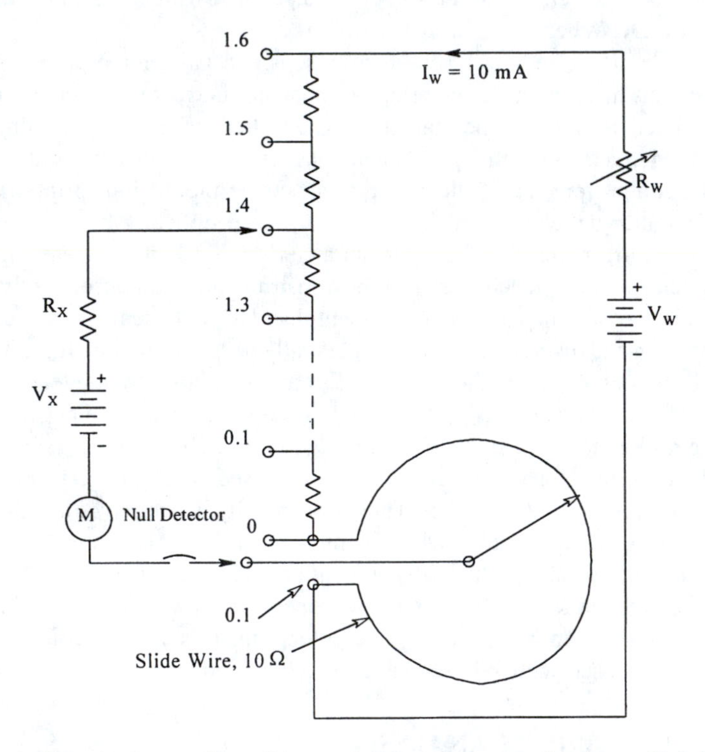

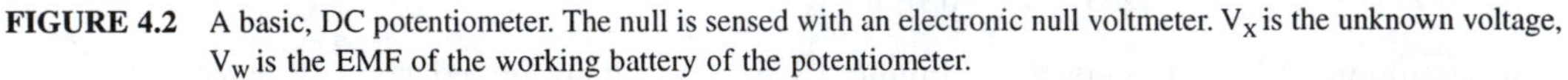

FIGURE 4.2 A basic, DC potentiometer. The null is sensed with an electronic null voltmeter. V_X is the unknown voltage, V_W is the EMF of the working battery of the potentiometer.

$$V_2 = V_b X/(P + X) \tag{4.2}$$

Hence, at null we have

$$V_o = V_b[X/(P + X) - N/(M + N)] = V_b \frac{XM - NP}{(P + X)(M + N)} \tag{4.3}$$

Hence, at null, $V_o = 0$, and we have the well-known relation for the unknown resistor, X:

$$X = NP/M \tag{4.4}$$

The limiting error in determining X at null may be found through application of the limiting error relation from Chapter 1 to Equation 4.4. In this case we have

$$\Delta X = \left|\frac{\partial X}{\partial M}\Delta M\right| + \left|\frac{\partial X}{\partial P}\Delta P\right| + \left|\frac{\partial X}{\partial N}\Delta N\right| \tag{4.5}$$

Algebraic substitution yields the simple relationship

$$\frac{\Delta X}{X} = \frac{\Delta M}{M} + \frac{\Delta P}{P} + \frac{\Delta N}{N} \tag{4.6}$$

Thus, for example, if the resistors M, P, and N are known to 100 ppm (0.01%), then the LE in X is $\Delta X/X = 0.03\%$.

The null detector resolution can also affect the LE in measuring X. Every null detector has a dead zone of $\pm\Delta V_o$ below which it is impossible to tell whether V_o is non-zero or not. This dead zone is caused by noise and/or the inability of a human operator to tell whether the physical indicator of the null, such as an analog meter pointer, has moved. To see how the dead zone, ΔV_o, affects the LE in X, we examine the Taylor's series for the Wheatstone bridge output relation;

$$\Delta V_o = (\partial V_o/\partial X)\Delta X_{MIN} \tag{4.7}$$

For the maximum bridge sensitivity case,

$$(\partial V_o/\partial X) = V_b S_V = V_b P/(X+P)^2 \to V_b/4X \tag{4.8}$$

Equation 4.8 for $\partial V_o/\partial X$ is substituted into Equation 4.7 and we solve for $\Delta X_{MIN}/X$:

$$\frac{\Delta X_{MIN}}{X} = \frac{\Delta V_o}{V_b}\left[(P+X)^2/PX\right] \to 4(\Delta V_o/V_b) \tag{4.9}$$

Equation 4.9 for $\Delta X_{MIN}/X$ can be added to Equation 4.6 for the LE in X. Generally, this term is numerically much smaller than the $\Delta R/R$ terms in Equation 4.6, demonstrating that the null detector is generally the smallest source of error in measuring resistances by Wheatstone bridge.

Often we are not interested in measuring X, per se, but rather wish to examine V_o for small changes in X around the value that satisfies Equation 4.4. For this case, we plot Equation 4.3 in Figure 4.3. We define a voltage sensitivity for the bridge in Equation 4.5:

$$S_V \equiv \frac{\partial(V_o/V_b)}{\partial X} = \frac{P}{(X+P)^2} \tag{4.10}$$

Clearly, S_V has a maximum when P = X. This leads to the conclusion that the maximum sensitivity bridge is one where P = X = M = N. $S_{VMAX} = 1/4X$ in this case.

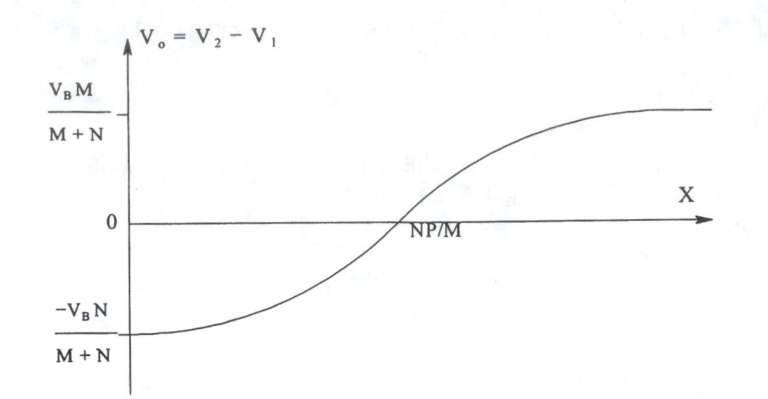

FIGURE 4.3 Output voltage of a Wheatstone bridge as a function of the unknown (X) arm.

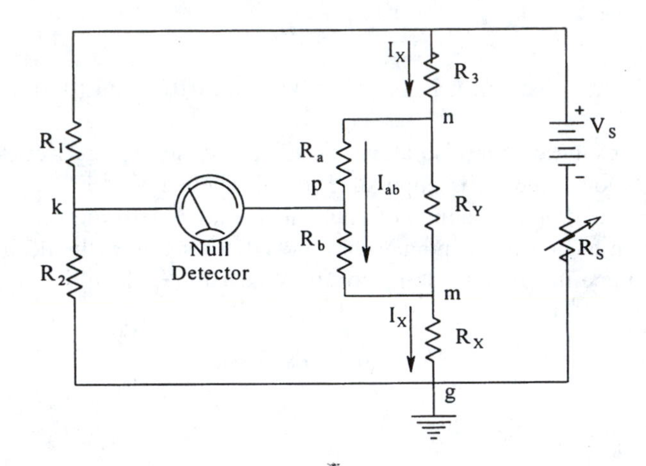

FIGURE 4.4 Current paths in a Kelvin bridge used to measure very low resistances, R_X.

4.2 THE KELVIN BRIDGE

The basic circuit for a Kelvin bridge is shown in Figure 4.4. Note that this bridge differs from a Wheatstone bridge by having four resistors in the right-half circuit. To derive the conditions for null, we note that at null, $V_k = V_p$. V_k can be written by the voltage divider expression:

$$V_k = V_S R_2/(R_1 + R_2) \tag{4.11}$$

and

$$V_p = V_{pm} + V_m = I_{ab} R_b + I_x R_x \tag{4.12}$$

where

$$I_{ab} = I_x R_y/(R_a + R_b + R_y) \tag{4.13}$$

and

$$I_x = \frac{V_S}{R_3 + R_x + R_y(R_a + R_b)/(R_a + R_b + R_y)} \tag{4.14}$$

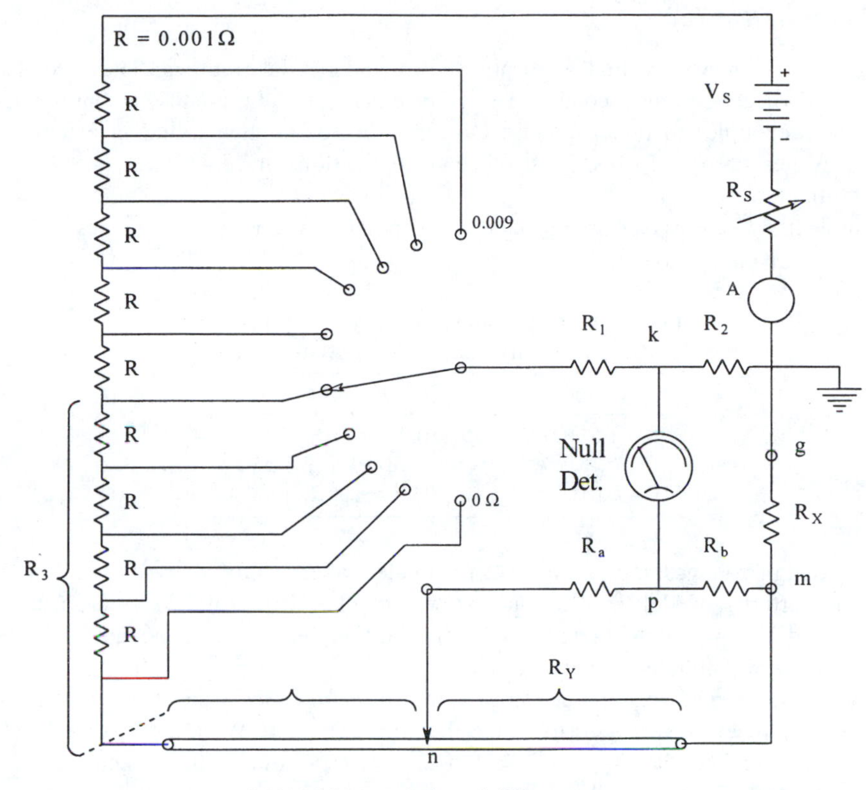

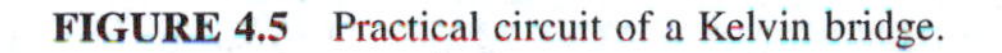
FIGURE 4.5 Practical circuit of a Kelvin bridge.

Thus, at null:

$$V_S R_2/(R_1 + R_2) = \left[\frac{V_S}{R_3 + R_x + R_y(R_a + R_b)/(R_a + R_b + R_y)}\right]\left[R_x + \frac{R_b R_y}{R_a + R_b + R_y}\right] \quad 4.15$$

Now it turns out that if we let

$$R_a/R_b = R_1/R_2 = \alpha \quad 4.16$$

and substitute α in Equation 4.15 above, terms cancel, we arrive at the Kelvin bridge null condition:

$$R_x = R_3/\alpha = R_3 R_2/R_1 \quad 4.17$$

Generally, the resistances in the Kelvin bridge are relatively small, so the battery or DC working voltage source, V_S, must be able to source considerable current. A practical version of a Kelvin bridge is shown in Figure 4.5. Note that the low resistance R_3 is set by a tap switch and 0.001 Ω resistors, with interpolation resistance, ΔR_3, being given by a sliding contact on a manganin bar resistor; the ΔR_3 resistance is proportional to the tap position along the bar.

4.3 POTENTIOMETERS

Potentiometers now generally find applications in standards laboratories. They are used as a precision means of calibrating secondary DC voltage standards. Originally, potentiometers were used with thermocouples to obtain precise temperature measurements, and in electrochemistry laboratories to measure cell EMFs. Most of these applications are now taken over by precision electronic voltmeters.

For example, the Leeds & Northrup Model 7556-1 potentiometer has three ranges. The accuracy of each range is given in Table 4.1.

Table 4.1 Ranges and Accuracy of the Leeds & Northrup 7556-1 potentiometer

Range	Accuracy
0 – 1.61100 V	±(0.0005% + 1 μV)
0 – 0.16110 V	±(0.0015% + 0.1 μV)
0 – 0.01611 V	±0.0025% + 20 nV)

To measure high voltages, a precision voltage divider called a *volt box* was used. For example, the Leeds & Northrup Model 7594-2 volt box was rated at 0 to 1500 V, with an accuracy of ±0.0025%. It had 25 taps, and was rated at 667 ohms/volt on the 1500 V range, quite a lossy device compared to modern voltmeter sensitivities.

Potentiometers are often calibrated using *Weston Standard Cells*, an electrochemical battery which until fairly recently was used to define the volt EMF. The Weston cell was developed in 1892 as an EMF standard. Two types of Weston cells exist, saturated and unsaturated. The more widely used saturated (normal) cell is illustrated in Figure 1.6. A Weston cell is generally used in a thermostatted oil bath to maintain its internal temperature to within 0.01°C. This is necessary because of the rather high EMF tempco of the normal cell, –40 μV/°C. The EMF of the saturated Weston cell is given by:

$$\mathrm{EMF} = 1.01830 - 4.06 \times 10^{-5}(\mathrm{T}-20) - 9.5 \times 10^{-7}(\mathrm{T}-20)^2 - 10^{-8}(\mathrm{T}-20)^3 \qquad 4.18$$

(This EMF is in International Volts; the EMF in absolute volts is 1.01864 at 20°C.) Thus the volt used to be defined as 1/1.01830 of the Weston normal cell at 20°C. Weston cells are very delicate and subject to instabilities if their EMFs are subjected to large or sudden temperature changes. They also should never have currents in excess of 100 μA drawn from them for any period of time. They also age, and should be recalibrated against freshly prepared cells every year, or against Josephson junction primary EMF standards.

Calibration of a potentiometer is done by first adjusting the potentiometer's reading to the calculated EMF of the standard cell, then adjusting R_W to obtain a null in the galvanometer or electronic null detector. Next, the unknown EMF is connected to the potentiometer and the null detector, and the calibrated potentiometer's output voltage is adjusted to obtain a null.

4.4 CHAPTER SUMMARY

In this chapter we discussed three basic measurement systems that make use of the principle of nulling or zeroing a DC voltage to obtain precision measurement of a resistance or an EMF. The accuracy attainable in a null measurement technique generally depends on the use of accurate resistors, rather than the accurate measurement of a DC voltage or current, per se.

5

AC Null Measurements

5.0 INTRODUCTION

In this chapter, we examine measurement techniques which make use of sensing an AC null. These include, but are not limited to, AC operation of (resistive) Wheatstone bridges, various types of AC bridges used to measure capacitance, capacitive dissipation factor (D), inductance, inductor quality factor (Q), mutual inductance, and the small-signal transconductance (g_m) of bipolar junction transistors (BJT) and field-effect transistors. As in the case of DC null methods, AC methods are used to obtain accurate measurements of component values based on the accuracy of the bridge's components.

It should be noted that in the real world of circuit components, there is no such thing as a pure or ideal resistor, capacitor, or inductor. All real-world devices have parasitic parameters associated with them. In some cases these parasitic components may be treated as lumped-parameter circuits, in other situations they are best described as distributed-parameter networks, similar to transmission lines. For example, depending on the frequency of the AC voltage across a resistor, the resistor may appear as having a pure inductance in series with a resistor in parallel with a capacitor. The resistor is an increasing function of frequency due to skin effect at high frequencies. The presence of parasitic components at high frequencies makes the operation of bridges and null circuits at high frequencies more subject to errors. To minimize the effects of parasitic components, most simple AC bridges operate at 1 kHz.

5.1 INDUCTOR EQUIVALENT CIRCUITS

Most practical inductors are made from one or more turns of conductor, wound either on an air core, or a ferromagnetic core made from ferrite ceramic or laminated iron. The use of ferromagnetic cores concentrates the magnetic flux and produces a higher inductance than would be attainable with an air core and the same coil geometry. Because the conductor used to wind an inductor has a finite resistance, the simplest, low-frequency equivalent circuit of a practical inductor is a resistor in series with a pure inductance. At very high frequencies, the small, stray capacitance between adjacent turns of the inductor's coil produces a complex, distributed-parameter, resistor, inductor, capacitor (RLC) circuit. If the coil is wound with several layers of turns, capacitance between the layers, as well as between adjacent turns produces an even more complex equivalent circuit of the inductor at very high frequencies.

One compromise to the problem of representing an inductor realistically at high frequencies is shown in Figure 5.1. Here the distributed capacitance between windings is modeled by a single, equivalent, lumped capacitor in parallel with the series R-L circuit. This model is probably reasonably valid in the low audio frequency range where most AC inductance bridges operate. To characterize and measure inductors at radio frequencies where the lumped model is not valid, we generally use a *Q-meter*, discussed below.

The quality factor or Q of an inductor is defined at a given frequency as the ratio of the inductor's inductive reactance to the real part of its impedance. The higher the Q, the "purer" the inductor and the lower are its losses. High Q inductors allow us to build more sharply tuned RLC frequency-selective circuits than do low Q inductors of the same inductance. For a given inductance, high Q inductors are generally more expensive. For a simple series R-L circuit, the Q is given by:

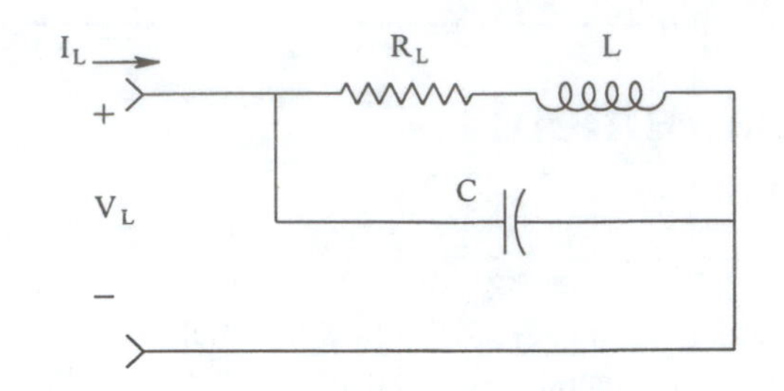

FIGURE 5.1 High-frequency, lumped-parameter model for an inductor.

$$Q \equiv \frac{X_L}{R_L} = \frac{\omega L}{R_L} \tag{5.1}$$

The Q of the RLC inductor model of Figure 5.1 may be found by finding its impedance:

$$\mathbf{Z}_L(j\omega) = \frac{(j\omega L + R)}{1 + j\omega RC + (j\omega)^2 LC} = \frac{R}{(1-\omega^2 LC)^2 + (\omega RC)^2} + j\frac{\omega L\left[(1-\omega^2 LC) - CR^2/L\right]}{(1-\omega^2 LC)^2 + (\omega RC)^2} \tag{5.2}$$

From the basic definition of Q given in Equation 5.1, and the impedance given by Equation 5.2, we can write Q for the series model inductor as:

$$Q_S(\omega) = \frac{\omega L\left[(1-\omega^2 LC) - CR^2/L\right]}{R} \tag{5.3}$$

Interestingly, $Q_S(\omega)$ rises to a maximum at $\omega = \omega_s(\sqrt{3})/3$, and then drops to zero at ω_s, the resonance frequency of the lumped RLC model of the inductor. At the resonance frequency, $\mathbf{Z}_L(j\omega)$ = Real. From Equation 5.2, this is seen to occur when:

$$\omega_s = \sqrt{\frac{1}{LC} - \frac{R^2}{L^2}} \text{ r/s} \tag{5.4}$$

Although the lumped RLC model for the inductor is crude, its behavior does mimic that observed in real inductors at high frequencies. Their Q does increase to a maximum, and then decreases to zero as the frequency is further increased. Some practical inductors exhibit multiple peaks in their Q curves at high frequencies due to the distributed nature of the stray capacitance. At 1 kHz, the frequency used in most AC bridges used to measure inductors, we are operating well to the left of the peak of the Q(ω) curve, and capacitive effects are generally second-order.

5.2 CAPACITOR EQUIVALENT CIRCUITS

There are many physical types of capacitors having a wide variety of geometries and using many different types of dielectrics, all depending on the selected application. The basic capacitor consists of a pair of parallel metal plates separated by a dielectric (insulating substance) which can be vacuum, air, sulfur hexafluoride, oil, oil-impregnated paper, glass, mica, metal oxides, or various plastics or ceramics. Each dielectric has its unique properties of DC leakage, dielectric constant, losses at high frequencies, temperature coefficient, etc. At high frequencies, capacitors can be modeled by a pure capacitance surrounded by parasitic inductors due to leads, and resistances modeling dielectric losses and leakage. Figure 5.2 illustrates a general, lumped-parameter, high-

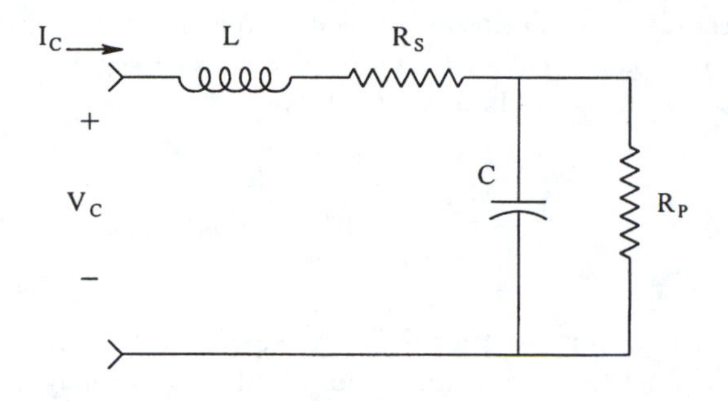

FIGURE 5.2 High-frequency, lumped-parameter model for a capacitor.

frequency equivalent circuit for a non-polarized (non-electrolytic) capacitor. As in the case of the inductor's equivalent circuit, we may write an expression for the impedance of the model:

$$\mathbf{Z}_C(j\omega) = R_S + j\omega L + \frac{R_P(1/j\omega C)}{R_P + 1/j\omega C} \tag{5.5}$$

$\mathbf{Z}_C(j\omega)$ may be put into the form, $\mathbf{Z}_C = \mathrm{Re} + j\mathrm{Im}$, using complex algebra:

$$\mathbf{Z}_C(j\omega) = \left[R_S + \frac{R_P}{1+(\omega R_P C)^2}\right] + j\frac{(\omega L - \omega R_P^2 C + \omega^3 R_P^2 L C^2)}{1+(\omega R_P C)^2} \tag{5.6}$$

This relation implies that the complex RLC circuit model for the capacitor at high frequencies can be reduced to a simple, equivalent model with a resistor given by the real term of Equation 5.6 being in series with a reactance given by the second term in Equation 5.6. It is possible to find an expression for the equivalent series capacitance, C_{sEQ}. From the second term in Equation 5.6 we can write:

$$X_c = \frac{-1}{\omega C_{S(EQ)}} = \frac{\omega(L - R_P^2 C) + \omega^3 R_P^2 C^2 L}{1+(\omega R_P C)^2} \tag{5.7}$$

From which it is easy to find

$$C_{S(EQ)} = \frac{1+(\omega R_P C)^2}{\omega^2\left[(R_P^2 C - L) - \omega^2 R_P^2 C^2 L\right]} \tag{5.8}$$

At low frequencies, where L is negligible, the capacitive reactance reduces to $-1/\omega C$. Note that the series high-frequency capacitor model exhibits resonance at a frequency where the reactance magnitude goes to zero. The resonant frequency of the capacitor is easily found to be

$$\omega_o = \sqrt{1/LC - 1/(R_P C)^2}\ \mathrm{r/s} \tag{5.9}$$

Thus, the reactance of the equivalent circuit appears as $-1/\omega C$ at low frequencies. The reactance magnitude decreases with ω until it reaches zero at ω_o, and then at $\omega >> \omega_o$, appears inductive as $+\omega L$.

Most laboratory bridges do not measure capacitive reactance; rather, they are calibrated in capacitance units and dissipation factor, D. D_S is defined as the ratio of the series equivalent resistance to the capacitive reactance. In algebraic terms,

$$D_S = \frac{R_{S(EQ)}}{X_{S(EQ)}} = \omega C_{S(EQ)} R_{S(EQ)} \tag{5.10}$$

Note that at any fixed frequency, a series R-C circuit, such as we have been discussing, has an equivalent parallel R-C circuit which has the same impedance. We may also define a dissipation factor for the parallel equivalent circuit. It is the ratio of the capacitive reactance to the equivalent parallel resistance. In algebraic terms,

$$D_P = \frac{X_{P(EQ)}}{R_{P(EQ)}} = 1/\left(\omega C_{P(EQ)} R_{P(EQ)}\right) \tag{5.11}$$

By equating the impedances or conductances of the series and parallel R-C models, it is possible to derive relations relating one circuit to the other:

$$R_{P(EQ)} = R_{S(EQ)} \left(\frac{D_S^2 + 1}{D_S^2} \right) \tag{5.12}$$

$$C_{P(EQ)} = C_{S(EQ)} / \left(1 + D_S^2\right) \tag{5.13}$$

$$R_{S(EQ)} = R_{P(EQ)} \left(\frac{D_P^2}{D_P^2 + 1} \right) \tag{5.14}$$

$$C_{S(EQ)} = C_{P(EQ)} \left(1 + D_P^2\right) \tag{5.15}$$

As you will see, certain capacitance bridge configurations make the assumption that the capacitance under measurement is represented by the series equivalent circuit model, while others use the parallel equivalent circuit. Since at a given frequency, the impedances of the two circuits are equal, the reason for using one model or the other lies in the practical derivation of the bridge balance equations in a form that allows two bridge elements to be uniquely calibrated in C_S and D_S or in C_p and D_p.

5.3 AC OPERATION OF WHEATSTONE BRIDGES

In this mode of operation, a low-frequency, AC excitation signal is used, usually ranging from 100 to 1000 Hz. The bridge arms are resistors, and the effects of stray capacitance to ground and inductance are generally negligible. Balance conditions are generally the same as for a DC Wheatstone bridge. However, one advantage of the AC excitation is that an electronic null detector with a lower uncertainty voltage, ΔV_o, can be used. This is probably not important, because bridge accuracy in measuring the unknown resistor, X, is largely due to the accuracy to which arms M, N, and P are known, as we demonstrated in the previous chapter. However, when the bridge is used in the voltage output mode, as with strain gauges, greater threshold sensitivity in measuring small changes in X are realized because amplification of V_o can be done with a low-noise, narrow-band,

AC amplifier working out of the 1/f noise region. Following amplification, the AC output voltage is generally converted into a DC signal proportional to ΔX through the use of a phase-sensitive rectifier (PSR). PSRs are discussed in detail in Section 2.7.

Another application for AC Wheatstone bridge operation is where the passage of DC through X will alter the magnitude of X. An example of this phenomenon is when we wish to measure the resistance of electrochemical electrodes (such as used in electrocardiography) attached to the body. The passage of direct current causes polarization of the electrodes due to ion migration in the DC electric field. Polarization increases electrode impedance. The use of AC generally avoids this phenomenon.

5.4 AC BRIDGES

The application and analysis of AC bridges is a large field with a rich literature. Most of the texts dealing with this topic in detail were written over 40 years ago. Great emphasis was placed on how an AC bridge approached its null. Vector (circle) diagrams were used to illustrate the interaction of the two variable bridge elements in reaching a null. In the following sections, we do not consider the details of how nulls are approached. Rather, we summarize the conditions at null, and discuss the applications of each type of bridge. Note that most modern AC bridges are designed to work at 1000 Hz, although other frequencies may be used for special applications.

Figure 5.3 illustrates a "general" AC bridge, in which the arms are impedances having real and imaginary parts. The AC bridge output voltage can in general be written:

$$V_o = V_s\left[\frac{\mathbf{Z_X}}{\mathbf{Z_X}+\mathbf{Z_P}} - \frac{\mathbf{Z_N}}{\mathbf{Z_N}+\mathbf{Z_M}}\right] \qquad 5.16$$

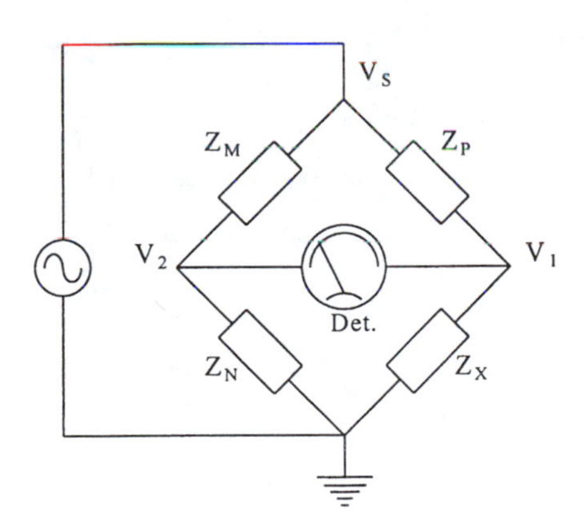

FIGURE 5.3 General configuration for an AC bridge. Det. is the null detector; it is a sensitive AC voltmeter assumed to have infinite input resistance.

From Equation 5.16 we find that at null, where $V_o = 0$, we can write:

$$\mathbf{Z_X Z_M} = \mathbf{Z_N Z_P} \qquad 5.17$$

which leads to the vector equation for $\mathbf{Z_X}$ in polar form:

$$\mathbf{Z_X} = |\mathbf{Z_X}|\angle\theta_X = \left|\frac{\mathbf{Z_N Z_P}}{\mathbf{Z_M}}\right|\angle\theta_N + \theta_P - \theta_M \qquad 5.18$$

Thus, finding unique expressions for the unknown L and Q, or C and D, requires solving vector equations of the form above. Generally this is done by equating the real terms on both sides of Equation 5.17, and independently, the imaginary terms.

It can be shown for all conventional, four-arm, AC bridges that the same conditions for null exist if the null detector is exchanged for the AC source; Equation 5.17 still applies.

We present below a summary of bridge designs used to measure capacitance, inductance, and mutual inductance using the null method.

5.4.1 Bridges Used to Measure Capacitance

In this section we examine the designs of bridges useful for the measurement of capacitance. Capacitance-measuring bridges can be subdivided into those designs suitable for low-loss (low D) capacitors and those giving best results for lossy (high D) capacitors.

5.4.1.1 The Resistance Ratio Bridge

The resistance ratio bridge is shown in Figure 5.4. It is best used to measure capacitors with low D, and it uses the series equivalent capacitor model. It is used in the well-known General Radio Model 1650A bridge. At null, $V_o = 0$, and we may write:

$$R_3(R_{XS} + 1/j\omega C_{XS}) = R_2(R_4 + 1/j\omega C_4) \qquad 5.19$$

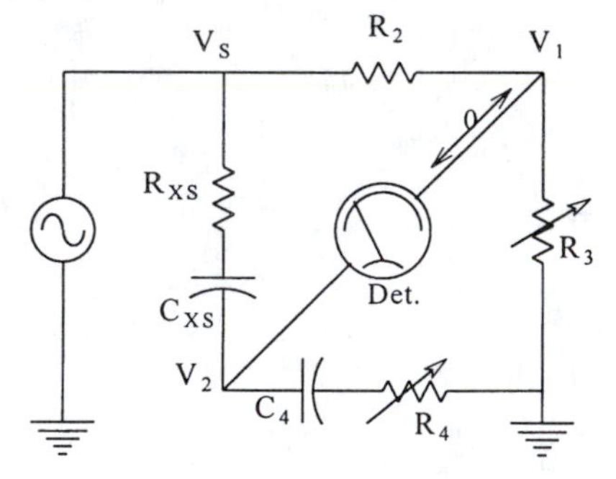

FIGURE 5.4 A resistance ratio bridge, used to measure capacitors with low D_S.

Equating the real terms, we find

$$R_{XS} = R_2 R_4 / R_3 \qquad 5.20$$

Equating the imaginary terms, we obtain

$$C_{XS} = R_3 C_4 / R_2 \qquad 5.21$$

The dissipation factor for the series capacitor model was shown to be $D_S = \omega C_{XS} R_{XS}$. If we substitute the relations for C_{XS} and R_{XS} into D_S, we obtain

$$D_S = \omega C_4 R_4 \qquad 5.22$$

Thus we see that R_4 can be uniquely calibrated in low D_S, and R_3 should be calibrated in C_{XS} units.

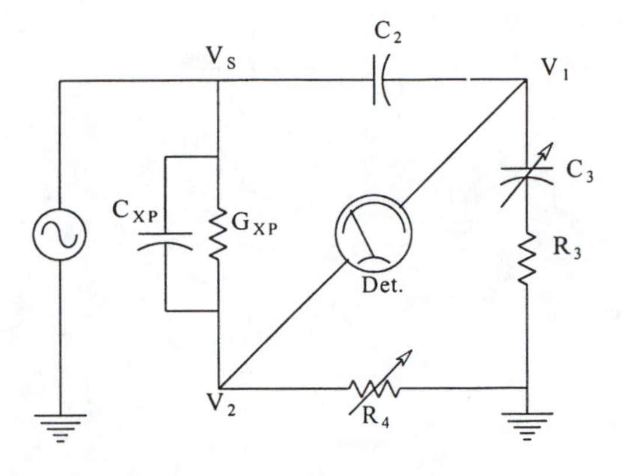

FIGURE 5.5 A Schering bridge, used to measure capacitors with high D_P.

5.4.1.2 The Schering Bridge

The Schering bridge is useful for measuring capacitors with high losses (high D). In finding the balance conditions for this bridge, it is expedient to use the parallel R-C equivalent circuit, as shown in Figure 5.5. At null, we can write, as usual:

$$\mathbf{Z}_X \mathbf{Z}_3 = \mathbf{Z}_2 \mathbf{Z}_4 \tag{5.23}$$

or

$$\mathbf{Z}_3 \mathbf{Y}_2 = \mathbf{Y}_X \mathbf{Z}_4 \tag{5.24}$$

Thus

$$(R_3 + 1/j\omega C_3) j\omega C_2 = (G_{XP} + j\omega C_{XP}) R_4 \tag{5.25}$$

Now by equating real terms and imaginary terms, we obtain the conditions at balance:

$$C_{XP} = R_3 C_2 / R_4 \tag{5.26}$$

$$R_{XP} = R_4 C_3 / C_2 \tag{5.27}$$

$$D_P = 1/(\omega C_{XP} R_{XP}) = \omega C_3 R_3 \tag{5.28}$$

Notice that to obtain uniqueness in finding C_{XP} and D_P, C_3 can be calibrated in D_P and R_4 can be calibrated in C_{XP}. The other components are fixed for a given range of C_{XP}.

5.4.1.3 The Parallel C Bridge

This bridge design, shown in Figure 5.6, is also used in the GR 1650A bridge to measure high D capacitors ($0.1 < D_P < 50$). At null, we can write $R_2 \mathbf{Z}_4 = R_3 \mathbf{Z}_X$. This leads to:

$$R_2 (G_{XP} + j\omega C_{XP}) = R_3 (G_4 + j\omega C_4) \tag{5.29}$$

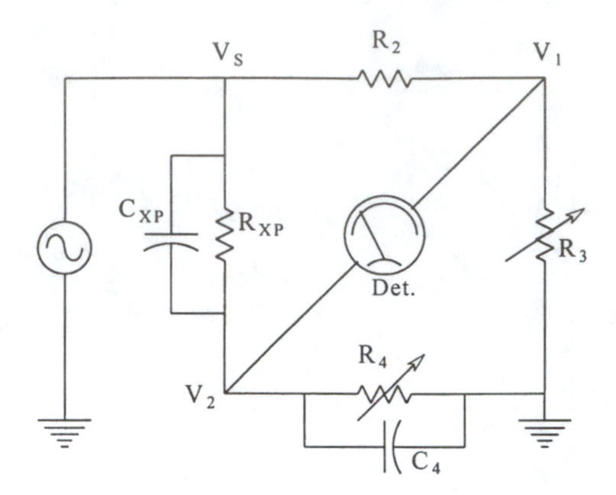

FIGURE 5.6 A parallel C bridge, used to measure capacitors with $0.1 < D_P < 50$.

from which we can easily find

$$C_{XP} = C_4 R_3 / R_2 \tag{5.30}$$

$$R_{XP} = R_2 R_4 / R_3 \tag{5.31}$$

$$D_P = 1/(\omega C_{XP} R_{XP}) = 1/(\omega C_4 R_4) \tag{5.32}$$

From the above equations, it is easy to see that R_4 should be calibrated in D_P, and R_3 or R_2 can read C_{XP}.

5.4.1.4 The De Sauty Bridge

The De Sauty bridge, shown in Figure 5.7, is a deceptively simple bridge which is often used to produce an output voltage which is proportional to a small change, δC, in one of the capacitors. The δC may be caused by a variety of physical phenomena, such as the deflection of a diaphragm due to a pressure change, a change in capacitor plate separation due to a change in material thickness, a change in capacitor plate separation due to applied force, etc.

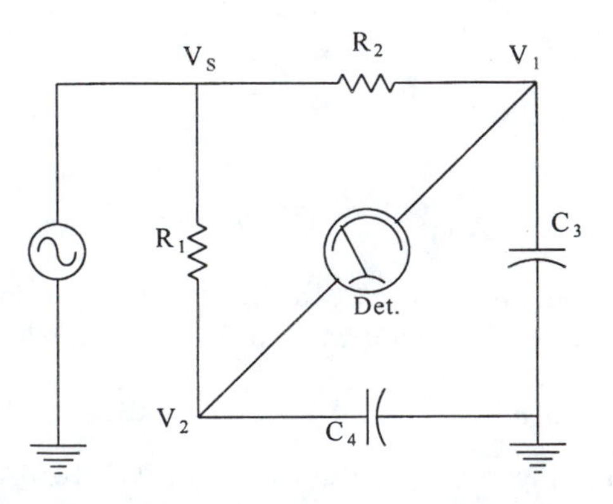

FIGURE 5.7 A DeSauty bridge, used to measure small changes in capacitance. $C_4 = C$, $C_3 = C + \delta C$, $R_1 = R_2 = R$, and $\omega = 1/RC$.

Because the output voltage is a nonlinear function of large capacitance changes, the De Sauty bridge is usually operated so that $\delta C/C_o << 1$, so output will be linearly proportional to the quantity under measurement. Also, the resistors R_1 and R_2 are made equal to R, and null is achieved by setting $C_4 = C_3 = C$. Now the output of the bridge as a function of R, C, ω, and δC can be easily written:

$$V_o = V_b \frac{j\omega RC(\delta C/C)}{(1 + j\omega RC)(1 + j\omega R(C + \delta C))} \quad 5.33$$

From Equation 5.33 we see that three approximate relations for V_o can be written, depending on the operating frequency. First, we let ω >> 1/RC. The transfer function reduces to

$$V_o/V_b = (\delta C/C)(1/j\omega RC)[1 - (\delta C/C)] \quad 5.34$$

Generally, the second-order term can be neglected. However, a square-law nonlinear distortion will occur for $\delta C/C \rightarrow 0.1$.

Second, we let the bridge be excited at a low frequency so that ω << 1/RC. Under this condition, we find

$$V_o/V_b = (\delta C/C)(j\omega RC) \quad 5.35$$

Third, if ω = 1/RC (tuned bridge condition), the output can be written:

$$V_o/V_b = \delta C/2C \quad 5.36$$

Maximum sensitivity generally occurs for the third case where ω = 1/RC. In this case, the just detectable δC can be estimated by using the series expansion:

$$\Delta V_o = \left(\frac{\partial V_o}{\partial C}\right)\delta C = V_b\, \delta C/(2C) \quad 5.37$$

Thus, the just detectable δC is found by assuming the bridge detector AC voltmeter resolution is $\Delta V_o = 0.1\ \mu V$, and the bridge excitation, V_b, is 5 V. If we let C = 100 pF, then δC_{MIN} is given by:

$$\delta C_{MIN} = (0.1 \times 10^{-6} \times 2 \times 100 \times 10^{-12})/5 = 4 \times 10^{-6}\ \text{pF} = 4\ \text{aF} \quad 5.38$$

This is an incredible theoretical sensitivity, generally not reachable in practice because of stray capacitances associated with the bridge arms.

5.4.1.5 The Wien Bridge

The Wien bridge, illustrated in Figure 5.8, is generally not used to measure capacitors because of the complexity of its solution at null. The Wien bridge is a frequency-dependent null network, and as such, finds application in the design of tuned bandpass and band-reject (notch) filters, and also oscillators. From Figure 5.8, we can write the general form of the bridge balance equation at null:

$$\mathbf{Z_1 Z_3 = Z_2 Z_X} \quad 5.39$$

This can be put as

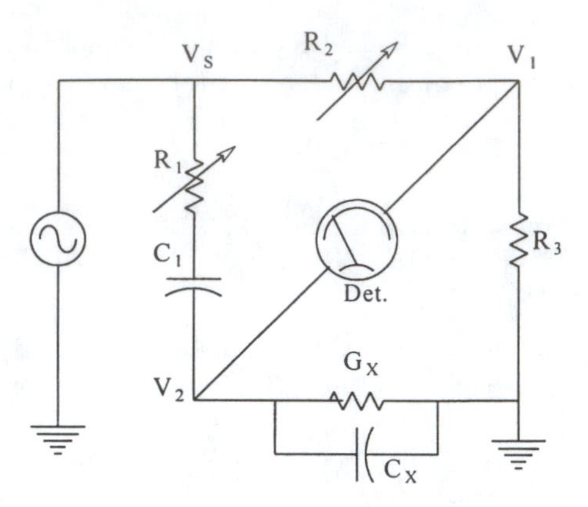

FIGURE 5.8 The Wien bridge is frequency-dependent. Here it is used to measure capacitance with $D_P << 0.1$. R_2 can be calibrated in C_X; R_1 in D_{PX}.

$$\mathbf{Z_1 Y_X = Z_2 Y_3} \quad \text{or} \quad (R_1 + 1/j\omega C_1)(G_X + j\omega C_X) = R_2 G_3 \tag{5.40}$$

To find the null conditions, we must solve the simultaneous equations derived from Equation 5.40 by setting real terms equal and imaginary terms equal.

$$G_X R_1 + C_X/C_1 = R_2 G_3 \tag{5.41}$$

$$-G_X/\omega C_1 + \omega C_X R_1 = 0 \tag{5.42}$$

Their solution leads to:

$$C_X = \frac{(R_2/R_3)C_1}{(\omega^2 C_1^2 R_1^2 + 1)} = \frac{(R_2/R_3)C_1}{(D_{1S}^2 + 1)} \to (R_2/R_3)C_1 \tag{5.43}$$

$$G_X = \frac{(\omega R_1 R_2/R_3)(\omega C_1^2)}{(\omega^2 C_1^2 R_1^2 + 1)} \tag{5.44}$$

$$D_{XP} = 1/(\omega C_X R_X) = \omega R_1 C_1 = D_{1S} \tag{5.45}$$

Providing $D_{XP} = D_{1S} << 0.1$, we can uniquely calibrate R_2 in C_X units, and R_1 in D_{XP} units.

The use of the Wien bridge as a frequency-sensitive null network can be demonstrated by writing its transfer function:

$$\mathbf{V_o/V_b} = \frac{1/(G_X + j\omega C_X)}{1/(G_X + j\omega C_X) + R_1 + 1/j\omega C_1} - \frac{R_3}{R_3 + R_2} \tag{5.46}$$

which reduces to:

$$\mathbf{V_o/V_b} = \frac{j\omega C_1 R_X}{1 + j\omega(C_1 R_X + C_1 R_1 + C_X R_X) + (j\omega)^2 C_1 R_1 C_X R_X} - \frac{R_3}{R_3 + R_2} \tag{5.47}$$

Now, if we tune the source frequency to

$$\omega = \omega_o = (1/C_1R_1C_XR_X)\ \text{r/s} \tag{5.48}$$

the transfer function becomes

$$\mathbf{V_o/V_b} = \frac{C_1R_X}{C_1R_X + C_1R_1 + C_XR_X} - \frac{R_3}{R_3 + R_2} \tag{5.49}$$

This real transfer function goes to zero at $\omega = \omega_o$ if we let

$$R_1 = R_X = R \tag{5.50}$$

$$C_1 = C_X = C \tag{5.51}$$

$$R_2 = 2R_3 = R \tag{5.52}$$

Then we have

$$\mathbf{V_o/V_b} = 1/3 - 1/3 = 0 \tag{5.53}$$

As was mentioned above, the frequency-dependent null of the Wien bridge can be exploited to create tuned filters and oscillators useful in instrumentation systems.

5.4.1.6 The Commutated Capacitor Bridge

A commutated capacitor bridge is shown in Figure 5.9. Unlike a conventional AC bridge, this bridge uses a DC V_s. Such bridges do not allow measurement of capacitor D. However, they are often used for high-accuracy, low-frequency applications. An MOS switch is driven by a square wave clock with period T. For T/2 seconds, the capacitor is connected to node "a" and charges toward V_s through R_2. For an alternate T/2 seconds, C_X is allowed to discharge through R_4 to ground. The waveforms at nodes "a" and "V_c" are shown in Figures 5.10A and 5.10B, respectively. We assume that the clock period and resistors R_2 and R_4 are chosen so $R_4C_X \ll T/2$ and $R_2C_X \gg T/2$. We also assume that the null detector responds to the average of $(V_2 - V_1)$. Hence, at null we find that:

$$V_sR_3/(R_1 + R_3) = (V_s/2)[1 + T/4R_2C_X] \tag{5.54}$$

From which we can solve for C_X:

$$C_X = \frac{T(R_3 + R_1)}{4R_2(R_3 - R_1)} \tag{5.55}$$

where obviously, $R_3 > R_1$. R_3 or R_2 can be varied to obtain null, and for a given range setting (fixed values of T, R_2, and R_1) can be calibrated according to Equation 5.55 above.

5.4.2 Bridges Used to Measure Inductance and Mutual Inductance

As in the case of bridges used for measuring capacitance, the inductance bridges can be subdivided into those optimal, in terms of reaching null, for measuring high-Q inductors and those best suited

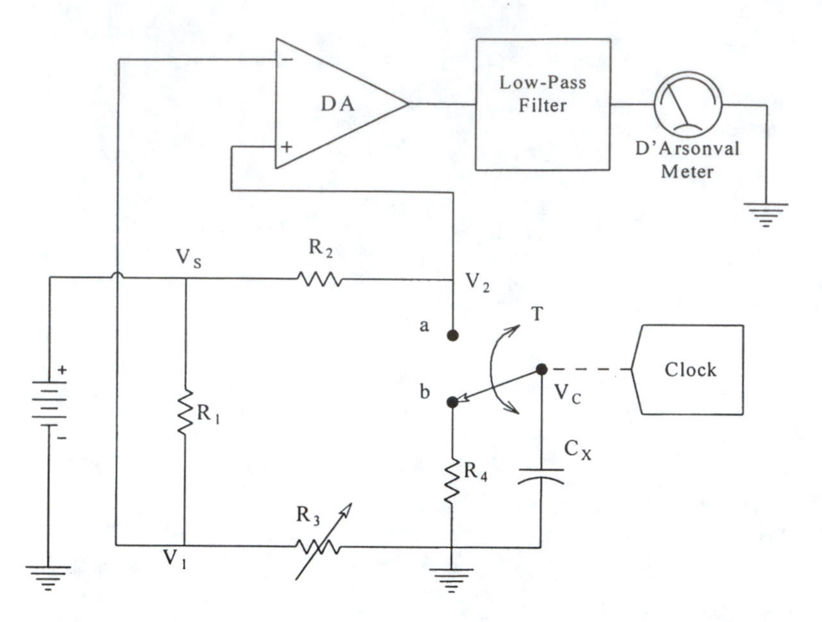

FIGURE 5.9 A commutated capacitor bridge. Unlike other AC bridges, the detector is a DC null meter which responds to the average voltage.

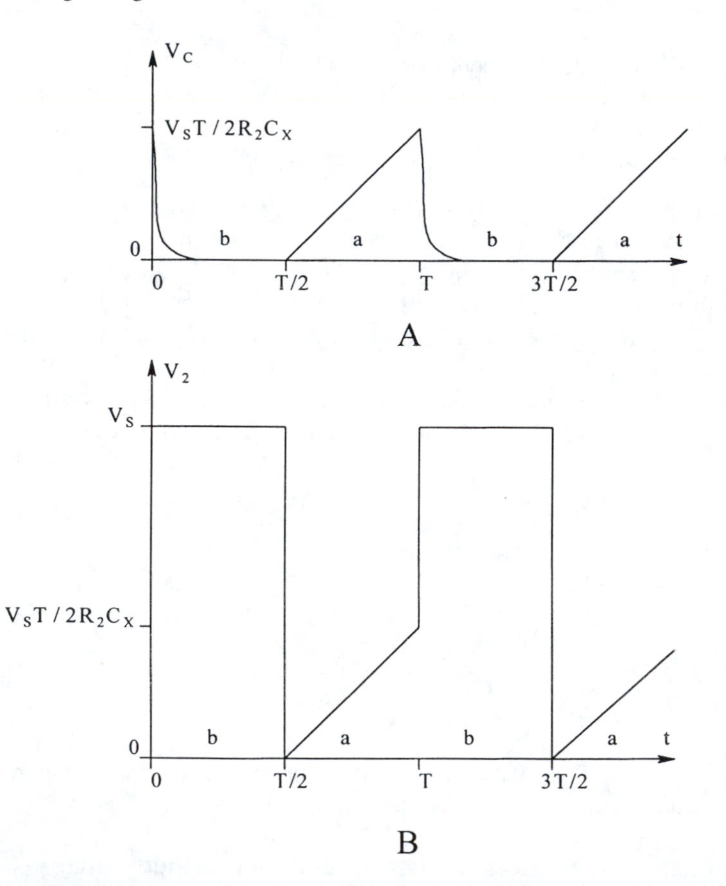

FIGURE 5.10 (A) Voltage waveform across C_X, $v_C(t)$. Note $RC_X << R_2C_X$. (B) Voltage waveform, $v_2(t)$.

for the measurement of low-Q inductors. In addition, we describe below the design of two specialized bridges used to measure the mutual inductance of power and audio frequency transformers.

5.4.2.1 The Maxwell Bridge

This is a well-known bridge which is used to measure low-Q inductors having Qs in the range of 0.02 to 10. The Maxwell bridge is used in the General Radio model 1650A multipurpose bridge; its circuit is shown in Figure 5.11. At null, we can write:

$$\mathbf{Z_X} = \mathbf{Z_2 Y_3 Z_4} \tag{5.56}$$

or

$$(R_X + j\omega L_X) = R_2(G_3 + j\omega C_3)R_4 \tag{5.57}$$

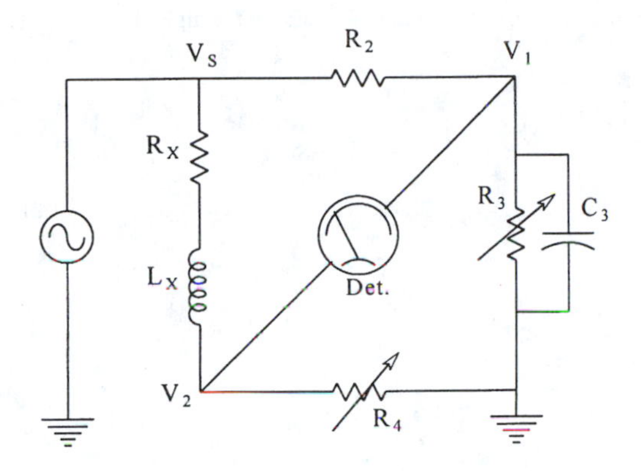

FIGURE 5.11 The Maxwell bridge for low-Q inductors (0.02 < Q < 10).

From Equation 5.54, we easily find

$$L_X = C_3 R_2 R_4 \tag{5.58}$$

$$R_X = R_2 R_4 / R_3 \tag{5.59}$$

$$Q_X = \omega L_X / R_X = \omega C_3 R_3 \tag{5.60}$$

Thus, R_4 is calibrated in inductance units, and the scale of R_3 scale reads Q. In the GR 1650A bridge, R_4 can be varied from 0 to 11.7 k ohms, and R_3 can be set from 0 to 16 k ohms.

5.4.2.2 Parallel Inductance Bridge

This bridge, shown in Figure 5.12, is also used in the GR 1650A bridge to measure high-Q inductors ($1 < Q_P < \infty$). It is somewhat unusual in that it uses a parallel equivalent circuit for the inductor. At a given frequency, any series R-L circuit can be made to be equal in impedance to a parallel R-L circuit. We describe these equivalences below.

At balance, we can write:

$$R_2 R_4 = (R_3 + 1/j\omega C_3)\frac{j\omega L_X R_X}{R_X + j\omega L_X} \tag{5.61}$$

From which we obtain

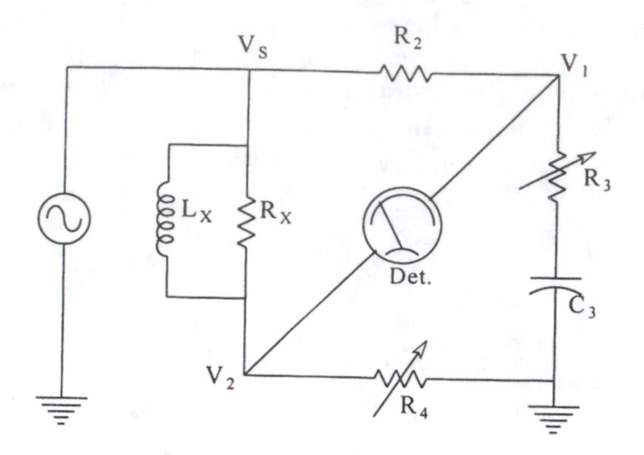

FIGURE 5.12 The General Radio parallel inductance model bridge for high-Q inductors ($1 < Q < \infty$). R_3 is calibrated in Q_P, R_4 in L_{PX}.

$$j\omega L_X R_2 R_4 + R_X R_2 R_4 = j\omega L_X R_X R_3 + L_X R_X / C_3 \tag{5.62}$$

Equating real terms and then imaginary terms in Equation 5.62, we finally obtain the expressions for L_{XP}, R_{XP}, and Q_P.

$$L_{XP} = C_3 R_2 R_4 \tag{5.63}$$

$$R_{XP} = R_2 R_4 / R_3 \tag{5.64}$$

$$Q_P = R_X / \omega L_X = 1/\omega C_3 R_3 \tag{5.65}$$

Thus, it is expedient to make R_4 the L dial, and R_3 the high-Q dial on this bridge.

5.4.2.3 The Hay Bridge

The Hay bridge, shown in Figure 5.13, uses the series R-L model for an inductor to measure the inductance and Q of high-Q coils. Assuming null, we find:

$$(j\omega L_X + R_X)(R_3 + 1/j\omega C_3) = R_2 R_4 \tag{5.66}$$

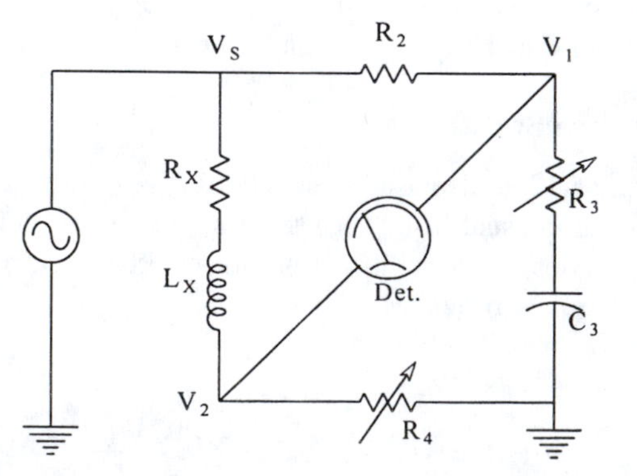

FIGURE 5.13 The Hay bridge is used to measure high-Q inductors. R_3 is calibrated in Q_S, and R_4 in inductance units.

The balance equation, Equation 5.66, can be broken down to two simultaneous equations:

$$L_X/C_3 + R_X R_3 = R_2 R_4 \qquad 5.67$$

$$\omega L_X R_3 - R_X/\omega C_3 = 0 \qquad 5.68$$

Their solution yields

$$L_X = \frac{R_4 R_2 C_3}{\left(\omega^2 R_3^2 C_3^2 + 1\right)} = \frac{R_4 R_2 C_3}{\left(Q^{-2} + 1\right)} \qquad 5.69$$

$$R_X = \frac{\omega R_3 R_2 R_4 \left(\omega C_3^2\right)}{\left(\omega^2 R_3^2 C_3^2 + 1\right)} \qquad 5.70$$

$$Q_S = \omega L_X/R_X = 1/\omega C_3 R_3 \qquad 5.71$$

Here, R_3 may be calibrated in Q_S, and R_4 in inductance units. Inductance calibration is substantially independent of $Q_S(R_3)$ as long as $Q_S > 10$.

5.4.2.4 The Owen Bridge

The Owen bridge, shown in Figure 5.14, uses the conventional, series inductance model, and is best used on large, low-Q inductors. This bridge is somewhat unique in that null may be obtained by varying both elements in the "4" arm, R_4 and C_4. At null we have:

$$\left(j\omega L_X + R_X\right)/j\omega C_1 = R_2\left(R_4 + 1/j\omega C_4\right) \qquad 5.72$$

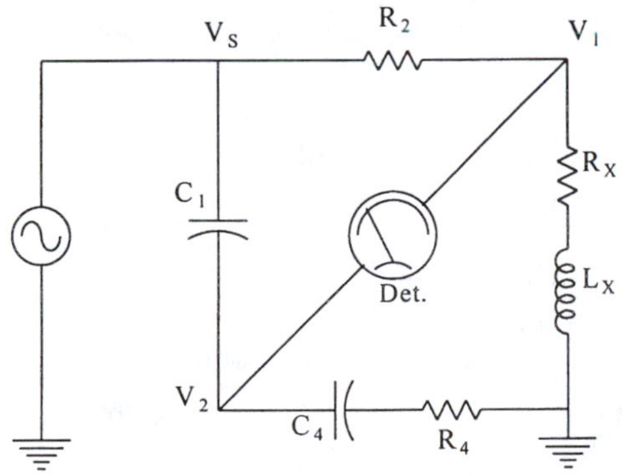

FIGURE 5.14 The Owen bridge is best used to measure large, low-Q inductors. R_4 can be calibrated in inductance, and C_4 in series equivalent resistance, R_{XS}.

By equating the real terms and then the imaginary terms in Equation 5.72, it is easy to find

$$R_X = R_2 C_1/C_4 \qquad 5.73$$

$$L_X = R_2 C_1 R_4 \qquad 5.74$$

Hence, R_4 may be uniquely calibrated in inductance, and C_4 in resistance (nonlinear scale). A major disadvantage of the Owen bridge is that it requires a precisely calibrated capacitor (C_4), and another is that for high-Q, low inductance coils, impractically large C_4 values may be required.

5.4.2.5 The Anderson Bridge

The Anderson bridge is illustrated in Figure 5.15. According to Stout (1950), the Anderson bridge gives the best convergence to nulls for low-Q coils. To analyze this bridge, we note that at null, $V_o = V_1 - V_2$, and voltage-divider relations can be used to find V_1 and V_2. Thus we have:

$$\frac{V_o}{V_s} = \frac{R_5}{R_5 + R_X + j\omega L_X} - \frac{R_3}{R_3 + R_2 + j\omega C[R_3R_4 + R_3R_2 + R_4R_2]} \quad 5.75$$

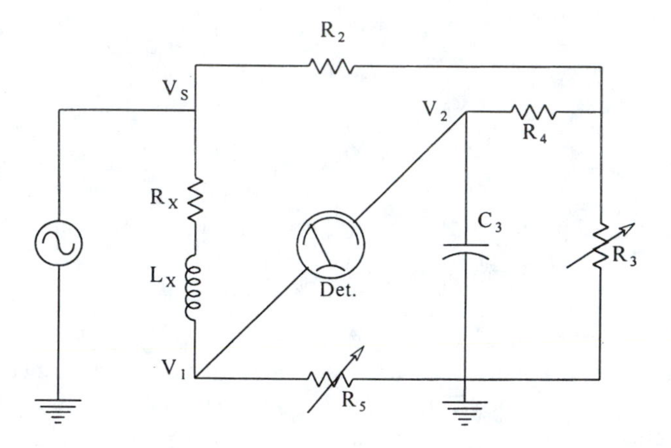

FIGURE 5.15 The Anderson bridge works best with low-Q coils. It has relatively complex balance equations.

Setting $V_o = 0$, we find:

$$R_X = R_2R_5/R_3 \quad 5.76$$

$$L_X = CR_5[R_4 + R_2 + R_4R_2/R_3] \quad 5.77$$

$$Q_X = \omega CR_3[1 + R_4/R_2 + R_4/R_3] \quad 5.78$$

It is easily seen that the price we pay in using this bridge, which is easily balanced with low-Q inductors, is a relatively complex set of balance equations. R_5 can be calibrated for L_X, and R_3 used to read Q_X.

5.4.2.6 The Heaviside Mutual Inductance Bridge

Before discussing the circuit and balance conditions for the Heaviside mutual inductance bridge, we review what is meant by the mutual inductance of a transformer. Figure 5.16 illustrates a simple equivalent circuit for a transformer operating at low frequencies. Under AC steady-state conditions we may write the loop equations:

$$V_1 = I_1(R_1 + j\omega L_1) + I_2(j\omega M) \quad 5.79$$

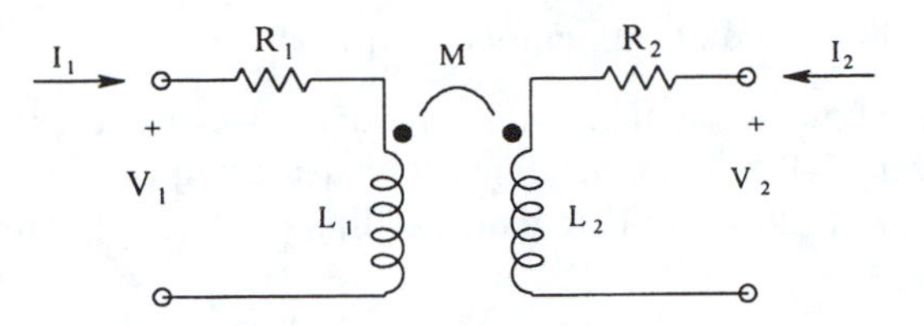

FIGURE 5.16 Circuit model for a transformer with mutual inductance at low frequencies. L_1 and L_2 are the self-inductances of the primary and secondary windings, respectively.

$$V_2 = I_1(j\omega M) + I_2(R_2 + j\omega L_2) \qquad 5.80$$

M is the mutual inductance in Henrys, given a positive sign for the dot convention shown in Figure 5.16.

Figure 5.17 illustrates the circuit of a Heaviside bridge. At null, $V_o = V_1 - V_2 = 0$, hence we can write:

$$I_1R_1 = I_2R_2 \qquad 5.81$$

and also,

$$V_3 = I_1(R_1 + R_3 + j\omega L_3) = I_2(R_2 + R_4 + j\omega L_4) + j\omega M(I_1 + I_2) \qquad 5.82$$

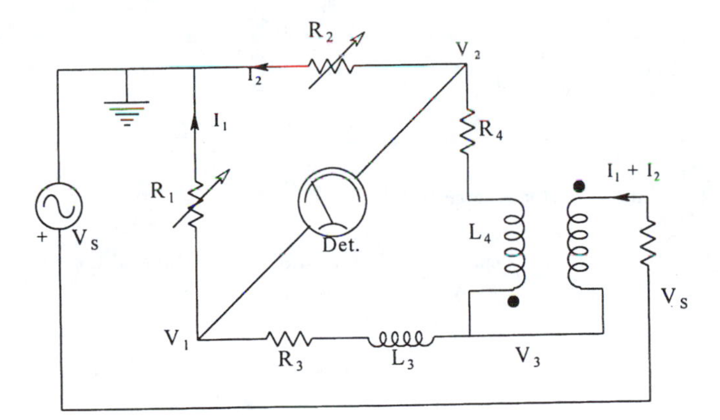

FIGURE 5.17 The Heaviside mutual inductance bridge.

If we solve for I_1 in Equation 5.81 and substitute in Equation 5.82, and then equate real terms and then imaginary terms, we find that at null,

$$R_4 = R_2R_3/R_1 \qquad 5.83$$

$$M = \frac{R_2L_3 - R_1L_4}{R_1 + R_2} \qquad 5.84$$

Note that M may have either sign, depending on the dots. The transformer's primary self-inductance, L_4, must have been measured with the secondary open-circuited. There is no need to know the secondary self-inductance, or the primary resistance when using the Heaviside mutual inductance bridge.

5.4.2.7 The Heydweiller Mutual Inductance "Bridge"

The Heydweiller means of measuring transformer mutual inductance uses a "bridge" with one arm a short-circuit (see Figure 5.18). Thus, the conventional method of examining $\mathbf{Z_1Z_3 = Z_2Z_4}$ will not work with this bridge! Instead, we note that at null, $V_2 = V_1 \rightarrow 0$, thus by use of KVL,

$$V_2 = 0 = I_1(R_3 + j\omega L_3) - j\omega M(I_1 + I_2) \tag{5.85}$$

and also,

$$I_2 = I_1(R_1 + 1/j\omega C_1)/R_2 \tag{5.86}$$

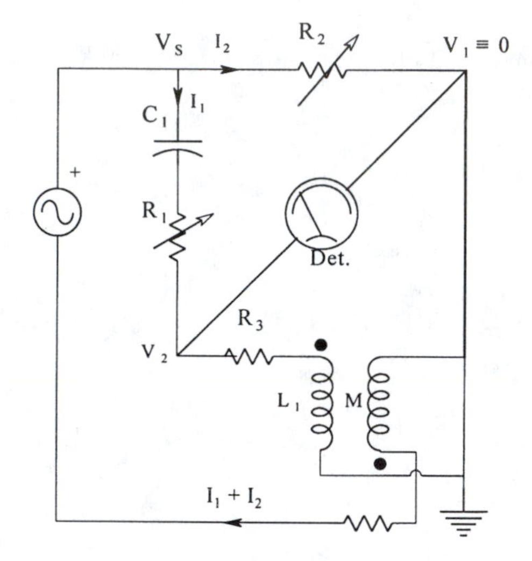

FIGURE 5.18 The Heydweiller mutual inductance null circuit.

If Equation 5.86 is substituted into Equation 5.85 and the sum of the real terms is set equal to zero, we find:

$$M = C_1R_2L_3 \tag{5.87}$$

Equating the sum of the imaginary terms to zero leads to:

$$L_3 = C_1R_3(R_1 + R_2) \tag{5.88}$$

Note that R_3 must be known, as it appears in both the expression for M and the primary self-inductance, L_3. R_1 and R_2 are manipulated to obtain the null.

5.4.3 Null Method of Measuring Transistor Small-Signal Transconductance and Feedback Capacitance

The null circuits discussed below are used to measure BJT or FET small-signal transconductance at a given DC quiescent operating point. An audio frequency AC signal, usually 1 kHz, is used. Measurement of small-signal transconductance is important to verify transistors are good, and to match pairs for like characteristics when building a discrete differential amplifier.

First, we illustrate the small-signal, modified hybrid-pi model for a BJT operating in its linear region (i.e., neither cut-off nor saturated). This model, shown in Figure 5.19, is generally accepted to be valid for frequencies up to the BJT's $f_T/3$. The small-signal transconductance, g_m, is defined as:

$$g_m = \left(\frac{\partial i_c}{\partial v_{b'e}} \right), \quad v_{ce} = 0 \tag{5.89}$$

g_m may be approximated by

$$g_m \approx h_{fe}/r_\pi = I_{CQ}/V_T \tag{5.90}$$

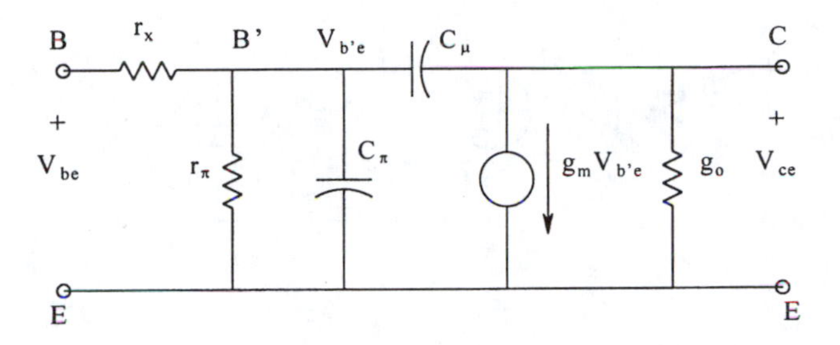

FIGURE 5.19 A simplified, hybrid pi, high-frequency, small-signal circuit model for BJTs.

The capacitance C_μ is the DC voltage-variable capacitance of the reverse-biased, collector-base junction evaluated at the DC operating point of the BJT. C_π is the capacitance of the forward-biased base-emitter junction, measured at the operating point; it is also a function of the DC voltage across the junction. r_π is the small-signal base input resistance. It is approximately equal to V_T/I_{BQ}. The base input spreading resistance, r_x, is generally less than 100 ohms, and we set it equal to zero to simplify analysis. The output conductance, g_o, is generally very small, but its exact value is not important in determining g_m.

Figure 5.20 illustrates the complete circuit of the transconductance bridge. Resistors R_B and R_C, and DC source V_{CC} are used to set up the DC quiescent operating (Q) point of the BJT. A is a milliameter to read I_{CQ}, and V is a DC voltmeter to read V_{CEQ}. A small, audio frequency, sinusoidal signal, v_1, is applied to the transistor's base through a large DC blocking capacitor, C_1. Simultaneously, the inverted AC signal, $-v_1$, is applied to the BJT collector. Capacitor C_N and resistor R_1 are adjusted to get a null or minimum AC signal at the v_c node. A node equation for v_c can be written for the small-signal equivalent circuit of Figure 5.21. Note that $v_{be} \cong v_1$. At null,

$$0 = V_c\left[G_1 + G_c + j\omega\left(C_\mu + C_N\right)\right] - V_1\left(j\omega C + G_1\right) - j\omega C_N\left(-V_1\right) + g_m V_1 \tag{5.91}$$

For $V_c \rightarrow 0$, it is evident that

$$g_m = G_1 \tag{5.92}$$

and

$$C_\mu = C_N \tag{5.93}$$

A similar g_m null circuit can be used to measure the small-signal transconductance of various types of FETs. Note that FET g_m at the DC operating point can be shown to be given by:

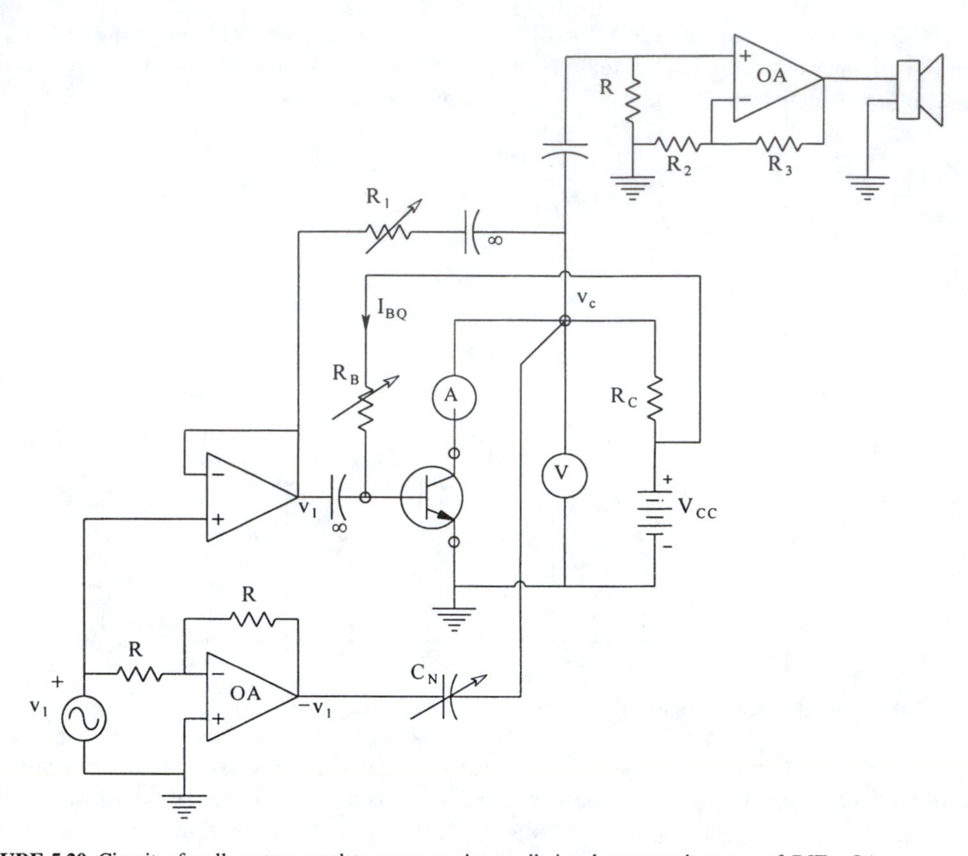

FIGURE 5.20 Circuit of null system used to measure the small-signal transconductance of BJTs. OAs are op amps. Capacitors marked "∞" are large and have negligible reactance at the operating frequency. A is a DC milliameter used to measure I_{CQ}. V is a DC voltmeter used to measure V_{CEQ}. The null is sensed acoustically.

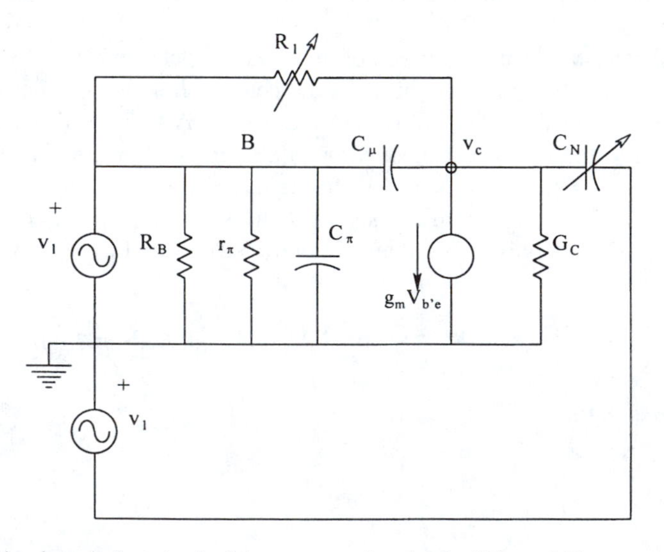

FIGURE 5.21 Small-signal, equivalent circuit of the gm measuring circuit of Figure 5.20.

$$g_m = \left(\frac{\partial i_d}{\partial v_{gs}}\right)_{v_{ds}\to 0} = g_{m0}\sqrt{\frac{I_{DQ}}{I_{DSS}}} = g_{m0}\left(1 - V_{GS}/V_P\right) \quad 5.94$$

where g_{m0} is the small-signal transconductance evaluated for $V_{GS} = 0$, V_P is the pinch-off voltage (for JFETs), and I_{DSS} is the JFET DC drain current measured for $V_{GS} = 0$ and $V_{DS} >> V_{GS} + V_P$.

5.5 CHAPTER SUMMARY

In this chapter we have reviewed the more important audio frequency AC bridges and null systems used for making precision measurements of capacitance, capacitor dissipation factor (D), inductance, inductor Q, mutual inductance, and transistor small-signal transconductance. General conditions for bridge null were derived, and the conditions at null for each circuit have been presented. No attempt has been made to discuss the effects of stray capacitances between bridge elements and the capacitance to ground offered by the two detector terminals. These are generally not significant sources of measurement error for most AC bridges operating at 1 kHz. A common means of compensating for detector capacitance to ground is by the Wagner Earth circuit (see Stout, 1950, Section 9-11 for details).

Measurement of circuit parameters at high frequencies (including video and radio frequencies) requires special apparatus such as the Q-Meter, or instruments such as the Hewlett-Packard (San Jose, CA) model 4191A RF Impedance Analyzer.

6

Survey of Input Sensor Mechanisms

6.0 INTRODUCTION

An input sensor or transducer is a device which permits the conversion of energy from one form to another. It is the first element in an instrumentation/measurement system. For example, it might convert temperature to voltage. A broader definition might substitute "information" for "energy". Its linearity, range, noise and dynamic response largely determine the resolution, sensitivity, and bandwidth of the over-all system. We will consider a transducer to be a device that in some way obeys reciprocity; e.g., the physical input quantity is converted to an output voltage, and if a voltage is applied to the output terminals, the input quantity is generated at the input interface of the transducer. Examples of transducers include piezoelectric crystals used to sense force or pressure, and electrodynamic devices such as loudspeakers and D'Arsonval meter movements. Obviously, not all sensors are transducers.

There are two approaches to categorizing sensors. One way, which we use in Chapter 7, is to group together all those different types of sensors used for a given application, such as the measurement of fluid pressure. The other method, which we use in this chapter, is to group sensors by the mechanism by which they work, such as the generation of an open-circuit voltage due to the input quantity under measurement (QUM), or a change in resistance proportional to the QUM.

6.1 CATEGORIES OF INPUT SENSOR MECHANISMS

In this section we present in outline form a comprehensive (yet incomplete) list of the categories of mechanisms by which sensors work. In the following sections we elaborate on details of representative sensors, and discuss their dynamic ranges, bandwidths, and the auxillary circuitry needed to produce a useful electrical (analog) output.

6.2. Resistive Sensors:
- 6.2.1 Resistive Temperature Sensors
- 6.2.2 Resistive Strain Gauges
- 6.2.3 Photoconductors
- 6.2.4 Resistive Relative Humidity Sensors
- 6.2.5 Direct Resistance Change Used to Sense Position or Angle

6.3. Voltage Generating Sensors:
- 6.3.1 Thermocouples and Thermopiles
- 6.3.2 Photovoltaic Cells
- 6.3.3 Piezoelectric Transducers
- 6.3.4 Sensors Whose Voltage Output is Proportional to the Rate of Change of Magnetic Flux, $d\phi/dt$
- 6.3.5 Sensors Whose Output EMF Depends on the Interaction of a Magnetic Field with Moving Charges

6.4. Sensors Based on Variable Magnetic Coupling:
- 6.4.1 The LVDT
- 6.4.2 Synchros and Resolvers

6.5. Variable Capacitance Sensors:

6.2 RESISTIVE SENSORS

The resistance of practically all resistive sensors varies around some baseline or average value, R_o, as the input quantity varies. Consequently, the most widely used means for converting the change in R due to the input to an output voltage is to include the resistive sensor as an arm of a Wheatstone bridge (see Section 4.1). As long as the ratio $\Delta R/R_o$ is $<< 1$, the bridge output voltage will be linear with ΔR. As $\Delta R/R_o \rightarrow 1$, the bridge output voltage vs. $\Delta R/R_o$ tends to saturate, hence sensitivity and linearity are lost. In some resistive sensors, such as unbonded strain gauge force sensors, we have available two resistors whose values increase linearly $(R_o + \Delta R)$ with the applied force, and two resistors whose values decrease with the input $(R_o - \Delta R)$. These resistors can be assembled into a four active-arm Wheatstone bridge which has an output linear in ΔR (see Figure 6.1).

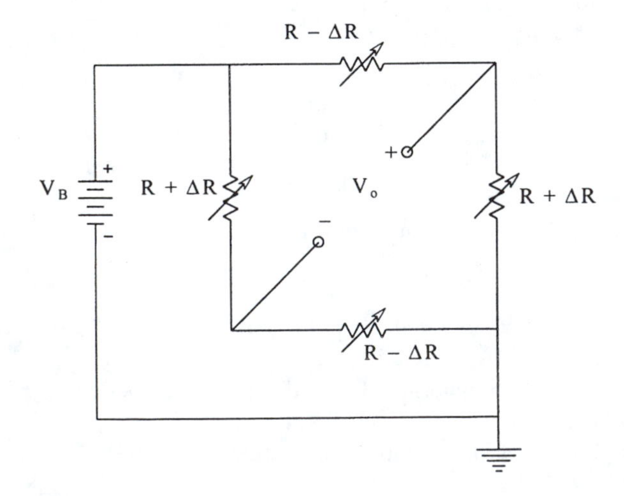

FIGURE 6.1 Wheatstone bridge with four active arms. Bridge sensitivity is four times that of a bridge with a single variable resistance.

6.2.1 Resistive Temperature Sensors

The electrical resistance of all metals and alloys increases with temperature. This increase can be modeled by a power series equation of the form:

$$R(T) = R(25°C) + \alpha(T - 25) + \beta(T - 25)^2 \quad 6.1$$

Here 25°C is taken as the reference temperature. The temperature coefficient (tempco) of a resistive conductor is defined as α:

$$\alpha \equiv \frac{dR(T)/dT}{R(T)} \quad 6.2$$

Table 6.1 below gives tempcos and useful temperature ranges for various metals used as resistance thermometers, or *resistance temperature detectors* (RTDs).

TABLE 6.1 Properties of Conductors Used for RTDs

Material	Tempco (α)	Useful Range (°C)	Relative Resistivity (vs. Cu)
Copper	0.00393	–200–+260	1
Platinum	0.00390	–200–+850	6.16
Nickel	0.0067	–80–+320	4.4
Thermistors	–0.05	–100–+300	—

Resistance temperature detectors (RTDs) are generally used in a Wheatstone bridge configuration. In order to compensate for lead lengths and thermoelectric EMFs at junctions with the RTD material, extra leads are run to the measurement site, as shown in Figure 6.2.

Another type of RTD material which has a much higher tempco than pure metals or metallic alloys is the thermistor, which is made from amorphous semiconductor material, generally sintered mixtures of oxides, sulfides, and silicates of such elements as Al, C, Co, Cu, Fe, Mg, Mn, Ni, Ti, U, and Zn. The resistance of negative tempco (NTC) thermistors generally follows the rule:

$$R(T) = R_o \exp\left[\beta\left(1/T - 1/T_o\right)\right] \quad 6.3$$

where T and T_o are Kelvin temperatures, and T_o is customarily taken to be 298 K. The NTC thermistor tempco is easily calculated to be:

$$\alpha = \frac{dR/dT}{R} = -\beta\frac{1}{T^2} \quad 6.4$$

The tempco evaluated at 298 K with $\beta = 4000$ is –0.045. When used in a Wheatstone bridge circuit, it is possible to resolve temperature changes of smaller than a millidegree Celcius.

If the power dissipation of the thermistor is sufficient to cause it to warm to above ambient temperature, then fluid at ambient tremperature moving past the thermistor will conduct heat away from it, cooling it and producing an increase in its resistance. Thus, this mode of operation may be used to measure fluid velocity when the fluid temperature is known.

PN junction devices (diodes) can also be used as temperature sensors. For example, the current in a forward-biased diode is given by the well-known approximation:

$$i_D = I_{rs}\left[\exp\left(v_D q/nkT\right) - 1\right] \quad 6.5$$

where v_D is the voltage across the diode (positive for forward bias), n is a constant ranging from 1 to 2, k is the Boltzmann constant, T is the junction temperature in Kelvin, and I_{rs} is the reverse

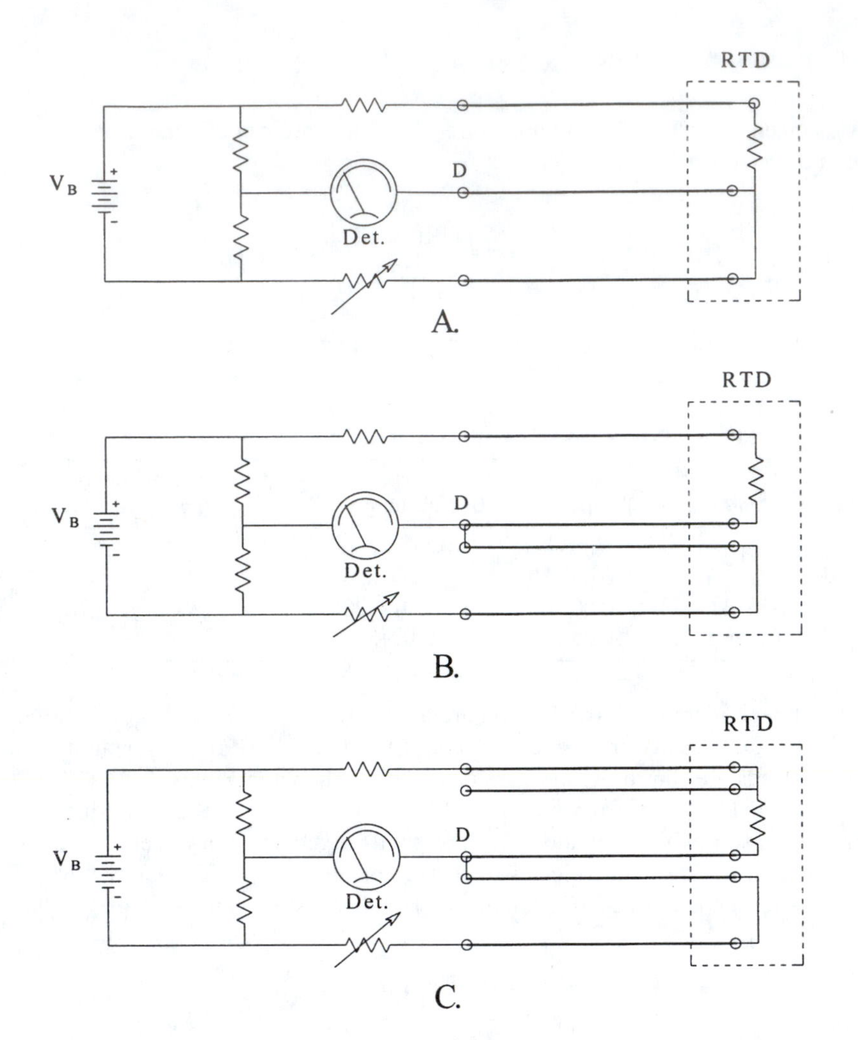

FIGURE 6.2 Three methods for compensating for lead resistance in a Wheatstone bridge used with an RTD. The RTD is the resistance element in the rectangular block. (Adapted from Figure 13.5 in Beckwith, T.G. and Buck, N.L., *Mechanical Measurements*, Addison-Wesley, 1961).

saturation current. Experimentally it has been observed that I_{rs} varies with temperature according to the relation (Millman, 1979):

$$I_{rs}(T) = I_{rs}(T_o)2^{(T-T_o)/10} \tag{6.6}$$

where $I_{rs}(T_o)$ is the reverse saturation current measured at temperature T_o.

The reverse saturation current approximately doubles for each 10°C rise in temperature. In general, the current through a forward-biased pn junction diode in a simple, series circuit with a resistor and a battery will increase, and the voltage across the diode will decrease, with increasing temperature. Figure 6.3 illustrates a simple Wheatstone bridge circuit which can be used to measure temperature using a diode. In this case, the diode forward voltage decreases approximately by 2 mV/°C.

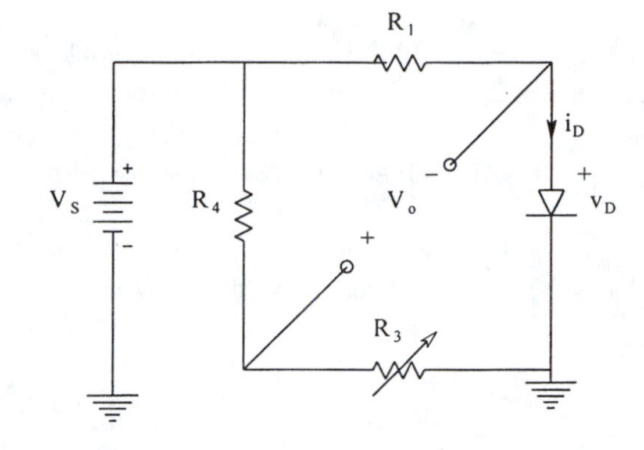

FIGURE 6.3 A forward-biased, Si, PN diode used as a nonlinear RTD in a Wheatstone bridge.

6.2.2 Resistive Strain Gauges

Resistive strain gauges fall into two, broad categories, bonded and unbonded gauges. Bonded gauges consist of fine wires or conducting films which are cemented to some structural beam or machine part in which we wish to measure the strain. The minute elongation or compression of the beam in response to a mechanical load causes a length change in the bonded gauge conductors. This length change, in turn, causes a small change in the resistance of the conductors which is usually sensed with a Wheatstone bridge circuit.

To examine how a strain gauge works, we consider a length of wire, L, with a circular cross section of area A, and a resistivity of ρ ohm cm. The resistance of this wire (at a given temperature) is given by:

$$R = \rho L/A \text{ ohms} \tag{6.7}$$

If the wire is stressed mechanically by a load W, a stress of s = W/A psi will occur. As the result of this stress, a strain, ε, will occur, given by:

$$\varepsilon = \Delta L/L = s/Y \text{ in./in.} \tag{6.8}$$

where Y is the Young's modulus for the material of the wire. Using a Taylor's series expansion on R, we can write:

$$\Delta R = \left(-L/A^2\right)\Delta A + (L/A)\Delta\rho + (\rho/A)\Delta L \tag{6.9}$$

To obtain the fractional change in R, we divide the equation above with Equation 6.7 for R:

$$\Delta R/R = -\Delta A/A + \Delta\rho/\rho + \Delta L/L \tag{6.10}$$

We now define the gauge factor, GF, as

$$GF = \frac{\Delta R/R}{\Delta L/L} = \frac{\Delta R/R}{\varepsilon} = 1 + \frac{\Delta\rho/\rho}{\Delta L/L} + \frac{-\Delta A/A}{\Delta L/L} \tag{6.11}$$

The last term in Equation 6.11 is Poisson's ratio, μ. From Equation 6.11 we can calculate the change in resistance, ΔR, for a given strain, ε. For example, if we load a 0.01" diameter steel wire with a one pound load, this causes a stress of $1/(\pi \times 0.0052) = 1.27 \times 10^4$ psi. The strain caused by this stress is $\varepsilon = s/Y = 1.27 \times 10^4/3 \times 10^7 = 4.23 \times 10^{-4}$ in./in. or 423 microstrains. Assume the wire has an unstrained resistance of 220 ohms, and the gauge factor, GF, is 3.2. ΔR is then given by:

$$\Delta R = GF\varepsilon R = 0.298 \text{ ohms} \qquad 6.12$$

The output voltage of an initially balanced, equal-arm, 220 ohm Wheatstone bridge with a 6 V excitation, V_s, which includes the wire described above as one arm would be:

$$V_o = V_s \frac{\Delta R}{4R} = 6 \frac{0.298}{4(220)} = 2.03 \text{ mV} \qquad 6.13$$

The output of a strain gauge bridge generally needs amplification. Sensitivity to temperature changes can be reduced by using two matched gauges on one side of the bridge; the lower one is active, the upper one is unstrained and acts as a thermal compensator for the active bottom gauge. Both gauges are at the same temperature. See Beckwith and Buck (1961) for an excellent description of bonded strain gauge applications. A good discussion of the sources of error in bonded strain gauge systems can be found in Lion (1959).

The frequency response of bonded strain gauges depends in large part on the mechanical properties of the structure to which they are bonded. It generally ranges from DC through the audio range.

Unbonded strain gauges are used in several applications: the direct measurement of forces, and the measurement of pressure (pressure acts on a diaphragm or piston to produce a force). Figure 6.4A shows the innards of an unbonded force sensing strain gauge. There are two sets of wires under tension. When force is applied, one set is strained more, and the other less. Due to this asymmetry, the resistance of one set of wires increases while that of the other set decreases a like amount. The two sets of wires connected in series is well suited to form one side of a Wheatstone bridge having inherent temperature compensation. Typically, the maximum $\Delta R/R$ attainable at full rated load is about 0.01. The bridge excitation voltage is practically limited, as in the case of bonded gauges, by power dissipation in the wires and the effects of heating on their resistances.

Unbonded strain gauges exhibit a mechanical resonance which limits their high-frequency response. The taut resistance wires and the moving armature form a mass and spring system which typically resonates at some low audio frequency. Some unbonded strain gauge transducers are designed to have a full bridge output, i.e., two of their four wire elements are resistors which increase like amounts, while two other wire elements decrease in resistance like amounts due to applied force. Balance of such a four-arm bridge at zero force is usually done with a potentiometer connected as shown in Figure 6.4B.

6.2.3 Photoconductors

Photoconductors are materials whose resistance decreases upon illumination with light. They should not be confused with photodiodes or solar cells, which can produce an EMF or short-circuit current in response to the absorption of light quanta. Uses of photoconductors include exposure meters for cameras, light sensors in spectrophotometers, light sensors in a variety of counting systems where an object interrupts a light beam hitting the photoconductor, systems which sense a decrease of overall ambient illumination and turn on outside lighting (of course, the lighting must not be sensed by the photoconductor), oil burner safety systems, and burglar alarms.

Absorption of light quanta with a certain energy range by semiconductor substances produces electron-hole pairs which enter the conduction band and drift in an applied electric field and contribute to the current passing through the semiconductor. Greatest photoconductor sensitivity occurs when the absorbed photons have energies equal to or slightly greater than the semiconductor's

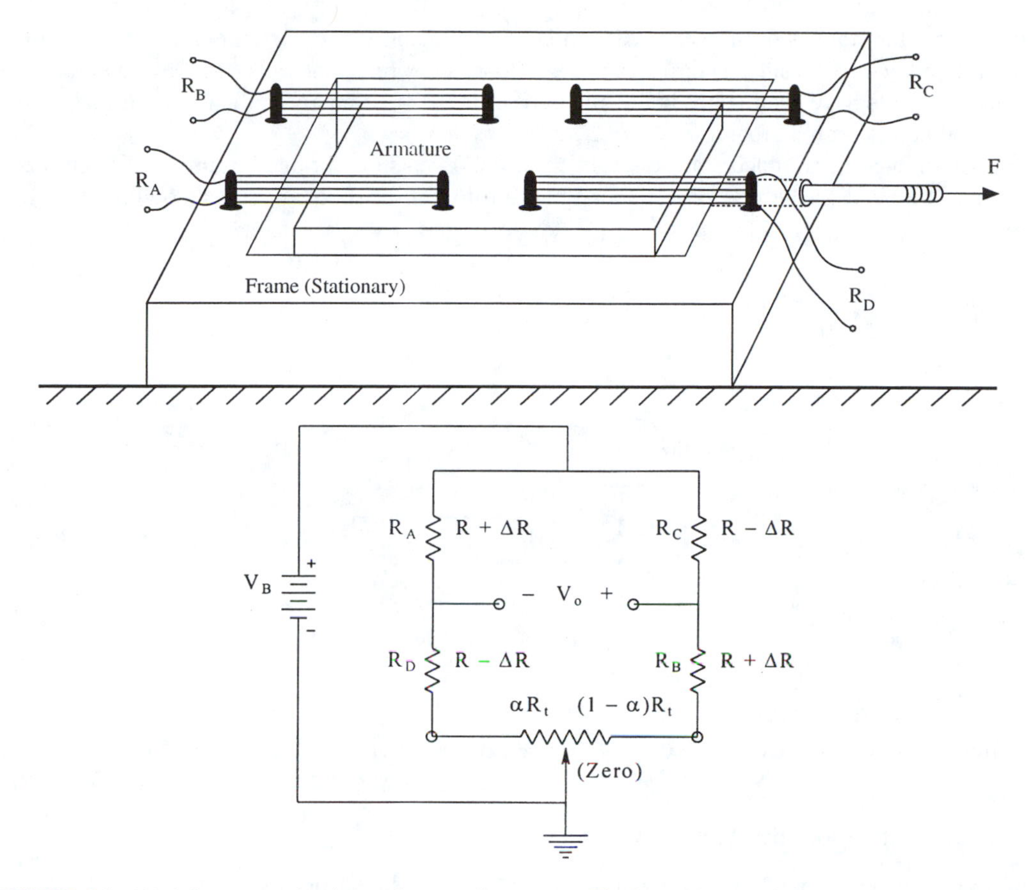

FIGURE 6.4 (A) A four-arm, unbonded strain gauge force sensor. The four resistance windings are under equal tension with zero applied force to the armature. When force is applied to the armature in the direction shown, strain increases in R_A and R_B, and decreases in R_C and R_D. (B) Schematic diagram of the four-arm unbonded strain gauge force sensor. R_t is used to obtain an output null at zero or reference force.

energy band gap. For example, cadmium sulfide (CdS) having an energy gap of 2.42 eV works best in the visible range of wavelengths. Germanium and indium antimonide having energy band gaps of 0.67 and 0.18 eV, respectively, are used in the near and far infrared wavelengths, respectively. EG&G-Judson offers a J15D series of mercury-cadmium-telluride photoconductors which operate in the 2 to 22 μm band (far infrared). These photoconductors are operated at 77 K, and are used for infrared spectroscopy, thermography, CO_2 laser detection, and missile guidance. Their response time constants range from 0.1 to 5 μsec, depending on the lifetime of electrons in the crystal, which in turn depend on material composition and temperature. HgCdTe photoconductive detectors generate 1/f noise which limits their resolution below 1 kHz.

CdS photoconductors are often made by depositing a thin layer of the material on an insulating (ceramic) substrate, often in a zigzag pattern to increase the effective length of the conductor's path, hence its resistance.

The frequency response of photoconductors (in response to modulated light intensity) depends on, among other factors, the intensity and wavelength of the incident light, the Kelvin temperature of the photoconductor, and the amount of metallic impurities in the semiconductor which can trap charge carriers and delay their spontaneous recombination. It can be shown that when the recombination time constant, τ_p, is large (corresponding to a low high-frequency response), the gain of the photoconductor is high. A short time constant, giving good high-frequency response, results in lower gain (Yang, 1988). Gain is defined by Yang as the ratio of carriers collected by the ohmic contacts to the carriers generated by photons per unit time. The range of frequency response of photoconductors ranges from over 1 MHz to below the audio range. Germanium and lead sulfide have relatively broadband responses (tens of kilohertz), while materials such as selenium and cadmium sulfide have bandwidths in the hundreds of hertz. A summary of the properties of various photoconductors may be found in Lion (1959).

Photoconductor transducers are typically used in Wheatstone bridge circuits, although an op amp may be used in the transresistor configuration to convert photocurrent to output volts (see Figure 6.5).

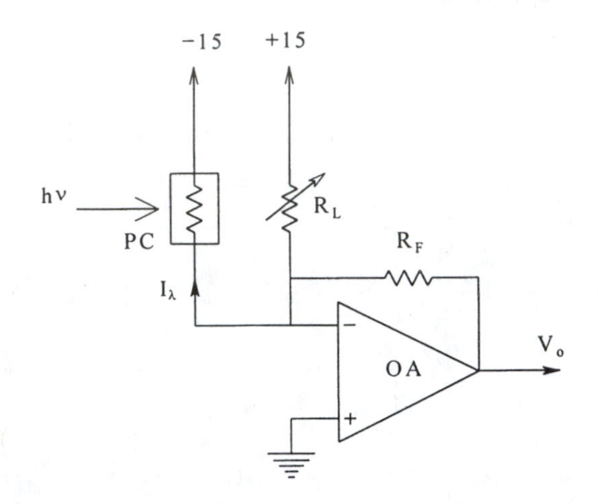

FIGURE 6.5 An op amp is used as an operational transresistor to condition the photocurrent from a photoconductor. The op amp should have low DC drift, such as found in a chopper design.

6.2.4 Resistive Relative Humidity Sensors

Humidity sensors respond to the amount of water vapor in the air or other gas. In this section we will consider only those humidity sensors that change their resistance in response to the partial pressure of water vapor to which they are exposed. In one embodiment, the Dunmore sensor, two noble metal wires are wound in the form of two nontouching helices on an insulating bobbin or form. A thin, hygroscopic coating consisting of an ionizable salt covers the wires. As the humidity increases, the salt in the coating reversibly absorbs water vapor and dissociates into ions. The ions conduct electric current which is an increasing function of the humidity. Alternately, one can say that the resistance of the sensor decreases with humidity. Because the sensor is an electrochemical system, the use of DC excitation for the bridge and sensor will lead to polarization and possible electrolysis and chemical breakdown of the hydrated salt. Polarization due to ion migration causes the apparent resistance of the sensor to increase, hence causes erroneous readings. The ionizable salt humidity sensors must be used with low-voltage, low-frequency, AC excitation to prevent these problems.

Lithium chloride, potassium dihydrogen phosphate, and aluminum oxide have been used in various resistive humidity sensor designs (Lion, 1959). The sensors using only lithium chloride appear to respond to a limited range of humidity, which is dependent on the LiCl concentration. The resistance of the sensor is strongly dependent on temperature, which affects water absorption and dissociation of the salt. At constant temperature, the LiCl sensor resistance decreases logarithmically over a three or four decade range in proportion to the relative humidity.

Figure 6.6A illustrates a wide-range relative humidity sensor described in Lion (1959), attributed to Cutting, Jason, and Wood, 1955. An aluminum rod was coated at one end with an aluminum oxide layer, and then a porous conducting layer such as graphite or vapor deposited metal. This sensor is used with AC excitation. Figure 6.6B, taken from Lion (1959), shows the resistance of the sensor varies from about 5 megohms at 0 RH to about 2 kilohms at 100% RH. Of special interest is the fact that there is a corresponding increase in sensor capacitance from about 500 pF at 0 RH to 10^5 pF at 100% RH. Hence an AC, capacitive-sensing bridge can also be used to obtain sensor output, as well as an AC Wheatstone bridge. This humidity sensor is unique in that it is claimed to be temperature insensitive over a –15° to 80°C range.

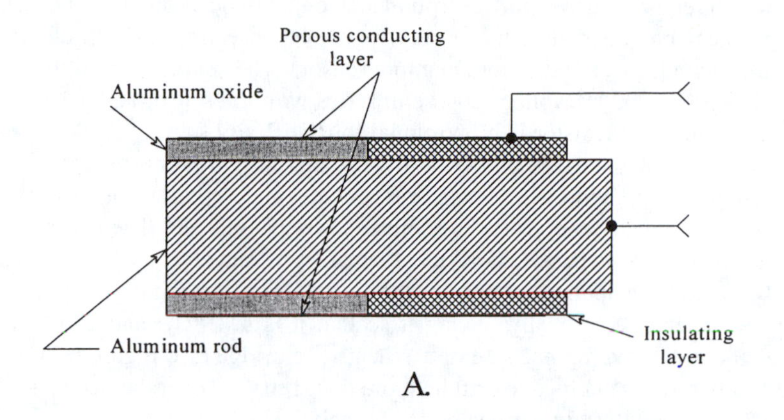

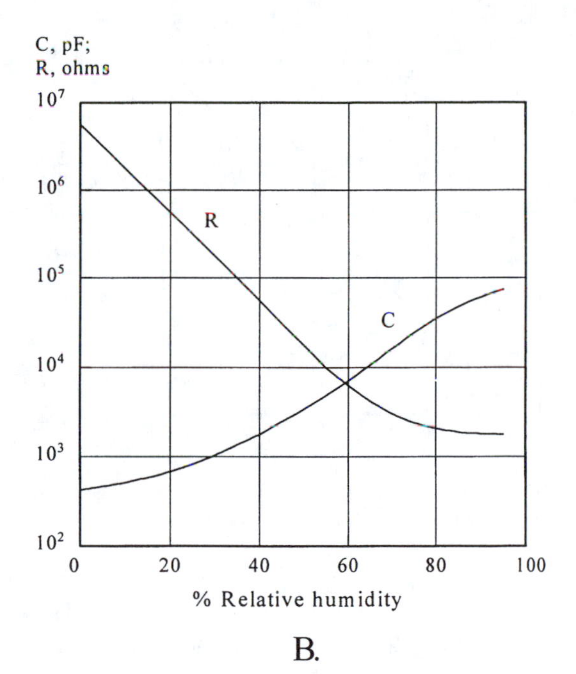

FIGURE 6.6 (A) Cross section through an aluminum oxide humidity sensor. (B) Plot of an aluminum oxide humidity sensor's equivalent parallel resistance and capacitance as a function of relative humidity. (Adapted from Lion, Figure (1-7)8, 1959.)

Commercial humidity measurement systems and sensors are offered by several companies. Omega Corp. markets a number of different humidity- and temperature-sensing systems; however, they use capacitance humidity sensors rather than conductive types. Ohmic Instruments Co. offers a series of conductive humidity sensors based on water vapor absorption by dissociable salts. They offer ten different Dunmore-type sensors, each with a narrow range of RH sensitivity. Sensor resistances vary from over 10 megohms to under 1 kilohm over their narrow ranges of RH sensitivity. A broad range instrument (model WHS 5-99L) covering from 5 to 90% RH uses four overlapping, narrow-range sensors in a weighted resistor network, which use 60 Hz excitation. Thunder Scientific Corp. has developed and marketed what they call the "Brady Array" humidity sensor. From published technical specifications, this sensor is a precise array of semiconductor crystals and interstitial spaces. Physical adsorption of water molecules from water vapor somehow causes an increase in conduction band electrons in the crystals, raising conductivity. Unlike the Dunmore-type sensors, which require water vapor adsorption and ionic dissociation, the Brady sensor operates on a purely physical basis and thus has a very rapid response time to step changes in RH (in hundreds of milliseconds) vs. minutes for Dunmore sensors. The simplest circuit for a Brady sensor consists of a 5 VRMS, 1000 Hz voltage source in series with the sensor and a 1 megohm resistor. The voltage across the 1 M resistor is proportional to RH. Brady sensors work over a range from nearly 0 to 100% RH. They have been shown to work normally in atmospheres of water vapor in air, oxygen, nitrogen, CO_2, hydrogen, natural gas, butane, and vapors from various hydrocarbons such as acetone, trichloroethylene, jet fuel, gasoline, kerosene, oils, etc. Their working temperature range has been given as –20° to +70°C. Typical accuracy is stated as ±2% of indicated RH, and resolution is 0.1% RH. Temperature compensation of a Brady sensor is easily accomplished by taking a like sensor and hermetically sealing it so that it is at 0 RH, and then using the sealed sensor as an element in the reference side of a Wheatstone bridge (see Figure 6.7). A temperature-compensated Brady array is claimed to drift no more than 1 mV/°F over the –20 to +70°C operating range. This is 0.02% of full-scale per degree Fahrenheit.

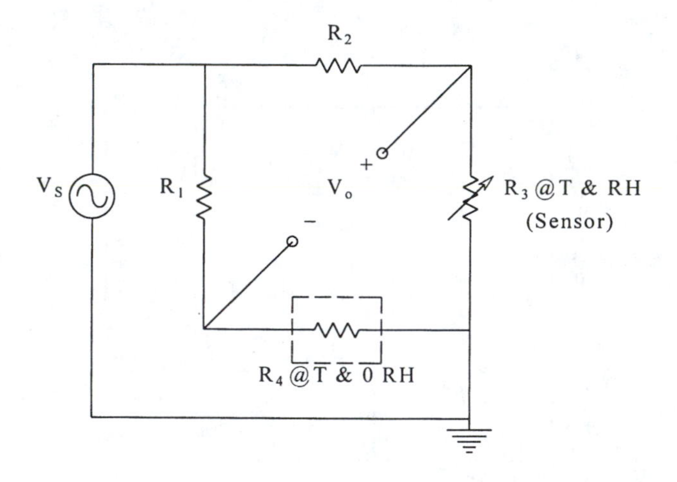

FIGURE 6.7 Two Brady array humidity sensors are used in a Wheatstone bridge to obtain temperature compensation over a wide range; sensor R4 is sealed at 0 RH, and responds only to temperature changes.

Calibration of resistive (and other kinds of) RH sensors is traditionally done by using saturated salt solutions at known temperatures. For example, at 25°C, the RH over a saturated aqueous $LiCl(H_2O)$ solution is 11%, over $MgCl_2(6H_2O)$ it is 33%, over $Na_2Cr_2O_7(2H_2O)$ it is 54%, over NaCl it is 75%, over K_2CrO_4 it is 86%, over $(NH_4)H_2PO4$ it is 93%, and over K_2SO_4 it is 97%.

6.2.5 Direct Resistance Change Used to Sense Position or Angle

Direct measurement of shaft angle in applications such as electromechanical, position feedback control systems is often done by coupling the shaft either directly, or through gears, to a single- or multi-turn potentiometer. The potentiometer can be one arm of a Wheatstone bridge, or used as a voltage divider. Often a feedback linearization circuit is used with a bridge, such as shown in Figure 6.8, to produce a voltage output which varies linearly with potentiometer resistance (shaft angle). The potentiometers used in such shaft angle transduction must be linear and have high resolution. The resolution of a standard wire-wound, three-, ten-, or fourteen-turn potentiometer is limited by the wiper connected to the shaft making contact with individual turns of the helically wound resistance wire. One way to avoid this problem is to use potentiometers that have continuous, cermet or conductive plastic resistance materials.

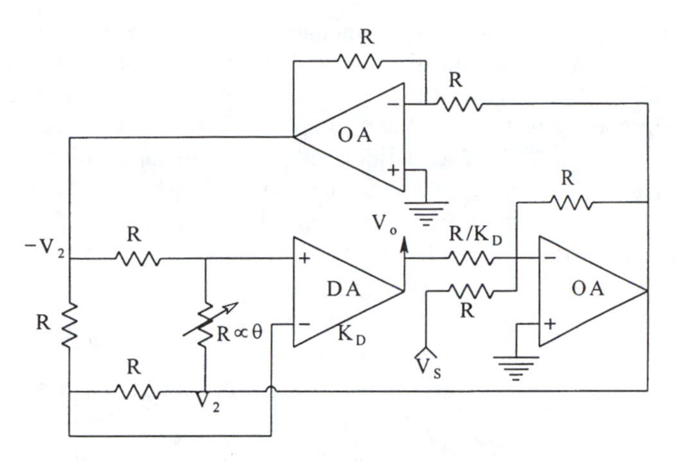

FIGURE 6.8 A Wheatstone bridge output linearization circuit, useful in applications where the variable resistance varies linearly over a wide range as a function of a parameter under measurement, such as shaft angle.

Potentiometers used as angular position sensors typically are specified to have independent linearity over their total range of rotation. The Bourns 3400 series of ten-turn, wire-wound potentiometers typically have ±0.15% independent linearity, and are available with ±0.05% independent linearity. The Bourns 3400 series has a resistance tempco of +0.007%/°C. Conductive plastic potentiometers, such as the Beckman single-turn, 6550 series, have independent linearities of ±0.5% and tempcos of –150 to +300 ppm. Cermet potentiometers, such as the single-turn, Beckman 6300 series also have independent linearities of ±0.5%, but have lower tempco ranges of ±100 ppm.

In addition to precision rotational potentiometers for sensing angles, linear displacements can be measured with linear potentiometers, i.e., pots whose resistance varies linearly with the linear position of the wiper. The Beckman series 400 rectilinear potentiometer series uses an infinite resolution, cermet resistance element. Wiper travel lengths range from 0.6" to 6.1" in various 400 series models. Linearity ranges from ±1.35% for the 0.6" model to ±0.25% in the 6.1" model.

One disadvantage of potentiometers as position sensors is that they require some extra torque or force to move their wipers. Another disadvantage is that because of the wiping action, they eventually wear out. Optical coding disks need no wipers, and are limited in resolution by the number of output bits available. A linear variable differential transformer (LVDT) can operate with negligible friction if a Teflon lining is used in its core. Still, potentiometers offer an inexpensive means of sensing angular or linear position in a number of applications.

6.3 VOLTAGE-GENERATING SENSORS

A wide variety of sensors generate an EMF in response to some physical input quantity. In most cases, the magnitude of the induced EMF is linearly proportional to the input quantity. Often, the induced voltage is very small, and special types of signal conditioning must be used.

6.3.1 Thermocouples and Thermopiles

Thermocouples are traditionally used to measure temperatures, often at extremes. Thermopiles are close arrays of thermocouples arranged in series to have high sensitivity and resolution, and are generally used to measure the energy of electromagnetic radiation at optical frequencies, such as from lasers; total absorption of the radiation causes a minute temperature rise, measured by the series thermocouples.

A thermocouple consists of junctions, often spot-welded, between two or more dissimilar metal wires. A basic thermocouple system must contain two couples (see Figure 6.9). However, thermocouple systems may contain three or more couples. In a two-couple system, there is a net EMF in the loop causing current to flow if the two couples are at different temperatures. In a simple, two-couple system, one couple junction is maintained at a reference temperature, typically 0°C, formed by melting ice in water. Often the thermocouple system (measuring and reference junctions), consisting of two different metals, must be joined to a third metal, such as copper, to connect to a millivoltmeter. The joining of the two thermocouple metals to copper produces one additional junction, as shown in Figure 6.9B. The addition of the third metal into the system will have no effect on the performance of the basic, two-metal, thermocouple system as long as the two junctions with the third metal are at the same (reference) temperature. This property is called the *law of intermediate metals* (Beckwith and Buck, 1961).

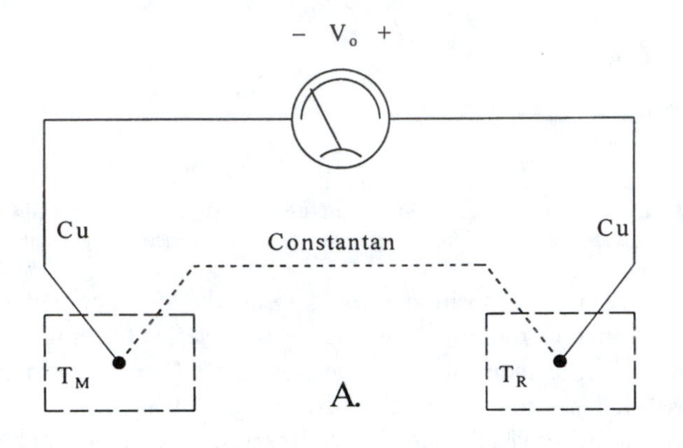

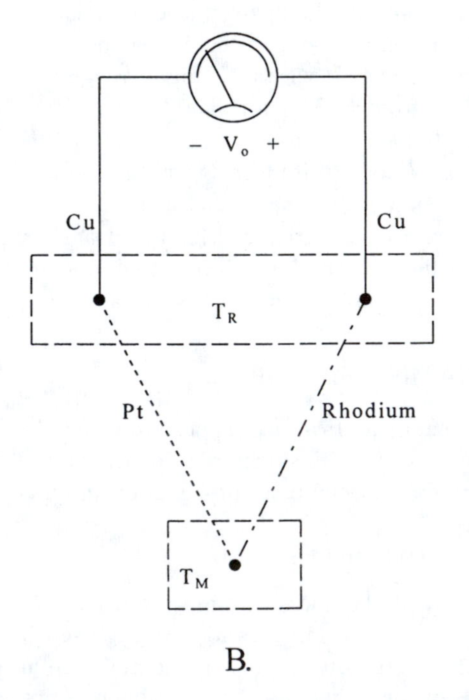

FIGURE 6.9 (A) A simple thermocouple circuit in which there are two couples (one for reference, one for measurement), and two metals. (B) A thermocouple system in which there are three metals. The two dissimilar junctions at the reference temperature behave as the single reference junction in (A).

Table 6.2 gives some common thermocouple materials and the temperature ranges over which they are effective. The net EMF in a thermocouple system is not a linear function of the temperature difference between the reference and measuring junction. In general, the net EMF can be expressed as a power series:

$$E_o = A(\Delta T) + B(\Delta T)^2/2 + C(\Delta T)^3/3 \quad 6.14$$

where ΔT is the temperature difference above the reference junction temperature, usually made 0°C. The thermoelectric sensitivity of a thermocouple pair is defined as $S_T \equiv dE_o/dT = A + B(\Delta T) + C(\Delta T)^2$, and is usually expressed in microvolts/°C. Lion (1959) gives a table of thermoelectric sensitivities of various metals vs. platinum at temperatures near 0°C. These are presented in Table 6.3 below. Lion points out that sensitivities of thermocouples made from any pair of metals can be found by subtracting the tabular values of S_T for platinum. For example, S_T for copper vs. constantan is (6.5 – [–35.5]) = 41.5 μV/°C. Of special note is the extremely high thermoelectric sensitivities of bulk semiconductor materials such as Ge and Si.

TABLE 6.2 Properties of Common Thermocouples

Useful Temp Materials	Max Temp for Range, °C	Sensitivity in Short Periods	μV/°C at °C
Copper/constantan	–300–350	600	15 @ –200 60 @ +350
Iron/constantan	–200–+800	1000	45 @ 0 57 @ +750
Pt/Pt90Rh10	0–1450	1700	0 @ –138 5 @ 0 12 @ +1500
Iron/Copnic	–200–+860	1000	60 @ 0
Chromel P/Alumel	–200–1200	1350	40 – 55 between 250 & 1000°C
W95Re5/W26Re7	0–2316		

TABLE 6.3 Thermoelectric Sensitivities of Thermocouples Made with Materials Listed vs. Platinum, in μV/°C, with Reference Junction at 0°C

Material	S_T vs. Pt	Material	S_T vs. Pt
Bismuth	–72	Silver	6.5
Constantan	–35	Copper	6.5
Nickel	–15	Gold	6.5
Potassium	–9	Tungsten	7.5
Sodium	–2	Cadmium	7.5
Platinum	0	Iron	18.5
Mercury	0.6	Nichrome	25
Carbon	3	Antimony	47
Aluminum	3.5	Germanium	300
Lead	4	Silicon	440
Tantalum	4.5	Tellurium	500
Rhodium	6	Selenium	900

Thermocouple EMFs were traditionally measured using accurate potentiometers (calibrated with standard cells). Modern thermocouple measurement makes use of high input impedance, solid-state, electronic microvoltmeters. Electronically regulated, solid-state temperature references for thermocouple systems are also used, replacing melting ice and water in a thermos flask. Wahl Instruments, Inc. offers electronically regulated, thermal calibration sources and an ice point reference chamber.

A *vacuum thermocouple* is used in conjunction with a sensitive DC microammeter to measure the true rms value of the current in its heater resistance element. A thermocouple junction is bonded to the heater wire and thus generates an EMF proportional to the temperature of the heater resistor. A reference junction is at room (ambient) temperature, so the net EMF of the thermocouple is proportional to ΔT, the temperature of the heater above ambient temperature. The temperature rise of the heater is given by

$$\Delta T = \overline{i^2} R_H \Theta_o \tag{6.15}$$

where $\overline{i^2}$ is the mean squared heater current, R_H is the heater resistance (assumed constant), and Θ_o is the thermal resistance of the heater *in vacuo*; its units are degrees Celcius/watt. The DC microammeter current is then given by Ohm's law:

$$I_M = E_o/(R_M + R_T) = A\Delta T/(R_M + R_T) = K_T \overline{i_H^2} R_H/(R_M + R_T) \tag{6.16}$$

where R_M is the microammeter's resistance, R_T is the thermocouple's resistance, and A is the thermocouple's EMF constant at room temperature (see Equation 6.14 above).

Typical parameters for a Western Electric model 20D vacuum thermocouple are: R_H = 35 ohms, R_T = 12 ohms, $K_T \cong 2.9$ V/W.

Note that vacuum thermocouples are used for true rms ammeters or voltmeters by giving the associated DC microammeter a square-root scale. Meter deflection is proportional to the power dissipated in R_H, hence $\overline{i_H^2}$. Thermocouple meters work at DC, and at frequencies from the tens of hertz up to the tens of megahertz. Because of their true mean squared deflection, they are widely used for noise voltage measurements.

6.3.2 Photovoltaic Cells

Photovoltaic cells are used to measure light intensity, as well as to generate DC electric power when used as solar cells. Photovoltaic cells include barrier-layer cells, and pn junction photodiodes. We first examine the photovoltaic effect in pn junction diodes. When light of energy hν is absorbed by an open-circuited pn junction device, the built-in electric field at the junction that results from the p,n doping acts to concentrate the hole-electron pairs generated by photon interaction with valence band electrons in such a manner that the holes are swept toward the p-side and the electrons toward the n-side of the junction. The net effect is the production of an open-circuit EMF given by the relation (Navon, 1975):

$$E_{oc} = v_T \ln\{[qA(L_p + L_n)G(\lambda)/I_{rs}] + 1\} \tag{6.17}$$

where $v_T = nkT/q = 0.026$ V at 300 K, q is the magnitude of the electron charge in coulombs, A is the cross-sectional area of the junction, n is a constant ranging from one to two, L_n is the electron diffusion length, L_p is the hole diffusion length, and $G(\lambda)$ is the hole-electron pair generation rate per unit volume, which is also a function of the light wavelength and intensity. This relation describes a logarithmic increase of the open-circuit photovoltage with light intensity, other factors being constant. Maximum E_{oc} in bright sunlight is typically 400 to 500 mV. Figure 6.10A illustrates the current-voltage curves for a photodiode or solar cell in the dark, and when illuminated. Figure 6.10B shows typical V_{OC} and I_{SC} curves vs. illumination for a photovoltaic device.

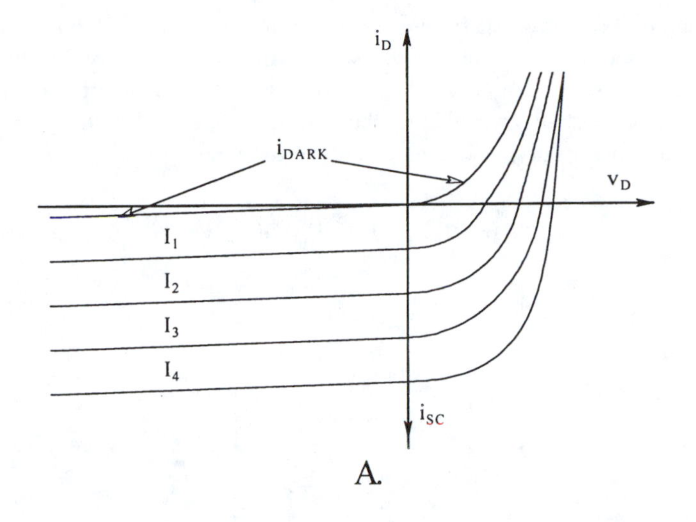

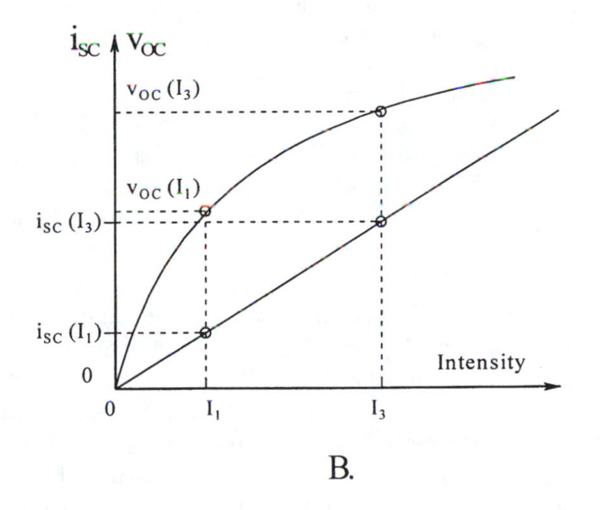

FIGURE 6.10 (A) Current-voltage curves for a photogalvanic (solar) cell or photodiode. This class of device is operated in the fourth quadrant. (B) Plots of the open-circuit voltage and short-circuit current of a photogalvanic device. Note that ISC is linear with illumination, and VOC varies logarithmically.

When illuminated, it can be shown (Nanavati, 1975) that the curent curve is given by:

$$i_D = I_{rs}\left[\exp\left(v_D/v_T\right) - 1\right] - qA\left(L_p + L_n\right)G(\lambda) \qquad 6.18$$

(The parameters in Equation 6.18 are defined in the previous equation.)

Photodiodes operated as photovoltaic cells are maximally sensitive to photons with energies near their band-gap energy, ϕ_o. If $h\nu$ is less than ϕ_o, the photons will not generate hole-electron pairs. On the other hand, if the photons have energies far in excess of ϕ_o, they will be absorbed near the surface of the exposed semiconductor layer where the recombination rate is high, decreasing efficiency. Semiconductor photodiodes made from mixed compounds (PbSe, InAs, InSb) with low band-gap energies are used to create far infrared detectors. A line of such detectors is offered by EG&G - Judson.

Germanium photodiodes have maximum sensitivity at a wavelength of 155 nm; however, their sensitivity ranges from visible wavelengths to 180 nm. An alternate means of using pn photodiodes is shown in Figure 6.11. An op amp is used as a current to voltage converter. Because the op amp's summing junction is at virtual ground, the photodiode is operated in the short-circuited mode, where $v_D = 0$ V. The op amp output is thus given approximately by

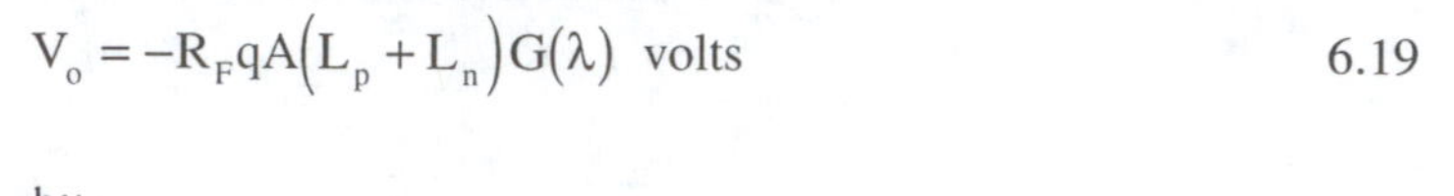

$$V_o = -R_F qA(L_p + L_n)G(\lambda) \text{ volts} \quad 6.19$$

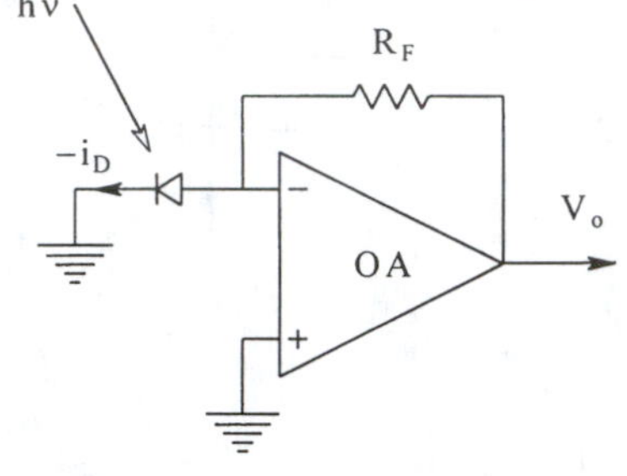

FIGURE 6.11 Op amp transresistor circuit in which a photodiode is operated in the short-circuit mode. Response speed is low due to large zero bias junction capacitance.

Under zero bias conditions, the diode's junction capacitance is quite large, and hence its speed of response to changes in the light flux is not as fast as it is for a reverse-biased photodiode operated as a photoresistor. EG&G - Judson's J16 series of germanium photodetectors are usable over an optical input intensity range of 2×10^{-12} to 10^{-2} W. Their −3 dB frequencies range widely, from 0.07 to 250 MHz. The faster photodiodes have smaller active junction areas and shunt capacitances, and higher shunt resistances.

A *solar cell* is operated in the fourth quadrant of its current-voltage curves. The open-circuit photovoltage causes a current to flow in a load, and thus the solar cell can deliver electrical power. A solar cell is made with a considerably expanded pn junction area, since it must do work. A typical solar cell of 2 cm^2 area might consist of a thin (0.5 mm), rectangular slab of n-type silicon, on top of which is diffused a thin, p-type (boron treated) surface layer, through which the light must pass. Such a cell can deliver about 10 mW in bright sunlight with about 12% conversion efficiency. Solar cells are designed to have maximum efficiency in delivering steady-state, DC power to a load, which might be charging batteries used for a portable instrumentation module. To obtain maximum efficiency, there is tradeoff between a number of design factors (Streetman, 1972). For example, the thickness of the top (p) layer of the solar cell, d, measured from its surface to the center of the junction, must be less than L_n to allow electrons generated near the surface to diffuse to the junction before they recombine. Similarly, the thickness of the n region must be such that the electrons generated in the region can diffuse to the junction before they, too, recombine. This requirement implies a match between the hole diffusion length, Lp, the thickness of the n region, and the optical penetration depth in the surface (p) layer. It is also desirable in solar cell design to have a large contact potential in order to realize a large photovoltage. This means that heavy doping must be used. The need for heavy doping must be compromised with the requirement for long carrier lifetimes, which are reduced by heavy doping. Efficient solar cells require a low internal (ohmic) resistance. There is no problem in making metallic electrode contact with the bottom (dark) layer of the solar cell, but low resistance also requires a large electrode contact area with the top layer. A large metallic area on the top semiconductor layer prevents light from reaching the cell, decreasing efficiency. The tops of solar cells are generally given an antireflection coating, similar to that used on camera lenses, to maximize light absorption.

6.3.3 Piezoelectric Transducers

There is a class of piezoelectric materials which can serve, among other applications, as mechanical input transducers, enabling the measurement of pressure, force, displacement, and other physical phenomena which can be related to them. Piezoelectic materials, when mechanically strained in a preferred manner, generate an open-circuit EMF. Actually, the mechanical strain causes a unidirectional separation of electric charges resident in the interior of the piezoelectric crystal structure. These displaced charges form an effective net charge on a capacitor formed by the bonding of metallic electrodes on the surface of the piezoelectric crystal. The transducer material itself, independent of piezoelectic activity, is an insulator having a very low conductivity and a high dielectric constant. Thus, the effectively separated charges produce an open-circuit voltage on the capacitor equal to Q/C. In practice, the situation is not that simple. The piezoelectric transducer material is not a perfect dielectric, and the charges leak off through the exceedingly small volume conductance. Of course, any voltage measuring system attached to the transducer will have a finite input impedance and perhaps a DC bias current, further altering the stress-caused EMF. In addition, the transducer has mechanical mass, stiffness, and damping, giving it mechanical resonance properties which can further complicate the transfer function of the transducer when the frequency of the mechanical input approaches the mechanical resonance frequency of the transducer. In the discussion below, we assume that the piezoelectric transducers are operated at frequencies well below their mechanical resonant frequencies.

Piezoelectric materials include a number of natural crystals and man-made ceramic materials. Quartz, Rochelle salt, and ammonium dihydrogen phosphate (ADP) are examples of naturally occurring piezo-materials; barium titanate, lead zirconate titanate, and lead metaniobate are synthetic piezoceramics. All piezoelectric materials have a *Curie temperature*, above which piezoelectric activity ceases to exist.

In order to understand how piezoelectric input transducers work, we will examine an example of a crystal responding to thickness compression. Figure 6.12 shows a rectangular crystal block with metallic electrodes vapor deposited on its top and bottom surfaces. The transducer has thickness t and top area A. The capacitance of the transducer is approximately:

$$C_X = \kappa\varepsilon_o A/t \text{ Farads} \tag{6.20}$$

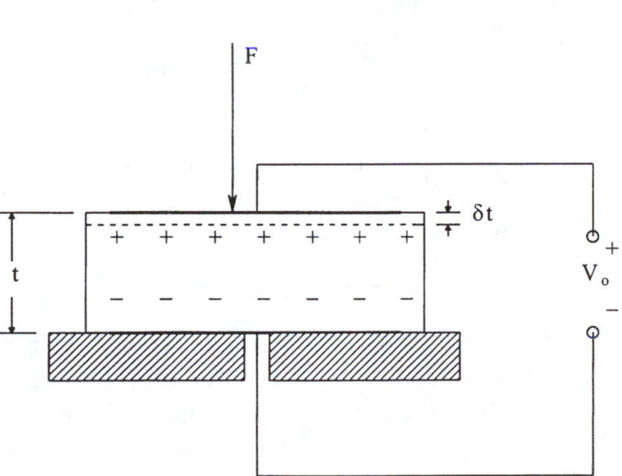

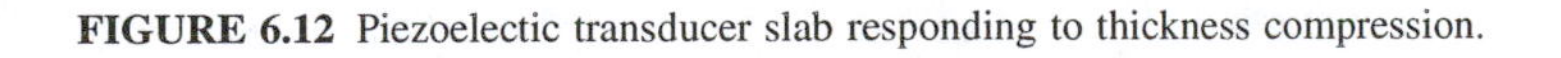
FIGURE 6.12 Piezoelectic transducer slab responding to thickness compression.

where κ is the dielectric constant of the piezoelectric material. A uniformly applied force F, will cause the crystal to compress a minute amount, δt. The piezo-material is assumed to have a Young's modulus, Y.

$$Y = \text{stress/strain} = (F/A)/(\delta t/t) \quad 6.21$$

The net charge displacement for the transducer is given by

$$Q = dF \;\text{ Coulombs} \quad 6.22$$

where d is the charge sensitivity of the material in (coulombs/m^2)/(newton/m^2). Thus we can write the open-circuit voltage across the transducer as

$$E_o = Q/C = \frac{dF}{\kappa\varepsilon_o A/t} = g(F/A) \;\text{ Volts} \quad 6.23$$

Here g is the voltage sensitivity of the transducer in (volts/m)/(newtons/m^2). F/A is the pressure on the transducer's active surface.

The axes along which a natural crystal of piezoelectic material is cut in order to make a transducer determine the type of mechanical input to which it will have maximum sensitivity. In addition to simple thickness compression, piezoelectric transducers also may be made to respond to length compression, thickness, and face shear, as well as twisting and bending. Synthetic piezo-materials may be cast in a variety of shapes not available with natural materials, such as thin-walled cylinders, half-cylinders, discs, hollow spheres, and hemispheres. Table 6.4 below summarizes some of the salient properties of certain piezo-materials.

TABLE 6.4 Properties of Piezoelectric Materials

Material	Orientation	Charge Sens. d in pCb/N	Voltage Sensitivity, g	Curie Temp.
Quartz	X cut; length along Y, length longitudinal	2.25	0.055	550°C
	X cut; thickness longitudinal	–2.04	–0.050	
	Y cut; thickness shear	4.4	0.108	
Rochelle salt	X cut 45°; length longitudinal	435	0.098	55°C
	Y cut 45°; length longitudinal	–78.4	–0.29	
Ammonium dihydrogen phosphate	Z cut 0°: face shear	48	0.354	125°C
	Z cut 45°; length longitudinal	24	0.177	
Barium titanate	Parallel to polarization	86–160	0.016	125°C
	Perpendicular to polarization	–56	0.005	
Lead zirconate titanate	Parallel to polarization	190–580	0.02–0.03	170–360°C
Lead metaniobate	Parallel	80	0.036	>400°C

For mechanical input frequencies well below piezoelectric transducer mechanical resonance, the equivalent circuit shown in Figure 6.13 may be used to model the electrical behavior of the transducer. Although charge is displaced internally by strain, we must use a current source in the model. Since current is the rate of flow of charge, the equivalent current source is

$$i_{eq} = dQ/dt = d(dF/dt) = d\dot{F} \quad 6.24$$

Thus, from Ohm's law, the open-circuit voltage across the transducer is given by the transfer function:

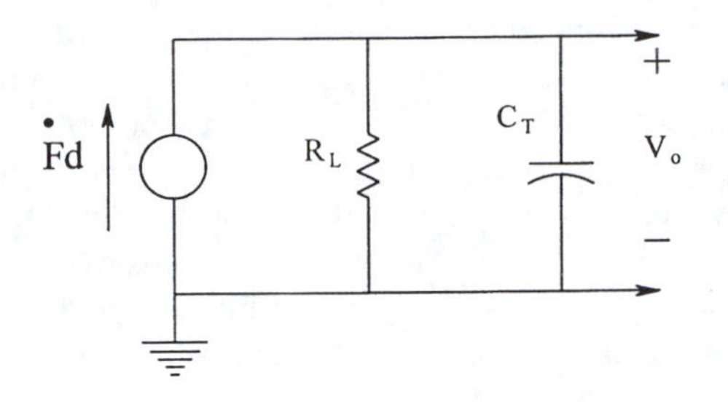

FIGURE 6.13 Electrical equivalent circuit of a piezoelectric transducer responding to applied force. Mechanical frequencies in the applied force are assumed to be well below the transducer's mechanical resonance frequency.

$$\frac{V_o}{F}(s) = \frac{sFdR_L}{1 + sC_T/G_L} \tag{6.25}$$

This transfer function is seen to belong to a simple high-pass filter. Thus, because the charge displaced by the strain caused by force F leaks off the crystal, true DC response of a piezoelectric transducer is impossible. G_L is the total leakage conductance of the piezoelectric crystal plus the input conductance of the meter or oscilloscope used to measure V_o, plus the leakage conductance of the cable used to connect the transducer with the measuring system. C_T is the capacitance of the transducer plus the input capacitance of the measuring device. If the transducer is connected to an oscilloscope with a 10 megohm input resistance, we can neglect the other leakage resistances because they are very large ($\geq 10^{13}$ ohms). C_T may be in the order of 1 nF for a piezoceramic transducer. Thus the time constant of the transducer/oscilloscope system is 10 msec, and the response to a step of force input will have decayed to zero in about 30 msec.

To control and improve low-frequency response of a piezoelectric transducer system, we connect the transducer to an op amp with an ultra-high input impedance. Such amplifiers are called *electrometer operational amplifiers*, and they generally have input resistances of 10^{14} to 10^{15} ohms. One common means of conditioning the output of a piezoelectric transducer is to use the charge amplifier configuration using an electrometer op amp (charge amplifiers are introduced in Section 2.6.1 of this text). Figure 6.14 shows one circuit for a charge amplifier. Here we neglect any series resistance between the transducer and the op amp's summing junction. If the op amp is considered to be ideal, it is easy to see that the transducer current, i_x = d(dF/dt), only flows through the amplifier's feedback conductance. Hence, the transfer function of the transducer and amplifier can be written as:

$$\frac{V_o}{F}(s) = \frac{-d/C_F s}{s + 1/C_F R_F} \tag{6.26}$$

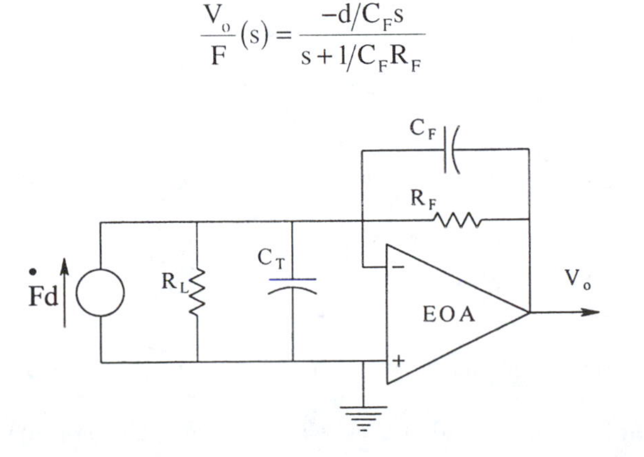

FIGURE 6.14 An electrometer op amp is used as a charge amplifier to condition the output of a piezoelectric force (or pressure) transducer.

We still have a high-pass system with no DC response, but now the effect of C_T and G_T are negligible, and the low frequency pole is controlled with the known feedback elements, R_F and C_F. For example, let R_F be 10^{10} ohms and C_F be 5 nF. The time constant is then 50 s, and the break frequency is 3.18 m Hz. The high-frequency gain is then $-d/C_F$ volts/newton. For a quartz transducer, d is about 2.3×10^{-12} coulombs/newton, thus the mid-band gain would be -4.6×10^{-4} V/N. The use of piezoceramics with larger d's obviously leads to greater sensitivities when used in the charge amplifier configuration. Because of their greater stability and higher Curie temperature, quartz pressure transducers are used in applications such as the measurement of internal combustion engine combustion chamber pressures.

Other applications of piezoelectric input transducers include measurement of minute, dynamic displacements (microphones, phonograph cartridges to translate the grooves in vinyl phonograph records into sound, measurement of minute pulsations on body surfaces, measurement of surface roughness, measurement of acceleration [with an attached mass], etc.).

Piezoelectric transducers used as sonar receivers, or in ultrasonic applications such as medical imaging are generally operated at their mechanical resonant frequencies for maximum sensitivity. At resonance, the electrical equivalent circuit of a transducer changes markedly. Figure 6.15A illustrates the Mason model for a piezoelectric input transducer valid for frequencies near the mechanical resonance frequency, f_o, of the transducer (Fox and Donnelly, 1978). η is the ideal transformer's turns ratio. It can be shown that $\eta = Ae_{33}t$, where A is the effective surface area of the transducer in m^2, e_{33} is the piezoelectric stress constant in the material in coulombs/m^2, and t is the transducer thickness in meters. (Here we assume a transducer responding in the thickness mode.) In addition, the characteristic impedance of the piezoelectric material is given by $Z_x = 2\rho Atf_o$, where ρ is the density of the piezo-material in kg/m^3, and f_o is the transducer's Hz mechanical resonant frequency. Parameter $\alpha = \pi f/f_o$. f is the frequency of the applied mechanical force. When the transducer is driven at exactly its mechanoresonant frequency, its electrical equivalent circuit (Thevenin source impedance) reduces to a series R-C circuit with capacitance C_o, and a real part of $4\eta^2/[(2\pi f_o C_o)^2 (Z_f + Z_b)]$ ohms. Here C_o is the transducer's clamped capacitance, and Z_f and Z_b are the acoustic source impedances faced by the front and back faces of the tranducer, respectively. As can be seen from the tranducer model of Figure 6.15, operation of the transducer even slightly off the resonant frequency leads to a far more complex model for the Thevenin source impedance of the transducer's electrical output. In general, the real part of the Thevenin impedance of the transducer's output is numerically small, typically ranging from the tens of ohms to the hundreds of ohms. Contrast this with the extremely high resistances presented by piezoelectric transducers operated well below their resonant frequencies.

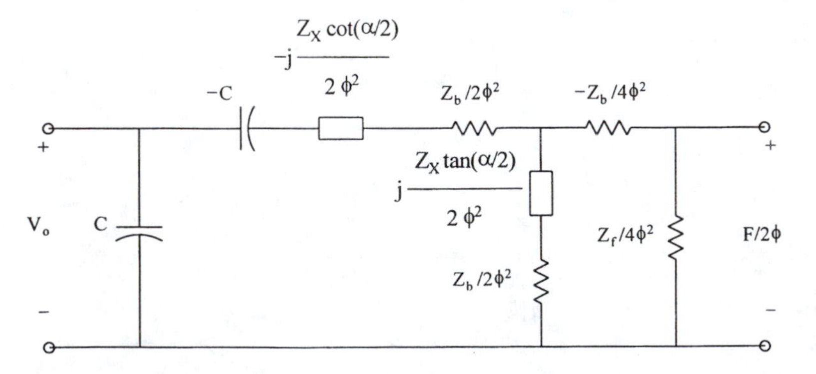

FIGURE 6.15 Electrical equivalent circuit of a piezoelectric transducer near its mechanical resonance frequency.

As a consequence of the low, real source impedance presented by a piezoelectric input transducer operated at its mechanical resonance frequency, a step-up transformer can generally be used to obtain optimum signal-to-noise ratio when measuring threshold signals, as is done in sonar systems. (Transformer optimization of output SNR is discussed in detail in Section 3.5 of this text.)

6.3.4 Sensors Whose Voltage Output is Proportional to dΦ/dt

It is well known that an EMF will be induced in a coil of N turns surrounding a magnetic flux, Φ, when the magnetic flux changes in time. This effect may be simply stated mathematically as

$$E_o = N\, d\Phi/dt \qquad 6.27$$

Many input transducers make use of this principle, including variable reluctance phonograph pickups, dynamic microphones, accelerometers, tachometers, etc. The source of the magnetic flux may be either a permanent magnet, or a DC- or AC-excited electromagnet (solenoid). The dΦ/dt may arise from a number of means, including modulation of the reluctance of the magnetic path by changing the size of an air gap in the path. Moving a coil perpendicualr to the flux density also gives a dΦ/dt, as does moving a permanent magnet in relation to a fixed coil.

6.3.4.1 The Variable Reluctance Phonograph Pickup

Magnetic reluctance, $\Re$, is defined as the ratio of magnetomotive force (MMF) to the magnetic flux, Φ, existing in a closed, magnetic "circuit." It may be thought to be analogous to electrical resistance in an electrical circuit where MMF is analogous to voltage and Φ is analogous to current. A simple variable reluctance phonograph pickup transducer is shown in Figure 6.16. The source of MMF is a permanent magnet (PM). Flux from the PM is split and passed through a right and left branch magnetic path so that the total flux in the magnet is the sum of left and right path fluxes.

$$\Phi = \Phi_R + \Phi_L \qquad 6.28$$

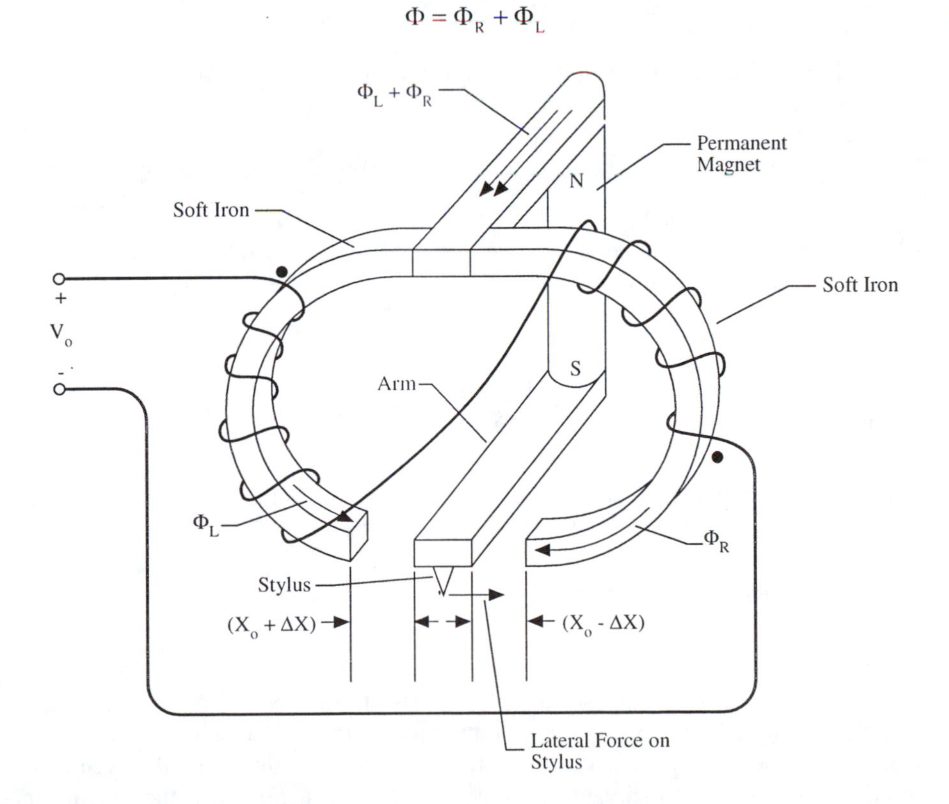

FIGURE 6.16 Schematic of a monophonic, variable-reluctance phonograph pickup.

The reluctances of the paths are modulated by the lengths of the air gaps caused to vary by the small lateral displacement of the movable arm in the air gap, δx. The arm is displaced by a lateral force, F, acting on the stylus, S, protruding downward from the arm. In playing an old-fashioned, non-stereo phonograph record, the stylus tip follows the helical groove of the record. Audio information is recorded on the record as lateral, side-to-side displacements of the groove. Because of its high inertia, the tone arm and cartridge do not respond to the small, audio frequency lateral forces the groove places on the stylus. Instead, the movable (stylus) arm of the cartridge follows the lateral modulations of the groove. The reluctance of the right- and left-hand magnetic paths may be written:

$$\Re_R = \Re_o - \delta x/\mu_o A \tag{6.29}$$

$$\Re_L = \Re_o + \delta x/\mu_o A \tag{6.30}$$

The total magnetic flux passing through the magnet is then

$$\Phi = \text{MMF}/(\Re_R + \Re_L) = \text{MMF}/2\Re_o \quad \text{(constant)} \tag{6.31}$$

An N-turn coil is wound with the dot polarity shown on the left arm of the magnetic circuit; this coil is connected in series with an N-turn coil on the right arm having the opposite dot polarity. The magnetic flux in the left arm is given by

$$\Phi_L = \frac{\text{MMF}/\Re_o}{(1+\delta x/\mu_o A\Re_o)} \cong \frac{\text{MMF}}{\Re_o}(1-\delta x/\mu_o A\Re_o) \tag{6.32}$$

Similarly, the flux in the right arm can be approximated by:

$$\Phi_R \cong \frac{\text{MMF}}{\Re_o}(1+\delta x/\mu_o A\Re_o) \tag{6.33}$$

The net EMF induced in the two series-connected coils is found by taking the time derivative of the flux in each arm, multiplying by the number of turns in each coil, and adding the EMFs algebraically:

$$E_o = \frac{2N(\text{MMF})}{\Re_o^2 \mu_o A}(d\delta x/dt) \tag{6.34}$$

Thus we see that the output EMF is proportional to the stylus velocity. The variable reluctance transducer does not respond to constant (DC) displacements.

6.3.4.2 Electrodynamic Accelerometer

Figure 6.17 illustrates a linear accelerometer made from a moving, cylindrical mass (which is also a permanent magnet), which moves inside a coil fixed to the accelerometer's case. The mass is constrained by a linear spring with stiffness k_s, and is surrounded by oil which gives viscous damping to the motion of the mass. The case of the accelerometer is attached to some structure of which we wish to measure the linear acceleration. The output voltage of the accelerometer is proportional to the relative velocity between the case (and structure) and the moving mass. In mathematical terms,

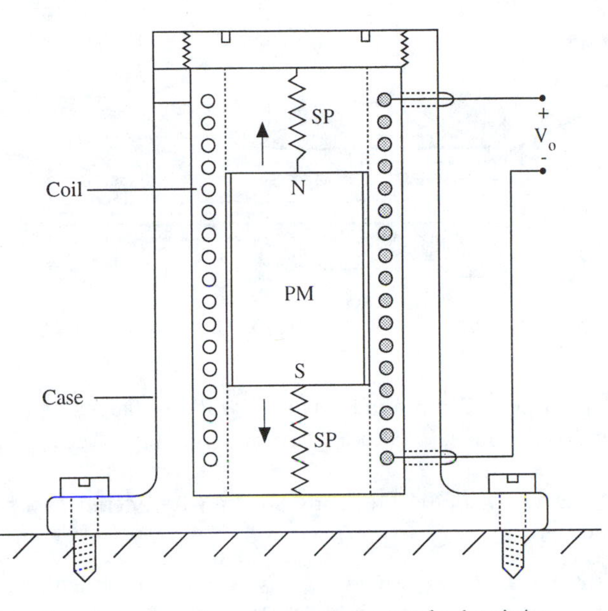

FIGURE 6.17 Cross section of an electrodynamic accelerometer. See text for description.

$$V_o = K_V(\dot{x}_c - \dot{x}_m) \tag{6.35}$$

The motion of the mass can be described by simple Newtonian mechanics:

$$M\ddot{x}_m + B(\dot{x}_m - \dot{x}_c) + k_s(x_m - x_c) = 0 \tag{6.36}$$

The acceleration, $\ddot{x}_c$, is considered to be the system input. If we separate terms of the force equation and take their Laplace transforms, we may write the transfer function:

$$\frac{X_m(s)}{X_c(s)} = \frac{sB + k_s}{s^2M + sB + k_s} = \frac{sB/ks + 1}{s^2M/k_s + sB/k_s + 1} \tag{6.37}$$

Now the relation for $\dot{X}_m(s) = sX_m(s)$ is substituted into the equation for the output voltage. After some algebra this yields:

$$V_o(s) = \ddot{X}_c(s)\frac{sM/k_s}{s^2M/k_s + sB/k_s + 1} \tag{6.38}$$

Thus we see that this transducer's response at low mechanical input frequencies is proportional to jerk, i.e., the rate-of-change of acceleration of the structure to which the case is attached. There is no output in response to constant acceleration, such as due to the Earth's gravitational field.

6.3.4.3 Linear Velocity Sensors

A linear velocity sensor (LVS), as the name suggests, produces an output voltage which is directly proportional to the rate of linear displacement of an object attached to its core. Figure 6.18 illustrates a section through an LVS. Note that the two coils are connected in series-opposing configuration to improve linearity over the full displacement range. Schaevitz Engineering offers LVSs covering ranges of motion from 12.5 to 500 mm. Sensitivities range from 4.8 mV per mm/s to 26 mV/mm/s in various models using Alnico-V magnetic cores. Linearity is ±1% of output over the nominal

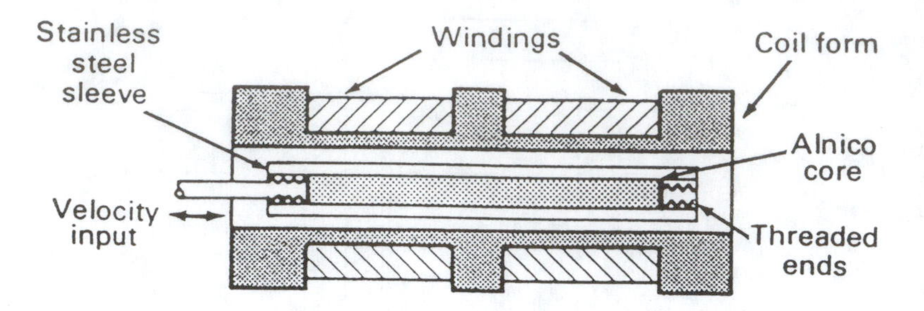

FIGURE 6.18 Cross section through a linear velocity sensor (LVS). (Figure courtesy of Lucas Schaevitz Controls, Hampton, VA.)

linear range. Robinson-Halpern also offers LVSs with working ranges of 12.5 to 225 mm; typical sensitivities are about 20 mV/mm/s. LVS case lengths are roughly twice the linear range of motion.

6.3.5 Sensors Whose Output EMF Depends on the Interaction of a Magnetic Field with Moving Charges

The fluid version of this class of generating sensor is generally used with a constant magnetic (B) field to measure the average velocity of a fluid flowing in an insulating pipe or conduit, including blood vessels. The solid-state version of this class of sensor is the well-known Hall-effect device, used to measure magnetic fields, or to compute average power as a wattmeter.

6.3.5.1 Faraday Effect Flowmeters

Figure 6.19 illustrates the basic geometry of a Faraday flowmeter. Two electrodes on opposite sides of the square cross section pipe make contact with a conductive fluid in the pipe moving with average velocity, $\bar{v}$. A uniform magnetic field of B Gauss is perpendicular to the **v** vector and the **L** vector between the centers of the electrodes. Faraday's law of induction predicts that the EMF between the electrodes is given by

$$E_F = \int_0^d (\mathbf{v} \times \mathbf{B}) \cdot \mathbf{dL} \text{ volts} \tag{6.39}$$

For orthogonal vector components, Equation 6.39 reduces to $E_F = BL\bar{v}$. For effective measurement, the conductivity of the fluid should be greater than 10^{-5} ohm^{-1} cm^{-1}. If a circular pipe of diameter d is used for the conduit, and d is also the separation of the electrodes, then the average volume flow, Q_{av}, of the conducting fluid may be shown to be given by (Lion, 1959):

$$Q_{av} = E_F d\mu / \left(4 \times 10^{-8} B\right) \text{ ml/s} \tag{6.40}$$

Here d is measured in centimeters, B in Gauss, and E_F in volts. Often the area of the conduit is not precisely known, hence flow estimates using the Faraday method are not accurate, since E_F is proportional to average velocity of the fluid.

In the Faraday flowmeter, the magnetic field is generally made to be a low-frequency sinusoid in order that very small E_Fs can be better resolved. An AC E_o can generally be amplified above the amplifier's 1/f noise frequency band.

The Faraday flowmeter is used to measure blood velocity and estimate blood flow. In one version, the electrodes make contact with the surface of the blood vessel, which is conductive itself. Due

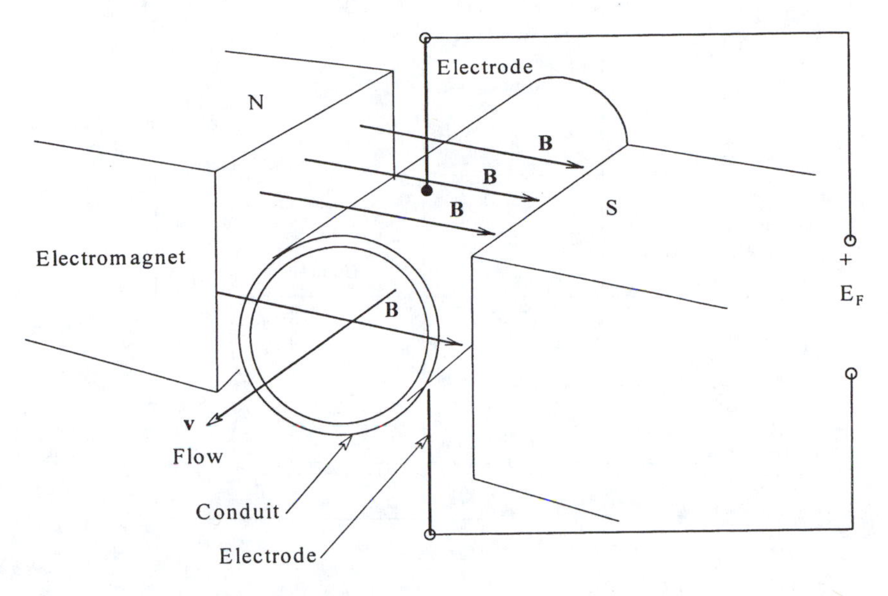

FIGURE 6.19 Schematic diagram of a Faraday-effect liquid velocimeter.

to geometrical errors in setting up the mutual orthogonality between B, $\overline{v}$, and L *in vivo*, a net time-varying magnetic flux passes through the one-turn coil formed by the electrodes used to measure the Faraday EMF. Thus, the total output EMF seen across the electrodes will be the sum of the Faraday EMF and $d\Phi/dt = A(dB/dt)$ volts which cuts the effective loop of area A formed by the E_F measurement circuit.

Webster (1978) describes several methods of cancelling the unwanted $d\Phi/dt$ term. One method makes use of a cancellation circuit formed by making one of the electrodes a twin electrode, as shown schematically in Figure 6.20A. Each circuit will have a transformer error voltage; one of the form $E_{T1} = k_1\, d\Phi/dt$, the other, $E_{T2} = -k_2\, d\Phi/dt$. Thus, for some potentiometer wiper position, the two transformer EMFs will cancel, leaving the pure Faraday voltage. Another means of eliminating the effect of the transformer voltage is based on the fact that the transformer voltage is 90° out of phase (in quadrature) with the Faraday voltage (assuming sinusoidal excitation). Thus, if we use a phase-sensitive, sampling demodulator, we can sample the peak values of the Faraday voltage sinewave at the times the $d\Phi/dt$ sinewave is going through zero. The sampled Faraday voltage sinewave is held and low-pass filtered to obtain a DC signal proportional to v. A third means of supressing the unwanted transformer voltage makes use of a feedback system, shown in Figure 6.21. This system makes use of two phase-sensitive rectifiers and low-pass filters; one detects the desired Faraday voltage, the other, using a quadtrature reference, detects the unwanted transformer voltage. A DC signal, V_T, proportional to the transformer voltage, acts through an analog multiplier to vary the level of a quadrature signal, V_{QF}, which is fed back through an auxillary transformer winding, n_2, to cancel the quadrature component, V_Q, which accompanies V_F. The quadrature detector loop acts to reduce the amount of transformer voltage contaminating the desired V_F signal, as shown in Equation 6.41.

$$V_{QD} = -V_Q\Big/\big[1 + (n_1/n_2)K_1 K_{DQ} K_2(10)\big] \qquad 6.41$$

If an integrator is put between the quadrature phase-sensitive detector/low-pass filter (PSD/LPF) and the analog multiplier, it is easy to show that there will be complete cancellation of V_{DQ}.

Webster (1978) describes several other configurations of electromagnetic blood velocimeters using the Faraday streaming effect.

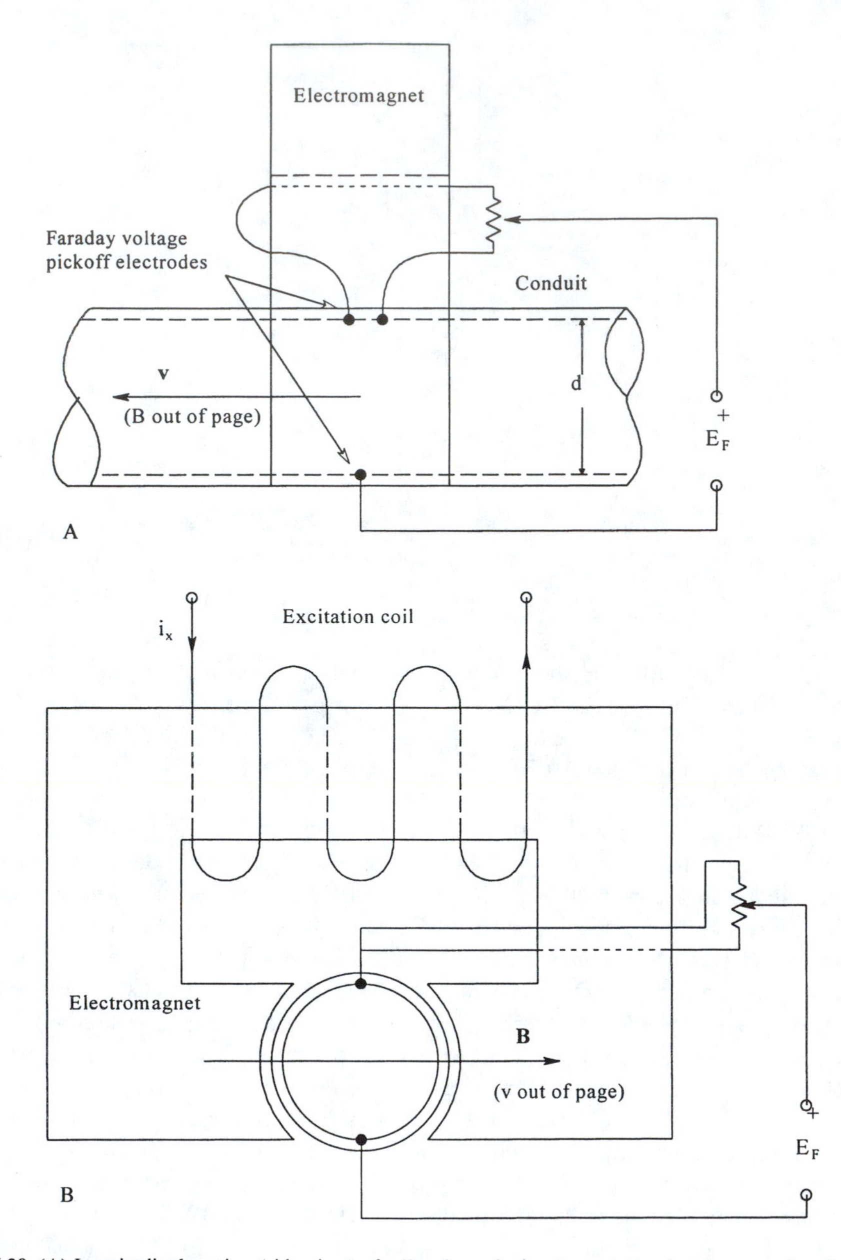

FIGURE 6.20 (A) Longitudinal section (side view) of a Faraday velocimeter system using three sensing electrodes. The upper two electrodes permit cancellation of the unwanted dΦ/dt (quadrature) term in the output voltage. (B) Axial (end-on) view of the three-electrode Faraday system.

6.3.5.2 Hall Effect Sensors

Hall effect sensors, unlike Faraday effect devices, use no moving, conductive fluid. Instead, majority charge carriers (electrons or holes) have some average drift velocity as they traverse a thin bar of doped semiconductor. Figure 6.22 illustrates the geometry of a typical hall sensor. Note that as in the case of the Faraday transducer, the magnetic field B, the drift velocity **v**, and the voltage measuring axis are ideally orthogonal. Each hole or electron with velocity **v** in the semiconductor will, in general, be subject to a *Lorentz force,* given by the vector equation

$$\mathbf{F} = q(\mathbf{v} \times \mathbf{B}) = \{|q||\mathbf{v}||\mathbf{B}|\sin(\theta)\}\mathbf{u} \text{ newtons} \tag{6.42}$$

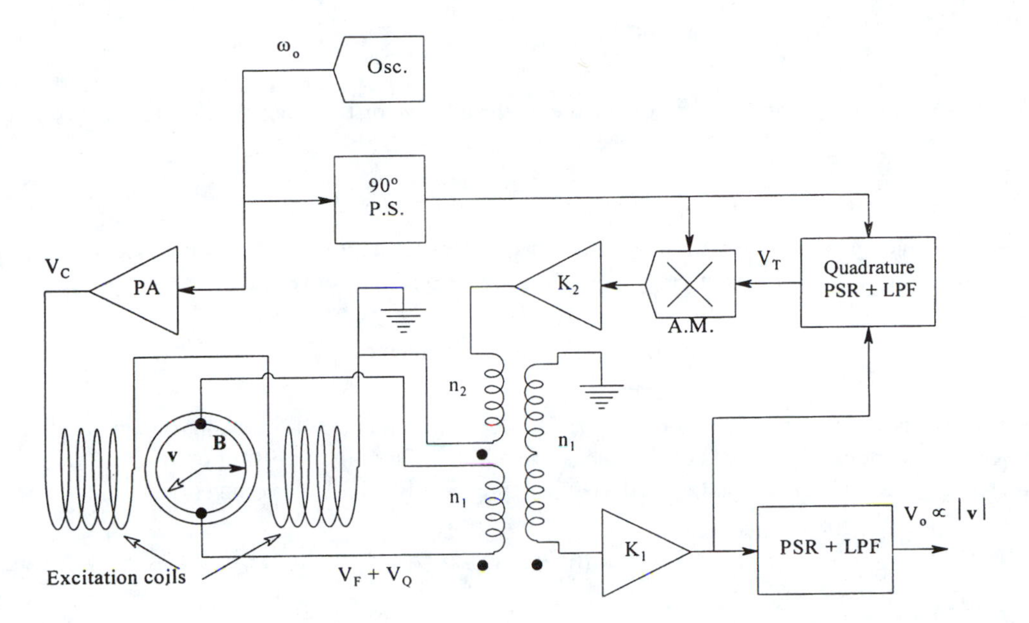

FIGURE 6.21 System to automatically null the unwanted quadrature output voltage from a Faraday liquid velocity sensor. See text for description.

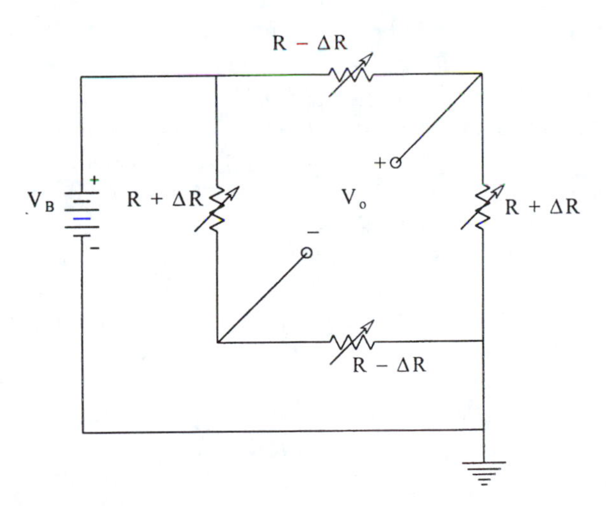

FIGURE 6.22 Schematic of a Hall sensor.

This force follows a right-hand screw rule in which its direction is found by rotating **v** into **B** and observing the direction a screw with right-hand (normal) threads would advance. **u** is a unit vector in this direction, q is the charge of an electron (or hole). If electrons are the carriers, then we use a – sign with q. As we see from Figure 6.22, the Lorentz force causes the moving electrons to crowd toward the front surface of the n semiconductor bar. An equilibrium density of electrons is reached when the average Lorentz force is balanced by the electrostatic force on the electron due to the electric field set up by their crowding. In other words, when

$$-qE_z = -qv_n By \quad \text{or} \quad E_z = v_n By \tag{6.43}$$

Now it is well known that in n semiconductor that the average drift velocity of electrons is given by:

$$v_n = -J_x/qn \tag{6.44}$$

where J_x is the current density in the semicon bar in the +x direction, and n is the electron doping density. This relation for v_n can be substituted in Equation 6.41 above, giving:

$$E_z = -J_x B_y/qn \ \text{volts/meter} \tag{6.45}$$

Now if we consider the geometry of the semicon bar, $J_x = I/2ah$, and the Hall voltage, developed across the two metallized faces of the bar is just $V_H = E_z\, 2a$ volts. Substitution of these relations into the equation above yields the well-known expression for the Hall voltage:

$$V_H = -IB_y/(qnh) = R_H IB_y/h \tag{6.46}$$

where h is the height of the semicon bar (in the y-direction) and R_H is defined as the Hall coefficient; $R_H = -1/qn$ for n semicon and $R_H = +1/qn$ for p semicon. Because the Lorentz force has the opposite sign for p semicon, all other vectors having the same direction, the Hall voltage for p material is $V_H = +IB_y/(qnh)$.

Hall sensors have many uses, most of which can be subdivided into either analog or switching applications. As analog sensors, Hall devices can be used to produce an output proportional to $|\mathbf{B}| \cos(\theta)$, where θ is the angle of the **B** vector with the tranducer's y-axis (axis of maximum sensitivity). Some commercial analog Hall sensors have built-in DC amplifiers to condition the V_H signal. Hall sensors listed by Sprague Electric Co. have sensitivities of about 1.4 mV/Gauss with high-frequency bandwidths of 20 to 25 kHz (on B). Other Hall sensors listed by F. W. Bell do not contain built-in amplifiers, and claim sensitivities ranging from 10 to 55 mV/kG. Bell also offers a Model BH-850 high sensitivity Hall sensor that uses a 9 in., high permeability bar to concentrate the magnetic flux. This sensor claims 18 mV/G, which is adequate sensitivity to measure the Earth's magnetic field and thus has applications as a solid-state compass.

One obvious application of analog Hall sensors is the measurement of magnetic fields. They also can be used to measure other physical or electrical parameters which can be made to be proportional to magnetic fields (AC or DC). A less obvious application of an analog Hall sensor is as an audio frequency wattmeter. The current through a reactive load is passed through a solenoid coil to produce a By proportional to I. The voltage across the load is conditioned and used to generate a current through the Hall sensor proportional to V. The phase between the I and V sinusoids is θ. It is easy to show that the DC component of the Hall voltage is proportional to the average power in the load; that is, it is given by

$$\overline{V_H} = kP_{AVE} = kVI\cos(\theta) \tag{6.47}$$

Here k is a proportionality constant, V is the RMS voltage across the load, I is the RMS current through the load, and $\cos(\theta)$ is the power factor.

Switching Hall sensors have many applications in counting and proximity (of a permanent magnet) detection. Their output is generally a TTL logic signal. An IC DC amplifier, Schmitt trigger, and open-collector logic output are mounted in the same package with the Hall chip. Hall switches have uses in automobile electronic ignitions, brushless DC motors (rotor position sensing), tachometers, keyboards, thermostats, pressure and temperature alarms, burglar alarms, etc.

6.4 SENSORS BASED ON VARIABLE MAGNETIC COUPLING

This class of sensors operates with AC excitation. Their outputs vary according to the degree of magnetic coupling between the excitation coil(s) and the output winding(s). Most sensors which operate on the principle of variable magnetic coupling between coils are mechano-transducers, i.e., they can be used to measure or generate linear or rotational displacements. Because many other physical phenomena can be converted to small linear rotations or displacements, this class of transducer also is the basis for certain pressure, temperature, and acceleration sensors. Lion's Section 1-22 (1959) gives a good summary of this class of transducer, which includes, among others, the linear variable differential transformer (LVDT) and the synchro transformer.

6.4.1 The LVDT

The *linear variable differential transformer* is one of the most widely used mechanical input sensors in modern instrumentation practice. LVDTs are used to measure linear mechanical displacement or position in control systems and in precision manufacturing gauging, and can be used indirectly to measure force, pressure, acceleration, etc., or any quantity that can be made to cause a small, linear displacement. Figure 6.23A illustrates a section through an LVDT showing the central excitation coil, symmetrically located pick-off coils, and the hollow, cylindrical core tube through which slides the high permeability magnetic core. A schematic representation of an LVDT is shown in Figure 6.23B. Note that the pick-off coils have opposite dot polarities, so that when the movable magnetic core is at magnetic center ($x = 0$), the induced EMFs in each coil sum to zero at the output. The AC excitation of an LVDT can range in frequency from about 50 Hz to 20 kHz. Peak output response for a given displacement, $\delta x > 0$, is maximum at some frequency which may range from 400 Hz to over 2 kHz. Some LVDTs contain built-in oscillators and phase-sensitive rectifiers so one has only to supply a DC power source (e.g., 24 V), and observe a DC output which has a voltage/length relationship similar to that shown in Figure 6.24. Several electronic IC manufacturers offer oscillator/phase-sensitive rectifier chips (e.g., Signetics' NE5520) to facilitate the use of LVDTs which do not contain this signal processing circuit internally.

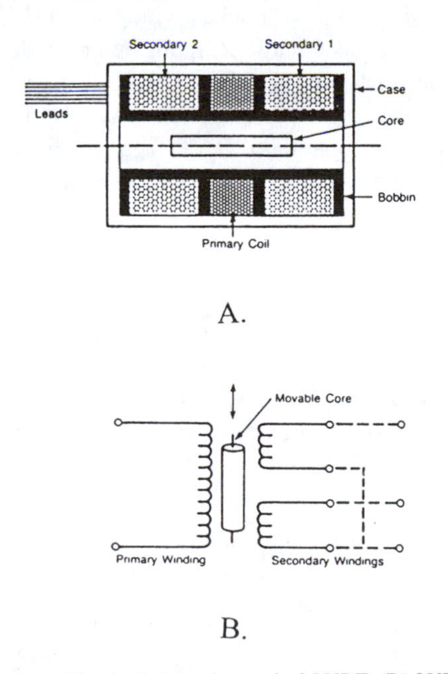

FIGURE 6.23 (A) Longitudinal section (side view) through a typical LVDT. (B) LVDT schematic.

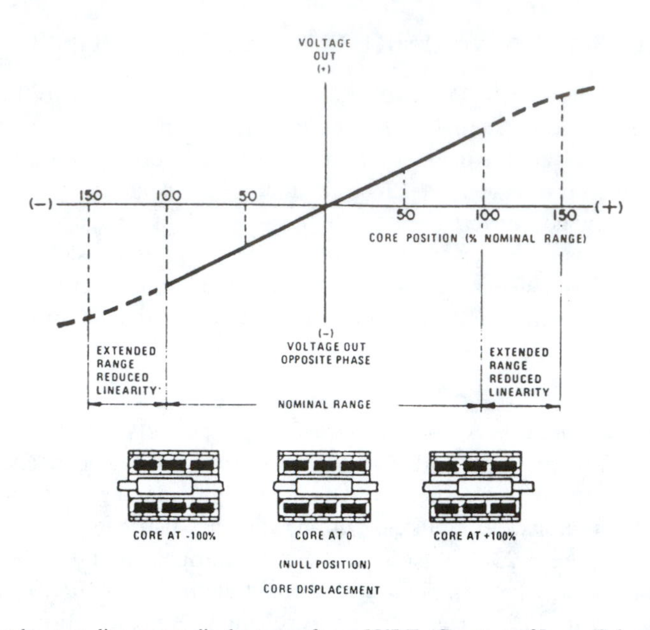

FIGURE 6.24 Output voltage vs. linear core displacement for an LVDT. (Courtesy of Lucas/Schaevitz Controls, Hampton, VA.)

LVDTs are not zero-force sensors. If their cores experience a lateral force, there will be increased friction with the inside of the core tube. In addition to friction forces, the solenoidal magnetic field from the excitation coil acts to pull the core toward the center ($x = 0$) position. The solenoid or axial force is proportional to the square of the excitation current. Typical peak axial force is less than 1 g, which is negligible in most applications.

Specifications for LVDTs vary widely according to application. Those used for gauging applications (e.g., measuring the contour of a cam) have infinite resolution, linearity of ±0.2% of full range and repeatability as small as 4 microinches. Gauge LVDTs generally have some sort of special tip to contact the work, such as a hardened, tungsten carbide ball, and also a helical spring to maintain constant static contact force with the work. Other versions of LVDT gauges use air pressure to obtain a variable gauge contact force which can be set for delicate (soft) materials. The Schaevitz PCA-230 series precision, pneumatic, LVDT gauges have gauging ranges from ±0.005" to ±0.100". Other models have ranges up to ±1.000", and some have built-in oscillator/demodulator ICs for DC operation.

6.4.2 Synchros and Resolvers

We first consider synchros, also called selsyns or autosyns. These are angular position sensors that work on the principle of variable mutual inductance. They generally have multiple windings. There are three main classes of synchros; transmitters, repeaters and control transformers. Figure 6.25A is a schematic diagram of a synchro. Rotor excitation is generally at 60 or 400 Hz. The three stator windings are arranged at 120° spacings, so the induced stator voltages may be written:

$$v_1(t) = V_{pk} \sin(2\pi f t)\sin(\theta) \quad 6.48A$$

$$v_2(t) = V_{pk} \sin(2\pi f t)\sin(180° + \theta) \quad 6.48B$$

$$v_3(t) = V_{pk} \sin(2\pi f t)\sin(240° + \theta) \quad 6.48C$$

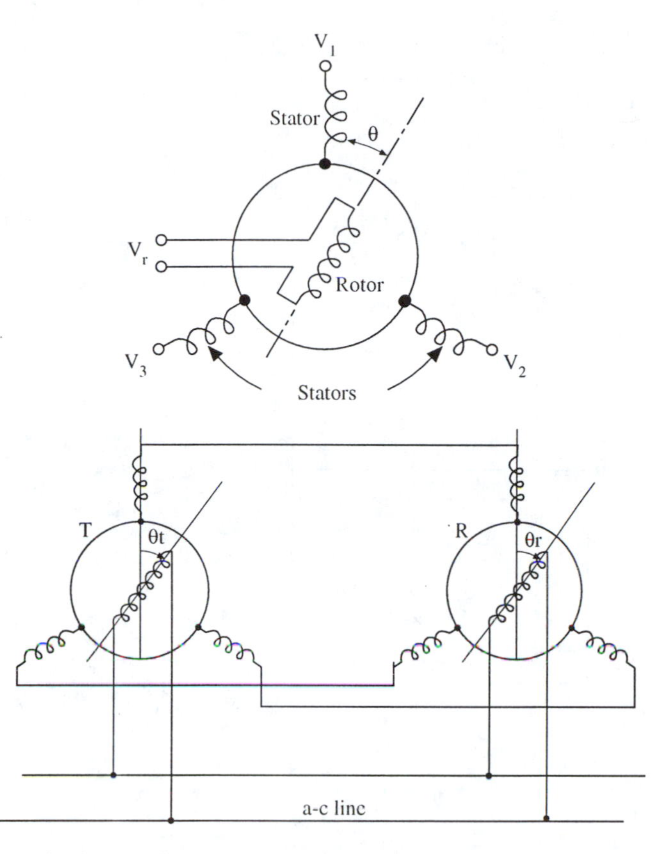

FIGURE 6.25 (A) Schematic diagram of a synchro transmitter. (B) Connection between a synchro transmitter (T) and repeater (R) enables the repeater rotor to follow the angle of the transmitter's rotor.

Synchros were first widely used in the 1940s in military systems as angular position sensors and components in servomechanisms used to train guns or gun turrets, sonar arrays, or radar antennas. Synchros come in all sizes, ranging from the massive Navy 7G developed in WWII, to modern, miniature synchros measuring 1" in diameter by 1.8" in length. Synchro accuracies range from ±0.1° to about ±1.0°.

The simplest application of two synchros is signaling a shaft rotation to a remote location. In this application, two synchros are used, a transmitter and a repeater, shown in Figure 6.25B. The rotors of the transmitter and repeater are connected in parallel to the AC line, and corresponding stator lines are connected. The rotor of the transmitter is at some angle θ_t. The rotor of the repeater is initially at some angle θ_r, and is free to rotate. The repeater rotor experiences a torque causing it to rotate so that $\theta_r = \theta_t$. The repeater rotor experiences zero torque when $\theta_r = \theta_t$. The repeater rotor generally has a light external inertial load, such as a pointer, and some viscous damping to prevent overshoots and oscillations when following sudden changes in the transmitter's θ_t.

Another major application of synchros is error signal generation in carrier-operated, angular position control systems. See Figure 6.26. A synchro transmitter and a synchro control transformer are used to generate a carrier frequency signal whose amplitude is proportional to $\sin(\theta_t - \theta_c)$, where θ_t is the rotor angle of the transmitter, and θ_c is the rotor angle of the control transformer. If we consider θ_t to be the input, reference, or command signal, and θ_c to be the output of a servomechanism designed to follow θ_t, then the output of the synchro transformer may be considered to be the servo system error for $\theta_e = \theta_t - \theta_c < 15°$, where $\sin(x) \approx x$ in radians. The synchro transformer output can be shown to be a double-sideband, suppressed-carrier modulated signal. When demodulated by a phase-sensitive rectifier, the output will be a voltage proportional to $\sin(\theta_t$

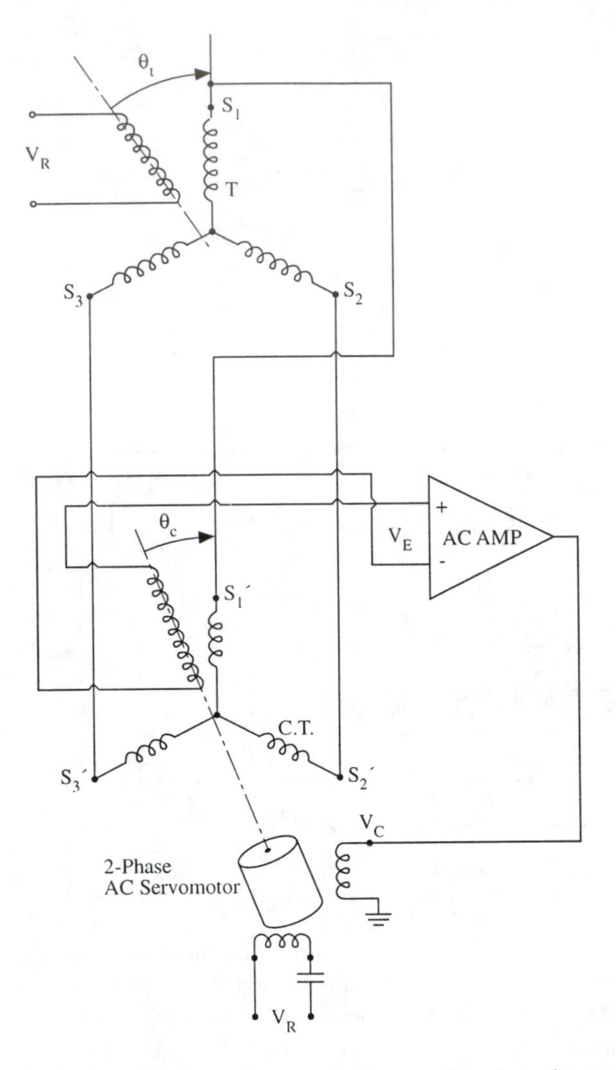

FIGURE 6.26 Schematic of a simple position control system which uses synchros. A synchro transmitter (T) and a synchro control transformer (CT) are used.

$- \theta_c$). Obviously, the synchro transmitter and tranformer can be located at some distance from one another, a distinct advantage.

Modern practice, however, is to replace the electromechanical synchro transformer with a commercially available, solid-state circuit system into which the input angle is read as a binary number, and the output of which is the same as would be seen from the rotor of a conventional synchro transformer. In spite of such solid-state conveniences, synchro systems are difficult to work with. Modern position control system design generally makes use of optical coding disks to sense angular position and directly code it into digital form for transmission to the control computer.

Resolvers are another rotational coordinate sensor. As shown schematically in Figure 6.27, resolvers differ from synchros in that they have two orthogonal stator windings, and two orthogonal rotor windings. The stator windings are generally energized so that:

$$v_{s1}(t) = V_{s1} \sin(\omega t) \qquad 6.49A$$

$$v_{s2}(t) = V_{s2} \sin(\omega t) \qquad 6.49B$$

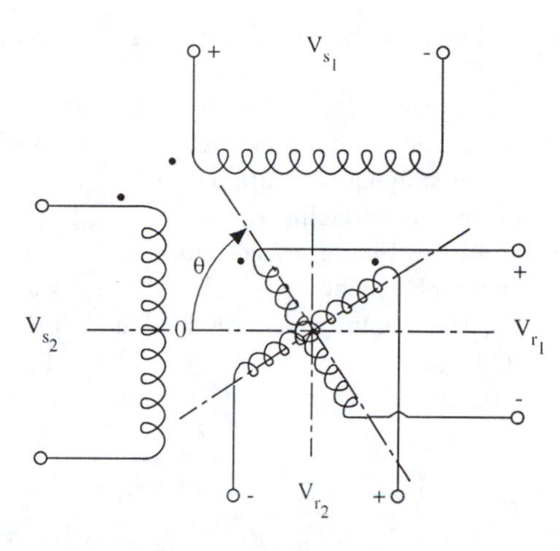

FIGURE 6.27 Schematic of a resolver.

Noting the dot convention shown in Figure 6.27, the induced output (rotor) voltages are:

$$v_{r1}(t) = v_{s1}(t)\cos(\theta) - v_{s2}(t)\sin(\theta) \tag{6.50A}$$

$$v_{r2}(t) = -v_{s1}(t)\sin(\theta) + v_{s2}(t)\cos(\theta) \tag{6.50B}$$

where θ is the rotor angle with respect to the reference shown. If just S1 is excited, the resolver may be used to do polar to rectangular coordinate transformation in an electromechanical system. In modern instrumentation systems, such transformations are better done by software in a computer.

6.5 VARIABLE CAPACITANCE SENSORS

Many mechanical input sensors, and some sensors which measure humidity or temperature, operate by the transduced quantity causing a change in capacitance, which in turn, is converted to an analog output voltage by an AC bridge circuit or other electronic system. Capacitive sensors have been designed to measure force by means of displacement of one or two capacitive electrodes. They also can be used to measure acceleration, thickness, depth of a dielectric liquid, and pressure. In its simplest form, the mechanical input quantity causes a change in the separation between two (or more) capacitor plates, as described below.

The capacitance between two, parallel, conducting plates, each of area A, separated by a dielectric of thickness d is given by (neglecting fringing effects):

$$C = \frac{\kappa\varepsilon_o A}{d} \tag{6.51}$$

where κ is the dielectric constant of the material separating the plates and ε_o is the permittivity of free space (8.85×10^{-12} farads/meter). Small variations in C, ΔC, which may be measured by several means, may occur if the plate separation is caused to change by some physical quantity, or if the dielectric constant, κ, changes due to pressure, temperature, mechanical strain, humidity, etc. The over-all dielectric constant may also be modulated by sliding a substance with a dielectric constant greater than one between the capacitor plates so that some fraction, α, of A has a dielectric constant of $\kappa = \kappa_1$, and the remaining area between the plates, $(1 - \alpha)$ A, has a dielectric constant $\kappa = 1$. The area of the capacitor plates may be effectively changed by sliding one plate over the other at constant d. An example of this latter means of effecting a ΔC is found in parallel-plate, rotary tuning capacitors used in radios.

Means of measurement of capacitance changes include various capacitance bridges or use of the capacitor in a tuned L-C circuit determining the frequency of an oscillator; the frequency of the oscillator is measured to obtain a signal proportional to $1/\sqrt{LC}$.

One of the most sensitive means of producing an output voltage proportional to ëC is through the use of a DeSauty bridge, shown in Figure 5.7 and discussed in Section 5.4.1.4 of this text. A capacitive pressure transducer in which a diaphragm deflects as the result of the pressure differential across it is shown in Figure 6.28. The diaphragm is grounded, and thus forms a common plate with two capacitors such that $C_1 = C_o - \Delta C$, $C_2 = C_o + \Delta C$, and $\Delta C = k(P_1 - P_2)$. If a De Sauty bridge is operated at a frequency $fo = 1/2\pi RC_o$ Hz, with peak excitation voltage V_s, it is easy to show that

$$\frac{V_o}{V_s}(s) = \frac{1/s(C_o - \Delta C)}{1/s(C_o - \Delta C) + R} - \frac{1/s(C_o + \Delta C)}{1/s(C_o - \Delta C) + R} \qquad 6.52$$

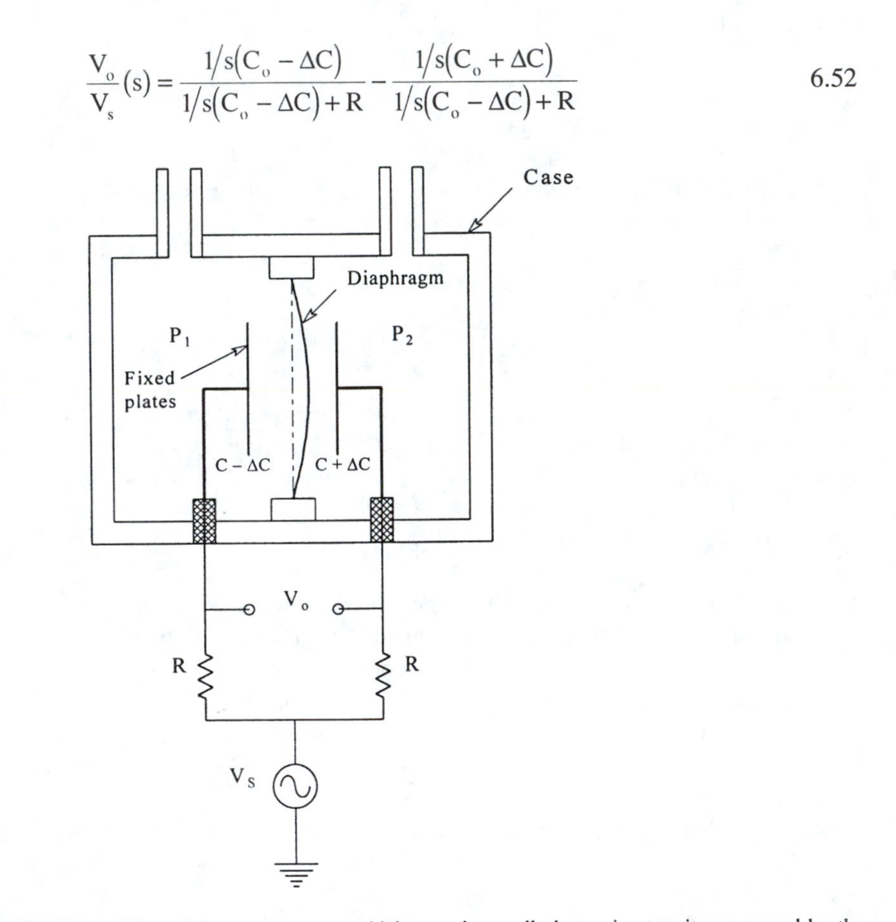

FIGURE 6.28 Section through a differential pressure sensor which uses the small change in capacitance caused by the deflection of its diaphragm to produce an AC output signal proportional to $(P_1 - P_2)$. A DeSauty bridge is used.

After some algebra, we can write

$$\frac{V_o}{V_s}(s) = \frac{s2R\Delta C}{(1 + sRC_o)^2 - s^2R^2C^2(\Delta C/C_o)^2} \qquad 6.53$$

Letting $s = j\omega = j/RC_o$, and noting that $(\Delta C/C_o)^2 << 1$, we obtain finally

$$\frac{V_o}{V_s}(j\omega) = \frac{\Delta C}{C_o} \tag{6.54}$$

$$v_o(t) = V_S\left(\frac{\Delta C}{C_o}\right)\sin(t/RC_o) \tag{6.55}$$

The bridge output, V_o, in this example is a double-sideband, suppressed-carrier signal whose amplitude is $V_o = V_s(\Delta C/C_o)$. Since it is possible to resolve a V_o as small as 1 μV, when V_s is typically 5 V pk, and C_0 is 100 pF, ΔC can be as small as 2×10^{-4} pF!

Pressure measurements can also be made using a capacitor with fixed geometry but with a dielectric constant which changes with pressure. Certain nonpolar liquids, gases, and piezoelectric materials can be used for this purpose. Lion (1959) reports that the dielectric constant of air at 19°C varies from 1.0006 at 1 atm to 1.0548 at 100 atm. Unfortunately, dielectric constants also change with temperature, so a null transducer can be used in a bridge circuit to obtain temperature compensation.

Dranetz, Howatt, and Crownover, cited by Lion, developed a titanate ceramic called Thermacon, in which the dielectric constant decreased linearly over the temperature range of –40 to +160°C. One specimen was described which had a dielectric constant of 490 at room temperature, and a dielectric constant tempco of –0.003/°C. However, the tempco increases dramatically as the Curie temperature is approached.

6.6 FIBER OPTIC SENSORS

Fiber optic (FO) dielectric waveguides were developed primarily for broad-band, long distance communications links. Their main advantages are low cost (silicon vs. copper), broad signal bandwidths, and immunity from interference caused by electromagnetic radiation, such as radio waves, pulses (EMPs) generated by lightning, or, perish the thought, nuclear explosions. They also may be used in harsh environments remote from their electrooptic signal conditioning systems.

Interestingly, fiber optic cables are also useful in a variety of sensor applications (Wolfbeis, 1991). They have been used in the measurement of electric current, magnetic fields, temperature, force, pressure, strain, acceleration, and pH. Several of their physical/optical properties have been used in realizing these applications. For example, (micro)bending of an optical fiber will result in the loss of light energy from the core into the cladding at the bend, decreasing the output light intensity. This phenomenon is used in sensing force, pressure, strain, etc. (Hochberg, 1986). pH may be sensed by measuring the emission of a fluorescent dye coating the end of the fiber; the emission characteristic of the dye changes with pH. pH has also been measured spectrophotometrically with fiber optics; a pH-sensitive indicator dye such as phenol red is immobilized inside a cuprophan membrane at the end of a pair of fiber optic cables. Light at two wavelengths is transmitted to the dye. Reflectance at one wavelength is pH-dependent, reflectance at the other wavelength is independent of pH and is used as a reference. Reflected light at both wavelengths is collected by the output fiber optic cable, and processed. System output is linear with pH over a limited range (Peterson and Goldstein, 1982).

Another effect that is used in fiber optics-based sensors is the modulation of the optical polarization properties of the fiber by mechanical strain, temperature changes, or magnetic fields. Fiber optic cables are also used in Sagnac-effect, fiber optic laser gyroscopes. Their use in fiber optic gyros is expeditious, however, and it should be noted that the Sagnac effect will occur whether the light travels in a closed path in air directed by mirrors, or in a multi-turn fiber optic coil. Sagnac-effect gyros are discussed in detail in Section 7.1.2.

6.6.1 Magneto-Optic Current Sensors

One of the major problems in measuring the electrical current in high voltage, high power transmission lines is that of measurement circuit isolation. One means of measuring heavy 60 Hz currents in a conductor is through the use of a current transformer. This transformer consists of a number of turns of wire wound on a high-permeability, toroidal magnetic core. The current carrying conductor is passed through the "hole" in the toroid, and its magnetic field induces a voltage in the toroid's winding proportional to the current in the conductor. This induced voltage is measured with an AC millivoltmeter. This means of measuring heavy AC currents works fine at lower voltages on the conductor. However, when the voltage on the conductor reaches values too high to safely isolate the toroid's winding (and meter circuit) from the conductor, some other means must be used to insure safety.

The *Faraday magneto-optic effect* offers one means to measure high currents on conductors at extremely high voltages above ground (50 kV and higher) with reasonable accuracy and excellent isolation. Magneto-optic current sensors (MOCSs) which make use of the Faraday magneto-optic effect, can use either solid glass pathways or fiber optic waveguides. When linearly polarized light is passed through a transparent, diamagnetic material through which a magnetic field is also passed in the same direction, there will be, in general, a rotation of the polarization vector of the emergent ray. It may be shown that the polarization rotation angle is given by:

$$\alpha = V\oint \mathbf{H} \cdot \mathbf{dl} \qquad 6.56$$

The rotation, α, is proportional to the line integral of the magnetic field intensity vector, H, along the light propagation path, l, which encloses the magnetic field. V is the Verdet constant for the material used. Note that by the Ampere circuital law, $\oint \mathbf{H} \cdot \mathbf{dl} = I$, where I is the current enclosed by one optical path loop. Hence, if N loops of a fiber optic cable are wound on a form so that all turns are aligned and have the same area, we may express the Faraday magneto-optical polarization rotation as:

$$\alpha = VNI \qquad 6.57$$

Figure 6.29 illustrates one version of a Faraday current transducer consisting of N turns of single-mode optical fiber wound around a current-carrying conductor. The optical rotation of the linearly polarized light entering the fiber is given by Equation 6.57. There are several means of sensing small changes of optical rotation which are discussed below. The optical rotation, in theory, will remain unchanged regardless of the position of the current-carrying conductor in the fiber optic coil. Also, by using a diamagnetic material such as glass fiber, the system has substantial temperature independence (Rogers, 1973).

Another version of the Faraday MOCS developed by Cease and Johnston (1990) is shown in Figure 6.30. Here, a single block of double extra dense flint (DEDF) glass is used. It acts both as the Faraday medium and as an electrical insulator for the bus bar. For this geometry, the optical rotation angle is given by:

$$\alpha = V'BL \qquad 6.58$$

where V′ is the Verdet constant, B is the magnetic flux density in Teslas, and L is the optical path length over which the Faraday effect occurs. Möller (1988) gives a table of Verdet constants. From this table we see that V′ for ethanol at 25°C is 11.12×10^{-3}, and V′ for dense lead glass at 16°C is 77.9×10^{-3} minutes of arc/gauss.cm.

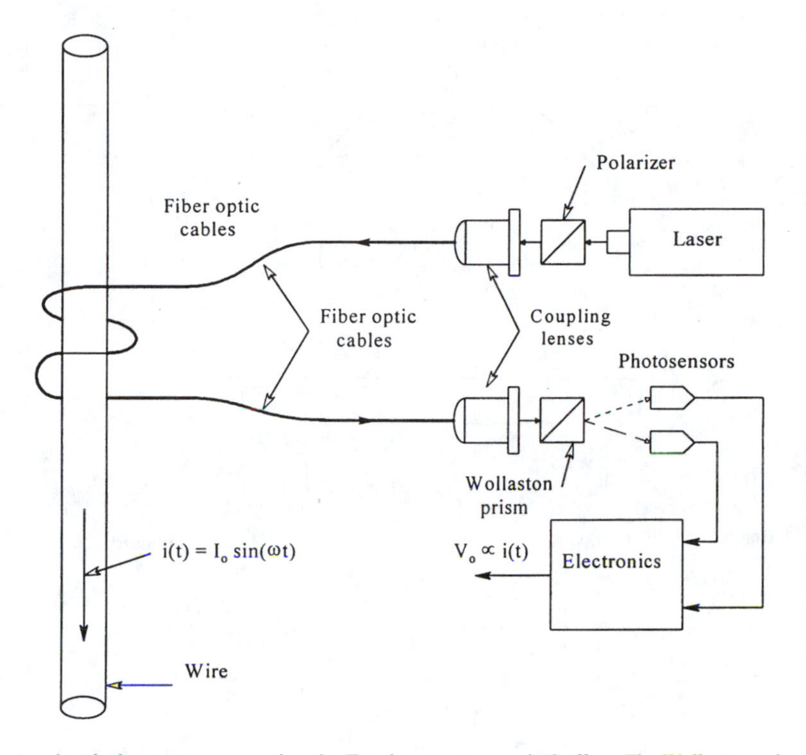

FIGURE 6.29 An electrical current sensor using the Faraday magneto-optical effect. The Wollaston prism separates the magneto-optically rotated E vector of the emergent light into two orthogonal components. The two orthogonal components are detected by photodiodes, and their signals are processed by a radio detector. See the text for analysis of the operation of this current sensor.

The magnetic field around a current-carrying conductor at a distance R, where R >> r, and r is the conductor's radius, may be shown by the Biot-Savart law to be given by:

$$B = \mu_o I/2\pi R \text{ webers/m}^2 \text{ (in MKS units)} \quad 6.59$$

B in gauss equals 10^4 w/m². B is everywhere tangential to a circle of radius R centered on the conductor. Such a magnetic field is called solenoidal, and if the light path consists of N optical fibers wound on a circular form of radius R, concentric with the conductor, then the Faraday rotation is given by

$$\alpha = V'(\mu_o I/2\pi R)(2\pi RN)(10^6) = V'\mu_o IN(10^6) \text{ minutes of arc} \quad 6.60$$

For example, a MOCS has 100 turns of a lead glass fiber with a Verdet constant $V' = 80 \times 10^{-3}$ minutes arc/(gauss cm), a current of 1 A will produce a B field which gives a Faraday rotation of 10.055 minutes, or 0.1676°/amp. The constant, 10^6, converts the Verdet constant, V′, in min/(gauss cm) to the mks V in units min/(weber/m² m); $\mu_o = 1.257 \times 10^{-6}$ webers/(amp meter).

As you will see in the next section, it is entirely possible to measure small angles of optical rotation to better than ±0.001°. This polarimeter resolution sets the ultimate resolution of a fiber optic MOCS.

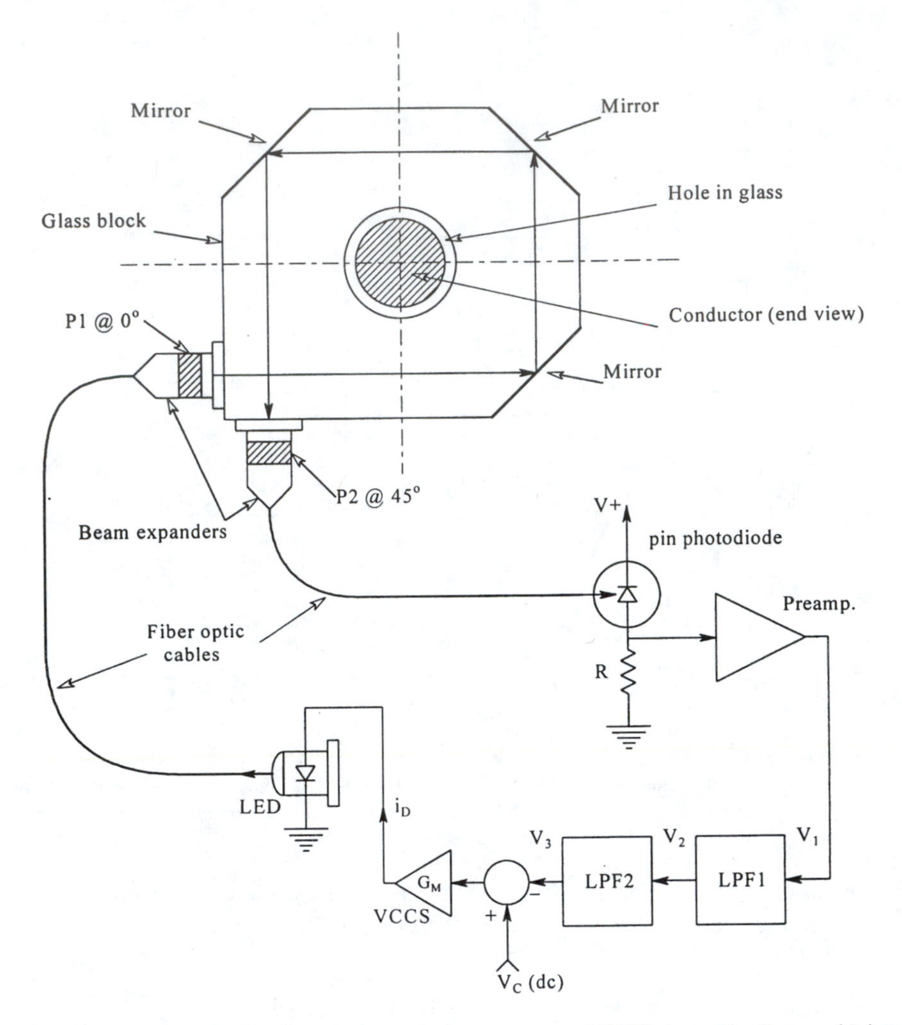

FIGURE 6.30 Diagram of another Faraday magneto-optical current sensor (MOCS) devised by Cease and Johnston, 1990.

6.6.2 Means of Measuring the Optical Rotation of the Linearly Polarized Light Output of Certain Optical Sensors

In order to consider the measurement of the rotation angle of linearly polarized light, we must first review the mathematical description of light as a propagating transverse electromagnetic wave in a dielectric medium. As is well known, the propagation of light can be described by Maxwell's equations. In the simplest case, we will assume a plane wave propagating in the z-direction in a source-free, isotropic medium with permittivity ε and permeability μ. From the Maxwell equations, we have

$$\nabla \times +\mathbf{E} = -dB/dt \qquad 6.61A$$

$$\nabla \times \mathbf{B} = (\mu\varepsilon)\, dE/dt \qquad 6.61B$$

$$\nabla \cdot \mathbf{E} = 0 \qquad 6.61C$$

$$\nabla \times \mathbf{B} = 0 \qquad 6.61D$$

The second-order wave equation may be derived from the Maxwell equations, 6.61:

$$\nabla^2\mathbf{E} - (\mu\varepsilon)d^2E/dt^2 = 0 \qquad 6.62$$

Assuming sinusoidal steady-state conditions, we can write

$$\nabla^2\mathbf{E} + \mathbf{k}^2\mathbf{E} = 0 \qquad 6.63$$

where k is the wavenumber defined as

$$k = 2\pi f/c = 2\pi/\lambda = 2\pi f(\mu\varepsilon)^{1/2} = \omega/c \qquad 6.64$$

Note that in an anisotropic medium, k can be expressed as a vector:

$$\mathbf{k} = \mathbf{a}_x k_x + \mathbf{a}_y k_y + \mathbf{a}_z k_z \qquad 6.65$$

Here we use $\mathbf{a}_x$ as a unit vector pointing along the positive x-axis, and k_x is the wavenumber in the x direction, etc. The anisotropy may be considered to be due to different propagation velocities in the x, y, and z directions (e.g., $c_x = 1/(\varepsilon_x\mu_x)^{1/2}$).

We note that the **B** and **E** field vectors are mutually perpendicular to each other, and to the direction of wave propagation, which we will take as the positive z axis.

The polarization of an electromagnetic wave is defined in accordance with the IEEE standard (Balanis, 1989). The polarization of a radiated electromagnetic wave is defined as "that property of a radiated electromagnetic wave describing the time-varying direction and relative magnitude of the electric field vector; specifically, the figure traced as a function of time by the extremity of the vector at a fixed location in space, and the sense in which it is traced, as observed along the direction of propagation." There are three categories of electromagnetic plane wave polarization: *linear, circular, and elliptical.* If the **E** vector at a point in space as a function of time is always directed along a line which is normal to the direction of wave propagation, the field is said to be *linearly polarized.* To illustrate this property mathematically, we assume that the **E** vector has x and y components and may be written:

$$\begin{aligned}\mathbf{E} = \mathbf{a}_x Ex + \mathbf{a}_y Ey &= \mathbf{Re}\left[\mathbf{a}_x\mathbf{E}_x^+ \exp(j\{\omega t - kz\}) + \mathbf{a}_y\mathbf{E}_y^+ \exp(j\{\omega t - kz\})\right] \\ &= \mathbf{a}_x E_{xr}\cos(\omega t - kz + \phi_x) + \mathbf{a}_y E_{yr}\cos(\omega - kz + \phi y)\end{aligned} \qquad 6.66$$

where $\mathbf{E}_\mathbf{x}^+$ and $\mathbf{E}_\mathbf{y}^+$ are complex, and E_{rx} and E_{yr} are real. Let us consider two cases of linear polarization: first, let $E_{yr} \equiv 0$, and $z = 0$ for convenience. This means that the locus of the instantaneous E field vector is a straight line which is directed along the x axis. The E field is given by:

$$\mathbf{E} = \mathbf{a}_x E_{xr}\cos(\omega t + \phi x) \qquad 6.67$$

In a more general case of linear polarization, we let $\phi_x = \phi_y = \phi$, $z = 0$, and we find that the **E** vector is given by:

$$\mathbf{E} = \left[\sqrt{(E_{xr})^2 + (E_{yr})^2}\right]\cos(\omega t + \phi) \qquad 6.68$$

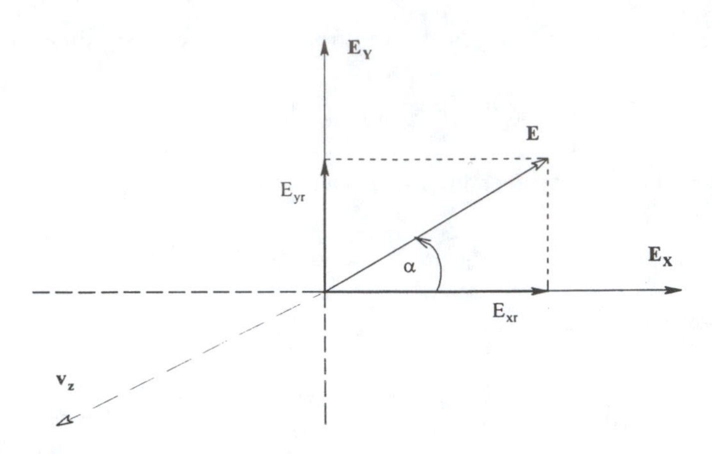

FIGURE 6.31 E vector of light linearly polarized along the α-direction.

This gives a straight line at an angle α with the x axis for all time, as shown in Figure 6.31. The polarization angle α can be written:

$$\alpha = \tan^{-1}\left(E_{yr}/E_{xr}\right) \tag{6.69}$$

Circular polarization is said to occur when the tip of the **E** vector traces a circular locus in the X-Y plane when viewed by an observer standing at the origin looking in the positive z direction. (The electromagnetic wave is assumed to propagate in the +z direction.) Circular polarization can be either clockwise (right-handed) or counterclockwise (left-handed), as viewed looking from the rear of the electromagnetic wave in the +z direction. It may be shown that circular polarization will occur when the **E** field has two, orthogonal, linearly polarized components with equal magnitudes, and the two orthogonal components must have a time-phase difference of odd multiples of 90°. As an example of circular polarization, let us assume in Equation 6.66 that $\phi_x = 0$, $\phi_y = -\pi/2$, $z = 0$, and $E_{xr} = E_{yr} = E_{CW}$. Thus:

$$\mathbf{E}_x = \mathbf{a}_x E_{CW} \cos(\omega t) \tag{6.70}$$

$$\mathbf{E}_y = \mathbf{a}_y E_{CW} \cos(\omega t - \pi 2) = \mathbf{a}_y E_{CW} \sin(\omega t) \tag{6.71}$$

The locus of the amplitude of the E vector in the X-Y plane is just

$$|\mathbf{E}| = \sqrt{E_x^2 + E_y^2} = \sqrt{E_{CW}^2\left[\cos^2(\omega t) + \sin^2(\omega t)\right]} = E_{CW} \tag{6.72}$$

At any instant of time, it is directed along a line making an angle α with the x axis;

$$\alpha = \tan^{-1}\left(\frac{E_{CW}\sin(\omega t)}{E_{CW}\cos(\omega t)}\right) = \tan^{-1}\left[\tan(\omega t)\right] = (\omega t) \tag{6.73}$$

Thus, the angle α rotates clockwise with angular velocity of ω. The right-hand, circularly polarized **E** vector can also be written (Balanis, 1989):

$$\mathbf{E} = E_{CW}\mathbf{Re}\{(\mathbf{a}_x - j\mathbf{a}_y)\exp[j(\omega t - kz)]\} \tag{6.74}$$

Using the development above, it is easy to demonstrate that a counterclockwise (left-hand) rotating circular polarization vector will occur when: $\phi_x = 0$, $\phi_y = \pi/2$, and $E_{xr} = E_{yr} = E_{CCW}$. In this case, $E = E_{CCW}$, and $\alpha = -\omega t$. The **E** vector may be written

$$\mathbf{E} = E_{CCW}\mathbf{Re}\{(\mathbf{a}_x + j\mathbf{a}_y)\exp[j(\omega t - kz)]\} \tag{6.75}$$

It may be shown that linearly polarized light can be resolved into right- and left-hand circularly polarized components. This can be demonstrated physically by passing a beam of linearly polarized light through a quarter-wave plate optical device. In an isotropic medium, both right- and left-hand circular components of linear polarized light rotate at the same angular velocity. In an optically active, anisotropic medium, the velocity of light (in the z direction) is different for left-and right-hand circularly polarized waves. This difference in the propagation velocity of the L- and R-hand circularly polarized waves is referred to as circular birefringence. Coté et al. (1990) show that in an optically active, circularly birefringent medium, the net rotation of linearly polarized light exiting the medium is

$$\alpha = (\pi z_o/\lambda)(n_R - n_L) \text{ radians} \tag{6.76}$$

where z_o is the distance the linearly polarized light propagates along the z axis through the circularly birefringent medium, and n_R and n_L are the indices of refraction in the medium for right- and left-hand circularly polarized light, respectively. (The index of refraction of a medium is defined as the ratio of the speed of light in vacuum to the speed of light in the medium; it is generally greater than unity.)

An electromagnetic wave is elliptically polarized if the tip of the E vector traces out an ellipse in the X-Y plane as the wave propagates in the +z direction. As in the case of circular polarization, ellipses traced clockwise are right-handed, and left-handed ellipses are traversed counterclockwise. The treatment of elliptical polarization is beyond the scope of this text. The interested reader who wishes to pursue this topic should consult Section 4.4 in Balanis (1989).

Polarimeters are optoelectronic systems designed to measure the optical rotation produced by an active medium. As we have seen above, the medium can be a piece of glass, an optical fiber, or a solution containing optically active molecules (such as D-glucose).

The first and most basic type of polarimeter design is shown in Figure 6.32. Monochromatic light from a source, S, is collimated and then linearly polarized at a reference angle of 0° by a polarizer. The linearly polarized beam is then passed through the optically active medium, M, of length L . The **E** vector of the emergent beam is still linearly polarized, but is rotated some angle α. The emergent beam is then passed through an analyzer polarizer whose axis is at 90° to the input polarizer. If $\alpha = 0$, then the detector, D, will sense (nearly) zero light intensity. If $\alpha > 0$, the light intensity will increase according to the relation:

$$I = (I_{max}/2)[1 - \cos(2\alpha)] \tag{6.77}$$

Thus, $\alpha = 90°$ will result in maximum light transmission through the system to the detector. Obviously, the $I = f(\alpha)$ function is nonlinear. This nonlinearity is true even for $\alpha < 7°$, where the intensity is given by $I \approx I_{MAX}(\alpha^2)$; α in radians. To eliminate system nonlinearity for small rotation angles, we can orient the polarization axis of the analyzer polarizer to 45°. Now the light intensity

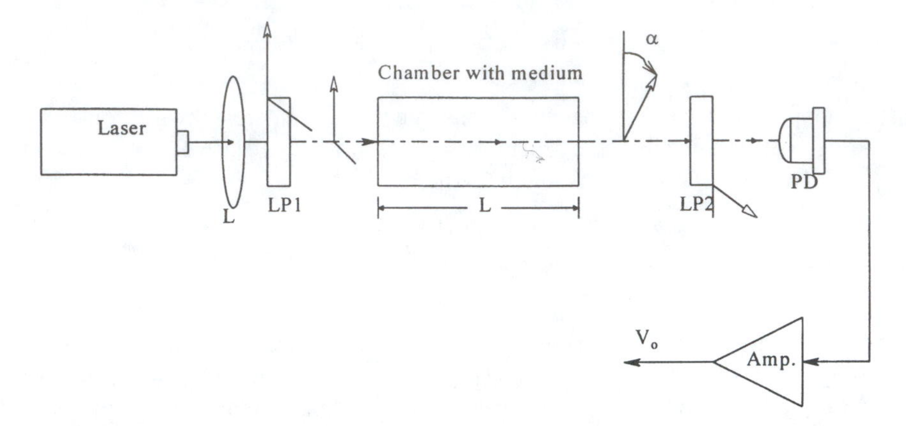

FIGURE 6.32 A basic polarimeter. The sample chamber length L is typically 10 cm.

at the detector will be maximum for $\alpha = 45°$, and a constant level will be detected for $\alpha = 0$. The intensity at the detector for a 45° analyzer can be written as:

$$I = (I_{MAX}/2)[1 + \sin(2\alpha)] \qquad 6.78$$

Now for $\alpha < 7°$, I may be approximated by

$$I \approx (I_{MAX}/2)[1 + 2\alpha] \qquad 6.79$$

where α is in radians. Thus, the detector output will be linear in α, and the DC bias may be subtracted from the rotation-proportional voltage. Limitations to the resolution and precision of the basic polarimeter are set by the noise in I_{MAX}, and the noise in the electronic photodetector.

A second type of polarimeter, described by Rogers (1973), was used in the dessign of a fiber optic, Faraday magneto-optic electric current sensor. The Faraday magneto-optic effect, described in the preceding section, causes a rotation in the polarization angle of linearly polarized light traversing a single-mode, fiber optic cable with turns around a current-carrying busbar. The emergent beam is collimated and passed through a Wollaston prism, WP, as shown in Figure 6.33. Two beams of light emerge from the prism and are directed to photodetectors D1 and D2. The Wollaston prism is made from two sections of calcite, a birefringent mineral having different refractive indexes in two different directions. As light travels through the first prism, both the ordinary and extraordinary rays travel colinearly with different refractive indexes. At the junction of the two prisms, the beams are interchanged so that the ordinary ray propagates with a lower refractive index and is refracted away from the normal to the interface, while the extraordinary ray is acted on by a higher refractive index, and is refracted towards the normal. The divergence angle between the two beams is increased upon exiting the prism; it depends on the wavelength of the light used. (The divergence angle is about 22° for HeNe laser light at 633 nm.) If the input beam to the Wollaston prism is linearly polarized at an angle α with the x-axis, the x- and y-components of the entering ray's **E** vector will be resolved into the emergent rays: the upper (ordinary) ray will have its **E** vector directed in the x-direction, and be proportional to E_{xr} of the input ray, and the lower (extra-ordinary) ray will have its **E** vector directed in the y-direction, and be proportional to E_{yr} of the input ray. The detector ouputs thus will be proportional to E_{xr}^2 and E_{yr}^2. In the Rogers detector, the output is formed by analog processing the detector outputs as described by the following equation:

$$V_o = K_D \frac{I_E - I_O}{I_E + I_O} \qquad 6.80$$

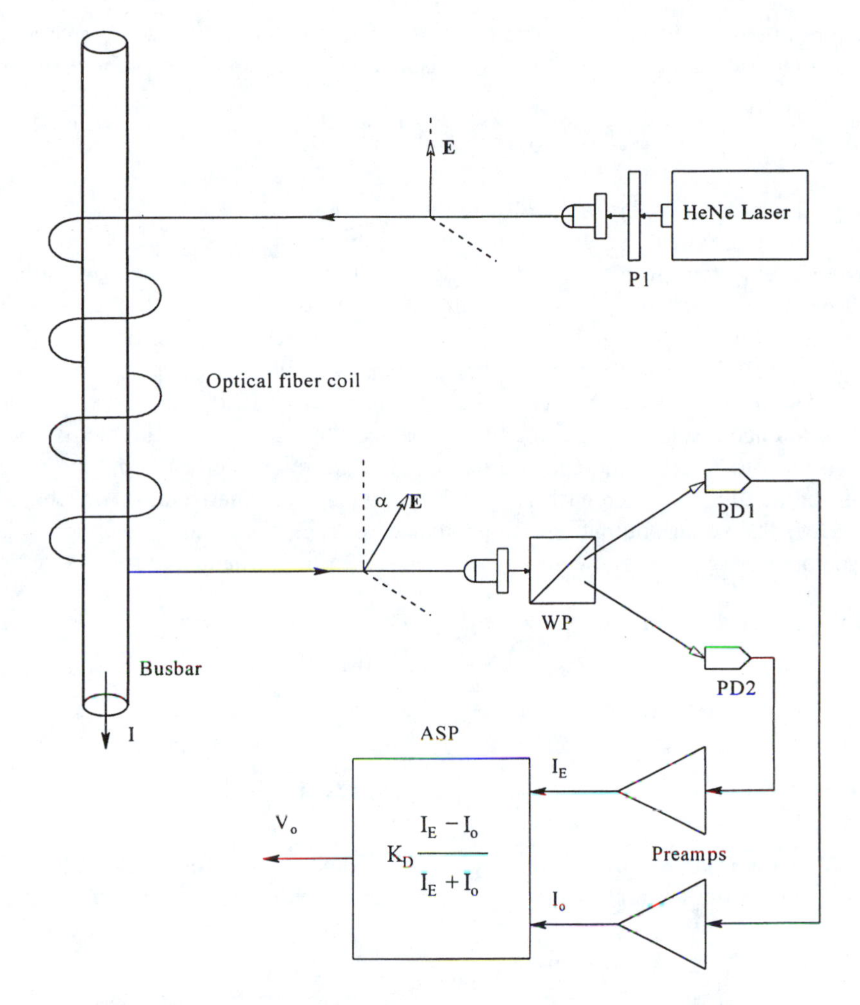

FIGURE 6.33 Rogers' (1973) polarimeter used in his MOCS system shown in Figure 6.29.

We note that if the analyzing Wollaston prism is rotated 45° clockwise with respect to the x-axis of the polarized input ray to the fiber optic coil, the intensity component of the extraordinary ray from the analyzer prism is given by:

$$I_E = KE^2 \cos^2(45° - \alpha) \quad 6.81$$

and the intensity of the ordinary emergent ray is

$$I_O = KE^2 \sin^2(45° - \alpha) \quad 6.82$$

When these intensity expressions are substituted into Equation 6.80, it is easy to show by trigonometric identities that the detector output is given by

$$V_o = K_D \sin(2\alpha) \quad 6.83$$

Equation 6.83 reduces to $V_o = K_D 2\alpha$, α being in radians, for $\alpha < 7°$. Thus, the Rogers ratio detector is substantially independent of noise affecting E, but still is affected by detector shot noise, and noise in the ratio circuit. It should be noted that AC current in the busbar will create a power line frequency B field, which in turn, will give a sinusoidally varying α. Thus, we will see for V_o,

$$V_o(t) = K_D 2\alpha_o \sin(2\pi ft) \tag{6.84}$$

It is this signal which is proportional to I(t) in the busbar.

A third means of resolving very small optical rotation angles has been developed by Coté et al. (1992). The Coté system, which is a true angle measuring system, is shown in Figure 6.34. A collimated beam of laser light is passed through a linear polarizer, then through a quarter-wave plate to produce circularly polarized light. The circularly polarized ray is then acted on by a rotating, linear polarizer (RP). The emergent beam is linearly polarized with the angle of its **E** vector rotating through 360° with the angular velocity of the rotating polarizer, $2\pi f_r$ r/s. The emergent beam is split into a reference beam and a beam which is passed through the optically active medium, M, in which we wish to measure the optical rotation, α_m. The reference beam and the measurement beam from the medium are then each passed through analyzer polarizers to two photodetectors, PDR and PDM. The voltage output of each photodetector is proportional to the magnitude squared of the **E** vectors emerging from the analyzer polarizers. In mathematical terms,

$$V_{oR} = \frac{K_D|\mathbf{E}_R|^2}{2}[1 + \cos(4\pi frt)] \tag{6.85A}$$

$$V_{oM} = \frac{K_D|\mathbf{E}_M|^2}{2}[1 + \cos(4\pi frt + 2\alpha_m)] \tag{6.85B}$$

$V_{oM}(t)$ and $V_{oR}(t)$ are passed through band-pass filters tuned to $2f_r$ in order to eliminate their DC components and restrict the noise bandwidth. A precision phasemeter is used to directly measure

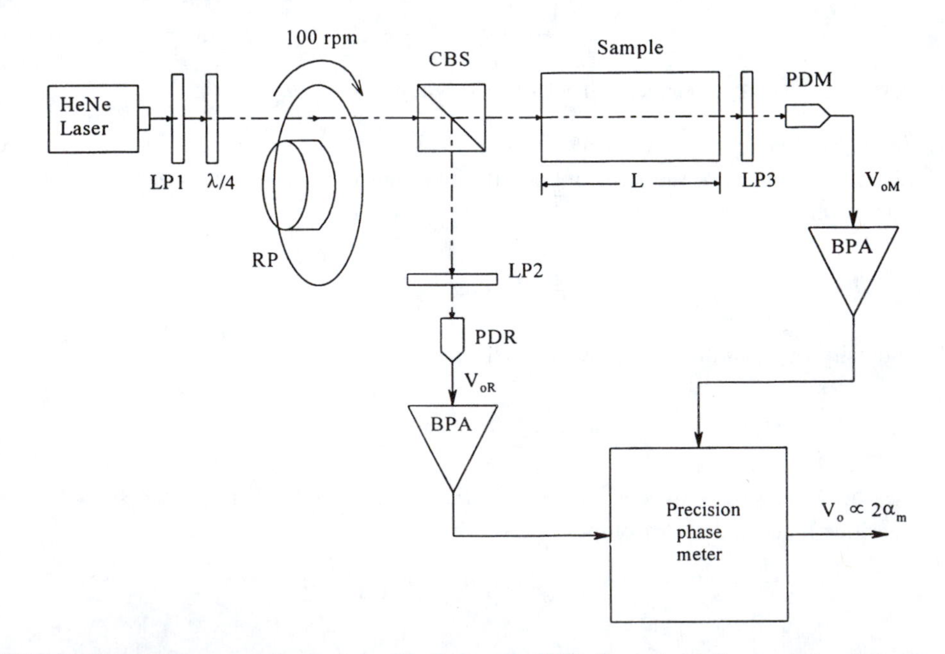

FIGURE 6.34 The millidegree polarimeter devised by Coté, et al. (1990). Symbols: LP, linear polarizer; λ/4, quarter wave plate; SM, synchronous motor; RP, rotating linear polarizer; PDR, reference photodetector; PDM, measurement photodetector; BPA, band-pass amplifier, tuned to $2\omega_r$.

the double optical rotation angle, 2 α_m, from the cosine terms in Equations 6.85. The magnitudes of E_R and E_M, and the noise associated with them have little effect on the measurement of 2 α_m. System resolution is better than ±0.0005° (Coté et al., 1990). Note that the rotating polarizer (RP) is effectively the "chopper" in the Coté et al. system, and effectively sets system bandwidth by the Nyquist criterion. It is apparent that this system will not work to measure α_m varying at power line frequencies unless the optical modulation frequency is made 600 to 1000 Hz. This system is best suited to measure small, static optical rotation angles.

A fourth type of polarimeter of interesting design described by Gillham (1957) had a noise-limited resolution of about ±0.0006°. A block diagram of the Gillham design is shown in Figure 6.35. Full-scale sensitivity was 0.02° for the open-loop version of the Gillham system, which makes use of a double-sideband, suppressed-carrier detection scheme. Referring to Figure 6.35, we see that monochromatic light passing through polarizer LP1 is linearly polarized at an angle of $-\theta_m$ (clockwise) with respect to the x (vertical) axis. A polarizer chopper was made by cementing four equally spaced quartz polarizers to a glass disk which was rotated at 1000 RPM. When the beam emerging from LP1 passes through the disk, the polarization vector is either unchanged as it passes through the glass, or is rotated +2 θ_m (counterclockwise) if it passes through the quartz. Thus, the beam emerging from the chopper wheel is alternately polarized at $-\theta_m$ and $+\theta_m$ at 66.7 cycles/s with a square-wave waveform. The polarization modulated beam next passes through a Faraday rotator (FR), whose DC coil current causes an additional rotation of the chopped beam so that the beam exiting the FR has the angles, $(-\theta_m + \alpha_F)$ and $(+\theta_m + \alpha_F)$. The beam next interacts with the rotatory sample, S, whose optical rotation, α_S, we wish to measure. The ray emerging from the sample thus has alternate polarization vectors of $(-\theta_m + \alpha_S + \alpha_F)$ and $(\theta_m + \alpha_S + \alpha_F)$. The beam next passes through polarizer LP2 on its way to the photodetector. The LP2 axis is put at −90° to the input (x) axis, so that the photodetector's output is a 66.7 Hz square wave with peak-to-peak amplitude given by

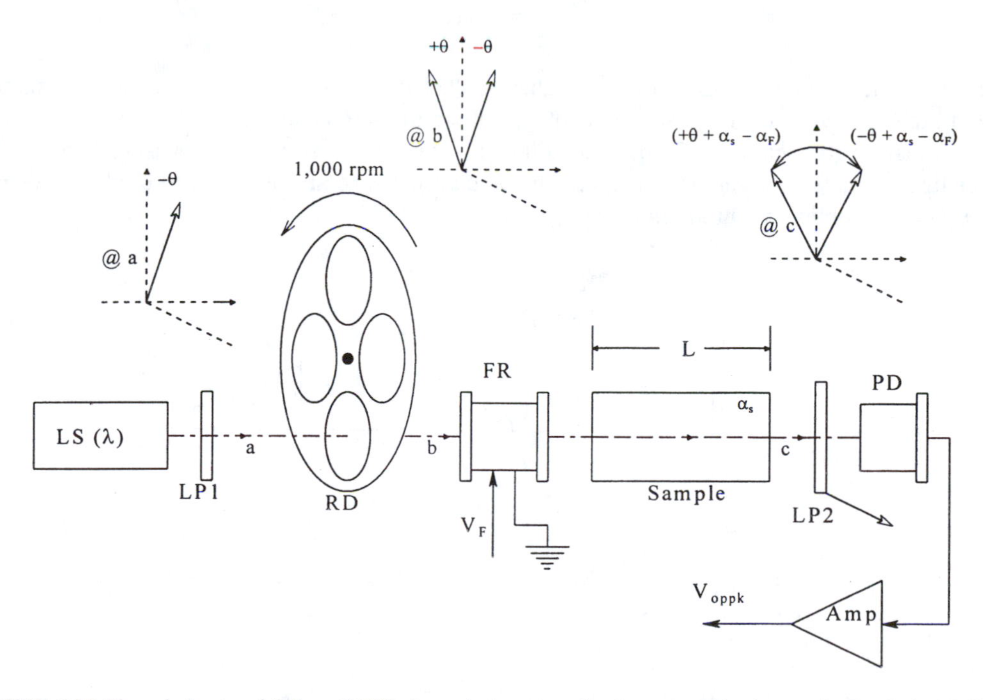

FIGURE 6.35 The polarimeter of Gilham (1957). An optical rotation of α_s is produced by the sample. Symbols: LP, linear polarizer; FR, Faraday magneto-optical rotator; PD, photodetector; RD, polarizing chopper disk rotating at 1000 RPM. RD has alternating segments of plain glass, and linear polarizer which rotates the beam's E vector 2θ degrees. LS, monochromatic light source.

$$V_{d(PPK)} = (KI_o/2)\{[1-\cos(2(\theta_m + \alpha_S - \alpha_F))] - [1-\cos(2(-\theta_m + \alpha_S - \alpha_F))]\} \quad 6.86$$

which reduces to:

$$V_{d(PPK)} = (KI_o/2)\{\cos(2(\theta_m + \alpha_F - \alpha_S)) - \cos(2(\theta_m - \alpha_F + \alpha_S))\} \quad 6.87$$

By trigonometric identity, we simplify the equation above to find:

$$V_{d(PPK)} = (KI_o)\sin(2\theta_m)\sin(2\alpha_S - 2\alpha_F) \quad 6.88$$

Assuming that the magnitudes of the arguments of the sine functions are less than 7°, we may finally write, noting that the angle values are in radians:

$$V_{d(PPK)} = 4KI_o\theta_m(\alpha_S - \alpha_F) \quad 6.89$$

From Equation 6.89, we see that if the Faraday rotator coil current is zero, the waveform observed at the detector is a true square-wave, double-sideband, suppressed carrier modulated signal. Gillham (1957) used $\theta_m = 3°$.

If the Faraday rotator is included in a feedback loop, as shown in the block diagram of Figure 6.36, it is easy to show that the steady-state output of the integrator, V_o, is exactly

$$V_o = \alpha_S/K_F \quad 6.90$$

where K_F is the Faraday rotator constant in radians/volt. This servo loop generates a Faraday rotation angle in the steady-state which is equal and opposite to the sample angle. Because of its relatively low modulation frequency (66.7 pps), the Gillham system is not suitable to measure 50 or 60 Hz power line currents with Faraday magneto-optical current sensors; it is intended to measure small (<0.03°), static optical rotation angles.

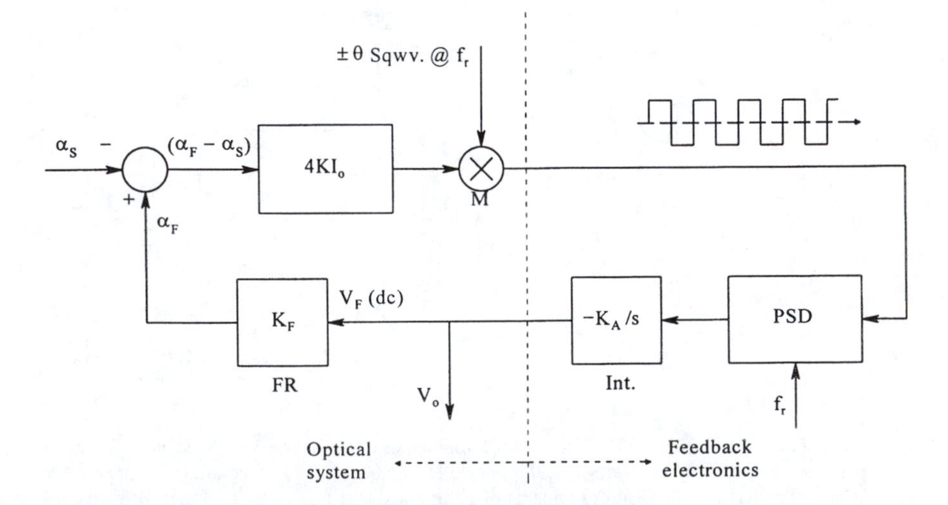

FIGURE 6.36 Block diagram of the feedback system used by Gilham (1957) to measure the optical rotation of the sample, α_S. The output of the phase sensitive demodulator, PSD, is integrated and the integrator output, V_o, is used to drive the Faraday rotator so its rotation is $\alpha_F = -\alpha_S$ in the steady state.

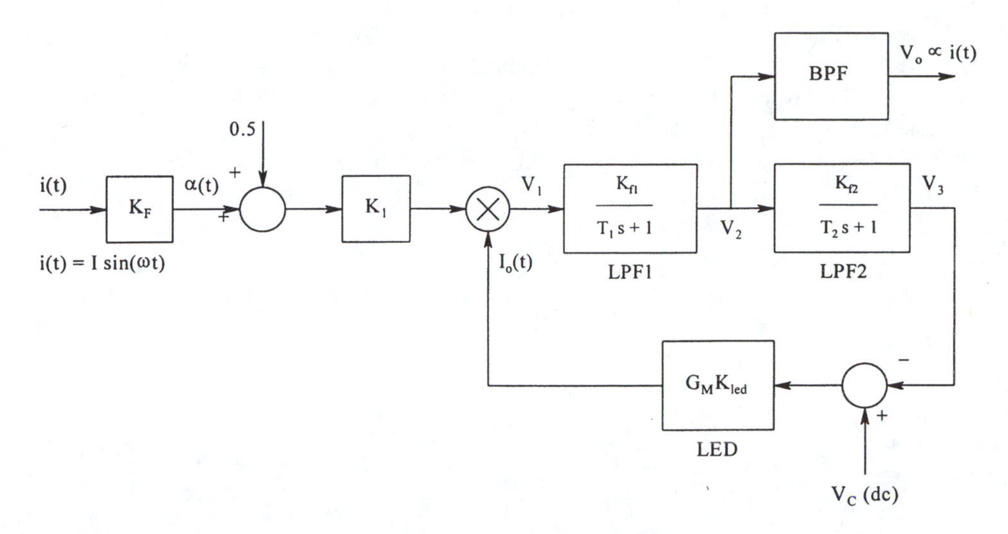

FIGURE 6.37 Block diagram of the polarimeter used in the MOCS described by Cease and Johnston, 1990.

A fifth type of polarimeter which also uses feedback was described by Cease and Johnston (1990). Their system is shown schematically in Figure 6.30. Instead of using multiple turns of fiber optic cable wrapped solenoidally around the current-carrying busbar, they used a single block of quartz glass as a single-turn optical pathway through which the busbar passes. The AC current in the the busbar causes a proportional optical rotation, α, by the Faraday magneto-optical effect. (Positive α is taken here as counterclockwise.) An analyzer polarizer, LP2, is assumed to have its axis at +45° (counterclockwise) to the vertical (0°) polarization axis of the input polarizer, LP1. Thus, the detector output voltage, V_1, may be expressed as:

$$V_1 = (K_1 I_o/2)[1 + \sin(2\alpha)] \quad 6.91$$

For $\alpha < 7°$, Equation 6.91 reduces to

$$V_1 \cong K_1 I_o[0.5 + \alpha] \quad 6.92$$

where, as expected, α is in radians. As is seen in the system block diagram, Figure 6.37, the photodetector output, V_1, is conditioned by two simple, low-pass filters. The output of the second filter, V_3, is subtracted from a DC level, V_c, and this difference is amplified and used to control the average light intensity, I_o, of the LED light source for the system. Thus, I_o is given by:

$$I_o = G_m K_{LED}(V_c - V_3) \quad 6.93$$

It is seen that this feedback system is nonlinear, and that a simple pencil and paper analysis of the transfer function for V_2 vs. α is out of the question. Accordingly, we may simulate the system's behavior using Simnon™, a computer program specialized for the simulation of complex nonlinear systems. The Simnon program is:

```
CONTINUOUS SYSTEM moct1
TIME t
STATE V2 V3
DER dV2 dV3
dV2 = Kf1*V1/T1 – V2/T1                "Output of first LPF.
```

```
dV3 = Kf2*V2/T2 – V3/T2                 "Output of second LPF.
ALPHA = Kf*I                            "Optical rotation angle from Faraday effect.
V1 = 0.5*Io*K1*(1 + sin(2*ALPHA))       "Output of PIN photodiode amp.
Io = GmKled*(Vc – V3)                   "Output intensity of LED.
I = Ipk*sin(377*t)                      "AC current in busbar.
"CONSTANTS: ................
K1: 2
Kf: 5.E-4
Kf1: 20
Kf2: 0.2
T1: 2.E-2
T2: 1.E-3
Vc: 1
GmKled: 0.5
Ipk: 1000
END
```

The original Cease and Johnston (1990) paper gave no constants for their system's components, so we arrived at the values above by reasonable estimates and trial and error. Figure 6.38 illustrates the results when the program above is run with the constants given above. Note that the photodetector output, V_1, is a sinusoid of line frequency whose peak-to-peak amplitude in the steady state is proportional to the current in the busbar. The DC level of V_2, however, must be removed by passing V_2 through a band-pass filter (not shown in the original Cease and Johnston paper). It is the AC voltage output of the band-pass filter which is actually measured to give the value of I(t) in the busbar. In our simulation of this system we noted that the linear proportionality between the peak-to-peak V_2 and I(t) is lost for $I_{pk} > 1200$ A, for the constants we used. An I_{pk} of 2000 A gives severely distorted peaks on the V_1 waveform.

In conclusion, we have seen that there are a number of ways that can be used to resolve optical rotation angles as small as 0.001°. Such small polarization angles may occur in instrumentation systems that make use of the Faraday magneto-optic effect, the electro-optic effect, and the electrogyration effect (Rogers, 1976).

6.6.3 Fiber Optic (FO) Mechanosensors

In this section we will review the operation of various mechanosensors using optical fibers and light (excluding fiber optic, Sagnac-effect gyroscopes). Fiber optic (FO) systems have been devised to measure pressure, force, displacement, acceleration, etc., using a variety of mechanisms. Hochberg (1986) places FO sensors in two classes: extrinsic FO sensors, in which the FO cable(s) is/are passive communications link(s) to an optically active terminating element, and intrinsic FO sensors in which some optical property of the FO cable itself is changed by the quantity under measurement (QUM).

Examples of extrinsic sensors are the intravascular pressure transducer developed by Hansen (1983) (see Figure 6.39), and the feedback pressure sensor developed by Neuman (1988) for measuring intracranial pressures. The Neuman sensor, shown in Figure 6.40A, is a feedback or null system. The microdiaphragm deflects proportional to the pressure difference across it. An increased external pressure, P_e, causes an inward deflection of the diaphragm, in turn causing less light to be reflected from the input optical fiber to fiber A, and more light to enter fiber B. Fibers A and B go to two photodiodes. The conditioned outputs of the photodiodes are subtracted to form an error signal that if negative, causes the pressure inside the probe, P_i, to increase, forcing the diaphragm outward until the error signal is zero. At this point the diaphragm is in its equilibrium position, and $P_i = P_e$. P_i is easily measured by a standard pneumatic pressure sensor, and in the

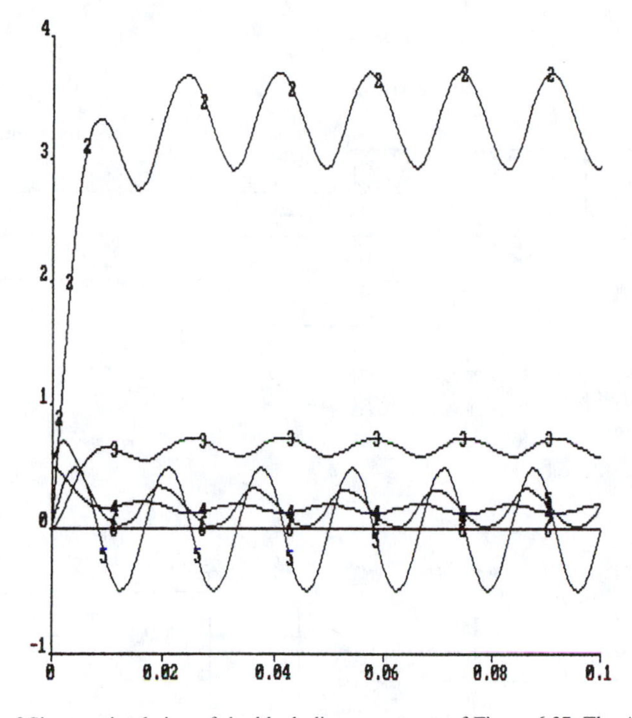

FIGURE 6.38 Results of Simnon simulation of the block diagram system of Figure 6.37. The AC component of voltage V_2 (trace 2) is proportional to i(t) in the busbar, and is isolated by a band-pass filter. Other traces are: #1, V_1 ; #3, V_3; #4, I_o; #5, alpha; #6, zero.

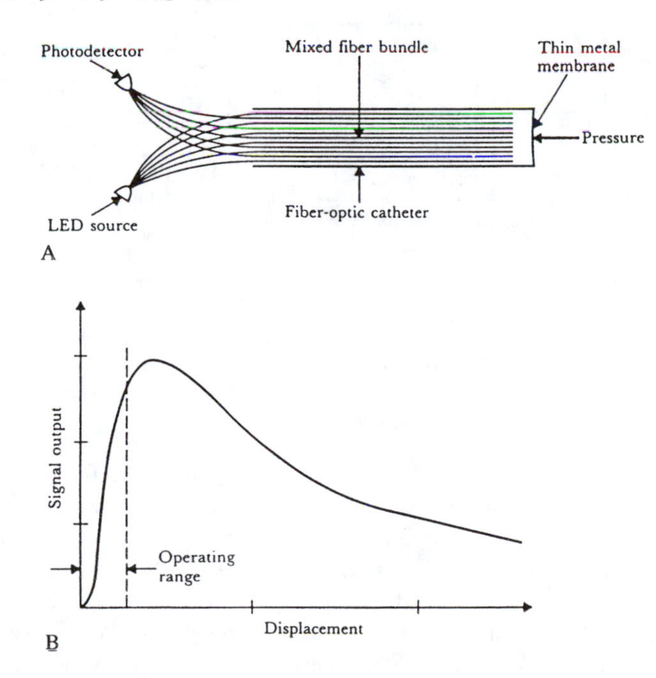

FIGURE 6.39 (A) Mixed fiber, fiber optic (FO) pressure sensor. The pressure difference across the diaphragm causes a change in the light output measured. (B) Relative sensor output vs. pressure difference for the Hansen (1983) FO pressure sensor.

steady-state, is equal to P_e. The block diagram for a simple controller for the Neuman probe is shown in Figure 6.40B. Other extrinsic FO sensors are shown in Figure 6.41.

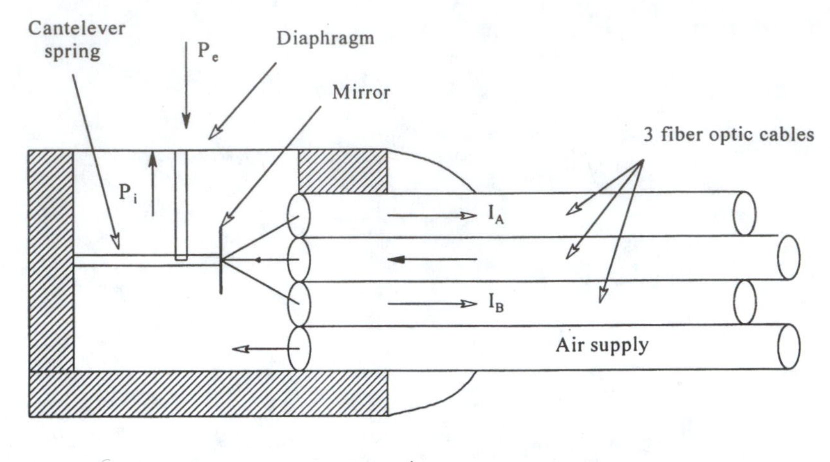

A.

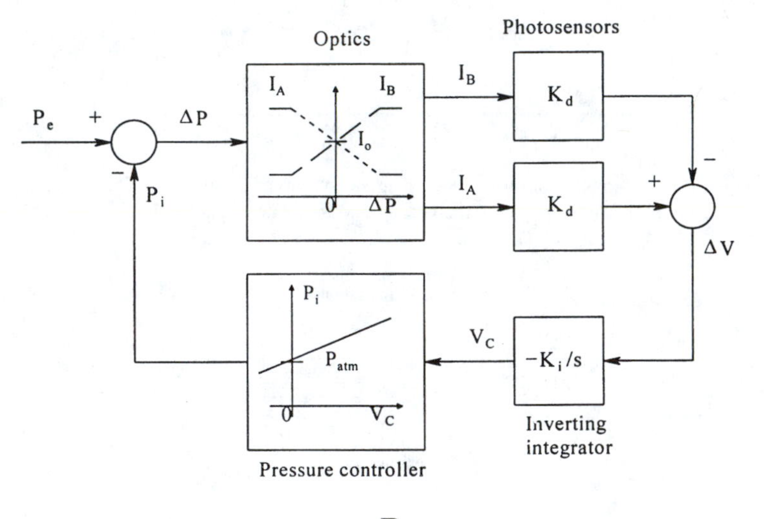

B.

FIGURE 6.40 (A) The Neuman (1988) feedback pressure sensor. When the internal pressure equals the external pressure being measured, there is no deflection of the diaphragm, and the light outputs are equal. (B) Block diagram of a simple controller for the Neuman pressure sensor. The system requires a voltage-to-pressure transducer to adjust the internal pressure of the sensor probe.

One intrinsic FO mechanosensor makes use of the property of optical fibers in which microbending causes a net decrease in the transmitted light intensity because of the loss of light power into the cladding; the bending angle prevents internal reflection of light back to the core (see Figure 6.42). The microbending obviously requires small deflections of the deformer relative to the fiber, and can be the result of the deflection of a diaphragm (pressure sensor) or elongation of a load cell (isometric force sensor). Another intrinsic FO mechanosensor design makes use of interferometry. Figure 6.43, after Hochberg (1986), illustrates an interferometric sensor system. The detector utilizes the light and dark interference patterns caused by combining the reference and measurement fiber light outputs. In this case, some net deformation of the measurement fiber causes a change in the refractive index of the fiber core, hence a phase change in the light output, and an interference fringe shift which can be sensed by the detector. The spatial distribution of light in an interference pattern at its center generally follows a function of the form:

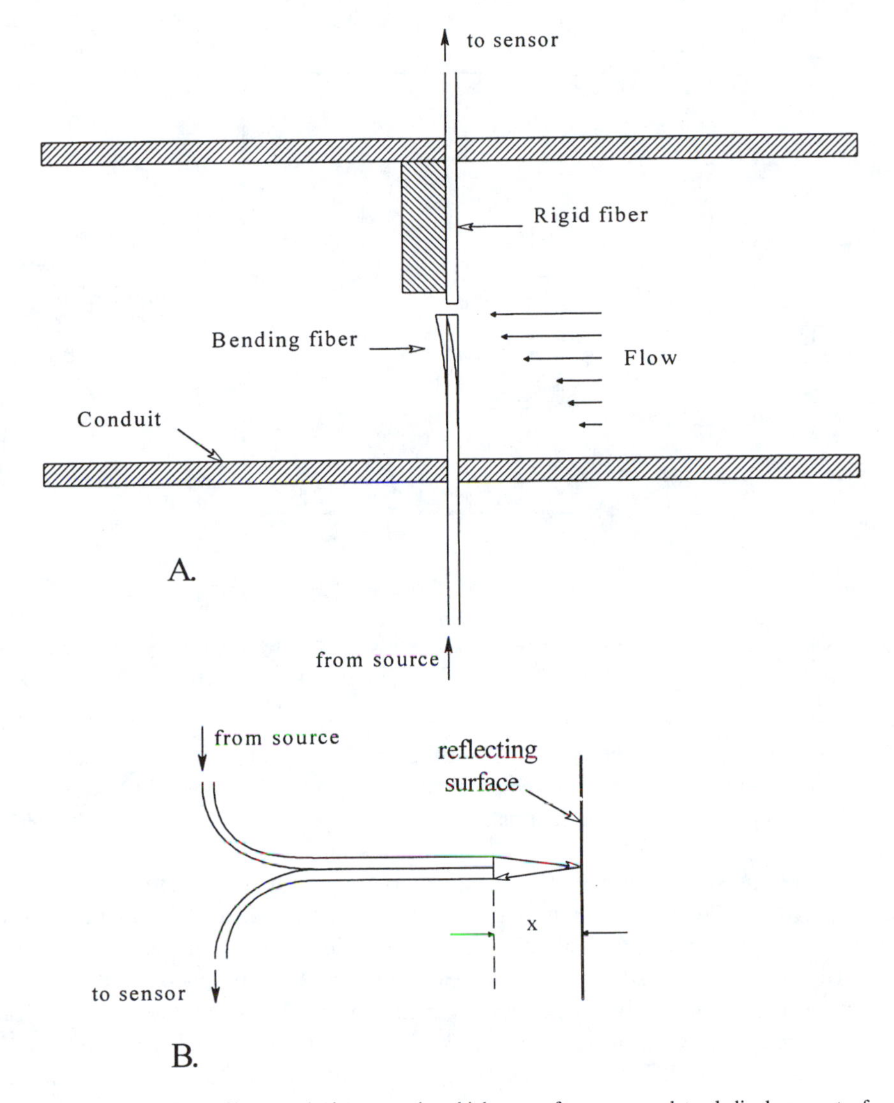

FIGURE 6.41 (A) Basic, single-fiber, extrinsic sensor in which some force causes lateral displacement of one or both fibers, decreasing light coupling. (B) The optical coupling between two OFs varies with the distance from a plane reflecting surface in a nonlinear manner.

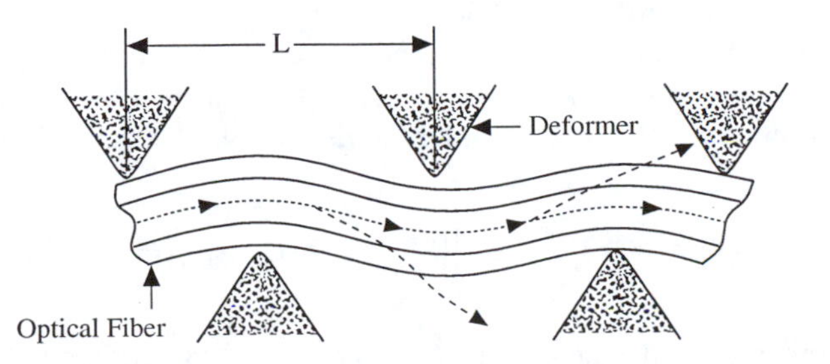

FIGURE 6.42 (A) Intrinsic fiber optic sensors may be based on microbending of a transmitting fiber. Microbending causes a decrease in the intensity of the exiting light because light is lost into the cladding as the result of bending. (B) A pressure sensor in which a membrane presses the OF into the microbending fixture.

$$I(x) = (I_o/2)\left[1 + \cos\left(2\{2\pi\delta x/X\}\right)\right] \quad 6.94$$

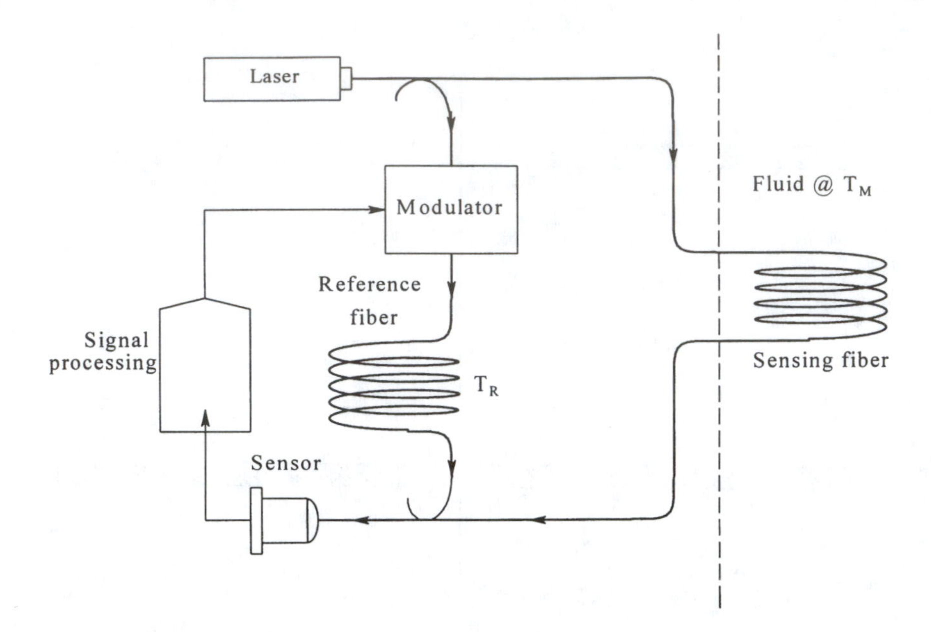

FIGURE 6.43 Schematic diagram of a two-fiber interferometer in which some physical quantity acting on the sensor fiber causes a phase change between the two light paths, hence a change in the detected intensity of the interference pattern.

where X is the spatial period of the fringes. The fringe shift occurs because of a small change (δx) in the spatial parameter, x. For maximum sensitivity, the fractional fringe shift should be sensed at the points of steepest slope of the cosinusoidal term in Equation 6.94. Thus, $I(\delta x) \cong (I_o/2)(1 + 4\pi\, \delta x/X)$.

An intrinsic, FO pressure sensor system using polarized light and interferometry was described by Bock et al. (1990). The Bock system is basically a one-fiber, Mach-Zehnder interferometer in which the sensing fiber carries two optical modes differing in polarization and whose phase delays change differentially in response to the applied external pressure. The pressure sensor of Bock et al. (1990) was shown to have almost perfect compensation for temperature changes. Figure 6.44 shows their experimental set-up, which used a HeNe laser as a source. The laser output was linearly polarized and launched parallel to one of the two principal axes of a highly birefringent (HB) fiber leading to the sensor. A bidirectional coupler couples light to the measurement fiber. The HB measurement fiber has a 45° axial splice (the reference section is rotated 45° with respect to the input segment). The reference section is of length L1. The sensing (terminal) section of HB fiber is axially spliced with a 90° rotation to the reference section. The terminal end of the sensing section of fiber has a gold mirror to reflect the light back down the fiber to the analyzer and detector. In another paper, Barwicz and Bock (1990) report further on their FO pressure sensing system. The photodetector output voltage was given by the rather complex formula:

$$v(p, \Delta T) = (V_o/2)\left\{1 + \cos\left[(2\pi L/\lambda)(\lambda/L_B - C_p \beta \Delta T + |C|\, ap)\right]\right\} \quad 6.95$$

where ΔT is the temperature difference from the reference temperature, p is the gauge pressure the FO sensor is exposed to, λ is the wavelength of light in the optical fibers, L_B is the beat length parameter of the fiber, L is the length of the fiber, a and β are material constants, C_p is the relative photoelastic constant, and C is a pressure-dependent photoelastic constant given by:

$$C = n^3(1 + \sigma/2E)(p_{12} - p_{11}) \quad 6.96$$

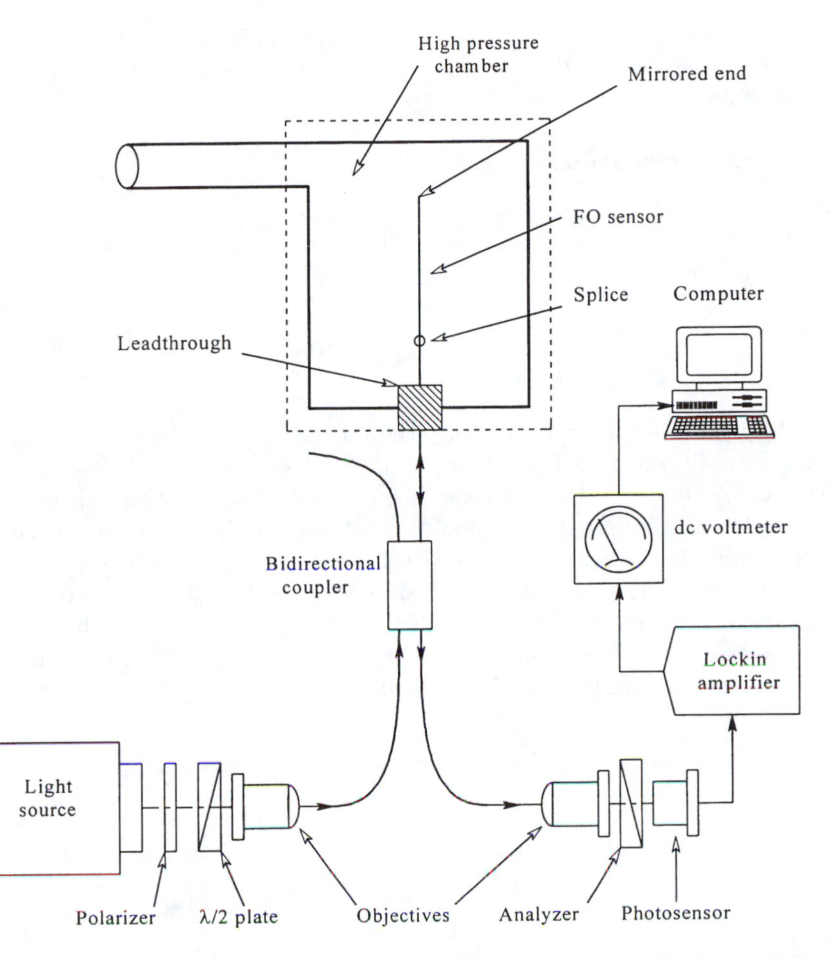

FIGURE 6.44 Schematic diagram of the interferometric, high-pressure sensor developed by Bock et al. (1990). It is a one-fiber, Mach-Zehnder interferometer which makes use of the pressure dependence of the sensing fiber's refractive index.

in which n is the refractive index of the fiber core, σ is the Poisson ratio, E is Young's modulus, and p_{11} and p_{12} are optical strain coefficients.

Barwicz and Bock (1990) showed that v(p) obtained experimentally follows Equation 6.95 quite closely; the period of v(p) varying inversely with sensing fiber length, L. For example, with L = 26 mm, the period in v(p) is 26 MPa; i.e., the cosinusoidal curve repeats itself every 26 MPa over a 100 MPa range. Note that 1 psi = 6895 Pa. By using three different lengths of FO sensor and computer processing of the sensor outputs, Barwicz and Bock were able to cover a 100 MPa range with 0.01 MPa resolution. This implies an 80 dB SNR for their system for high pressure measurements.

Additional examples of FO mechanosensors may be found in the texts by Culshaw (1984), Marcuse (1981), Davis (1986), and Cheo (1990). The design of FO mechanosensors is a rapidly growing field which has an active literature in instrumentation and in optoelectronics journals.

6.7 ELECTROCHEMICAL SENSORS

This class of sensor is generally used in a "wet" environment, i.e., one in which the substance to be measured is dissolved in water or another solvent. The outputs of such sensors are generally currents or voltages which are functions of the concentration of the substance to be measured. Electrochemical sensors may be subdivided into three categories; those which generate EMFs as the result of the sum of two half-cell potentials (i.e., they are electrochemical cells), those which operate polarographically, and those which are basically fuel cells. All three categories involve electrochemical reactions.

6.7.1 pH and Specific Ion Electrodes

The traditional pH measurement system is composed of two half-cells forming a "battery" whose EMF is a function of pH. pH is defined as the logarithm to the base 10 of the reciprocal of the hydrogen ion concentration. In terms of an equation:

$$\mathrm{pH} \equiv \log_{10}\left(1/\left[\mathrm{H}^{+}\right]\right) \tag{6.97}$$

A traditional pH cell is shown in Figure 6.45. The pH electrode itself is made from a special glass membrane across which the EMF is a function of hydrogen ion concentration. The pH electrode is filled with 0.1 *N* HCl, and electrical contact is made internally through a silver chloride-coated silver wire (Ag/AgCl). The resistance of the glass pH electrode is quite high, around 50 megohms or more; this limits the current which may be drawn from the pH cell so that an electrometer amplifier must be used to measure the cell's EMF without voltage drop errors due to current flow. The reference half-cell electrode is traditionally a saturated calomel design, shown in Figure 6.45. Undesolved potassium chloride crystals saturate the internal solution of the reference half-cell. The total pH cell's EMF is given by summing the half-cell EMFs:

$$\mathrm{E_{GL}} = \mathrm{E_{0GL}} - (\mathrm{RT}/\Im)\ln\left(\mathrm{a_{H^+}}\right) = \mathrm{E_{0GL}} + (2.3026\,\mathrm{RT}/\Im)(\mathrm{pH}) \tag{6.98A}$$

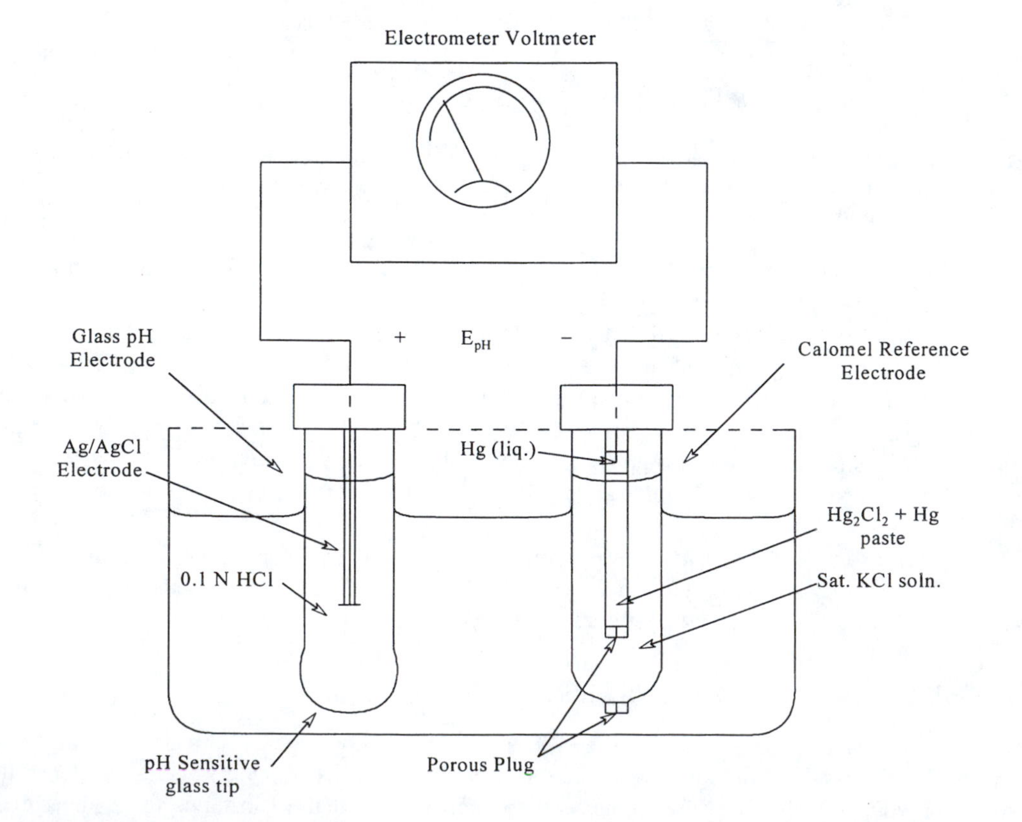

FIGURE 6.45 A basic pH measuring system.

$$E_{CAL} = E_{0CAL} - 7.6 \times 10^{-4}(t - 25°C) \qquad 6.98B$$

$$E_{NET} = E_{GL} - E_{CAL} = (E_{0GL} - 0.2415) + 7.6 \times 10^{-4}(t - 25) + (2.3026\,RT/\Im)(pH) \qquad 6.98C$$

where a_{H^+} is the activity of the hydrogen ions in solution; activity may be considered to be equal to concentration at low concentrations. E_{GL} is the glass half-cell electrode's EMF, E_{CAL} is the calomel electrode's half-cell EMF, R is the MKS gas constant (8.3143 joules per degree Kelvin), T is the temperature in Kelvin, t is the Celcius temperature, $\Im$ is the Faraday number (96,500), E_{0CAL} is the standard potential for the calomel half-cell (0.2415 V), and E_{0GL} is the standard potential for the glass electrode. E_{0GL} varies between glass electrodes and is compensated for when the individual pH measuring system is calibrated. Note that acidic solutions have pHs ranging from 0 to 7.0, and pHs for basic solutions range from 7.0 to 14. Also note that the pH potential, E_{NET}, is highly temperature dependent, and therefore all pH meters use a temperature probe of some sort also immersed in the solution under measurement. The signal from the temperature probe is used to null the (t – 25°) term in the pH cell EMF, E_{NET}, and to correct for changes in the (2.3026 RT/$\Im$) term if t of the measurement solution differs from the temperature of the standard pH buffer solutions used to calibrate the pH meter. It should be noted that there are many physical variations on the standard glass pH electrode and the calomel reference electrode. Designs exist which have been optimized for the measurement of blood pH, the pH of high-temperature solutions, pH measurements in the stomach, etc. In addition to the common glass pH electrodes, under limited conditions, other half-cells can be used whose half-cell potentials are proportional to pH. Tungsten metal, quinhydrone, and palladium oxide electrodes are responsive to pH (Maron and Prutton, 1959; Grubb and King, 1980). (Grubb and King (1980) are cited in Liu and Neumann's article in *Diabetes Care*, 5(3): 1982.)

Specific ion electrodes are used in conjunction with a reference electrode to measure the logarithm of the activity (concentration) of a specific anion or cation in solution. Specific ion electrodes are temperamental in that most have ranges of pH for optimum operation, and they also may require the complete absence of certain ion(s) which compete in the electrode's half-cell chemical reaction, and therefore cause errors. We mention a few, representative, specific ion electrodes and their properties,

The Orion model 94-17 chloride electrode has a log-linear half-cell potential which increases by 59.2 mV/decade change in concentration over a 10^{-4} to 1 *M* range. This electrode tolerates a pH range of 0 to 14, as well as nitrate, sulfate, phosphate, fluoride, and bicarbonate ions. It will not function in strongly reducing solutions, nor in the presence of sulfide, iodide, or cyanide ions. The chloride level must be at least 300 times the bromide level, 100 times the thiosulfate level, and 8 times the ammonia level for proper operation.

The Orion model 94-06 cyanide ion electrode has a log-linear half-cell potential over a 10^{-6} to 10^{-2} *M* range. The cyanide electrode does not respond to cations except silver, and most common anions, including nitrate, fluoride, and carbonate do not interfere with normal operation. The electrode will not work with sulfide ions in the solution. The chloride concentration may not be more than 10^6 times the cyanide concentration, the bromide level may not be more than 5000 times the cyanide level, and the iodide level may not be more than 10 times the cyanide before error will occur in measuring the cyanide level.

The Orion model 92-19 potassium ion electrode produces a log-linear, 59.6 mV/decade concentration change of K^+ ions over a 10^{-6} to 1 *M* concentration range. It operates from pH 1 to 12. Selectivity constants, given as the ratio of the electrode's response to the interfering ion to the response to K^+, are given by Orion as: Cs^+(1.0), NH_4^+(0.03), H^+(0.01), Ag^+(0.001), Na^+(2×10^{-4}), Li^+(10^{-4}). Thus, we see that the potassium electrode is relatively insensitive to other metal ions, but has a 1:1 ambiguity with cesium ions, and is affected at the 1% level by pH.

The concentration of cadmium ions may be measured with the Orion model 94-48 electrode. The EMF of this half-cell is log-linear from 10^{-7} to 10^{-1} *M* $[Cd^{2+}]$. The pH range is 1 to 14, and silver, mercury, and copper must not be present in the solution. The level of free lead or ferric ions must not exceed the cadmium ion concentration for absence of error.

The *Severinghaus* pCO_2 *electrode* described by Wheeler (in Webster, 1978) makes use of the property that the pH of a buffer solution is proportional to the $\log[pCO_2]$ over the range of 10 to 90 mmHg partial pressure. A Teflon membrane separates the solution in which CO_2 gas is dissolved (blood), and the buffer solution. The membrane is permeable to CO_2 gas, but blocks the passage of ions such as H^+ and HCO_3^-. The pH of the Severinghaus cell may be shown to be equal to:

$$pH = \log[HCO_3^-] - \log(k) - \log(a) - \log[pCO_2] \qquad 6.99$$

where k is the equilibrium constant for CO_2 gas in water going to bicarbonate ion and hydrogen ion, a is the constant relating the CO_2 gas pressure above blood to the concentration of CO_2 dissolved in blood, and $[HCO_3^-]$ is the concentration of bicarbonate ions in the buffer solution around the pH electrodes. The Severinghaus pCO_2 electrode requires calibration with two known partial pressures of CO_2, and, like any chemical cell, is affected by temperature.

6.7.2 Polarographic Electrodes

Polarographic electrodes differ from pH and ion-specific cells in that a polarographic cell is run at a constant potential to force two electrochemical reactions to take place. The current which flows in the cell is determined by and is proportional to the limiting concentration of one reactant. The *Clark cell* for the measurement of the partial pressure of oxygen dissolved in water, blood, or of oxygen in the air, is shown schematically in Figure 6.46. The Clark cell has four components: a platinum or gold cathode, an Ag/AgCl anode, an electrolyte solution, and a plastic membrane (polypropylene, Teflon, or Mylar) which is permeable to O_2 gas and little else. The electrolyte is generally a buffered, saturated KCl solution. At the noble metal cathode, a reduction reaction takes place, which is described as:

$$O_2(g) + 2H_2O + 4e^- + 4KCl(s) \rightarrow 4KOH(s) + 4Cl^- \qquad 6.100$$

At the Ag/AgCl anode, the oxidation reaction below takes place:

$$4Cl^- + 4Ag(s) \rightarrow 4AgCl(s) + 4e^- \qquad 6.101$$

It is found experimentally that when the anode is held at 0.7 V positive with respect to the cathode, a current flows which is described by the relation:

$$I = I_o + \beta[pO_2] \qquad 6.102$$

β is typically about 10 nA/mmHg pO_2. Both I_o and β are increasing functions of temperature. The response time of the Clark electrode to step changes in pO_2 is limited by the ease of O_2 diffusion through the membrane to the cathode surface. Typically in the tens of seconds, the response time

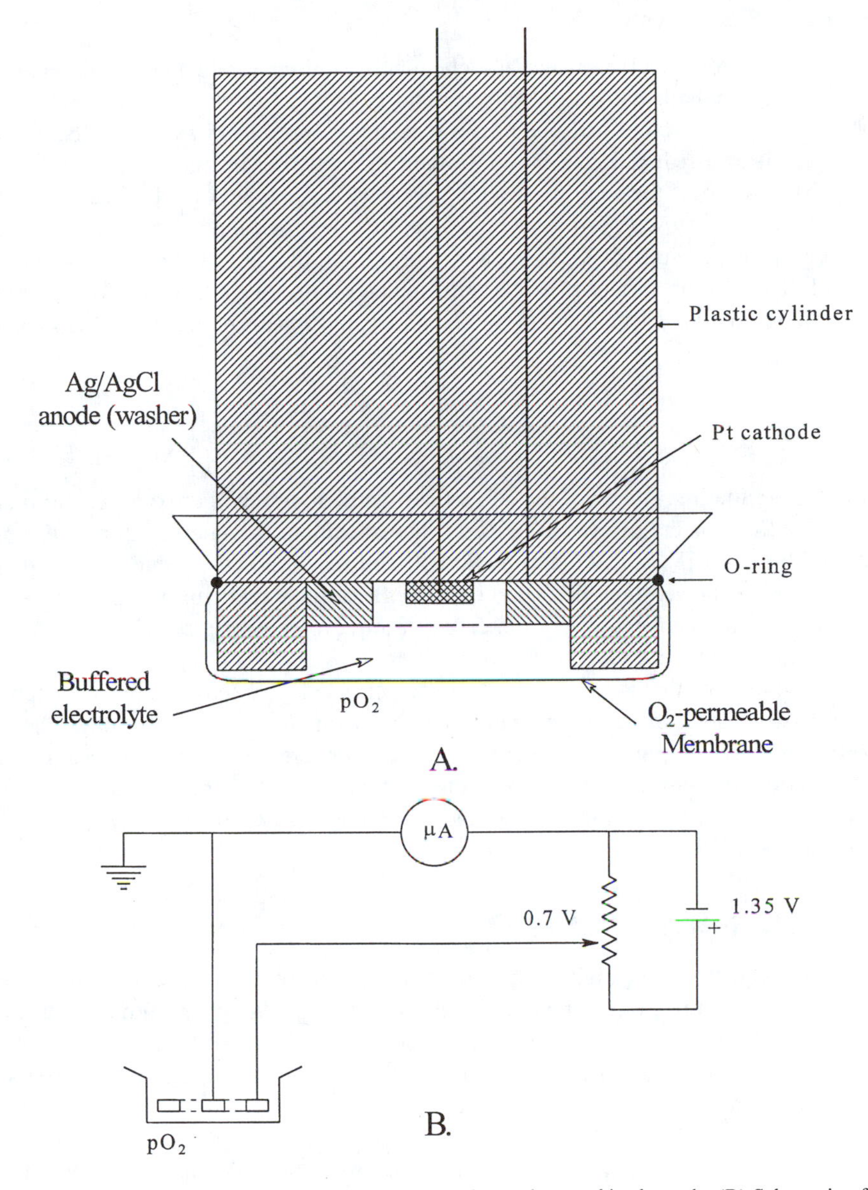

FIGURE 6.46 (A) Cross section through a Clark, oxygen sensing, polarographic electrode. (B) Schematic of basic circuit used with a Clark O_2 probe. No temperature compensation is shown.

of a fast Clark O_2 electrode has been cited by Geddes and Baker (1968) as 0.3 msec. According to Wheeler (in Webster, 1978), the settling time for a Clark cell is proportional to the pO_2 being measured.

The Clark cell has two major sources of error in measuring pO_2. One is the effect of temperature. The Clark cell must be calibrated and operated at a controlled, fixed, temperature which varies no more than ±0.1°C, or, knowing its tempco, we can measure its temperature and calculate compensation terms for Equation 6.102 above.

6.7.3 Fuel Cell Electrodes

A fuel cell is a battery (cell) in which the chemical reactions producing the EMF and output current occur at one or both half-cell electrodes and consume two or more reactants which are broken down to reaction products. If one of the reactants is present in excess concentration, and the other is present at a reduced concentration, then the cell's EMF may be proportional to the logarithm of the activity of the limiting reactant. A specific example of a fuel cell system is the catalyzed oxidation of glucose to form gluconic acid, hydrogen peroxide, and heat. A glucose fuel cell described by Wingard et al. (1982) used a mixture of the the bioenzymes glucose oxidase and catalase immobilized on the surface of a platinum screen electrode. Glucose oxidase catalyzes the oxidation of glucose to gluconic acid and H_2O_2. Catalase catalyzes the breakdown of H_2O_2 according to the reaction:

$$H_2O_2 \xrightarrow{\text{catalase}} O_2 + 2H^+ + 2e^- \qquad 6.103$$

The authors state that the source of the cell's EMF is the hydrogen peroxide reaction, although firm evidence for this contention was not given. They speculate that the reaction at the platinum anode is the oxidation of H_2O_2, and the cathodic half-cell reaction is the reduction of AgCl to Ag. Wingard et al. (1982) showed that the open-circuit EMF of their fuel cell was proportional to the logarithm of the glucose concentration from 5 to 500 mg/dl. Another type of glucose fuel cell is described in Chapter 8 of Wise (1989).

It should be pointed out that there are many other electrochemical cells which have been devised to measure glucose. Some use polarographic methods in which the current through the cell is a function of the glucose concentration, others measure the decrement of pO_2 across a membrane caused by the oxidation with a Clark O_2 cell, others measure the temperature rise in a chamber resulting from the oxidation. The interested reader should see the review of glucose sensors by Peura and Mendelson (1984).

6.8 IONIZING RADIATION SENSORS

Ionizing radiation can be either high-energy, electromagnetic radiation which includes photons, gamma radiation, and X-rays, or can be a directed beam of high energy, subatomic particles such as beta particles (electrons), alpha particles (helium nucleii, less the two orbital electrons), or certain ions, moving at high velocity. Ionizing radiations are given that name because as they pass though various media which absorb their energy, additional ions, photons, or free radicals are created. In a biological context, such ions or free radicals can, under certain circumstances, lead to genetic damage, mutations, or cancer. On the other hand, the production of ions, electrons, or photons under controlled circumstances can lead to a means of counting the incident ionizing particles and determining their energy. We describe below several designs for radiation sensors, including the familiar gas-filled Geiger-Müller tube, solid-state crystal detectors, and scintillation counters. The basic unit of radioactivity is the Curie, defined as 3.7×10^{10} disintegrating nucleii per second. The unit of exposure is the Roentgen, defined as producing 2.58×10^{-4} coulombs of free charge per kilogram of dry air at STP. The biologically absorbed dose is the rad, defined as 100 ergs of absorbed radiation per gram of absorber. The rem is an equivalent dose of radiation that has the same "effect on a man" as does one rad of 200 kV X-rays.

6.8.1 Geiger-Müller Tubes

A basic Geiger-Müller tube (GMT) with coaxial symmetry is shown in Figure 6.47A. The outer wall of the tube is generally a thin metal such as aluminum. The outer wall is the negative electrode or cathode of the GMT. The anode is a thin wire of tungsten or platinum run axially at the center of the tube and is, of course, insulated electrically from the cathode. The other major type of GMT

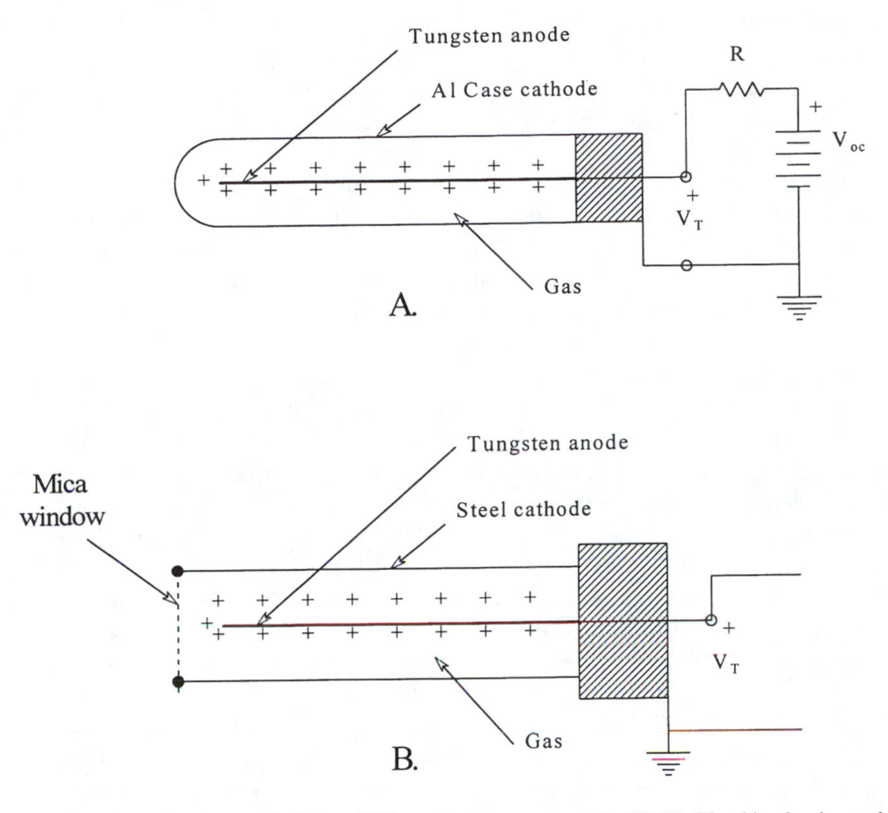

FIGURE 6.47 (A) Conventional, coaxial, Geiger-Müller radiation counting tube (GMT). The thin aluminum shell passes most ionizing radiation particles except alpha and low-energy electrons. (B) GMT with thin mica end window to enable counting of alpha particles.

is the end-window design, shown in Figure 6.47B. The end window is a thin sheet of mica or Mylar, and serves to admit particles which have low penetration such as α and β particles. (Although α particles have very high energies in the range of 4 to 10 MeV, their penetration in air is relatively short; a 3 MeV α particle travels about 1.7 cm in air, and α particles are stopped by a sheet of paper or aluminum foil. β particles (electrons) also arise from the decay of certain natural and man-made radioisotopes. An electron with 3 MeV energy will travel about 13 m in air before being stopped, and may penetrate as much as 1 mm of aluminum.)

Geiger-Müller tubes are generally filled with an inert gas such as helium, argon, or nitrogen at pressures ranging between a few mmHg and atmospheric. Most modern GMTs are self-quenching; i.e., they also contain a polyatomic gas such as methane or alcohol vapor which prevents energetic photons emitted from the activated inert gas molecules from reaching the cathode and stimulating the release of secondary electrons by photoelectric emission. The polyatomic gas is either dissociated or decomposed by the high-energy photons in the process of absorbing their energy. Because the polyatomic gas is decomposed, the self-quenching GMT gradually looses its quenching ability with its total number of counts. The effective life of a GMT may be as high as 10^{10} counts. Methane as a quenching gas gives a shorter count life (10^{8}) than does alcohol vapor.

The DC potential difference across a GMT determines its properties as radiation counter. Lion (1959) describes five distinct ranges of applied voltage to the GMT (the voltage ranges given are for example; every type of GMT will have numerical values peculiar to its own design). The first region (100 to 400 V) is called the ionization chamber region. In this region, an ionizing particle entering the GMT causes the generation of one or more electron–ion pairs. These charged particles drift toward the electrodes in the E-field of the tube. If the field is high enough so that they reach their destinations before an appreciable number of them recombine, and not so strong that additional ionization by collision occurs, the number of charges moving to and arriving at the electrodes is equal to the number initially produced by the collision(s) of the ionizing particle with the inert gas molecule(s). Each surge of ions from an ionizing particle collision thus causes a very small current pulse, the time integral of which is proportional to the number of ions, n_j, produced by the jth collision.

The second region (400 to 700 V) is called the proportionality region. Because of the higher electric field strength, ions and electrons produced in the GMT by an energetic particle are accelerated to higher kinetic energies, and thus cause the generation of new ions and electrons by inelastic collisions. This process lacks the energy to become self-sustaining, but a current pulse produced has an area proportional to kn_j. The multiplication factor, k, depends on many factors, including the type and pressure of inert gas used. k may range from 10 to 10^4. Because the current pulses caused by the collisions of ionizing particles with inert gas in the GMT are larger, proportional operation of the GMT allows the measurement of nj and thus the energy of the ionizing particles.

The third region (700 to 1000 V) is the region of limited proportionality. In this region, the sizes of the pulses are no longer proportional to the initial number of ions formed by the collision, and hence not proportional to ionizing particle energy. This region has little practical use.

The fourth region (1000 to 1300 V) is the well-known Geiger counter region. Here a collision of an ionizing particle with an inert gas molecule in the GMT causes output pulses that are all of the same size, independent of n_j and the energy of the particle. The mechanisms by which this mode of operation occurs have been described by Lion (1959): the initial + ion and electron formed in the collision are accelerated in the strong E field and gain sufficient kinetic energy to cause further ionizations through collisions with inert gas molecules. The additional ions and electrons so produced form an "avalanche" of charge. Photons are emitted from the excited gas atoms in the avalanche, and the discharge forms along the center wire anode where the field strength is highest. According to Lion, "The light [photon] emission procedes in every direction, but only in the vicinity of the central wire is the field strength high enough to facilitate the formation of new avalanches. The discharge spreads along the wire, therefore, until the entire anode is surrounded by a narrow cylinder of ions, and the discharge becomes self-sustaining."

Lion continues: "The termination of the discharge comes about in the following manner. The positive ions surrounding the central anode move toward the cathode with a velocity which is considerably smaller (because of their mass) than that of the electrons. These ions cause a positive space charge surrounding the positive wire; as they move toward the cathode, the positive space-charge cylinder ('the virtual anode') increases in diameter and, therefore, the field strength in the counter decreases." The recovery time in GMTs following a "count" ranges from 30 to 300 μsec (Lion, 1959), thus the maximum counting rate can be from 3,000 to 30,000 counts per second. The action of the quenching gas is described above.

The fifth region (over 1300 V) is called the unstable region. Here the field strengths are so large that the quenching gas is ineffective, and one ionizing particle collision with an inert gas molecule can lead to multiple pulses, or in extreme cases, a sustained plasma discharge which may destroy the GMT.

It should be stressed that the current pulses from a GMT operated in the Geiger or plateau region are about 1000 times the size of the pulses observed when the GMT is operated in the proportionality region. The circuit of a simple, portable Geiger counter is shown in Figure 6.48. Several output modes are commonly used: an audio output through headphones or a loudspeaker, an analog meter output, and a digital counter output reading in counts per minute, or other time unit. A GMT counter system does not provide information as to what the ionizing particle or radiation is. This information must be inferred from an *a priori* knowledge of the source of radiation.

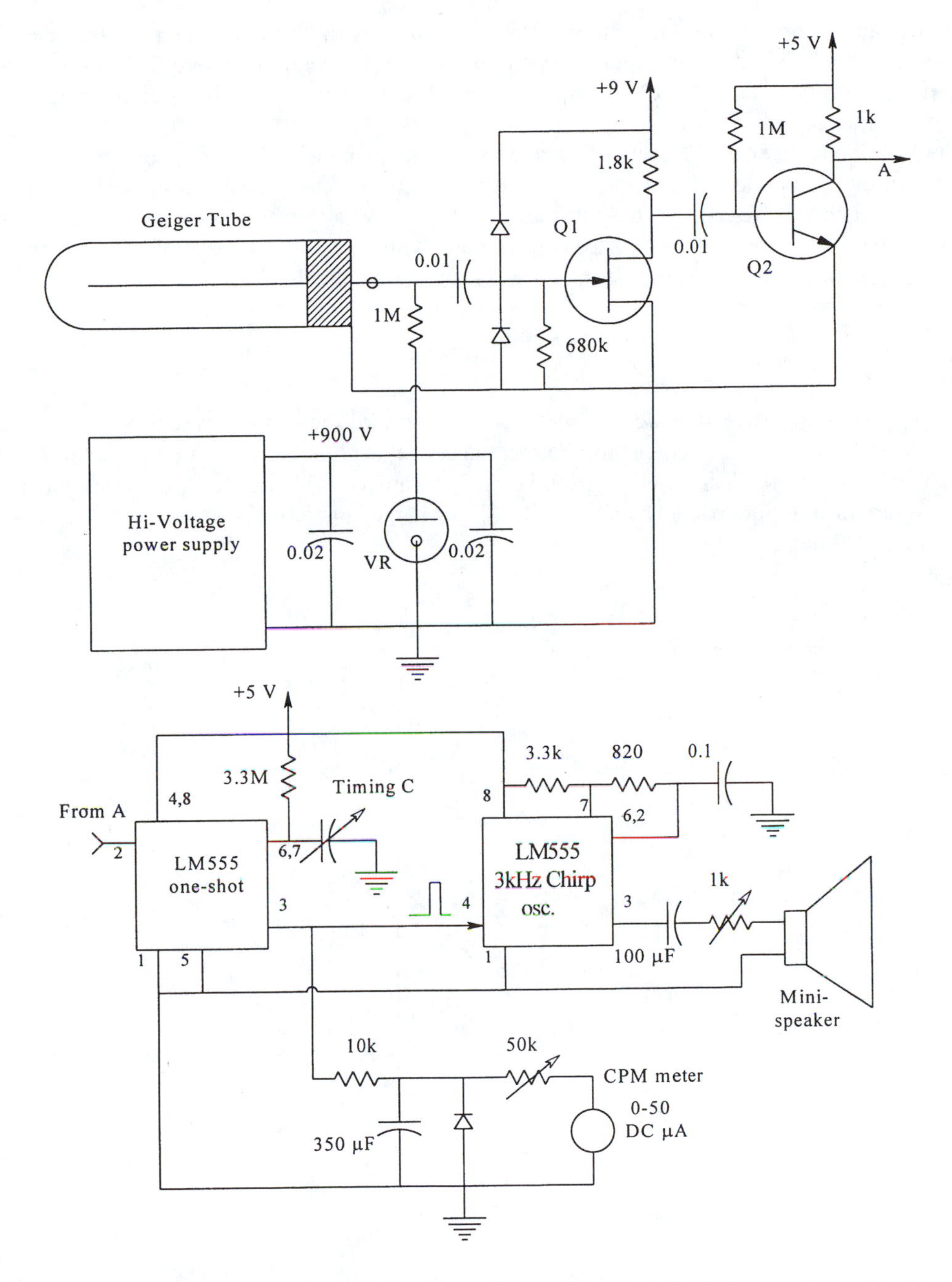

FIGURE 6.48 Circuit of a simple, analog, radiation counter. Voltage pulses from the GMT are conditioned by transistors Q1 and Q2, and used to trigger LM555-1 which acts as a one-shot multivibrator to produce pulses of standard width and height. The output pulses from LM555-1 are averaged by a low-pass filter preceding a DC microameter. The DC current in the microameter is proportional to the average number of GMT pulses per minute. The output pulses from LM555-1 also gate a second LM555 acting as a chirp oscillator. The chirp is sent to headphones or a small loudspeaker to give the operator a subjective reading.

6.8.2 Solid-State Crystal Radiation Sensors

Crystal radiation sensors have used crystals of materials such as lightly chlorine-doped cadmium telluride (a p-semiconductor), thallium chloride-thallium bromide, silver chloride, silver bromide, and diamond on which metallic electrodes have been deposited on opposite surfaces. A DC potential is applied to the electrodes, generating an internal electric field in the crystal of several kilovolts/centimeter. If an ionizing particle enters the crystal and collides with its atoms, a burst of free electrons is released which then drift through the crystal to the positive electrode where their arrival causes a brief current pulse. Only 1.0 eV is required in a Cl-doped CdTe crystal to put an electron into the conduction band.

Let us assume the total electron charge released by the ionizing event is $n_o q$ coulombs (n_o is the number of electrons released, q is the charge on one electron), and these electrons are created near the cathode and drift, en masse, the thickness, d, of the crystal to the positive electrode. In their travel to the anode, some of the electrons recombine with positive charges (holes) or are trapped by impurities in the crystal such that the charge reaching the anode is given by

$$Q_A = n_o q \exp(-d/\lambda) \tag{6.104}$$

Thus, $n_o q$ coulombs start to move at some velocity, v, toward the anode, decreasing in number as they travel. Because the applied electric field in the crystal is high (about 5 kV/cm), and d is small, the travel time of the electrons, $T_d = d/v$, is small compared with the RC time constant of the external circuit in Figure 6.49. The magnitude of the charge bundle in the crystal may be found as a function of time:

$$\begin{aligned} Q(t) &= n_o q \exp(-tv/\lambda) \quad \text{for } 0 \le t \le T_d \\ Q(t) &= 0 \quad\quad\quad\quad\quad\quad \text{for } t > T_d \end{aligned} \tag{6.105}$$

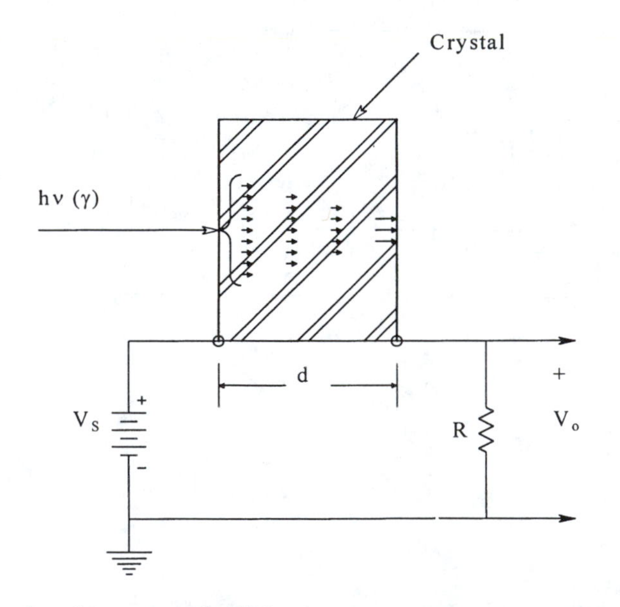

FIGURE 6.49 Schematic of a solid-state, crystal radiation detector.

We note that the peak voltage on the capacitor formed by the crystal and its electrodes is given by $V_c = \overline{Q}/C$. $\overline{Q}$ is the average charge flowing over the interval, $[0, T_d]$. Integrating Equation 6.105 from 0 to T_d, and dividing by CT_d yields the peak value of the voltage induced by the moving electrons:

$$V_c = (n_o q \lambda / Cd)[1 - \exp(-d/\lambda)] \tag{6.106}$$

V_c decays exponentially as $v_c(t) = V_c e^{-t/RC}$. In the equations above, the constant λ is known as the *Schubweg* of the crystal; this is the space constant for electron decay in the crystal, as seen from Equation 6.104 above. The value of the Schubweg depends on the crystal material, and is proportional to the applied field strength. Lion (1959) states that the crystal will behave as a good proportional counter when the Schubweg lies between a few tenths of a millimeter and one centimeter. He states: "If the incident particle penetrates deeply into the crystal and causes the release of electrons uniformly throughout the crystal volume,..." The peak voltage is then given by:

$$V_c = (n_o q\lambda/Cd)\left[1-(d/\lambda)\left(1-e^{-d/\lambda}\right)\right] \tag{6.107}$$

As in the case of GMTs, there is a critical range of applied DC voltage where the pulse size is proportional to the energy of the ionizing particle. The particle energy for single ion-pair production in a crystal is of the order of 5 to 10 eV. Lion states that a 1 MeV beta particle causes the release of about 1.2×10^5 electrons in an AgCl crystal. The lowest detectable energy for an incident particle is in the order of 1 keV; this produces from 100 to 200 electrons. With a 5 kV/cm internal DC field, an AgCl crystal can resolve about 10^6 cps, and a diamond can count about 10^7 events/second. According to Lion, space charges may accumulate in the crystal due to the accumulation of immobile holes in the crystal lattice, or from electrons trapped near the anode. Such space charges may seriously limit the counting life of a crystal, but their effect may be overcome by irradiating the crystal with infrared light, or periodically reversing the polarity of the DC applied to the crystal.

6.8.3 Scintillation Counters

A scintillation counter has three main components: the scintillation crystal, the photomultiplier tube used to count flashes (scintillations) from the crystal, and electronic pulse forming and pulse height discriminating circuits. The scintillation crystal is generally sodium iodide, about one-half inch thick. The energy from ionizing radiation which collides with the crystal lattice is converted in part to a flash of light (secondary photons), the intensity of which is proportional to the energy of the absorbed radiation. These flashes or scintillations are very weak and must be converted to electrical pulses by a photomultiplier vacuum tube (PMT). The PMTs used in scintillation counters have flat, transparent photocathodes. Photons from the scintillation crystal strike the photocathode of the PMT and cause the release of a number of electrons proportional to the intensity of the flash, the intensity of which is proportional to the energy of the ionizing particle. The few photoelectrons released at the photocathode are amplified by the dynode electrodes of the PMT to form a large output current pulse. The current amplification of a PMT (anode current/photocathode current) depends on the supply voltage used, the geometry of the tube, etc., and can range from 10^2 to 5×10^6 (Lion, 1959). Overall photosensitivity of PMTs can range from 0.01 to 2×10^4 amps/lumen. The PMT current amplification increases proportional to the square root of the supply voltage. With liquid nitrogen cooling to reduce noise, PMTs can resolve scintillations as small as 2×10^{-16} lumen. The response time of a PMT is in general less than 10 nsec (Lion, 1959), hence such tubes should be capable of counting as fast as 10^7 cps. The output pulses from PMT circuits are often reshaped to make them sharper. One such sharpening circuit, taken from Webster (1978), uses a shorted delay line to cancel the long tail on the primary output pulse, making the pulse narrow (see Figure 6.50).

Scintillation systems are widely used in nuclear medicine in conjunction with gamma-emitting isotopes to image tissues with cancerous growth. The patient swallows or is injected with a small amount of an isotope which is selectively taken up metabolically by the organ in question. The presence of a tumor generally means that more of the radioisotope will be taken up than by normal tissue. A scintillation system with a lead collimator is used to scan over the tissue in question and to detect areas of abnormal (high) radioactivity. An example of such a system is the use of the isotope ^{131}I, which has a half-life of 8.1 d. ^{131}I is selectively taken up by cells in the thyroid gland

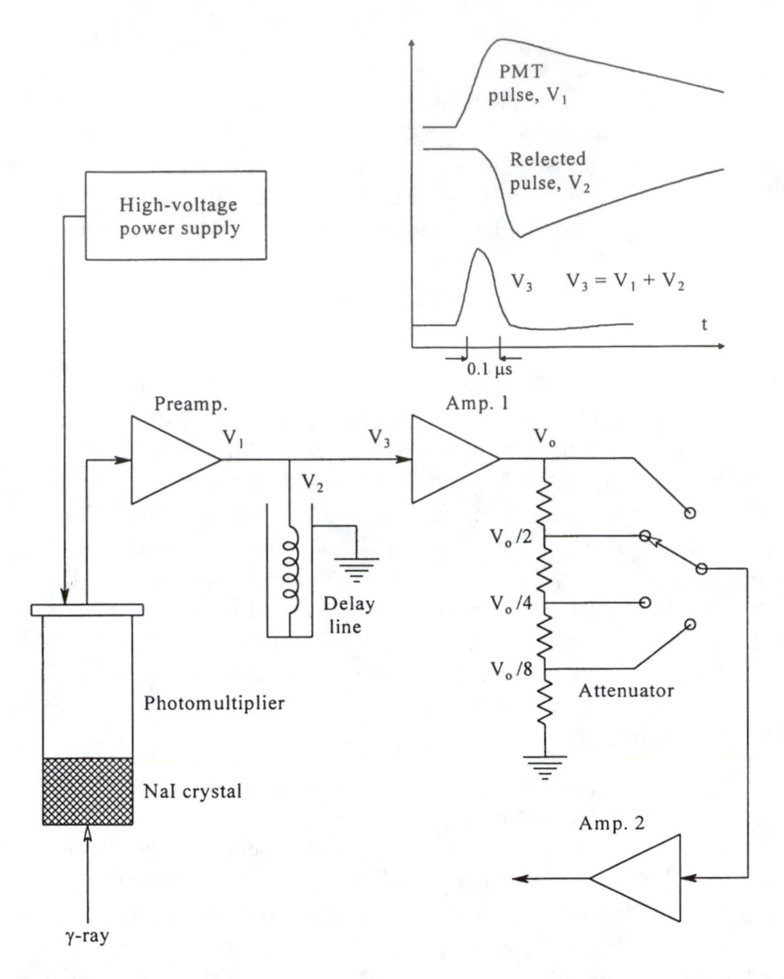

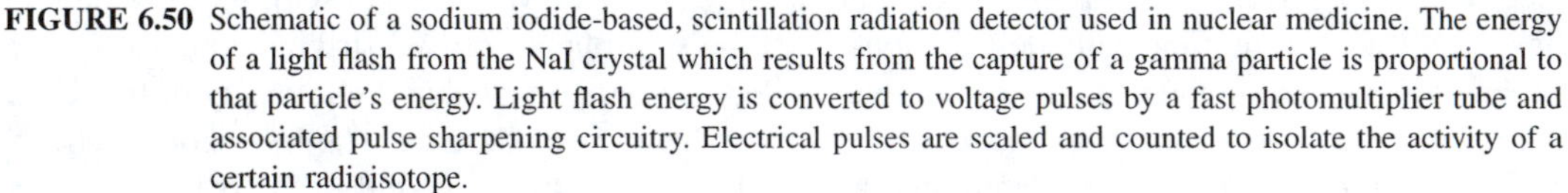
FIGURE 6.50 Schematic of a sodium iodide-based, scintillation radiation detector used in nuclear medicine. The energy of a light flash from the NaI crystal which results from the capture of a gamma particle is proportional to that particle's energy. Light flash energy is converted to voltage pulses by a fast photomultiplier tube and associated pulse sharpening circuitry. Electrical pulses are scaled and counted to isolate the activity of a certain radioisotope.

in the process of the formation of the hormone thyroxin. Planar arrays of scintillation crystals and their associated PMTs are called gamma cameras, and are used in nuclear medicine to visualize parts of the body that have selectively absorbed a radioisotope. This type of gamma camera provides relatively coarse resolution because of the physical size of the scintillation crystals and PMTs. In the Anger camera, a single, large scintillation crystal is used with an array of 19 PMTs. The location of a single radioactive event in the crystal is estimated by a formula involving the intensity of the flash produced at each PMT, and the location of each PMT. Another configuration of scintillation camera used in nuclear medicine is the pinhole collimator. In this mode of operation, pixel resolution is traded off for sensitivity (Macovski, 1983).

6.9 MECHANO-OPTICAL SENSORS

In this section, we discuss certain sensor mechanisms, excluding fiber optics, in which mechanical parameters such as displacement and velocity cause changes in the intensity of transmitted or reflected light, either directly, or as the result of induced changes in the linear polarization angle of the output light, or from optical interference. In the simplest case, shaft angle is sensed with an optical encoder disk. Shaft velocity can be found by differentiating the position signal.

Another means of measuring angular velocity is through the use of the Sagnac effect, discussed below. The displacement of interference fringes is proportional to the angular velocity. The Sagnac effect forms the basis for modern optical gyroscopes which are discussed in the following chapter. While their primary response mode is to angular velocity, gyroscopes can be combined with electromechanical feedback systems in order to sense angular position; such systems are covered in Chapter 7. The linear velocity of a fluid medium can be measured optically by laser Doppler velocimetry (LDV), the basics of which are described below.

6.9.1 Optical Coding Disks

Optical coding disks are used to convert analog shaft rotation angle to a digital word proportional to the angle. Optical coding disks may be subdivided into two categories: incremental position encoders, and absolute position encoders. In both types of sensors, light is either transmitted to or blocked from a linear phototransistor array, depending on the angle of the shaft to which the photoetched coding disk is attached. Thus, coding disk output is digital, and may be input directly into a computer controlling the shaft angle or recording its value. Figure 6.51A shows a simple, three-bit, absolute position encoding disk etched for a straight binary output code and Figure 6.51B illustrates a gray code etched disk. Practical resolution of an absolute position disk is 14 bits. The most significant bit (MSB) (2^0) track is the innermost track; higher-order bits are encoded radially outward, with the least significant bit (LSB) (2^{N-1}) lying on the outside edge of the disk. The arc length of the smallest (outer) sectors can be shown to be given by:

$$I \equiv R \tan\left(360/2^{N+1}\right) \tag{6.108}$$

Thus, for N = 14, and R = 4", the radius to the outer track, the resolvable sector length l is 0.000767" or 19.5 μm. This also implies that one LSB of shaft revolution is $\theta = 360/2^{N+1} = 1.099 \times 10^{-4}$ degrees. To obtain finer resolution, one only has to connect a second coding disk to the primary shaft through a step-up gear ratio of 2K, where K is 1, 2, 3, etc., and use the outer K tracks of the second disk as the K LSBs.

Incremental coding disks are far simpler in their construction. On their outer edges they have photoetched two, simple, light/dark square wave patterns, one of which is 90° out of phase with the other. Depending on the application, from 100 to 1000 or 1024 cycles may be used. Once a reference angle is established, total shaft angle can be obtained by counting the output pulses with up/down counters, such as the 74LS192 or 74LS193. Comparison of the phase of the two square wave outputs of the incremental coding disk gives rotation direction and sets the up/down control on the counters.

It should be pointed out that angular velocity information can be obtained from the digital signals from optical coding disks by simple digital differentiation, in the case of absolute position encoders, and by simply counting the pulses per unit time for the output of an incremental disk, or doing digital differentiation on the position counts stored in the up/down counters.

Optical mask techniques similar to those used in angle coding disks can be used to realize direct linear analog displacement to digital coding sensors. Linear, charge-coupled photodetector arrays can also be used with an opaque mask and lens optics to measure linear displacements with a precision on the order of 1/1024. The output of such a charge-coupled device (CCD) system is basically analog, however, and the array element number at which the light/dark transition occurs gives the desired displacement. Still another method of optical detection of linear displacement, described in Barney (1985, Section 7.4.1), makes use of the moving interference fringes caused by the superposition of two linear Moiré patterns. An optical detection scheme is used that uses the relative motion of the fringe pattern to obtain direction and displacement information.

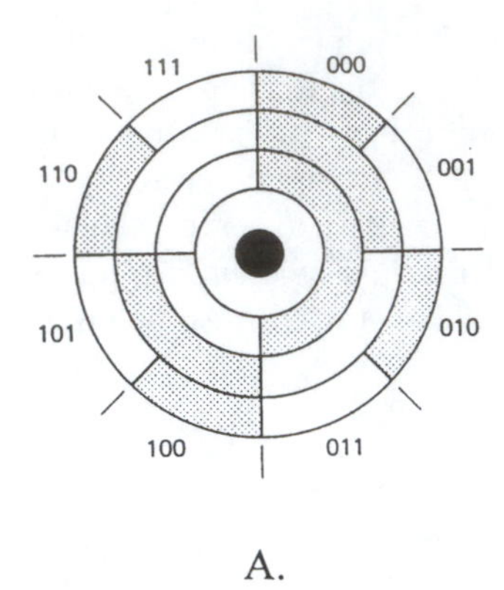

A.

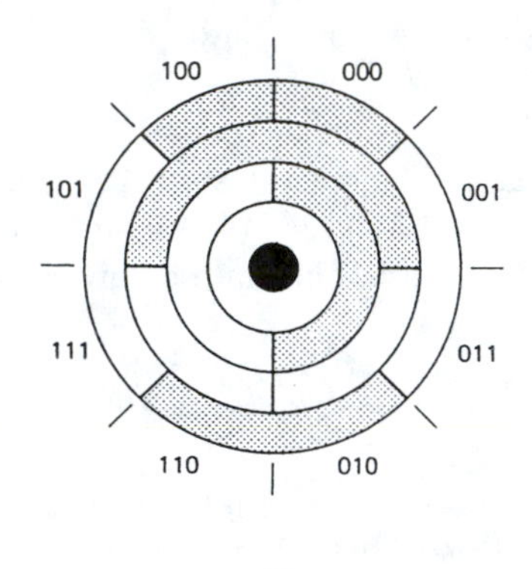

FIGURE 6.51 (A) A three-bit, straight binary, absolute position shaft coding disk. (B) A three-bit, gray-code disk.

6.9.2 Sagnac Effect Sensing of Angular Velocity

The Sagnac effect is the principle underlying the modern fiber optic gyroscope (Post, 1967). In this section, we will describe the basic Sagnac effect. Fiber optic gyros will be discussed in detail in the next chapter. It has been observed that the Sagnac effect occurs whether the light traverses a closed path in air or in glass optical fiber loops.

The Sagnac effect was first observed by Harress in 1911. Harress constructed a polygonal ring interferometer using glass prisms. His purpose was to measure the dispersion properties of glasses. When the ring was rotated, a shift in the interference fringe pattern was noted. Harress assumed that this shift was due to Fresnel-Fizeau drag. This drag effect is observed when light propagates through a linearly moving optical medium.

Sagnac performed his experiments in 1914. A simplified diagram of his apparatus is shown in Figure 6.52. A mercury green line source was collimated and directed at a beam splitter (a half-silvered mirror at 45°). The two beams were directed in opposite but collinear paths in air around an optical table. The half-silvered mirror served to direct the beams to a screen where the incident beams produced interference fringes. Sagnac's loop area was 866 cm^2, and he rotated his interferometer at 2 revolutions/second. Sagnac observed a fringe shift of 0.07 fringes as the entire optical table, containing mercury lamp, lenses, mirrors, and screen, was rotated about an axis perpendicular to the table. Sagnac was able to show that the fringe shift was not due to Fresnel-Fizeau drag, and he derived a relation for the fractional fringe shift given by Equation 6.109. In 1926 Pogany repeated Sagnac's experiment with improved apparatus in which $\lambda_o = 546$ nm, A = 1178 cm^2, and $\dot{\phi} =$ 157.4 r/s. He measured $\Delta z = 0.906$; the theoretical value was 0.900. Pogany repeated his experiment in 1928 and obtained results within 1% of theoretical values.

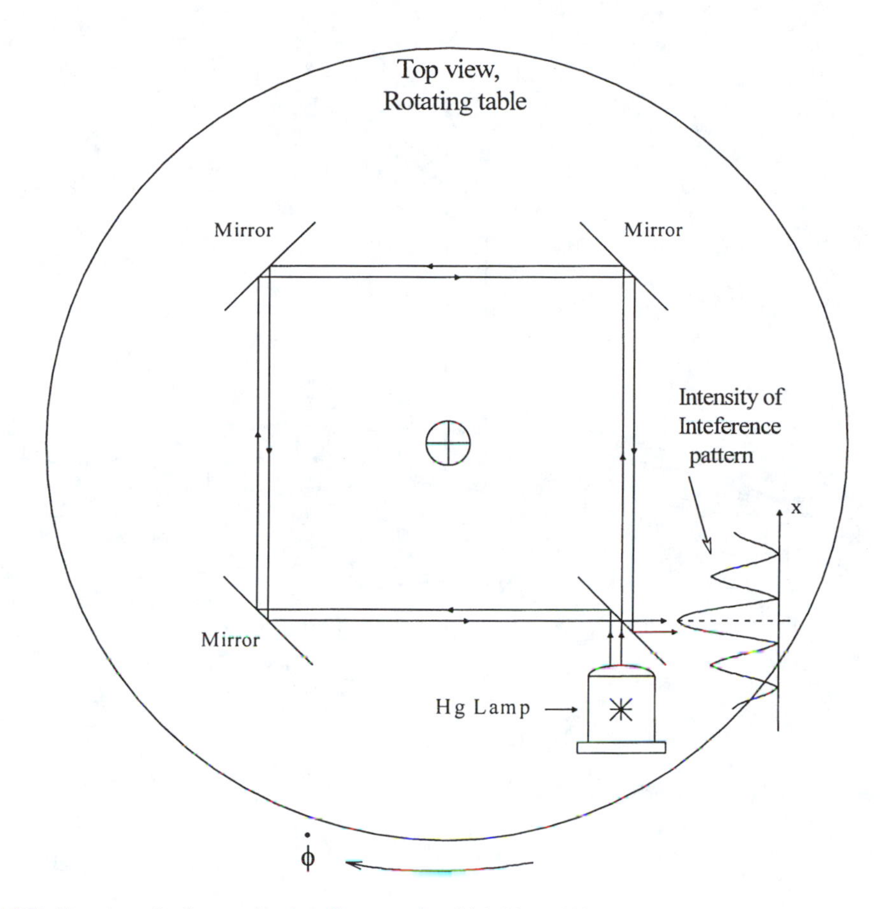

FIGURE 6.52 Top view of a Sagnac ring interferometer in which fringe shifts are proportional to the angular velocity of the apparatus.

From the early work on the Sagnac effect, four principles emerge: (1) Equation 6.109 applies for a one loop system. (2) The fringe shift does not depend on the shape of the closed light path, only on its area. (3) The fringe shift does not depend on the axis of rotation; it does depend on θ, however. (4) The fringe shift does not depend on the propagation medium.

The Sagnac equation for fringe shift is:

$$\Delta z = \left[4\dot{\phi}A\cos(\theta)\right] / (\lambda_o c) \qquad 6.109$$

Here Δz is the fractional fringe shift, λ_o is the wavelength of light, A is the area of the light path, θ is the angle the rotation axis makes with the perpendicular to the plane of the light path, c is the velocity of light around the path, and $\dot{\phi}$ is the angular velocity of path rotation around an axis at an angle θ to a perpendicular to its plane, in radians/second.

It is easy to derive Equation 6.109 above if we consider a circular light path rather than a rectangular one. Such a circular path could be the limiting result of using a very large number of prisms or mirrors, as shown in Figure 6.53 (polygonal approximation), or of using a single circular turn of a glass optical fiber, as shown in Figure 6.54. Assume that at t = 0, the input/output point of the ring is at 0°. First, we assume that the ring is stationary, i.e., $\dot{\phi} = 0$. In this case it takes $\tau_o = 2\pi R/c$ seconds for a light wave to circle the ring, either clockwise or counterclockwise. Now, let the ring rotate clockwise at a constant angular velocity, $\dot{\phi}$. In the time, τ_{cw}, a fixed point on the ring rotates an arc length ΔL″ to point P″. This may be expressed as:

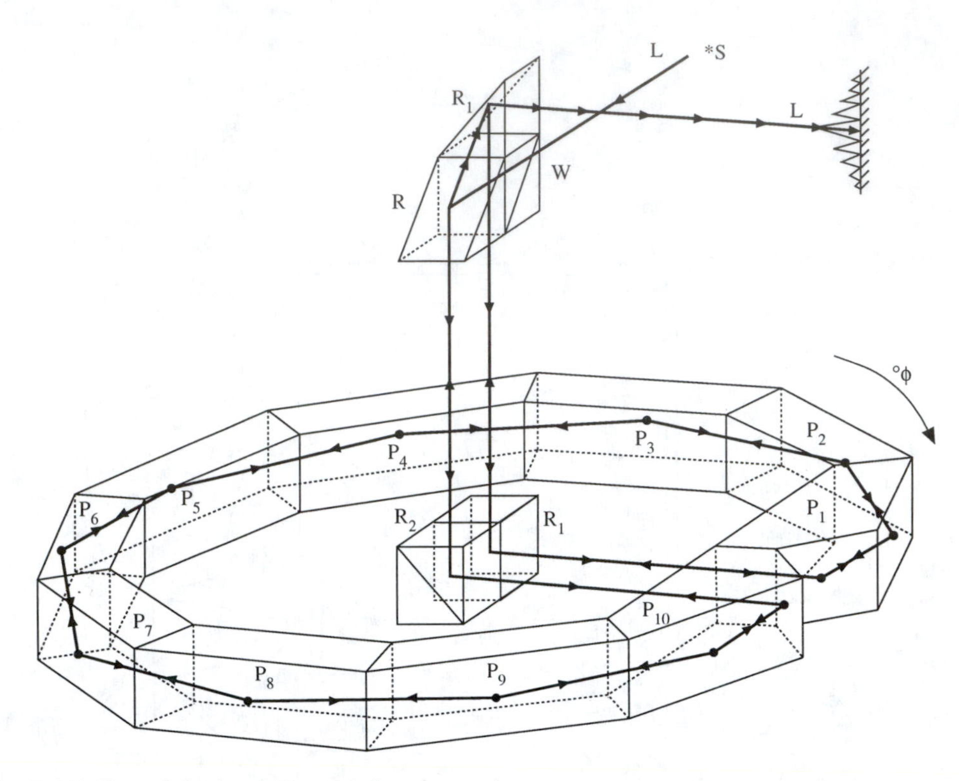

FIGURE 6.53 Harress' ring interferometer (a Sagnac system).

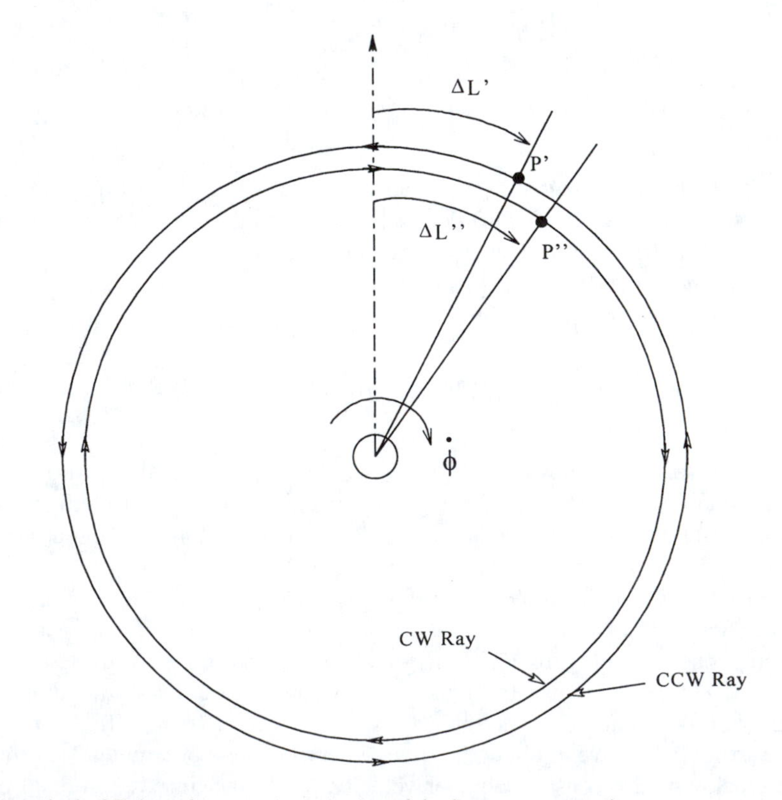

FIGURE 6.54 A single OF ring relevant to the derivation of the Sagnac gyro equation.

$$\tau_{cw} = \Delta L''/(R\dot{\phi}) \tag{6.110}$$

In the same time, the clockwise lightwave travels

$$\tau_{cw} = (2\pi R + \Delta L'')/c \tag{6.111}$$

Obviously, $\tau_{cw} > \tau_o$.

A wave propagating counterclockwise takes less time to reach point P′:

$$\tau_{ccw} = (2\pi R - \Delta L')/c = \Delta L'/(R\dot{\phi}) \tag{6.112}$$

The fractional fringe shift, Δz, is given by:

$$\Delta z = c\,\Delta\tau/\lambda_o \tag{6.113}$$

where $\Delta\tau$ is given by:

$$\Delta\tau = \tau_{cw} - \tau_{ccw} = (2\pi R + \Delta L'')/c - (2\pi R - \Delta L')/c = (\Delta L'' + \Delta L')/c \tag{6.114}$$

Now it is easy to show that:

$$\Delta L'' = 2\pi R^2\dot{\phi}/(c + R\dot{\phi}) = (2\pi R^2\dot{\phi}/c)/(1 + R\dot{\phi}/c) \approx 2\pi R^2\dot{\phi}/c \tag{6.115}$$

Similarly,

$$\Delta L' = 2\pi R^2\dot{\phi}/(c - R\dot{\phi}) = (2\pi R^2\dot{\phi}/c)/(1 - R\dot{\phi}/c) \approx 2\pi R^2\dot{\phi}/c \tag{6.116}$$

So the fractional fringe shift can be written finally as:

$$\Delta z = (\Delta L'' + \Delta L')/\lambda_o = 4\pi R^2\dot{\phi}/(c\lambda_o) = 4A\dot{\phi}/(c\lambda_o) \tag{6.117}$$

which is the same as Equation 6.109, with $\cos(\theta) = 1$.

As an example, the basic sensitivity of a one-turn Sagnac system can be calculated from Equation 6.109: θ is assumed to be 0°.

$$\Delta z/\dot{\phi} = 4A/(c\lambda_o) = 4 \times 1\ \text{m}^2/(3 \times 10^8\ \text{m/s} \times 633 \times 10^{-9}\ \text{m}) = 0.021\ \text{fringe/r/s} \tag{6.118}$$

Clearly, sensitivity will be raised if multiple optical paths are used, a situation addressed in the discussion of fiber optic gyroscopes in the next chapter.

6.9.3 Laser Doppler Velocimetry

Laser Doppler velocimetry (LDV) provides a "no touch" means of measuring the linear velocity of fluids (and particles). As the name suggests, LDV makes use of the *Doppler effect* at optical frequencies. In this section, we will review the basics of LDV, and discuss certain key applications.

First, we describe the Doppler effect. We are all familiar with the Doppler effect on sound. A car moving towards us blows its horn. As it passes, there is a perceptable downward shift in the pitch of the horn. Johann Christian Doppler gave a paper, "On the Colored Light of Double Stars and Some Other Heavenly Bodies" in 1842 before the Royal Bohemian Society of Learning. Doppler was professor of elementary mathematics and practical geometry at the Prague State Technical Academy. He apparently got little recognition for his work, and died in 1854 at the age of 45, of consumption. A contemporary of Doppler, Buys Ballot, in 1844 contested Doppler's theory as an explanation for the color shift of binary stars. Ballot actually did an experiment using sound waves where a trumpet player played a constant note while riding on a flatcar of a train moving at constant velocity. A musician with perfect pitch, standing at trackside, described the trumpet tone as a half-tone sharp as the train approached, and a half-tone flat as it receded. In spite of this obvious evidence, Ballot continued to object to Doppler's theory. Ballot's publications apparently served to cast doubt on Doppler's theory for a number of years.

To derive the Doppler effect for either electromagnetic or sound waves, we assume a moving, reflecting target and a stationary source/observer, as shown in Figure 6.55. Assume that the sinusoidal waves leaving the stationary transmitter (TRX) propagate at velocity, c, over a distance, d, to the target, T. The target is moving at velocity, **V**, toward the source at an angle θ. Velocity **V** can thus be resolved into a component parallel to the line connecting TRX and the reflecting target, and a component perpendicular to the TRX-T line. These components are $|\mathbf{V}|\cos(\theta)$ and $|\mathbf{V}|\sin(\theta)$, respectively. The reflected wave from T propagates back to the stationary receiving transducer, RCX, along path d. The receiving sensor output waveform can be written as:

$$V_o = B\sin[\omega_o t + \psi] \qquad 6.119$$

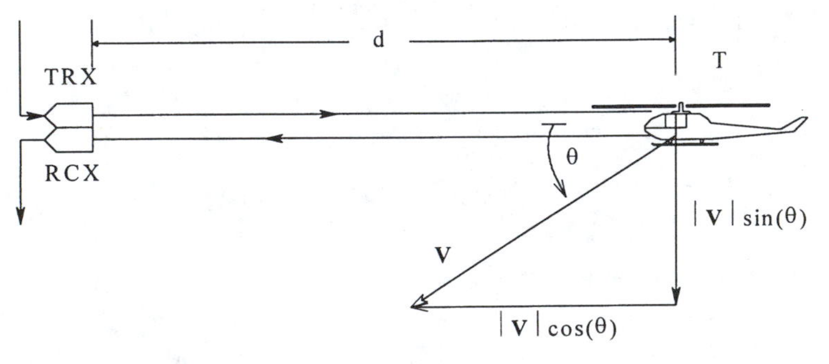

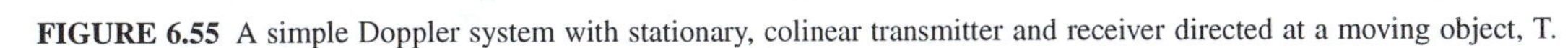
FIGURE 6.55 A simple Doppler system with stationary, colinear transmitter and receiver directed at a moving object, T.

The transmitted radian frequency is ω_o, and ψ represents the phase lag between the transmitted signal and the received signal. In general, the phase lag, ψ, is given by:

$$\psi = 2\pi 2d/\lambda = 2(2\pi)d/(c/fo) = \omega_o 2d/c \text{ radians} \qquad 6.120$$

However, the distance 2d is changing because of the target velocity component along the line from the tranducers to the target. Thus, the frequency of the received signal, ω_r, is the time derivative of its phase:

$$\omega_r = d[\omega_o t + \omega_o 2d/c]/dt = \omega_o(1 + [2/c]\dot{d}) = \omega_o(1 + [2|\mathbf{V}|/c]\cos(\theta)) \qquad 6.121$$

The frequency, ω_D, is the Doppler frequency shift which contains the velocity information:

$$\omega_D = \left(\omega_o 2|\mathbf{V}|/c\right)\cos(\theta) \text{ r/s} \qquad 6.122$$

Note that $\omega_r > \omega_o$, because in this example, the target is approaching the source/sensor.

In LDV systems, the wave return path is seldom colinear with the path from the source to the reflecting object. A basic LDV system geometry is shown in Figure 6.56. The source, S, emits light with wavelength, λ_s, that propagates toward reflecting particle, P. Unit vector, **i**, is directed from S to P. P moves with velocity, **V**. It is assumed that $(|\mathbf{V}|/c)^2 << 1$, so that relativistic effects are negligible. If **V** = 0, the number of wave-fronts striking the particle per second are: $f_s = c/\lambda_s$. (f_s for a HeNe laser is $3 \times 10^8/632.8 \times 10^{-9} = 4.741 \times 10^{14}$ Hz.) The number of wavefronts hitting a *moving* particle, P, per second are:

$$f_p = (c - \mathbf{V}\cdot\mathbf{i})/\lambda_s \text{ Hz} \qquad 6.123$$

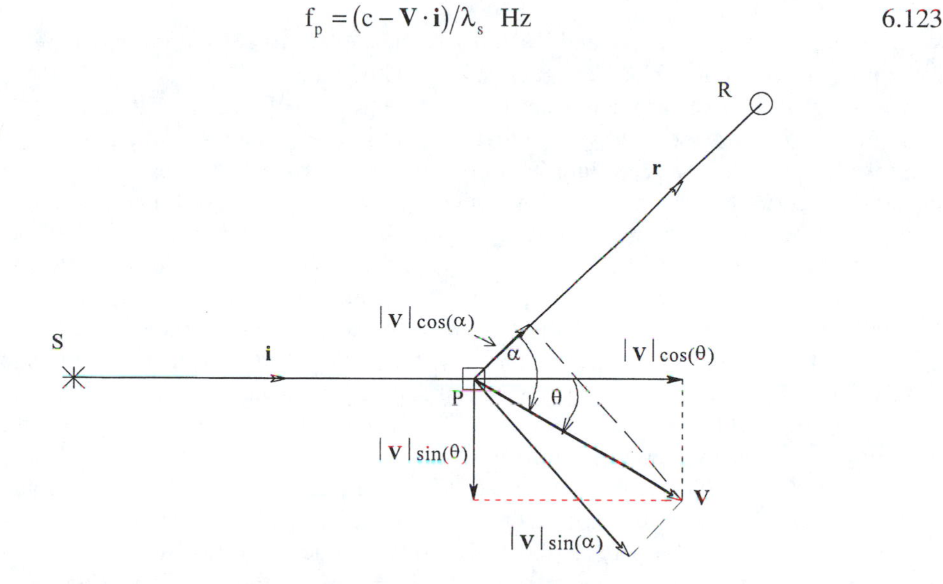

FIGURE 6.56 Two-dimensional geometry of a Doppler system used in LDV. Note that the transmitter (laser) (S) and the receiver (R) are at different angles with the object's velocity vector, V.

Note that the vector dot product gives the particle velocity component parallel to **i**. The wavelength λ_p apparent to the particle is:

$$\lambda_p = c/f_p = (\lambda_s c)/(c - \mathbf{V}\cdot\mathbf{i}) = (\lambda_s c)/[c - V\cos(\theta)] \qquad 6.124$$

Now the unit vector from the instantaneous particle position, P, to the stationary receiver, R, is **r**. An observer at R sees a scattered wavelength, λ_r, from P:

$$\lambda_r = (c - \mathbf{V}\cdot\mathbf{r})/f_p = \left[c - |\mathbf{V}|\cos(\theta)\right]/f_p \qquad 6.125$$

Here again, **V** · **r** represents the velocity component of the particle in the **r** direction. Hence, the frequency of the received, scattered radiation is:

$$f_r = c/\lambda_r \text{ Hz} \qquad 6.126$$

Substituting from above, we find:

$$f_r = c(c - \mathbf{V} \cdot \mathbf{i})/[\lambda_s(c - \mathbf{V} \cdot \mathbf{r})] \text{ Hz} \quad 6.127$$

Now the observed Doppler frequency shift is:

$$f_D = f_r - f_s = (c/\lambda_s)\left[\frac{c - \mathbf{V} \cdot \mathbf{i}}{c - \mathbf{V} \cdot \mathbf{r}} - 1\right] \text{ Hz} \quad 6.128$$

For $|\mathbf{V}| << c$, Equation 6.128 reduces to:

$$f_D \cong \mathbf{V} \cdot (\mathbf{r} - \mathbf{i})/\lambda_s = |\mathbf{V}|[\cos(\theta) - \cos(\alpha)]/\lambda_s \text{ Hz} \quad 6.129$$

where θ is the angle between **V** and **i**, and α is the angle between **V** and **r**. Note that if the reflected radiation is directed back along **i** to R coincident with S, the angle $\alpha = 180° + \theta$. Because $\cos(180 + \theta) = -\cos(\theta)$, f_D given above reduces to the value given by Equation 6.122.

Figure 6.57 shows the basic system first used for LDV by Yeh and Cummins in 1964. A HeNe laser was used with $\lambda_s = 632.8$ nm. 0.5 μm diameter polystyrene spheres were used in water at a density of 1:30,000 by volume to obtain scattering. Sphere velocity is assumed to be water velocity. The laser reference beam was frequency-shifted by 30 MHz using a Bragg cell acousto-optic modulator. The measurement beam was directed parallel to the fluid velocity vector, **V**, hence $\theta = 0°$. The scattered, Doppler-shifted beam was taken off at 30° relative to the angle of **V** and passed through a half-silvered mirror and directed to the photocathode of a photomultiplier tube (PMT), colinear with the frequency-shifted reference beam. Optical mixing occurs at the surface of the PMT photocathode. Optical mixing is a process analogous to square-law detection in a radio, where two sinewaves of different amplitudes and frequencies (f_s and f_r) are added together, then squared. This process produces sinusoidal terms at the PMT output with frequencies $2f_s$, $2f_r$, $(f_s + f_r)$, $(f_r - f_s)$, and 0 (DC). Of interest in LDV is the difference frequency term. In the Yeh and Cummins (1964) system, $f_s' = f_s + 30$ MHz, so the difference term equals 30 MHz for $\mathbf{V} = 0$. The Doppler frequency shift, f_D, is added to the 30 MHz center frequency in the Yeh and Cummins system, which was used to measure the parabolic laminar flow profile in water. They could measure velocities as low as 0.007 cm/s, corresponding to a Doppler frequency shift of 17.5 Hz. The maximum velocity they measured was 0.05 cm/s, corresponding to a Doppler shift of 125 Hz. The minimum detectable f_D was 10 Hz, or one part in 10^{14}!

Many architectures for LDV systems have been described. Figure 6.58 illustrates a "single-beam laser anemometer" described in Durst et al. (1974). The laser is focused on a microregion of the fluid whose velocity is to be measured. Part of the light is scattered and forms the signal beam; part is transmitted to form the reference beam. Through the use of lenses, slits, mirrors, and a half-silvered mirror, the signal and reference beams are combined at the surface of the photocathode in the PMT. This system was effectively used to measure both laminar flow in a liquid and the flow of air. No particles needed to be added to the liquid, but smoke particles were required to obtain adequate signal strength in the air measurements. Air velocities of up to 1000 fps were measured, corresponding to $f_D = 33$ MHz. It was found that the signal energy due to scattering was several hundred times stronger for small scattering angles in the forward direction, rather than for light scattered back along the input path. It should be noted that the Foreman system design which picks off the signal beam at a shallow forward angle suffers from the disadvantage that the laser and the PMT are on opposite sides of the fluid flow path.

A dual-beam LDV design, shown in Figure 6.59A, is described in Durst et al. (1974). Note that the laser output (reference) beam is perpendicular to V, and passes directly to the PMT through aperture A. The signal beam irradiates the scattering particle at angle θ. Some of the scattered signal light also passes through aperture A and mixes with the reference beam at the PMT photocathode. Observe that the dual-beam LDV is similar to the single-beam system except that

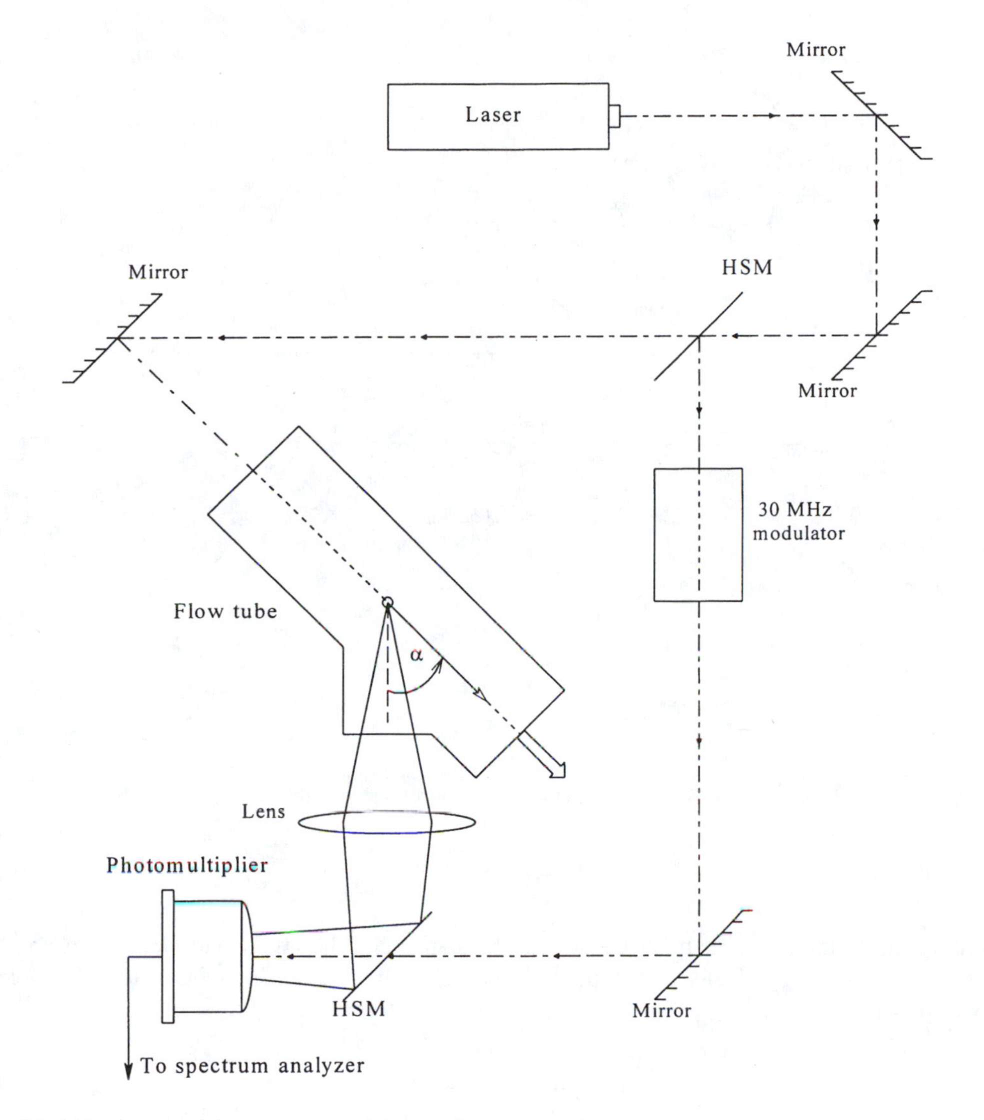

FIGURE 6.57 Diagram of the system used by Yeh and Cummins (1964) to first measure fluid velocity by Doppler shift. In their system, $\theta = 0$ and $\alpha = 30°$.

the laser and the PMT are interchanged. This system was used to measure the air flow profile at the inlet of a square duct (Figure 6.59B).

There are many other considerations in the design of LDV systems which we will only touch upon; space considerations do not permit us to treat them in detail. First, we have noted that the detected LDV signal originates from the interaction of particles suspended in the moving fluid with the laser beams. Particle size, density, and concentration in the fluid medium are important considerations in LDV system design. Particles are generally assumed to be randomly distributed in the sampling volume. Under conditions of low Reynolds number flow (laminar flow, no turbulence) one can generally assume that the particle velocity is the same as the fluid velocity. However, in oscillating or turbulent flows, particles generally do not follow the fluid without some lag, creating measurement errors. As an example of the error caused by particle lag, consider the force on a smoke particle in moving air. From the Stokes drag formula (Durst et al. 1974), we can write the drag force on the particle in one dimension as:

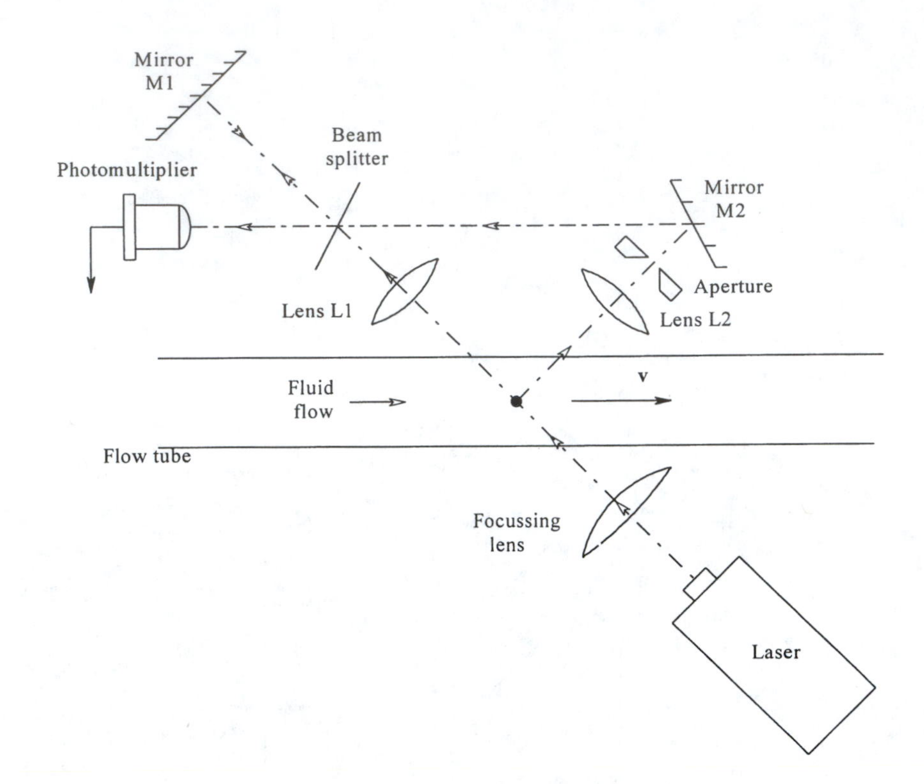

FIGURE 6.58 The LDV system of Foreman et al. (1966): a so-called single-beam anemometer.

$$F_d = 3\eta d(v_p - v_f) \quad 6.130$$

where η is the dynamic viscosity of the gas, d is the diameter of the smoke particle, v_p is its absolute velocity, and v_f is the velocity of the fluid. The Newtonian equation of motion for a particle of density ρ is thus

$$F = ma = (\pi d^3/6)\rho\dot{v}_p = -3\eta d(v_p - v_f) \quad 6.131$$

Equation 6.131 is an ordinary, first-order, linear differential equation in v_p, and if there is a step change in fluid velocity of size v_{f0}, then the ODE's solution is:

$$v_p(t) = v_{f0}[1 - \exp(-t/\tau)] \quad 6.132$$

The particle time constant is

$$\tau = (\pi d^2\rho)/(18\eta) \text{ seconds} \quad 6.133$$

This implies that the smoke particles lag behind gas velocity changes with first-order dynamics (the situation is actually far more complicated). A 1 μm diameter smoke particle with a density of 2.5 will have a τ of 20 μsec, and will follow velocity fluctuations well at frequencies up to about 8 kHz. Some other particles used in LDV are silicone oil aerosols in air (density difference 900), magnesium oxide in flame (density difference 1.8×10^4), PVC spheres in water (density difference

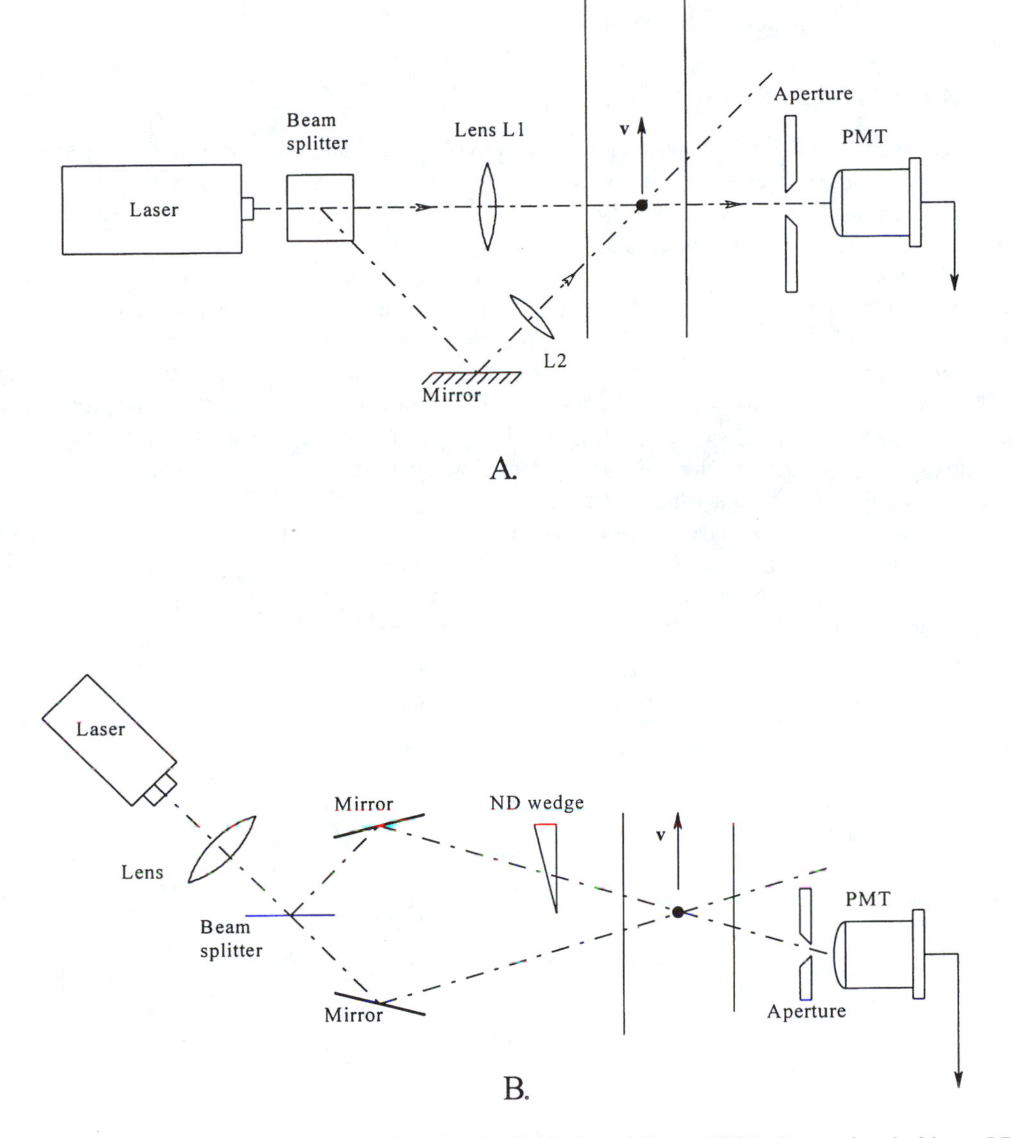

FIGURE 6.59 (A) A dual-beam LDV system described by Goldstein and Hagen (1967). (B) Another dual-beam LDV system, described by Goldstein and Kreid (1967).

1.54), polystyrene spheres in water (density difference 1). An 0.8 μm diameter particle with a relative density of 10^3 will follow 10 kHz velocity fluctuations with 1% error.

The calculation of the minimum laser power to achieve a given signal-to-noise ratio (SNR) in an LDV system is ordinarily quite complex, and we refer the interested reader to the many texts on LDV. Durst et al. (1974) derive expressions for the SNR at the photosensor as a function of incident power. For example, a PMT photosensor will give a 40 dB SNR for 10 μW power incident on the detector. The same 10 μW incident power gives a 50 dB SNR for PIN and avalanche photodiode sensors. At incident light powers below 10^{-7} W, the PMT generally has a better SNR than the photodiodes.

LDV systems are used in a wide variety of applications ranging from meteorology (studies of wind shear, clear air turbulence, tornadoes, etc.), studies of air flow in automotive streamlining, respiratory physiology, wind tunnel studies in aircraft design, investigation of laminar flows with recirculation and turbulent flow in ducts, studies of combustion dynamics in furnaces and gas turbines, and investigations of blood flows.

6.10 CHAPTER SUMMARY

In this chapter, we have examined various basic physical mechanisms underlying the operation of sensors and sensor systems used in making mechanical, chemical, physical, and electrical measurements. In some cases, we have also described certain sensor applications in order to give a better understanding of the sensor mechanisms.

Many sensors are seen to work through the input modality causing changes in resistance or capacitance. Other sensors involve the generation of voltages through the separation or release of charges (electrons, holes, ions); these include piezo-electric, Faraday, and Hall effect sensors. Still other sensors make use of certain physical wave effects, e.g., the Doppler and the Sagnac effects. Electrochemical sensors are seen to involve ion-specific chemical half-cells (batteries) and polarographic electrochemical reactions, such as used in the Clark oxygen cell. Radiation sensors are basically event-indicating devices, where the radiation event generates ionized atoms or molecules which are then detected by a variety of means.

Our survey of sensor mechanisms in this chapter has focused on the more important principles whereby physical quantities are converted to electrical signals. Our listing is by no means exhaustive. Chapter 7 deals with the various means by which we can measure specific physical quantities, using various types of transducers, devices, and systems.

7

Applications of Sensors to Physical Measurements

7.0 INTRODUCTION

In this chapter, we examine various means available to measure certain physical quantities. In Chapter 6 we reviewed various transducer mechanisms underlying measurement system function, such as changes in resistance. (Many physical quantities cause resistive changes, such as temperature, strain [caused by force or pressure], light intensity, humidity, etc.). Our purpose in this chapter is to examine the most commonly employed sensor types used to measure a given quantity, such as temperature. The quantities which we shall consider here are: acceleration, velocity and position (angular and rectilinear), temperature, pressure, force, and torque.

7.1 MEASUREMENT OF ANGULAR ACCELERATION, VELOCITY, AND DISPLACEMENT

There are many applications where it is important to know information about the attitude of a vehicle in inertial space. In the case of aircraft, instrumentation of roll, pitch, and yaw are critical for the design of robust, closed-loop, fly-by-wire autopilots and flight stabilization systems. Similar knowledge is important for the control of high-speed hydrofoils, hovercraft, and submarines. Guided missiles and "smart" bombs also require feedback on their roll, pitch, and yaw to perform their tasks effectively. In addition to vehicle stabilization and guidance, measurement of angular acceleration, velocity, and position is also important in the design of constrained mechanical systems which face variable loads and inputs, for example, automobile and truck suspension systems.

In the following sections, we describe some of the common sensor systems used to measure angular acceleration, angular velocity, and angular position.

7.1.1 Angular Acceleration Measurement

In this section, we describe several means by which angular acceleration can be measured. One is tempted to obtain angular acceleration by simply differentiating the output of a rate gyroscope. While theoretically correct, this procedure will generally result in a poor signal-to-noise ratio and in poor resolution because the process of differentiation enhances the noise present in the angular velocity signal. It is generally better to measure angular acceleration directly, as described below.

7.1.1.1 Angular Acceleration Measurement with a Constrained Mechanical Gyro

The inner gimbal of a mechanical gyro (see Section 7.1.2 for the development of the linear ordinary differential equations (ODEs) describing the behavior of mechanical gyroscopes) is restrained by a stiff, piezoelectric twister crystal as shown in Figure 7.1A. We may assume $\dot{\theta}$ and $\dot{\theta}$ are zero, and the torsional moment for the inner gimbal is given by the gyro equation, 7.26A, as:

$$M_\theta = -H_r\dot{\phi} \qquad 7.1$$

The twister piezoelectric transducer has an equivalent circuit given by Figure 7.1B at mechanical frequencies well below its mechanical resonance frequency. A net charge is displaced by the mechanical torque, $q = M_\theta d$ (d is a constant). Current is simply the rate of flow of charge, so the

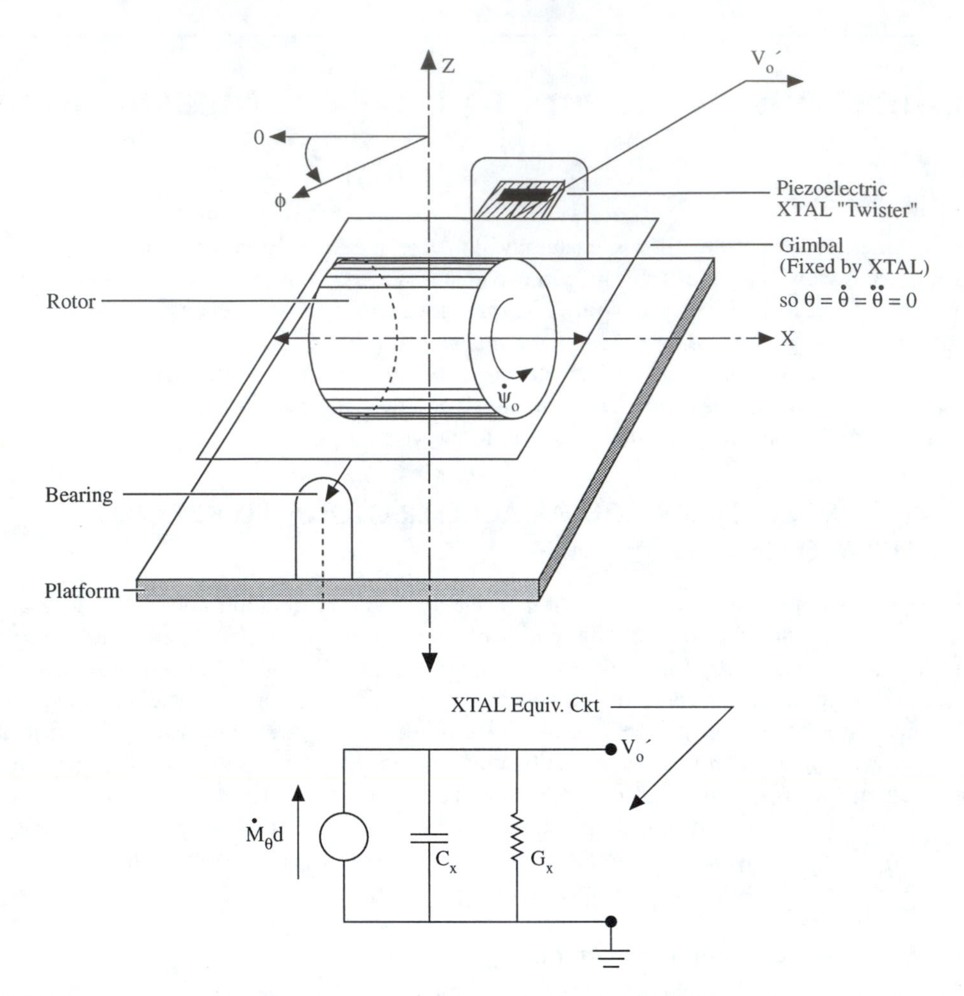

FIGURE 7.1 (A) A simple mechanical gyroscope with restrained gimbal. Gimbal torque is proportional to the input angular velocity, $\dot{\phi}$. A piezoelectic "twister" transducer senses the gimbal torque. (B) Equivalent circuit of a piezoelectric transducer at mechanical frequencies well below the mechanical resonance frequency of the transducer.

equivalent current source in the model for the transducer is $i_x = -H_r \ddot{\phi} d$, and the output voltage of the transducer is

$$V_o(s) = \frac{H_r R_x d\ddot{\Phi}(s)}{(sC_x R_x + 1)} \qquad 7.2$$

Thus, the steady-state output of the transducer is proportional to the angular acceleration of the gyro around the ϕ-axis. If a charge amplifier is used (see Section 2.6), the time constant in the output transfer function becomes $C_F R_F$, the feedback C and R of the charge amplifier. A far simpler means of measuring $\ddot{\phi}$ is shown below.

7.1.1.2 Simple Inertia Wheel-Spring-Dashpot Angular Accelerometer

Figure 7.2 illustrates a simple inertia wheel angular accelerometer in which the output voltage is proportional to the relative position of the wheel in the case, i.e.,

$$V_o = K_o(\phi_R - \phi_C) \qquad 7.3$$

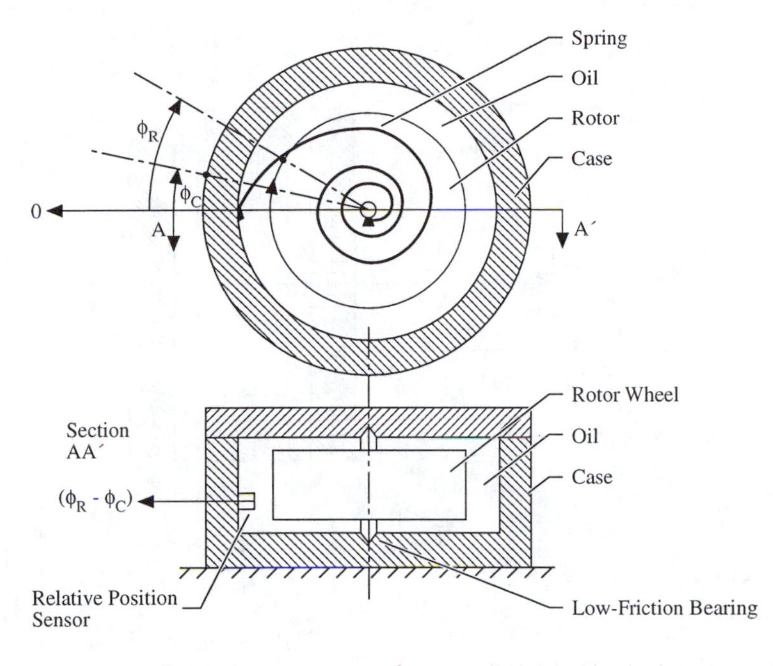

FIGURE 7.2 Two views of an inertia-wheel angular accelerometer. Wheel position in the case can be sensed photo-optically, or by a potentiometer. The inertia wheel is restrained by a helical spring.

The torque equation is

$$J\ddot{\phi}R + B(\dot{\phi}_R - \dot{\phi}_C) + K_S(\phi_R - \phi_C) = 0 \qquad 7.4$$

Here B is the viscous friction between the wheel and the case, and K_S is the torsion spring constant. If we Laplace transform the ODE of Equation 7.4 and solve for ϕ_R, then substitute into the expression for V_o, we obtain

$$\frac{V_o}{\ddot{\Phi}_C}(s) = \frac{K_o J/K_S}{s^2 J/K_S + sB/K_S + 1} \qquad 7.5$$

Thus, the steady-state output of this simple angular accelerometer is $V_o = \ddot{\phi}_C K_o J/K_S$. Such an accelerometer should be well damped so that the rotor wheel will not oscillate. The accelerometer's damping factor is easily seen to be $\zeta = B/(2\sqrt{JK_S})$, and should be made to be about 0.5 by adjusting B, to obtain a rapid transient response without excessive overshoots. The natural frequency of this accelerometer is $\omega_n = \sqrt{K_S/J}$ r/s. Thus, extending the accelerometer's bandwidth decreases its sensitivity, a good example of a gain-bandwidth trade-off. To see how it is possible to maintain high sensitivity and bandwidth, we will examine an active accelerometer design.

7.1.1.3 Servo Angular Accelerometer

In this design, shown in Figure 7.3, a torque motor is used as an active spring. The torque motor has a flat torque-speed characteristic over most of its speed range, the torque being proportional to the motor's input current, I_M. The accelerometer output voltage is made proportional to I_M. The torque balance equation for the servo accelerometer is

$$T_M = K_T I_M = J\ddot{\phi}_R + B(\dot{\phi}_R - \dot{\phi}_C) \qquad 7.6$$

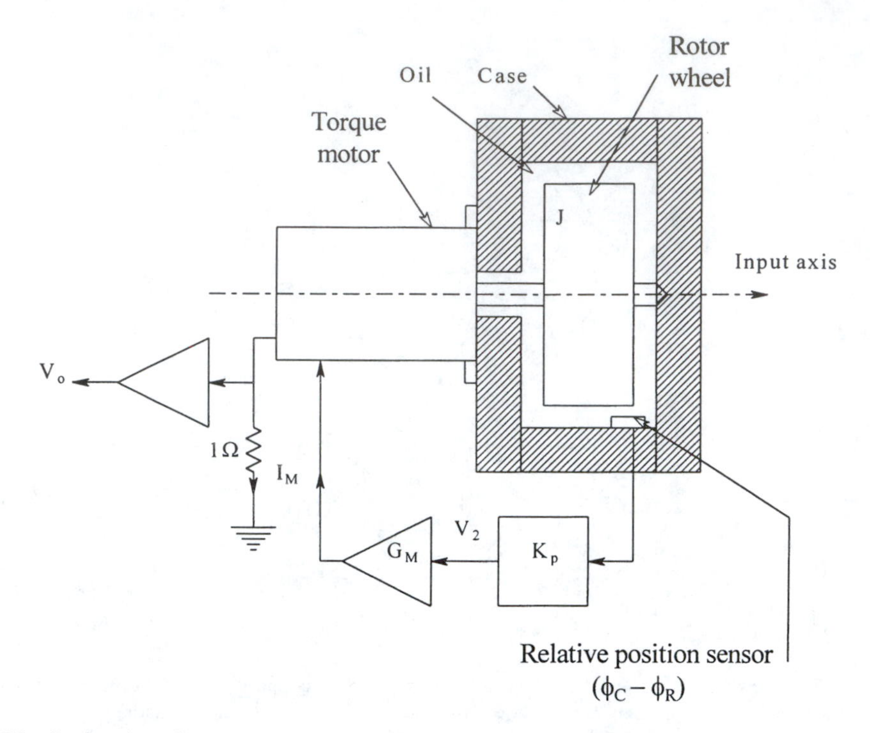

FIGURE 7.3 Section through a servo angular accelerometer. The torque motor current, I_M, may be shown to be proportional to the angular acceleration. G_M = transconductance amplifier. K_P, K_o = gains.

Also, we have

$$V_2 = K_P(\phi_C - \phi_R) \tag{7.7}$$

$$I_M = G_M V_2 \tag{7.8}$$

$$V_o = K_O I_M \tag{7.9}$$

Combining the equations above, it can be shown that

$$\frac{V_o}{\ddot{\Phi}_C}(s) = \frac{K_O/K_T}{s^2 J/(K_T K_P G_M) + sB/(K_T K_P G_M) + 1} \tag{7.10}$$

Now the natural frequency governing the high-frequency response is set by the electrical parameters K_T, K_P, and G_M. In this design, K_O can be made large to compensate for a high K_T, thus preserving sensitivity and bandwidth.

7.1.2 Angular Velocity Measurement with Gyroscopes

The angular velocity of vehicles, such as associated with the roll, pitch, or yaw rate of aircraft, missiles, boats, submarines, or spacecraft needs to be measured in order to control vehicle attitude in inertial space, and in the case of certain modern hypersonic aircraft, stabilize their flight under all conditions. The principal means of sensing angular velocity is by *rate gyro.* Rate gyros can be either electromechanical with spinning rotors, or optoelectronic, using the Sagnac effect, introduced in Section 6.9.2. In this section, we first discuss conventional mechanical gyroscopes, and then describe the design and sensitivity limitations of fiber optic, Sagnac effect gyros.

The innards of a conventional mechanical gyroscope are illustrated schematically in Figure 7.4. The rotor is spun electrically at a high, constant angular velocity, typically as high as 24,000 rpm for 400 Hz excitation. The rotor is held in an inner frame or gimbal, which is in turn suspended inside an outer gimbal. In early gyroscopes used in autopilots, etc., the angle between the inner and outer gimbal, θ, was sensed by a wiper on a wire-wound resistor potentiometer. Resolution therefore was limited by the diameter of the resistance wire vs. the area of the wiper. In our analysis which follows, we will assume infinite resolution in sensing the angle θ.

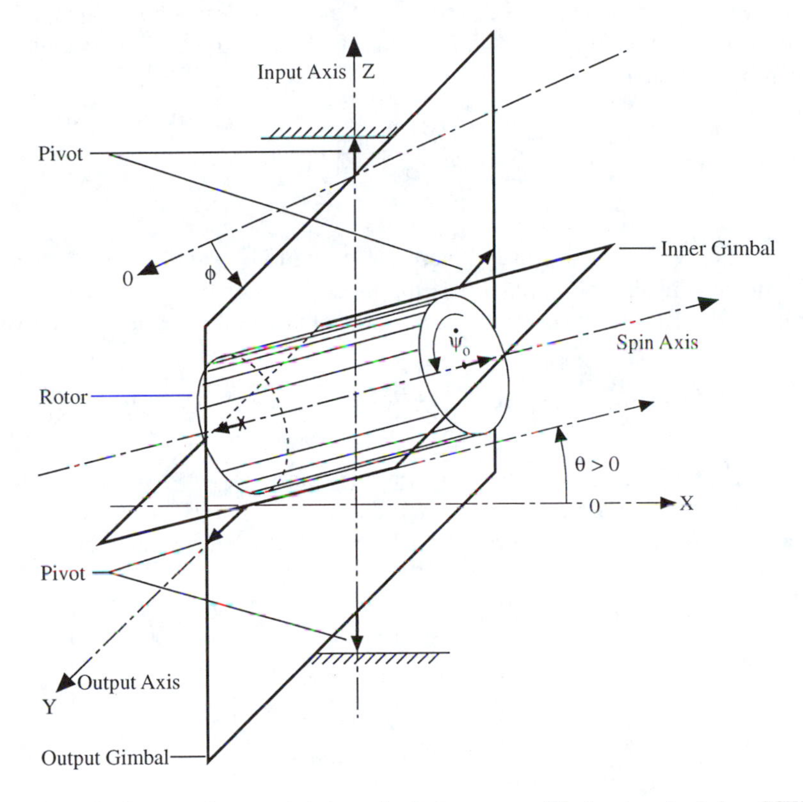

FIGURE 7.4 Schematic diagram of a two-gimbal, mechanical gyroscope. The input angle, ϕ, is a CCW rotation of the outer gimbal around the Z axis.

Referring to Figure 7.4, we will define the angle, ϕ, as the *input angle*. If the rotor is spinning in the direction shown with angular velocity $\dot{\psi}$, a positive (CCW) increase in ϕ will result in the gyroscopic effect of having the inner gymbal and rotor tip up to the right, giving a θ > 0 (CCW rotation). The detailed analysis of the dynamics of a mechanical gyro results in three coupled, highly nonlinear differential equations. Analysis proceeds smoothly if we define three mutually orthogonal coordinates for the gyro system; these are in fact the angular momentum axes of the rotor and inner gimbal. $\mathbf{H}_1$ is the angular momentum vector component perpendicular to the rotor spin axis and the plane of the inner gimbal. $\mathbf{H}_2$ is the angular momentum vector of the inner gimbal and rotor as it pivots on the outer gimbal. $\mathbf{H}_2$ is in the plane of the inner gimbal, and perpendicular to the rotor spin axis and $\mathbf{H}_1$. $\mathbf{H}_3$ is the angular momentum vector of the rotor, along the rotor axis and perpendicular to the other two momentum vectors. The direction of the angular momentum vectors is given by the right-hand rule; the fingers curl in the direction of rotation and the thumb points in the vector direction.

A good way to describe the dynamics of a mechanical gyroscope is by the Euler-Lagrange equations. These equations can be derived from Newton's laws, and provide a general means of describing the motion of complex mechanical systems having rotating or fixed coordinate systems. The Lagrange equations are of the form:

$$\frac{d}{dt}\left[\frac{\partial U}{\partial \dot{x}_j}\right] - \left[\frac{\partial U}{\partial x_j}\right] = M_j, \quad j = 1, 2, 3, \ldots N \qquad 7.11$$

where U is the total kinetic energy of the system, x_j is the jth general coordinate, and M_j is the jth generalized reaction force or torque. The differential work done in causing a small change in the coordinates can be written as

$$\delta W = \sum_j M_j \delta x_j \quad j = 1, 2, 3, \ldots N \qquad 7.12$$

In the case of the gyroscope shown in Figure 7.4, x_1 will be taken as the rotor angle, ψ; x_2 is defined as the inner gimbal angle, θ, and x_3 is the input angle, ϕ. The total kinetic energy of the gyro is the sum of the KEs associated with each orthogonal coordinate, and can be written as

$$U = (1/2)J_r\left[\dot{\psi} + \dot{\phi}\sin(\theta)\right]^2 + (1/2)J_1\dot{\theta}^2 + (1/2)J_1\left[\dot{\phi}\cos(\theta)\right]^2 \qquad 7.13$$

Here J_r is the moment of inertia of the rotor around its spin axis, and J_1 is the moment of inertia of the rotor and inner gimbal around the axes normal to the spin axis.

From Equation 7.11, we can write the first nonlinear ODE for the system. Consider j = 1, so $x_1 \equiv \psi$. It is easy to show that:

$$\frac{\partial U}{\partial \psi} = 0 \qquad 7.14$$

$$\frac{\partial U}{\partial \dot{\psi}} = J_r\dot{\psi} + J_r\dot{\phi}\sin(\theta) \qquad 7.15$$

$$\frac{d}{dt}\left[\frac{\partial U}{\partial \dot{\psi}}\right] = J_r\ddot{\psi} + J_r\ddot{\phi}\sin(\theta) + J_r\dot{\phi}\dot{\theta}\cos(\theta) \qquad 7.16$$

Hence the Lagrange equation for the gyro rotor spin axis is

$$M_\psi = J_r\left[\ddot{\psi} + \ddot{\phi}\sin(\theta) + \dot{\phi}\dot{\theta}\cos(\theta)\right] \qquad 7.17$$

The rotor friction moment, M_ψ, is usually taken as zero. The rotor is run at constant (synchronous) velocity, so $\ddot{\psi} = 0$. Now the second Lagrange equation is for $x_2 \equiv \theta$, the inner gimbal angle. In this case we have:

$$\frac{\partial U}{\partial \theta} = J_r\dot{\psi}\dot{\phi}\cos(\theta) + J_r\dot{\phi}^2\sin(\theta)\cos(\theta) - J_1\dot{\phi}^2\cos(\theta)\sin(\theta) \qquad 7.18$$

$$\frac{\partial U}{\partial \dot{\theta}} = J_1 \dot{\theta} \tag{7.19}$$

$$\frac{d}{dt}\left[\frac{\partial U}{\partial \dot{\theta}}\right] = J_1 \ddot{\theta} \tag{7.20}$$

Hence the Lagrange equation for the θ coordinate is

$$M_\theta = J_1 \ddot{\theta} - J_r \dot{\psi} \dot{\phi} \cos(\theta) - J_r \dot{\phi}^2 \cos(\theta) + J_1 \dot{\phi}^2 \cos(\theta)\sin(\theta) \tag{7.21}$$

The third Lagrange equation is written for $x_3 \equiv \phi$, the input angle.

$$\frac{\partial U}{\partial \phi} = 0 \tag{7.22}$$

$$\frac{\partial U}{\partial \dot{\phi}} = J_r \dot{\psi} \sin(\theta) + J_r \dot{\phi} \sin^2(\theta) + J_1 \dot{\phi} \cos^2(\theta) \tag{7.23}$$

$$\frac{d}{dt}\left[\frac{\partial U}{\partial \dot{\phi}}\right] = J_r \ddot{\psi} \sin(\theta) + J_r \dot{\psi} \dot{\theta} \cos(\theta) + J_r \ddot{\phi} \sin(\theta) + 2J_r \dot{\phi} \dot{\theta} \sin(\theta)\cos(\theta) - 2J_1 \dot{\phi} \dot{\theta} \cos(\theta)\sin(\theta) + J_1 \ddot{\phi} \cos^2(\theta) \tag{7.24}$$

The Lagrange equation for ϕ is just

$$M_\phi = \frac{d}{dt}\left[\frac{\partial U}{\partial \dot{\phi}}\right] \tag{7.25}$$

To linearize the gyro ODEs, Equations 7.21 and 7.25, we note again that $\ddot{\psi} = 0$, and we assume that system operation is such that $|\theta| < 15°$ so $\sin(\theta) = \theta$ in radians. Also, $\cos(\theta) \approx 1$, and terms containing products of the small quantities, $\dot{\phi}$, $\ddot{\phi}$, $\dot{\theta}$, and $\ddot{\theta}$ are negligible. Under these approximations, we finally obtain the two simultaneous, *linear* ODEs which describe the dynamic behavior of a mechanical gyro.

$$M_\theta = J_1 \ddot{\theta} - H_r \dot{\phi} \tag{7.26A}$$

$$M_\phi = H_r \dot{\theta} + J_1 \ddot{\phi} \tag{7.26B}$$

H_r is the angular momentum of the rapidly spinning rotor, $J_r \dot{\psi}$.

Under most conditions, the outer gimbal angle, ϕ, is considered to be a rotation of the vehicle about a principal axis, such as its roll axis. The response of the roll gyro having its outer gimbal axis pointing in the direction of the roll axis is considered to be the inner gimbal angle, θ, or in some cases, the torque, M_θ. The linear gyro differential equations above can be Laplace transformed and written as

$$M_\theta(s) = J_1 s^2 \Theta(s) - H_r s \Phi(s) \quad 7.27A$$

$$M_\phi(s) = H_r s \Theta(s) + J_1 s^2 \Phi(s) \quad 7.27B$$

We shall use these equations to describe the dynamics of the rate gyro, and later, the behavior of the rate integrating gyro (RIG).

7.1.2.1 The Rate Gyro

In the rate gyro, the motion of the inner gimbal is restrained by a spring with torque constant, K_s, in parallel with a viscous damper (dashpot) with rate constant B, as shown in Figure 7.5. In this gyro, we have dispensed with the outer gimbal, and assume the gyro base rotates an angle ϕ, as shown, as the result of the vehicle turning around the input axis. The rotor gimbal moment is thus:

$$M_\theta = -\Theta(s)[Bs + K_s] \quad 7.28$$

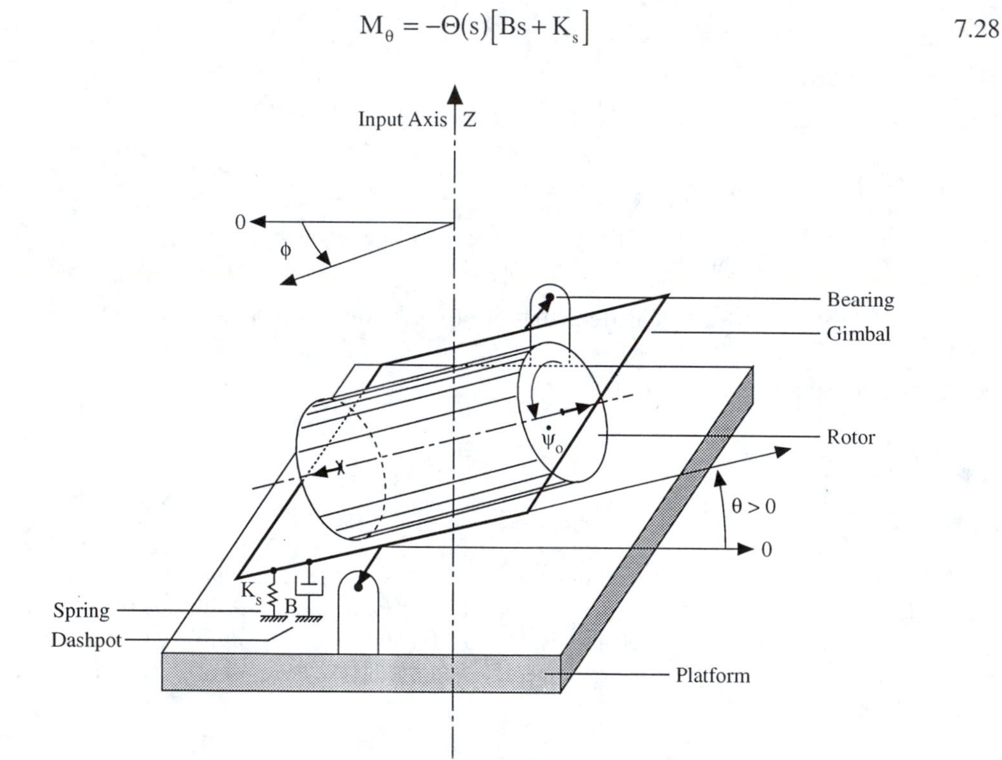

FIGURE 7.5 Schematic diagram of a one-gimbal, mechanical, rate gyro. In the steady state, the gimbal angle $\theta = \dot{\phi} H_r / K_s$.

The gyro equations can now be written:

$$0 = \Theta(s)[s^2 J_1 + sB + K_s] - \Phi(s) H_r s \quad 7.29A$$

$$M_\phi = \Theta(s) s H_r + \Phi(s) J_1 s^2 \quad 7.29B$$

Using Cramer's rule, we can solve Equations 7.29A and B for the relationship between θ and ϕ.

$$\frac{\Theta}{\Phi}(s) = \frac{\begin{vmatrix} 0 & -sH_r \\ M_\phi & s^2J_1 \end{vmatrix}}{\begin{vmatrix} s^2J_1 + sB + K_s & 0 \\ sH_r & M \end{vmatrix}} = \frac{M_\phi sH_r}{M_\phi\left(s^2J_1 + sB + K_s\right)} = \frac{sH_r/K_s}{s^2J_1/K_s + sB/K_s + 1} \quad 7.30$$

This transfer function is easily seen to be a relation between the inner gimbal tilt angle, θ, and the angular velocity of the gyro around the input axis, $\dot{\phi}$.

$$\frac{\Theta}{\dot{\Phi}}(s) = \frac{H_r/K_s}{s^2J_1/K_s + sB/K_s + 1} \quad 7.31$$

For a constant velocity (step) input,

$$\dot{\Phi}(s) = \dot{\phi}_o/s \quad 7.32$$

and the output angle is given by:

$$\Theta(s) = \frac{\dot{\phi}_o\left(H_r/K_s\right)}{s\left(s^2J_1/K_s + sB/K_s + 1\right)} \quad 7.33$$

In the time domain, the response θ(t) is shown in Figure 7.6. It is a classic second-order low-pass system step response, with steady-state value, $\theta_{SS} = \dot{\phi}_o H_r/K_s$. Thus, in response to a constant angular velocity input, the rate gyro rotor tips an angle θ_{SS}. θ can be measured by a low-friction potentiometer wiper or by a frictionless optical coding disk.

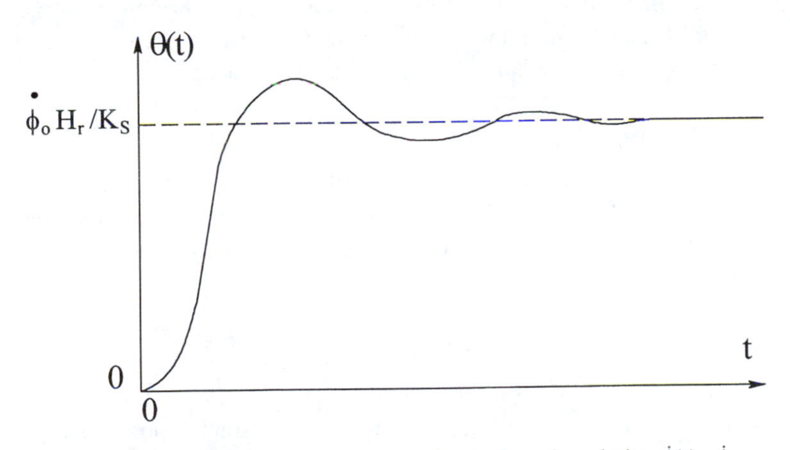

FIGURE 7.6 Second-order response of the rate gyro to a step input of angular velocity, $\dot{\phi}(t) = \dot{\phi}_o U(t)$.

Another approach to rate gyro design is shown in Figure 7.7. Here, the rotor gimbal is constrained not to rotate by a stiff torsion spring. The spring has strain gauges cemented to it arranged so that torque exerted on the spring produces an output voltage. Referring to the linearized gyro equations, 7.27A and B, for practical purposes we assume that, $\dot{\theta} = \ddot{\theta} = 0$. Thus, Equation 7.27A reduces to

$$M_\theta(s) = -H_r s\Phi(s) \quad 7.34$$

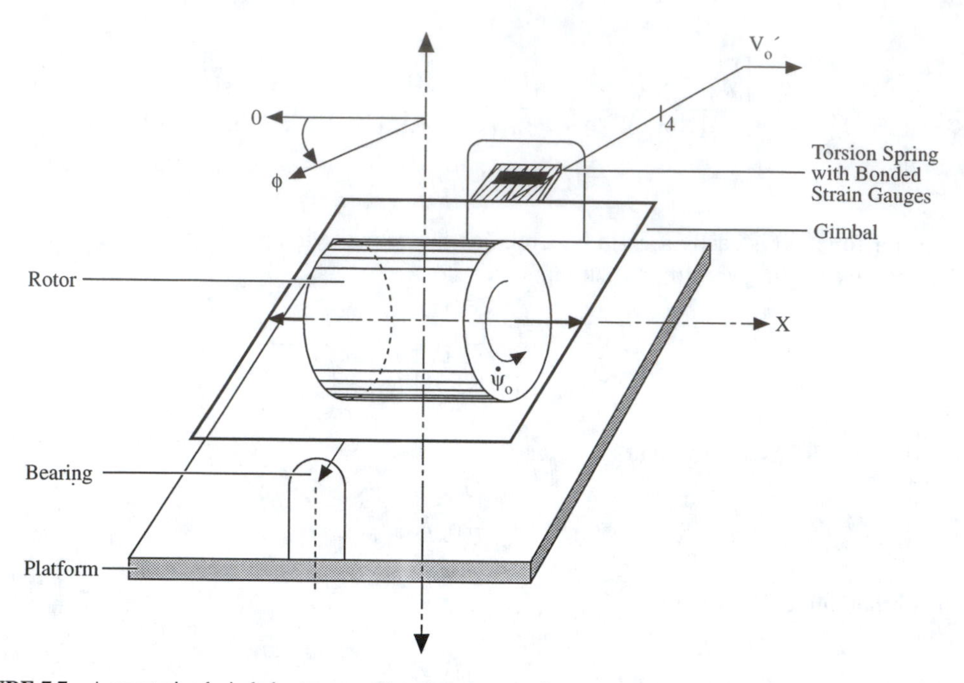

FIGURE 7.7 A constrained gimbal rate gyro. The gimbal torque is measured by a strain gauge, and is proportional to $\dot{\phi}$. Compare this system to the constrained gyro angular accelerometer shown in Figure 7.1.

Hence, the output of the strain gauge bridge attached to the torsion spring can be written as

$$v_o(t) = K_B H_r \dot{\phi} \tag{7.35}$$

Thus, we see that the inner gimbal does not need to move appreciably to sense $\dot{\phi}$. Note that the $\dot{\phi}$ term is multiplied by a $\cos(\beta)$ term in the relations above when the gyro is rotated around some axis at an angle β with the input or ϕ-axis.

7.1.2.2 Sagnac Effect Fiber Optic Gyroscopes

In Section 6.9.2 we derived the relation for interference fringe shifts as a function of rotation rate for a Sagnac interferometer. The fractional fringe shift, Δz (generally measured electro-optically), was shown to be related to the input rotation rate by Equation 6.115. When we include the cosine term for off-axis rotation, the relation for Δz becomes:

$$\Delta z = \dot{\phi}\cos(\beta)[4A/c\lambda] \tag{7.36}$$

Equation 7.36 above describes the fractional fringe shift for a single optical circuit. Sensitivity of the Sagnac gyroscope can be increased by increasing the number of optical circuits; a convenient way of doing this is by using multiple turns of an optical fiber waveguide. A very simple fiber optic (FO) gyro is shown in Figure 7.8. Since N turns of FO cable are used, the fringe shift equation is easily seen to become:

$$\Delta z = \dot{\phi}\cos(\beta)[4NA/c\lambda] = \dot{\phi}\cos(\beta)[2(N2\pi R)R/c\lambda] = \dot{\phi}\cos(\beta)[2L_F R/c\lambda] \tag{7.37}$$

where R is the radius of the FO coils, N is the number of turns, c is the speed of light on the optical fiber, λ is the wavelength of the light, and L_F is the total linear length of the FO cable used. Practical

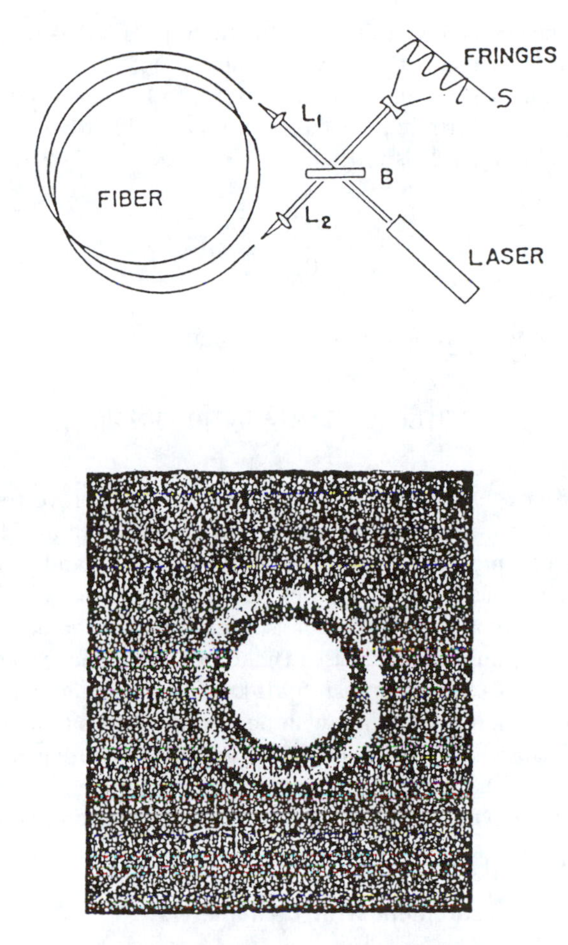

FIGURE 7.8 (A) A simple Sagnac effect, DC fiber optic gyroscope. (B) Interference fringes seen on the screen S under steady-state conditions. The fringes shift radially for $|\dot{\phi}| > 0$.

considerations concerning signal-to-noise ratio limit the maximum number of turns used, or L_F. Light intensity from the laser or laser diode source is attenuated by the optical fiber as a function of path length. This attenuation is given by the well-known exponential relation,

$$P = P_o 10^{-\alpha L_F} \tag{7.38}$$

Here α is the attenuation constant which is wavelength and fiber dependent, P_o is the input power in watts, P is the output power, and L_F is the length of the fiber. The noise associated with the photodetector used to sense fringe shifts may be expressed in terms of an rms uncertainty in the fringe shift, Δz_{RMS}. This noise is given by

$$\Delta zRMS = (1/2\pi)\sqrt{\frac{h\nu}{\eta P}}\sqrt{B} = (1/2\pi)\sqrt{\frac{h\nu}{\eta P_o 10^{-\alpha L_F}}}\sqrt{B} \tag{7.39}$$

Hence the fringe shift rms signal-to-noise ratio (SNR) can be written as

$$SNR = \frac{\Delta z}{\Delta z_{RMS}} = \frac{\phi(2L_F R/c\lambda)}{(1/2\pi)\sqrt{(h\nu B/\eta)P_o 10^{-\alpha L_F}}} = \phi \frac{4\pi R\sqrt{P_o}}{c\lambda\sqrt{h\nu B/\eta}}\left(L_F 10^{-\alpha L_F/2}\right) \tag{7.40}$$

where h is Planck's constant, ν is the frequency of the light in Hz, B is the Hz noise bandwidth, and η is the photodetector's quantum efficiency (typically 1/2).

Regardless of ϕ, the SNR has a maximum with respect to L_F. This optimum L_F can be found by differentiating the right-hand term in parentheses in Equation 7.40 with respect to L_F and then setting the derivative equal to zero. This procedure yields

$$L_{F(OPT)} = \frac{2}{\alpha IN(10)} = (2/\alpha)(0.4343) \tag{7.41}$$

The optimum number of turns, N_{OPT}, can be found from:

$$N_{OPT} = INT\left[L_{F(OPT)}/2\pi R\right] = INT[0.13824/(\alpha R)] \tag{7.42}$$

Other forms of the Sagnac FO gyro have been developed to improve resolution by allowing more precise measurement of the fringe shift, and carrier operation to avoid 1/f noise associated with DC and low-frequency measurements. By introducing AC modulation of the light phase traversing the FO coil, the intensity of the interference fringes is also AC modulated, and fringe shift detection can be made more precise and noise-free because photodetector outputs can be amplified out of the range of amplifier 1/f noise. The use of single-mode, polarization-preserving optical fibers with attenuations (α) of less than 5 dB/km and integrated coherent optical systems has permitted the design of Sagnac FO gyros with noise-limited sensitivities of greater than 10^{-2} degrees/hour using a 1 Hz noise bandwidth centered at the modulating frequency (10^{-2} degrees/hour is 5.56×10^{-6} degrees/second). By way of contrast, the earth's rotation rate is 4.167×10^{-3} degrees/second. Hence an FO gyro with its input axis pointing north can easily measure the earth's axial spin.

7.1.3 Angular Velocity Measurement with Tachometers

A tachometer is a sensor/system which converts the angular velocity of a rotating shaft to an electrical output, generally a voltage. Tachometers are typically used to measure the angular velocity, generally in RPM, of rotating machines such as electric motors, generators, turbines, pumps, internal combustion engines, etc.

A commonly found form of tachometer is the DC, permanent magnet generator, in which the output voltage linearly follows the shaft velocity. Shaft velocity is read by a calibrated, DC voltmeter. Response time of the DC generator tachometer is basically set by the mechanical natural frequency of the analog voltmeter. In the case of a digital voltmeter, the response time is limited by the low-pass filtering at the input analog-to-digital converter.

Other tachometers avoid the expense of the DC generator, and use either optical or magnetic means to generate a train of pulses whose repetition rate is proportional to the shaft speed. A light beam is bounced off one or more reflecting strips cemented to the rotating shaft. Every time the shaft comes around, the light beam is reflected onto a photosensor, producing a train of pulses which is shaped by a comparator and one-shot multivibrator. The average value of this conditioned pulse train is seen to be proportional to the RPMs of the shaft. A low-pass filter averages the pulse train, and an op amp is used to correct for DC offset and set the output calibration, as shown in Figure 7.9. An alternate means of generating a pulse train whose rate is proportional to shaft speed is to cement permanent magnets to the rotating shaft, and place a digital Hall sensor in a fixed location near the path of the magnet. Every time the magnet comes close to the Hall sensor, its output changes state, generating the pulse train. The disadvantage of the pulse averaging tachometer is that the low-pass filter time constant needs to be long to suppress output voltage ripple at low RPM. The long time constant means a slow response to changes in RPM.

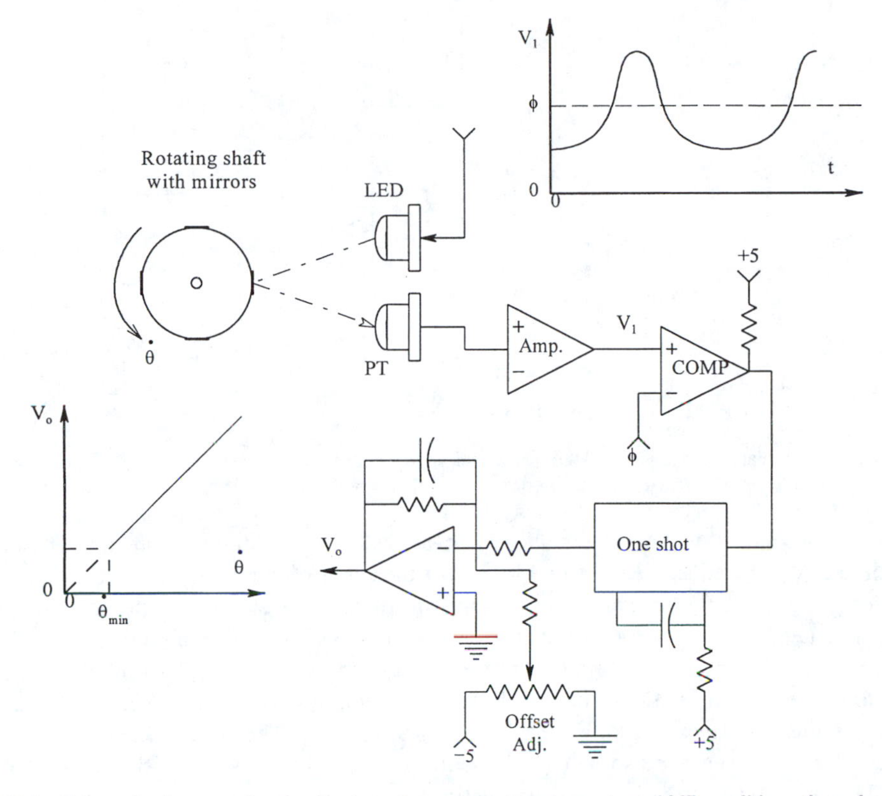

FIGURE 7.9 Schematic diagram of a simple photoelectronic tachometer system. AMP conditions the pulses from the phototransistor, PT. The conditioned pulses V_1 are made into standard-width TTL pulses by the amplitude comparator COMP and the one-shot multivibrator, OS. Op amp OPA low-pass filters the pulse train from OS and adjusts the DC level. The DC output of OPA, V_o, is proportional to $\dot{\phi}$; however, at very low $\dot{\phi}$ the low frequency of the pulse train causes excessive ripple on V_o, setting a practical lower limit to tachometer resolution determined by the feedback time constant, RC.

One way to get around the problem of a long filter time constant being required for ripple-free output at low RPMs is to use a phase-lock loop (PLL) to convert the rate of the pulse train to a DC output proportional to RPM. The block diagram of a PLL tachometer is shown in Figure 7.10. A 4046 CMOS PLL is used to convert input frequency to a proportional output voltage, V_o. In the block diagram, the first integrator is required to convert the input signal's frequency to phase, the quantity the PLL operates on. In the steady state, $\theta_i = \theta_o$, and it is easy to show that the DC output voltage is given by

$$V_o = 2\pi f_i N/K_V \tag{7.43}$$

where N is a digital frequency divider ratio (N = 1, 2, 4, …, 64, …, etc.), and K_V is the PLL VCO constant in (r/s)/V. Reduction of the block diagram yields the transfer function:

$$\frac{V_o}{f_i}(s) = \frac{2\pi N/K_V}{s^2C(R_1 + R_2)N/(K_PK_V) + s(1 + CR_2K_PK_V/N)N/K_PK_V + 1} \tag{7.44}$$

where K_P is the phase detector gain. The PLL tachometer is seen to be a second-order system with a natural frequency of $\omega_n = \sqrt{K_PK_V/(C(R_1 + R_2)N)}$ r/s. In this design, the VCO runs at N times the input frequency at lock. This allows the measurement of very low frequencies. All VCOs have

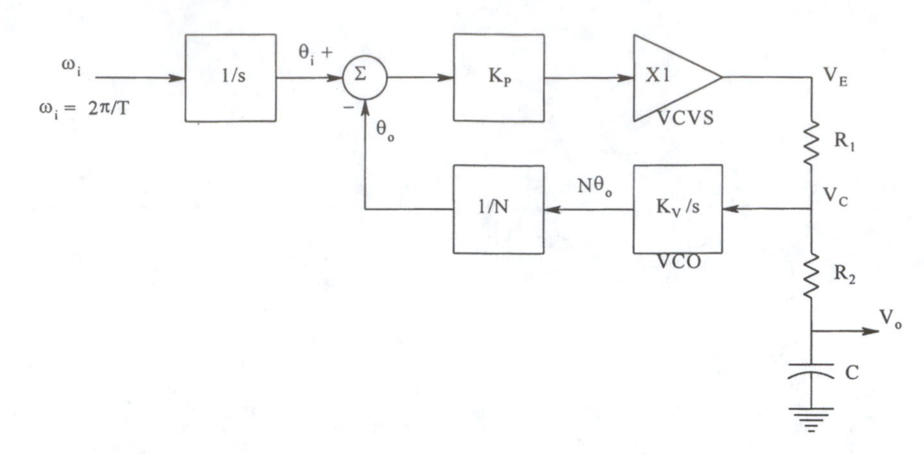

FIGURE 7.10 Block diagram of a 4046 CMOS phase-lock loop used as a tachometer. The input integrator converts the input event rate (frequency) to the phase input, θ_i, of the phase detector.

a minimum frequency at which they will oscillate, as well as a maximum frequency. The minimum and maximum VCO frequencies, divided by N, set the input range of this PLL tachometer.

The final means for the measurement of angular velocity which we will describe can be used with turbine wheels as well as rotating shafts. A system known as an *instantaneous pulse frequency demodulator* (IPFD) operates on a pulse sequence obtained by the rotation of magnets attached to the turbine wheel or shaft passing fixed Hall switch, or alternately, by the rotation of illuminated, alternating white and black markings past a fixed photosensor. The IPFD measures the period or time interval between each pulse pair in the sequence, and then calculates the reciprocal of that interval. The reciprocal of the ith interval is defined as the ith element of instantaneous frequency of the pulse sequence, r_i, and is proportional to the angular velocity of the shaft or wheel. If, for example, each revolution produces 36 pulses, then 2400 RPM would produce 86,400 ppm, or 1440 pps. Since the IPFD is responsive to the reciprocal of the length of each period, any change in the speed of rotation from 2400 RPM will be sensed within about 0.7 msec. This is a considerably faster response time than is afforded by pulse averaging or PLL tachometers. The IPFD is thus a useful system to study subtle, short-term changes in angular velocity. A block diagram for an analog IPFD tachometer is shown in Figure 7.11A. The processing time, i.e., the time required to charge the capacitor to the maximum voltage, sets the maximum possible instantaneous frequency (shortest input period). The maximum input period (lowest input frequency) is determined by the lowest capacitor voltage at which the decay waveform follows the curve given by:

$$v_c(t) = \frac{C/\beta}{t + \tau_o} \tag{7.45}$$

where V_o is the peak voltage the capacitor is charged to at time t = 0, when the voltage begins decaying hyperbolically. Time τ_o is the processing time between the (i – 1)th pulse and the beginning of the hyperbolic decay, $v_c(t)$, which is sampled at time t_i when the ith pulse occurs. $\beta = C/(\tau_o V_o)$. The hyperbolic decay of voltage from the capacitor occurs because the capacitor is discharged through a nonlinear resistor with the voltampere relation:

$$i_{nl} = \beta v_c^2 \tag{7.46}$$

Hence the node equation for the capacitor discharge is:

$$\dot{v}_c + (\beta/C)v_c^2 = 0, \quad t \geq 0 \tag{7.47}$$

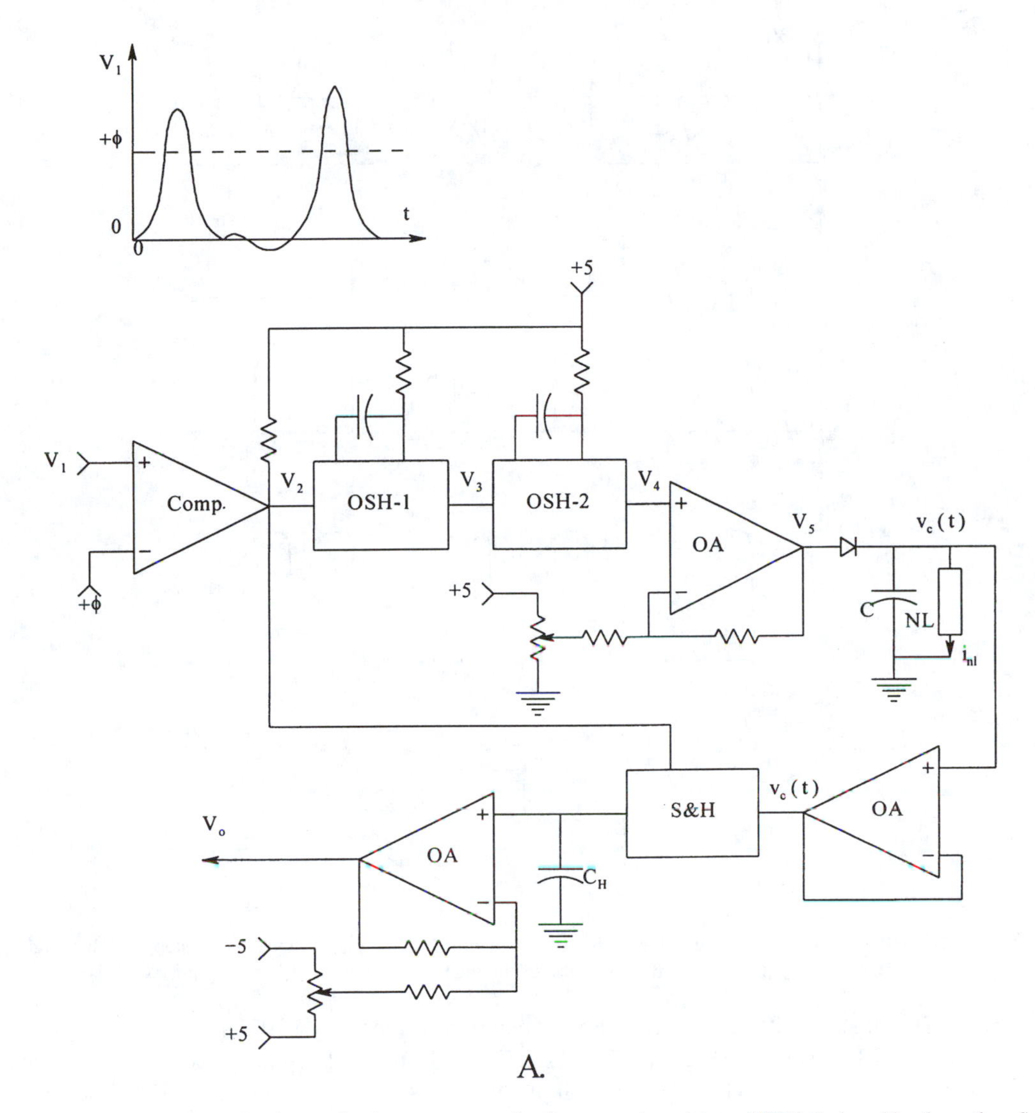

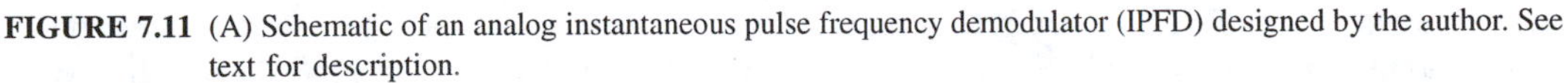

FIGURE 7.11 (A) Schematic of an analog instantaneous pulse frequency demodulator (IPFD) designed by the author. See text for description.

This is a form of Bernoulli's equation, which has a solution for the initial condition of $v_c = V_o$ at $t = 0$ given by Equation 7.45 above. The value of the hyperbola is sampled and held at the occurence time of the (i + 1)th pulse, generating a stepwise voltage waveform, q(t). The height of each step is the instantaneous frequency of the preceding pulse pair interval, and is proportional to RPM. In mathematical terms, the IPFD output can be written

$$q(t) = k\sum_{i=1}^{\infty} r_i\left[U(t - t_i) - U(t - t_{i+1})\right] \qquad 7.48$$

Here U(t − a) is the unit step function, zero for $t < a$, and 1 for $t \geq a$. r_i is the ith element of instantaneous frequency, defined as

$$r_i \equiv 1/(t - t_{i+1}) \text{ second}^{-1} \qquad 7.49$$

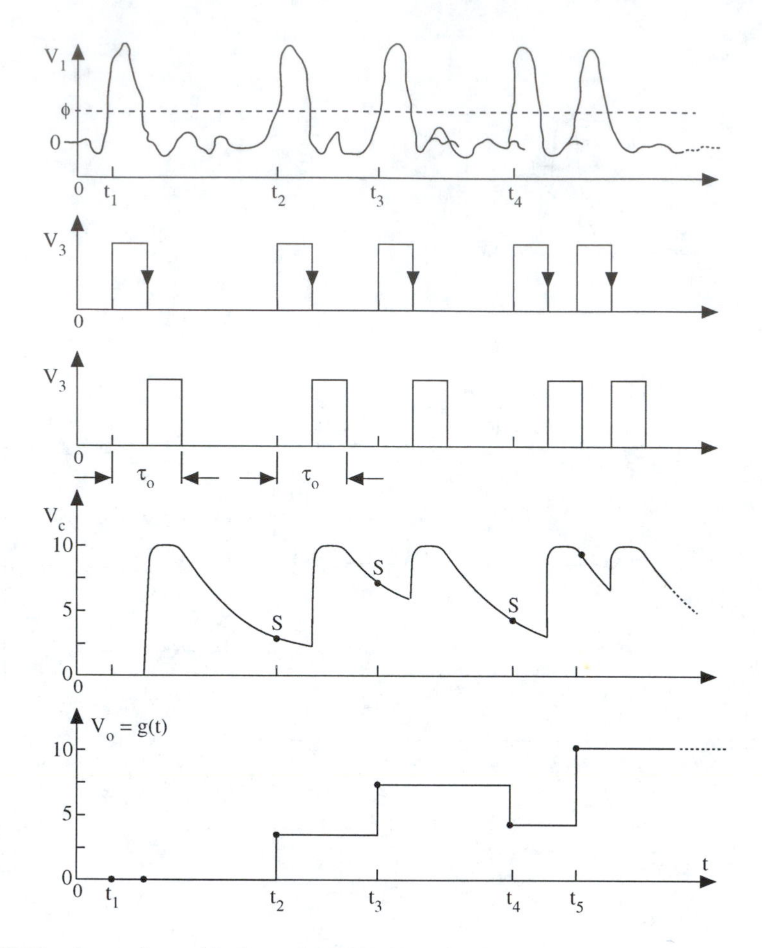

FIGURE 7.11 (B) Waveforms observed in the analog IPFD. Note that the output voltage appears stepwise, the height of a step being proportional to the reciprocal of the time between the two preceding pulses.

Figure 7.11B illustrates key waveforms in the operation of the analog IPFD.

Webster (1978) describes a digital IPFD which is used as a cardiotachometer (to measure heart rate). The Webster sytem has an analog output which is also given by Equation 7.48 above. Illustrated in Figure 7.12, the reciprocal relation between interpulse interval and r_i is ingeniously generated by the use of a 12-bit digital-to-analog converter (DAC) whose output resistance increases linearly with the size of its binary input, N_i. N_i is proportional to $(t_i - t_{i-1})$. A timing diagram for the Webster (1978) IPFD is shown in Figure 7.13. The op amp's output is thus $V_o = V_R R_F / R_{DAC} = Kr_i$ over the interval (t_i, t_{i+1}).

In summary, we have seen that tachometers can have a wide variety of embodiments, ranging from simple DC generators whose output voltage is proportional to shaft angular velocity, to systems which generate pulse trains whose frequencies (and instantaneous frequencies) are proportional to angular velocity. The pulse trains can be processed by averaging phase-lock-loops, or instantaneous pulse frequency demodulators to obtain DC signals proportional to angular velocity.

7.1.4 Angular Position Measurement with Gyroscopes

Modern inertial navigation systems and attitude control systems such as those used by space vehicles, helicopters, aircraft, submarines, etc., need estimates of angular position for their operation. There are several simple modifications of mechanical and Sagnac FO gyroscopes which will allow them to sense angular position, rather than angular velocity. These include the rate integrating gyro (RIG) and various servo platform gyroscope configurations.

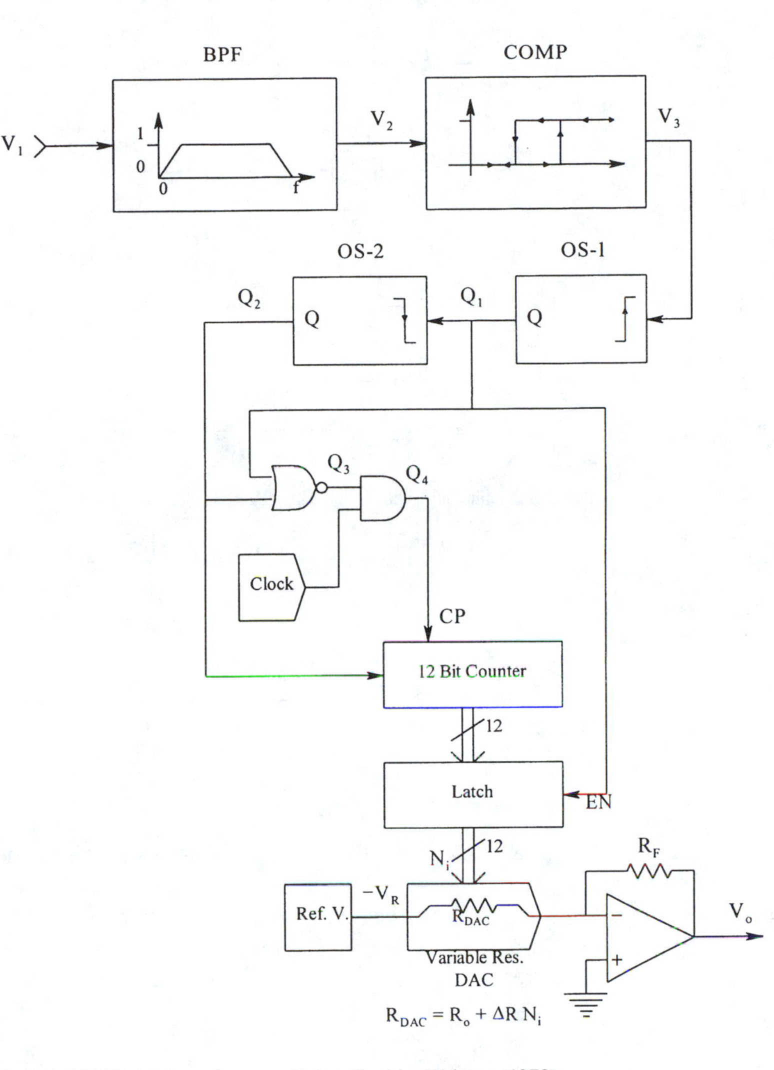

FIGURE 7.12 A digital IPFD (with analog ouput) described by Webster (1978).

The Rate Integrating Gyro (RIG) is the same as the rate gyro shown in Figure 7.5, except that the spring is removed. The linearized gyro equations now become

$$M_\theta = -B\dot{\theta} = J_1\ddot{\theta} - H_r\dot{\phi} \qquad 7.50$$

$$M_\phi = H_r\dot{\theta} + J_1\ddot{\phi} \qquad 7.51$$

Solution of Equations 7.50 and 7.51 yields

$$\frac{\Theta}{\Phi}(s) = \frac{H_r/J_1}{s + B/J_1} \qquad 7.52$$

Thus, in the steady state, a rotation ϕ_o around the input axis will produce an inner gimbal deflection of $\phi_o H_r/J_1$ radians. The RIG is seen to be approximately a first-order, low-pass sensor system with a time constant of J_1/B seconds.

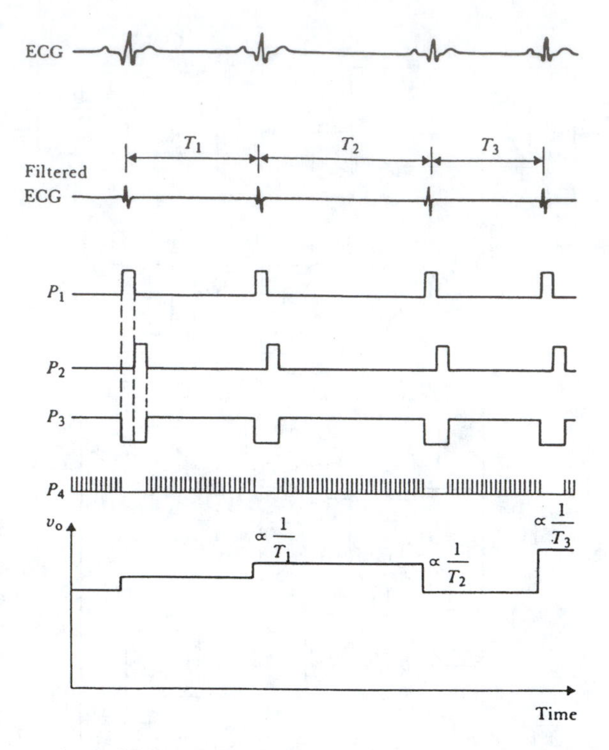

FIGURE 7.13 Timing diagram for the Webster IPFD of Figure 7.12. Here the IPFD is used to measure the beat-by-beat heart rate.

Figure 7.14 illustrates the use of a feedback platform to obtain an output signal proportional to the steady-state input rotation, ϕ_{io}. In this case, we use a servomotor-driven platform to counter-rotate the RIG as the assembly rotates with the vehicle. The inner gimbal angle, θ, is sensed by a potentiometer or optical sensor. The servogyro equations are as follows:

$$\phi_e = \phi_i - \phi_m \tag{7.53}$$

$$M_\theta = 0 = B\dot{\theta} + J_I\ddot{\theta} - H_r\dot{\phi}_e \tag{7.54}$$

$$M_\phi = H_r\dot{\theta} + J_1\ddot{\phi}_e \tag{7.55}$$

The platform position is driven by the θ signal through a proportional plus derivative compensated servomotor. The platform position may be related to θ by the relation

$$\frac{\Phi_m}{\Theta}(s) = \frac{K_M K_A(\tau_c s + 1)}{s(\tau_M s + 1)} \tag{7.56}$$

The block diagram describing the servo RIG is shown in Figure 7.15. System output is proportional to the platform angle, ϕ_m. From the block diagram, we can write the transfer function of the servo RIG:

$$\frac{V_o}{\Phi_i}(s) = \frac{K_o(\tau_C s + 1)}{s^3 J_1 \tau_M/(H_r K_A K_M) + s^2(B\tau_M + J_1)/(H_r K_A K_M) + s(\tau_C H_r K_A K_M)/(H_r K_A K_M) + 1} \tag{7.57}$$

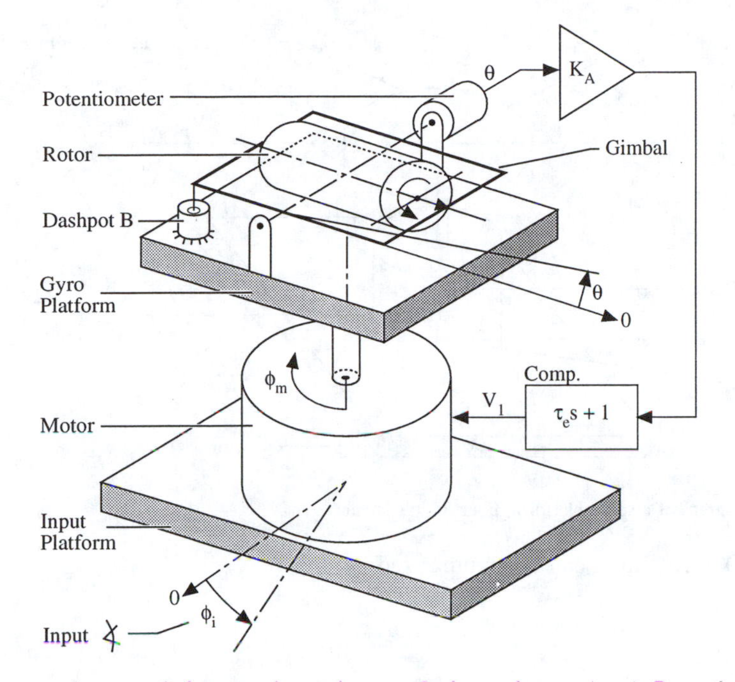

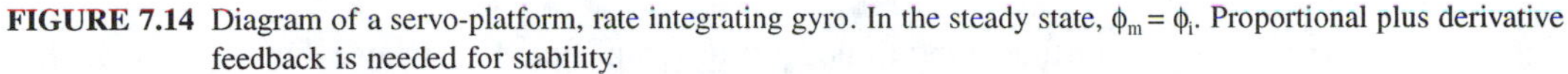

FIGURE 7.14 Diagram of a servo-platform, rate integrating gyro. In the steady state, $\phi_m = \phi_i$. Proportional plus derivative feedback is needed for stability.

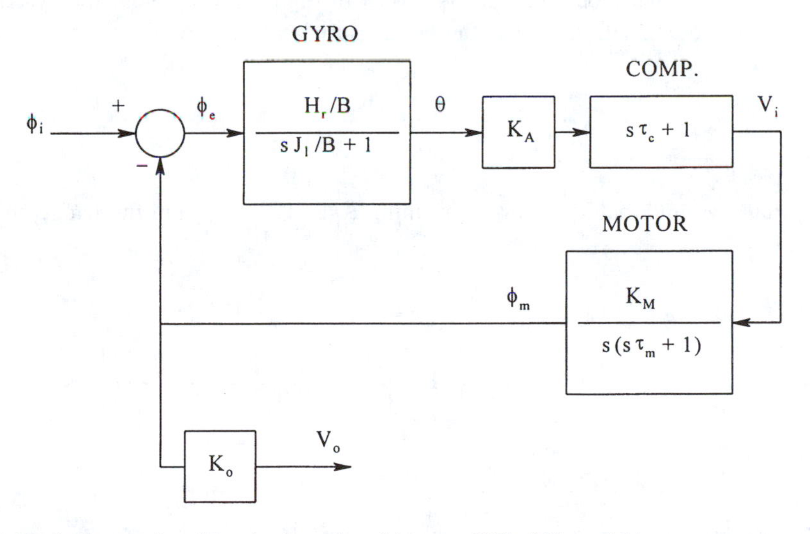

FIGURE 7.15 Block diagram describing the dynamics of the servo-RIG of Figure 7.14.

In the steady state, $V_o = K_o\Phi_i$. The advantage of the servo-RIG configuration is that sensitivity is set by the parameter, K_o, and the dynamic range is not limited by inner gimbal tilt angle, θ. In fact, it is easy to show that in the steady state, $\theta/\Phi_i \rightarrow 0$. The natural frequency of the servo-RIG is seen to depend on the gains K_A and K_M.

A *Sagnac FO gyro* can also be placed on a servo platform to realize a RIG. Figure 7.16 illustrates such a system. As in the example above, the platform counter-rotates against the input angle so we have:

$$\phi_e = \phi_i - \phi_m \qquad 7.58$$

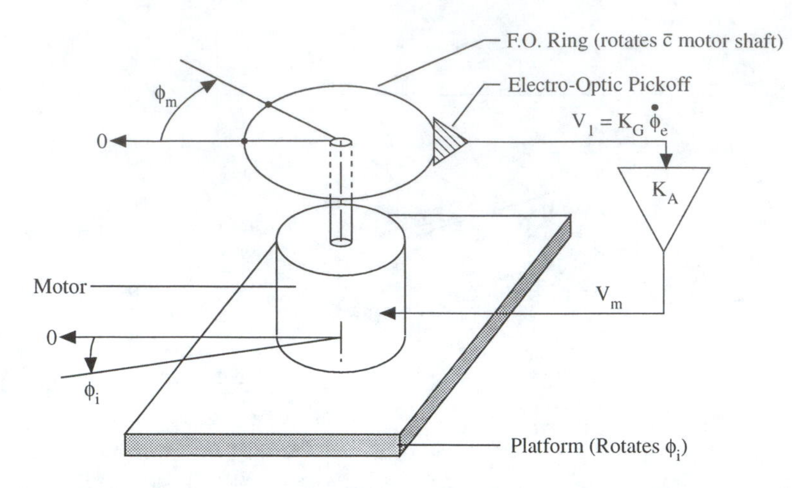

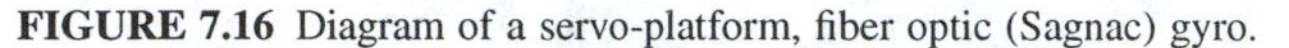

FIGURE 7.16 Diagram of a servo-platform, fiber optic (Sagnac) gyro.

The Sagnac FO gyro output can be summarized by

$$V_1 = K_G \dot{\phi}_e \tag{7.59}$$

The Sagnac gyro output is further conditioned by an amplifier, K_A, whose output acts on the servomotor system. The net block diagram is shown in Figure 7.17. Again, the system output is proportional to the motor rotation, which can be measured with a potentiometer or a coding disk.

$$\frac{V_o}{\Phi_i}(s) = \frac{K_O}{s\tau_M/(K_G K_A K_M)+1} \tag{7.60}$$

In the steady state we find that $V_o = K_o \phi_{io}$, assuming $K_G K_M K_A >> 1$, and the fringe shift is nulled to zero.

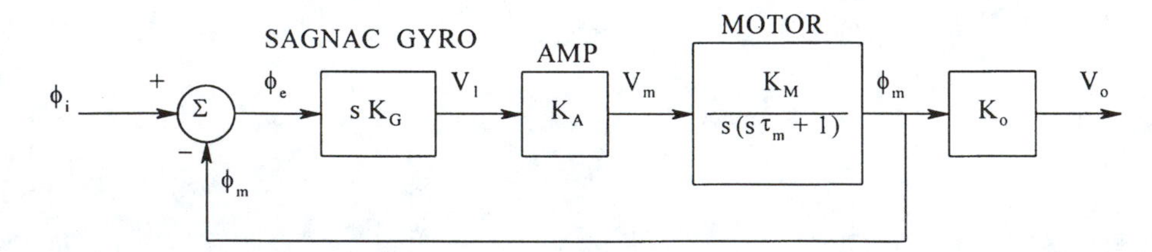

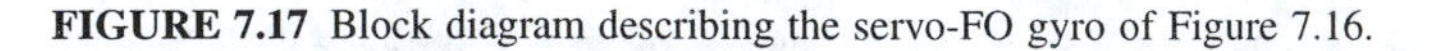

FIGURE 7.17 Block diagram describing the servo-FO gyro of Figure 7.16.

Placing a gyroscope on a servo-controlled platform is seen to be an effective means of increasing the gyro's dynamic range of response and its natural frequency. In designing such a system, it is clear that there is a trade-off between the considerable extra cost of the servo platform and the increase in performance obtained.

7.1.5 Angular Position Measurement with Clinometers

Gyroscopes, discussed in the previous two sections, were seen to operate either in a gravitational field, or in a gravitational field-free environment, such as in an orbiting vehicle. Clinometers, on the other hand, require gravity to work. Clinometers are used on ships to sense roll and pitch angle, and on heavy earth-moving machinery to sense tilt. Obviously, excessive tilt or roll can lead to capsizing, and loss of life and/or the equipment. A clinometer is considerably simpler in design than is a mechanical gyroscope, and consequently is a far less expensive instrument. A schematic design of a clinometer is shown in Figure 7.18. The roll axis is perpendicular to the plane of the page. A pendulum of total mass, M, and moment of inertia, J, is suspended in the center of a cylindrical case. The case is filled with light oil for viscous damping, and the position of the clinometer pendulum relative to the case is measured by a potentiometer wiper with negligible friction. A torsion spring acts to restore the pendulum to the center position of the case. Summing the torques around the pivot point, we can write

$$0 = -J\ddot{\theta}_P + K_S(\dot{\theta}_C - \dot{\theta}_P) + B(\theta_C - \theta_P) - MgR\sin(\theta P) \quad 7.61$$

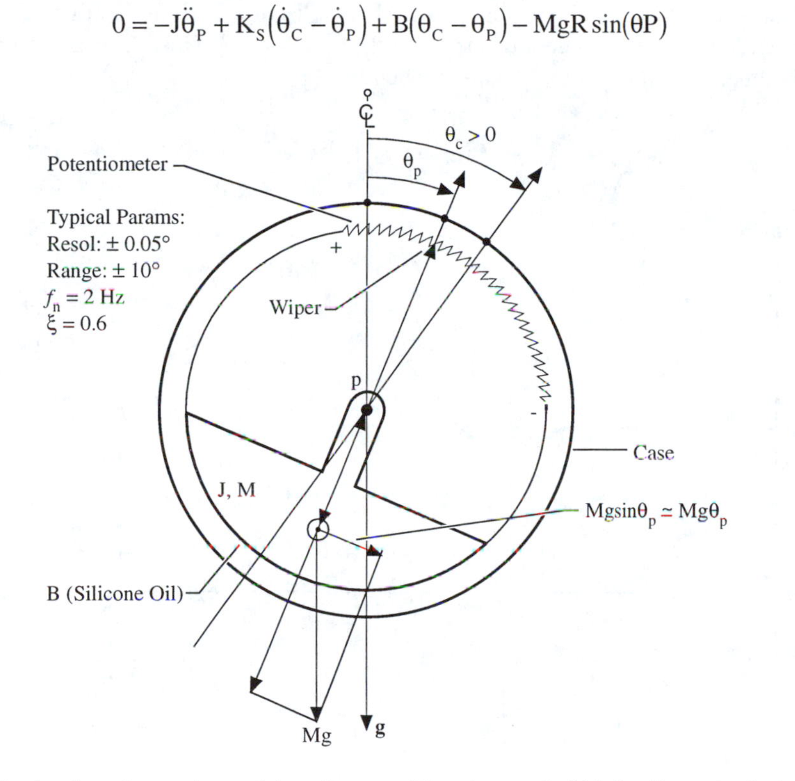

FIGURE 7.18 Section through a passive pendulum clinometer. The axis around which the clinometer tilts is perpendicular to the plane of the paper.

This ODE can be linearized by assuming $|\theta_P| < 15°$, so $\sin(\theta_P) \approx \theta_P$ (in radians). Laplace transforming and separating Θ terms, we can write:

$$\frac{\Theta_P}{\Theta_C}(s) = \frac{(sB + K_S)}{s^2J + sB + (K_S + MgR)} \quad 7.62$$

Now the clinometer output is assumed to be given by:

$$V_o = K_o(\theta_C - \theta_P) \quad 7.63$$

Substituting Equation 7.62 in the relation for V_o, we obtain finally

$$\frac{V_o}{\Theta_C}(s) = \frac{K_o(s^2J/MgR + 1)MgR}{[s^2J/(K_S + MgR) + sB/(K_S + MgR) + 1](K_S + MgR)} \quad 7.64$$

The input/output transfer function of the pendulum clinometer is that of a mechanical notch filter. The transfer function has a pair of conjugate zeros on the jω axis in the s-plane at $\omega_0 = \sqrt{MgR/J}$ r/s, and a pair of complex-conjugate poles with $\omega_n = \sqrt{(K_S + MgR)/J}$ r/s. The steady-state frequency response of the pendulum clinometer to $\theta_C(t) = \theta_o\sin(\omega t)$ is shown in Figure 7.19. The response is flat from DC to just below the notch frequency, ω_o. In order that the pendulum follow rapid changes in θ_C, $\omega_o = \sqrt{MgR/J}$ should be made as large as possible. In the limiting case, the pendulum can be considered to be a mass point, M, at a distance R from the pivot. It is well known that such a pendulum has a moment of inertia given by $J = MR^2$. Thus the notch frequency of the pendulum is independent of its mass, and is given by

$$\omega_o = \sqrt{g/R} \text{ r/s} \quad 7.65$$

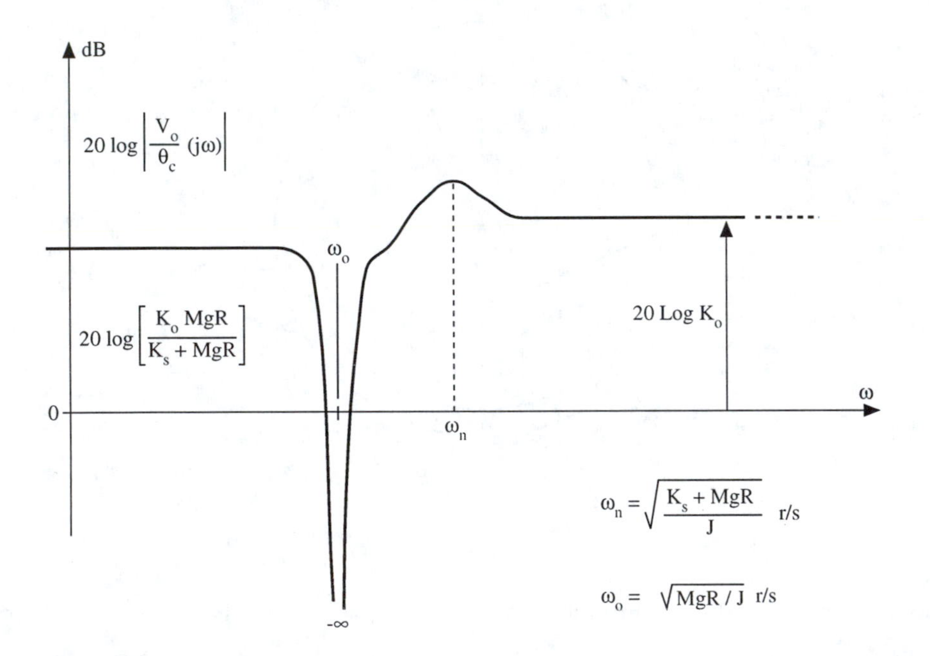

FIGURE 7.19 Steady-state frequency response of the pendulum clinometer subject to a sinusoidal tilting of its case. Note that it is a mechanical notch filter.

Hence a 10 cm pendulum in the earth's gravitational field will have an f_o of about 1.6 Hz, a 2 cm pendulum has an f_o of 3.5 Hz. The measurement of most vehicle roll angles is done under almost static conditions, so the dynamic performance of the clinometer angle sensor is generally not an issue. Even the roll rate of large ships is slow enough so that most clinometers can follow the roll angle accurately.

7.1.6 Angular Position Measurement of Shafts

We have already discussed the use of optical coding disks in Section 6.9.1, and the use of precision potentiometers in Section 6.2.5 in the measurement of shaft angles. Another magnetic sensor used in angular position sensing is the Inductosyn™, manufactured by Farrand Controls, Valhalla, NY. Inductosyns have been used in the inertial navigation systems, the periscopes, and the fire control systems of Polaris, Poiseidon, and Trident submarines. They have also been used as position sensing transducers in the Remote Manipulator System Arm installed in the cargo hold of the U.S. space shuttle. Physically, the rotational inductosyn consists of two opposing, flat, toroidal cores, each with a flat printed circuit coil with repeated parallel hairpin turns attached to one flat surface of the flat toroidal cores. The length of one complete cycle of the hairpin coil is called the pitch, P. Inductive coupling between the two flat coils is used to measure the angular displacement between the two coils. An AC current, generally between 2.5 and 100 kHz in frequency, is used as the excitation to one coil. The voltage induced in the pickup coil is given by

$$v_i(t) = kV_s \cos(2\pi x/P)\sin(2\pi ft) \qquad 7.66$$

where k is the coupling coefficient, V_s is the input voltage, x is the angular displacement between the inductosyn disks, P is the angular period of one hairpin turn, and f is the frequency of the excitation voltage. If a second pickup winding is located adjacent to the first, but given an angular dispacement of P/4 with respect to the first pickup coil, the voltage induced in the second coil is given by:

$$v_i'(t) = kV_s \sin(2\pi x/P)\sin(2\pi ft) \qquad 7.67$$

Only the amplitudes of v_i and v_i' change with x/P, their phase and frequency remain the same.

If the space quadrature windings of an inductosyn are excited by constant amplitude carriers in time quadrature, the resulting output signal has a constant amplitude, but its phase undergoes a 360° shift for an angular rotation of x = P.

The available nominal outer diameters of inductosyn "rings" are 3", 4", 7", and 12". Limiting accuracies for these sizes are ±2, ±1.5, ±1.0, and ±0.5 seconds of arc, respectively. Initial alignment and concentricity of the inductosyn rings is essential in order to realize the high angular precision cited above. Care also must be taken to avoid electrostatic coupling and ground loops to avoid artifact errors in the operation of these sensors. Inductosyns are used with custom electronics packages to give a DC voltage output proportional to angular position between the coils. Because there is no physical contact required between the inductosyn stator and rotor, there are no brushes or wipers to wear out, similar to the case of optical coding disks.

7.2 MEASUREMENT OF LINEAR ACCELERATION, VELOCITY, AND DISPLACEMENT

In this section we examine certain means to measure the linear acceleration of structures and vehicles with sensors known as accelerometers, as well as certain techniques for measuring linear velocity and position. The output of a linear accelerometer can be integrated to obtain a velocity estimate, a second integration yields a position or deflection estimate. There are also other means for finding the velocity of structures and vehicles, including the use of the Doppler effect with light, microwaves, or sound. Knowing the precise position of a vehicle is really a navigation problem, and if done from within a closed vehicle such as a submerged submarine or an aircraft flying under instrument flight rules (IFR), it is an inertial navigation problem. On the other hand, the relative position of small objects and structures can be measured with a wide variety of position sensors, including ones which do not physically touch the object. This latter class of no-touch position sensors include the use of ultrasound, microwaves, and light in various schemes that utilize the waves' propagation velocity, interferometry, or phase-lock techniques.

7.2.1 Linear Accelerometers

A basic linear accelerometer is shown in Figure 7.20. A mass, M, is restrained to slide inside a cylindrical case. There is viscous damping, B, between the case and the mass, and a spring capable of compression or stretch is used to return the mass to the center of the case when there is no acceleration. The relative position between the moving mass and the case is measured by a linear variable differential transformer (LVDT). The body of LVDT is attached to the case, and its moving core is attached to the movable mass. The case itself is firmly attached to the object whose acceleration is to be measured. A typical application of an accelerometer might be to measure the acceleration of a vibration table used to test the mechanical stability of components, or to measure the up/down acceleration of an automobile body while driving over a bumpy road. For a mass and spring accelerometer moving in the x-direction, we may write, using Newton's law:

$$0 = M\ddot{x}_2 + B(\dot{x}_2 - \dot{x}_1) + K_S(x_2 - x_1) \quad 7.68$$

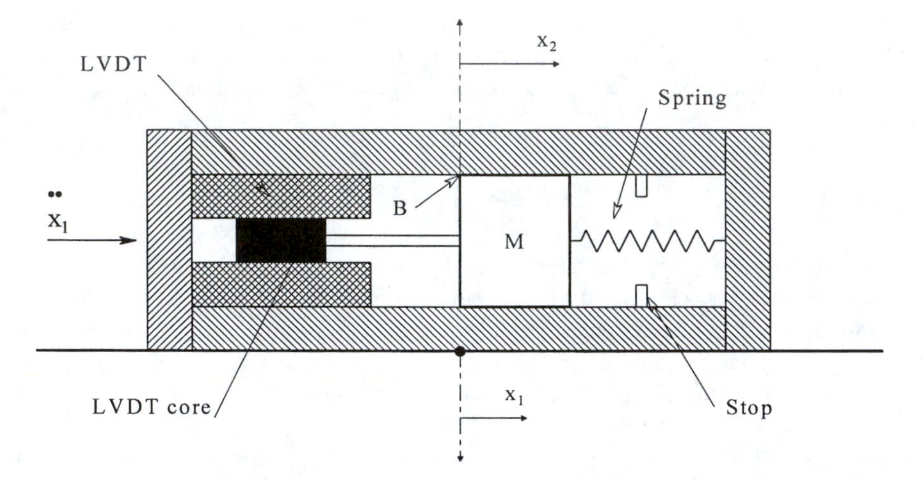

FIGURE 7.20 Section through a passive linear accelerometer. Mass position is sensed by the LVDT.

Here x_1 is the position of the case (and the structure whose acceleration we wish to measure), and x_2 is the position of the mass inside the case with respect to an external inertial reference point (x = 0). This ODE may be Laplace transformed and put in the form of a transfer function.

$$\frac{X_2}{X_1}(s) = \frac{sB/K_S + 1}{s^2M/K_S + sB/K_S + 1} \quad 7.69$$

Now we assume that the output of the LVDT is demodulated by a phase-sensitive rectifier and filtered to give a voltage proportional to the relative position of the mass in the accelerometer case.

$$V_o = K_O(X_1 - X_2) \quad 7.70$$

Substitution of Equation 7.69 into Equation 7.70 gives us

$$\frac{V_O}{X_1}(s) = \frac{K_O(M/K_S)}{s^2M/K_S + sB/K_S + 1} \quad 7.71$$

Thus, for a constant acceleration, $V_o = \ddot{X}_1 K_o(M/K_S)$. Linearity of this simple mechanical accelerometer depends on spring linearity over the working range of mass displacement, as well as LVDT linearity. The natural frequency of this sensor is $f_n = (1/2\pi)\sqrt{K_S/M}$ Hz, and the sensor's damping constant is given by $\zeta = B/(2\sqrt{MK_S})$. If the magnitude of $\ddot{x}_1$ becomes too large, the mass will hit the stops in the case and the output voltage will saturate. On the other hand, if $\ddot{x}_1$ is very small, small amounts of static friction between the mass and case will prevent the mass from moving relative to the case, and there will be an output deadzone. The latter problem can be mitigated by lubrication, by precision machining of the mass/case assembly, or by dithering the mass with an AC solenoidal magnetic field; the former problems are overcome by the use of the servo-accelerometer design, described below.

A *servo accelerometer* is shown in cross section in Figure 7.21. We see that the same components are present as for the simple design described above, except the spring has been replaced with a force coil, not unlike the voice coil of a standard loudspeaker. The force coil moves in the air gap of a strong, coaxial permanent magnet, also like that used for loudspeakers. The Newtonian force balance equation can be written

$$F_C = K_F I_C = M\ddot{x}_2 + B(\dot{x}_2 - \dot{x}_1) \qquad 7.72$$

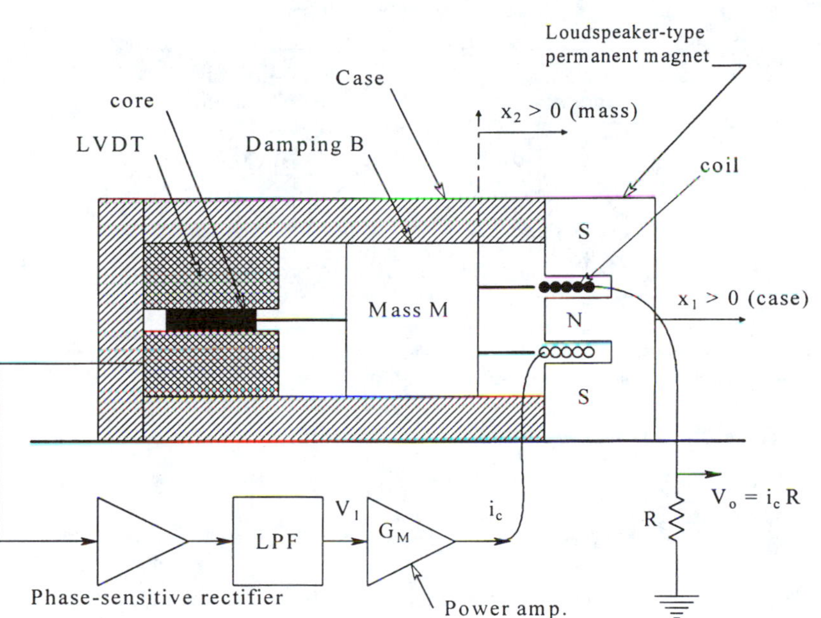

FIGURE 7.21 Section through an active, servo, linear accelerometer. The servo design allows the natural frquency to be raised while preserving sensitivity. See text for analysis.

Here we assume the force produced by the coil and magnet is proportional to the current through the coil. In addition, the LVDT system output is $V_1 = K_D(x_1 - x_2)$, and the current is produced by a voltage-controlled voltage source (VCCS) with transconductance, G_M. We assume further that the servo-accelerometer output voltage is proportional to i_c. Combination of these relations after Laplace-transforming yields:

$$\frac{V_O}{\ddot{X}_1}(s) = \frac{K_O(M/K_F)}{s^2 M/(K_F G_M K_D) + sB/(K_F G_M K_D) + 1} \qquad 7.73$$

Now the system's natural frequency is $f_n = (1/2\pi)\sqrt{K_F G_M K_D/M}$ Hz, and its damping factor is $\zeta = B/(2\sqrt{MK_F G_M K_D})$. The feedback design of this active sensor allows the natural frequency to be increased using $G_M K_D$, without lowering sensitivity, as happens with the passive accelerometer. Of course, the complexity of the servo-accelerometer design makes it more expensive than the simple moving-mass design.

One way to eliminate the problem of coulomb friction which accompanies any moving-mass design is to not have the mass move relative to the case. Figure 7.22 illustrates two single-mass, compression-type accelerometers in which the mass is in contact with the surface of a piezoelectric crystal such as quartz, barium titanate, lead zirconate-titanate, etc. The mass is often preloaded with respect to the case with a very stiff spring. Acceleration of the case in the ±x-direction causes a corresponding acceleration of the mass, M. Hence a force $F_x = \pm M\ddot{x}$ is exerted on the piezoelectric material which is compression-sensitive. Thus an equivalent current, $i_x = \pm\dot{F}d$, is generated in parallel with the below-resonance, R_x-C_x parallel equivalent circuit for the piezo material. It is easy to show that:

$$V_o = \frac{\dddot{X}(s)MdR_x}{sR_xC_x + 1} \quad 7.74$$

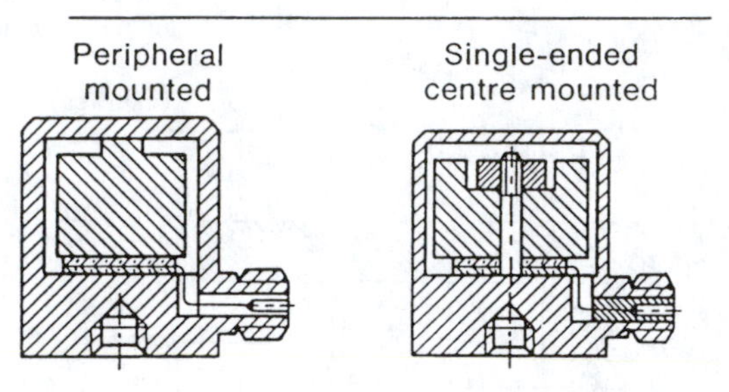

FIGURE 7.22 Two piezoelectric crystal accelerometer designs are shown in cross section. (Courtesy of Brüel & Kjaer Instruments, Inc., Marlborough, MA.)

Thus for the accelerometer, we may write:

$$\frac{V_O}{\ddot{X}}(s) = \frac{sMdR_x}{sR_xC_x + 1} \quad 7.75$$

Equation 7.75 is a simple high-pass transfer function. It tells us that the piezoelectric accelerometer will not measure constant or low-frequency accelerations (such as due to gravity), but will work well for accelerations containing frequencies above $f_b = 1/(2\pi R_xC_x)$ Hz. Thus, piezo-accelerometers are widely used for vibration testing, where the frequencies of the applied displacement are above f_b. If a charge amplifier is used, the break frequency is now set by the feedback R_F and C_F of the electrometer op amp. The transfer function becomes:

$$\frac{V_o}{\ddot{X}}(s) = \frac{-sMdR_F}{sR_FC_F + 1} \quad 7.76$$

The break frequency, $1/(2\pi R_FC_F)$, will generally be lower than that for the piezo-element alone, given above. The high-frequency gain is $-Md/C_F$ volts/meter/second^2.

Simple compression-type, piezo-accelerometers suffer from two drawbacks: one, a high temperature sensitivity (temperature equivalent acceleration output caused by rapidly changing transducer temperatures); and two, sensitivity to base bending (under the piezo sensor) and off-axis accelerations (linear and angular) which cause the mass to exert force asymmetrically on the piezo material, producing erroneous outputs. The Delta-Shear piezo-accelerometer design developed by Brüel & Kjaer Instruments, Inc. (Marlborough, MD) minimizes these problems (see Figure 7.23).

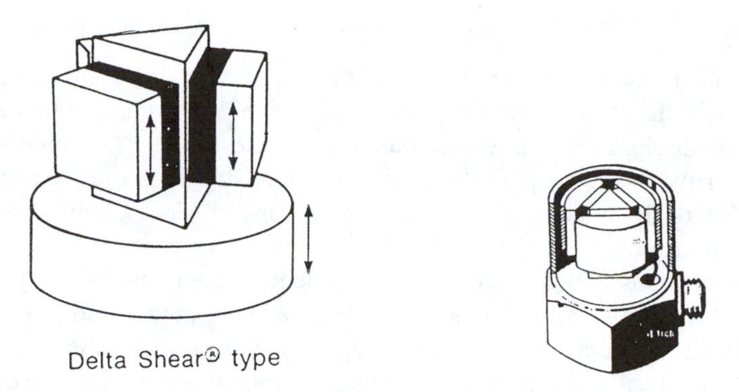

FIGURE 7.23 Sections of a Delta Shear™ piezo-accelerometer designed by B&K Instruments to minimize output sensitivity to base bending and off-axis acceleration components. (Courtesy of Brüel & Kjaer Instruments, Inc., Marlborough, MA.)

Because there is no relative motion between the mass and the case, there is no coulomb friction to cause a dead zone, and the dynamic range of piezoelectric accelerometers can be an enormous 160 dB. Because of noise and output saturation, amplifiers do not have this dynamic range, and so signal conditioning gains would have to be switched to accommodate an $\ddot{X}$ which varies 160 dB. The mechanical resonance frequency of piezo-accelerometers is orders of magnitude higher than that of moving mass-and-spring accelerometers. This difference is because the stiffness of the piezo material is orders of magnitude larger than that of the spring. Because of their high-frequency bandwidth, piezo-accelerometers are well suited for vibration testing applications in the kilohertz range.

7.2.2 Linear Velocity Measurement

Measurement of linear velocity may be subdivided into two classes: velocity measurements on solid or mechanical objects, and velocity measurements of fluids. In some cases, the velocity observer is inside the moving object (e.g., an automobile, boat, or airplane), in other cases, the observer is without.

An obvious means of estimating the linear velocity of a mechanical (solid) object is to integrate the output of a linear accelerometer attached to it. This method works well when the SNR at the accelerometer output is high, and the integrator does not drift. If the accelerometer output is integrated digitally, a 16-bit analog-to-digital converter (ADC) is needed and the integration routine and sampling interval must be carefully specified.

Generally, other means of velocity measurement depend on what velocity is being measured. We have already discussed the use of laser Doppler velocimetry (LDV) to measure the velocity of fluids containing small reflecting particles in Section 6.9.3. LDV can be used with any moving, reflecting surface to measure velocity. LDV has recently been adapted to work with a pulsed infrared (IR) laser diode to measure vehicle velocity, thus competing with the conventional microwave Doppler radar detector. Not only is the IR laser beam highly focused, but it is invisible; the IR radar speed "gun" has to be aimed with a telescopic sight. Within the limitations of propagation range, ultrasound can also be used to measure vehicle velocity in a Doppler system. A major problem with all Doppler systems for measuring velocity of moving, reflecting objects is the necessity to know θ, the angle between the velocity vector and the line connecting the transmitting and receiving transducers with the moving object. This problem has been largely overcome by the cross-beam techniques developed by George and Lumley (1973) for LDV, and by Fox (1978) for ultrasound velocimetry.

We have already described systems to measure fluid velocity by laser Doppler velocimetry, and by the Faraday effect. We now consider several other means.

7.2.2.1 Hot Wire and Hot Film Anemometers

The hot wire or hot film anemometer makes use of the fact that the resistance of a metallic wire or film changes with its temperature. The temperature of the wire, on the other hand, depends on the electric power dissipated in the wire (Joule's law), and the rate at which the heat from the power is removed by the surrounding medium. The rate at which the heat is dissipated is an increasing function of the velocity of the fluid surrounding the wire or film. Several representative hot wire and hot film probes are illustrated in Figure 7.24.

Hot wire sensors are used in gas velocity and turbulence measurements. They are generally very fine wires (e.g., 13 μm diameter, 1.3 mm length) stretched between two gold-plated needle supports. Depending on the gas temperature and velocity range, materials used for the wires can be platinum-coated tungsten, which has a tempco of 0.0042 ohms/°C, pure platinum (tempco of 0.003 ohms/°C), useful up to 300°C, and platinum-iridium alloy (tempco of 0.0009 ohms/°C), useful up to 750°C. Less fragile are metal film sensors. A high-purity platinum film is bonded to a fused quartz substrate, and then covered with alumina for gas velocity measurements or with quartz for measuring the velocity of conducting liquids. Metal film sensors can be configured as wires (by bonding the platinum to quartz filaments as small as 0.001" in diameter), or as wedges (the film is bonded to the end of a quartz "screwdriver"), or as a flat stip on a flat quartz disk, or as a thin ring around the circumference of the tip of a quartz cone, etc. Film sensors can be used to about 300° C on high-temperature probes.

All hot wire and hot film probes have minimum and maximum fluid velocities over which they work effectively. The minimum velocity detectable is set by the amount of free convection from the heated sensor. Velocities down to 0.15 m/s (0.5 ft/s) can be measured in air, and down to 0.003 m/s (0.01 ft/s) in water. The maximum velocities which can be reliably measured are primarily limited by mechanical considerations, i.e., the strength of the fine wires. Special hot film sensors have been used in gases with supersonic velocities, and in liquids with velocities up to 15 m/s.

Although most fluid velocity measurements are made with single probes or pairs of probes, true three-dimensional vector flow now can be measured with a special 3-wire probe developed by TSI, Inc. (St. Paul, MN). This probe is used with a special computer which solves "Jorgenson's equations" for the three-wire probe to give the exact, three-dimensional velocity vector.

A heuristic, static analysis of a single, hot wire probe is based on several simplifying assumptions. First, we assume that the probe's resistance, R_W, is linearly related to its temperature. Mathematically, this may be expressed as

$$R_W = R_o + \alpha \Delta T R_o \tag{7.77}$$

where R_o is the probe's resistance at some reference temperature, T_o, and α is the first-order, positive temperature coefficient of the metal wire. Second, we assume that the temperature rise of the probe is given by the heat sink equation:

$$\Delta T = P\Omega \tag{7.78}$$

where P is the average electrical power dissipated in the probe, and Ω is the thermal resistance of the probe, in degrees Celsius per watt. Ω is determined by the geometry of the hot probe, the heat conducting properties of the surrounding medium, whether the medium is moving or not, and the velocity of the medium relative to the probe's geometry. For the case where the fluid velocity is perpendicular to the axis of the wire, the thermal resistance is given approximately by

$$\Omega = \Omega_o - \beta\sqrt{|\mathbf{v}|} \tag{7.79}$$

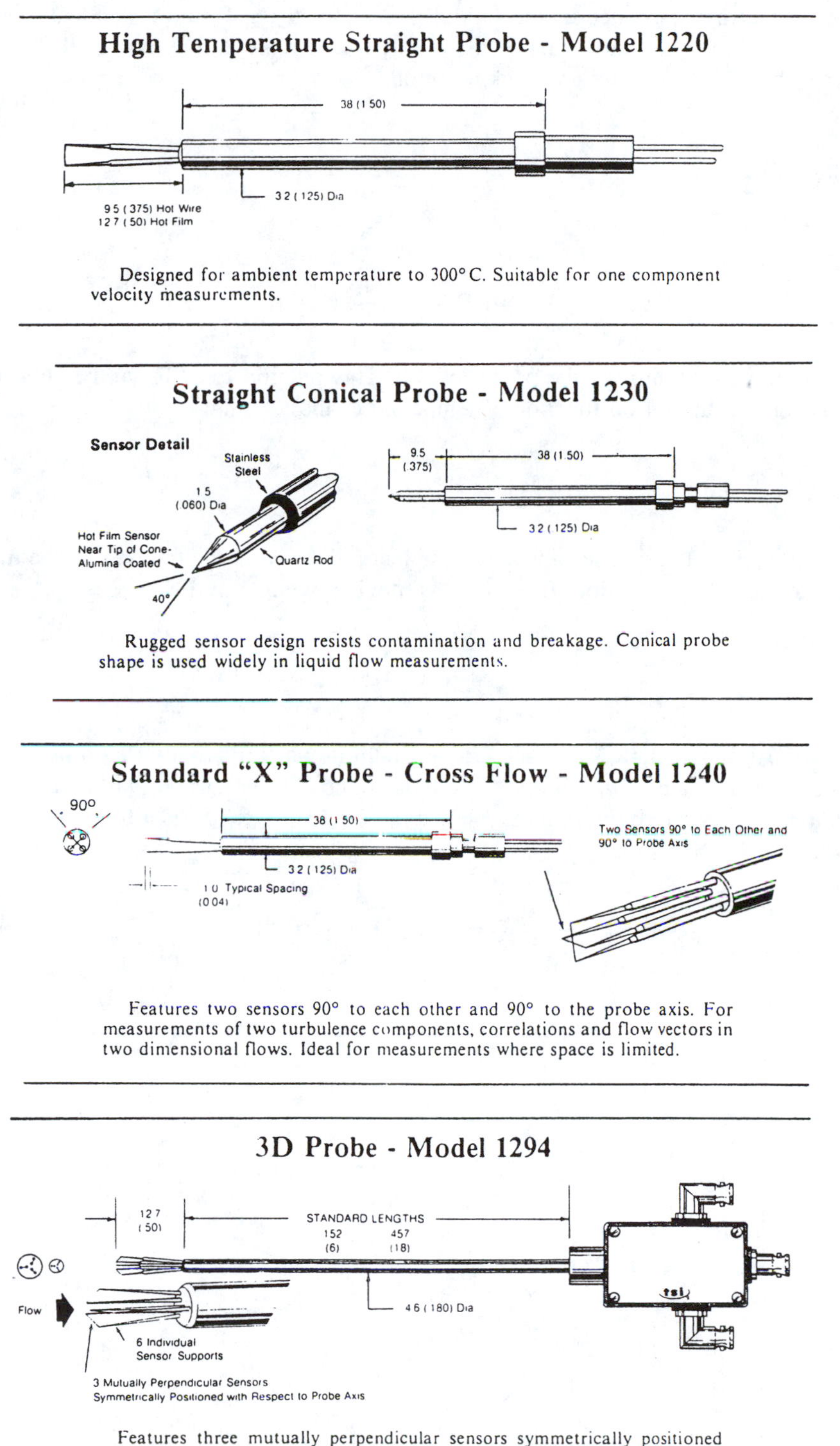

FIGURE 7.24 Representative hot-wire and hot-film fluid velocity probes. (Figure courtesy of TSI, Inc., St. Paul, MN.)

Here Ω_o is the thermal resistance in still fluid, β is a geometry and fluid-dependent constant, and $|\mathbf{v}|$ is the magnitude of the fluid velocity, assumed perpendicular to the axis of the wire. In still fluid, ΔT_o can be found from the heat sink equation:

$$\Delta T_o = P_o\Omega_o = I_o^2 R_o(1+\alpha\Delta T_o)\Omega_o \qquad 7.80$$

Solving for ΔT_o, we obtain:

$$\Delta T_o = I_o^2 R_o \Omega_o / (1-\alpha I_o^2 \Omega_o) \qquad 7.81$$

I_o is the average current through the probe resistor. This relation for ΔT_o can be substituted into the equation for resistance as a function of temperature, and we find:

$$R_{Wo} = R_o / (1-\alpha I_o^2 \Omega_o) \qquad 7.82$$

R_{Wo} is the new, elevated, resistance above the ambient resistance of the hot wire probe in still fluid due to electrical power dissipation. If the fluid is moving, we may write:

$$R_W(I, v) = R_o \Big/ \left[1-\alpha I^2\left(\Omega_o - \beta\sqrt{|\mathbf{v}|}\right)\right] \qquad 7.83$$

Figure 7.25 illustrates a simple, feedback bridge circuit in which the hot wire probe is maintained at nearly constant resistance, hence operating temperature. In the circuit, we assume that $R_1 \gg R_W(I,v)$ so that the current through the probe is given by $I = V_B/R_1$. V_o is a DC source used to set R_{Wo}. With the switch SW1 open, $V_B = V_o$, and $V_1 = V_{1o}$. Thus:

$$V_{1o} = V_o R_{Wo}/R_1 = V_o R_o \Big/ \left[R_1\left(1-\alpha V_o^2 \Omega_o / R_1^2\right)\right] \qquad 7.84$$

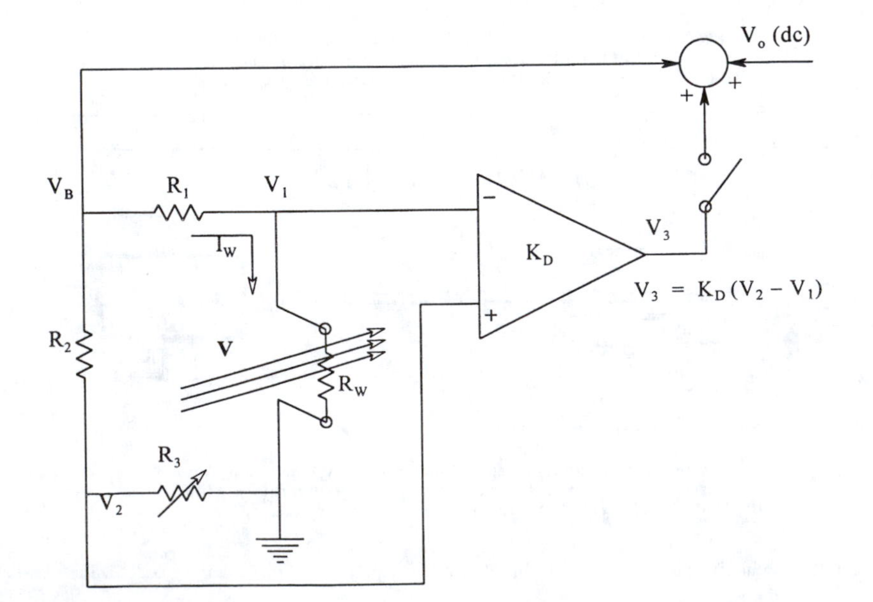

FIGURE 7.25 Schematic of a simple feedback bridge circuit useful with hot wire probes. See text for analysis.

V_2 is adjusted to have this same value, i.e., $V_{2o} = V_{1o}$, so $V_3 = 0$. Exact analysis of the closed-loop system is tedious and involves the algebraic solution of a cubic equation. It is useful to assume that the high gain of the amplifier, K_D, forces the closed-loop $R_W(I,v)$ to equal the R_{wo} in still fluid. Thus:

$$\frac{R_o}{1-\alpha V_o^2\Omega_o/R_1^2} = \frac{R_o}{1-\alpha V_B^2\left(\Omega_o - \beta\sqrt{|\mathbf{v}|}\right)/R_1^2} \tag{7.85}$$

From Equation 7.85 above, we find that

$$V_B^2 = V_o^2 / \left(1 - \beta\sqrt{|\mathbf{v}|}/\Omega_o\right) \tag{7.86}$$

If we assume that $\beta\sqrt{|\mathbf{v}|}/\Omega_o << 1$, then it is easy to show that

$$V_B = V_o\left(1 + \beta\sqrt{|\mathbf{v}|}/2\Omega_o\right) \tag{7.87}$$

and finally,

$$V_3 = V_o\beta\sqrt{|\mathbf{v}|}/2\Omega_o \tag{7.88}$$

Thus, the electrical output of this system for conditioning the hot wire anemometer response is proportional to the square root of the fluid velocity. Further linearization can be obtained by passing V_3 through a square-law circuit. Fluid flows at an angle to the hot wire probe axis give a reduced response, the exact form of which depends on the probe geometry, including the supports.

7.2.2.2 Other Means of Measuring Fluid Velocity and Flow

Fluid flow through a uniform conduit may be subdivided into two major categories; laminar flow, where the velocity vectors have a parabolic distribution, shown in Figure 7.26A, and turbulent flow, which is no longer characterized by a parabolic velocity profile, and which no longer contains only parallel streamlines but also eddies and whorls. The transition between laminar and turbulent flow occurs over a narrow range of (average) velocities. It has been found that the transition between laminar and turbulent flow occurs when a diagnostic variable called the Reynolds number exceeds a threshold value. The Reynolds number is given by:

$$N_R = \overline{V}\rho D/\eta \tag{7.89}$$

where $\overline{V}$ is the average velocity (LT^{-1}) in the pipe, ρ is the mass density (ML^{-3}) of the fluid, η is the absolute viscosity of the fluid ($ML^{-1}T^{-1}$), and D the internal diameter of the pipe (L). Combining the dimensions above, N_R is seen to be dimensionless. If English units are used, (lbs, ft, s), a velocity above that which will produce $2000 < N_R < 2300$ will result in turbulent flow. Needless to say, the mathematics describing turbulent flow are beyond the scope of this text, so we will concentrate on flow and velocity measurements under laminar conditions ($N_R < 2000$). Laminar flow velocity in a pipe with circular cross section may be shown to be described by:

$$V(r) = \frac{(p_1 - p_2)}{4\eta L}\left(R^2 - r^2\right), \quad 0 \le r \le R \tag{7.90}$$

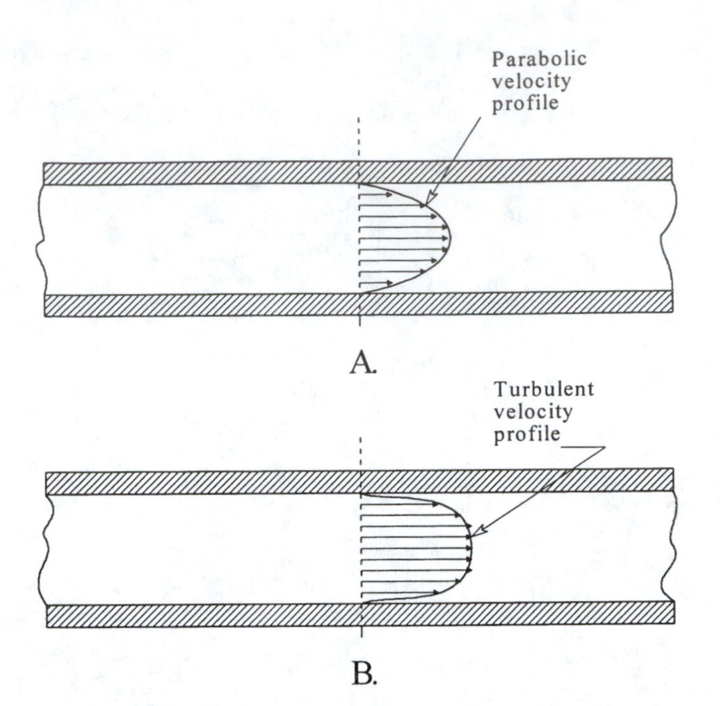

FIGURE 7.26 (A) Axial section through a pipe carrying fluid under laminar flow conditions. Note the parabolic distribution of axial fluid velocity with near zero velocity at the pipe walls. (B) This pipe is carying fluid under turbulent flow conditions. Note that the axial velocity component is no longer parabolic with the pipe's radius.

Here $(p_1 - p_2)$ is the pressure difference across a length L of pipe, R is the radius of the pipe, and r is the radial distance from the center of the pipe. The total volume flow across the cross-sectional area of the pipe may be found by integrating the parabolic velocity profile over the cross-sectional area of the pipe.

$$Q = \int v(r)\, dA = \int_0^R V_m\left(1 - r^2/R^2\right) 2\pi\, dr = V_m A/2 \ \ m^3/s \qquad 7.91$$

Here V_m is the peak velocity, equal to $(p_1 - p_2)R^2/(4\eta L)$, and A is the cross-sectional area, πR^2. Thus, under laminar flow conditions, if we can measure the peak velocity (at the center of the pipe), we can calculate the net volume flow rate, Q.

A common means of measuring the volume flow of incompressible fluids (generally, liquids) is by obstruction meters (Beckwith and Buck, 1961). These include venturi tubes, flow nozzles, and orifice nozzles. Figure 7.27 illustrates these obstruction meters. What is actually measured is the pressure differential across the nozzle. Beckwith and Buck show that the volume flow through a venturi tube can be expressed as

$$Q_v = CMA_2\sqrt{2g(P_1 - P_2)/\alpha} \ \ m^3/s \qquad 7.92$$

and the volume flow though nozzles and orifices is given by

$$Q_n = KA_2\sqrt{2g(P_1 - P_2)/\alpha} \ \ m^3/s \qquad 7.93$$

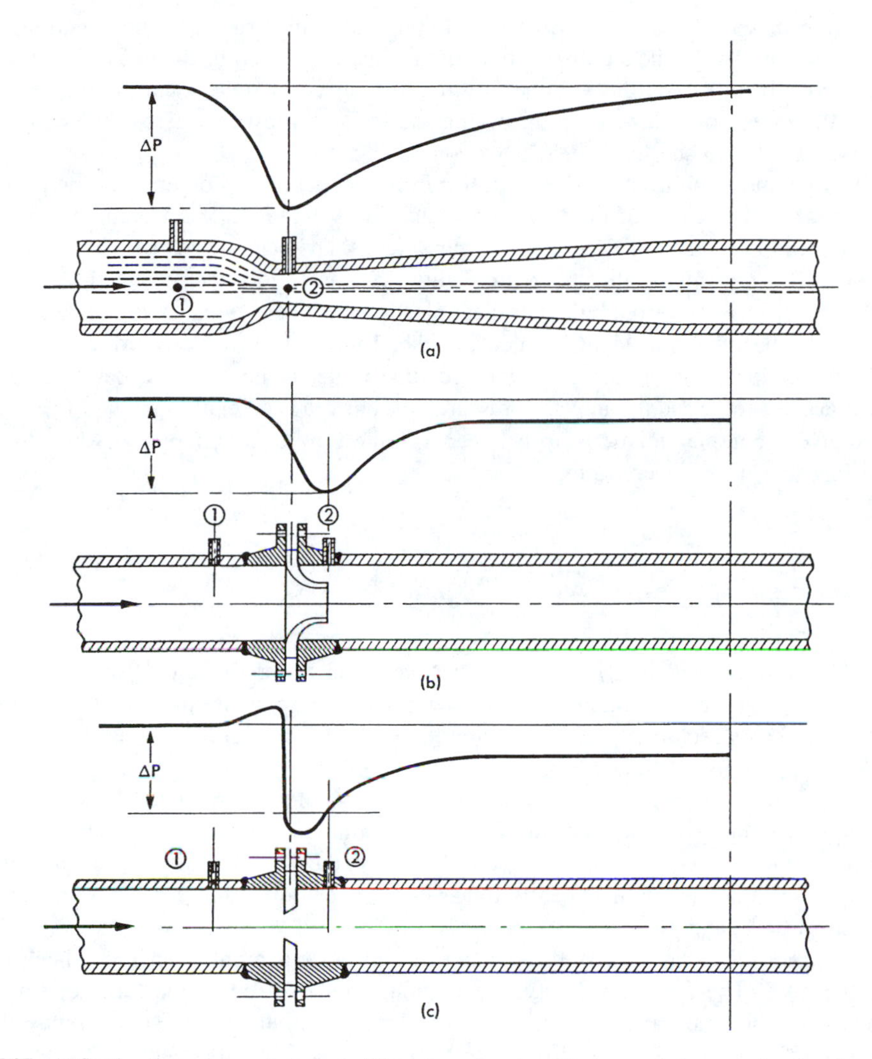

FIGURE 7.27 (A) Section through a venturi nozzle. The volume flow through the nozzle is proportional to the square root of the pressure difference for incompressible liquids. (B) A flow nozzle. (C) An orifice flowmeter. (From Beckwith, T.G. and Buck, N.L., *Mechanical Measurements*, 1st ed., 1961. Reprinted by permission of Addison-Wesley Longman Publishing Company, Inc.)

where C and K are constants which depend on, among other factors, the Reynolds number and the geometry of the obstruction. M is given by

$$M = 1\Big/\sqrt{1-\left(A_2/A_1\right)^2} \qquad 7.94$$

K = CM, g is the acceleration of gravity, α is the specific weight of the liquid, A_1 is the upstream pipe cross-sectional area, and A_2 is the minimum area of the obstruction. Since the pressure difference, $P_1 - P_2$, is measured, it is clear that the actual volume flow is proportional to its square root. Because of the turbulence losses associated with obstruction flowmeters, there is always a steady-state downstream pressure loss, in some cases up to 30% for an orifice flowmeter. There is significantly less steady-state pressure loss with a venturi system because of its streamlining and reduced turbulence.

Obstruction meters can also be used with compressible fluids (gases). The situation here is far more complex, because of the adiabatic compression and expansion of the gas as it passes through the obstruction. In this case, the volume flow is no longer a simple square root function of the pressure difference; power-law terms involving the ratio of pressures across the obstruction are involved (see Beckwith and Buck, 1961, Section 12-15).

The fixed geometry obstruction meters suffer from the common problem that the pressure drop is proportional to the square of the flow rate. Thus a very wide range differential pressure meter is required. If a large range of flow is to be measured, then accuracy may be low at the low end of the range. The *rotameter*, illustrated in Figure 7.28, is a flow measuring device of extreme simplicity. It consists of a vertical glass or clear plastic tube whose inner diameter increases toward the top. Inside the tube is a "float," which is pushed upward by the upward flowing fluid. The equilibrium position of the float in the tapered tube is determined by the balance between the downward gravity force and the upward pressure, viscous drag force, and buoyant force. Beckwith and Buck give a complex relationship between volume flow of a noncompressible fluid and the height of the float in the tapered tube:

$$Q = (\pi/4)\left[(D + by)^2 - d^2\right]\frac{\sqrt{2gv_f(\rho_f - \rho_w)}}{\sqrt{A_f\rho_w}}C \qquad 7.95$$

where D is the effective diameter of the tube at the height of the float, y; v_f is the volume of the float; b is the change in tube diameter (taper) with y, d is the maximum diameter of the float; A_f is the effective cross-sectional area of the float, ρ_f is the float's density; ρ_w is the density of the liquid; and C is the discharge coefficient.

Rotameters are used for gases as well as incompressible liquids. A major disadvantage in their use is that they cannot be used with opaque fluids or fluids with large amounts of particles in suspension. Rotameters must also be read out manually; they are ill-suited for electronic readout. Advantages of rotameters are their nearly linear flow scale and that their pressure loss is constant over the entire flow range.

Another type of flow or velocity meter makes use of a free-spinning turbine wheel in the pipe carrying the fluid. The flow or velocity is a function of the speed of the turbine, which can be measured by a tachometer generator which must be coaxial with the turbine in the pipe. Alternately, if small magnets are placed in the tips of the turbine's vanes, a Hall effect switch sensor can be used outside the pipe to measure the rotating vanes passing under it. The number of pulses/second is proportional to the turbine's angular velocity, which is in turn an increasing function of volume flow. Because no analytical expression exists for turbine velocity as a function of flow or velocity, turbine meters must be calibrated for a given application.

Pressure probes, which include Pitot tubes, are also useful in measuring fluid velocity in ducts or in free space. The Pitot static probe is used to estimate the relative airspeed of aircraft through the air. Here again, we see that the velocity is approximately proportional to the square root of the pressure difference for air. A differential pressure transducer may be used to effect readout. A typical Pitot tube geometry is illustated in Figure 7.29. Pitot tubes lose calibration when the fluid flow is no longer coaxial to the probe axis. A yaw angle of as small as 8° can produce significant errors in Pitot tube calibration. The Kiel ram tube, shown in Figure 7.30, operates similarly to the Pitot tube, except the reference pressure is not taken from the probe. The Kiel probe is evidently far more insensitive to off-axis air velocity, maintaining its calibration up to yaw angles of 40° (Beckwith and Buck, 1961).

In medicine and physiology, it is important to measure blood flow and velocity in various blood vessels and in the heart. One means of estimating blood flow is by the Faraday effect, already discussed in Section 6.3.5.1. Faraday flow probes have been designed which clamp on to the blood vessel in question. They do not actually contact the blood, but do require surgery to expose the

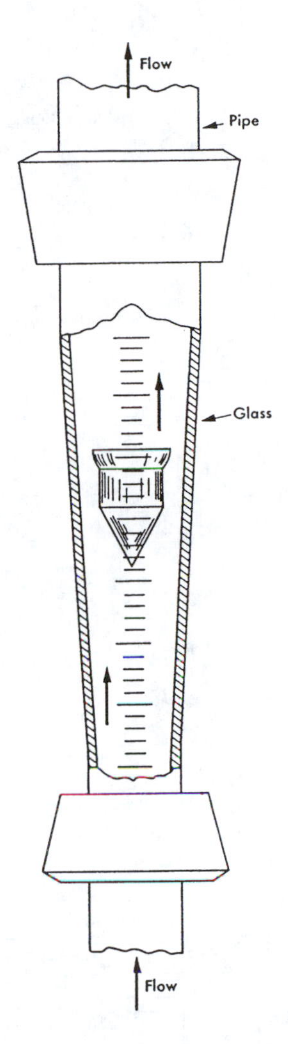

FIGURE 7.28 Section through a typical rotameter. (From Beckwith, T.G. and Buck, N.L., *Mechanical Measurements*, 1st ed., 1961. Reprinted by permission of Addison-Wesley Longman Publishing Company, Inc.)

vessel in question. Their use is obviously invasive and not without risk. Webster (1978) describes the designs of several Faraday blood flowmeters. It should be stressed that the Faraday system is responsive to the average velocity of the conducting fluid (blood), and that flow is a derived quantity based on a knowledge of the internal diameter of the blood vessel.

The use of Doppler ultrasound transmitted through the skin to the blood vessel of concern offers a risk-free, noninvasive means to estimate blood velocity and flow. (See Webster, 1978, Section 8.4.) All ultrasonic blood velocity/flow systems make use of the fact that blood contains erythrocytes or red cells suspended in it. Erythrocytes are typically found in densities of about 5×10^6 per mm^3, and are biconcave disks about 8 μm in diameter, 2 μm thick at their edges, and 1 μm thick at their centers. The velocities of individual erythrocytes follow the laminar flow stream lines of velocity in the larger blood vessels (arteries, veins). Ultrasound incident on the moving blood in a vessel is therefore reflected off myriads of tiny scatterers whose velocities range from zero to maximum, and whose cross-sectional areas vary randomly in time and space. Hence, the return signal is composed of two major components: the largest component is at the carrier frequency and is reflected from all non-moving tissues in the beam path. The information-carrying component is much smaller in energy, and is made up from the superposition of many amplitudes and frequencies reflected from the ensonified, moving red cells. The exact nature of the Doppler return signal

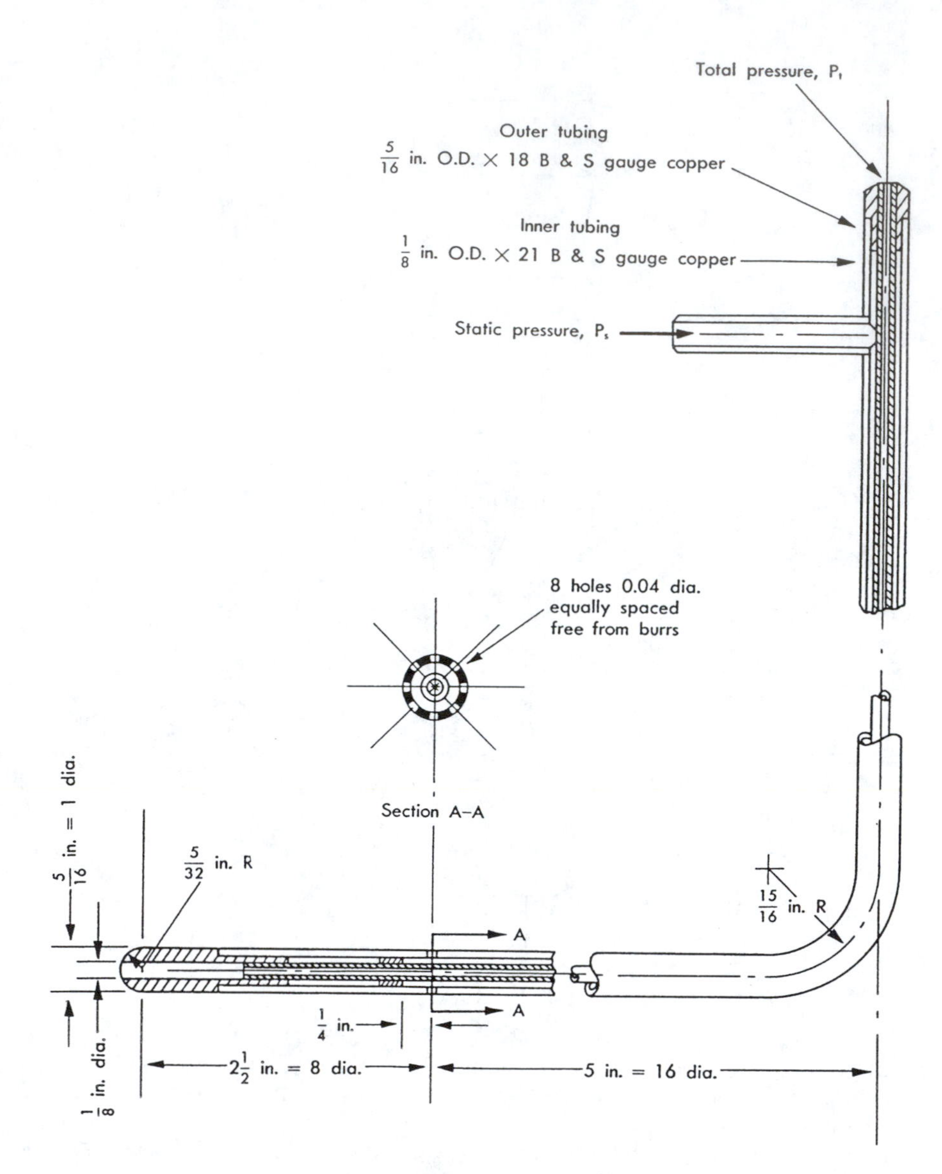

FIGURE 7.29 A pitot airspeed tube. (From Beckwith, T.G. and Buck, N.L., *Mechanical Measurements*, 1st ed., 1961. Reprinted by permission of Addison-Wesley Longman Publishing Company, Inc.)

depends on how tightly the ultrasound beam can be focused in the interior of the vessel to discriminate individual regions of velocity.

The most basic form of Doppler ultrasound system is the CW configuration, where the return signal is heterodyned with a sine wave at the transmitted frequency, then low-pass filtered and amplified. The frequency of this output signal is proportional to the average velocity of the ensonified blood. The output signal is listened to by the operator, and may be used to detect foetal heartbeats, or pulsating aneurisms in artery walls, in addition to blood velocity in vessels such as the common carotid arteries. In the latter case, obvious asymmetries in carotid blood velocity may indicate an asymmetrical obstruction, such as by atherosclerotic plaque in one carotid sinus. CW Doppler ultrasound systems are generally more qualitative in their medical applications than are the more sophisticated pulsed Doppler systems. Pulsed Doppler systems can be range-gated to enable a blood velocity profile to be constructed for large vessels such as the descending aorta.

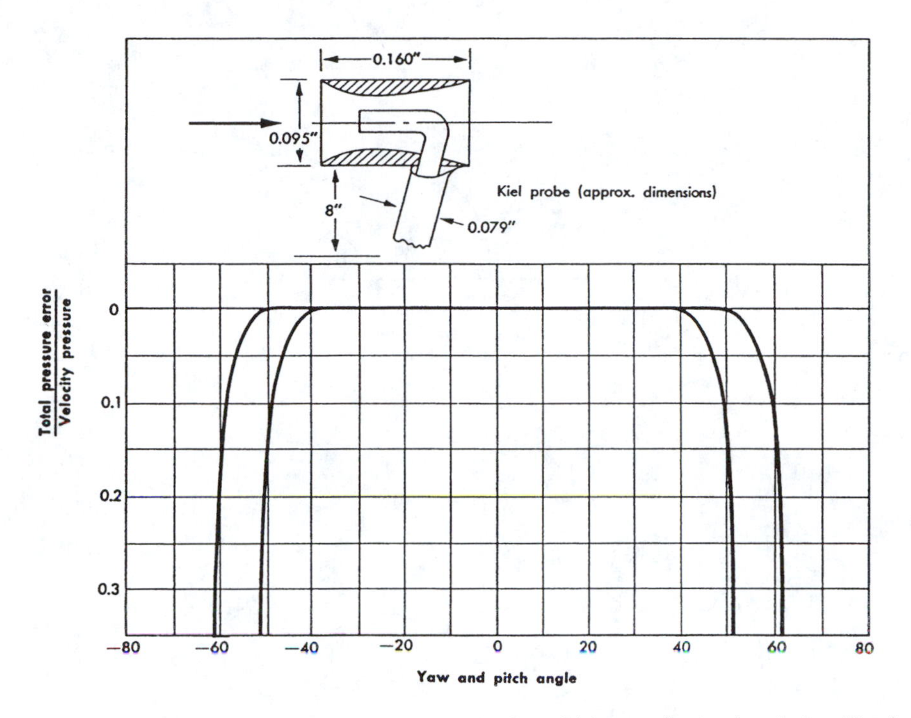

FIGURE 7.30 A Kiel-type air velocity probe. Note its relative insensitivity to off-axis air velocity. (Reprinted with permission from The Airflo Instrument Company, Glastonbury, CT.)

To be used quantitatively, all conventional Doppler blood velocity systems require a precise knowledge of θ, the angle between the blood velocity vector and a line connecting the transmitting/receiving transducers with the small volume of blood whose velocity is being sensed. Recall that the frequency of the Doppler-shifted return signal is given by the relation:

$$f_r = f_t\left(1 + \frac{2v\cos(\theta)}{c}\right) \qquad 7.96$$

where c is the average velocity of sound in blood, v is the velocity of the small blood volume ensonified, and f_t is the transmitted (carrier) frequency. The Doppler shift, here taken as positive because the velocity of the reflector has a component toward the transducers, is detected electronically and is given by:

$$f_d = \left[f_t 2v\cos(\theta)\right]/c \qquad 7.97$$

Obviously, v is proportional to $f_r/\cos(\theta)$, and an error in θ will give an error in v. In 1985, Fox presented a closed-form solution to the two-dimensional Doppler situation which utilizes the outputs of two independent transmit-receive probes. Fox's solution yields the velocity magnitude, $|\mathbf{v}| = \sqrt{v_x^2 + v_y^2}$, and the angle θ_1 between the velocity vector and a line from number 1 probe (see Figure 7.31). Note that the probes lie in the XY plane with the velocity vector. The probes are separated by an angle ψ, and their beams converge on the moving reflecting object at P. The Doppler frequency returned to each probe is given by:

$$f_{d1} = 2f_1|\mathbf{v}|\cos(\theta_1)/c \qquad 7.98$$

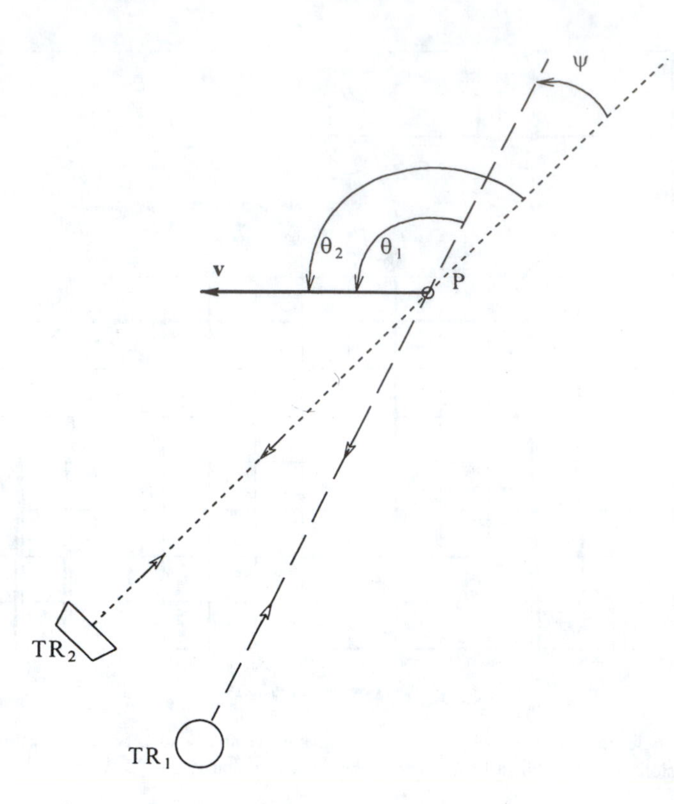

FIGURE 7.31 Geometry relevant to Fox's angle-independent Doppler system. TR1, ultrasound transducer 1 transmits at frequency f_1, transducer TR2 transmits at f_2. See text for analysis.

$$f_{d2} = 2f_2|\mathbf{v}|\cos(\theta_2)/c \qquad 7.99$$

Now the angle θ_2 can be written in terms of θ_1:

$$\theta_2 = \theta_1 + \psi \qquad 7.100$$

After some trigonometry and algebra, Fox shows that:

$$|\mathbf{v}| = \sqrt{v_x^2 + v_y^2} = (cf_{d1}/2f_1)\sqrt{(1 - 2R\cos(\psi) + R^2)/\sin^2(\psi)} \qquad 7.101$$

and

$$\theta_1 = \tan^{-1}[(\cos(\psi) - R)/\sin(\psi)] \qquad 7.102$$

where R is the Doppler shift ratio corrected for carrier frequency:

$$R = (f_{d2}/f_2)/(f_{d1}/f_1) \qquad 7.103$$

In order to calculate $|\mathbf{v}|$ and θ_1, one must measure f_{d1} and f_{d2}, knowing f_1, f_2, and ψ. The accuracy of the method is limited by the accuracy with which one can determine f_{d1} and f_{d2}. Because of the parabolic (laminar) flow profile in blood vessels, the finite size of the ensonified volume (typically

3 mm³ for a 2.25 MHz carrier; Fox, 1978), and the random scattering nature of moving red blood cells, there typically is a bell-shaped distribution of f_ds, rather than a single sharp peak. The mode of the distribution is generally taken as the desired f_d. The Fox two-probe method of determining the velocity vector in two dimensions is best implemented with a computer system which algorithmically processes the Fourier-transformed Doppler return signals to determine their modes to estimate f_{d1} and f_{d2}, and then calculates $|\mathbf{v}|$ and θ_1 using the relations above and the known parameters, Ψ, f_1, and f_2.

Fox and Gardiner (1988) have extended the two-dimensional closed-form solution for v to three dimensions. Their equations are too long to include here, but they have the same general form as the simpler two-dimensional case described above. Their results showed that the calculated $|\mathbf{v}|$ remained within 5.6% of the theoretical value for Doppler angles up to 50°. Also, their angle estimate agreed with the theoretical values with a correlation coefficient, r = 0.99937.

The two- and three-dimensional Doppler flow velocimetry technique developed by Fox and colleagues is of course not restricted to the ultrasonic measurement of blood velocity. The Fox technique can be extended to the other Doppler modalities (lasers and microwaves), when a two- or three-dimensional estimate of object velocity is required.

The final system that we will dicuss which is used for the measurement of gas velocity and flow and which is used in modern pulmonary diagnostic instruments is the *pneumotachometer*, or pneumotach. Basically, a pneumotach is a pneumatic analogue of a resistor. A resistor obeys Ohm's law; i.e., I = V/R. In the case of laminar gas flow through the pneumotach,

$$Q = (P_1 - P_2)/R_P \qquad 7.104$$

where Q is the volume flow of gas, $(P_1 - P_2)$ is the pressure drop across the pneumotach, and R_P is the equivalent pneumatic resistance of the pneumotach. The pneumotach itself can have one of several forms. The body or duct of a pneumotach is typically 1.5" in diameter, and it uses either a fine-mesh screen or hundreds of parallel capillary tubes, about 1 to 2" long to obtain the desired acoustic resistance. Unfortunately, a capillary tube does not exhibit a pure frequency-independent resistance, but appears reactive at high frequencies. The acoustic impedance of a single capillary tube is given by (Olsen, 1943):

$$\mathbf{Z}_A(j\omega) = \frac{8L\eta}{\pi R^4} + j\frac{4L\omega\rho}{3\pi R^2} \text{ acoustic ohms} \qquad 7.105$$

where η is the viscosity of air at 20°C (1.86×10^{-4} poise), R is the radius of the tube, L is its length, ρ is the density of air (1.205×10^{-3} g/cm³), and ω is the radian frequency of pressure variation across the pneumotach elements. If N identical capillary tubes are packed in parallel, the resultant acoustic impedance is

$$\mathbf{Z}_{A(NET)} = \frac{8L\eta}{N\pi R^4} + j\frac{4L\omega\rho}{3N\pi R^2} \text{ acoustic ohms} \qquad 7.106$$

The acoustic impedance of a capillary tube pneumotach appears resistive for frequencies from DC up to about 1/10 of the frequency where $R_P = X_P$, or

$$f_{max} \approx 0.3\eta/(\pi\rho R^2) \text{ Hz} \qquad 7.107$$

f_{max} is about 50 Hz for R = 0.01715 cm and L = 1 cm. This means that as long as the frequency components in the pressure applied across the pneumotach impedance are below 50 Hz, the pneumotach will appear resistive, and the pressure drop will be proportional to volume flow. A broader bandwidth pneumotach can be made from thin, parallel, rectangular slits. The acoustic impedance of one such slit is given by (Olsen, 1943):

$$\mathbf{Z}_A(j\omega) = \frac{12\eta L}{t^3 d} + j\frac{\omega 6\rho L}{5td} \quad 7.108$$

where η is the viscosity of air, ρ is the density of air, L is the length of the slit in the direction of flow, t is the height of the slit perpendicular to the direction of flow, and d is the width of the slit perpendicular to the flow direction. The critical frequency for this type of pneumotach is again taken as 1/10 the frequency where $X_P = R_P$, or:

$$f_{min} \approx \eta/(t^2\rho) \text{ Hz} \quad 7.109$$

For slits with t = 0.01 cm, the acoustic impedance of the pneumotach would appear real up to about 240 Hz.

We see that the pneumotach can be used to measure volume flow under laminar conditions by sensing the pressure drop across it with a differential pressure sensor. Average air velocity can easily be derived by multiplying the volume flow by the pipe's area. The pneumotach can also be used to measure the acoustic impedance of a series acoustic load at frequencies where the pneumotach appears resistive itself. Figure 7.32 illustrates such a system. A pressure source (a loudspeaker) is coupled to the pneumotach. Two independent pressure sensors are used to measure the pressure difference across the pneumotach ($P_1 - P_2$), and the pressure across the unknown acoustic impedance (P_2). The unknown acoustic impedance is then just

$$\mathbf{Z}_{XA}(jw) = \frac{P_2}{(P_1 - P_2)/R_P}(j\omega) \quad 7.110$$

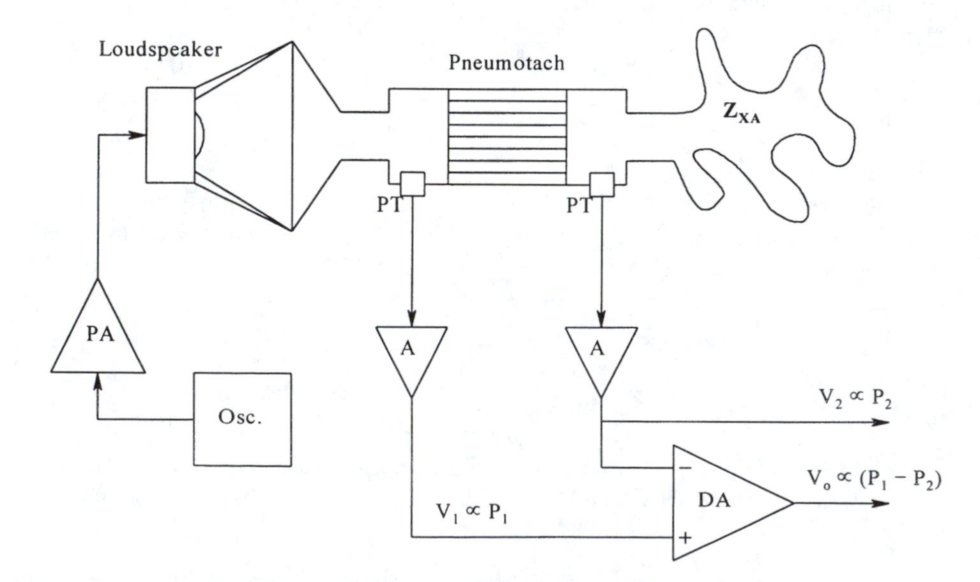

FIGURE 7.32 Diagram of pneumotach system used to measure acoustic impedance. OSC, oscillator; PA, power amplifier; A, voltage amplifier; LS, loudspeaker; PT, pressure transducer; DA, differential amplifier.

Such means have been used by Pimmel et al. (1977) to investigate the impedance of the human respiratory system over a frequency range of 1 to 16 Hz, and by Peslin et al. (1975) over a frequency range of 3 to 70 Hz.

7.2.3 Measurement of Linear Position

The technology of linear position measurement is a very broad area, ranging from position measurements of vehicles from within (navigation and inertial navigation) to position measurements of mechanical objects, radiation sources, estimation of vehicle position from without (as in the location and tracking of aircraft around an airport, or the tracking of submarines). It is not within the scope of this text to consider vehicle location problems by radar or sonar, nor will we consider the specialized topics of inertial navigation, LORAN and satellite navigation. Instead, we will focus on the measurement of mechanical position.

Most mechanical systems operate under coordinate constraints, i.e., they may only move in a path determined by their associated connections. For example, the position of the printing head of a dot-matrix printer at any instant is located somewhere along the length of its guide bar; this is a bounded one-dimensional position problem which has been solved cleverly by not measuring the position of the print head as the printer operates, but by locating the zero or reference position at printer power-up by use of a simple photoelectric position indicator at the left-hand side of the guide bar. Once the zero position is set, a stepping motor is used to position the print head along the guide bar as required. Because the stepping motor rotates in discrete angular increments with every applied pulse sequence, all the printer CPU has to do is use the net count of stepping motor command pulses stored in an up/down counter to "know" the location of the print head. The discrete nature of printing and symbols allows this simple indexing system to work effectively. However, there are many cases where there must be continuous feedback on position to optimize position control system performance.

In Section 6.4.1 we described the operation of the linear variable differential transformer (LVDT), a sensor widely used to give precise information in bounded linear displacement. LVDTs come in all sizes, and cover the range of displacements from millimeters to ten centimeters. An LVDT is an AC, carrier-operated sensor with an AC, double-sideband, suppressed carrier output which requires demodulation to a voltage proportional to core displacement by a phase-sensitive rectifier (PSR) and low-pass filter (LPF). In some LVDT models, the carrier oscillator and PSR/LPF are built into the case of the LVDT. In all cases, the response speed of the LVDT is limited by the carrier frequency and the time constants of the LPF.

One-dimensional mechanical position can also be sensed photoelectrically. One scheme we have used makes use of a red LED and two photodiodes or phototransistors connected as shown in Figure 7.33. An opaque vane attached to an axially moving shaft blocks the amount of light impinging on the photodiodes. Near the center of its range, the voltage output is linear with shaft position. (This system was used by the author to provide position feedback to a servomechanism used to provide controlled stretches to small muscles in physiological studies.) The United Detector Technology Co. (UDT) describes single-axis, position-sensing photodiodes which can sense the linear position of a collimated spot of light. The UDT sensor can be operated differentially as a fourth-quadrant (photovoltaic) solar cell; the ends of the linear sensor are connected to a DC differential amplifier, the center to ground. The spectral response of the UDT sensors peaks at about 850 nm. The sensor can also be used in the reverse-biased photodiode mode. Rise times range from 0.5 to 7 μsec, depending on the model. The UDT linear position sensors have maximum displacement ranges of 0.21" or 1.18". A 1% output linearity in the central 75% of the length is claimed for the displacement of a 1 mm diameter light spot.

In some cases, linear position can be converted to rotational position by use of a cable passed around a pulley to a weight. The weight keeps the cable taut, and the pulley drives one of many

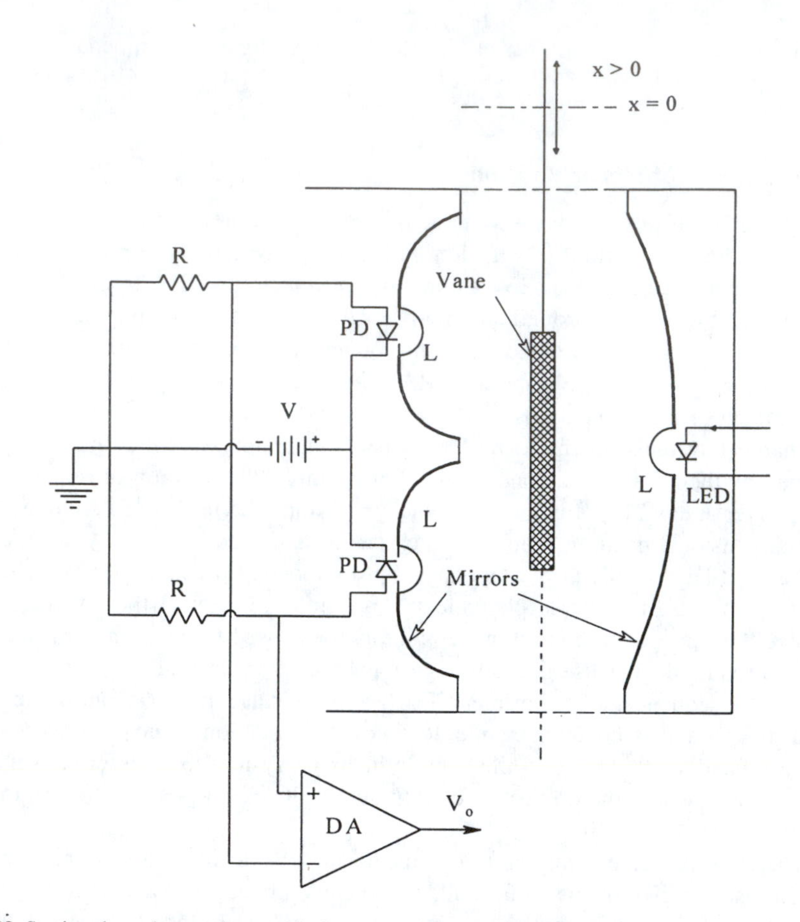

FIGURE 7.33 Section through a mechano-optical system using photodiodes used to measure linear displacements in the millimeter range. PD, photodiode; L, lens; V, opaque vane; LED, light emitting diode.

angular position measuring sensors. Probably the simplest angular position measuring sensor for this application is a 10-turn potentiometer. The potentiometer can be used as a voltage divider, or be one arm of a Wheatstone bridge. This type of system has been used to measure the level of liquid in a tank or well.

Linear position of distant objects can also be measured by means of reflected laser light, ultrasound waves, or microwaves. In the simplest implementation, as developed for the Polaroid camera rangefinder, a short ultrasound pulse is transmitted toward the object whose position is to be sensed, and the time delay for the echo to return is used as a measure of the distance. The same principle applies to laser ranging, used in modern surveying and geophysics, and with radar, whose uses should be obvious. All of the above pulse-echo systems rely on a precise knowledge of the propagation speed of the energy in the propagating medium. Unfortunately, the speed of sound is affected by air temperature, pressure, and moisture content. The speed of sound in water is also affected by temperature, pressure, density, and salinity. The speed of light in various media is only slightly affected by factors that affect the speed of sound. Also critical in any pulse-echo system design is the need for a precise, stable clock to measure the time intervals between pulse transmission and echo reception. Such clocks are often derived from thermostated quartz crystal oscillators using frequency multiplication techniques.

Often, we do not wish to measure the absolute position of an object, but rather a small change in its position. In these cases, we can make use of interferometric techniques; coherent light is commonly used. Fox and Puffer (1976) have described the use of a holographic interferometric system to measure the three-dimensional growth of plants in response to applied stimuli such as short flashes of light. Resolution of displacement was on the order of 1/10 of a fringe, or 0.16 μm, using a HeNe laser. Growth movement velocities under constant illumination of 0.4 to 0.05 μ/min were observed using the cactus *Stapelia.*

In Section 11.3, we describe a phase-lock technique using 900 kHz, air-coupled ultrasound developed by Northrop and Nilakhe (1977) to measure the ocular pulse (minute displacements of the cornea of the eye due to blood pressure) with resolution of over 1 μm.

Drake and Leiner (1984) have described the use of a single-mode, fiber optic, Fizeau interferometer to measure the displacement of the tympanic membrane of the cricket in response to a sinusoidal sound stimulus of 90 dB sound pressure level (SPL) over a range of 1 to 20 kHz. Their system could resolve peak displacements ranging from 0.1 to 300 Å. The 0.1 Å limit was set by system noise, and the 300 Å limit by linearity considerations in the interferometer. A HeNe laser was used. Drake and Leiner's system is illustrated schematically in Figure 7.34. Note that the distal end of the single-mode optical fiber is brought to within a millimeter or so of the reflecting tympanic membrane, this distance not being exceptionally critical except if too large; the power of the return signal picked up by the optical fiber is too small for interferometer operation. The specimen is mounted on a thickness-mode piezoelectric crystal which acts as displacement vernier to adjust the exact working distance between the end of the optical fiber and membrane for maximum interferometer sensitivity.

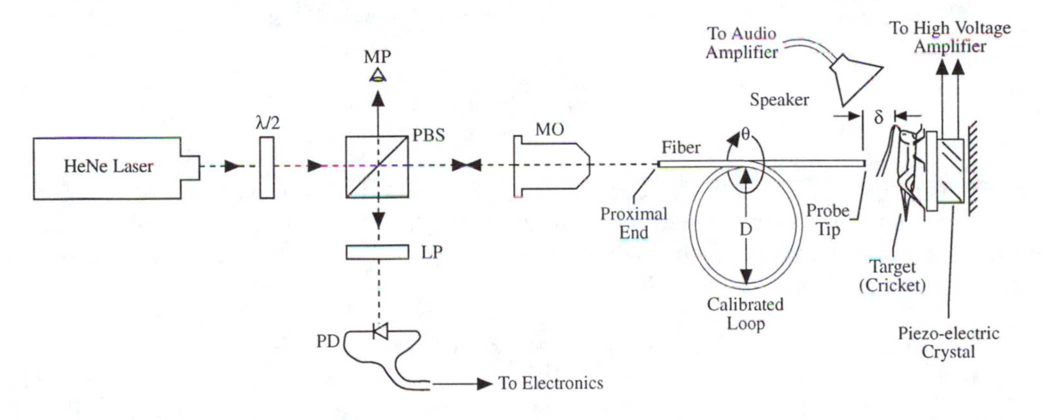

FIGURE 7.34 Diagram of the single optical fiber Fizeau interferometer used by Drake and Leiner (1984) to measure displacements of 0.1 to 300 Å in a sound-excited, insect tympanic membrane. MO, microscope objective lens; λ/2, halfwave plate; LP, linear polarizer; PBS, polarizing beamsplitter; MP, (visual) monitoring point; PD, photodiode.

Northrop (1980), and Northrop and Decker (1977) have described a 40 kHz ultrasonic system used as an incremental motion detector. The application of Northrop's system was a prototype, no-touch, infant apnea monitor and convulsion alarm for at-risk new-born babies. The system's transmit and receive transducers were installed in the top of a conventional incubator system, and the CW ultrasound beam was directed down to ensonify the surface of the baby's skin, a good reflecting surface. The return signal was amplified and processed electronically, as shown in the block diagram of Figure 7.35. The apnea monitor system output is responsive to the small changes in the ultrasound pathlength. The frequency of the closed-loop system voltage-controlled oscillator (VCO) is changed by feedback so that at any instant, there is a constant number of wavelengths in the air in the round-trip airpath (TRX to BABY to RCX). That is, the controlled variable in this feedback system is the total phase lag between TRX and RCX. It may be shown that the analog output of the apnea monitor is proportional to incremental changes in the total airpath length. Such changes in airpath length would normally be from chest and abdominal movements caused by normal breathing of the sleeping infant. Changes in the normal amplitude and rate of these displacements are what are used to trigger the apnea alarm to summon medical help. The analog output is given by:

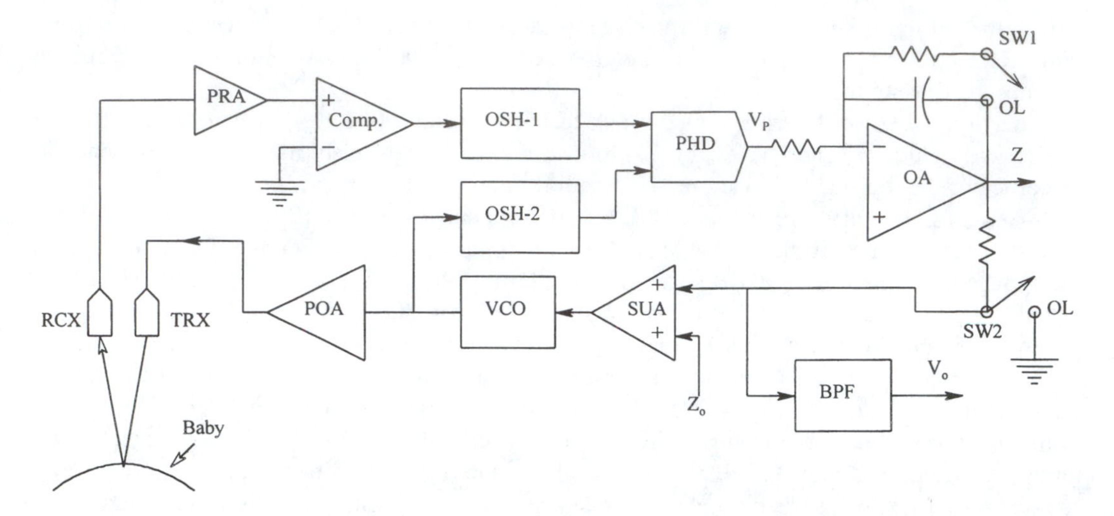

FIGURE 7.35 Block diagram of the 40 kHz, ultrasonic, infant apnea monitor system of Northrop (1980). Symbols: TRX, transmitting ultrasound transducer; RCX, receiving transducer; PRA, preamplifier; POA, power amplifier; COM, analog comparator; OSH, one-shot multivibrator; INT, integrator; BPF, bandpass filter; SUA, summing amplifier; VCO, voltage-controlled oscillator; z_o, DC reference voltage.

$$V_o = K_o \frac{2\pi(N + 1/2)c}{2X^2 K_V} \Delta x \quad 7.111$$

where K_o is a constant, K_V is the VCO constant in (r/s)/volt, c is the speed of sound in air, X is the average pathlength from the transducers to the baby, N is the integer number of ultrasound wavelengths in 2X, the total airpath length, and Δx is the (incremental) displacement of the baby's chest and abdomen. N may be given by:

$$N = \mathrm{INT}\left[\frac{2XK_V}{2\pi c}\right] V_C \quad 7.112$$

Here, V_C is a constant (DC) voltage. The ultrasonic apnea monitor was calibrated using an 18" woofer loudspeaker whose DC displacement sensitivity was known. Limiting resolution of the Northrop system was found to be about ±12 μm, due to system noise and the sound wavelength at 40 kHz (0.86 cm). Output sensitivity for X = 21 cm, was 0.92 V/mm. Once lock is established, the distance X may vary slowly over a wide range, limited only by the working range of the VCO and its ability to keep the steady-state relation, Equation 7.113,

$$f_o = (N + 1/2)c/(2X) \quad 7.113$$

satisfied. Here f_o is the VCO output frequency. Large, rapid changes in X may cause the system to loose lock. This event can be detected electronically (Northrop and Decker, 1977) and used to relock the system to a new N value. Note that the phase-lock design principle of this system can be extended to microwaves. Such a microwave system might be useful in detecting motion of living persons buried in opaque but otherwise microwave-transparent media, such as snow, or rubble from collapsed buildings.

7.3 MEASUREMENT OF FORCE AND TORQUE

Measurement of force is important in many branches of science, engineering, and biomedicine. For example, the magnetic attractive or repulsive force between current-carrying conductors is used in the definition of ampere as the unit of electric current. The measurement of forces in a uniform gravitational field are how we weigh things. The thrust of a jet engine is measured as a force. In biomechanics, force plates are used to study the forces involved in various gaits (walking, etc.). Physiologists and physical therapists are concerned with the contractile force produced by muscles under various conditions. Aerospace engineers and naval architects are concerned with the lift and drag forces on airframes, wings, hulls, rotors, etc. A locomotive or tractor-trailer exerts force on a load to accelerate it to speed, and then keep it rolling. Traction force times velocity is the power expended in moving the load (horsepower). The range of forces which are commonly measured is from micrograms to hundreds of thousands of kilograms. Obviously, no one sensor can cover this dynamic range, so there are many types and sizes of force sensors (load cells) available to the instrumentation engineer, depending on the application.

Torque measurements are generally made on rotating shafts connecting a power source (electric motor, Diesel engine, etc.) to a device doing work (pump, rolling mill, conveyor belt, etc.). Torque measurements are used to calculate rotatory output power (torque × angular velocity). In general, direct torque measurements are more difficult to make than are force measurements because the shaft is rotating, and slip rings must be used to couple the output signal to the stationary, outside world. There are means, as we will see below, of measuring torque optically or magnetically which do not require direct contact with the rotating shaft.

7.3.1 Load Cells for Force Measurement

Curiously, most force measurements involve the small displacement of a stressed, linearly elastic member. Displacement or strain is proportional to the force load-produced stress. The transduction of the small displacement may be done optically by interferometry, or electrically by a sensitive LVDT, a bonded or unbonded wire strain gauge bridge, a semiconductor strain gauge bridge, a circuit which senses small changes in capacitance, or by a piezoelectric crystal (crystal force transducers only respond to time-varying forces). Any force applied to a piston acting on a fluid will produce a rise in pressure, which, if there are no leaks, will remain constant and can be measured with a pressure transducer. A basic type of load cell uses four bonded strain gauges in a Wheatstone bridge configuration, illustrated in Figure 7.36. The four-arm bridge is inherently temperature-compensated, and the circular load cell configuration works for compressive as well as tension loads.

Interface, Inc., of Scottsdale, AZ offers a large line of bonded strain gauge bridge load cells with ranges from 5 to 200,000 lbs full-scale in tension or compression. The Interface load cells are quite precise and linear. Nonlinearity is ±0.03% of rated output, hysteresis is ±0.02% of rated output, and nonrepeatability is ±0.01% of rated output.

Schaevitz Engineering, Pennsauken, NJ markets LVDT-based load cells with ranges from 10 g full-scale, to as much as 500,000 lbs full-scale. Frequency response of an LVDT-based load cell is about one quarter the carrier frequency used with the LVDT, which can be as high as 10 kHz. The Schaevitz load cells have linearities of 0.2% of full range, resolutions of 0.1% of full range, and repeatability of 0.1% of full range. Deflection at full-scale force may be 0.005" if the Schaevitz 004-XS-B sub-miniature LVDT is used in the cell.

Still another type of load cell is the capacitive cell marketed by Kavlico Corp., Chatsworth, CA. In this cell, the applied force causes the deflection (5×10^{-4} inch at full-scale force) of a ceramic diaphragm, one surface of which is metallized, which in turn causes a proportional capacitance change which is sensed electronically. Kavlico load cells have full-scale ranges of 100 g to 60 lbs. System accuracy is ±5% of full-scale, and repeatability and hysteresis are typically ±0.36%.

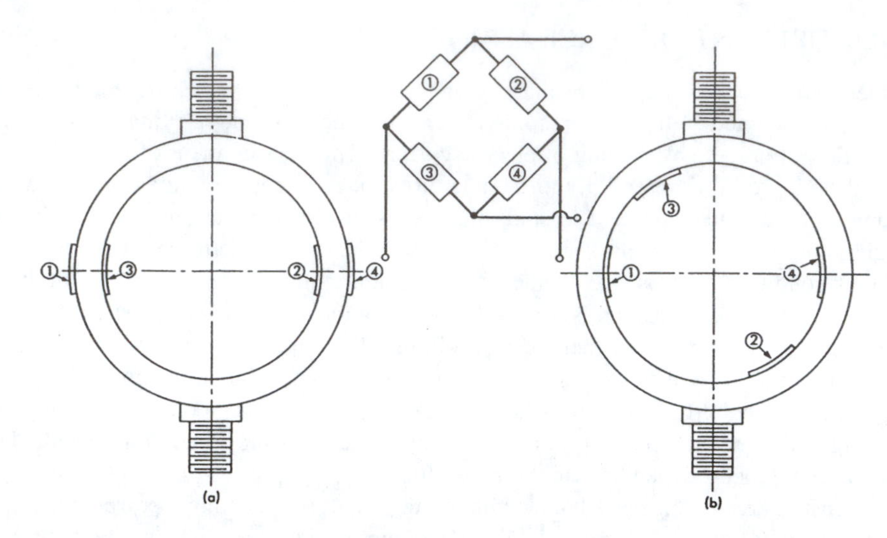

FIGURE 7.36 Two ring-type load cells used to measure tensile or compressive forces. In (a), the strain gauges respond only to the bending strains; the axial strain components cancel out in the bridge. In (b), the Wheatstone bridge has greater sensitivity because gauges 1 and 4 respond to both axial and bending strain components. (From Beckwith, T.G. and Buck, N.L., *Mechanical Measurements*, 1st ed., 1961. Reprinted by permission of Addison-Wesley Longman Publishing Company, Inc.)

Grass Instruments, Quincy, MA, makes a physiological force sensor which is basically a cantilever to which are bonded four strain gauges that make a temperature-compensated Wheatstone bridge. The Grass force sensors have sensitivities which can be decreased by the addition of insertable spring sets. The Grass FT-03C has a working range of ±50 g with no insertable springs, but can be given a full-scale range of 2 kg with the stiffest spring set. Deflection at full rated load is 1 mm. The mechanical natural resonant frequency is 85 Hz for the 50 g range and 500 Hz for the 2 kg range. The strain gauge bridge output is 1.5 mV/mm/V applied to the bridge. Obviously, differential amplification is needed.

Accurate, linear load cells for extreme compressive forces can be made using hydraulic technology. Figure 7.37 illustrates such a hydraulic cell. Beckwith and Buck (1961) state that capacities as high as 5×10^6 lbs with accuracies of ±0.1% of full-scale have been attained with this design. Pneumatic load cells offer the advantage over hydraulic fluid designs over a lower force range in terms of temperature insensitivity. Air is compressible, and does not freeze or boil over a very wide temperature range. Pneumatic load cells are generally made using a compliant diaphragm, rather than a piston. In addition, they use a feedback design in which the internal air pressure is adjusted by a needle valve attached to the moving diaphragm such that if the external, downward load force exceeds the upward pneumatic force on the diaphragm, the diaphragm drops downward and decreases the outflow of air by decreasing the pneumatic bleed conductance. This causes the internal air pressure to rise, raising the diaphragm and increasing the outflow conductance. An equilibrium is reached in which the internal pressure times the effective diaphragm area equals the applied load. Also, the volume inflow must equal the volume outflow at equilibrium. Refer to Figure 7.38. Assume the needle valve has a pneumatic conductance which is a function of diaphragm displacement given by:

$$G_o(x) = G_{oMAX} \qquad x < 0 \tag{7.114A}$$

$$G_o(x) = G_{oMAX}\left(1 - x/x_{MAX}\right) \qquad 0 \le x \le x_{MAX} \tag{7.114B}$$

$$G_o(x) = 0 \qquad x \ge x_{MAX} \tag{7.114C}$$

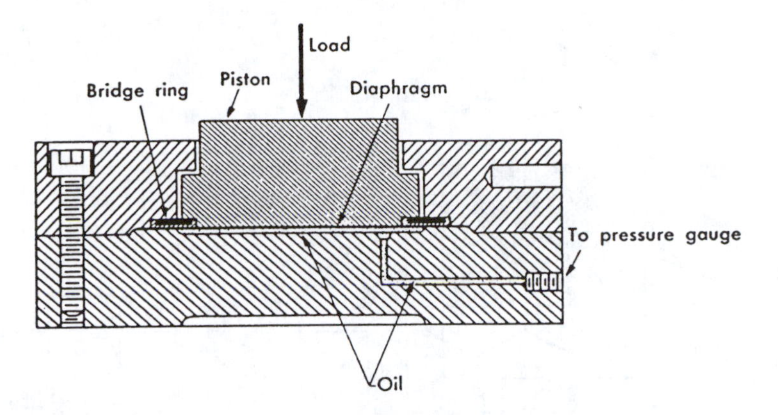

FIGURE 7.37 Section through a hydraulic load cell. This system responds only to compressive forces. (From Beckwith, T.G. and Buck, N.L., *Mechanical Measurements*, 1st ed., 1961. Reprinted by permission of Addison-Wesley Longman Publishing Company, Inc.)

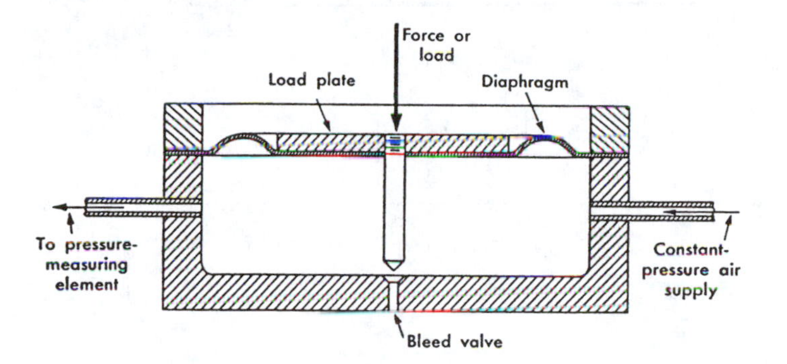

FIGURE 7.38 Section through a pneumatic load cell. See text for mathematical analysis. (From Beckwith, T.G. and Buck, N.L., *Mechanical Measurements*, 1st ed., 1961. Reprinted by permission of Addison-Wesley Longman Publishing Company, Inc.)

Here negative x is upward displacement of the membrane from x = 0. It is not difficult to show that in the steady state,

$$P_i = F/A = P_o G_i/(G_i + G_o) \qquad 7.115$$

where G_i is the fixed, pneumatic input conductance, A is the effective area of the diaphragm, F is the external load force, and P_o is the pressure in the air reservoir supply. From Equation 7.115 above we see that the practical bounds on the measurement of F are set by $F_{MAX} = P_o A$ at the upper end, and $F_{MIN} = P_o A G_i/(G_i - G_{oMAX})$ at the low end. One problem with simple pneumatic load cells is that they may become unstable under certain conditions. If the force on the pneumatic load cell is a mass, the air in the chamber behaves like a nonlinear spring, and oscillations and instability in the operation of this sensor can occur unless pneumatic damping is used in its design. Beckwith and Buck (1961) report that 40 ton capacity pneumatic load cells are commercially available.

Closed-loop, electromagnetic force cells are used for weighing light objects in the range of 0.1 mg to 100 g. In this design, an upward force is generated electromagnetically to equal the external, downward force, as shown in Figure 7.39A. Equilibrium position can be sensed either electro-optically or with a high-resolution LVDT. The electromagnetic force is generated by a coil and permanent magnet assembly similar to that used in a loudspeaker, except that there is no spring required. The coil current is made proportional to the integral of the downward

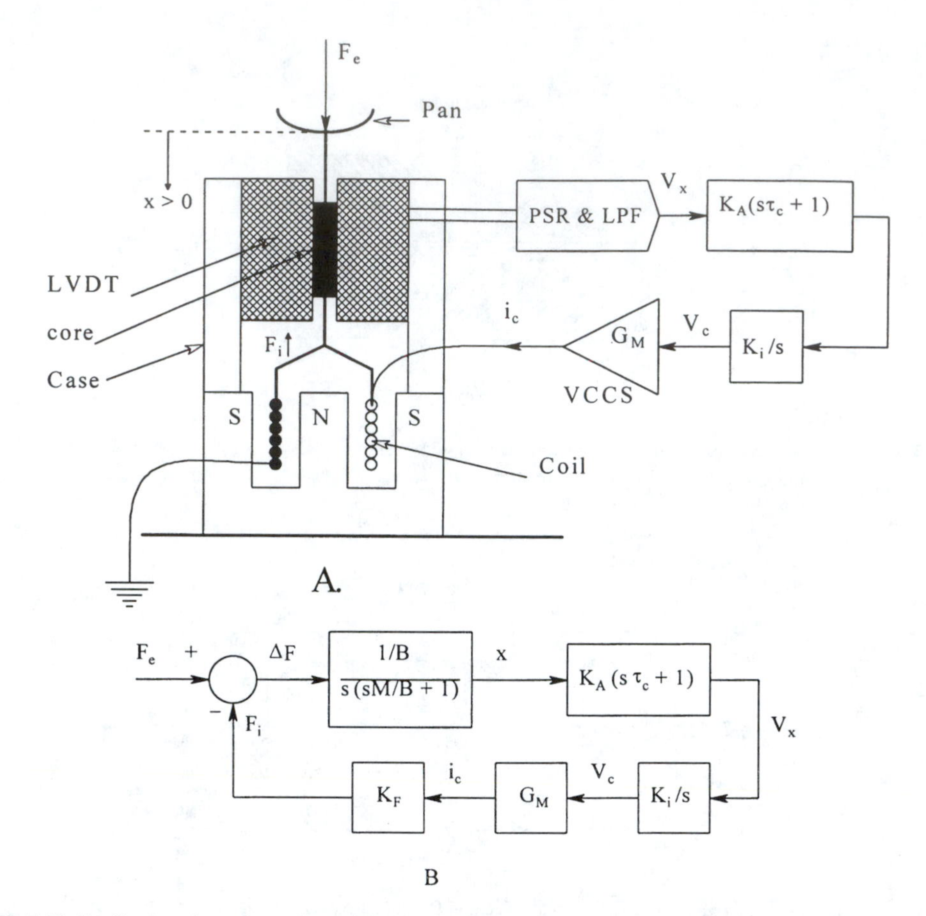

FIGURE 7.39 (A) Section through an electrodynamic feedback load cell. Position is sensed by an LVDT. The LVDT output is conditioned by a phase-sensitive rectifier and low-pass filter (PSR & LPF), then by a proportional plus derivative compensator (COMP) and integrator (INT). The integrator's output, VC, is then conditioned by a transconductance amplifier (VCCS) whose output, i_C, is the input to the force coil. See text for analysis. (B) Block diagram of the electrodynamic feedback load cell.

displacement. A systems block diagram describing this system is shown in Figure 7.39B. There is zero steady-state displacement error, and it can be shown that:

$$\frac{V_c}{Mg} = \frac{1}{G_M K_F} \text{ volts/newton} \tag{7.116}$$

Other obvious, and far less technical means for measuring forces less than several hundred kilogram include spring scales which make use of Hooke's law and a calibrated scale; displacement of the moving end of the spring is proportional to the applied force. Steelyards, laboratory beam balances, platform balances, etc., all make use of calibrated weights, and lever arms pivoted on knife-edge bearings. A human operator adjusts the position of one or more known weights on the long arm of the balance until the arm is level and stationary, signifying that the unknown force times its lever arm equals the sum of the known forces times their lever arms.

7.3.2 Torque Measurements

Torque measurements on rotating shafts are made with more difficulty than are those on stationary shafts. The reason for this difficulty is that the wires from sensor elements, such as a strain gauge bridge cemented to the shaft, must be brought out to the nonrotating world through slip rings. Noise introduced by the brushes making contact with the rotating slip rings can reduce the resolution of a rotating, torque measuring system below that of an identical system used on a stationary shaft without slip rings. Figure 7.40 illustrates the configuration of bonded strain gauges on a shaft in order to measure torsional strain caused by a torque couple applied to the ends of the shaft. Such a configuration may be shown to be insensitive to temperature changes, axial load, and bending of the shaft.

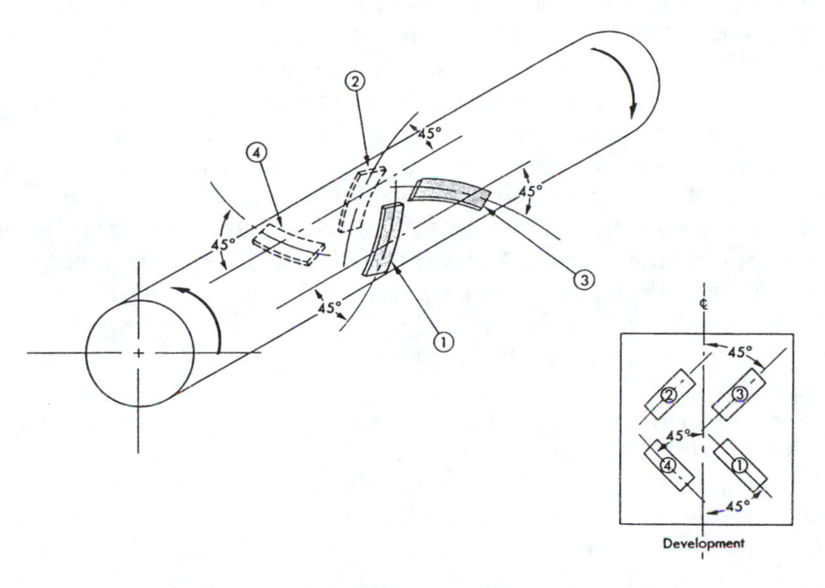

FIGURE 7.40 Four bonded strain gauges cemented to a rotating torque shaft. This configuration is insensitive to temperature, axial load, or bending. Slip rings must be used to make electrical connections to the gauges (not shown). (From Beckwith, T.G. and Buck, N.L., *Mechanical Measurements*, 1st ed., 1961. Reprinted by permission of Addison-Wesley Longman Publishing Company, Inc.)

There are other means of measuring torque on a rotating shaft that do not require electrical contact with the shaft. In one approach, alternating black and white markings are painted around the circumference of each end of the shaft under measurement. A photosensor consisting of an IR LED and a phototransistor is placed above both rotating, black/white bands. The output of each photosensor is a square wave with frequency proportional to the shaft's angular velocity. However, the phase between the square waves is determined by the torque-produced displacement angle between the black/white bands. This angle is given by $\phi = KTd$

$$\phi = KTd \qquad 7.117$$

where d is the distance between the bands, T is the torque on the shaft, and K is a constant determined by the material and dimensions of the shaft. The electrical phase angle between the two square wave trains, θ, can be measured, and is related to ϕ by:

$$\theta = N\phi \qquad 7.118$$

where N is the integer number of black/white cycles inscribed around the circumference of the torque shaft at each end. The frequency of each square wave is just:

$$f = NRPM/60 \qquad 7.119$$

where RPM is the shaft's revolutions per minute. One way of measuring the phase angle, θ, is to use a phase detector IC, such as the Motorola (Phoenix, AZ) MC4044. The MC4044 has a working range of ±360°, which generally can be used to resolve angles as small as ±0.1°. Obviously, to measure low values of torque, a phase detector is needed which can resolve hundredths of a degree or better, and work over a wide range of frequencies.

A frequency-independent, millidegree resolution phase detector design was devised by Northrop and Du (Du, 1993), based on a design by Nemat (1990). The Northrop and Du system is illustrated schematically in Figure 7.41. The periodic reference signal (A) is converted to a transistor-transistor logic (TTL) waveform by a fast comparator, the output of which is the input to a phase-lock loop in which the VCO output is made to follow the input frequency and multiply it by a factor of 360,000. Thus, every input cycle produces 360,000 PLL clock cycles, and the input phase is effectively divided into millidegrees, i.e., there is one PLL clock pulse for every 0.001° of input, regardless of the exact input frequency. The reference and complimentary, phase-shifted square wave inputs act through a NAND gate (G1) to enable a 24-bit, high-frequency binary counter which counts the high-frequency output of the PLL for N cycles of the input wave. For purposes of discussion, assume that the input signals are 100 Hz, thus the PLL high-frequency output will be 36 MHz. The number of high-frequency pulses counted depends, of course, on the phase shift between the 100 Hz input square waves, as well as N.

If the phase shift is zero, no pulses are counted. On the other hand, if, for example, the phase shift is 3.6°, then 3600 PLL clock cycles will be counted for every one of the N gated input cycles. After N input waveform cycles, the cumulative count in the binary counter will be 9.216×10^5, if $N = 256$ and $\theta = 3.6°$. In order to obtain an average count for N input cycles, the cumulative count in the counter is loaded in parallel to 24-bit shift registers (3, 74F299s). Division by N is easily accomplished by shifting the register contents to the right by k clock cycles from the controlling computer. We make $N = 2^k$ in this system, with k = 0, 1, 2, ... 8. The 16 least significant bits of the shifted binary number in the register is then loaded into the computer and displayed as a decimal angle in the format, 88.888°. For our 3.6° phase shift example, this would be 3600 or 3.600°. The PLL allows precise measurement of the phase difference between the input signals A and B even though their frequency may not be fixed. In fact, it will vary as the shaft speed varies.

A problem with the Northrop and Du system occurs when the phase angle, θ, is large. Now the counter must count upwards of 360,000 pulses per cycle. The cumulative count for N = 256 would be about 92.2 million before division by 256. Thus, if it is intended to work with large phase angles, the Northrop and Du system needs counters capable of holding raw counts of over 2^{26}. A practical upper bound on the input signal frequency is set by the maximum operating frequency of the PLL, and the maximum counting rate of the counter. A Signetics NE564 PLL will operate up to 50 MHz. The 74F579 decade counters work up to 115 MHz.

An alternate to the use of photoelectronic measurement of the shaft's torque angle, ϕ, is the use of a variable reluctance pickup, such as shown in Figure 7.42. Here, a voltage pulse is induced in the coil every time a steel gear tooth passes under the permanent magnet. Two gears, concentric with the torque shaft, are used, as in the case of the photoelectronic pickup described above. Because the output pulses are biphasic, fast analog comparators, such as LT1016s, are used to convert them to TTL pulse trains for input to the phase detector.

Other schemes of no-touch torque measurement in rotating shafts have been described which make use of the magnetoelastic (Villari) effect (Lion, 1959). Here the magnetic permeability of ferromagnetic materials is seen to change in a nonlinear manner when the material is subjected to mechanical stress, such as torsion. Lion reports that the Villari effect has been used to measure forces in load cells. As the force is increased, there is a nonlinear decrease in the inductance of a coil wound around the ferromagnetic core under compression. Villari sensors may be adaptable to the measurement of torque (NASA Tech Briefs 15(3): 1991, p. 50). Besides their obvious nonlinearity, the Villari effect has a strong temperature dependence which would need compensation in a field-grade instrument. The inductance change resulting from mechanical strain of the core must

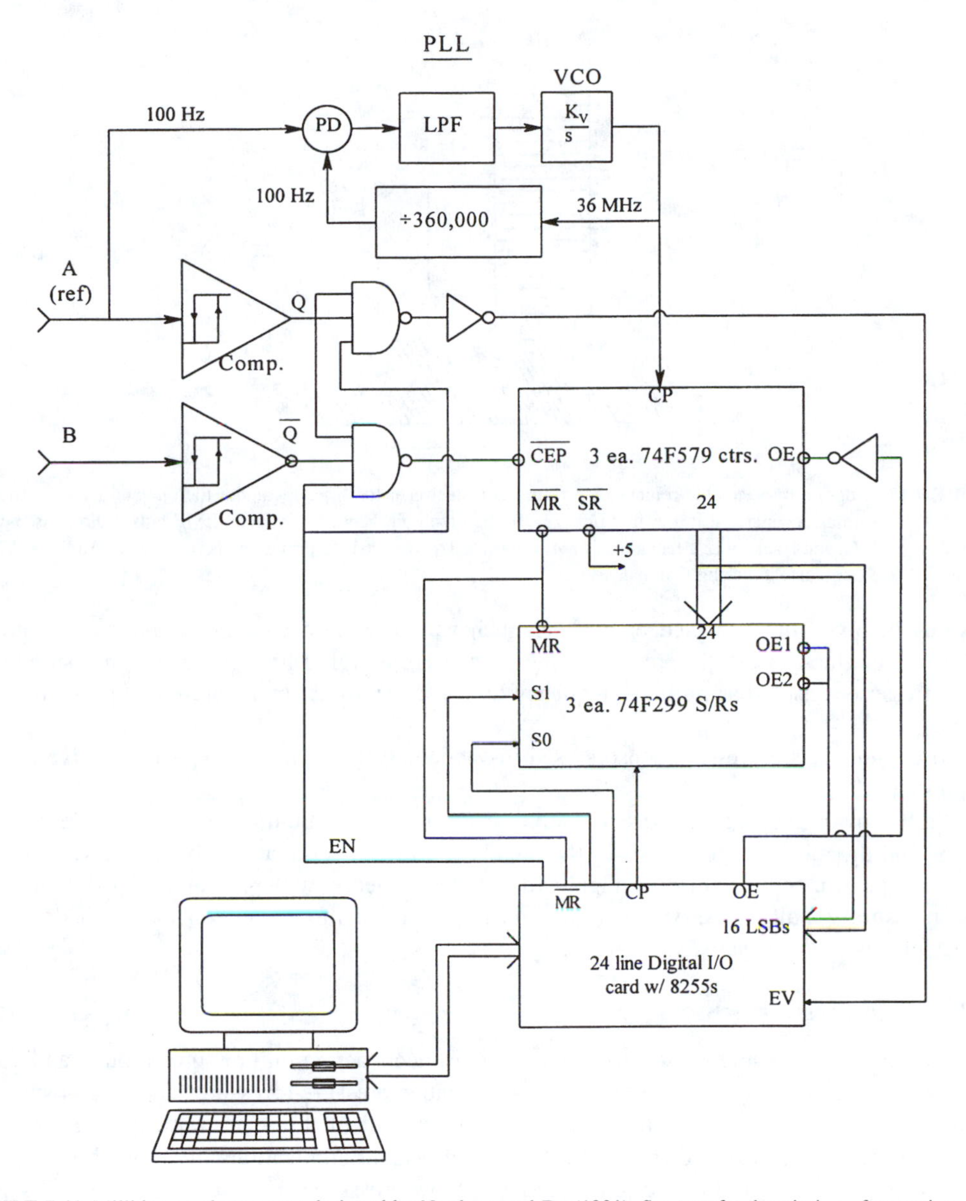

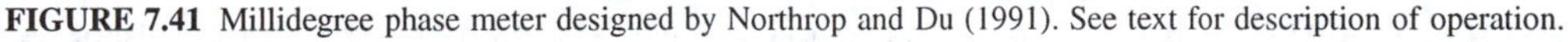
FIGURE 7.41 Millidegree phase meter designed by Northrop and Du (1991). See text for description of operation.

be detected with some form of inductance bridge or by sensing the change in the output frequency of an oscillator in which the inductance is a frequency-determining component.

7.4 PRESSURE MEASUREMENTS

Many interesting transducer systems have been developed to measure fluid pressure, as gauge pressure, absolute, or differential pressure. Applications of pressure sensors are broad, including but not limited to aircraft altimeters, submarine depth meters, pressure meters for gas tanks, sensors for aircraft rate-of-climb and airspeed indicators, medical/physiological pressure sensors for blood, cerebrospinal fluid, intraocular pressure, intrathoracic pressure, kidney dialysis pressure, etc., hydraulic system pressure, oil pressure in bearings for turbines, generators, diesel engines, etc., liquid level indicators for storage tanks, purge liquid pressure in oil well reconditioning, etc.

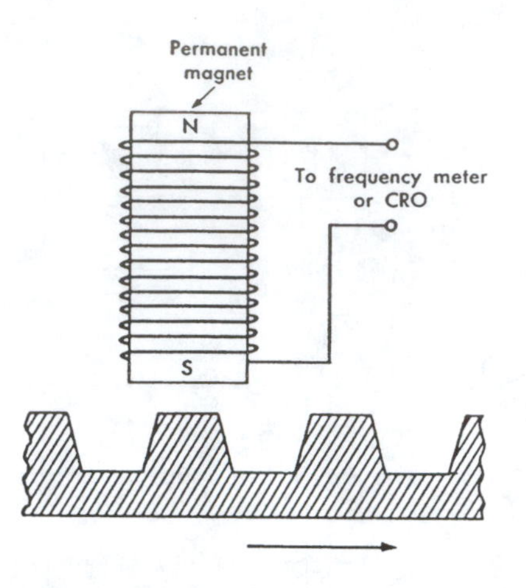

FIGURE 7.42 Diagram of a variable-reluctance pickup. As a steel gear tooth passes under the magnet and coil, there is a transient increase in the flux linkages through the coil, hence an induced EMF pulse. (From Beckwith, T.G. and Buck, N.L., *Mechanical Measurements*, 1st ed., 1961. Reprinted by permission of Addison-Wesley Longman Publishing Company, Inc.)

Pressures involved in controlled explosions, such as combustion chamber pressures in internal combustion engines, or the gas pressure in the chamber or barrel of firearms, can be measured with spatial pressure sensors having high-frequency response, and high temperature and pressure capabilities.

Also of interest are pressure sensors used in vacuum systems, i.e., sensors responsive to very low pressures.

In the following sections we discuss some of the important, commercially available pressure sensors, and the mechanisms by which they work. It will be seen that nearly all pressure sensors use some type of membrane or diaphragm which deflects linearly with pressure, relative to the zero gauge pressure condition. As you will see, measurement of this deflection has been done in a number of ingenious ways.

7.4.1 High-Pressure Sensors

In this category of pressure sensor, we place all transducers acting on pressures around and above atmospheric pressure, including those that measure gauge pressure (pressures above atmospheric), absolute pressure, and pressure differences. At the low end of the measurement range, we find pressure sensors specialized for biomedical applications and applications in automobile fuel and emissions control systems. One design approach for low-pressure sensors used by Schaevitz Engineering (Hampton, VA) is to use a compliant bellows attached to the core of an LVDT. Bellows deflection is proportional to the pressure difference across its walls, and models are available which use vacuum, vented gage, or differential pressure references. Such designs can be made quite sensitive (0 to 2 in. H_2O in the Schaevitz Model P-3000 series). Because of the high compliance (low stiffness) of the bellows, and the mass of the LVDT core, such sensitive pressure sensors have limited frequency response, and are generally suited for measuring pressures which vary slowly.

Another approach to pressure measurement is seen in the design of pressure sensors made by Conal Precision Instruments, Inc., Franklin Lakes, NJ. Conal pressure sensors use the pressure of a bellows against a quartz crystal used in a 40 kHz oscillator to reduce the oscillator's frequency. Unfortunately, the relation between oscillator period and input pressure is nonlinear, and a calibration equation must be used to convert period to a function proportional to pressure. Conal claims resolution of 0.0001% of FS, and hysteresis of 0.01% of FS. Full-scale pressure ranges from 20 to 1000 psi are available.

Many of the pressure sensors described below use the deflection of a thin diaphragm, or the strain produced in this diaphragm by pressure as the primary step in converting pressure to an electrical output. When pressure is applied to a flat diaphragm on the side opposite the gauges (see Figure 7.43), the central gauge or gauges experience tension, while the gauge cemented to the edge of the diaphragm sees compression. In terms of the deflection of the center of the diaphragm, deflection is linear with applied pressure for deflections up to about 30% of the diaphragm thickness (Beckwith and Buck, 1961). The central deflection of a diaphragm is given approximately by:

$$\delta y = \frac{3\Delta P R^4 (1-\mu^2)}{16 Y t^3} \tag{7.120}$$

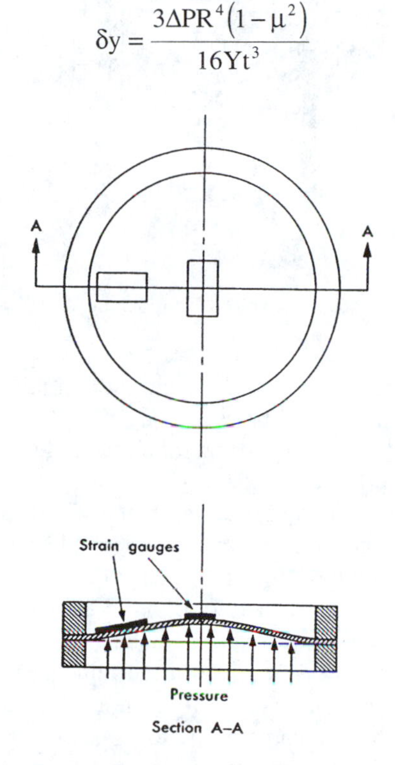

FIGURE 7.43 Diagram of a diaphragm pressure sensor in which the sensing elements are a pair of strain gauges bonded to the diaphragm. The strain gauges are put in one arm of a Wheatstone bridge, giving temperature compensation. (From Beckwith, T.G. and Buck, N.L., *Mechanical Measurements*, 1st ed., 1961. Reprinted by permission of Addison-Wesley Longman Publishing Company, Inc.)

where ΔP is the pressure difference across the diaphragm, R is the radius of the diaphragm, t is its thickness, Y is its Young's modulus, and μ its Poisson's ratio. The radial stress in the diaphragm at its edge is:

$$\sigma_r = (3/4)(R/t)^2 \, \Delta P \tag{7.121}$$

The tangential stress at the center of the diaphragm is given by:

$$\sigma_t = (3/8)(R/t)^2 (1+\mu) \Delta P \tag{7.122}$$

The parameters are the same as for Equation 7.120 above (Beckwith and Buck, 1961, Chapter 12).

Stow Laboratories, Inc., Hudson, MA, make miniature, Pitran™ pressure sensors. In the Stow design, illustrated in Figure 7.44, pressure acting on a diaphragm causes a pointed, insulated stylus to bear mechanically on the "layer cake" of an NPN silicon transistor. This transferred pressure produces a large, reversible, effective decrease in the transistor's β, lowering I_C and thus raising V_{CE}. Stow claims that Pitrans have 0.5% linearity and hysteresis, and a 65 dB dynamic range.

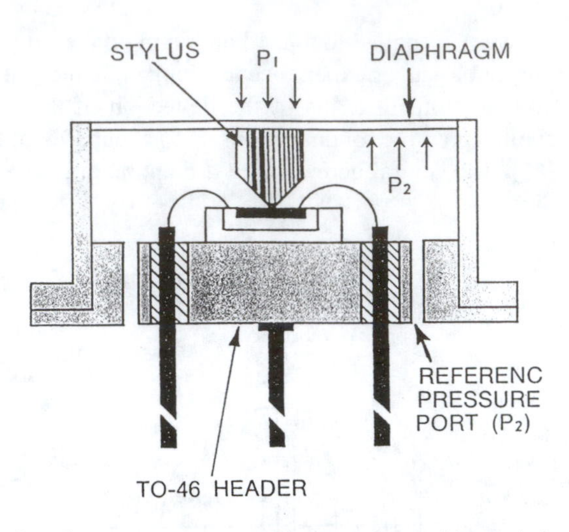

FIGURE 7.44 Section through a Pitran pressure sensor. (Figure reprinted with permission of Stow Laboratories, Inc., Hudson, MA.)

Maximum membrane deflection is 2 microinches, and the Pitran's mechanical resonant frequency is about 150 kHz. Pitran models have FS pressure ranges of 0.1 to 20 psi.

A family of pressure sensors based on a differential, variable reluctance transducer principle is sold by Validyne Engineering Corp, Northridge, CA. A cross-sectional view of the Validyne pressure sensor is shown in Figure 7.45. The Validyne sensor design uses a diaphragm of magnetically permeable stainless steel clamped between two identical blocks. Embedded in each block are identical "E" core and coil assemblies. With zero deflection of the diaphragm, both coil and core assemblies see an identical air gap, and therefore both coils have identical inductances. When a pressure difference exists across the diaphragm, its deflection causes an increase in the air gap or reluctance of one coil assembly, and a decrease in the reluctance of the flux path of the second coil. Coil inductance, L, is inversely proportional to the reluctance of the magnetic flux path, so we see that a small bowing of the diaphragm due to ΔP will cause a corresponding $+\Delta L$ in one coil and a $-\Delta L$ in the other. This inductance change is symmetrical and can be sensed with a simple AC bridge, such as shown in Figure 7.46. It is left as an exercise to show that when the bridge sinusoidal excitation frequency is $\omega_o = (R + R_L)/L_o$ r/s, the bridge output voltage is:

$$\frac{V_1 - V_2}{V_B} = (\Delta L/L_o)[R/(R + R_L)] \qquad 7.123$$

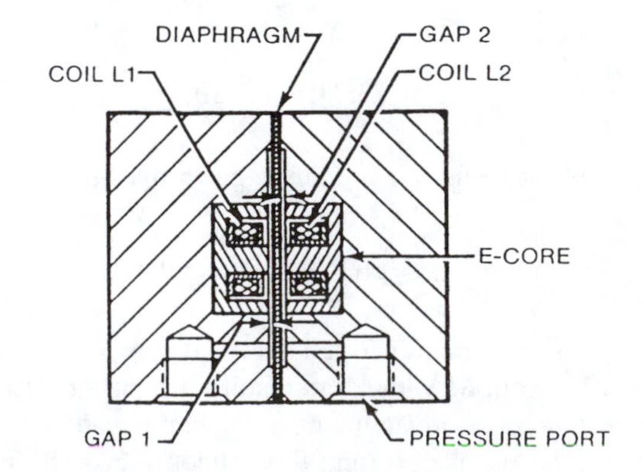

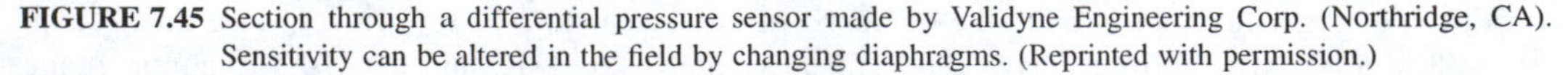
FIGURE 7.45 Section through a differential pressure sensor made by Validyne Engineering Corp. (Northridge, CA). Sensitivity can be altered in the field by changing diaphragms. (Reprinted with permission.)

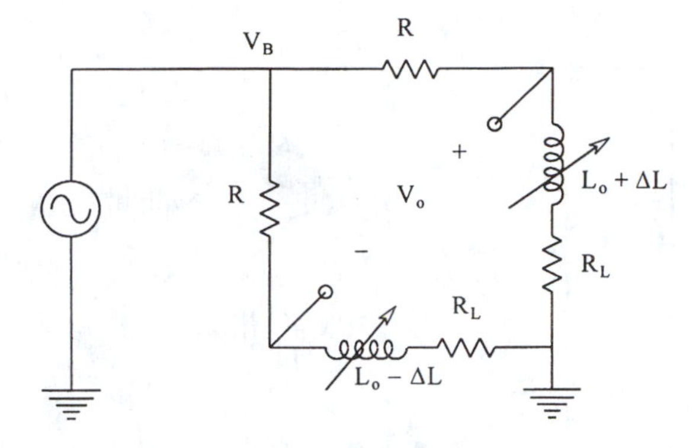

FIGURE 7.46 A simple AC inductance bridge which can be used to give an output proportional to ΔL, which in turn is proportional to ΔP in the sensor shown in Figure 7.45.

The bridge output is a double sideband, suppressed carrier (DSBSC) signal, and can be demodulated by a phase-sensitive rectifier (PSR) and low-pass filter. The output of the PSR/LPF is seen to be linearly proportional to ΔP across the transducer diaphragm. For greater range, stiffer (thicker) diaphragms are used. Diaphragms are available to give full-scale ranges from 0.0125 to 12,500 psi. Excitation voltage is typically 5 Vrms, 3 to 5 kHz. Linearity and hysteresis are typically ±0.5%.

A pressure sensor design that is more commonly encountered than the types described above uses four strain gauges bonded to the protected side of a deflecting diaphragm. In such a full Wheatstone bridge design, two of the bridge arms increase resistance with diaphragm strain, and two decrease resistance. The close proximity of the strain gauges affords nearly complete temperature compensation. In other designs, only two gauges are active, increasing resistance with strain. The other two are inactive and in close proximity to the active gauges, so a temperature-compensated output can be realized.

There are a number of miniature pressure sensor designs that make use of integrated circuit fabrication techniques. In the design described by ICSensors, Inc., Milpitas, CA, four piezoresistive strain gauges are diffused into the surface of a single silicon crystal diaphragm to form a fully active Wheatstone bridge. The single crystal diaphragm has negligible hysteresis. Entran Devices, Fairfield, NJ, makes an EPI series of ultra-miniature pressure sensors using a fully active, four arm Wheatstone bridge of piezoresistive material bonded to a silicon diaphragm. In the EPI-050 series of sensors, the diameter of the body is 0.05" and the body length is 0.25"; full scale pressure ranges from 2 psi to 300 psi, and the corresponding mechanical resonance frequencies range from 100 kHz to 1.7 MHz. Useful, mechanical frequency response is given as 20% of the resonance frequency. Ultra-miniature sensors such as the Entran EPI-050 series can be mounted in the tips of catheters, and used invasively to measure pressures in the chambers of the heart, blood vessels, etc., in physiological studies.

Another approach to pressure sensor design has been devised by Motorola Semiconductors, Phoenix, AZ. Motorola's MPX series of pressure sensors do not use a conventional, four active-arm Wheatstone bridge. Instead, a single, p-type, diffused silicon strain resistor is deposited on an etched, single-crystal silicon diaphragm, as shown in Figure 7.47. A second resistor is deposited to form an X-shaped, four-terminal resistor with two current taps (1 and 3) and two output voltage taps (2 and 4). When current is passed through terminals 1 and 3, and pressure is applied at right angles to the current flow, a transverse electric field is generated, producing an EMF between pins 2 and 4. Motorola suggests that this effect can be considered to be an electromechanical analog of the Hall effect, with pressure replacing the magnetic field. In the Motorola MPX3100 sensor series, a complete signal conditioning system with four op amps and laser-trimmed resistors has been built on the margin of the silicon wafer holding the diaphragm and the transverse voltage strain gauge element (see Figure 7.48). The output of the MPX3100 is quite linear, ±0.2% FS. Temperature compensation of full-scale span is ±2% FS over a 0 to 85°C range. In the design of the MPX2000

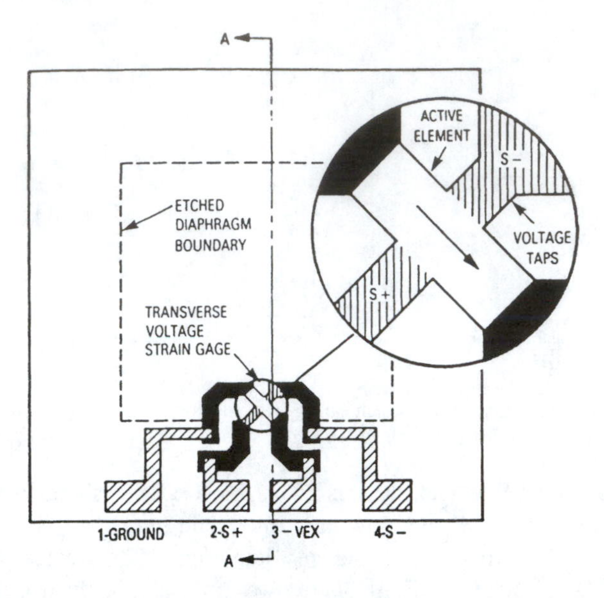

FIGURE 7.47 Diagram of the etched microcircuit diaphragm of a Motorola MPX series pressure sensor. The patented X-shaped structure is a four-terminal resistor with two voltage and two current taps. Motorola claims that the sensor is the electromechanical analog of a Hall effect device. (Courtesy of Motorola Semiconductor Products, Inc., Phoenix, AZ.)

series pressure sensors, five laser-trimmed resistors and two thermistors are deposited on the margin of the silicon chip around the pressure diaphragm. These resistors and thermistors are used to achieve a ±1% FS temperature effect over a 0 to 85°C span. A similar temperature range on one of the non-thermally compensated sensor models, such as the MPX10, would be a whopping ±16.2% FS.

Measurement of extremely high pressures in the range of 10,000 to over 250,000 psi requires special transducer designs. The simplest is a cylindrical pressure cell, illustrated in Figure 7.49. Two active bonded strain gauges respond to the hoop stress of the outer surface caused by pressure difference across the wall of the cylinder. Two reference strain gauges are located on the top of the cylinder. It may be shown that the hoop strain of the outer surface is given by

$$\varepsilon_H = \frac{P_i d^2 (2-\mu)}{Y\left(D^2 - d^2\right)} \text{ inch/inch} \qquad 7.124$$

where Y is Young's modulus, D is the cylinder's outer diameter, d is its inner diameter, μ is Poisson's ratio, and P_i is the internal pressure (Beckwith and Buck, 1961).

A different sensor mechanism which is used to measure extremely high pressures makes use of the fact that the electrical resistance of a conductor changes if it is compressed by an applied, external pressure. Such sensors are called bulk compression gauges (see Figure 7.50). The resistance of the compressed conductor may be approximated by:

$$R(P) = R_o(1 + bP) \qquad 7.125$$

where R_o is the resistance of the conductor at 1 atm and at standard temperature, and b is the pressure coefficient of resistance.

Two metals commonly used for bulk compression gauges are an alloy of gold and 2.1% chromium, and manganin. The gold alloy is preferred because of a much lower tempco than manganin ($\Delta R/R$ = 0.01% for a 70 to 180°F change in temperature for the gold alloy vs. 0.2% for

A.

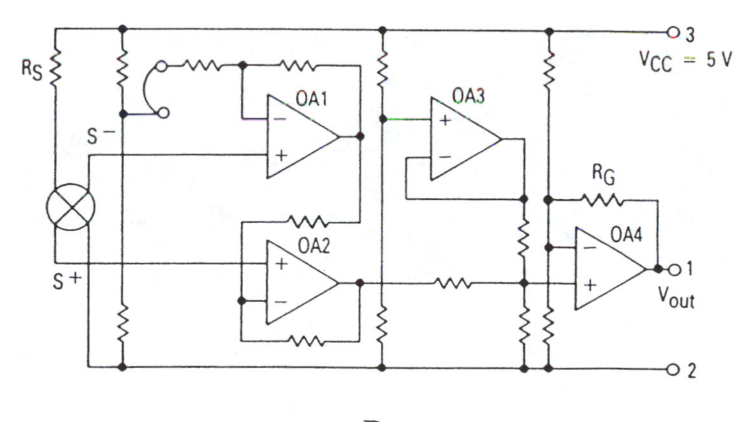

B.

FIGURE 7.48 (A) Top view of the integrated circuit of a Motorola MPX3100 pressure sensor. (B) Schematic of the temperature-compensated MPX3100 pressure sensor. (Courtesy of Motorola Semiconductor Products, Inc., Phoenix, AZ.)

manganin). The $(\Delta R/R)/P$ for the gold alloy is 0.673×10^{-7} ohm/ohmú psi, and is 1.692×10^{-7} ohm/ohm/psi for manganin. Although the gold alloy has a lower pressure sensitivity, its lower tempco makes its use preferred in bulk compression gauges (Beckwith and Buck, 1961, Chapter 12). A bulk compression pressure sensor which works up to 1.5×10^6 psi was described by Hall (1958). Bulk compression pressure sensors must be temperature-compensated. Technical difficulties are encountered at very high pressures in bringing the leads of the insulated resistance element out of the case without leaks or plastic flow problems.

The final type of pressure sensor that we will describe in this chapter is the quartz crystal, piezoelectric gauge. As we have seen in Section 6.3.3, quartz transducers are only suitable for

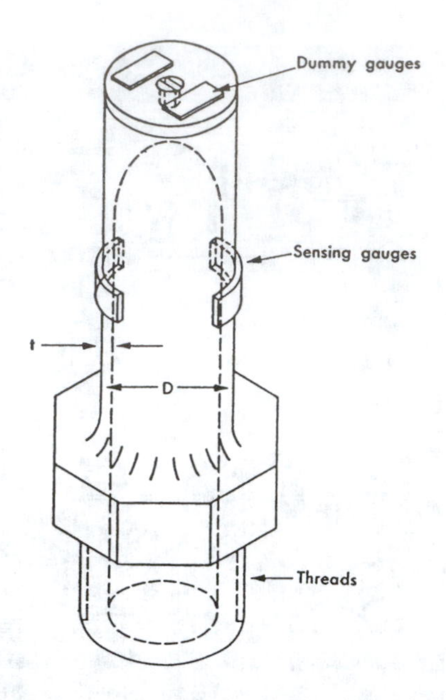

FIGURE 7.49 Cylindrical high-pressure sensor cell which uses four bonded strain gauges in a Wheatstone bridge; the two gauges on the end are inactive and are used for temperature compensation. (From Beckwith, T.G. and Buck, N.L., *Mechanical Measurements*, 1st ed., 1961. Reprinted by permission of Addison-Wesley Longman Publishing Company, Inc.)

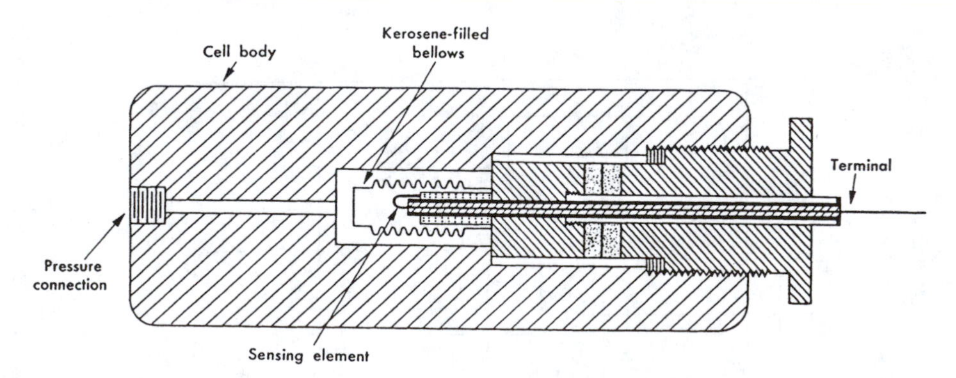

FIGURE 7.50 Section through a very high-pressure, bulk-modulus compression, pressure sensor. The sensing resistor is made one arm of a Wheatstone bridge. (From Beckwith, T.G. and Buck, N.L., *Mechanical Measurements*, 1st ed., 1961. Reprinted by permission of Addison-Wesley Longman Publishing Company, Inc.)

measuring transient and time-varying pressure changes which contain frequencies above their low-frequency (high-pass) pole. Quartz transducers do not respond to static (constant) pressures. Because quartz pressure sensors work well at elevated temperatures, and respond to transient changes in pressure with frequency responses which typically extend to hundreds of kilohertz, they are often used to measure the transient pressure changes associated with combustion in internal combustion engines, and the transient pressure associated with internal ballistics of firearms (peak chamber pressures in military and hunting rifles are typically 50 to 65,000 psi). Some quartz dynamic pressure sensors are water-cooled to prevent the quartz crystal from reaching its Curie temperature and failure in certain high-temperature applications such as measurement of cylinder pressure in an internal combustion engine. The output of a quartz piezoelectric pressure sensor is generally conditioned by a charge amplifier (see Section 2.6.1). Kistler Instrument Corp., Amherst, NY, makes a wide selection of quartz pressure sensors. For example, the Kistler model 6205 sensor is specifically designed for ballistics research and ammunition testing. It has a working range of 0 to 75,000 psi, a charge sensitivity of –0.09 picocoulomb/psi, a resonant (upper) frequency of 300 kHz, a rise time of 1.5 μsec, nonlinearity and hysteresis of ≤1% FS, operating temperature range of –50 to 200°C. The Kistler model 6213 piezoelectric pressure sensor has a maximum range of 145,000 psi, and their special Z-series of sensors operate over an extended temperature range of –196 to 400°C. PCB Piezotronics, Inc., Buffalo, NY also makes a full line of quartz pressure sensors, accelerometers, and load cells.

7.4.2 Low-Pressure Sensors

Here we define low-pressure sensors as those sensors designed to measure pressures significantly lower than atmospheric. Their primary application is in vacuum system measurements. Two means of measuring low pressures do not have electrical outputs: the McLeod gauge, shown in Figure 7.51, is a mercury-filled glass system which is operated manually and read out visually. Its useful range is typically 5 to 0.005 mmHg pressure. The McLeod gauge makes use of Boyle's law: $P_1 = P_2(V_2/V_1)$. P_1 is the pressure we desire to measure, P_2 is the increased pressure in the McLeod gauge obtained by compressing the gas from volume V_1 to a smaller volume, V_2. The gas whose pressure is under measurement should not contain water vapor, or the compression process will cause condensation. Beckwith and Buck (1961) describe the operation of a McLeod gauge:

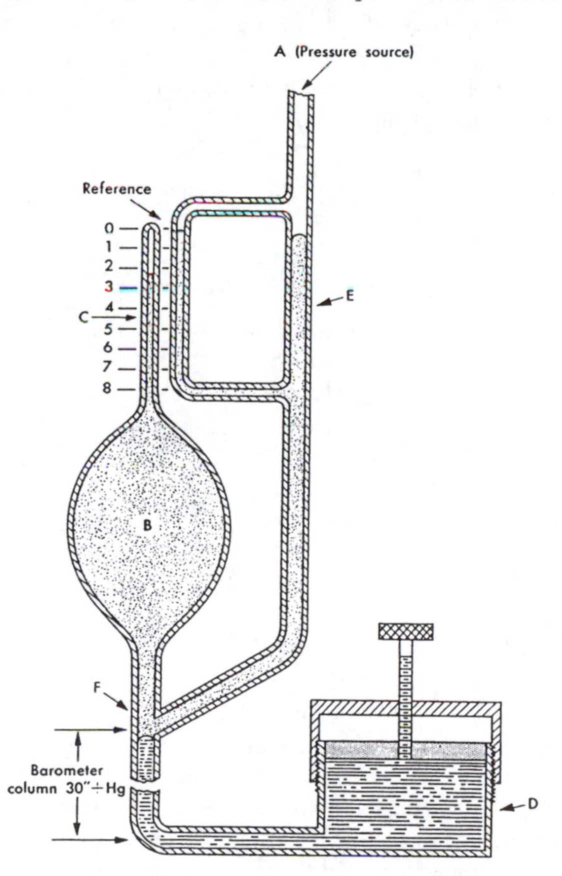

FIGURE 7.51 Section through a McLeod vacuum gauge. (From Beckwith, T.G. and Buck, N.L., *Mechanical Measurements*, 1st ed., 1961. Reprinted by permission of Addison-Wesley Longman Publishing Company, Inc.)

Measurement is made as follows. The unknown pressure is connected to the gauge at point A, and the mercury level is adjusted to fill the volume represented by the darker shading. Under these conditions the unknown pressure fills the bulb B and capillary C. Mercury is then forced out of the reservoir D, up into the bulb and reference column E. When the mercury level reaches the cutoff point F, a known volume of gas [V1] is trapped in the bulb and capillary. The mercury level is then further raised until it reaches the zero reference point in E. Under these conditions the volume remaining in the capillary is read directly from the scale, and the difference in heights of the two columns is the measure of the trapped pressure [P2]. The initial pressure may then be calculated by use of Boyle's law.

Another directly indicating pressure gauge is the General Electric (Fairfield, CT) "*Molecular Vacuum Gauge*". The GE gauge is calibrated over a 0.002 to 20 mmHg range. In a cylindrical chamber which is at the pressure under measurement, a synchronous motor drives a fan. The gas in the chamber is given an angular momentum by the fan, and its molecules strike the blades of an adjacent fan connected to a pointer and a helical spring. The higher the gas pressure in the chamber, the higher the density of molecules striking the pointer's fan, and the more torque produced on it, hence the greater the deflection of the pointer. The scale of the GE gauge is nonlinear, with the highest resolution at the low end of the scale (0 to 0.1 mmHg covers the first 90° of meter deflection, while 0.1 to 20 mmHg cover the remaining 150° of scale).

It is of course important to have low-pressure sensors which can be read out electrically. The first and simplest electrical vacuum gauge is the *Pirani gauge*. The Pirani gauge consists of a filament of platinum, tungsten, or nickel in a glass bulb connected to the vacuum system. Various gases have different heat conductivities, so the Pirani gauge must be calibrated for the gas or gases being used. A simple operating mode of the Pirani gauge is to place it in one arm of a Wheatstone bridge circuit, as shown in Figure 7.52. An identical gauge *in vacuo* is used as a temperature reference. Recall from Section 7.2.2.1 on hot wire anemometers that in a vacuum there will be a rise in the Pirani gauge's filament temperature, given by:

$$\Delta T = I_o^2 R_o \Omega_o / \left(1 - \alpha I_o^2 \Omega_o\right) \qquad 7.126$$

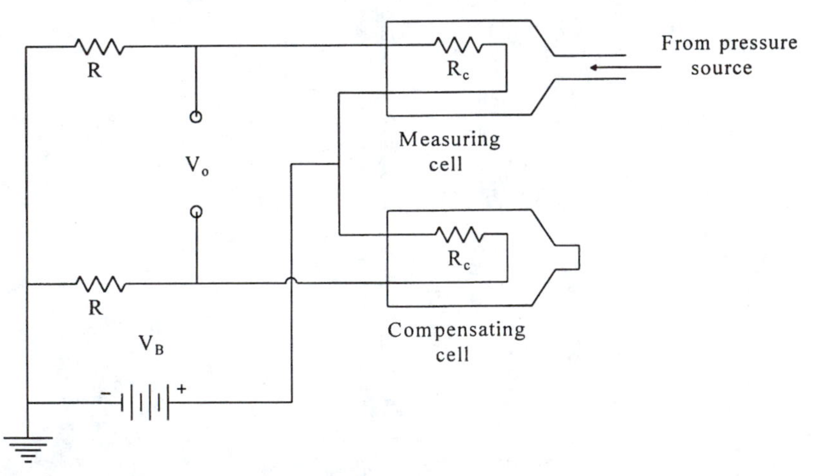

FIGURE 7.52 Schematic diagram of a Pirani low-pressure sensor.

where I_o is the rms current through the Pirani filament, R_o is the resistance of the filament at ambient (reference) temperature, α is the tempco of the wire, ΔT is the equilibrium increase in temperature of the wire above ambient temperature, and Ω_o is the thermal resistance of the filament *in vacuo*, in degrees Celcius per watt. We also showed in Section 7.2.2.1 that the resistance of the hot filament *in vacuo* is given by:

$$R(\Delta T, 0) = R_o / \left(1 - \alpha I_o^2 \Omega_o\right) \qquad 7.127$$

Now as the gas pressure is increased from zero, filament heat is no longer only dissipated by radiation and conduction through the leads, but also by heating gas molecules and giving them increased kinetic energy. Thus as the pressure increases, Ω decreases from its vacuum value. At low pressures (below 10 mmHg; Lion, 1959), the thermal resistance may be given by:

$$\Omega(P) = \Omega_o(1 - \sigma P) \tag{7.128}$$

where σ is dependent on system geometry, filament material, and the heat conductivity of the gas. If Equation 7.128 above is substituted into Equation 7.127, then we see that the Pirani gauge filament resistance may be approximated by

$$R(P) = R(\Delta T, 0)\left[1 - \frac{\alpha I_o^2 \Omega_o \Delta P}{1 - \alpha I_o^2 \Omega_o}\right] \tag{7.129}$$

The linear working range of most Pirani gauges is from 10^{-5} to 1 mmHg, although modifications have allowed measurements of pressures as low as 5×10^{-9} mmHg. Of interest is the closed-loop, feedback operation of the Pirani gauge. In this mode of operation, the Wheatstone bridge is balanced at some reference pressure, P_r. As P increases from P_r, the thermal resistance decreases, and the Pirani filament cools and its temperature and resistance decrease. Decreasing resistance causes the bridge output to increase. This increase in V_o is applied to increase the voltage across the bridge, thereby increasing the power dissipation in the Pirani filament, thereby increasing its temperature and resistance, restoring bridge balance. The other resistors in the bridge are assumed to have very low thermal resistances and tempcos.

Another low-pressure (vacuum) sensor makes use of the heater/vacuum thermocouple device introduced in Section 6.3.1 as a means of measuring true rms current. In the vacuum version of this sensor, the thermal resistance of the heater decreases with increasing gas pressure. This is because at increased gas pressure, there are more gas molecules present to absorb heat energy from the heater and convey it to the walls of the glass envelope where it is lost by radiation and convection. Hence the heater temperature drops (at constant power input) as the gas pressure increases, and thus the EMF of the associated thermocouple also decreases with the pressure increase. Thermocouple vacuum sensors work in the range from 10^{-3} to 10 mmHg. Below pressures of 10^{-3} mmHg, heat loss through the heater wire supports exceeds that from the gas molecules, and the thermocouple pressure sensor loses linearity and sensitivity. Thermocouple pressure sensors are ideally suited for feedback operation; one such closed-loop system is shown schematically in Figure 7.53. In this system, feedback is used to nearly match the DC EMF from the pressure-responsive thermocouple to the DC EMF from the reference (vacuum) thermocouple. For the reference thermocouple we have a heater temperature rise given by:

$$\Delta T_r = I^2 R_W \Omega_o \tag{7.130}$$

The thermocouple EMF is proportional to ΔT_r:

$$V_r = K_T\left(I^2 R_H \Omega_o\right) \tag{7.131}$$

For the pressure-responsive thermocouple we have

$$V_P = K_T\left[I_o^2 R_W \Omega_o (1 - \sigma P)\right] \tag{7.132}$$

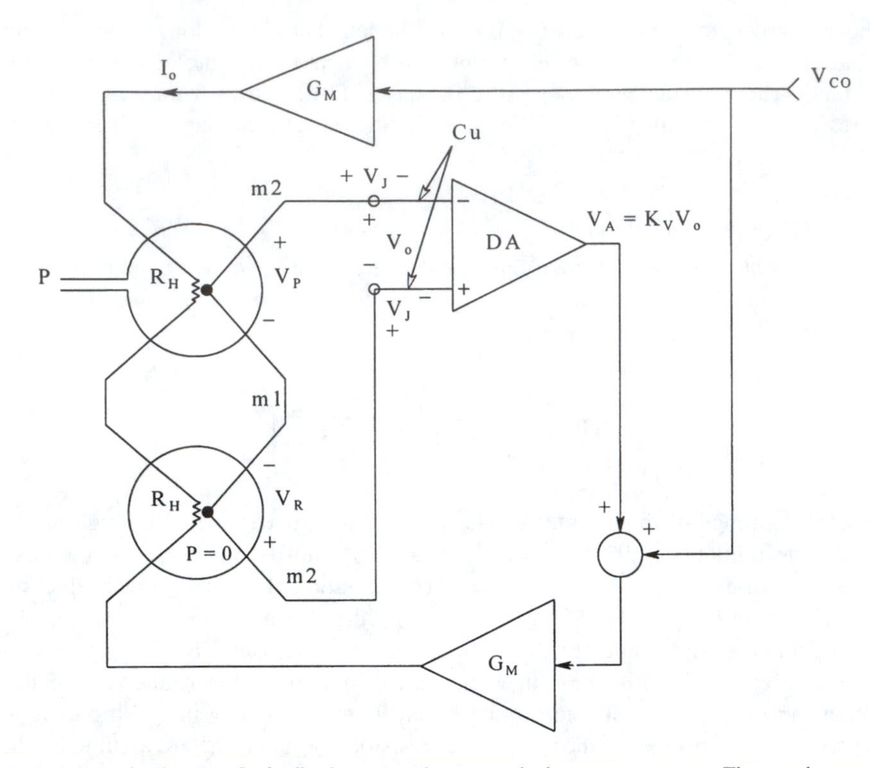

FIGURE 7.53 Schematic diagram of a feedback vacuum thermocouple, low-pressure sensor. The two thermocouples must have matched characteristics. Amplifiers Gm are voltage-controlled current sources, DA is a differential amplifier (VCVS).

Here we assume that the thermal resistance of the heater decreases linearly with increasing pressure, P, at very low pressures. The current in the pressure-responsive thermocouple heater is constant.

$$I_o = G_M V_{co} \tag{7.133}$$

G_M is the transconductance of the two voltage-controlled voltage sources (VCCSs), and V_{co} is a DC voltage. Thus, the current in the reference thermocouple heater is:

$$I = G_M\left(V_{co} - K_V V_o\right) \tag{7.134}$$

where K_V is the differential amplifier gain, and V_o is the difference between the reference thermocouple's EMF and the pressure-responsive thermocouple's EMF. Thus:

$$V_o = V_r - V_P \tag{7.135}$$

If the relations above are substituted into Equation 7.135 above, we obtain a quadratic equation in V_o:

$$0 = V_o^2 - V_O \frac{1 + 2V_{co}\mu}{\mu K_V} + \frac{P\sigma V_{co}^2}{K_V^2} \tag{7.136}$$

where

$$\mu = K_V K_T R_W \Omega_o G_M^2 \tag{7.137}$$

If we assume that $2V_{co}K_VK_TR_H\Omega_oG_M^2 \gg 1$, then one root of the quadratic equation can be shown to be:

$$V_o = P\sigma V_{co}/2 \qquad 7.138$$

and the differential amplifier output is:

$$V_A = P\sigma K_V V_{co}/2 \qquad 7.139$$

Thus, the proposed design for the feedback thermocouple, low-pressure sensor system is directly proportional to the pressure. Note that the calibration depends on the value of the constant, σ, which will depend on the composition of the gas. Hydrogen would have a larger σ than would air, for example.

The *ionization vacuum gauge* is routinely used to measure very low pressures, in the range of 10^{-8} to 10^{-3} mmHg. Special designs have been used down to 10^{-10} mmHg (Lion, 1959, Section 1-57). The ionization gauge is constructed similar to a triode vacuum tube, as shown in Figure 7.54. A glass envelope is kept at the pressure under measurement. In the envelope are a heated cathode which emits electrons. Surrounding the cathode is a grid, and outside the grid is a plate. Unlike a vacuum tube triode, the grid of the ionization detector is maintained at a high (100 to 250 V) positive potential with respect to the cathode. The plate is kept at a negative potential (–2 to –50 V). Electrons from the hot cathode are accelerated by the field from the grid; they collide with gas molecules and ionize them. The + ions in the space between the grid and the plate are collected by the negative plate; those formed in the space between the cathode and the grid are accelerated toward the cathode. The free electrons and negative ions are collected by the positive grid. The rate of ion production is proportional to the density of gas molecules in the tube (or to the gas pressure), and to the number of electrons available to ionize the gas. Lion (1959) shows that the gas pressure is proportional to the ratio of the positive ion current at the plate to the grid current. Thus:

$$P = K\left(i_p^+/i_g^-\right) \qquad 7.140$$

where K is about 10^{-2} mmHg(mA/μA). K depends on the gas composition, the geometry of the tube, and the voltages used. The emission of photoelectrons from the plate, caused by light and

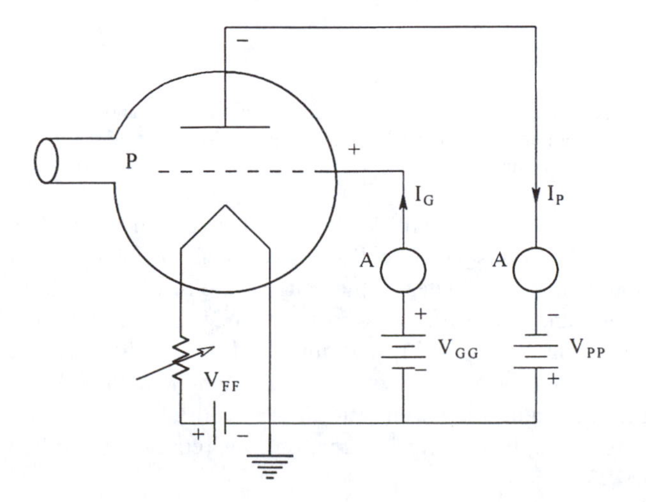

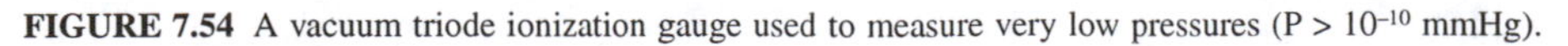

FIGURE 7.54 A vacuum triode ionization gauge used to measure very low pressures ($P > 10^{-10}$ mmHg).

soft X-rays from the grid, sets the lower bound of ion gauge sensitivity. One mode of operation of the ion gauge is to keep the grid current constant (1 to 20 mA) and to measure the plate current. A log ratio I_C (see Section 2.5.7 of this text) can also be used to measure P over the wide range of this sensor.

Instead of creating an ion current by collisions with energetic electrons, a radioactive, α particle emitter can be used to make a radioactive ionization gauge, illustrated in Figure 7.55A. The number of + ions formed is proportional to the gas pressure as long as the range of the α particles is longer than the dimensions of the chamber. The useful range of pressures is 10^3 to 10^{-3} mmHg. Below 10^{-3} mmHg, the mean free path of the α particles increases beyond the dimensions of the chamber so that the probability of an ionizing collision with a gas molecule is reduced and the ion current drops off sharply. One advantage of the radioactive ionization gauge is that it is not damaged if it is turned on at atmospheric pressure, and it requires no degasing.

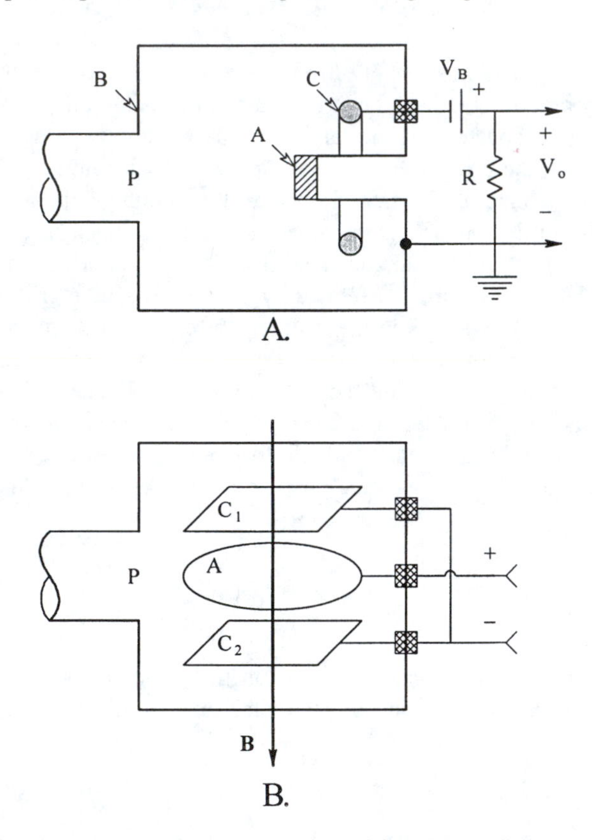

FIGURE 7.55 (A) A radioactive ionization vacuum sensor. A, α particle-emitting radioisotope; C, + ion collector (cathode); B, metal housing (anode). (Diagram after Lion, 1959.) (B) Diagram of a Philips-Penning vacuum sensor. C1, C2, cathodes; A, anode.

The final low-pressure sensor we will describe in this section is the *Philips-Penning gauge*, illustrated in Figure 7.55B. In this gauge, collisions between gas molecules and moving electrons creates an ion plasma, which contributes to the gauge current. Electrons are emitted as the result of the high electic field, and bombardment of the electrode surfaces by ions. The working range of the Philips-Penning gauge is 4×10^{-7} to 10^{-3} mmHg (Lion, 1959). The electrodes of this gauge are run at potentials of 1 to 3 kV, and a transverse, DC magnetic field 300 to 8000 oersteds is used to decrease the mean free path of electrons and thus extend the low-pressure range. One version of the Philips-Penning gauge, the inverted magnetron, was reported to work at pressures as low as 10^{-12} mmHg (Lion, 1959).

7.5 TEMPERATURE MEASUREMENTS

Many physical and chemical phenomena and physical "constants" are found to be functions of temperature, and thus can be used to measure temperature. Temperature-dependent properties and constants include resistance, dielectric constant, and the magnetic permeability and susceptiblity (of paramagnetic salts). Other temperature-sensitive phenomena include linear and volume expansion of solids and gases, generation of the Seebeck (thermoelectric EMF) by thermocouples, and the generation of Johnson (thermal) white noise by resistors.

Scientific temperature measurements are generally made using the Celsius (centigrade) or Kelvin scales. Absolute zero (thermodynamic zero) occurs at 0 K, or –273.15°C. That is, degrees K = degrees C + 273.15. While most of the civilized world uses the Celsius scale for such mundane things as cooking and weather reports, in the United States, use of the Farenheit scale is dominant for these applications. The nominal boiling and freezing temperatures of water were originally taken as the two calibration points for linear temperature scales; 100° and 0° are those respective temperatures in the Celcius scale, and 212° and +32° are boiling and freezing in the Fahrenheit scale. It is easy to derive a conversion formula between degrees Fahrenheit and degrees Celsius:

$$°C = 0.55556(°F - 32) \qquad 7.141$$

In the following discussions, we will examine the details of some of the common means of temperature measurement, and secondary standards for temperature sensor calibration.

7.5.1 Temperature Standards

A primary standard for temperature is the *triple point* of pure water. The triple point of a pure substance is defined as that temperature and pressure at which all three phases (solid, liquid, vapor) are in equilibrium in a closed vessel. The triple point of pure water occurs at +0.0098°C and 4.58 mmHg pressure. This is a single, unique point in the P,T phase diagram for H_2O. The transition (melting) temperature between solid (ice) and liquid water above 4.58 mmHg pressure is a decreasing function of pressure, hence the melting point of ice is slightly pressure-dependent at atmospheric pressures (see Figure 7.56). The pressure dependence is sufficiently small so that an ice bath made with double-distilled water (both ice and water) can serve as a secondary standard for 0°C.

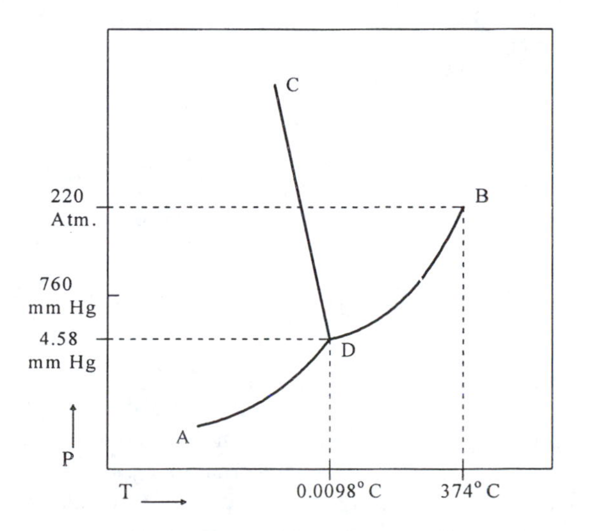

FIGURE 7.56 Phase diagram for water, showing the triple point.

The phase transition temperatures of other pure substances can be used as secondary temperature standards. In some cases, these transitions can be observed through a microscope focused on a single crystal in a capillary tube. A change in the index of refraction generally accompanies the phase change; often there is a change of color, or of shape. When the liquid to solid phase change of a large amount of a pure calibrating substance is used, the cooling curve has a plateau at the freezing temperature. This effect is illustrated in Figure 7.57 for the metal bismuth. As a substance freezes, it releases isothermally, at the freezing temperature, a heat of fusion which prevents the temperature of the substance from continuing to drop until the phase change is complete. This process is reversed in a melting curve (not shown); the rising temperature plateaus because of the isothermal absorption of the heat of fusion. Some substances that can be used for secondary temperature calibration at atmospheric pressure are listed below with their melting points:

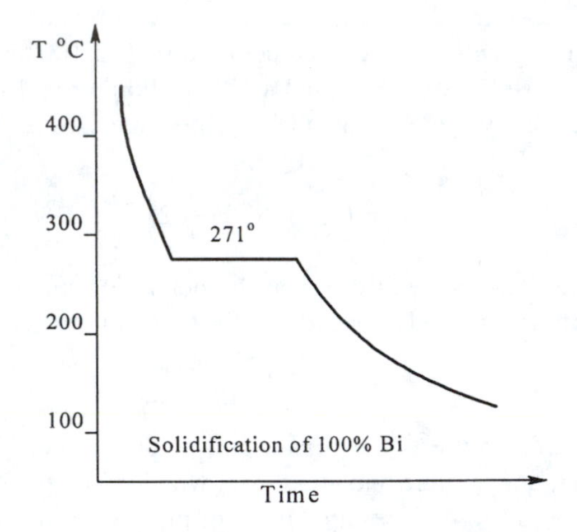

FIGURE 7.57 Cooling diagram for liquid bismuth. The plateau occurs at the freezing temperature of bismuth.

TABLE 7.1 Melting Points of Some Pure Substances that Can Be Used for Temperature Calibration

Substance	Melting Temperature (°C)
Oxygen	–218.4
Mercury	–38.87
Water	0
Sulfur	112.8
Bismuth	271.3
Cadmium	320.9
Silver	961.9
Gold	1064.4

7.5.2 Some Common Means of Temperature Measurement

7.5.2.1 Mechanical Temperature Sensors

The bending bimetallic strip is one of the more commonly encountered mechanical temperature-sensing systems. This device is used in many common household thermostats and thermometers, and is illustrated in Figure 7.58. At reference temperature, T_O, the strip is perfectly straight and has a length, L. At some higher (or lower) temperature, T, the strip is seen to bend in a section of an arc with radius R, as shown. Beckwith and Buck (1961) have shown that R is given by:

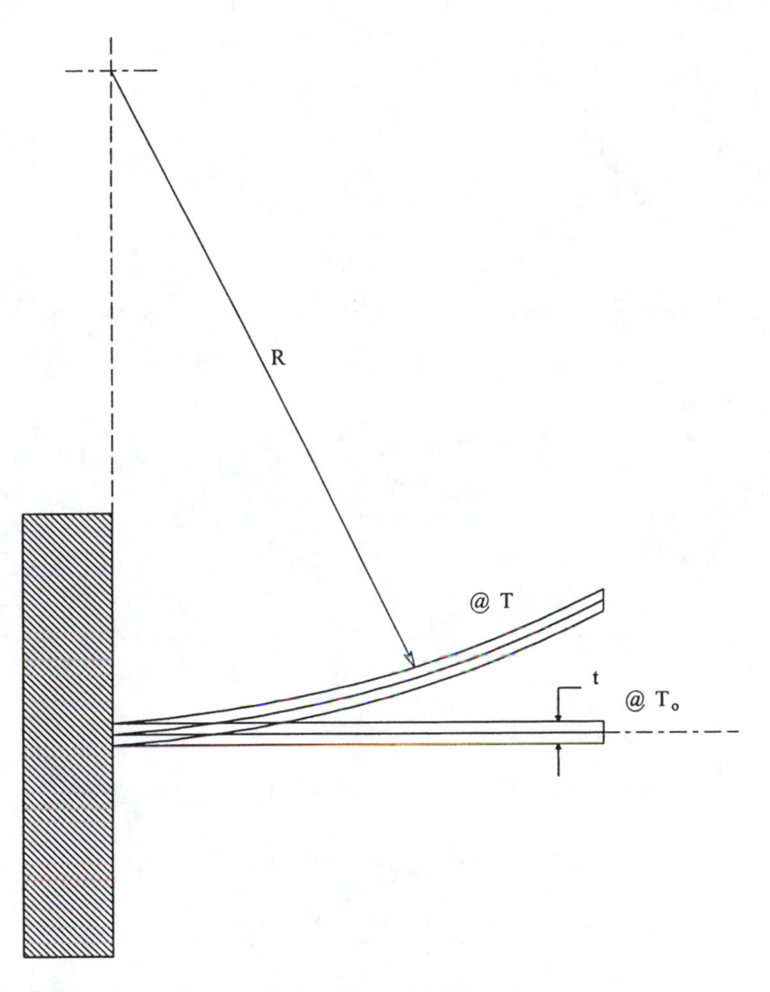

FIGURE 7.58 Bending of a bimetallic strip with temperature change.

$$R = \frac{t\{3(1+m)^2 + (1+mn)[m^2 + 1/(mn)]\}}{6(\alpha_B - \alpha_A)(T - T_O)(1+m)^2} \qquad 7.142$$

where t is the total thickness of the strip, m is the ratio of thickness of the top to the bottom strip, n is the ratio of Young's moduli of the top to the bottom strip, α_A and α_B are the coefficients of linear expansion of the top and bottom strips, respectively. We assume $\alpha_B > \alpha_A$. If the strips have the same thickness, and their Young's moduli are the same, Equation 7.142 reduces to:

$$R = \frac{2t}{3(T - T_O)(\alpha_B - \alpha_A)} \qquad 7.143$$

In general, R will be large compared to L. From the system geometry, and a knowledge of R(T), we can predict the deflection of the tip of the bimetallic strip, and the angle the tip makes with the equilibrium (horizontal) position of the strip. Refer to Figure 7.59. From simple trigonometry, it can be shown that the angle β that the tip of the bimetallic strip makes with its equilibrium position is given by:

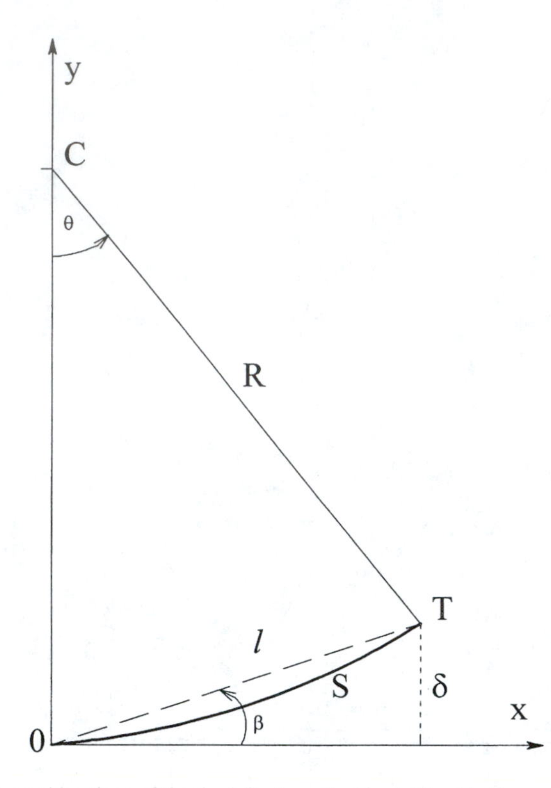

FIGURE 7.59 Geometrical considerations of the tip deflection of a bimetallic strip thermometer.

$$\beta = \theta/2 = L/2R \tag{7.144}$$

and the linear deflection, δx, of the tip T from the X-axis is

$$\delta x = 2R\sin^2(L/2R) \tag{7.145}$$

And if $\beta < 15°$, then, more simply,

$$\delta x = L\beta = L^2/2R \tag{7.146}$$

Thus, we see that both the linear tip deflection and the deflection angle are linear functions of the temperature difference, $T - T_O$. In some simple mechanical thermometers, the deflection of the tip of the bimetallic strip is converted to rotary motion of a pointer by means of a linkage. In some home thermostat designs, the tip moves to close a set of contacts which will start the furnace, or the tilt angle acts on a mercury-filled switch to do the same thing.

Another commonly encountered mechanical temperature sensor is the mercury or colored alcohol-filled, sealed glass reservoir and capillary tube. In this common design, the filling liquid expands at a greater rate than the glass, so the column of liquid rises. The space above the expanding liquid in the sealed capillary tube is often filled with an inert gas, or is under vacuum. Most mercury-filled glass laboratory thermometers are designed for 76 mm immersion in the medium whose temperature is being measured. Other thermometers may require total immersion. If a mercury-filled glass capillary thermometer calibrated for 76 mm immersion depth is used at a different immersion depth, then there will be a small error between the true and indicated temperatures. Beckwith and Buck (1961) discuss means for correction of this error.

Another type of mechanical thermometer often used for industrial applications, including steam boilers, refrigeration systems, lubricant cooling systems, etc., is the pressure thermometer, illustrated in Figure 7.60. Three types of pressure thermometer exist: gas-filled, liquid-filled, and vapor-filled. In the latter type, there is an equilibrium between liquid and vapor in the system. A large volume reservoir is immersed in the environment of which we wish to know the temperature. Because of thermal expansion, the fluid acts through the capillary tube to the Bourdon gauge. The movable tip of the C-shaped Bourdon tube deflects radially outward with increasing temperature and fluid pressure. The tip motion is converted to rotary motion of a pointer by a linkage and sector gear acting on the pointer's pinion gear. When liquid- or gas-filled pressure tubes are used, both the reservoir and the capillary tube are temperature sensitive. Temperature sensitivity of the capillary tube is minimized by minimizing the ratio of the volume of the capillary tube to that of the immersed reservoir. When a two-phase, vapor-filled pressure tube is used, the temperature of the capillary tube does not affect the gauge reading. This is because of Dalton's law, which states that if both phases are present in the closed system's reservoir, only one pressure is possible for a given temperature. When dry nitrogen is the filler, a temperature span of –200°F to +800°F is possible for a pressure thermometer, although a typical span for a thermometer used for a boiler might be 0°F to 800°F. A variety of organic liquids can also be used for vapor-phase units; ethane and propane being useful in refrigeration systems (Patranabis, 1983).

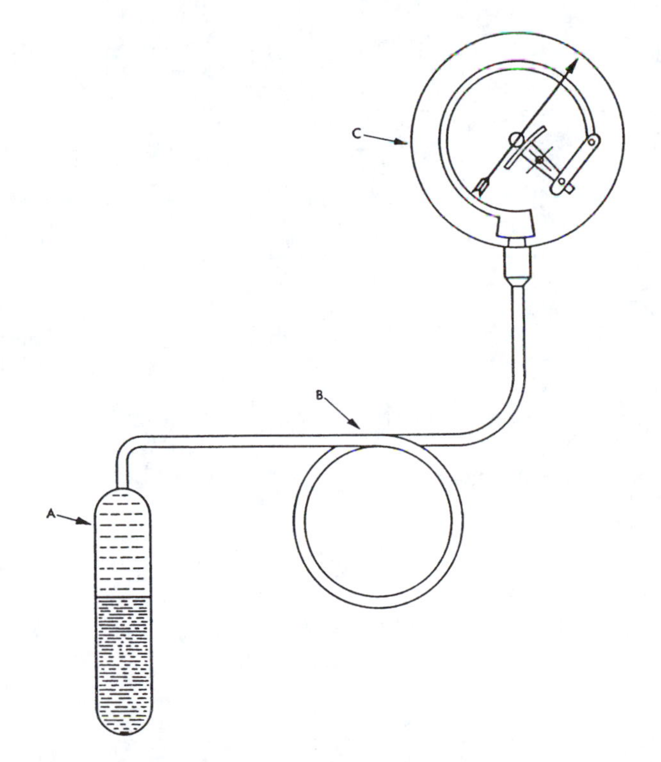

FIGURE 7.60 A pressure thermometer. (From Beckwith, T.G. and Buck, N.L., *Mechanical Measurements*, 1st ed., 1961. Reprinted by permission of Addison-Wesley Longman Publishing Company, Inc.)

7.5.2.2 Electrical and Electronic Temperature Sensors

There are several types of electronic temperature-sensing systems. We have already discussed resistance temperature detectors (RTDs) in Section 6.2.1. Thermocouples and thermopiles were covered in Section 6.3.1. In the following paragraphs, we will describe some of the less well-known electronic temperature measuring systems.

The *resistance noise thermometer* is a system which uses the fact that the power density spectrum of Johnson (thermal) noise from a resistor is proportional to the Kelvin temperature of the resistor (see Section 3.1.3.1). That is,

$$S(f) = 4kTR \ \text{MSV/Hz} \tag{7.147}$$

Figure 7.61 illustrates the schematic diagram of a resistance noise thermometer. AMP is a low-noise, high-gain voltage amplifier, BPF is a unity-gain band-pass filter with an equivalent noise bandwith of B. TRMS is a sensitive, broadband, true rms voltmeter. R1 is the probe resistor at the unknown temperature, T_1. R_2 is the variable reference resistor at known reference temperature, T_2. With the switch in position A, the mean squared meter voltage is:

$$N_A = \left(4kT_1R_1 + e_{na}^2\right)K_V^2B \ \text{MSV} \tag{7.148}$$

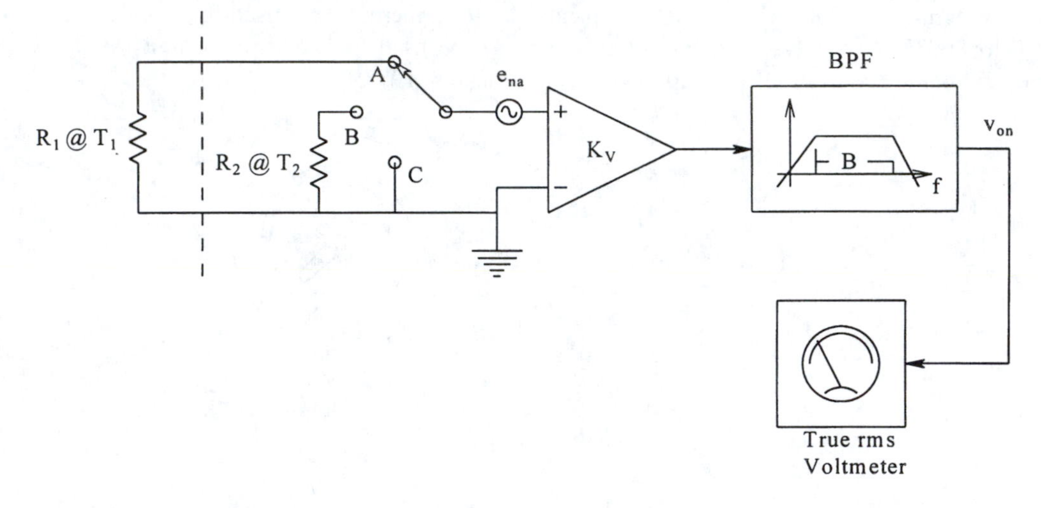

FIGURE 7.61 Schematic of a resistance noise thermometer. See text for description.

With the switch in position B the mean squared meter voltage is

$$N_B = \left(4kT_2R_2 + e_{na}^2\right)K_V^2B \ \text{MSV} \tag{7.149}$$

In practice, the resistor R_2 is adjusted until $N_B = N_A$. Equating N_A and N_B, we easily find that at equality,

$$T_1 = T_2R_2/R_1 \tag{7.150}$$

The lower limit of this system's range is set by the e_{na} of the amplifier. The T_1 at which the test resistor noise is equal to e_{na} is given by:

$$T_{1L} = e_{na}^2/(4kR_1) \ \text{K} \tag{7.151}$$

If we let $e_{na} = 3$ nV/ $\sqrt{\text{Hz}}$, $R_1 = 3000\ \Omega$, and $k = 1.38 \times 10^{-23}$, then $T_{1L} = 54.3$ K. The rms output voltage of the filter when the switch is in position A, $T_1 = 54.3$ K, and $K_V = 10^4$, is 13.4 mV, well within the range of a modern, true rms voltmeter. Since the noise thermometer works by adjusting

R_2 so that the thermal noise from R_2 at T_2 equals the noise from R_1 at T_1, and there is uncertainty in the matching process, system accuracy is improved when the matching is repeated N times, and the mean R_2 is used in Equation 7.150. The resistance noise thermometer should be useful at temperatures as low as 0.5 K. A noise thermometer made with fine tungsten wire has been used to temperatures as high as 1700 K with an accuracy of 0.1% (Lion, 1959).

Because the Johnson noise from a resistor is a linear function of resistance (at a given temperature), and because the resistance of metals increases with temperature, the noise output of a noise thermometer increases more rapidly with temperature than given by the simplified relations for N_A and N_B above. This nonlinearity in response makes it necessary to calibrate a noise thermometer at several temperatures when it is to be used over a wide range of temperatures. However, the use of special, low tempco alloys for the resistor R_1, such as manganin, eliminate the noise thermometer's nonlinearity problem in limited ranges of measurement.

The *resistance thermometer* (RTD) makes direct use of the fact that resistance of a metal increases with temperature. This effect was discussed in Section 6.2.1, and will be expanded on here. In general, R(T) is a nonlinear function which can be approximated around a given reference temperature, T_O, by a power series:

$$R(T) = R_O\left(1 + \alpha\Delta T + \beta\Delta T^2 + \gamma\Delta T^3 + \ldots\right) \qquad 7.152$$

where $\Delta T = T - T_O$, and α, β, etc., are temperature coefficients.

Generally, when ΔT is small, we neglect the higher-order coefficients and use only α to describe R(T). In general, α is defined by:

$$\alpha \equiv \frac{dR(T)/dT}{R(T)} \qquad 7.153$$

We note that the first-order temperature coefficient, α, defined by Equation 7.153, is in fact, a function of temperature. Resistance thermometers are called resistance temperature detectors, or RTDs. They are generally used in a Wheatstone bridge circuit which is nulled at the reference temperature (e.g., 0°C), and the bridge unbalance voltage, V_O, is an almost linear function of ΔT. Table 7.2 below lists the first-order tempco of some of the common metals used for RTDs measured at room temperature, and the useful range of temperatures over which they can be used.

Table 7.2 Some Metals Used for RTDs

Metal	Tempco (α)	Working Range of RTD (°C)
Platinum	+0.00392	–190–800 –264–1000 with corrections
Nickel	0.0067	–100–300
Tungsten	0.0048	–100–400
Copper	0.0043	–100–250

Platinum is the preferred RTD metal because of several factors: it has a high melting point (1775.5°C), it does not oxidize in air at high temperatures, it is relatively chemically inert, and its R(T) characteristic is quite linear from –190 to +400°C, and has a slight negative second derivative above 400°C. In other words, the high-order coefficients of the power series, Equation 7.152, for platinum are very small. Of course, pure platinum wire is expensive.

Sources of error in using an RTD include self-heating, and the effect of lead resistances. Self-heating can be avoided by keeping the power dissipation in the RTD very small, and constructing the RTD so there is low thermal resistance between the wire of the RTD and the medium whose temperature is being measured. A low thermal resistance also means a rapid response time to changes of temperature in the medium. Means of compensating for lead resistances are illustrated in Figure 6.2.

In the biological temperature range, platinum RTDs are very precise; resolution of 0.0001°C is not uncommon. Accuracy decreases with increasing temperature, however. At 450°C, accuracy is several hundredths of a degree, and at around 1000°C, accuracy is around 0.1°C (Lion, 1959).

Thermistors are negative temperature coefficient (NTC), sintered, amorphous semiconductor devices whose resistance decreases with increasing temperature according to the relation:

$$R(T) = R_O \exp\left[\beta\left(1/T - 1/T_0\right)\right] \qquad 7.154$$

where T is in degrees K. From the tempco definition of Equation 7.153, for an NTC thermistor, α is given by

$$\alpha = -\beta/T^2 \qquad 7.155$$

β is typically 4000 K, and for T = 300 K, $\alpha = -0.044$. This relatively large tempco enables the precise measurement of temperatures in the biological range with 0.0001°C resolution. Consequently, thermistor bridges are used in many applications in ecology and environmental studies, in physiological measurements, and in physical chemistry. Thermistors come in a wide variety of sizes and shapes, including 0.01" diameter, glass-covered balls. As in the case of RTDs, the power dissipation of a thermistor must be kept low enough so that self-heating does not disturb the temperature measurement. Thermistors can also be used as fluid flow probes. Two matched thermistors are used; one in still fluid for temperature compensation, the other in the moving stream. Both devices are operated at average power dissipations high enough to raise their temperatures well above ambient. The moving fluid extracts more heat from the measurement thermistor, cooling it and causing its resistance to increase. This increase of resistance is a nonlinear function of fluid velocity (see Section 7.2.2.1 for a description of conventional hot wire anemometers). We often see thermistors used as frequency-determining resistors in oscillators, the frequency of which is a function of temperature. Such oscillators have been used in biotelemetry applications. The thermal response time constant of a thermistor depends on its mass, its insulation, and the medium in which it is immersed. Thermistor time constants range from 100 msec to several minutes.

Thermocouples and *thermopiles* are discussed in depth in Section 6.3.1. Traditionally, a thermocouple system is used with a precision potentiometer to read its EMF, and the temperature is found from a look-up table for the particular couple materials used, such as copper and constantan. Potentiometers are expensive, slow to use, require a standard EMF and a reference temperature source. Consequently, where possible, it is more economical and timely to use RTDs or thermistors for temperature measurements formerly done with thermocouples.

Electronic IC temperature sensors are specialized integrated circuits used for sensing temperatures in the –55 to +150°C range. Physically, they can be packaged in the form of a small metal (TO-52) can, a flat pack (F-2A), or a plasic (TO-92) case. Analog Devices (Norwood, MA) offers the AD590 and AD592, two-terminal, temperature to current sensors. The AD sensors operate from a 4 to 30 V DC supply, and provide an output current which is a linear function of the sensor's Kelvin temperature. That is, at 0°C, the output current is a nominal 273.15 μA, and at 25°C, the output current is 298.15 μA, etc. Over the entire operating temperature range, the AD590M has a maximum nonlinearity of ±0.3°C, maximum repeatability and long-term drift are ±0.1°C, and the absolute error with the sensor calibrated for zero error at 25°C is no more than ±1.0°C. The AD temperature to current sensors are generally used with an op amp current to voltage amplifier which allows generation of a voltage output proportional to the Fahrenheit or Celsius scales. Figure 7.62

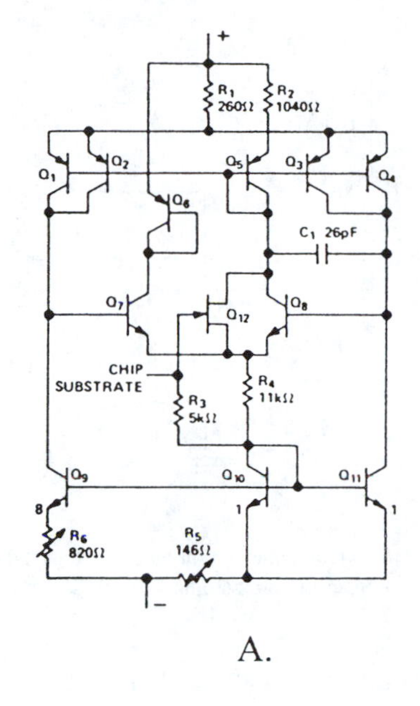

A.

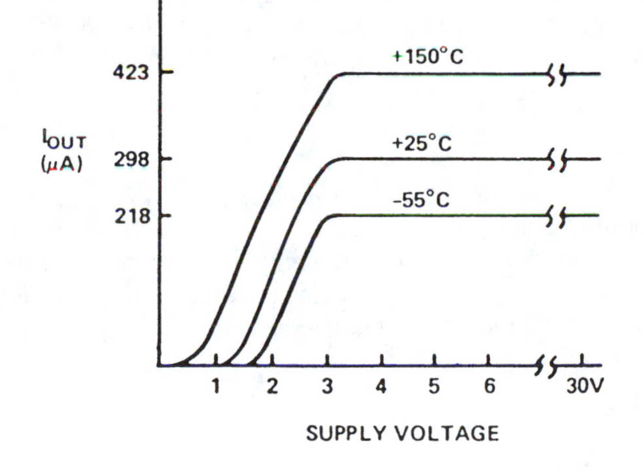

B.

FIGURE 7.62 (A) Schematic of the innards of an Analog Devices AD590 electronic temperature sensor. (B) Volt-ampere-temperature curves of the AD590 sensor. (Figures courtesy of Analog Devices, Inc., Norwood, MA.)

shows the internal schematic of an AD590 sensor, and its current/voltage/temperature curves. An op amp conditioning circuit for an AD590 is shown in Figure 7.63.

National semiconductor Corp. (Santa Clara, CA) also makes IC temperature sensors; the LM34, LM35, LM134, LM135, LM234, LM235, LM335 series. These sensors are three-terminal devices which produce an output voltage proportional to the temperature. For example, the AD34CA has a 10 mV/°F output voltage gain, a range from −50 to +300°F, and a nonlinearity of ±0.3°F. The LM35 series has a 10 mV/°C output voltage gain over a −55 to +150°C range, and ±0.15°C nonlinearity. IC temperature sensors are ideally suited for environmental monitoring, and control applications within their temperature ranges.

Optical pyrometers provide a non-contact means of estimating the surface temperatures of hot objects in the range of 775 to 4200°C, such as metals being hot-worked, molten metals, gas plasmas,

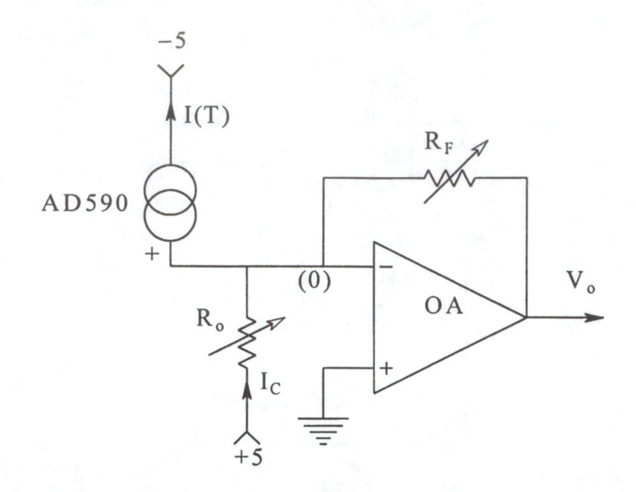

FIGURE 7.63 A simple op amp circuit to condition the current output of the AD590 temperature sensor. For DC stability, the op amp should be a chopper or auto-zeroing type. R_o sets $V_o = 0$ at 0°C, and R_F sets the output scale so 100°C gives $V_o = 10$ V.

and furnace interiors. Optical pyrometers make use of the fact that all objects at temperatures above 0 K radiate heat in the form of broad-band, electromagnetic energy. The range of the electromagnetic spectrum generally considered to be thermal radiation lies from 0.01 to 100 μm wavelength. Objects that are radiating heat are charaterized by three parameters which describe what happens to long-wave, electomagnetic radiation (heat) at their surfaces. This relation is:

$$e = a = 1 - r \qquad 7.156$$

where e is emissivity of the surface, which is always equal to its absorbtivity; r is the reflectivity of the surface. An ideal blackbody radiator has e = a = 1, and r = 0 (that is, all radiant energy striking its surface is absorbed, and none is reflected). For a non-ideal radiator, a finite fraction of the incident energy is reflected. The same properties exist for emitted radiation from a blackbody. For a hot object which is not an ideal radiator, the radiated heat output is given by:

$$W = eW_{bb} \qquad 7.157$$

where W is the total radiant emittance (in MKS units, watts/m^2) from the hot surface, e is the emissivity, and W_{bb} is the radiant emittance from an ideal blackbody.

Max Planck, using the newly developed quantum theory in the early 1900s, developed an expression for the power spectrum of heat radiation from a blackbody (Sears, 1949):

$$W_f = \frac{2\pi h f_3}{c^2 \exp(hf/kT) - 1} = \frac{dW_{bb}}{df} \qquad 7.158$$

where W_f is in watts/(m^2 Hz), c is the speed of light, k is Boltzmann's constant, h is Planck's constant, and W_{bb} is the radiant emittance of the blackbody in watts/m^2. From an experimental and practical point of view, it is useful to write Planck's equation in terms of wavelength instead of frequency. Substituting f = c/λ, we find:

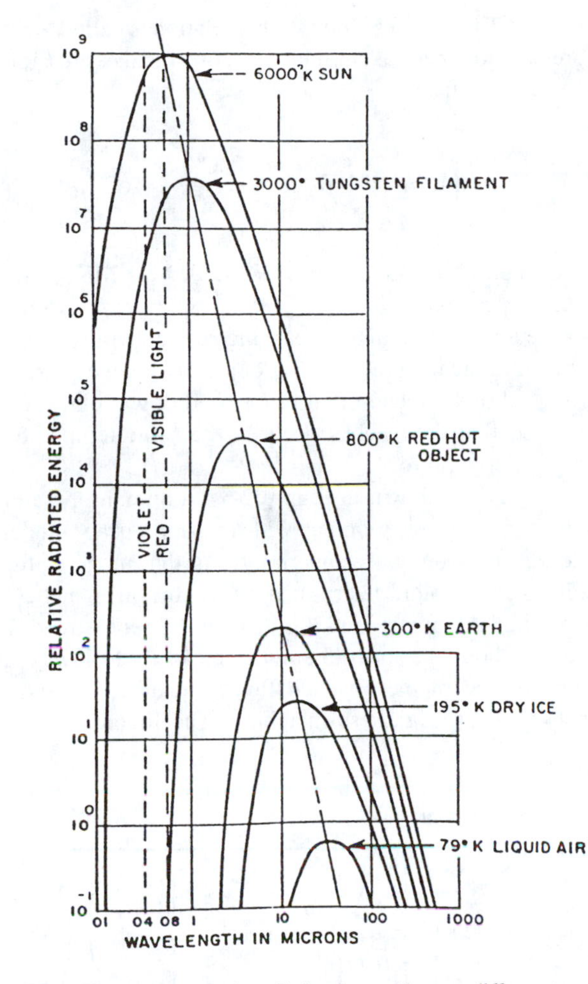

FIGURE 7.64 Spectral characteristics of blackbody radiation from objects at different temperatures. (Courtesy of Barnes Engineering, Inc., Colorado Springs, CO.)

$$W_\lambda = \frac{C_1\lambda^{-95}}{\exp(C_2/\lambda T)-1} = \frac{dW_{bb}}{d\lambda} \tag{7.159}$$

where $C_1 = 2\ \pi c^2 h = 3.740 \times 10^{20}$, $C_2 = hc/k = 1.4385 \times 10^7$, and W_λ is in watts/(m^2 µm). A plot of Equation 7.159 is shown in Figure 7.64. There are two important observations to make about this figure:. (1) as the temperature of the blackbody increases, the peak of maximum spectral emittance shifts systematically to shorter wavelengths; (2) as the temperature increases, the area under the spectral emittance curves increases. Subjectively, this means that when we look at a hot object, the color of the object's surface shifts from a dim, dull red to a bright orange to a brighter yellow, etc., as its temperature increases. The peaks of the spectral emittance curves can be found by differentiating the expression for W_λ and setting the derivative equal to zero. This gives a transcendental equation which can be solved numerically to yield the wavelength at the peak, λ_{PK}.

$$\lambda_{PK} = 2.8971 \times 10^6/T \tag{7.160}$$

Here λ_{PK} is in millimicrons, and T in degrees Kelvin. This relation is called Wien's displacement law.

If we integrate the expression for spectral emittance, we obtain an expression for the total radiated power/m^2 from the blackbody's surface:

$$W_{bb} = \int dW_{bb} = \int W_\lambda d\lambda = \frac{\pi^4 C_1 T^4}{15 C_2^4} = \sigma T^4 \qquad 7.161$$

The constants have been defined above, and $\sigma = 5.672 \times 10^{-8}$ for MKS units. Equation 7.161 is known as the *Stefan-Boltzmann equation.*

Practical hot surfaces have non-unity emissivities which are generally a function of wavelength. Thus, the practical spectral emittance curve for a hot object may have many peaks and valleys. If these irregularities are averaged out, we can often fit a scaled-down, blackbody spectral emittance curve to the practical curve so that their peaks occur at the same temperature, T. Such a scaled blackbody W_λ is called a graybody curve.

In one form of optical pryometer, shown in Figure 7.65, a human operator makes a subjective color comparison of a glowing tungsten filament with the hot surface under measurement. The color comparison is made easier by optically superimposing the image of the filament on that of the hot object. When the filament's color, determined by its current, matches that of the object, it disappears on the background of the object, and the filament current is read. Because the filament may be considered to be a blackbody, its spectral emittance closely follows that of the object when its temperature is the same as that of the object. Often a red filter is used to convert the color matching task to one that involves brightness matching. The filament ammeter is calibrated in temperature.

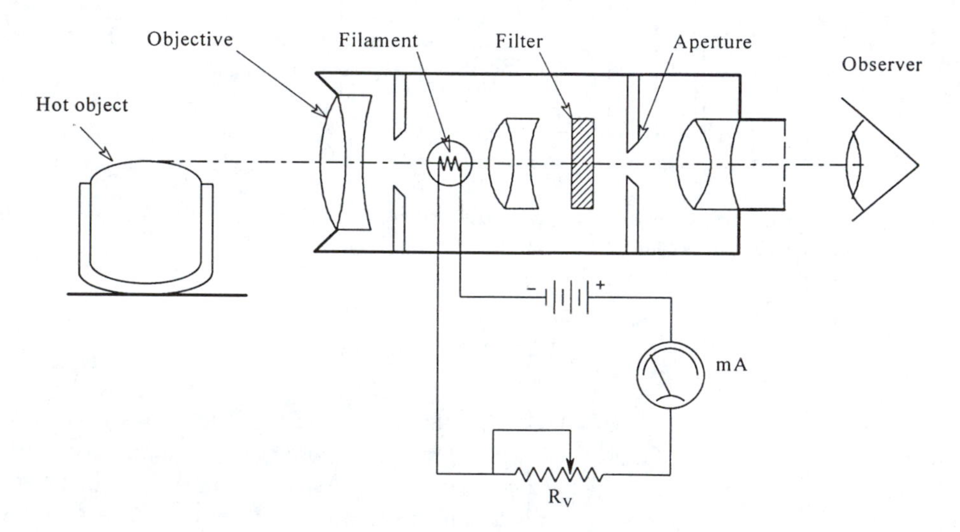

FIGURE 7.65 Diagram of an optical pyrometer. The operator adjusts R_V until the color temperature of the filament matches that of the object. The human operator must have reliable color vision in the red-yellow end of the spectrum.

In another form of optical pyrometer, shown in Figure 7.66, the filament is run at a constant current and brightness. The intensity of the image of the hot object is then varied with a neutral density wedge. Again, a red filter is used to make the null process an exercise in monochromatic intensity matching. The wedge position is calibrated in object temperature, assuming blackbody emission.

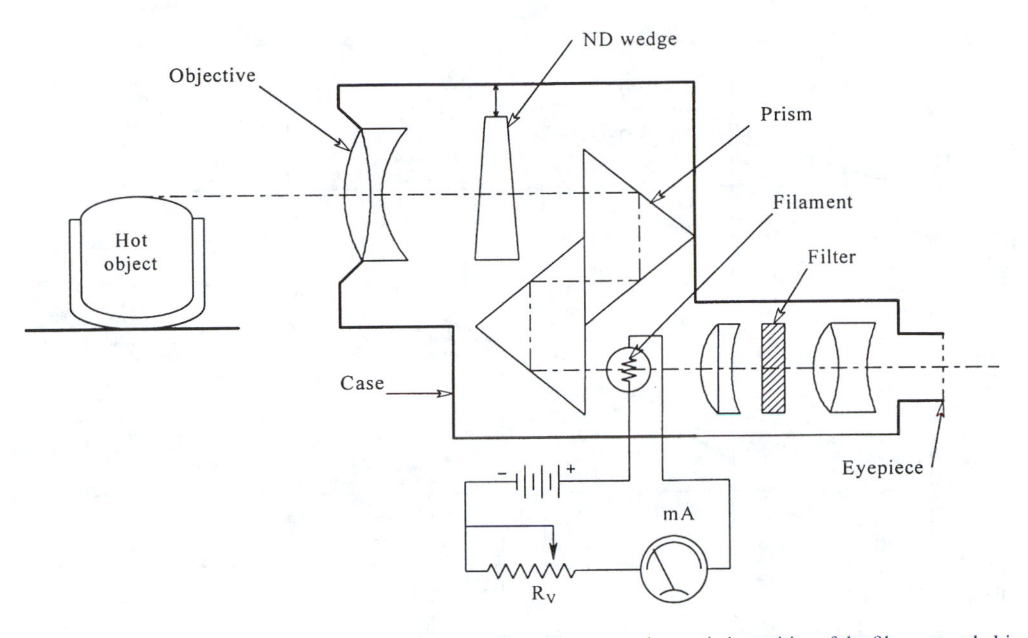

FIGURE 7.66 In this form of optical pyrometer, the operator matches monochromatic intensities of the filament and object. The neutral density wedge position is calibrated in object temperature.

Total radiation pyrometers are also known as radiometers, and are electronic instruments that measure the integral of the W_λ curve. That is, their output is proportional to the integral of W_{bb} over the area of the instrument's aperture. In one form of radiometer, the detector is a thermopile in which half of the thermocouple junctions must be kept at a reference temperature. The thermopile is designed to be a nearly 100% blackbody absorber. Thus, regardless of the shape of $W(\lambda)$, all of the incoming energy is captured and converted to a temperature rise of the sensing junctions. Total radiation pyrometers can be used with blackbody radiation as well as coherent sources such as lasers.

Other detectors used in radiation pyrometers may include photoconductors, photodiodes, and *pyroelectric detectors*. Pyroelectric detectors absorb thermal energy and generate electrical signals. One example of a pyroelectic detector material is the polarized polymer film, polyvinylidene difluoride (PVDF). PVDF film absorbs strongly in the IR, and a free mounted, 28 μm PVDF film has a response time constant of about 5 s to a step temperature change. All radiation detectors have spectral response characteristics. The broadest response characteristic is that of the blackbody thermopile; this sensor responds to radiation from UV to far IR (0.25 to 20 μm). PVDF pyroelectric sensors respond primarily from 6.6 to 66 μm. $LiTaO_3$ crystals are also used in faster-responding pyroelectric sensors. Most photodiodes and photoconductors respond maximally in the near IR, and are relatively narrow-band devices compared with PVDF and thermopiles. Figure 7.67 illustrates the spectral sensitivities of some common photoconductive IR detectors, pyroelectric detectors, thermopiles, and thermister bolometers. Note that several of the photoconductive sensors are operated at cold temperatures (e.g., HgCdTe at 77 K) to reduce noise.

The optics of pyrometers also present a problem, as conventional glasses do not transmit effectively at wavelengths beyond 2.6 μm. Thus, special materials must be used in IR radiometers for windows, mirrors, and lenses. A broadband, IR radiometer/pyrometer uses front surface mirrors coated with aluminum or gold to focus the radiation on the sensor. A chopper wheel is used to 0,1 modulate the radiation and permit AC amplification of the sensor output signal out of the 1/f noise band. Phase-sensitive rectification is then used to recover the amplified signal. A typical IR radiometer system is shown in Figure 7.68. When lenses and windows are used in radiometer design,

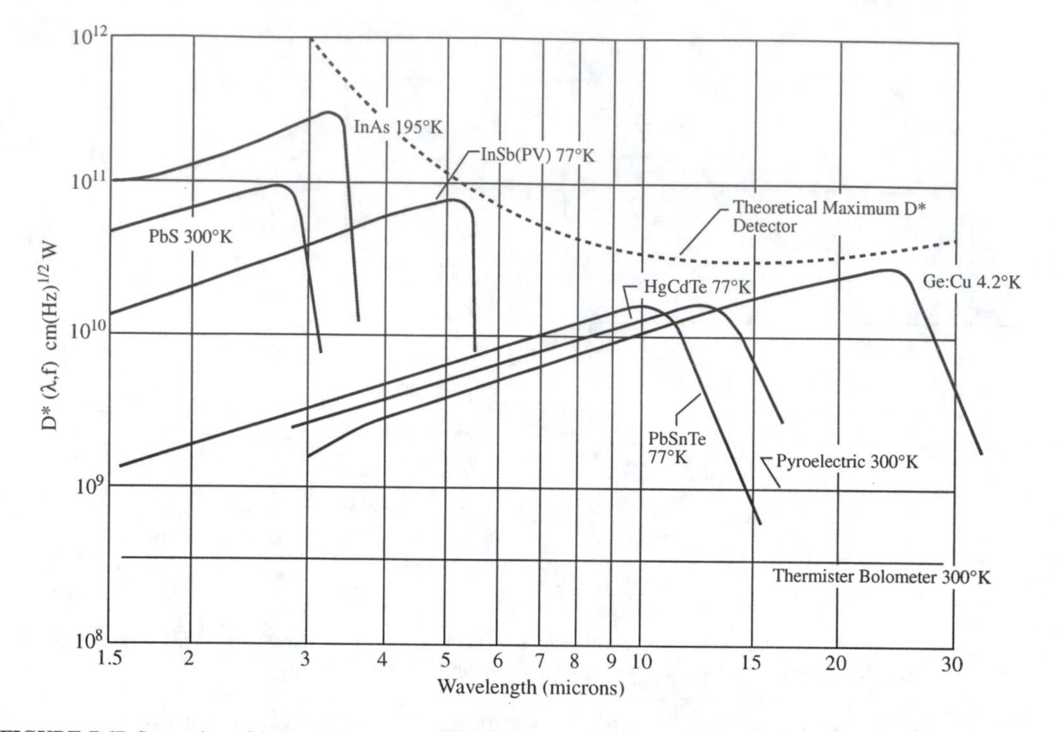

FIGURE 7.67 Spectral sensitivities of some common IR sensor materials. Sensitivity can generally be improved by chilling the sensor with liquid nitrogen. Vertical axis is D*, the specific detectivity. (Courtesy of Barnes Engineering, Inc., Colorado Springs, CO.)

they must be able to transmit in the IR. Figure 7.69 summarizes the IR transmission characteristics of commonly used IR optic materials. Note that pure germanium is now widely used for IR optics (Barnes Engineering, 1983). The passband of germanium is from about 2 to 30 μm. Pure germanium can be considered to be a dielectric, and has a high refractive index of $n \cong 4$. The reflection coefficient of germanium, r, is given by the Fresnel reflection equation:

$$r = \frac{(n-1)^2}{(n+1)^2} = 0.36 \qquad 7.162$$

Thus, assuming zero absorption for a Ge window, only $(1 - 0.36)^2 = 0.41$ of the incident IR would be transmitted (Barnes Engineering, 1983).

The optics of some radiometers make use of a chopping scheme in which the IR sensor is alternately exposed to the radiation under measurement, dark, and a standard blackbody source of known temperature. Thus calibration and correction for dark response can be done simultaneously by a computer attached to the radiometer.

7.6 CHAPTER SUMMARY

In this chapter we have presented a survey of some of the common (and not so common) means of measuring certain physical quantities, including angular acceleration, velocity and position, linear acceleration, velocity and position, force, torque, pressure and vacuum, and temperature.

Various kinds of linear and rotational accelerometers were described. Mechanical and Sagnac gyroscopes were presented as systems for sensing the inertial angular velocity and position of

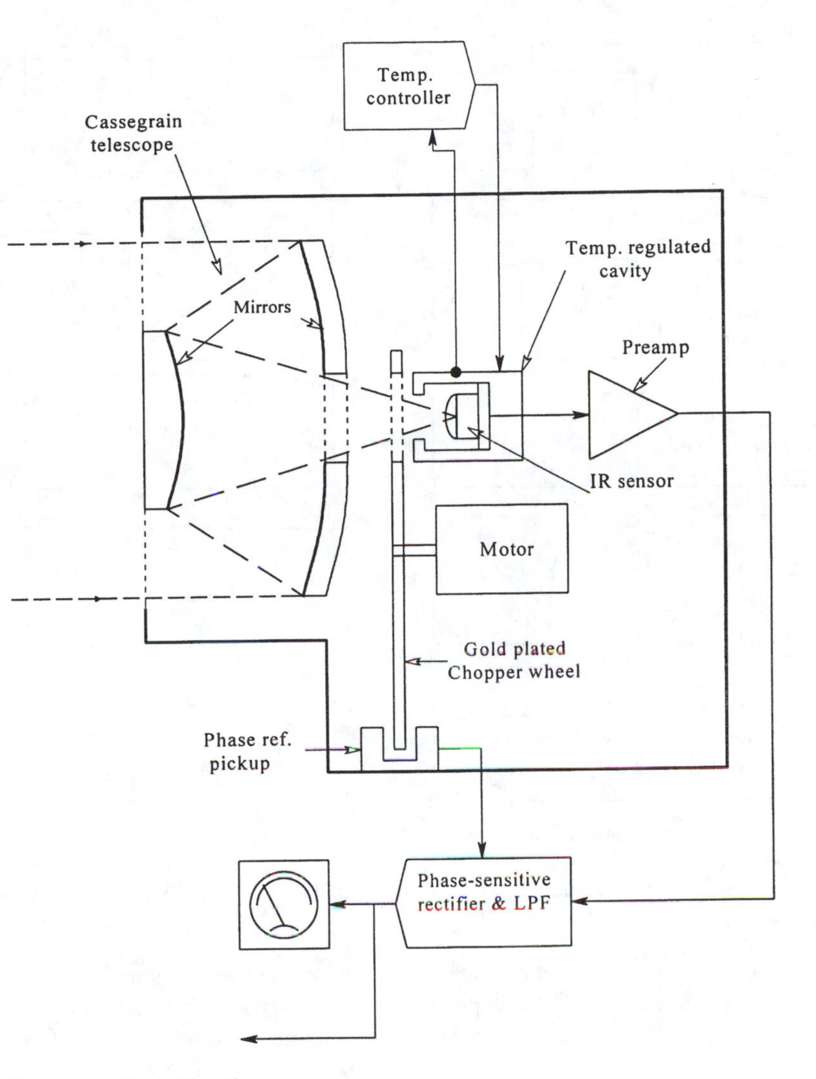

FIGURE 7.68 Diagram of a basic IR radiometer.

vehicles. The Doppler effect was introduced as the basis for many instruments using sound or electromagnetic radiation to measure the linear velocity of vehicles or particles suspended in a moving fluid. Laser Doppler velocimetry was described.

Sensors used to measure force and pressure were seen to be related in that a pressure acting on the area of a diaphragm or piston produces a force. Forces can produce changes in resistance, capacitance, inductance, charge displacement (in piezoelectric transducers), as well as the optical transmission properties of fiber optic waveguides. These physical changes can be sensed as functions of force or pressure. Some torque sensors also make use of sensed mechanical strain caused by the torque.

We saw that many physical processes respond to changes in temperature. These include chemical reactions, conduction in semiconductors, resistance of conductors, dielectric constant, speed of sound, refractive index of optical fibers, etc. In fact, many other physical measurements require that we compensate for temperature changes. Optical pyrometry was introduced as a means of remote sensing the temperature of objects.

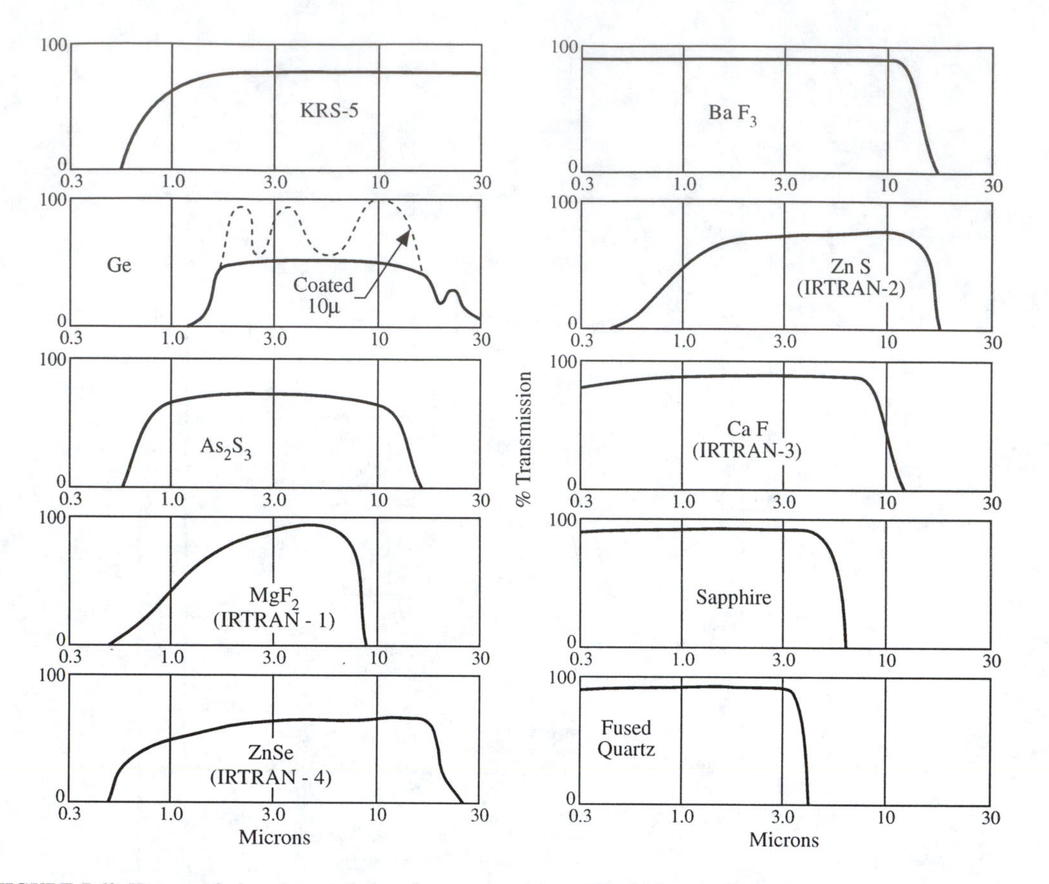

FIGURE 7.69 IR transmission characteristics of some materials used in LIR optical systems. (Courtesy of Barnes Engineering, Inc., Colorado Springs, CO.)

The basis for many measurement systems and sensors has been found to lie in several basic physical laws. These laws include the Doppler effect, the Sagnac effect, the piezoelectic effect, the Hall effect, the piezoresistive effect (strain gauges), to name a few.

8

Basic Electrical Measurements

8.0 INTRODUCTION

Electrical measurements are defined in this chapter as being measurements on the traditional electrical parameters of voltage, current, electric field strength, charge, magnetic fields, resistance, capacitance, inductance, and the steady-state AC parameters of impedance and admittance, power, frequency, and phase.

8.1 DC VOLTAGE MEASUREMENTS

DC voltage measurements can be made over an enormous range; from nanovolts to thousands of kilovolts. The practical limits to low DC voltage measurement are noise and thermoelectric EMFs, and the practical limits to high-voltage measurements involve circuit loading, insulation, and measurement system isolation. Of course, the specialized voltmeters that allow measurement of nanovolt potentials are generally unsuitable for high-voltage measurements. All DC voltmeters, regardless of their specialization, can be represented by an equivalent circuit consisting of a parallel impedance by which the voltmeter loads the circuit under test (CUT), and an "ideal" infinite impedance voltmeter, or alternately, a zero impedance microammeter in series with the loading resistance. Thus we see that every voltmeter takes some power from the CUT. An ideal voltmeter would take zero power and not load the CUT. The CUT itself, by Thevenin's theorem, can be represented by an open-circuit voltage and an equivalent series impedance. Obviously, connecting a practical voltmeter to the Thevenin equivalent circuit will result in current flowing in the circuit, and the (indicated) voltage at the voltmeter's terminals being less than the open-circuit voltage by the amount of the voltage drop across the Thevenin resistor. This situation is illustrated in Figure 8.1. In mathematical terms, assuming resistors rather than impedances, the voltmeter reads:

$$V_M = V_{OC} R_M / (R_M + R_{TH}) \tag{8.1}$$

Under normal circumstances, $R_M >> R_{TH}$.

Voltmeter loading of the CUT is related to a parameter called the *voltmeter sensitivity,* η. η is defined as the full-scale voltage of the meter divided by the power dissipated in the meter at full-scale DC voltage, V_{FS}.

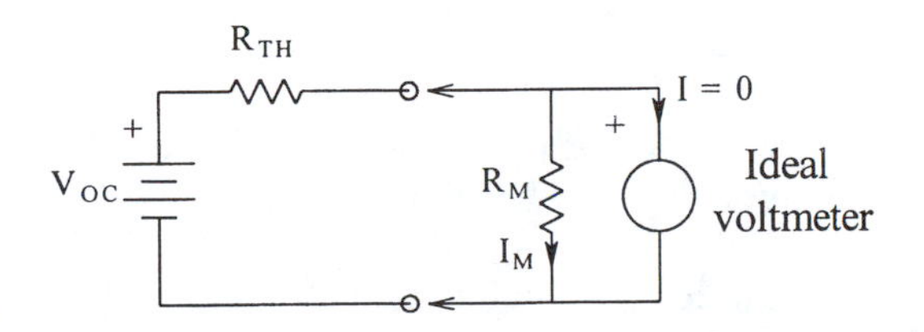

FIGURE 8.1 Simple circuit showing a non-ideal voltmeter attached to a Thevenin equivalent source.

$$\eta \equiv V_{FS}/P_{FS} = \frac{V_{FS}}{V_{FS}^2/R_M} = \frac{R_M}{V_{FS}} = 1/I_{MFS} \text{ ohms/volt} \qquad 8.2$$

where I_{MFS} is the current drawn by the voltmeter when at its full-scale voltage. Typical analog DC voltmeters, such as found in multimeters, have ηs of 20,000 ohms/volt, which means that the voltmeter draws 50 μA from the CUT at full-scale voltage. The equivalent resistance of any DC voltmeter can be found by mutiplying V_{FS} by η. Some electronic voltmeters have a fixed input resistance, regardless of scale. This resistance is typically 10 or 11 megohms, but some modern electronic voltmeters have R_Ms as large as 10^{10} ohms.

Figure 8.2 illustrates the role of Johnson (thermal) noise from R_{TH} in limiting the resolution of DC voltage measurements. Recall from Section 3.1.3.1 that the Johnson noise power density spectrum from a resistor is given by S(f) = 4kTR mean squared volts/hertz. The effective Hz bandwidth of a DC voltmeter may be considered to be the reciprocal of the time required to take a reading of an applied step of voltage. This response time generally ranges from 0.1 to 10 s, hence the effective bandwidth ranges from 10 to 0.1 Hz. If we assume that the DC open-circuit voltage, V_{OC}, must be greater than the Johnson noise from R_{TH}, $R_M >> R_{TH}$ and R_{TH} is the sole source of noise in the circuit, then we can write

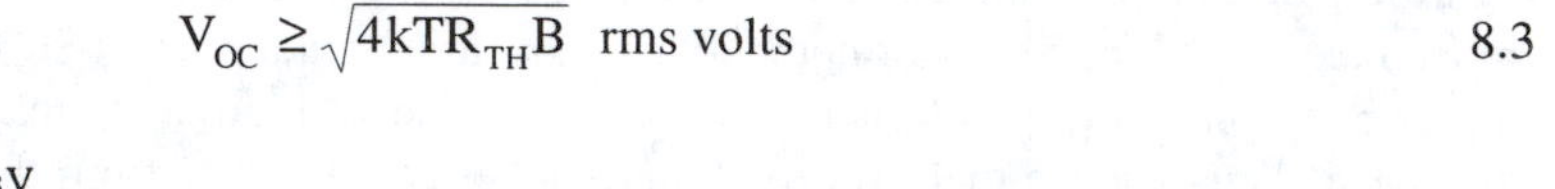

$$V_{OC} \geq \sqrt{4kTR_{TH}B} \text{ rms volts} \qquad 8.3$$

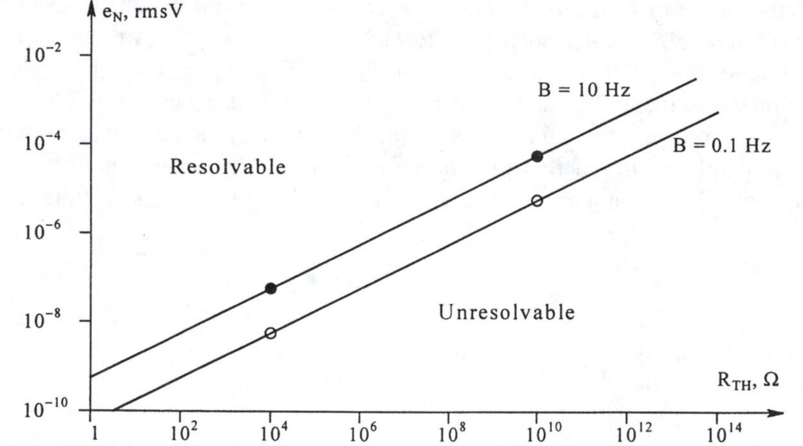

FIGURE 8.2 Graph illustrating the role of thermal noise from the Thevenin source resistor in limiting the resolution of low-level DC voltage measurements. 10 Hz and 0.1 Hz noise bandwidths are considered.

This function is plotted for B = 0.1 and 10 Hz in Figure 8.2. Note that DC voltages below the lines are unresolvable because of noise.

8.1.1 Types of Electromechanical DC Voltmeters

There are many types of DC voltmeters; some in fact are effective as power line frequency AC voltmeters as well. In this section, we examine the designs of some DC analog voltmeters. Included are D'Arsonval meters, dynamometer meters, capacitor voltmeters, electrometer voltmeters, chopper-type nanovoltmeters, and thermocouple voltmeters.

D'Arsonval DC voltmeters, or permanent magnet-moving coil voltmeters, are the most common design of electromechanical DC voltmeter. A typical D'Arsonval microammeter meter movement, illustrated in Figure 8.3, is basically a current-to-position transducer. Current flowing through the

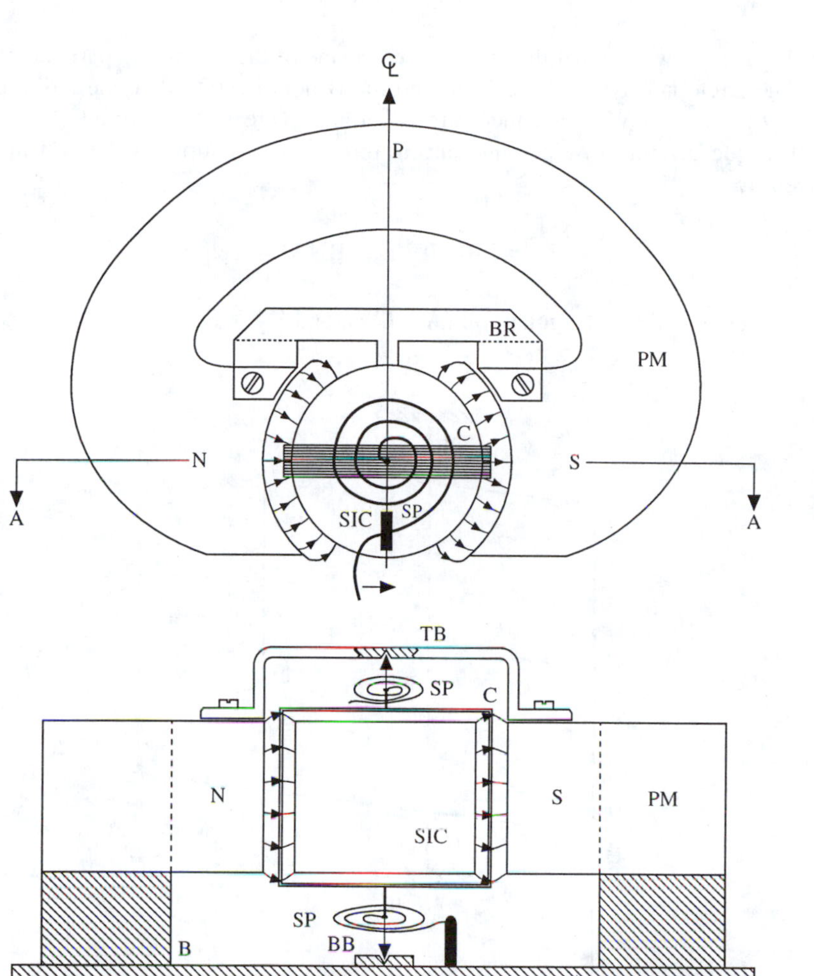

FIGURE 8.3 (A) Top view of the innards of a D'Arsonval meter movement. P, pointer; PM, permanent magnet; C, pivoted rectangular coil; SP, helical spring; SIC, soft iron core to concentrate magnetic flux. (B) Vertical section (AA′) through D'Arsonval meter. TB, BB, top and bottom bearings; B, base. Cross-hatched material is bakelite.

turns of the rectangular coil suspended in the uniform magnetic field in the airgap generates an orthogonal component of force on each conductor on each side (but not the top or bottom) of the coil. The magnetic field is generated by a strong permanent magnet. The resultant torque acts against a helical spring, or in some meters, a flat torsion spring, used to suspend the coil. When the magnetically produced torque equals the spring torque in the steady state, the meter pointer stops moving.

The force per unit length of a current-carrying conductor in a uniform magnetic field is given by the vector equation:

$$\mathbf{dF} = \mathrm{I}(\mathbf{dI} \times \mathbf{B}) \tag{8.4}$$

If the vector **IdI** is rotated into the vector **B**, the direction of **dF** is given by the right-hand screw rule. Accordingly, the magnitude of **dF** can be written:

$$\mathrm{dF} = |\mathbf{I}||\mathbf{B}|\sin(\varphi)\,\mathrm{dI} \tag{8.5}$$

Now from Figure 8.4 we see that the length of each side of the coil in the perpendicular magnetic field is L. The angle φ between the current and the B field is 90°. Also, the radial distance from each side to the pivot is W/2, and there are N turns of wire in the pivoted coil. Thus the total mechanical torque generated by passing current through the N turns of the coil in the magnetic field is given by

$$T_M = 2BLNIW/2 = BNAI = K_T I \qquad 8.6$$

where A is the area of the rectangular coil (A = LW), and $K_T \equiv BNA$ is the torque constant of the meter.

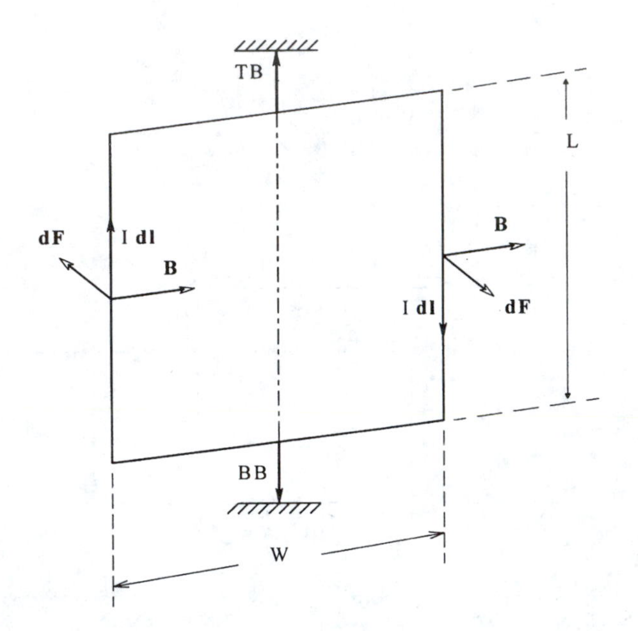

FIGURE 8.4 Magnetic (Faraday) forces on one turn of the rectangular, current-carrying coil of a D'Arsonval meter.

Unfortunately, the dynamic behavior of a D'Arsonval meter movement is not simply described by the balance between T_M and the spring torque. If the coil is moving, an EMF is induced in it, given by

$$E_B = 2NLB\left(\dot{\theta}W/2\right) = NBA\dot{\theta} = K_B\dot{\theta} \qquad 8.7$$

where θ is the angle of rotation (pointer angle) of the coil, $\dot{\theta}$ W/2 is the tangential (linear) velocity of a side of the coil, and K_B is the coil's back EMF constant. A complete circuit of the D'Arsonval meter coil is shown in Figure 8.5. L_C is the self-inductance of the coil which is generally negligible at low frequencies. From Figure 8.5 we see that the current in the coil is given by:

$$I_C = \left(V_C - E_B\right)/\left(R_C + R_I\right) = \left(V_C - E_B\right)/R_T \qquad 8.8$$

The Newtonian torque balance equation can be written:

$$T_M = K_T I_C = J\ddot{\theta} + D\dot{\theta} + K_S\theta \qquad 8.9$$

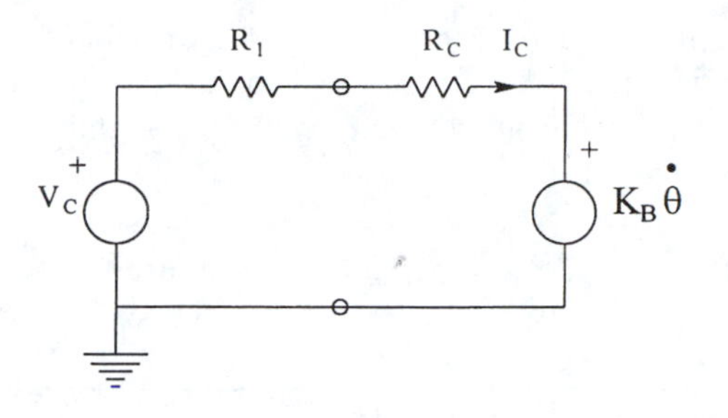

FIGURE 8.5 Equivalent circuit of the D'Arsonval meter coil. The EMF due to coil motion in the magnetic field is $V_B = K_B\dot{\Phi}$.

$$K_T\left[V_C - K_B\theta/R_T\right] = J\ddot{\theta} + D\dot{\theta} + K_S\theta \qquad 8.10$$

which reduces to:

$$V_C(t) = \ddot{\theta}JR_T/K_T + \dot{\theta}\left(DR_T/K_T + K_B\right) + \theta K_S R_T/K_T \qquad 8.11$$

where D is the viscous damping torque constant, J is the moment of inertia of the coil and pointer, and K_S is the torque constant of the torsion spring.

When this second-order, linear ordinary differential equation (ODE) is Laplace transformed, we finally obtain a transfer function relating pointer angle, θ, to applied voltage:

$$\frac{\Theta(s)}{V_C(s)} = \frac{K_T/(R_T K_S)}{s^2 J/K_S + s\left[\left(B + K_T K_B/R_T\right)/K_S\right] + 1} \qquad 8.12$$

Note that the natural frequency of the D'Arsonval meter movement is:

$$\omega_n = \sqrt{K_S/J}\ \text{r/s} \qquad 8.13$$

and its damping factor is given by:

$$\zeta = \frac{D + K_T K_B/R_T}{2\sqrt{JK_S}} \qquad 8.14$$

Equation 8.12 above is a classic example of the transfer function of a second-order linear electromechanical system. The step response of the D'Arsonval meter is easily found by setting

$$V_C(s) = V_C/s \qquad 8.15$$

and using a table of Laplace transforms. The meter deflection as a function of time is found to be:

$$\theta(t) = V_C \frac{K_T}{R_T K_S}\left[1 - \frac{e^{-\omega n \zeta t}}{\sqrt{1-\zeta^2}} \sin\left(\left\{\omega_n\sqrt{1-\zeta^2}\right\}t + \tan^{-1}\frac{\sqrt{1-\zeta^2}}{\zeta}\right)\right] \qquad 8.16$$

The mechanical response of a D'Arsonval meter to a step of applied voltage is shown in Figure 8.6. Note that as the damping factor, ζ, varies from 1 to 0, the response becomes more and more oscillatory. Obviously, it is inconvenient to have to wait for a highly underdamped meter to settle down to its steady-state reading of

$$\theta_{SS} = V_C K_T / (R_T K_S) \tag{8.17}$$

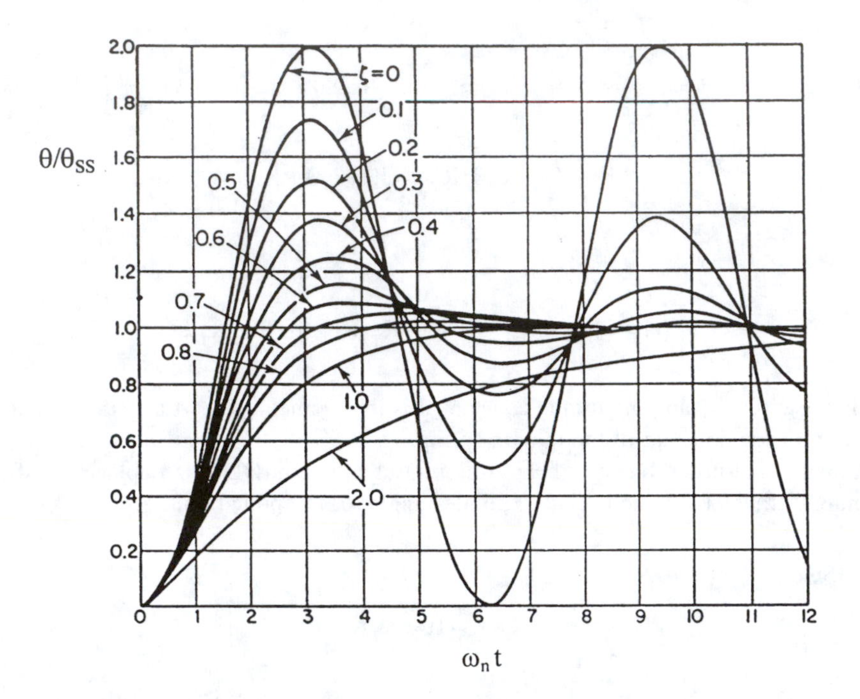

FIGURE 8.6 Mechanical response of a D'Arsonval meter to an applied step of voltage.

Experience tells us that the best dynamic performance at a fixed ω_n for a second-order system's step response occurs when the damping factor lies somewhere between 0.5 and 0.707.

From the analysis above we observe several important properties of D'Arsonval meters used as voltmeters. First, they require a steady-state current to produce a reading. This current must come from the circuit under test. Practical full-scale current is typically 20 to 50 μA, although special, sensitive D'Arsonval meter movements with torsion spring suspensions have been built with full-scale currents of 1 μA or less. To obtain greater voltmeter sensitivity, we see from Equation 8.17 that we must increase K_T = BNA. Meter size and cost are design considerations. B is set by the state of the art and cost of permanent magnets available. A is set by size considerations, and N by the maximum tolerable mass and moment of inertia of the coil. Too high a mass will give problems with bearings and coil suspension, a high J will result in an underdamped transient response and long settling time. Likewise, increased sensitivity reached by making the torsion spring constant, K_S, small will also give low damping. Second, the total resistance in the circuit, R_T, is set by the voltmeter's sensitivity and the desired full-scale voltage. From the relations above, we see that:

$$V_{FS} = (R_S + R_C)/\eta = R_T/\eta = R_T I_{MFS} \tag{8.18}$$

Obviously, for any given full-scale voltage, circuit loading by a D'Arsonval analog DC voltmeter will be less for a meter with a high η. Also, for very large V_{FS}, R_T will be very large. For example,

R_T for a 500 VFS with a 20,000 ohms/volt movement will be 10 megohms. Multirange DC D'Arsonval voltmeters have a switch that switches various R_Ss in series with the meter, satisfying Equation 8.18 above.

The capacitor, or electrostatic voltmeter, is an unusual type of voltmeter which is best suited for DC or line frequency AC, high-voltage measurements. Capacitor voltmeters have been used from about 1 kV, to over 200 kV with special insulation. The schematic of a capacitor voltmeter is shown in Figure 8.7. The design is very much like a parallel-plate, radio tuning capacitor. Rotor plates pivot on low friction bearings and are restrained by a linear torsion spring. The capacitance between rotor and stator plates is a function of the rotation angle, θ (pointer angle), and in a linear capacitor, assumed here for convenience, can be written as;

$$C(\theta) = C_O + K_C\theta \tag{8.19}$$

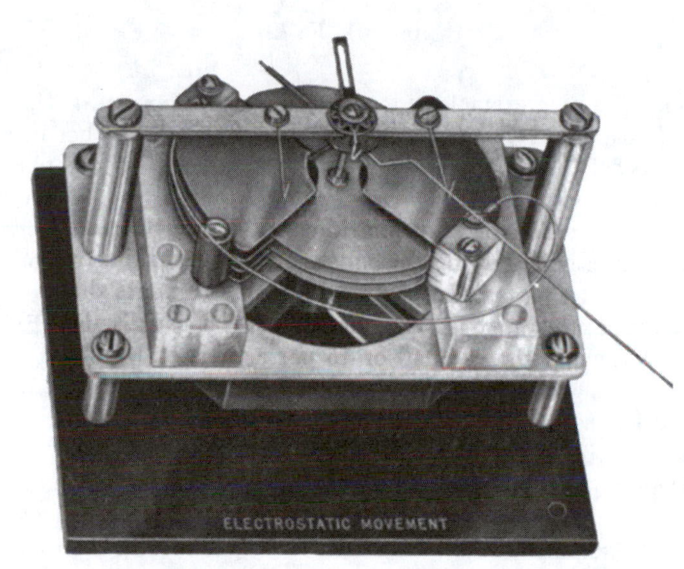

FIGURE 8.7 Cutaway view of an electrostatic capacitor voltmeter. (Courtesy of Electrical Instrument Service, Inc.)

Here C_O is the capacitance between the plates at $\theta = 0°$, and K_C is the capacitance constant. $C(\theta_{MAX})$ ranges from about 20 pF up to several hundred pF, depending on design. It can be shown that at a given voltage, there is an attractive torque acting to rotate the rotor into the stator to produce maximum capacitance and maximum stored electric field energy in the capacitance. The energy stored in the capacitor at steady-state is:

$$W = \frac{1}{2}C(\theta)V_C^2 \tag{8.20}$$

The torque acting to rotate the rotor into the stator plates may be shown to be given by:

$$T_M = \left.\frac{\partial W}{\partial \theta}\right|_{V_C} = \frac{V_C^2}{2}\frac{dC(\theta)}{d\theta} = \frac{V_C^2}{2}K_C \tag{8.21}$$

The electrostatic-derived torque must equal the Newtonian torques:

$$T_M = J\ddot{\theta} + D\dot{\theta} + K_S\theta \quad 8.22$$

After substituting the expression for T_M into the second-order differential equation above and Laplace transforming, we obtain the transfer function:

$$\frac{\Theta}{V_C^2}(s) = \frac{K_C/2K_S}{s^2J/K_S + sD/K_S + 1} \quad 8.23$$

The mechanical properties of the capacitor voltmeter, similar to those of the D'Arsonval meter movement, act as a mechanical low-pass filter above $\omega_n = \sqrt{K_S/J}$ r/s. Hence, for steady-state conditions:

$$\theta_{SS} = \overline{V_C^2}K_C/2K_S \quad 8.24$$

The capacitor voltmeter is an example of a *square-law meter movement.* The scale has nonlinear calibration from Equation 8.24, and also because $C(\theta)$ is in general not linear. After the meter reaches steady-state deflection, no current flows in the circuit for an applied DC voltage step. However, if a capacitor voltmeter is suddenly placed directly across an energized, high-voltage circuit, it appears initially as a short circuit and heavy current flows into the meter. This may damage the CUT, so often a series resistor is used with the electrostatic voltmeter to limit the initial charging current.

A typical electrostatic voltmeter, such as the Sensitive Research model ESD, has a scale calibrated from 1 to 5 kV, with the most sensitive (expanded) scale at the center of the meter ($\theta_{MAX}/2$) at about 2.75 kV. This meter has a guaranteed accuracy of 1% of full-scale voltage, a capacitance of 18 pF at full-scale, and an insulation resistance of 10^{15} ohms. θ_{MAX} is about 65° for this meter, which is less than the 90° typical for most D'Arsonval meter movements.

The electrodynamometer meter movement is another voltmeter design, which as we will see below, is square-law, and can be used for the measurement of DC or powerline frequency AC voltage, current, and power. Here we will examine the electrodymamometer as a DC voltmeter. Figure 8.8 illustrates the basic electrodynamometer movement. Note that there are two symmetrically placed stator coils around a rotor coil which, as in the case of other analog meters, is constrained by a torsion spring. When an electrodynamometer meter is connected as a voltmeter, the three coils are wired in series, as shown. An equivalent circuit of the electrodynamometer voltmeter is shown in Figure 8.9. Here, as in the case of the D'Arsonval DC voltmeter, a series resistor is used to limit the meter's full-scale current. The energy stored in the magnetic field of the electrodynamometer meter coils is given by:

$$W = i_M^2\left[L_S/2 + L_R/2 + M_{SR}(\theta)\right] \quad 8.25$$

where i_M is the current in the series coils, L_R is the self-inductance of the rotor coil, L_S is the self-inductance of the two stator coils together, and $M_{SR}(\theta)$ is the mutual inductance between the stator coils and the rotor coil. Note that $M_{SR} = M_{RS}$. Because of meter coil geometry, $M_{SR}(\theta)$ may be approximated by:

$$M_{SR}(\theta) = K_M(\theta - \theta_{MAX}/2) \quad 8.26$$

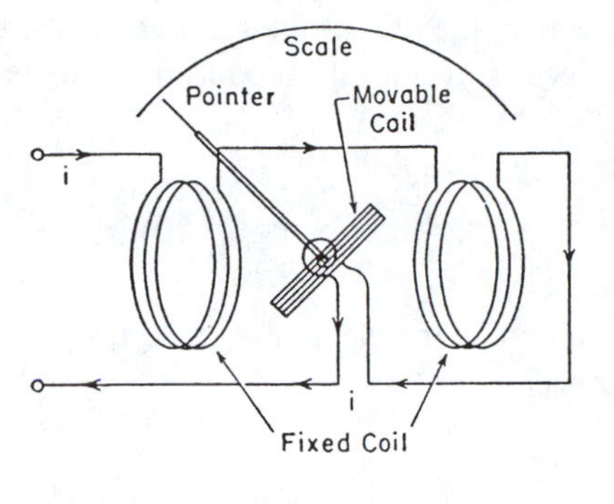

A.

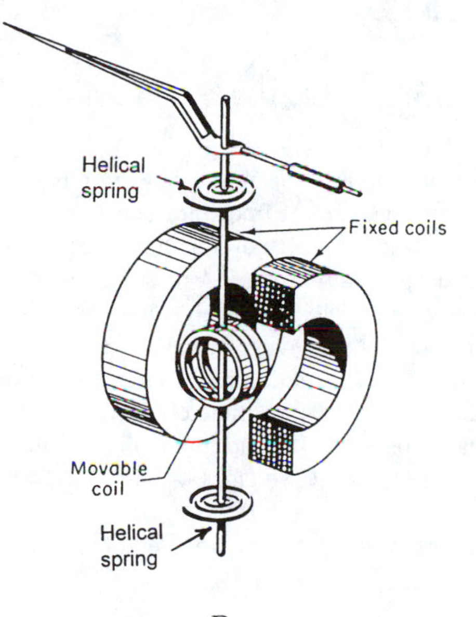

B.

FIGURE 8.8 (A) Schematic of an electrodynamometer meter movement connected as a voltmeter. (From Stout, M.B., *Basic Electrical Measurements*, 2nd ed., Prentice-Hall, Englewood Cliffs, NJ, 1960. Reproduced with permission.) (B) Cut-away view of an electrodynamometer meter movement.

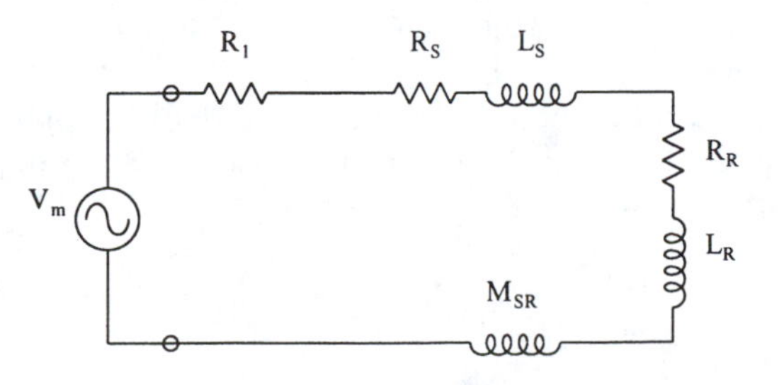

FIGURE 8.9 AC equivalent circuit of an electrodynamometer voltmeter.

Note that at half maximum deflection, the axis of the rotor coil is perpendicular to the axis of the stator coils, and $M_{SR} = 0$. The electromagnetically produced torque is given by:

$$T_M = \left.\frac{\partial W}{\partial \theta}\right|_{i_M} = i_M^2 \frac{dM_{SR}(\theta)}{d\theta} = i_M^2 K_M = \left[V_M/(R_S + R_R + R_1)\right]^2 K_M \qquad 8.27$$

As in the previous examples, we equate the electromagnetic torque with the Newtonian reaction torques, and form a low-pass transfer function:

$$\frac{\Theta}{\overline{V_M^2}}(s) = \frac{K_M/\left[(R_S + R_R + R_1)^2 K_S\right]}{s^2 J/K_S + sD/K_S + 1} \qquad 8.28$$

In the steady state, the meter deflection is:

$$\theta_{SS} = \overline{V_M^2} K_M/\left[(R_S + R_R + R_1)^2 K_S\right] \qquad 8.29$$

Here, as in the case of the electrostatic voltmeter, the deflection is square-law. Electrodynamometer voltmeters are seldom used above power line frequency because the coil inductive reactance begins to become more than 1% of (R_S + R_R + R_1), producing a frequency-dependent calibration error. Electrodynamometer voltmeters generally require far more current for steady-state, full-scale deflection than do D'Arsonval voltmeters. Typical electrodynamometer voltmeter sensitivities range from 10 to 50 ohms/volt, or full-scale currents range from 0.1 to 0.02 A. It appears that this type of meter is truly a "wattsucker," and best suited for measurements on power systems rather than on electronic circuits. Dynamometer voltmeters, ammeters, and wattmeters generally have 80° to 90° arc scales with scale lengths of 6.5" to 7.0", and a mirror to eliminate parallax when reading the pointer. They are calibrated for DC or AC voltages or currents ranging from about 25 to 500 Hz, depending on design. Accuracies range from 0.1 to 0.25% of full-scale reading. Because of the mass of the moving coil, the bearings must be rugged and have low friction.

8.1.2 Electronic DC Voltmeters

In this section we will consider the design, applications, and limitations of various types of electronic DC voltmeters, beginning at the low end of the voltage scale with nanovoltmeters, which are designed to work with extremely low DC input potentials seen through Thevenin resistors on the order of tens of ohms. Figure 8.10 illustrates the design of a chopper-type nanovoltmeter. A special, low-noise electromechanical chopper running at 60 Hz is used to convert the low-level DC potential, V_S, to an AC signal, V_2, which is amplified by an RC amplifier, A1, with a gain of -10^4. The AC signal's frequency, 60 Hz, is above the 1/f noise portion of the RC amplifier's input voltage noise spectrum, $e_{na}^2(f)$. The equivalent short circuit voltage noise spectrum of the input amplifier is of the form:

$$e_{na}^2(f) = \eta + b/f \ \text{msv/Hz} \qquad 8.30$$

In the flat (white) part of the noise power density spectrum, η can be as low as 49×10^{-18} msv/Hz (Toshiba 2SK146 JFET). Following amplification by A1, V_3 is synchronously demodulated by a second chopper. The output of the second chopper, V_4, has a DC component which is conditioned by the op amp low-pass filter, A2. A block diagram summarizing these operations is shown in Figure 8.11. With the feedback resistors shown, the nanovoltmeter's transfer function can be written:

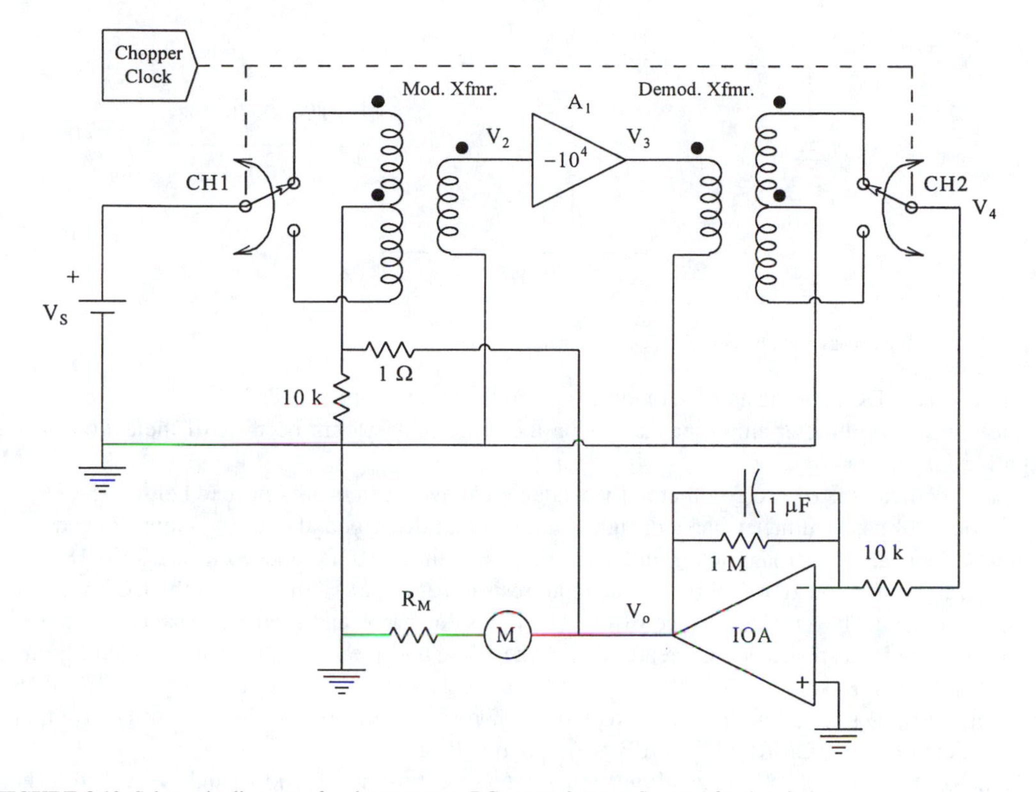

FIGURE 8.10 Schematic diagram of a chopper-type, DC nanovoltmeter. See text for description.

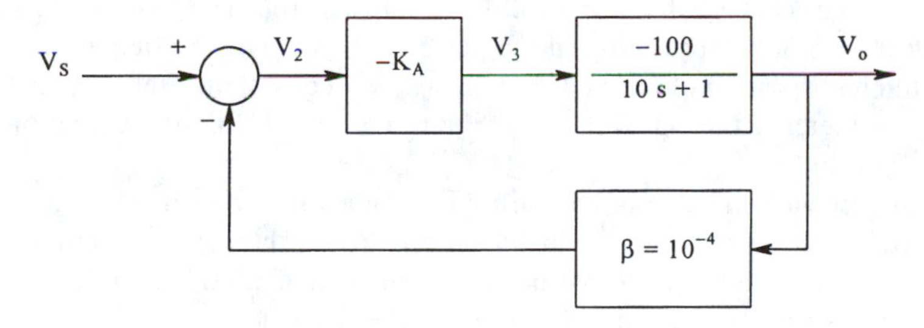

FIGURE 8.11 Block diagram of the chopper nanovoltmeter.

$$\frac{V_O}{V_S}(s) = \frac{(-10^4)[-100/(10s+1)]}{1-[-(10^6)(10^{-4})/(10s+1)]} = \frac{9.901\times 10^3}{s10/101+1} \qquad 8.31$$

Thus there is a low-noise amplification of the DC input voltage by 79.9 dB, and single-pole, low-pass filtering with a break frequency at 1.61 Hz. When operating on the lowest range (typically 10 nV full-scale) it is necessary to further amplify V_O to a level where an analog D'Arsonval microammeter can serve as an output indicator. An additional DC gain stage of 10^2 will boost a 10 nV input to a 10 mV output. Now if a 20 μA full-scale, DC, D'Arsonval meter is used, it will read full scale if its internal resistance is 500 ohms (or less).

The input resistance of the nanovoltmeter can be calculated using the equivalent input circuit of Figure 8.12. At DC, R_{IN} can be shown to be approximately:

$$R_{IN} \cong (R_T + 1\Omega)101.01 \text{ ohms} \qquad 8.32$$

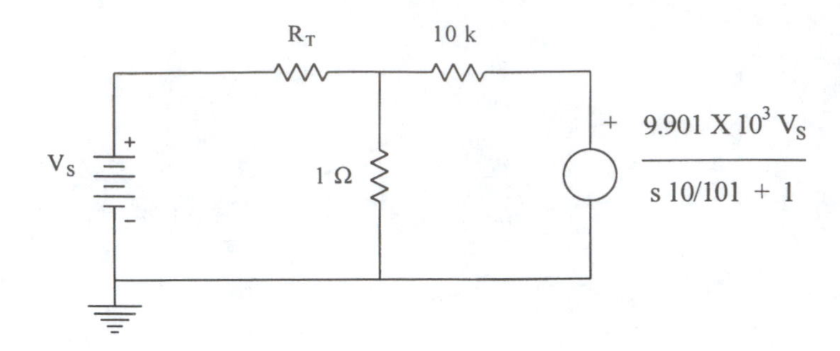

FIGURE 8.12 Equivalent input circuit of the chopper nanovoltmeter.

where R_T is the DC series resistance of either side of the input chopper coil. The DC input resistance of the nanovolt voltmeter amplifier may be on the order of 1 kilohm because of the action of the negative feedback.

The specifications of two commercially available nanovoltmeters are outlined below. The Keithley model 148 nanovoltmeter has a design similar to that described above. Its output is read on a center-scale analog microammeter. Full-scale ranges begin at 10 nV and go to 0.1 V in 18 overlapping ranges. On the 10 nVFS range, the input resistance is over 1 kilohm, and the DC nV source must not have a Thevenin resistance of over 10 ohms for noise and loading reasons. On the 0.1 V_{FS} range, the input resistance is greater than 1 megohm, and the source resistance should be less than 10 kilohms. Resolution is better than 1 nV on the 10 nV range, and accuracy is ±2% of full scale on all ranges. The line frequency rejection is 3000:1 on the 10 nV range, and the common-mode rejection ratio (CMRR) is 160 dB at 60 Hz on all ranges.

The Keithley model 181 nanovoltmeter has a 6½ digit, digital readout, and is IEEE-488 bus compatible. It uses a special low-noise JFET headstage to obtain an input resistance of over 10^9 ohms on the low ranges. There are seven full-scale ranges, from 2 mV to 1 kV. Meter noise is claimed to be less than 30 nV peak-to-peak on the 2 mV scale with the output low-pass filter on. The low-pass filter is a three-pole digital design, with an equivalent time constant of 0.5, 1, or 2 s, depending on meter range. This meter can be run with 5 1/2 digits resolution when inputs are noisy.

Because of the special challenge in measuring DC voltages in the nanovolt range, meters must be allowed to warm up at least an hour before use to reach thermal equilibrium. DC nanovolt measurements are confounded by factors which are normally neglected in high-level measurements. These include the generation of thermal (Seebeck) EMFs by junctions between dissimilar metals at different temperatures, the induction of low-frequency EMFs by time-varying magnetic fields whose flux passes through the circuit from the meter to the source, low-frequency and DC potentials due to ground loops, low-frequency EMFs due to triboelectric effects on connecting coaxial cables, and piezoelectric EMF artifacts. A good discussion of sources of error in low-level, DC measurements can be found in Keithley (1992).

In order to operate either an analog meter or a digital interface, the DC voltage under measurement must be amplified to a level useful to drive the analog indicating meter (about 100 mV) or the analog-to-digital converter (1 to 10 V). This means that microvolt level signals need to be amplified by factors of 10^5 to 10^7. Inherent in all DC amplification is the addition of noise from the amplifier, generally consisting of a sum of white noise and 1/f noise, DC offset and drift due to temperature changes. It was seen in Section 3.6 that the noise performance (as well as the DC drift performance) of a high-gain amplifier is set by the headstage, which must have a gain greater than five. Thus, the amplifiers used in any sensitive, low-level, DC voltmeter must have low noise, and also have very low DC drift and offset voltage. Op amps suitable for high-gain, DC signal conditioning include chopper-stabilized and commutating autozero amplifiers, the latter being described in Section 2.4.3. The Intersil ICL7600/ICL7601 commutating autozero op amp is well suited for DC voltage amplification system headstages. The ICL7600 has a low input offset voltage of 2 μV, a very low input offset voltage drift of 0.2 μV/year, and a very low input offset voltage tempco of ±0.005 μV/°C, and a DC input bias current of 300 pA. The switching (commutating) frequency of the ICL7600 system capacitors can be set from 160 to 5200 Hz.

Another approach to low-noise, drift-free DC amplification is to use a chopper-stabilized op amp, such as the ICL7650. This op amp has a DC input offset voltage of 1 μV over the operating temperature range, an offset voltage tempco of ±0.01 μV/°C, a DC input bias current of 35 pA, a DC open-loop gain of 134 dB, and a gain bandwidth product (GBWP) of 2 MHz. The internal chopping frequency is 200 Hz. Many other manufacturers offer integrated circuit, chopper-stabilized or auto-zero DC amplifiers. Tempcos of 50 nV/°C are pretty much state of the art, and input DC offset voltages range from 1 to 5 μV.

A low-pass filter is generally used at the output of the DC amplifier stages, before the analog-to-digital converter (ADC) or the digital voltmeter (DVM) module. If a D'Arsonval analog meter is used, the meter movement itself acts as an electromechanical low-pass filter (see Equation 8.11). However, additional low-pass filtering may be required to improve the DC microvoltmeter's resolution. The schematic of a DC voltmeter with electronic amplification is shown in Figure 8.13. If an analog meter is used, precision is seldom more than ±0.25% of full scale because of the visual uncertainties of reading the pointer position. Properly designed digital meters can commonly reach precisions of ±0.005%, or 50 ppm. Noise, DC drift, and quantization error are still the ultimate determinors of resolution in instruments with digital readouts, however.

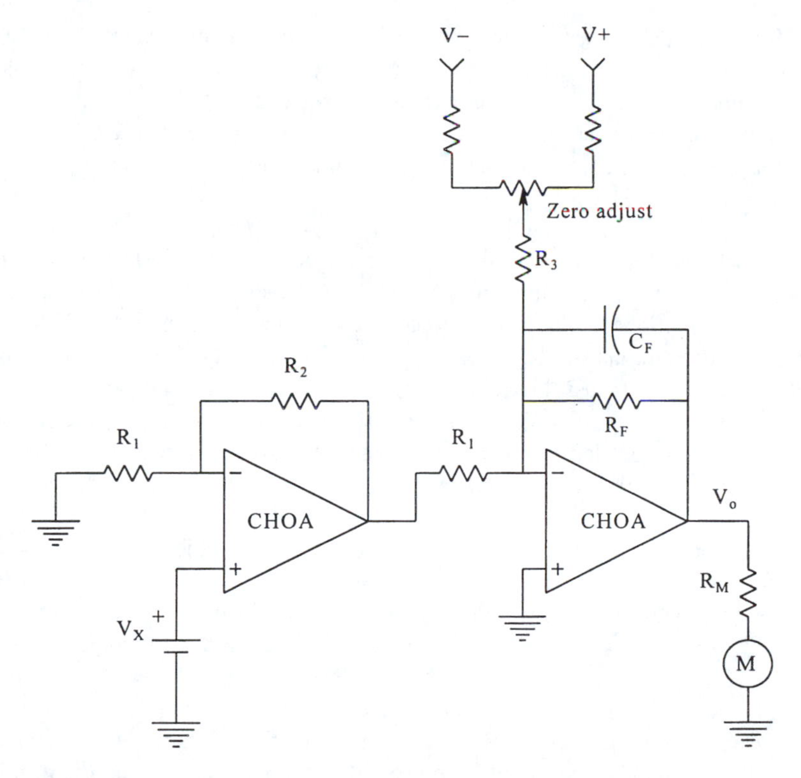

FIGURE 8.13 Schematic diagram of an analog DC millivoltmeter. Two chopper-stabilized op amps are used for amplification. The second op amp is used as a low-pass filter and for zeroing.

Electronic DC voltage measurements above the single volt range generally do not require amplification, either for an analog indicating meter or a digital output indicator. In fact, electronic voltmeters used in the range of 10 to 1000 V generally make use of internal input voltage dividers to attenuate the voltage under measurement to a value useful to the meter amplifier or ADC. An external resistive voltage divider probe is generally used to measure DC voltages from 0.5 to

40 kV. Such a probe makes an external voltage divider with the DC voltmeter's input resistance. For example, the Fluke model 80K-40 high-voltage probe is designed to be used with any DC voltmeter having an input resistance of 10 megohms. It makes a 1000:1 voltage divider, so 1 kV applied to the probe gives 1 V at the meter. The probe and meter present a 1000 megohm load to the high-voltage circuit under test, hence the probe and meter draw 1 μA/kV from the CUT.

At DC voltages above 40 kV, special measurement techniques must be used because of insulation problems associated with the resistive voltage divider. At very high-voltages, resistors may behave erratically due to the generation of corona discharge, or leakage from the absorption of water from the atmosphere. In fixed installations, these problems can be overcome by using proper insulators and couplings, and housing the resistors in an inert, insulating gas such as SF_6 or Freon, or in high-pressure (≥10 atm) dry air or nitrogen.

In order to measure DC voltages in excess of 200 kV electronically, use is often made of electrometer-based, electric field measuring instruments. Such techniques make use of the fact that the circuit geometry is fixed in super high-voltage systems, hence the electric field geometry is fixed, and at any point in the field, the field strength is proportional to the DC voltage generating the field. We discuss instruments used to measure electric fields in the next section.

8.2 MEASUREMENT OF STATIC ELECTRIC FIELDS AND THE POTENTIAL OF CHARGED SURFACES

Static (DC) electric fields can arise from a variety of physical and electrical causes which include the presence of bound or mobile surface charges (electrons, ions) on conductors or insulators. The surfaces can become charged by being bombarded with charged particles such as sand or raindrops, or from triboelectric effects where charges become separated from the surfaces of insulators when they rub on conductors or other insulators. Of course, static fields exist around conductors operating at high DC voltages with respect to ground.

Static charges and fields can create problems ranging from annoying ("static cling") to extreme danger (static discharge sparks ignite solvent vapors or combustible dust, causing an explosion). Other problems caused by static electricity include the attraction of fine dust particles to clean, charged surfaces, attraction of charged dust particles to clean, grounded surfaces, the physical attraction of charged materials such as paper or plastic film to other objects, and when the electric field strength exceeds a critical value (about 3×10^6 V/m in air at STP), there can be an abrupt discharge of charged persons or objects to static-sensitive semiconductor devices or electronic equipment, causing damage. Clearly, it is important to be able to measure the DC voltages associated with the production of DC fields, to give warning of potentially damaging or dangerous conditions (pun intended).

The first method of measuring the DC potential of a charged object field makes use of a DC, electrometer voltmeter connected to a special probe as shown in Figure 8.14A. To measure the potential of the object in question, which may be a charged insulating surface or conductor, the probe is brought to a standard distance from the object. The standard distance is required to set the capacitance, C_2, to a known value. Figure 8.14B shows the equivalent circuit for the DC electrometer field meter. To measure V_S, capacitance C_1 is initially shorted to ground, and a grounded metal plate is used to cover the proximal plate of capacitor C_2. (C_1 is the input capacitance of the electrometer voltmeter plus the capacitance to ground of the wire connecting the proximal plate of C_2 to the electrometer input. C_2 is the capacitance between the proximal plate of C_2 and the charged object.) Next, the metal plate is slid aside, fully exposing the plate of C_2. C_2 and C_1 are in series, and acquire charge from the field from the potential on C_3. Initially, if C_3 is isolated and is at potential V_S, some charge flows from C_3 into C_2 and C_1, causing the voltage on C_3 to drop to V_S'. In the steady state, this can be written:

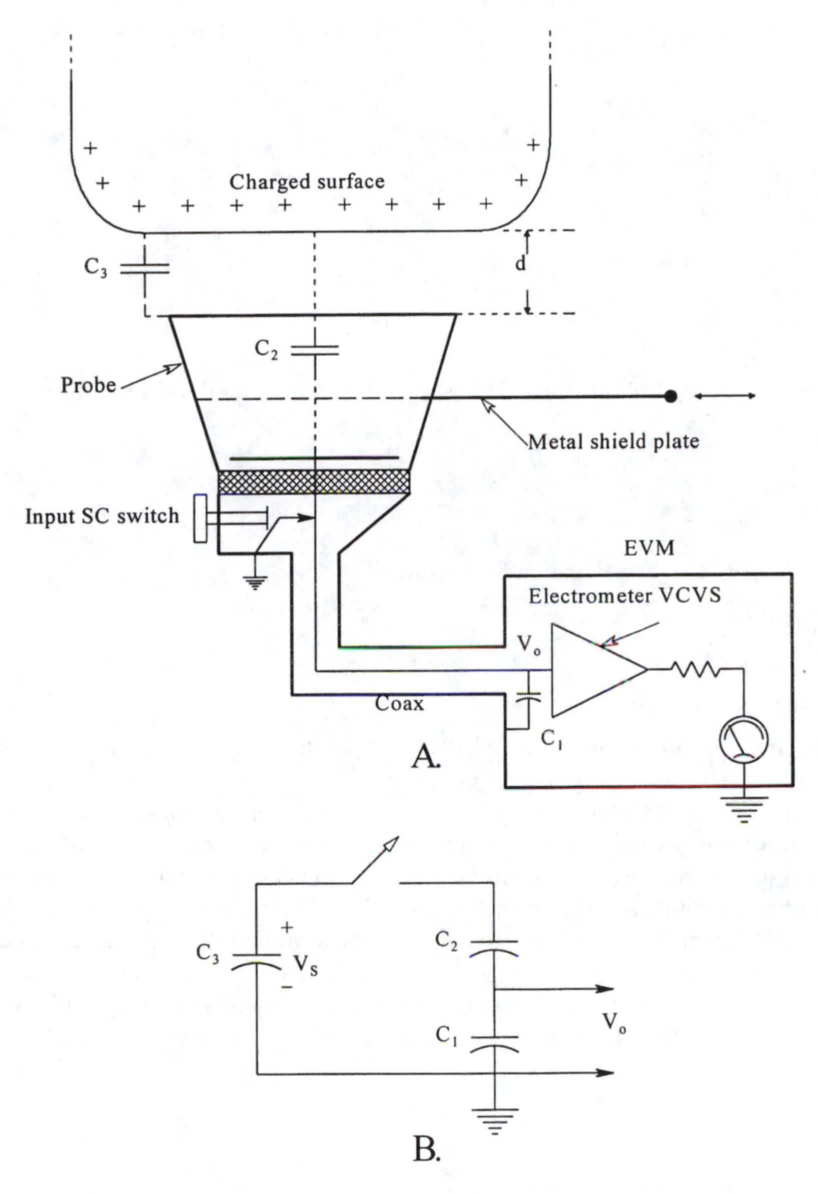

FIGURE 8.14 (A) Section through a capacitance probe used to measure high dc, electrostatic potentials on charged objects. The probe acts as a voltage divider. (B) Equivalent circuit of the high-voltage capacitance probe. See text for analysis.

$$(V_S - V_S')C_3 = V_S' \frac{C_1 C_2}{C_1 + C_2} \tag{8.33}$$

The new voltage on the isolated object with capacitance C_3 is found to be:

$$V_S' = \frac{V_S C_3}{\dfrac{C_1 C_2}{C_1 + C_2} + C_3} \tag{8.34}$$

Now because the same current flows in C_3, C_2, and C_1, the charge that resides in C_2 also is present in C_1. Thus, from

$$V_O C_1 = (V_S' - V_O) C_2 \qquad 8.35$$

we find that

$$V_O = \frac{V_S' C_2}{C_1 + C_2} \qquad 8.36$$

If the expression for V_S' in Equation 8.34 is substituted in Equation 8.36 above, we finally obtain:

$$\frac{V_O}{V_S} = \frac{C_2 C_3}{C_1 C_2 + C_3 (C_1 + C_2)} \qquad 8.37$$

If $C_3(C_1 + C_2) >> C_1 C_2$, then Equation 8.37 reduces to

$$\frac{V_O}{V_S} \cong \frac{C_2}{C_1 + C_2} \qquad 8.38$$

For example, if $C_2 = 10^{-11}$ F, and $C_1 = 10^{-7}$ F, then $V_O = 10^{-4}\ V_S$. Thus, an initial 10 kV charge on C_3 will result in a 1 V reading on the electrometer.

In the system described above, the value of C_1 depends on the capacitance to ground of the cable connecting the probe with the electrometer, as well as the input capacitance of the electrometer. In order to eliminate the system's requirement for a fixed input geometry to determine C_1, the charge amplifier configuration shown in Figure 8.15 can be used. Now C_1 appears between the summing junction, which is at virtual ground, and actual ground. Hence negligible current flows through C1, and C_F determines the output voltage of the system. R_2 limits the initial charging current through C_2 and C_F so that the electrometer op amp will not go into current saturation. Now it is easy to show that the charge amplifier output is:

$$V_Q = -V_S (C_2 / C_F) \qquad 8.39$$

We wish C_2/C_F to be 10^{-3} or 10^{-4} so the op amp will not saturate for large V_S.

Another approach to the problem of measuring DC voltages in the kilovolt range through their fields is to use a time-modulated C_2, and to measure the AC current which flows in resistor R, shown in Figure 8.16. In this form of DC high-voltage meter, the modulation of C_2 may be expressed mathematically as

$$C_2(t) = C_{2O} + \Delta C \sin(\omega t), \quad \Delta C \ll C_{2O} \qquad 8.40$$

There are several ways in which C_2 can be modulated. One common method is to have the blades of a grounded, rotating chopper wheel alternately cover and uncover the proximal plate of C_2. Another method physically modulates the distance between the proximal plate and the charged surface. As a result of the modulation of C_2, an AC current of frequency ω will flow through resistor R to ground. This current may be found by writing a loop equation for the system of Figure 8.16:

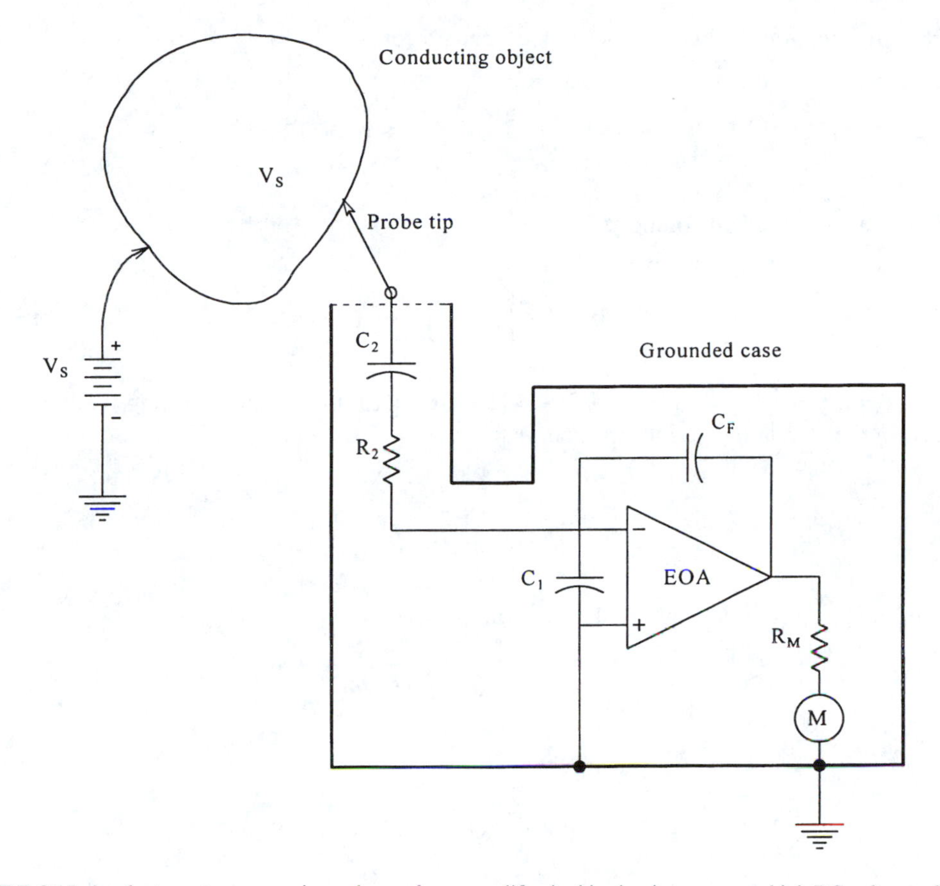

FIGURE 8.15 An electrometer op amp is used as a charge amplifier in this circuit to measure high DC voltages. See text for analysis.

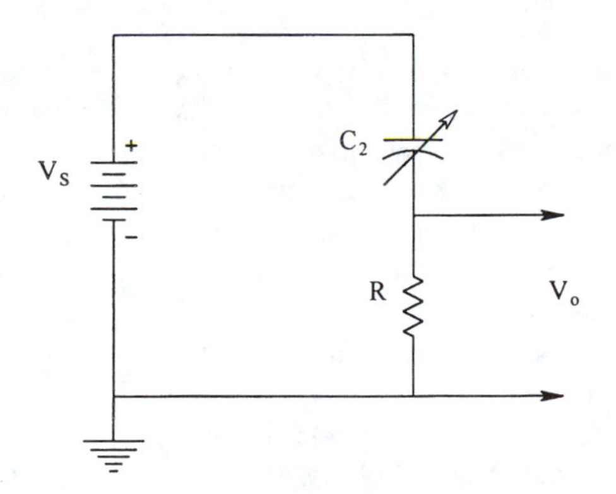

FIGURE 8.16 Equivalent circuit of a field mill, DC voltmeter. The value of C_2 is modulated sinusoidally. See text for analysis.

$$V_S = \dot{q}R + q/C_2(t) \qquad 8.41$$

or

$$V_S = \dot{q}R + q(1 - \Delta C/C_2)/C_{2O} \qquad 8.42$$

The ODE of Equation 8.42 can be put in the standard form:

$$\dot{q} + q\frac{1}{C_{2O}R}\left(1 - \frac{\Delta C \sin(\omega t)}{C_{2O}}\right) = V_S/R \tag{8.43}$$

This ODE has the standard solution:

$$y \exp\left(\int P\,dx\right) = \int\left(\exp\left(\int P\,dx\right)\right)Q\,dx + c \tag{8.44}$$

in which x = t, y = q, Q = V_S/R, and P(t) = $[1 - \Delta C \sin(\omega t)]/RC_{2O}$. Performing the integrations indicated in the standard solution, we find that:

$$q(t) = \frac{V_S C_{2O}}{1 - [\Delta C \sin(\omega t)]/C_{2O}} + c \tag{8.45}$$

The sinusoidal steady-state sinusoidal current in the series circuit is:

$$i_{SS}(t) = \dot{q} = V_S \Delta C \omega \cos(\omega t) \tag{8.46}$$

Thus, the AC voltage across R is proportional to V_S:

$$v_O(t) = V_S R \Delta C \omega \sin(\omega t) \tag{8.47}$$

For example, if ω = 100 r/s, V_S = 100 kV, R = 10^5 ohms, and $\Delta C = 10^{-11}$ F, then the peak value of v_O is 10 V. Of course, the input resistance of the AC voltmeter used to measure v_O should be over 10^8 ohms.

The two methods of measuring the high DC voltage, V_S, described above do not require direct contact with the charged surface. However, their accuracy is limited by the accuracy in determining either C_2 or ΔC, which in turn depends on accurate determination of the distance from the proximal plate of C_2 to the charged surface. At least one commercially available "digital field meter" designed to have a working distance of 4.0" from the charged surface, uses an ultrasonic ranging system similar to that used in popular cameras which use self-developing, "instant" film, to permit precise adjustment of the working distance. The Semtronics model EN 235 Auto-DFM reads $0 \le V_S \le 20$ kV with ±5% accuracy of the reading at 4" distance. Correct distance is indicated by LEDs. Other field voltmeters require a physical measurement of their working distance, presumably with an insulated ruler. Voltage ranges of some portable, battery-operated field meters can span from 0–500 V to 0–200 kV. Such meters are typically used to measure static charges accumulated in industrial processes such as paper and plastic sheet manufacturing.

Another severe static electricity problem occurs when helicopters fly through charged particles such as sand or water droplets. The metal airframe of the helicopter can acquire a potential of thousands of volts. If not discharged before landing, this charge can be dangerous in fueling and cargo-handling operations. Bradford (1975) described a system which automatically senses the magnitude and sign of the helicopter's charge, and automatically discharges the airframe by spraying charged droplets of isopropyl alcohol from the helicopter's fuselage. The helicopter's fuselage has a capacitance of about 2200 pF when on the ground. Its body capacitance when hovering at 25 m is about 500 pF. To measure the helicopter's body potential, Bradford used a corona field sensor manufactured by Dayton Aircraft Products. A block diagram of the Dayton sensor is shown in Figure 8.17A. A 400 Hz AC voltage is amplified to 4000 peak volts and applied to a wire brush

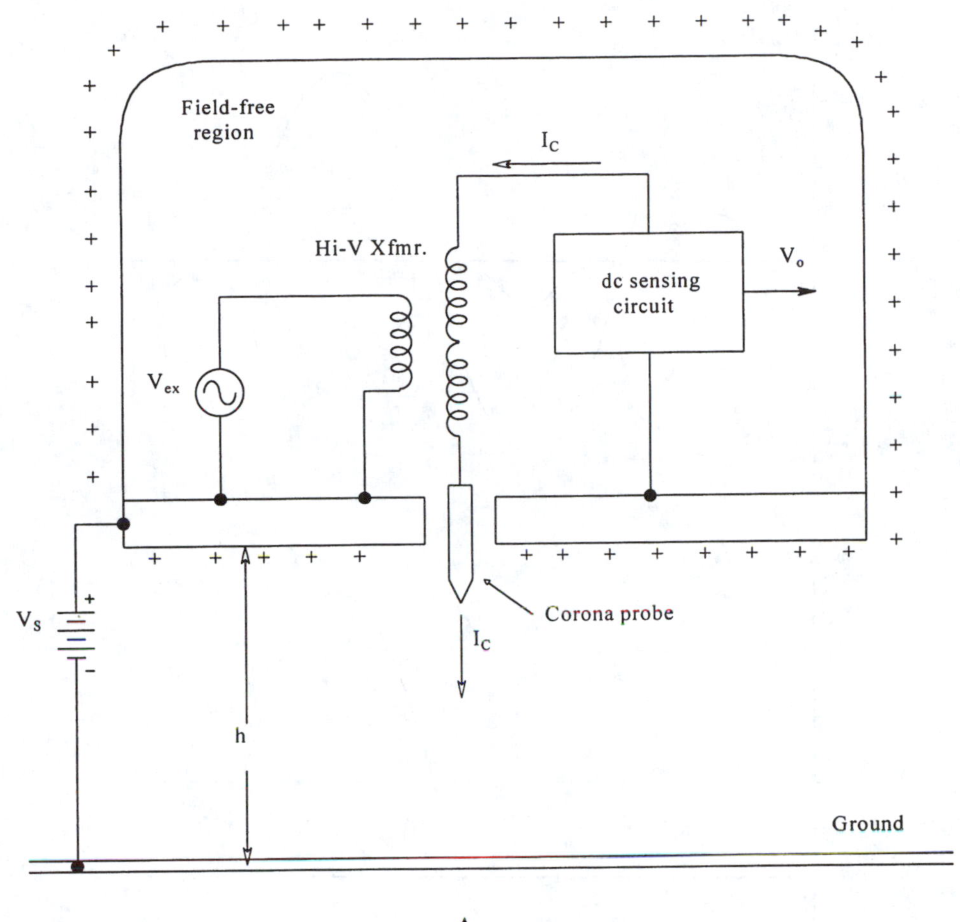

FIGURE 8.17 (A) Diagram of a Dayton Aircraft corona field sensor, used to measure the potential accumulated on helicopter fuselages.

corona probe which protrudes through a hole in the fuselage into the high DC field region next to the fuselage. As seen in Figure 8.17B, if the fuselage is at zero potential with respect to ground, the AC current due to corona flowing from the probe is symmetrical, and has zero average. If the helicopter body is at some positive potential, V_S, the DC field from V_S causes extra corona current to flow on the positive cycles, and less to flow during the negative cycles, giving the corona current an average (DC) component which is found to be proportional to E, hence V_S. The sensor electronics and static charge discharge controller were located inside the fuselage where, of course, the electric field is zero.

Calibration of the field sensor was made difficult by the fact that helicopter body capacitance decreases nonlinearly with hover height. As this decrease occurs, the body potential increases if the charge remains constant. Offsetting this increase in V_S is the fact that the field strength in the vicinity of the corona probe tends to decrease from geometrical considerations as altitude increases. Bradford (1975) used a large, parallel-plate capacitor (plate separation of 1 foot) to generate a uniform field to calibrate the corona probe field sensor. He obtained a conditioned sensor (DC) output of about 1 V per kilovolt on the capacitor.

It is generally difficult to measure the strength of an actual electric field in volts/meter without disturbing the field with the measurement apparatus. This is true of field measurements in conducting media as well as in free space. If the measurement sensor is a dielectric with a low dielectric constant, and not grounded, the disturbance of the field will be minimal.

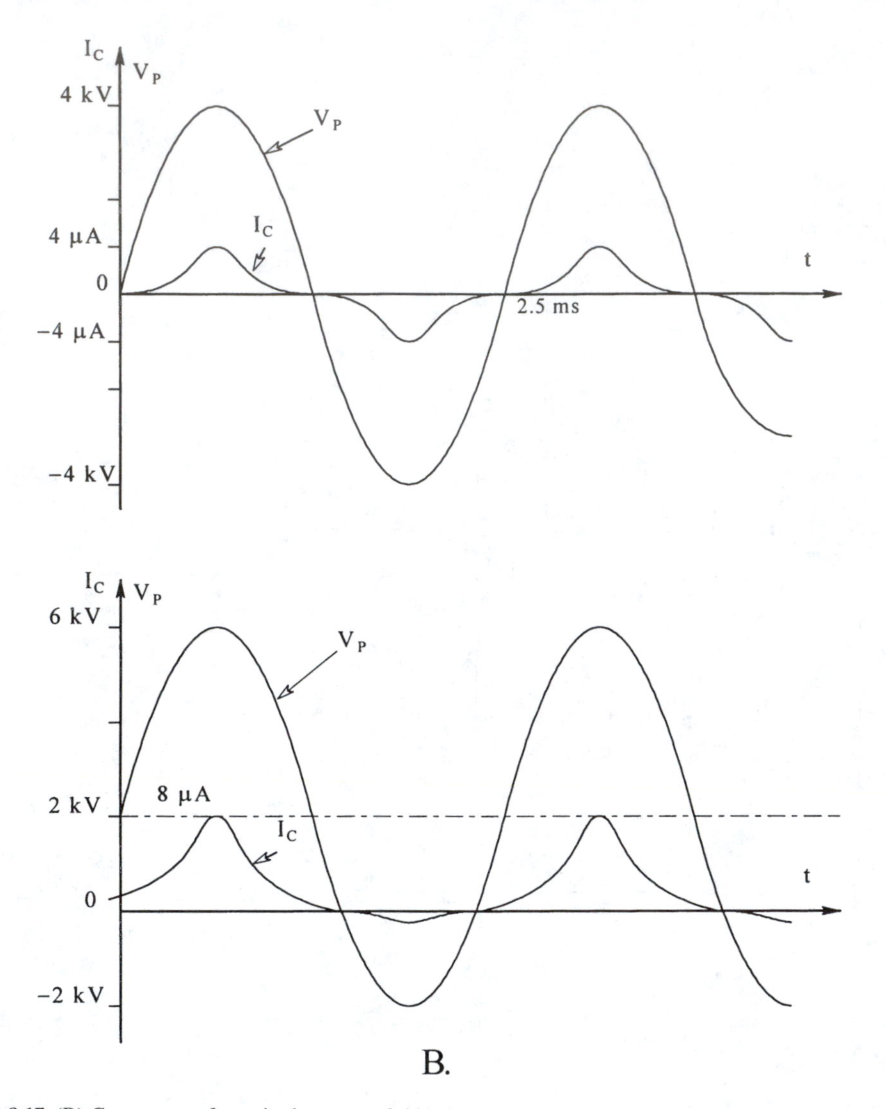

FIGURE 8.17 (B) Current waveforms in the corona field sensor.

There are several possible electro-optical approaches to the problem of measuring DC electric fields. One is to measure the change in the linear polarization of light transmitted through a quartz crystal subject to an electric field applied across a preferred axis. This phenomenon is called the *electrogyration effect* (Rogers, 1977). Another electro-optical method proposed for measuring DC electric field strength makes use of the *Pockels effect.* Certain crystalline substances, such as ammonium dihydrogen phosphate (ADP), potassium dihydrogen phosphate (KDP), cuprous chloride (CuCl), cadmium telluride (CdTe), and gallium arsenide (GaAs) exhibit the Pockels effect, in which application of an electric field along the privileged axis of the crystal causes changes in the refractive indices along two orthogonal axes (x and y).

To examine how the Pockels effect might be used to measure an electric field, we orient a Pockels crystal so that the privileged axis is perpendicular to the gradient of the field to be measured. An internal E field is set up in the crystal which is proportional to the external field we wish to measure. The internal electrical field, if correctly aligned, causes a symmetric change in the optical index of refraction along the x- and y-axes. Circularly polarized light is directed at the input surface of the crystal, and propagates within the crystal along the z-axis. The light is acted on by the birefringence induced by the internal E field, and emerges elliptically polarized. The emergent light

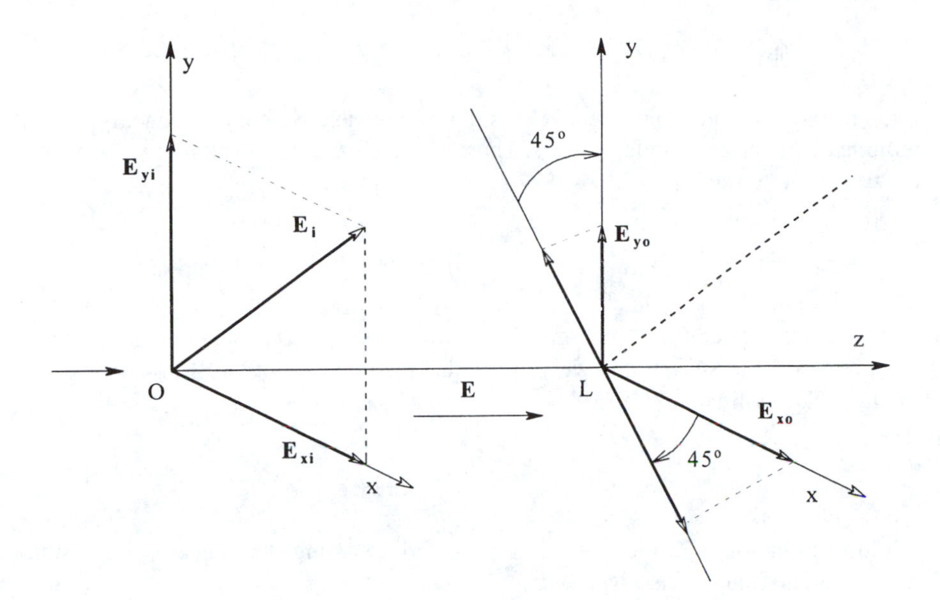

FIGURE 8.18 Vector diagram relevant to the use of a Pockel's cell to measure electric fields.

is passed through a linear polarizer analyzer plate which is oriented at 45° with respect to the x- and y-axes. The intensity of light from the output of the linear polarizer can be shown to be a linear function of the crystal's internal **E** field. Referring to Figure 8.18, we see that the x- and y-components of the electric vector of the circularly polarized input light at the entrance face of the crystal can be written as:

$$E_{Xi} = E_i \cos(\omega t) \quad \text{8.48A}$$

$$E_{Yi} = E_i \cos(\omega t - \pi/2) \quad \text{8.48B}$$

Now the refractive indexes of the Pockels crystal are affected by its internal electrical field, **E**:

$$n_X = n_o + K_P E \quad \text{8.49A}$$

$$n_Y = n_o - K_P E \quad \text{8.49B}$$

After propagating a distance L meters through the crystal, the x- and y-components of the exiting light ray's **E** vector can be written:

$$E_{XO} = E_i \cos(\omega t + \phi_X) \quad \text{8.50A}$$

$$E_{YO} = E_i \cos(\omega t + \phi_Y - \pi/2) \quad \text{8.50B}$$

The phase lags due to orthogonal changes in propagation velocity are assumed to be linearly proportional to the internal **E** field, and can be written as:

$$\phi_X = 2\pi L n_X/\lambda = 2\pi L n_o/\lambda + 2\pi L K_P E/\lambda = \phi_o + 2\pi L K_P E/\lambda \quad \text{8.51A}$$

$$\phi_Y = 2\pi L n_Y/\lambda = 2\pi L n_o/\lambda - 2\pi L K_P E/\lambda = \phi_o - 2\pi L K_P E/\lambda \qquad 8.51B$$

Propagation through the linear polarizer selects those lightwave **E** field components along the line, LAN, defining the polarizer's preferred axis. Thus, the lightwave **E** field along the LAN axis exiting the polarizer can be written:

$$E_{AN} = E_{XO}/\sqrt{2} - E_{YO}/\sqrt{2} = \left(E_i/\sqrt{2}\right)\left[\cos(\omega t + \phi_X) - \cos(\omega t + \phi_Y - \pi/2)\right] \qquad 8.52$$

We note that the second term in Equation 8.52 is of the form $\cos(\alpha - \pi/2)$ which by trigonometric identity can be written as $\sin(\alpha)$. It is well known that the intensity of an EM plane wave is related to its **E** field peak amplitude by:

$$I_{AN} = E_{AN}^2(\varepsilon_o c/2) \qquad 8.53$$

By substituting Equation 8.52 into Equation 8.53, and assuming that $\sin(x) \approx x$ for small x, we find that the average intensity is given by:

$$I_{AN} = \frac{E_i^2}{2}\left[1 + \frac{4\pi L K_P E}{\lambda}\right](\varepsilon_o c/2) \qquad 8.54$$

In the foregoing analysis, we have assumed that there was zero neutral density and reflection losses in E_i, and that $2\pi L K_P E/\lambda << 0.25$ radians. Other detection schemes are possible, but the important result is that output intensity varies linearly with the Pockel crystal's internal **E** field. For a numerical example, consider a 1.0 cm crystal of potassium dihydrogen phosphate (KDP, KH_2PO_4): we evaluate the sensitivity of KDP to be:

$$\frac{4\pi L K_P}{\lambda} = \frac{4\pi \times 0.01 \text{ m} \times 29 \times 10^{-12} \text{ m/V}}{632.8 \times 10^{-9} \text{ m}} = 5.7576 \times 10^{-6} \text{ m/V} \qquad 8.55$$

Thus, an internal field on the crystal of 10^4 V/m or 10 V/mm would produce a 5.76% increase in the detected intensity. A DC field measuring instrument based on the Pockels effect would need a means of correcting for intensity changes in the laser light source.

8.3 DC CURRENT MEASUREMENTS

The technology of analog DC current measurements may be traced back to the early 19th century (Drysdale, et al., 1952). Oersted discovered in 1819 that a wire carrying DC current, when held near a compass needle, caused the compass to deflect — implying that the current-carrying wire generated a magnetic field. This effect was put to use by Lord Kelvin, who, in 1858, developed a four-coil mirror galvanometer. In Kelvin's instrument, shown in Figure 8.19, four coils were arranged in two opposing pairs; their fields acted on two arrays of permanent magnets with opposing polarities attached to a vertical torsion spring suspension, to which was also attached a small mirror. As the magnetic fields of the current-carrying coils interact with the magnetic fields of the suspended permanent magnets, torque is developed causing rotation of the suspension and mirror. A collimated beam of light directed at the mirror is reflected onto a distant scale. Even a slight rotation of the suspension is seen as a linear deflection of the spot of light. Galvanometers of this type were used more to detect very small currents (on the order of nA), rather than to measure large currents accurately. Such galvanometers found a major application as null detectors for DC Wheatstone and

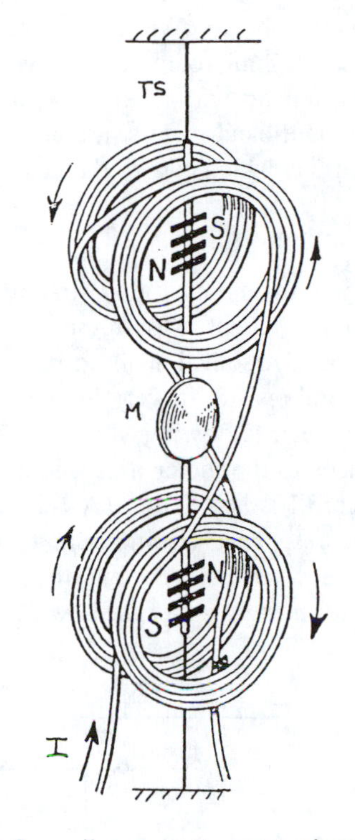

FIGURE 8.19 Diagram of Lord Kelvin's four-coil, permanent magnet, mirror galvanometer. (From Drysdale, C.V. et al., *Electrical Measuring Instruments*, Part 1, 2nd ed., John Wiley & Sons, New York, 1952. With permission.)

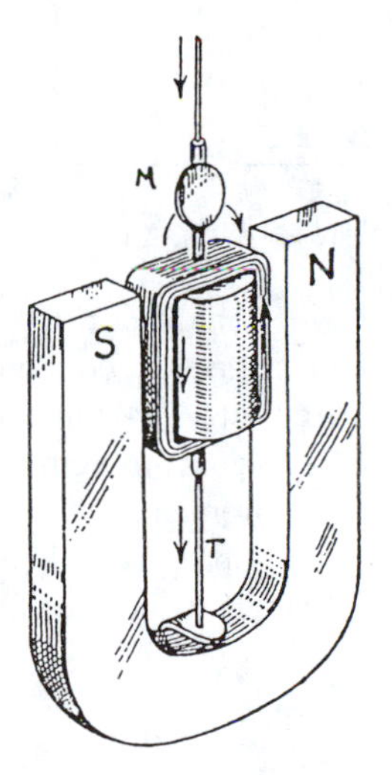

FIGURE 8.20 Diagram of D'Arsonval's mirror galvanometer. M, mirror; T, torsion spring. (From Drysdale, C.V. et al., *Electrical Measuring Instruments*, Part 1, 2nd ed., John Wiley & Sons, New York, 1952. With permission.)

Kelvin bridges, and for DC potentiometers. In 1836, Sturgeon employed a coil suspended in the field of a permanent magnet as a galvanometer. This design evolved, and in 1882, D'Arsonval built a moving-coil, permanent magnet, mirror galvanometer with a torsion spring suspension (see Figure 8.20). Notable in the D'Arsonval design was the use of a soft iron cylinder inside the rectangular coil to concentrate the magnetic flux perpendicular to the sides of the coil. In 1888, Weston developed a similar permanent-magnet, moving-coil microammeter movement in which the coil was suspended by pivot bearings, and the restoring torque was provided by a pair of helical springs.

A pointer was attached to the moving coil. The Weston version of the D'Arsonval meter is what is widely used today as the universal, DC microammeter movement which has applications in all analog multimeters, and in most instruments with analog readouts. D'Arsonval-type DC microammeters using torsion springs, called taut-band meters, are also widely used because of their ruggedness; they have no jewelled bearings which can be damaged by mechanical shocks.

8.3.1 Electromechanical DC Ammeters

Analog DC ammeters are generally of two types, the D'Arsonval permanent magnet-moving coil design or the electrodynamometer movement. Measurement of currents in the range of 10 to 10,000 A with either dynamometer or D'Arsonval meter movements generally requires an *external shunt* resistance. A shunt is a precision resistor of very low value designed so that by Ohm's law, when the full-scale current for the meter is flowing, the voltage drop across the resistor and the meter is 50 mV. The lead resistances to the meter are generally not important when measuring I_{FS} in excess of 10 A. In cases where I_{FS} is less than 10 A, and an insensitive dynamometer meter movement is used, the meter lead resistance (from shunt to meter) may be important, and a specified set of meter leads with known resistance must be used to obtain readings with rated accuracy. The circuit for an ammeter using a shunt is shown in Figure 8.21.

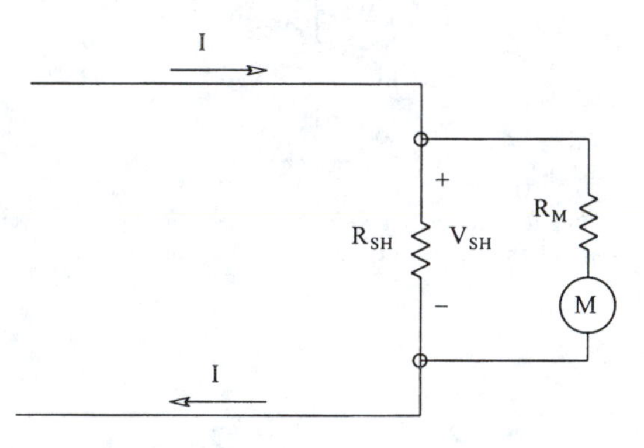

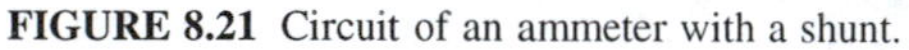

FIGURE 8.21 Circuit of an ammeter with a shunt.

Analog DC ammeters can be subdivided into benchtop (precision) meters and panel meters which are generally used in instruments, systems, and communications equipment. Some D'Arsonval microammeters have been miniaturized into 1" diameter bodies for use as tuning and signal strength meters in radio equipment, and as signal strength meters in portable tape recorders, etc.

We note that the same basic DC meter movements which can be used for analog DC voltmeters are suitable for use as DC ammeters with appropriate shunts. When full-scale DC currents less than 20 μA are to be measured, it is generally necessary to employ electronic amplification, discussed in the following section.

8.3.2 Electronic DC Ammeters

To make measurements in the range from 1 fA (10^{-15} A) to 50 μA, it is necessary to employ DC amplification of some sophistication. In general, current measurements in the range from 10^{-8} to 10^{-15} A are made with an electrometer picoammeter. Electrometer picoammeters find application in measuring the low current outputs from photomultiplier tubes and ion chambers, as well as specialized measurements in the semiconductor area, etc. As in the case of low-level voltage measurements, one fundamental factor limiting precision is the thermal noise from the Norton equivalent conductance seen in parallel with the DC current source being measured. Noise from the electrometer amplifier also confounds resolution. Both e_{na} and i_{na} input noise root power spectrums can be significant in reducing instrument resolution, as well. In addition to problems

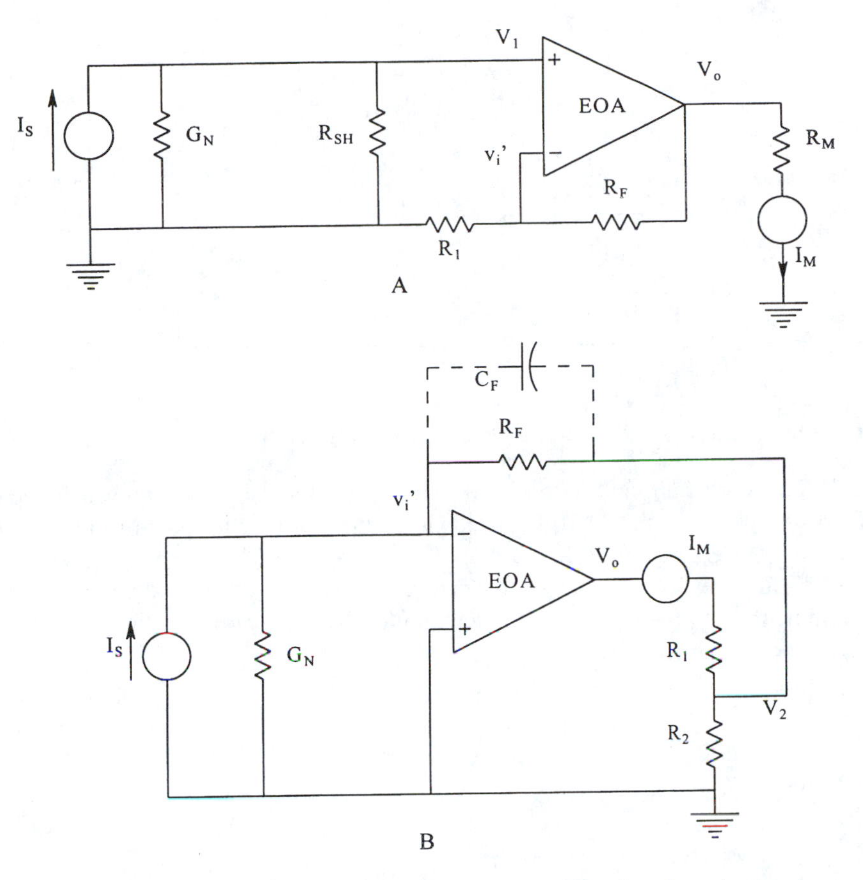

FIGURE 8.22 (A) Circuit of an electronic DC ammeter. The op amp amplifies the voltage developed across the shunt. (B) An op amp used as a transresistor. The summing junction appears at virtual ground.

with noise, there can be measurement errors caused by instrument DC bias current and offset voltage.

Figure 8.22 illustrates the two types of picoammeter circuits commonly used. In Figure 8.22A we see a simple shunt picoammeter in which most of the current to be measured flows through R_{SH}. Ordinarily, $R_{SH} << 1/G_N$, so $V_1 = I_S R_{SH}$. V_1 is amplified by the electrometer op amp (EOA) connected as a noninverting, DC millivoltmeter. In Figure 8.22B, the DC current under measurement is the input to an operational transresistor circuit, the output of which can easily be shown to be $V_O = -I_S R_F$. To obtain greater sensitivity, it would appear that R_F should be increased. There is a practical limit to the size of R_F set by its cost and accuracy. Resistors above 10^8 ohms are expensive, and as their size increases, their accuracy decreases. Also, when $R_F > 10^9$ ohms, the effects of shunt capacitance act to slow the overall response time of the operational transresistor (the parallel combination of R_F and C_F act like a low-pass filter).

8.3.2.1 Error Analysis of the Shunt Picoammeter

Three major sources of error in the shunt picoammeter circuit are discussed in this subsection. First, errors arise because of calibration errors in R_{SH}, R_1, and R_2. Second, errors arise because of the EOA's DC bias current, I_B, and offset voltage, V_{OS}. V_{OS} can be nulled to zero at one temperature, but V_{OS} will drift with temperature. I_B cannot be nulled in solid-state EOAs, and we must live with it; fortunately, I_B is very small for modern, IC EOAs, about 40 fA in the AD549L (Analog Devices, Norwood, MA). The V_{OS} tempco is 5 μV/°C for the AD549L. The third source of measurement errors comes from amplifier noise, and from Johnson (thermal) noise in associated resistors. The short-circuit input voltage noise, e_{na}, for an AD549L electrometer op amp is 4 μV peak-to-peak in

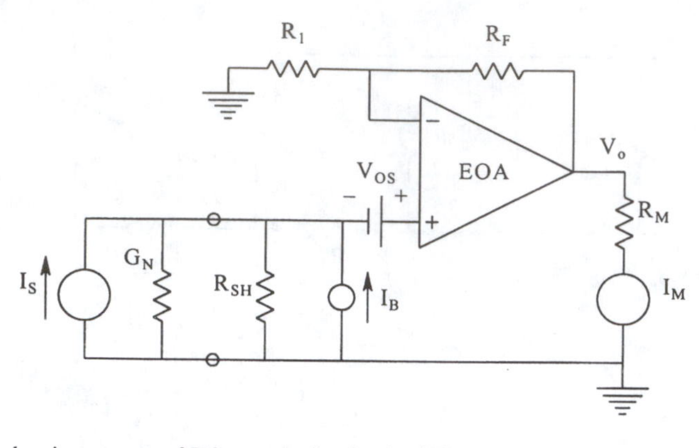

FIGURE 8.23 Circuit showing sources of DC error in the circuit of Figure 8.22A.

the 0.1 to 10 Hz bandwidth, and 35 nV/$\sqrt{\text{Hz}}$ above about 1 kHz. The equivalent input current noise is 0.36 fA ppk in the 0.1 to 10 Hz bandwidth, and can be considered to be 0.11 fA/$\sqrt{\text{Hz}}$ white noise from 0.1 Hz to 100 MHz.

Figure 8.23 illustrates the DC sources that can cause errors in reading the unknown DC current, I_S. Let us assume that $R_N >> R_{SH}$. The amplifier gain for the noninverting input is $K_V = (1 + R_F/R_1)$. Now the amplifier output voltage due to I_S is just:

$$V_{OM} = I_S R_{SH} K_V \tag{8.56}$$

The ouput voltage due to V_{OS}, I_B, and I'_B is:

$$V_{OE} = V_{OS} K_V + I_B R_{SH} K_V + I'_B R_F \tag{8.57}$$

Thus we can write an expression for the DC output signal to DC output voltage error ratio:

$$\frac{V_{OM}}{V_{OE}} = \frac{I_S}{V_{OS}/R_{SH} + I_B + I'_B R_F/(R_{SH} K_V)} \tag{8.58}$$

Clearly, the meter should be zeroed with input open-circuited to null out V_{OE}.

To examine the effect of resistor thermal noise and amplifier noises on the measurement of the DC I_S, we calculate the shunt picoammeter's signal-to-noise ratio. This calculation is made more difficult by the fact that the $e_{na}(f)$ root spectrum is in its 1/f region. Thus, to evaluate the total rms output noise, we must assume that all noise sources are independent and uncorrelated, and we must integrate the 1/f spectrum over the nominal frequency range used to make DC measurements; from 0.1 to 10 Hz. From the published $e_{na}(f)$ curve on the AD549L data sheet, we can write the short-circuit input, noise voltage power density spectrum as:

$$e_{na}^2(f) = \eta + b/f = 1.2250 \times 10^{-15} + (4.0833 \times 10^{-17})/f \ \text{MSV/Hz} \tag{8.59}$$

In a JFET-input amplifier, $i_{na}(f)$ generally has a flat power density spectrum (PDS) at low frequencies, and can be considered to be white.

Now the total mean-squared noise output voltage can be found in the 0.1 to 10 Hz bandwidth:

$$N_O = K_V^2 \int_{0.1}^{10} e_{na}^2(f)df + K_V^2 4kT \frac{R_{SH}R_N}{R_{SH}+R_N} B + K_V^2 \frac{(R_{SH}R_N)^2}{(R_{SH}+R_N)^2} i_{na}^2 B \quad 8.60$$

$$+ 4kTR_1(R_F/R_1)^2 B + 4kTR_F B \text{ MSV}$$

In Equation 8.60, $B = f_H - f_L = 10 - 0.1 = 9.9$ Hz, $i_{na} = 0.11$ fA/$\sqrt{\text{Hz}}$, and $K_V = (1 + R_F/R_1)$. Note that we have included Johnson noise from R_1 and R_F. The mean-squared output signal is:

$$V_{OS}^2 = I_S^2 \frac{(R_{SH}R_N)^2}{(R_{SH}+R_N)^2} K_V^2 \text{ MSV} \quad 8.61$$

The mean-squared signal-to-noise ratio can be written algebraically:

$$SNR_{OUT} = \frac{I_S^2/B}{\eta(G_{SH}+G_N)^2 + (b/B)\ln(f_H/f_L)(G_{SH}+G_N)^2 + i_{na}^2 + 4kT(G_{SH}+G_N) + 4kTR_F K_V^{-1}(G_{SH}+G_N)^2} \quad 8.62$$

A numerical value for the rms signal-to-noise ratio can be found from the expressions above. For simplicity, assume $R_N >> R_{SH}$, $R_{SH} = 10^6$ ohms, $R_F = 100$ K, $R_1 = 1$ K, and $T = 300$ K. The rms signal-to-noise ratio is found to be:

$$SNR_{(RMS)} = \frac{I_S}{4.2184 \times 10^{-13}} \text{ A/rmsA} \quad 8.63$$

Thus, noise limits the resolvable I_S to $>3 \times 4.22 \times 10^{-13}$ A for the parameters given above, to avoid excess random errors.

8.3.2.2 Error Analysis of the Feedback Picoammeter

As in the case of the shunt picoammeter, the EOA's offset voltage and bias current cause uncertainty in the measurement of I_S. From the equivalent circuit of Figure 8.24, we can write an expression for the worst-case, DC output signal to DC output error ratio:

$$\frac{V_{OM}}{V_{OE}} = \frac{I_S}{I_B \pm V_{OS}(G_N + G_F)} \quad 8.64$$

Here again, the systematic DC error due to V_{OS} and I_B can generally be observed with the input open-circuited, and be compensated for.

The error due to random noise includes the thermal noise from R_N and R_F, as well as the amplifier's e_{na} and i_{na}. The output mean-squared signal-to-noise ratio can be written as:

$$SNR_O = \frac{I_S^2 R_F^2}{[i_{na}^2 + 4kT(G_N + G_F)]R_F^2 B + [\eta B + b\ln(f_H/f_L)](R_N + R_F)^2/R_N^2} \quad 8.65$$

This equation can be reduced to a more interesting form:

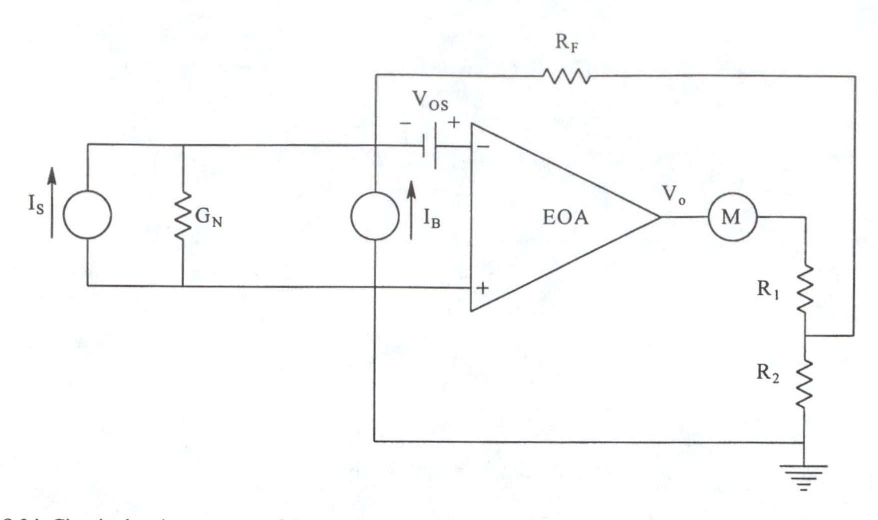

FIGURE 8.24 Circuit showing sources of DC error in the feedback picoammeter of Figure 8.22B.

$$SNR_O = \frac{I_S^2}{B\left\{i_{na}^2 + 4kT(G_N + G_F) + [\eta + (b/B)\ln(f_H/f_L)](G_N + G_F)^2\right\}} \quad 8.66$$

It is instructive to evaluate Equation 8.66 above for typical system parameters of: $\eta = 1.2250 \times 10^{-15}$ MSV/Hz, $b = 4.0833 \times 10^{-17}$ MSV, $R_F = 10^9$ ohms, $f_H = 10$ Hz, $f_L = 0.1$ Hz, $i_{na} = 0.11$ fA/$\sqrt{Hz}$, $4kT = 1.656 \times 10^{-20}$, and $R_N = 10^6$ ohms. We find:

$$SNR_O = \frac{I_S^2}{1.643 \times 10^{-25}} \quad 8.67$$

So for reliable, low-noise measurements, the DC I_S should be over three times the equivalent input rms noise current, or $I_S > 1.216 \times 10^{-12}$ A.

A problem that often arises in the design of feedback picoammeters is the loss of high-frequency response due to the parasitic shunt capacitance which appears in parallel with the very large R_F used. For example, if $R_F = 10^{11}$ ohms, a 1 pF stray capacitance in parallel with R_F will produce a transresistance given by:

$$\frac{V_O}{I_S}(s) = \frac{-R_F}{sR_F C_F + 1} \quad 8.68$$

The time constant is $\tau = 0.1$ s in this case. This would give little problem in the measurement of DC currents, but would be unsuitable for the measurement of AC currents. A circuit to speed up the response of the feedback picoammeter is shown in Figure 8.25. It is posed as a problem for the reader to find the condition on R_1, C_1 that will cancel the low-pass effect of C_F in parallel with R_F.

8.3.3 Magneto-Optic Current Sensors

The Faraday magneto-optic effect has been used in several designs for the measurement of high currents in power distribution systems. Typically, currents can range from tens of amperes to kiloamperes, and the conductors can be at kilovolts above ground potential, rendering the use of a shunt and millivoltmeter quite impractical. The Faraday magneto-optic effect makes use of the

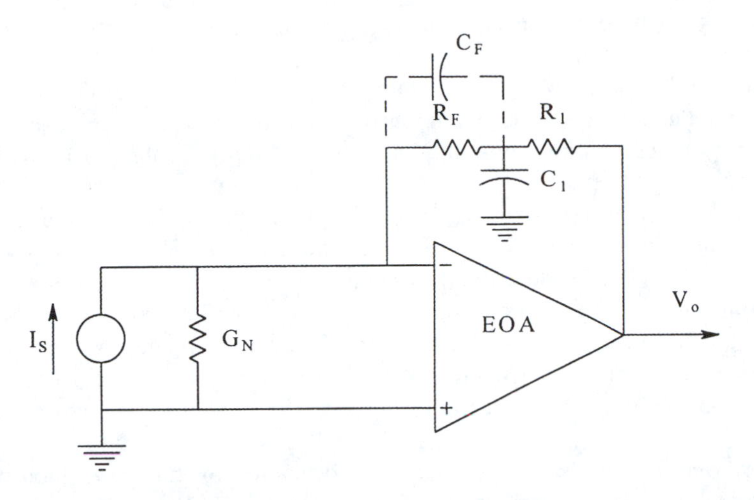

FIGURE 8.25 A circuit to speed up the response of the feedback picoammeter.

fact that a current-carrying conductor is surrounded by a solenoidal magnetic field whose magnitude is proportional to the current. The magnetic field, in turn, induces changes in the optical polarization properties of glass or fiber optic cable on which it acts. Thus by measuring the amount of optical rotation induced on linear polarized light emerging from the optical path in the B field, we can indirectly sense the current in the conductor. Details of several Faraday magneto-optic effect current sensors are described in Section 6.6.1 of this text.

8.4 AC VOLTAGE MEASUREMENTS

Any discussion of AC voltage measurements necessarily must cover a wide range of applications in which the amplitude and frequency of the voltage to be measured are of critical importance. AC voltage measurements can be divided into a number of frequency ranges: audio (including power line frequencies), ranging from around 10 Hz to about 30 kHz, and radio frequencies, including the following bands: low frequency (LF) 30 to 300 kHz, medium frequency (MF) 300 to 3000 kHz, high frequency (HF) 3 to 30 MHz, very high frequency (VHF) 30 to 300 MHz, ultra high frequency (UHF) 300 to 3000 MHz, super high frequency (SHF) 3 to 30 GHz, and extra high frequency (EHF) 30 to 300 GHz. Resolvable amplitudes range from nanovolts to hundreds of kilovolts. AC voltages are generally sinusoidal, but situations arise in which one must measure other periodic waveforms, including squarewaves, triangle and sawtooth waves, pulse trains, etc. We also have need to measure completely random (noise) voltages with various spectral contents and amplitudes. In the following sections, we will first describe various nonelectronic, electromechanical AC voltmeters, and then discuss the designs for electronic AC voltmeters.

8.4.1 Electromechanical AC Voltmeters

Most electromechanical AC voltmeters are designed to work at powerline or low audio frequencies, and at voltages ranging from about 1 V to tens of kilovolts. As in the case of DC voltmeters, basic electromechanical AC voltmeter design uses a series resistor to limit meter current to give full-scale deflection at full-scale applied voltage at the meter terminals. Almost all AC voltmeters have their scales calibrated in root mean square volts of a sinewave, regardless of the deflection mechanism of the meter movement. Only true rms-reading voltmeters will remain calibrated when a periodic voltage waveform other than a sinewave is the input to the meter. There are five common types of AC, powerline frequency voltmeter movements, some of which we have already described in Section 8.1.1. These include the dynamometer movement, the capacitor (electrostatic) movement, the iron vane movement, the rectifier-D'Arsonval design, and the vacuum thermocouple-D'Arsonval design. Of these AC voltmeters, only the thermocouple meter is also useful at radio frequencies, up to 50 MHz in some designs.

We defined voltmeter sensitivity for DC voltmeters as the reciprocal of the current required for full-scale meter deflection. A similar definition is useful for AC voltmeters, except it is the total meter impedance, which is found by multiplying the sensitivity in ohms per volt by the meter's full-scale voltage.

The *dynamometer AC voltmeter* is generally a large, insensitive, bench-top instrument used to measure sinusoidal voltages with frequencies from about 20 to 133 Hz. Accuracy is typically 0.2% of full-scale voltage, and meter deflection, assuming a linear spring whose torque is proportional to the deflection angle, and a mutual inductance which varies linearly with deflection angle, is proportional to the mean-squared current in the coils. Typical dynamometer voltmeters can be found with full-scale ranges from 30 to 600 Vrms. About 73.34 mA is required for full-scale deflection of the GE type P-3 electrodynamometer voltmeter. See Figure 8.8 for diagrams of a dynamometer meter.

Capacitor voltmeters are also large, insensitive, bench-top instruments which can be used to measure sinusoidal voltages with frequencies from 15 Hz to 300 kHz. Accuracy is typically 1% of full-scale voltage, and meter deflection, assuming linear spring torque with deflection angle and a linear increase of meter capacitance with deflection angle, is proportional to the mean-squared voltage across the meter. Typical electrostatic voltmeters can be found with full-scale deflections of 300 Vrms to 50 kVrms. It should be noted that the current at a given deflection of an electrostatic meter is given by Ohm's law, i.e., $I_M = V_M 2\pi f C_M$, where the meter capacitance, C_M, is itself an increasing function of meter deflection. The innards of an electrostatic voltmeter are shown in Figure 8.7. Note that while the deflection angle of the movement may be proportional to the MSV, the scale is calibrated in rms volts.

Iron vane voltmeters are found as large, insensitive, benchtop instruments which can be used to measure sinusoidal voltages with frequencies from about 15 to 133 Hz, with accuracies of about 0.5% of full-scale voltage. Iron vane AC panel voltmeters are also found. Iron vane meters can be of the repulsion type, as shown in Figure 8.26, or be of the attraction type, as shown in Figure 8.27. Typical meter efficiency is about 29 ohms/volt. This means that the meter draws about 34 mA at full-scale deflection. Because of their inefficiency, iron vane voltmeters are generally used to monitor powerline voltages, etc. Typical full-scale voltages for these meters range from 8 to 750 Vrms, depending on the application. Iron vane meter scales are generally nonlinear because of the geometry of the vanes and/or coils; the lower quarter of the scale is generally crowded.

Rectifier/D'Arsonval AC voltmeters generally find application in benchtop and pocket volt-ohm-milliameters (multimeters). Full-scale voltages typically range from 3 Vrms to over 1 kVrms. The circuit of a rectifier/D'Arsonval AC voltmeter is extremely simple, as shown in Figure 8.28. A series resistor, R_1, limits the meter current for a given full-scale voltage. A series capacitor, C_1, blocks DC from the meter circuit. A bridge rectifier converts AC current in R_1 to a full-wave rectified current with a DC average value. The D'Arsonval microammeter movement deflects proportional to the average current through its coil, hence deflection of the rectifier/D'Arsonval meter is proportional to the full-wave rectified, average value of the input voltage. AC voltmeter calibration is, of course, in terms of the rms of a sinewave input voltage. The D'Arsonval meter movement acts as a mechanical low-pass filter (see Equation 8.12), its deflection being proportional to the average current through its coil. The average value of a full-wave rectified sinewave is easily shown to be

$$\overline{i_M} = \overline{|I_{M(PK)} \sin(\omega t)|} = \frac{2}{\pi} I_{M(PK)} = \frac{2\sqrt{2}}{\pi} I_{M(RMS)} \qquad 8.69$$

If we assume at high input voltages that the rectifier diodes are ideal (i.e., they have zero forward resistance and voltage drop, and infinite reverse resistance), then we may write for full-scale meter deflection:

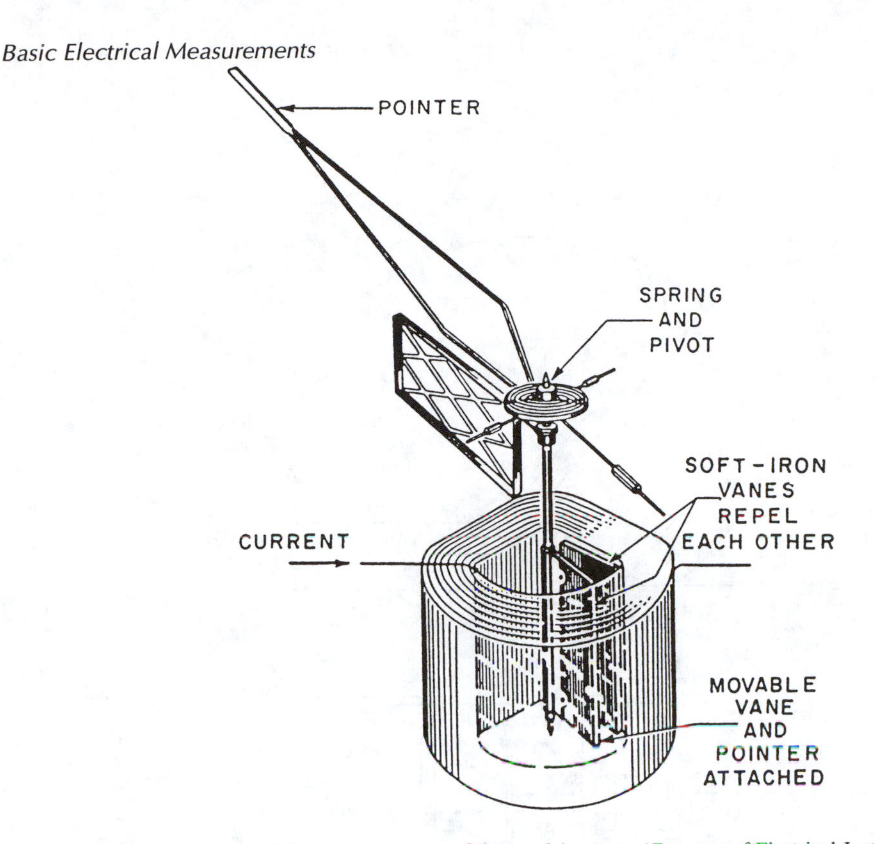

FIGURE 8.26 An iron vane AC meter movement of the repulsion type. (Courtesy of Electrical Instrument Service, Inc.)

$$I_{M(FS)} = \frac{2\sqrt{2}V_{S(RMS,FS)}}{\pi(R_1 + R_M)} \tag{8.70}$$

For a D'Arsonval microammeter with resistance R_M and full-scale current, $I_{M(FS)}$, we may solve for the R_1 required to make an AC voltmeter with a full-scale voltage reading of $V_{S(RMS,FS)}$, assuming ideal diodes:

$$R_1 = \frac{2\sqrt{2}V_{S(RMS,FS)}}{\pi I_{M(FS)}} - R_M \tag{8.71}$$

The AC, full-wave rectifier voltmeter's efficiency can be shown to be:

$$\eta_{ac} = \frac{1}{I_{M(RMS)}} = \frac{R_1 + R_M}{V_{S(RMS,FS)}} = \frac{\left(2\sqrt{2/\pi}\right)\left(V_{S(RMS,FS)}\right)/I_{M(FS)}}{V_{S(RMS,FS)}} = 0.9\,\eta_{dc} \tag{8.72}$$

The relations above are generally valid for AC input voltages above 10 Vrms. The scales of meters reading below 10 Vrms full-scale generally must be corrected for the finite forward and reverse resistances of the real diodes used in the meter rectifier. Also, practical diodes exhibit a voltage-dependent, depletion capacitance which is effectively in parallel with their reverse resistance. Thus, at high audio frequencies, considerable reverse diode current can flow, decreasing the average current through the meter movement, and causing frequency-dependent low meter readings. At powerline frequencies and low audio frequencies, diode capacitance is generally not a problem.

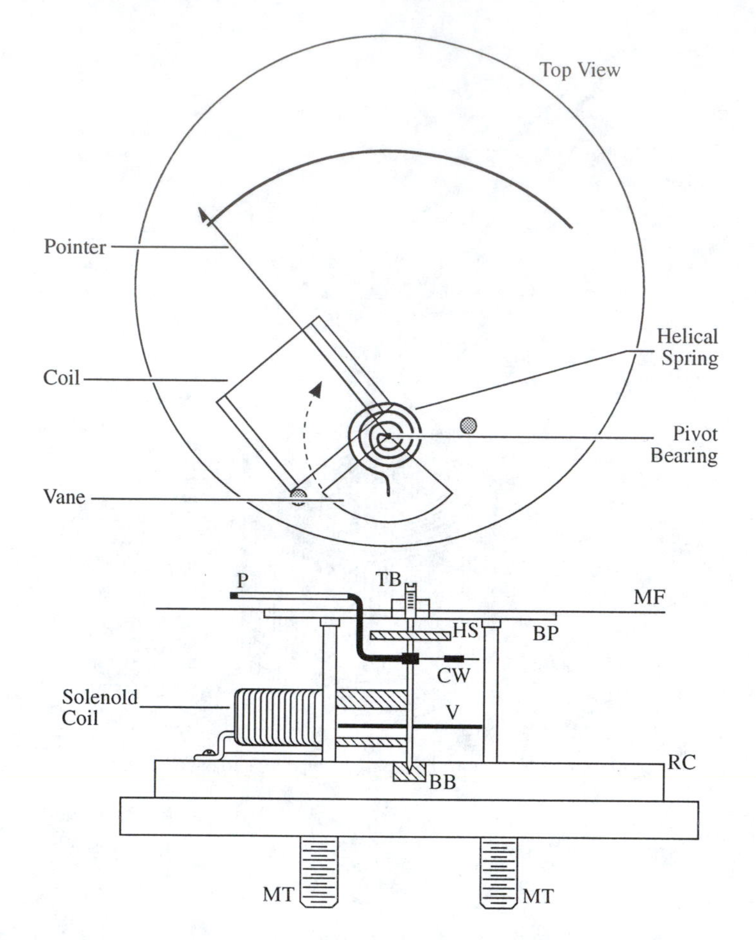

FIGURE 8.27 (A) Face view of an iron vane AC meter of the attraction type. (B) Side view of the attraction-type AC iron vane meter. Magnetic force pulls the vane into the current-carrying coil.

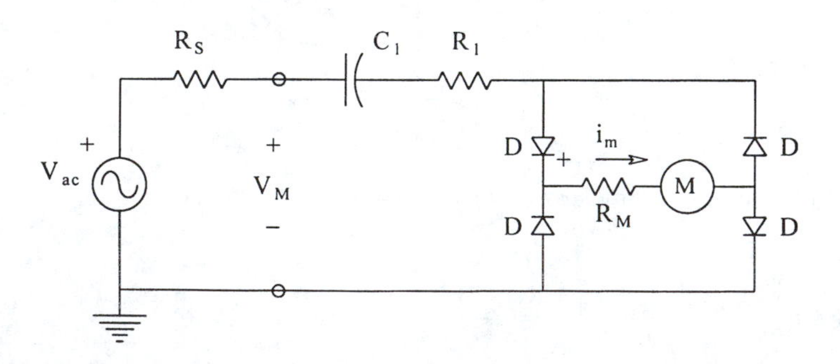

FIGURE 8.28 Schematic of a bridge rectifier AC voltmeter. A DC D'Arsonval meter movement is used.

A problem commonly encountered in using AC voltmeters of this sort is, how do they respond to nonsinusoidal voltages, such as square waves, triangle waves, etc.? To analyze this problem, we note first that any average or DC component in V_S is removed from the voltmeter input current, I_S, by the series capacitor, C_1. The average (DC) current through the D'Arsonval movement can be written as:

$$\overline{i_M(t)} = \frac{\overline{|v_S(t)|}}{R_1 + R_M} = \frac{\overline{|v_S(t)|}}{\left(2\sqrt{2/\pi}\right)\left(V_{S(RMS,FS)}\right)/I_{M(FS)}} \quad 8.73$$

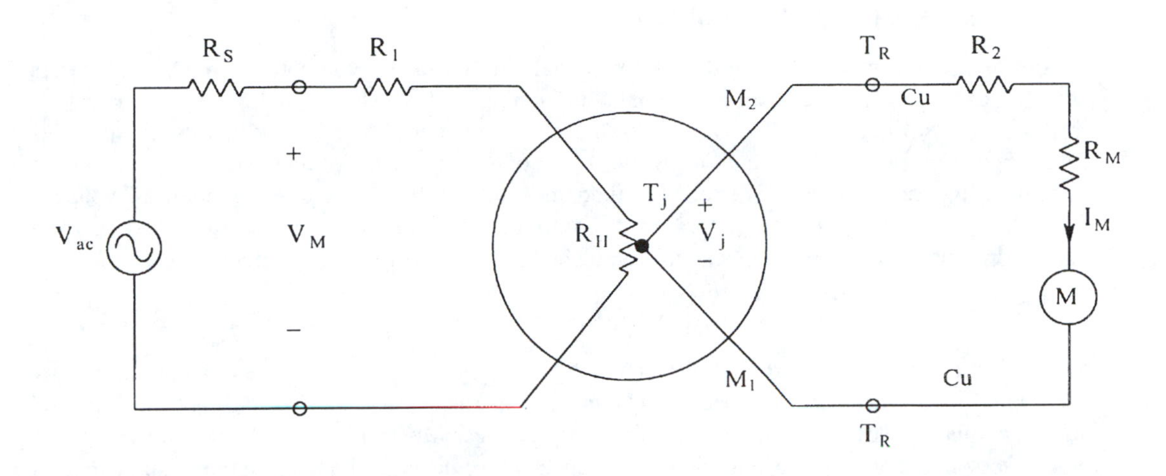

FIGURE 8.29 Schematic of a vacuum thermocouple AC voltmeter. Meter deflection is proportional to the mean-squared heater current.

The actual meter reading can be expressed as a fraction of the full-scale voltage:

$$V_{READ} = \frac{\overline{i_M(t)}}{I_{M(FS)}} V_{S(RMS,FS)} = \frac{\overline{|v_S(t)|}}{2\sqrt{2}/\pi} = 1.111\overline{|v_S(t)|} \tag{8.74}$$

Vacuum thermocouple/D'Arsonval AC voltmeters are the only type of electromechanical AC voltmeter which will remain calibrated for input voltage frequencies from 15–20 Hz to about 50 MHz. The reading of a vacuum thermocouple AC voltmeter is proportional to the true rms of the voltage being measured.

A schematic of a vacuum thermocouple AC voltmeter is shown in Figure 8.29. The input circuit consists of a series resistor, R_1, and a heater filament with resistance, R_H. For example, R_H for a Western Electric type 20K vacuum thermocouple is given as 750 ohms. The maximum safe heater current is given as 5 mArms. A Western Electric type 20D vacuum thermocouple has a heater resistance of 35 ohms, and a safe current of 16 mArms. An over-current of as little as 50% is likely to burn out the heater wire, so extreme care must be taken to prevent over-range inputs when using vacuum thermocouple/D'Arsonval AC voltmeters. The maximum rated EMF for the Western Electric vacuum thermocouples mentioned above is 15 mV (DC), and the thermocouple resistance is 12 ohms. If the thermocouple is connected to a 0 to 20 μA D'Arsonval meter, and the total series resistance in the circuit ($R_{TC} + R_M + R_2$) equals 750 ohms, then the meter will read full-scale when the heater current is at its rated maximum (3.6 mArms in the 20K vacuum thermocouple heater).

The DC current from the thermocouple, hence meter deflection, is proportional to the mean-squared heater current (or heater voltage). This is easy to see from Joule's law; the thermocouple EMF is proportional to heater temperature. Heater temperature is proportional to the electrical power dissipated in the heater, and the power, in turn, is proportional to the mean-squared heater current times the heater resistance. In order that the meter be calibrated in rms volts, a square root scale is painted on the meter face, rather than the linear one commonly found on D'Arsonval meters used in DC applications. Thus the lower end of the scale is crowded, and calibrations are seldom given below 20 V on a 100 VFS meter. Laboratory quality thermocouple meters commonly have accuracies of 0.5% of full-scale voltage.

8.4.2 Analog Electronic AC Voltmeters

In this section we will review the common analog types of AC voltmeters, i.e., AC voltmeters which make use of analog electronic signal conditioning, and which use analog output meters. AC voltages can be measured effectively from the microvolt range to tens of kilovolts. Frequencies can range from powerline to tens of gigahertz. No one type of AC voltmeter is effective over the ranges of frequency and voltage cited. The fundamental limit to the resolution of small AC voltages is noise; thermal noise from resistors, noise from amplifiers, and coherent noise picked up from the environment in spite of our best efforts to shield and guard the circuits being used.

8.4.2.1 AC Amplifier-Rectifier AC Voltmeters

In the first class of AC electronic voltmeter, the AC voltage under measurement is conditioned by a flat band-pass, RC amplifier of accurately known gain. The amplified AC voltage is then converted to DC which causes meter indication of the AC voltage. Most electronic AC electronic analog voltmeters make use of the common D'Arsonval movement as the output indicator, either in conjunction with some type of electronic rectifier to convert amplified AC voltage to DC current, or with a vacuum thermocouple to convert the mean-squared value of the amplified AC voltage to a DC current to drive the meter. A block diagram of the AC amplifier-rectifier type of AC voltmeter is shown in Figure 8.30. The amplifier's gain is generally fixed and large (e.g., 3.33×10^3), and a precision, frequency-compensated attenuator is used to reduce input signals which are above the minimum full-scale voltage (e.g., 3 mV) to a level that will not saturate the amplifier. Input resistance is thus set by the attenuator, and is typically 10 or 11 megohms. Because the amplifier's AC output is rectified to drive the D'Arsonval meter, the rectifier's frequency response generally sets the upper useful frequency of this type of meter. Such meters are typically used at audio frequencies, and some versions are useful up to 10 Mhz.

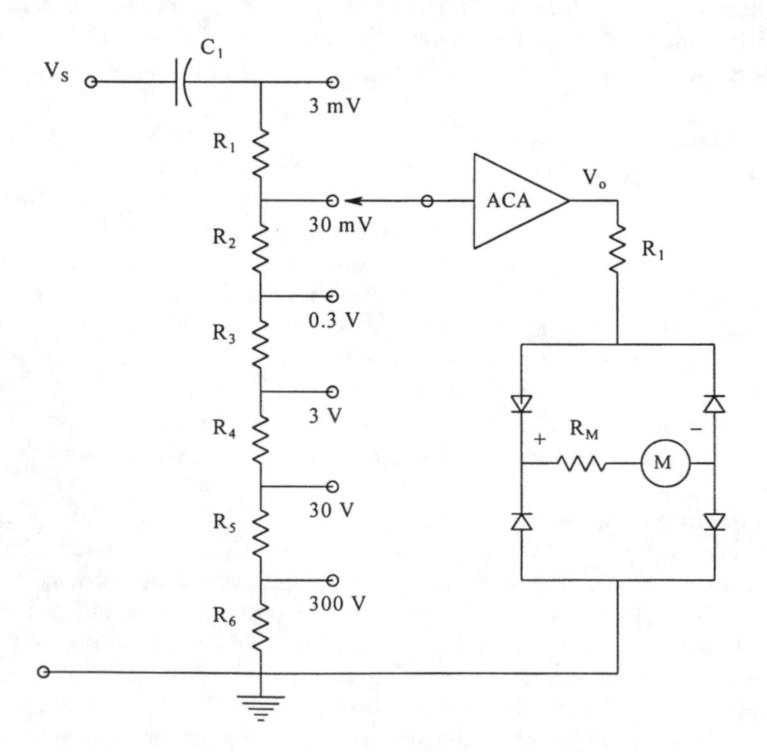

FIGURE 8.30 Schematic of an AC amplifier-rectifier type AC voltmeter.

To overcome the nonlinearity of the rectifier diodes at low output voltage levels, it is possible to use an operational rectifier circuit, shown in Figure 8.31, instead of a passive diode bridge. Such a circuit will improve meter linearity at the low end of the meter scale; however, the reverse capacitance of the diodes, and the finite slew rate and gain-bandwidth product of the op amp provide fundamental limitations to the operating frequency of the operational rectifier. Graeme (1974) gives several circuit modifications for operational rectifiers which improve their performance.

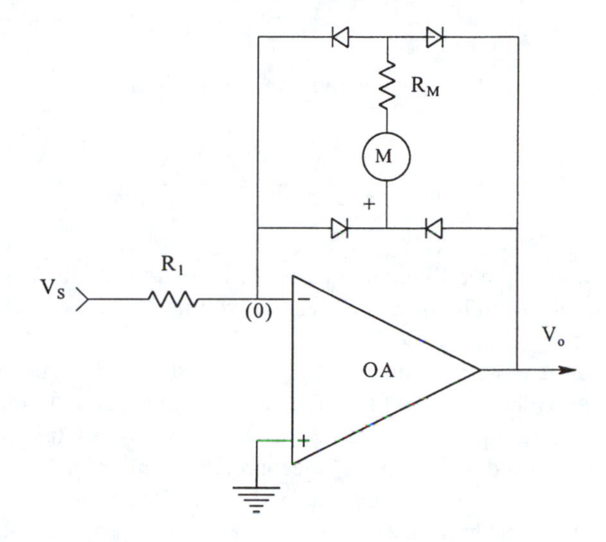

FIGURE 8.31 An operational rectifier circuit that provides linear rectification of the audio frequency AC input current even at low values of V_o.

8.4.2.2 Peak-Reading Electronic AC Voltmeters

The second type of AC electronic voltmeter (peak reading) uses a low-capacitance diode to half-wave rectify the AC voltage under measurement. The rectified AC is smoothed by a capacitor low-pass filter, and then is amplified by a drift-free, DC amplifier with known gain, such as a chopper-stabilized or commutating auto-zero circuit. The DC output of the amplifier causes the deflection of the voltmeter's D'Arsonval meter. The rectifier used in the peak-reading AC voltmeter can be a vacuum thermionic diode, although low-capacitance semiconductor diodes can also be used. The advantage of this design lies in the measurement of RF voltages. Because the AC voltage under measurement is converted to DC at the front end, there is no need for an expensive, high-gain, high-bandwidth linear amplifier to condition the AC voltage. There is need for a low-gain, drift-free, DC amplifier, however. Because of the inherent nonlinearity of semiconductor and thermionic diodes at low voltages, peak-reading AC voltmeters are generally used for AC input voltages above 1 Vpk. Below 1 Vpk, special scale calibration must be used to compensate for diode nonlinearity, which can be shown to be square-law for input voltages below 50 mV (Oliver and Cage, 1971). Most peak-reading AC voltmeters have input ranges of 1 Vpk full-scale to 300 Vpk full-scale. Their frequency response ranges from powerline to over 300 MHz. Indeed, some special microwave diodes have been used to measure peak voltages in waveguides at frequencies up to 40 GHz (Oliver and Cage, 1971).

There are two common forms of peak-detecting AC voltmeter, illustrated in Figure 8.32. In Figure 8.32A, we see a DC peak detector in which the capacitor charges through the diode to a DC voltage equal to the positive peak value of $v_S(t)$. Thus, the output voltage across the diode is $v_1(t) = [v_s(t) - V_{PK+}]$. For the case of a sinewave input with zero mean, the average voltage across the diode is $v_1 = V_{S(PK)}$. This DC level is recovered by low-pass filtering, amplified, and used to drive the D'Arsonval indicating meter. Thus meter deflection is proportional to the peak value of the zero mean sinewave input. In more general terms, for an arbitrary input $v_S(t)$, the mean value of $v_1(t)$ can be shown to be $\overline{v_1(t)} = -(+\text{peak } v_S - \overline{v_S})$. To demonstrate this property, consider the pulse train waveform in Figure 8.32C. Its peak value is V_P, and its average value is:

$$\overline{V_S} = \frac{V_P T_1 - V_2 T_2}{T_1 + T_2} \quad 8.75$$

So the average value of $v_1(t)$ is the – peak-above-average value of $v_S(t)$:

$$\overline{V_1} = -\left[V_P - \frac{V_P T_1 - V_2 T_2}{T_1 - T_2}\right] = -\frac{T_2(V_P + V_2)}{T_1 + T_2} \quad 8.76$$

Peak-reading voltmeters are generally calibrated in terms of the rms of a sinusoidal input. The input impedance of the diode probe of a peak-reading voltmeter can be shown to be $Z_{IN} = R_L/3$, where in general, $R_S << R_L$ (Angelo, 1969).

A second type of peak-reading electronic AC voltmeter is shown in Figure 8.32D. In this simple circuit, the capacitor acts as the low-pass filter. The capacitor charges up to the positive peak value of $v_S(t)$. Hence, the average output of the detector is $\overline{v_1(t)} = V_{S(PK)}$. Charge leaks off C through R_L with a time constant adjusted to be about 1/2 second. The input impedance of the peak rectifier can be shown to be $Z_{IN} = R_L/2$ at frequencies where $1/\omega C << R_L$ (Angelo, 1969). A peak-to-peak-reading AC voltmeter can be made from a positive and a negative peak rectifier connected to a DC differential amplifier, as shown in Figure 8.33.

8.4.2.3 True rms AC Voltmeters of the Feedback Type

The circuit for this AC voltmeter is shown in Figure 8.34. In the heart of the circuit are two matched vacuum thermocouples. The heater current of the input thermocouple is derived by AC amplification (or attenuation) of the AC input signal, V_S. The heater current of the feedback thermocouple comes from the DC current through the D'Arsonval meter movement. We will show below that the DC meter current is proportional to the true rms value of $v_S(t)$ when certain conditions on circuit parameters are met.

After attenuation and/or amplification by the linear AC amplifier, an AC signal proportional to $v_S(t)$, v_{HS}, is applied to the input thermocouple heater, causing a rise in its temperature, and a resultant DC output voltage, V_{TS}, from the thermocouple junction attached to the input heater. The input to the high-gain DC differential amplifier is the difference between V_{TS} and V_{TR}, V_{TD}. V_{TD} is amplified and produces a DC output voltage, V_{OD}. The DC meter current is thus $I_M = V_{OD}/(R_O + R_M + R_H)$. From inspection of the circuit, we can write the following equations:

$$R = R_O + R_M + R_H \quad 8.77$$

$$V_{TS} = K_T \overline{v_{HS}^2} \quad 8.78$$

$$V_{TR} = K_T V_{HR}^2 = K_T (R_H I_M)^2 \quad 8.79$$

In the equations above, R_O is the output resistance of the DC differential amplifier, K_D is its gain, R_H is the thermocouple heater resistance, R_M is the DC microammeter's resistance, and K_T is the

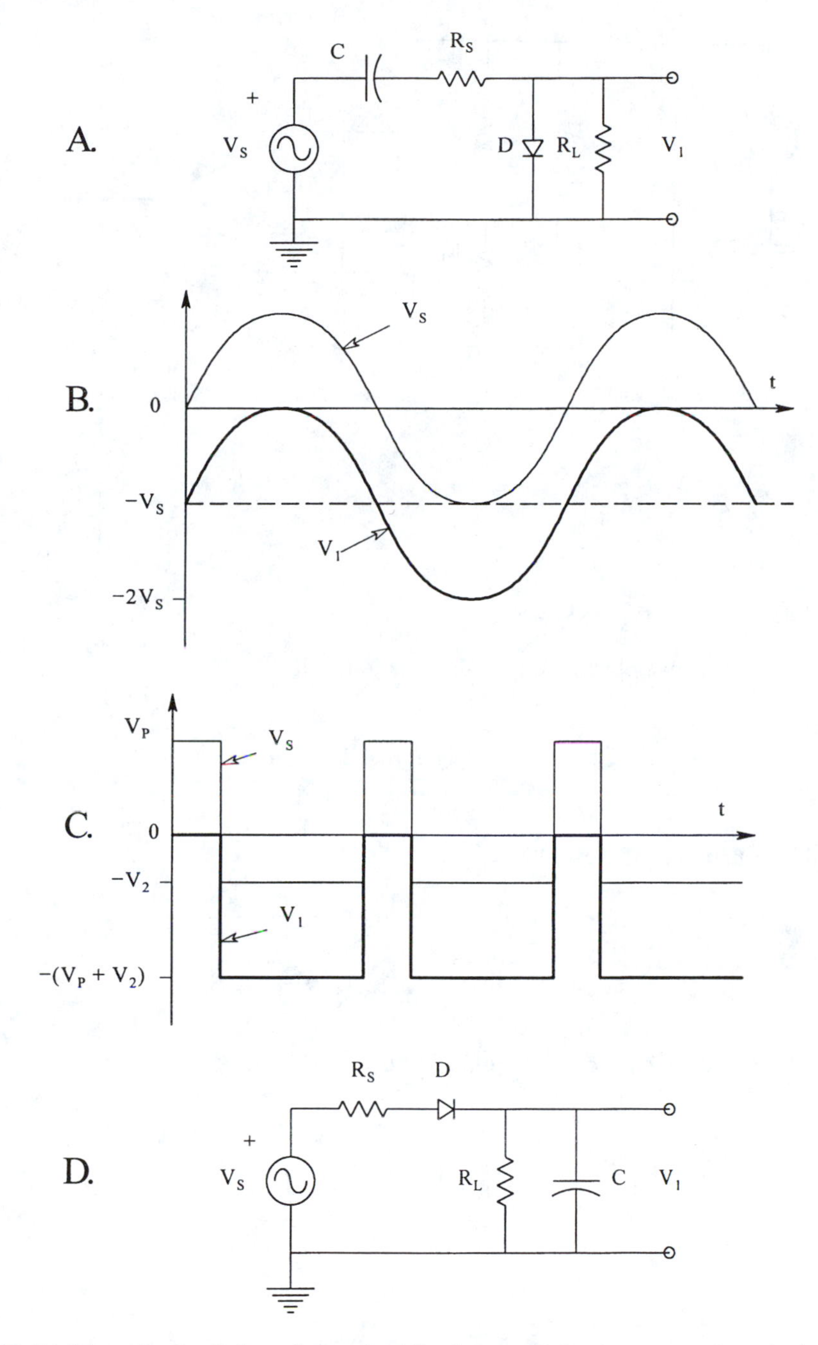

FIGURE 8.32 (A) Schematic of a diode, peak-detecting AC voltmeter. (B) Steady-state waveforms in the peak detector given a sinewave input. (C) Steady-state waveforms in the peak detector given a pulse train input. (D) A second type of peak-reading AC voltmeter.

thermocouple's transfer gain in volts/mean-squared volt. The DC meter current can be written from Ohm's law:

$$I_M = \frac{K_D\left(V_{TS} - V_{TR}\right)}{R} = \frac{K_D K_T\left(\overline{v_{HS}^2} - V_{HR}^2\right)}{R} \qquad 8.80$$

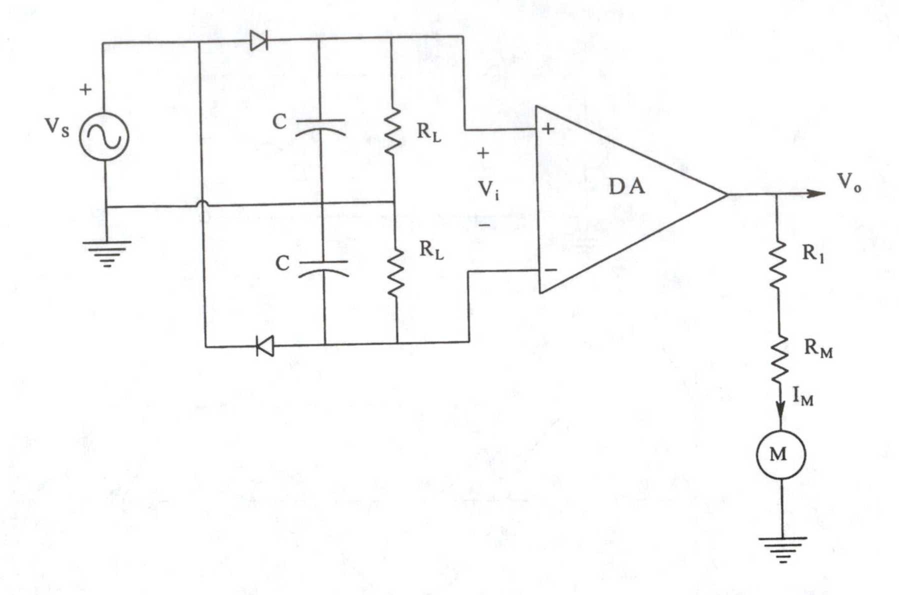

FIGURE 8.33 Schematic of a peak-to-peak reading AC voltmeter.

If we substitute for V_{HR}, we obtain a quadratic equation in I_M:

$$I_M^2 + I_M \frac{R}{K_D K_T R_H^2} - \frac{\overline{v_{HS}^2}}{R_H^2} = 0 \qquad 8.81$$

Solution of this equation is of the form:

$$I_M = -\frac{R}{2K_D K_T R_H^2} \pm \frac{1}{2}\sqrt{\frac{R}{\left(K_D K_T R_H^2\right)^2} + \frac{4\overline{v_{HS}^2}}{R_H^2}} \qquad 8.82$$

We extract the $4/R_H^2$ factor from Equation 8.82 above, and write,

$$I_M = -\frac{R}{2K_D K_T R_H^2} \pm \frac{1}{R_H}\sqrt{\frac{R_2}{\left(2K_D K_T R_H\right)^2} + \overline{v_{HS}^2}} \qquad 8.83$$

Now if we make

$$V_{HS(RMS)} \gg \frac{R}{2K_D K_T R_H}$$

I_M is approximately given by:

$$I_M \cong V_{HS(RMS)}/R_H \qquad 8.84$$

and V_{OD} is

$$V_{OD} = \left(V_{HS(RMS)}/R_H\right)\left(R_M + R_H\right) \text{ DC volts} \qquad 8.85$$

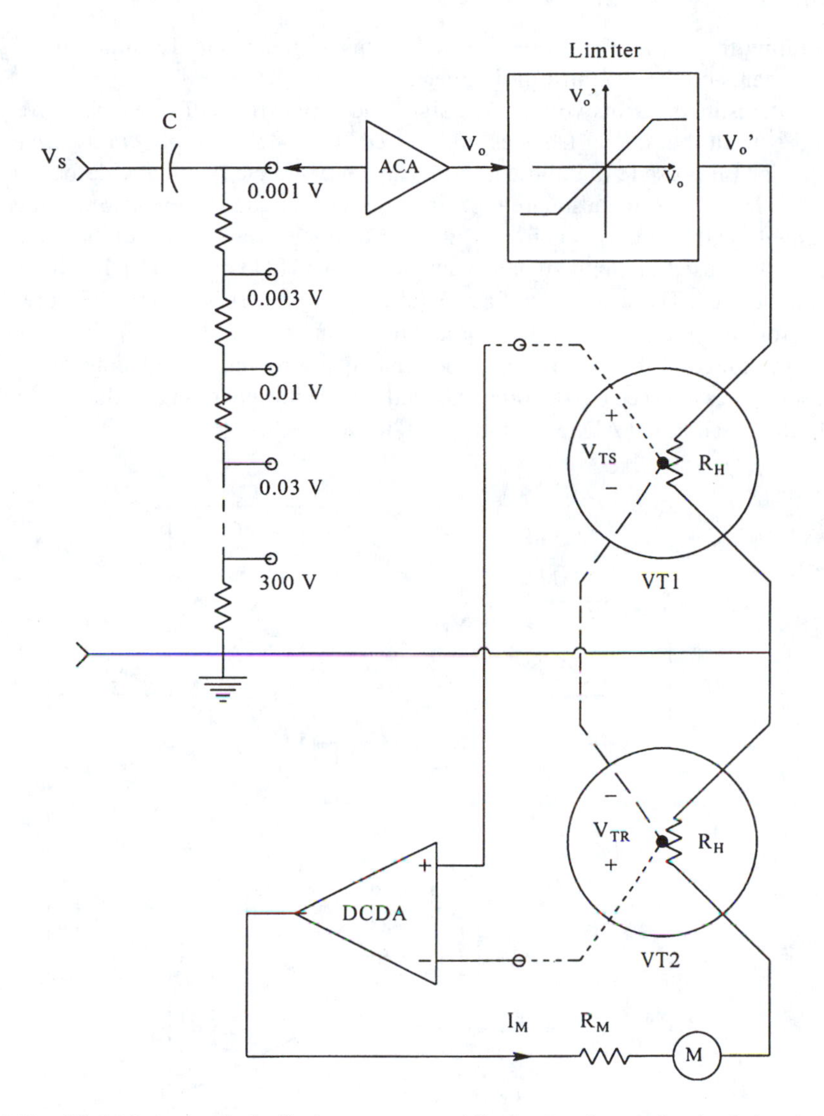

FIGURE 8.34 Simplified schematic of a feedback-type, true rms AC voltmeter. See text for circuit description.

In a typical vacuum thermocouple, $I_{M(MAX)}$ through the heater is 16 mA, and the heater resistance is 35 ohms. Thus, $V_{SH(RMS,MAX)}$ is 0.56 V. Now if we let $K_D = 10^6$, $K_T = 3 \times 10^{-2}$, $R_M = 250$ ohms, and $R_O = 50$ ohms, then $V_{HS(RMS,MIN)}$ would be around 1 mV. This gives the meter around a 560 to 1 dynamic range over which meter deflection is proportional to the true rms value of $v_S(t)$.

The Hewlett-Packard (San Jose, CA) venerable model 3400A true rms voltmeter uses a design similar to the one analyzed above.

8.4.2.4 True rms AC Voltmeters Using the Direct Conversion Approach

In Section 2.5.4, we saw that there are several dedicated ICs available that will perform true rms conversion on AC signals. The DC output of these ICs can be conditioned and used to drive a D'Arsonval DC microammeter. The maximum frequency response for a specified percent error in conversion is generally obtained for maximum AC input. For example, The AD536 rms converter (Analog Devices, Norwood, MA) has less than 1% error at frequencies up to about 140 kHz when its input is a 7 Vrms sinewave, and when the input is a 10 mVrms sinewave, less than 1% error at frequencies up to 6 kHz. Thus, to make a true rms voltmeter with ICs of this type, we need to adjust the input gain or attenuation such that the input to the IC is always as large as possible without saturating the circuit. One way of ensuring this accuracy and dynamic range is to always make measurements on the top third of the meter scale.

For some interesting true rms voltmeter designs and applications of true rms converter ICs, the reader should consult the Analog Devices' *RMS to DC Conversion Application Guide*, 1983.

Note that it is also possible to build a true rms analog voltmeter from a wide-bandwidth analog multiplier, a low-pass filter, and a square root circuit. Such a circuit is capable of a wider bandwidth than is obtainable with a true rms converter IC. For example if a four-quadrant analog multiplier such as an AD834 is used, an input signal bandwidth from 5 Hz to over 20 MHz with peak voltages up to 10 V can be used. The schematic of a squarer-averager-rooter is shown in Figure 8.35. Again, maximum performance in terms of low error and maximum bandwidth is obtained with the use of the full dynamic range of the analog multiplier and square rooter. The analog multiplier used in the square-rooter does not need outstanding bandwidth since it is operating on dc, so a high-accuracy analog multiplier such as the AD534 can be used here.

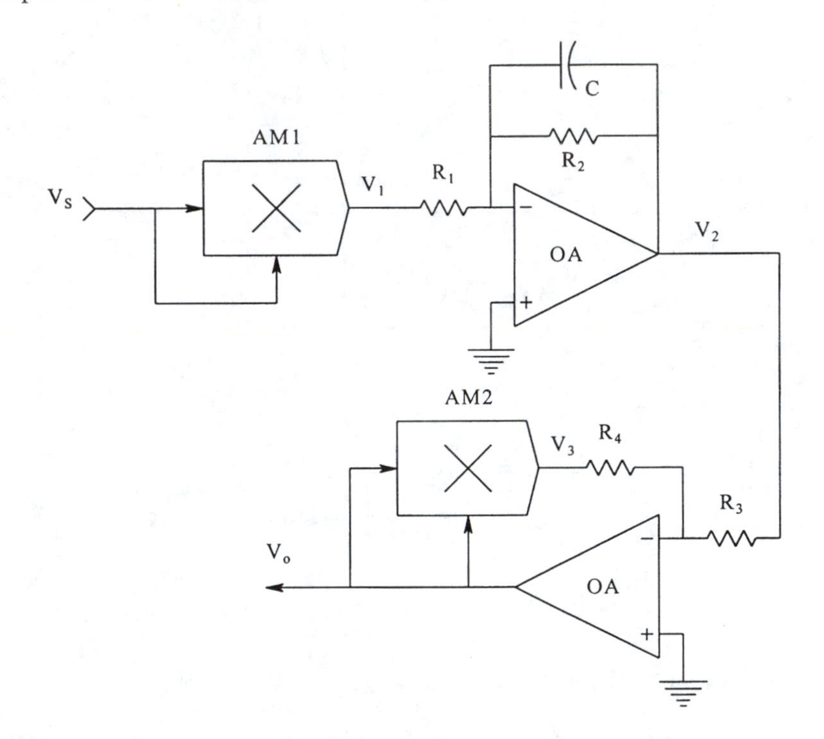

FIGURE 8.35 Schematic of a true rms converter using an analog multiplier as a squarer, an op amp low-pass filter averager, and a square root circuit. A DC output is obtained.

8.4.3 Measurements of Amplifier Noise Voltages, Noise Factor and Figure

In making noise voltage measurements, one must be certain that the noise being measured is indeed uncontaminated by coherent interference or hum from sources external to the system under measurement. The best way to gain this assurance is to visualize the noise spectrum in question on a modern, sampling, fast fourier transform (FFT) spectrum analyzer. Such instruments generally display a one-sided, root power spectrum whose units are rms $\text{volts}/\sqrt{\text{Hertz}}$. The presence of any spike-like peaks generally indicates the presence of contamination of the desired spectrum by coherent interference. Obviously, steps must be taken to remove extraneous coherent voltages from the desired spectrum in order to obtain valid noise measurements.

The quantitative measurement of noise and noise figure requires the use of true rms AC voltmeters, such as we have described above. Noise measurements also require a knowledge of amplifier gain, and the equivalent hertz noise bandwidth of band-pass filters conditioning the noise from the noise source(s). Noise measurements can be subdivided into two categories: the use of narrow

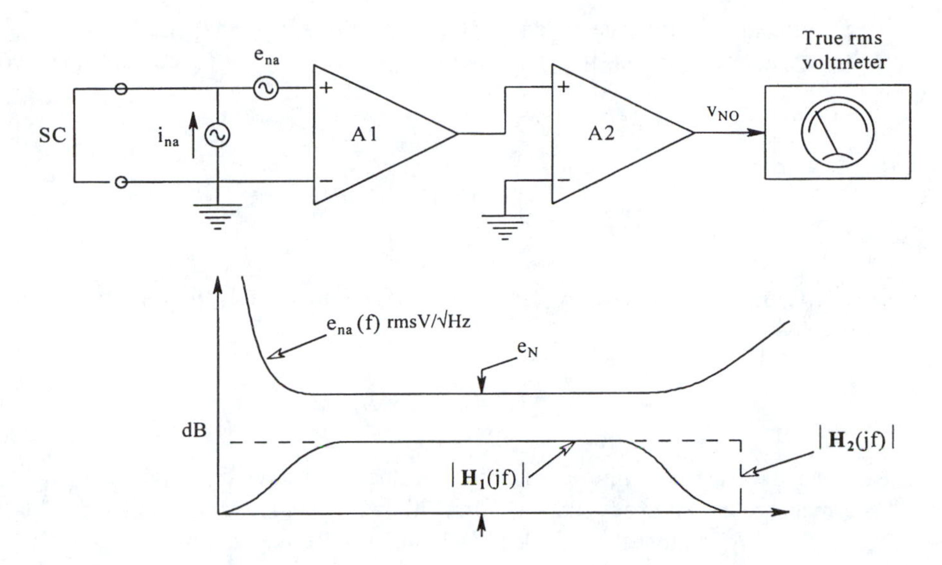

FIGURE 8.36 Circuit used to measure the white noise portion of e_{na} from A1 The noise passband excludes the 1/f region of the e_{na}^2 power density spectrum.

band-pass filters to characterize the power density spectrums of noise sources, and broadband noise measurements over the entire bandwidth of a signal conditioning system. We are often interested in measuring the equivalent input noise sources of an amplifier, $e_{na}(f)$ and $i_{na}(f)$. $e_{na}(f)$ is the equivalent, short-circuited input noise of the amplifier. It is a root power spectrum, and its units are rms volts per root Hz. $e_{na}(f)$ generally increases at low frequencies, is flat in a midrange, and rises again at high frequencies. Ideally, in order to obtain a maximum signal-to-noise ratio (SNR) at the amplifier's output, we would like to condition the input signal over a range of frequencies where $e_{na}(f)$ is a minimum.

Figure 8.36 illustrates a system which can be used to measure the flat white noise portion of the $e_{na}(f)$ root spectrum. It is often necessary to follow amplifier A1 with a second, high-gain, low-noise amplifier, A2, which has a broader bandwidth than A1 That is, the band-pass characteristics of A1 dominate the system, and determine B, the equivalent system Hz noise bandwidth. It can be shown (Northrop, 1990) that as long as the midband gain of A1 is greater than 5, any noise from A2 is negligible in the measurement process. If we assume $e_{na}(f)$ is flat over B, then we can write an expression for the noise power density spectrum at the system output, $S_{NO}(f)$.

$$S_{NO}(f) = e_N^2 \left|\mathbf{H}_{12}(j2\pi f)\right|^2 \tag{8.86}$$

The rms meter must have a flat frequency response which is wider than that of A1. The rms noise read by the meter is given by:

$$v_{NO(RMS)} = \sqrt{\int_0^\infty e_N^2 \left|\mathbf{H}_{12}(j2\pi f)\right|^2 df} = e_N K_{V12} \sqrt{B} \tag{8.87}$$

where

$$B = \int_0^\infty \left|\mathbf{H}_{12}(j2\pi f)\right|^2 df \Big/ K_{v12}^2 \ \text{Hz} \tag{8.88}$$

and K_{V12} is the midband gain of the cascaded amplifiers, A1 and A2, and $\mathbf{H}_{12}(j2\pi f)$ is the frequency response of the cascaded amplifiers. If $e_{na}(f)$ is not flat over B, then the rms output noise voltage is given by:

$$v_{NO(RMS)} = \sqrt{\int_0^{\infty} e_{na}^2(f)\left|\mathbf{H}_{12}(j2\pi f)\right|^2 df} \qquad 8.89$$

To measure $e_{na}(f)$ in its low-frequency, 1/f region, we note that the e_{na} power density spectrum can be described by:

$$e_{na}^2(f) = b/f + e_N^2 \ \text{MSV/Hz} \qquad 8.90$$

Now we use a narrow band-pass filter to examine the spectral power of $e_{na}^2(f)$. We assume that the filter has a rectangular band-pass characteristic with peak gain K_F and noise bandwidth B Hz. Refer to Figure 8.37. We see that the mean-squared output noise is given by:

$$\begin{aligned} v_{NO}^2 &= \int_{fo-B/2}^{fo+B/2} \left(\frac{b}{f}\right) K_F^2 K_{V12}^2 \, df + e_N^2 K_F^2 K_{V12}^2 B \\ &= K_F^2 K_{V12}^2 \left[b \ln\left(\frac{2Q+1}{2Q-1}\right) + e_N^2 B \right] \end{aligned} \qquad 8.91$$

where $Q = f_o/B$. From Equation 8.91 above, we can solve for the b coefficient:

$$b = \frac{\left[v_{ON}^2/\left(K_F^2 K_{V12}^2\right) - e_N^2 B\right]}{\ln\left(\frac{2Q+1}{2Q-1}\right)} \ \text{Hz} \qquad 8.92$$

In certain modern, low-noise op amps, e_N is about $3\text{nV}/\sqrt{\text{Hz}}$, and b is about 27×10^{-18} MSV.

Although in many applications, the amplifier's equivalent input noise current, $i_{na}(f)$, has negligible effect on the output SNR, there are instances where its value does have an effect. To measure the effect of $i_{na}(f)$, we replace the input short-circuit with a resistor, R, at temperature T K. Now there are three sources of noise voltage at the A1 input: e_{na}, the Johnson noise from R (assumed white), and the noise $i_{na}R$. Using the procedures from Chapter 3, we can write an expression for the mean-squared output noise:

$$v_{NO}^2 = \int_0^{\infty} \left[e_{na}^2(f) + 4kTR + R^2 i_{na}^2(f)\right] \left|\mathbf{H}_{12}(j2\pi f)\right|^2 df \qquad 8.93$$

Now if we assume both e_{na} and i_{na} are white over B, Equation 8.93 can be simplified to:

$$v_{NO}^2 = \left[e_{na}^2 + 4kTR + R^2 i_{na}^2\right] K_{V12}^2 B \qquad 8.94$$

Since we know K_{V12}, B, e_{na}, R, T, and B, it is easy to calculate i_{na}^2:

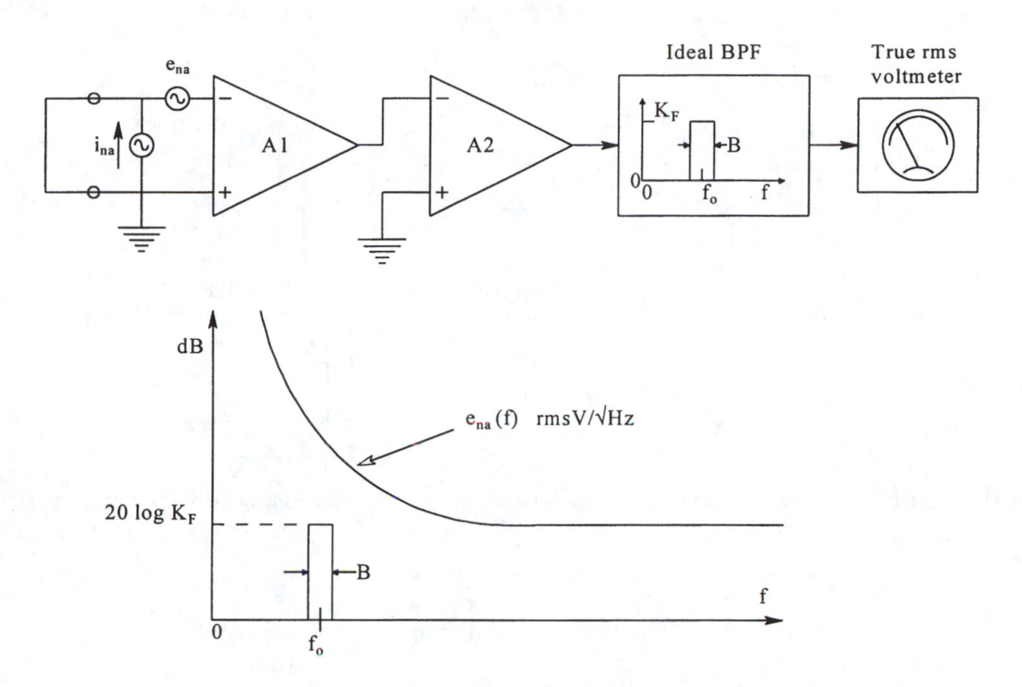

FIGURE 8.37 System used to measure the 1/f portion of the e_{na}^2 power density spectrum. See text for explanation.

$$i_{na}^2 = \left(\frac{v_{NO}^2}{K_{V12}^2 B} e_{na}^2 + 4kTR \right) \Big/ R^2 \qquad 8.95$$

Our final topic in this section will be to describe how we can measure an amplifier's *spot noise figure*, NF. Spot noise factor, F, is defined as the ratio of the mean-squared signal-to-noise ratio at the amplifier's input to the mean-squared signal-to-noise ratio at the amplifier's output. An ideal, noiseless amplifier would have an F = 1. An amplifier's noise figure is defined as 10 $\log_{10}$ (noise factor). An amplifier's spot noise factor is a useful measure of its noise performance under various conditions of source resistance and frequency.

A circuit which can be used to measure spot noise figure of amplifier A1 is shown in Figure 8.38. A sinusoidal voltage source is used at the input to enable the measurement of F_{SPOT}. A narrow band-pass filter (NBPF) with peak gain, K_F, and equivalent Hz noise bandwidth, B, is used to select the range of noise frequencies for the measurement. A broad-band, true rms responding meter is used to read the total rms voltage at the amplifier output. With the switch in position A, the meter reading squared can be written

$$\begin{aligned} v_{NOA}^2 &= \int_0^\infty K_{V12}^2 \left|\mathbf{H}_F(j2\pi f)\right|^2 \left[e_{na}^2(f) + i_{na}^2 R_S^2 + 4kTRS\right] df \\ &= \int_0^\infty K_{V12}^2 \left|\mathbf{H}_F(j2\pi f)\right|^2 e_{na}^2(f)\, df + K_{V12}^2 K_F^2 B\left(i_{na}^2 R_S^2 + 4kTRS\right) \end{aligned} \qquad 8.96$$

Note that the bandwidth of the meter and the amplifier are assumed to be greater than and overlap that of the NBPF, so that the filter's frequency response curve is dominant. Now the switch is set to position B, and the sinusoidal source is adjusted to the center frequency of the NBPF. Its amplitude is adjusted until:

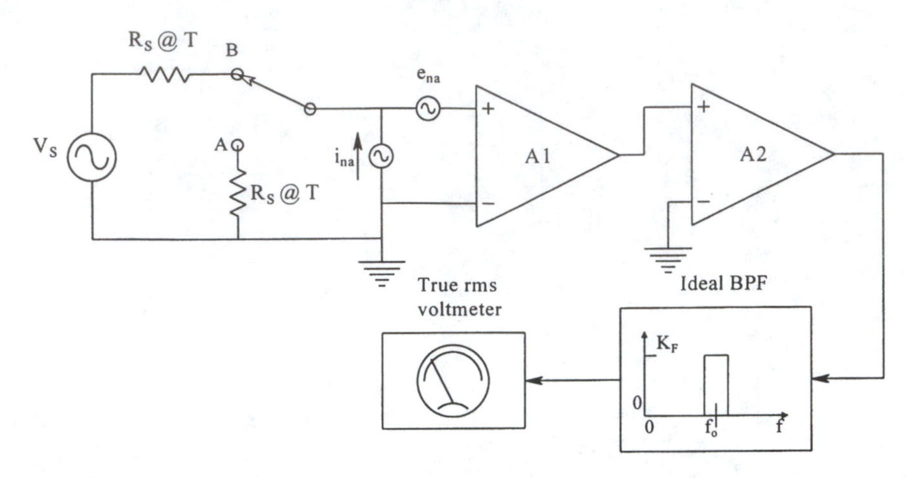

FIGURE 8.38 Circuit used to measure the spot noise figure of amplifier A1 over a wide range of frequencies. Amplifier A2 is assumed noiseless.

$$v_{NOB}^2 = 2v_{NOA}^2 = K_F^2 K_{V12}^2 (V_S^2/2) + \int_0^\infty K_{V12}^2 |\mathbf{H}_F(j2\pi f)|^2 e_{na}^2(f)\,df + K_{V12}^2 K_F^2 B(i_{na}^2 R_S^2 + 4kTR_S) \tag{8.97}$$

Now the mean-squared SNR in bandwidth B at the amplifier's input is:

$$SNR_{IN} = \frac{\overline{v_S^2}}{4kTR_S B} \tag{8.98}$$

The SNR_{OUT} is:

$$SNR_{OUT} = \overline{v_S^2} \Big/ \left[\int_0^\infty K_{V12}^2 |\mathbf{H}_F(j2\tilde{a}f)|^2 e_{na}^2(f)\,df + K_{V12}^2 K_F^2 B(i_{na}^2 R_S^2 + 4kTR_S) \right] \tag{8.99}$$

Now the noise factor,

$$F_{SPOT} \equiv \frac{SNR_{IN}}{SNR_{OUT}} \tag{8.100}$$

can be written, after some algebra,

$$F_{SPOT} = \frac{\int_0^\infty |\mathbf{H}_F(j2\pi f)|^2 e_{na}^2(f)\,df / K_F^2 + B(i_{na}^2 R_S^2 + 4kTR_S)}{4kTR_S B} \tag{8.101}$$

If we combine Equations 8.96 and 8.97, it is easy to show that the numerator of Equation 8.101 is equal to $V_S^2/2$, which is the mean-squared sinusoidal input signal of frequency f_o required to make $v_{NOB}^2 = 2v_{NOA}^2$. Thus, we can finally write the expression for the spot noise factor as:

$$F_{SPOT} = \frac{V_S^2/2}{4kTR_SB} \tag{8.102}$$

Note that V_S is the peak value of the sinusoid of frequency f_o required to make the true rms meter reading increase by $\sqrt{2}$ in the "B" switch position, B is the NBPF's equivalent Hz noise bandwidth, and T is the temperature in degrees Kelvin of R_S. F_{SPOT} is often presented as a plot of $10 \log_{10}(F_{SPOT})$ vs. R_S and f_o values by manufacturers of low-noise amplifiers and transistors.

An alternate means of measuring F_{SPOT} makes use of the same circuit as in Figure 8.38, except the sinusoidal source is replaced with a calibrated, broad-band, white noise source with a flat spectrum in the 1/f region of e_{na}. The same procedure is followed: first we measure v_{NOA}^2, and then adjust the noise source level to e_{NS}^2 MSV/Hz so that $v_{NOB}^2 = 2v_{NOA}^2$. It is left as an exercise for the reader to show that for the noise source method

$$F_{SPOT} = \frac{e_{NS}^2}{4kTR_S} \tag{8.103}$$

Note that the center frequency of the NBPF and its noise bandwidth do not appear in this simple expression, but they should be specified in presenting the calculated values for F_{SPOT}. In general, a different value of e_{NS} will be found for each point in f_o,R_S space.

Often we wish to measure the rms voltage of a noise source in bandwidth B when the noise is of the same order of magnitude or smaller than the noise arising in our signal conditioning amplifiers. Figure 8.39 illustrates a means of measuring such low-level noise voltages. The noise voltage under measurement, e_{NS}, appears in series with a Thevenin source resistance, R_S. This source is connected to two identical amplifier channels with gain K_V and noise bandwidth B Hz. Both amplifiers have short-circuit input equivalent voltage noise sources which are equal statistically, have zero means, but otherwise are independent and uncorrelated. The outputs of the two channels are multiplied together using a precision, four-quadrant, broad-band, analog multiplier. The output of the multiplier is then passed through a low-pass filter to average it. For simplicity, let us assume that the noises have finite bandwidths less than B. Thus, the multiplier output can be written:

$$\begin{aligned} v_M(t) &= (0.1)\left\{K_V\left[v_{NS}(t) + e_{na1}(t)\right]\right\}\left\{K_V\left[v_{NS}(t) + e_{na2}(t)\right]\right\} \\ &= (0.1)K_V^2\left[v_{NS}^2(t) + v_{NS}(t)e_{na2}(t) + e_{na1}(t)v_{NS}(t) + e_{na1}(t)e_{na2}(t)\right] \end{aligned} \tag{8.104}$$

Low-pass filtering $v_M(t)$ effectively computes $\overline{v_M}$. Because the three voltages have zero means and are statistically independent and uncorrelated with one another, the average of each cross-term product is the product of the averages, which by definition, are zero. Thus, the output of the filter is a DC voltage, V_F:

$$V_F \cong (0.1)K_V^2\overline{v_{NS}^2} \tag{8.105}$$

V_F is thus proportional to the mean square of the noise voltage under measurement, taken over bandwidth B. The effects of the amplifier voltage noise drops out; we have assumed the current noise to be negligible (i.e., $i_{na} R_S << e_{na}$). Also, because of R_S,

$$\overline{v_{NS}^2} = \left(4kTR_S + e_{NS}^2\right)B \tag{8.106}$$

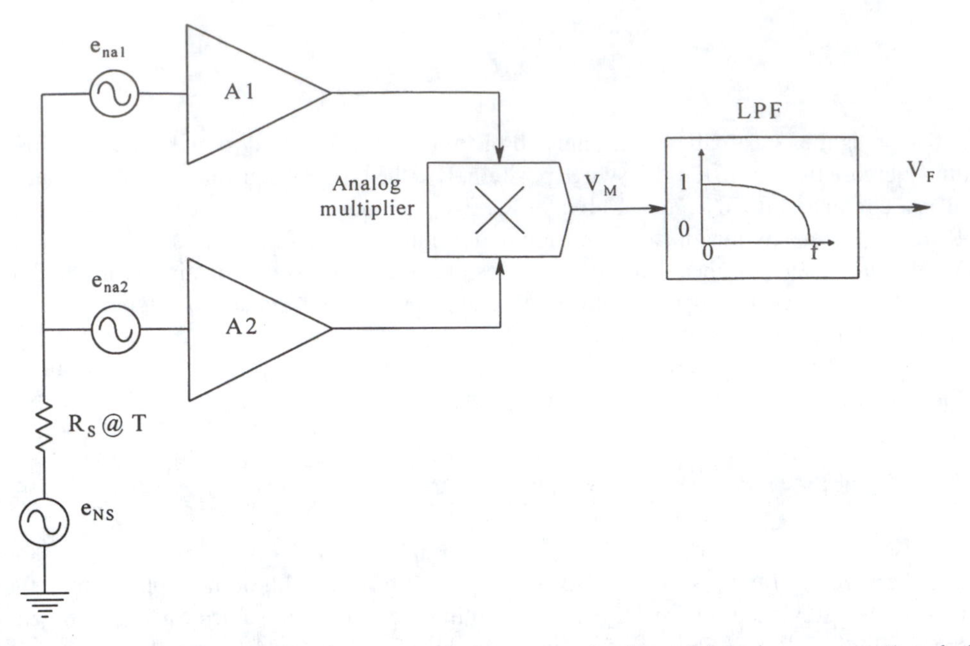

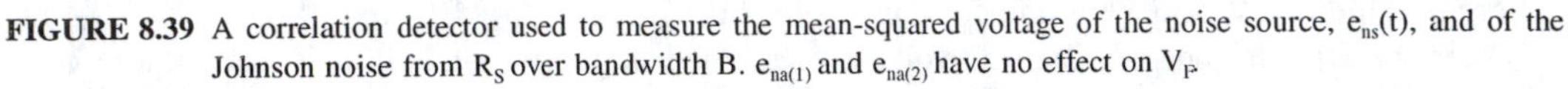

FIGURE 8.39 A correlation detector used to measure the mean-squared voltage of the noise source, $e_{ns}(t)$, and of the Johnson noise from R_S over bandwidth B. $e_{na(1)}$ and $e_{na(2)}$ have no effect on V_F.

An approximate means of estimating the rms value of a broadband, Gaussian noise source was described by Hewlett-Packard. The *tangential noise measurement method* is extremely simple. It makes use of a dual-beam oscilloscope. The noise voltage is applied to both channels with the same gain. At wide trace separation, we see two "grassy" noise traces separated by a dark band. The vertical position control of one trace is then adjusted slowly to bring the grassy traces together to a critical separation so there is no dark band between the traces, i.e., there is a uniform illumination of the CRT screen between the traces. It can be shown that this condition corresponds to a trace separation of two standard deviations of the Gaussian noise. Referring to Figure 8.40, we see that:

$$\sigma_x = \frac{V_U - V_L}{2} \qquad 8.107$$

If the measurement is repeated five times and the results averaged, the $\overline{\sigma_x}$ is claimed to be accurate to about ±0.5 dB, or ±6%.

8.5 AC CURRENT MEASUREMENTS

AC current measurements can be made over a wide range of frequencies and amplitudes. At audio and power frequencies, currents can be measured from nanoamperes to hundreds of kiloamperes. Different ranges require different techniques, which we shall describe below. At radio frequencies, the techniques available are considerably fewer. Regardless of the meter mechanism employed, AC ammeters are generally calibrated in terms of rms of a sinewave.

8.5.1 Electromechanical (Analog) AC Ammeters

Most of the electromechanical meter movements we have already discussed in considering AC voltmeters can be used for AC ammeters, with attention to their range and input current frequencies. To measure currents above the full-scale sensitivity of the meter movement, a shunt must be used. For currents in excess of about 5 A, a 50 mV external shunt is generally used. Recall that a 50 mV

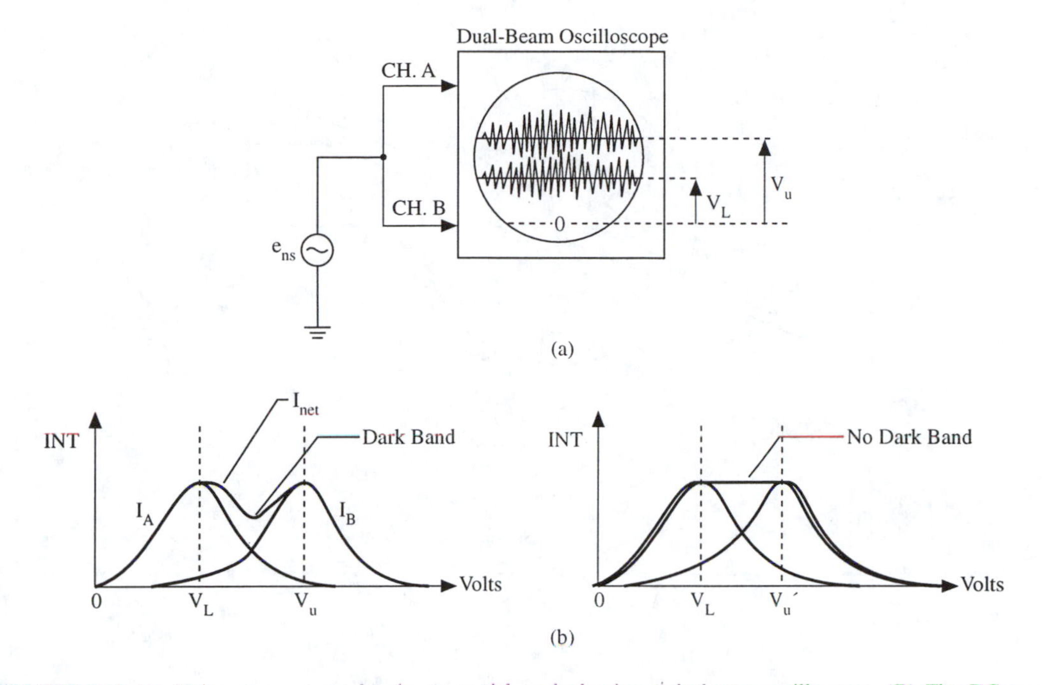

FIGURE 8.40 (A) Noise measurement by the tangential method using a dual-trace oscilloscope. (B) The DC trace separation is adjusted until the noise appears as a solid band. See text for details.

shunt has a resistance such that when the full-scale current enters it, 50 mV is developed across the parallel combination of the shunt and the 50 mV voltmeter which reads the current. Shunts are typically used for high current measurements at audio or powerline frequencies using thermocouple, electrodynamometer, or iron vane millivoltmeters. They are not used with electrostatic or D'Arsonval/rectifier-type AC meters. The thermocouple ammeter can be used at radio frequencies up to 50 MHz or so. The other meter movements, using coils, are only useful at powerline frequencies.

Another means used at powerline frequencies to measure high currents on conductors at high potentials is the *current transformer.* One or more turns of the alternating current-carrying conductor are passed through the center of a high-permeability, toroidal iron core, inducing a sinusoidally changing magnetic flux. Also wound on the toroidal core are a number of turns of wire which are connected to an AC ammeter, generally having a dynamometer or iron vane movement. Figure 8.41A illustrates a typical current transformer circuit. In a Midwest Electric Products, Inc. current transformer, the secondary (meter) winding was found to have a resistance of 0.04 ohms, an inductance of 3.66 mH, and a turns ratio of 75/5. This transformer was rated for 2 VA, and works over a 25 to 400 Hz range. Stout (1950) gives a current transfer function for the current transformer:

$$\frac{I_1}{I_2} \cong \frac{n_2}{n_1}\left[1 + \frac{I_O}{I_1}\sin(\theta + \phi)\right] \qquad 8.108$$

where n_2/n_1 is the ratio of secondary (output) winding turns to primary (input) turns, I_O is the excitation or magnetizing component of the primary current, I_1, θ is the power factor angle between V_1 and I_1, and includes both the effect of the load (burden) and the current transformer. I_2 is the AC current in the secondary ammeter, and ϕ is the angle between the net magnetic flux in the current transformer core, Φ_o, and the equivalent sinewave of I_O. A phasor diagram showing currents and flux in a typical current transformer is shown in Figure 8.41B. Current transformers should never be operated with their secondaries open-circuited; dangerous high-voltages can exist there.

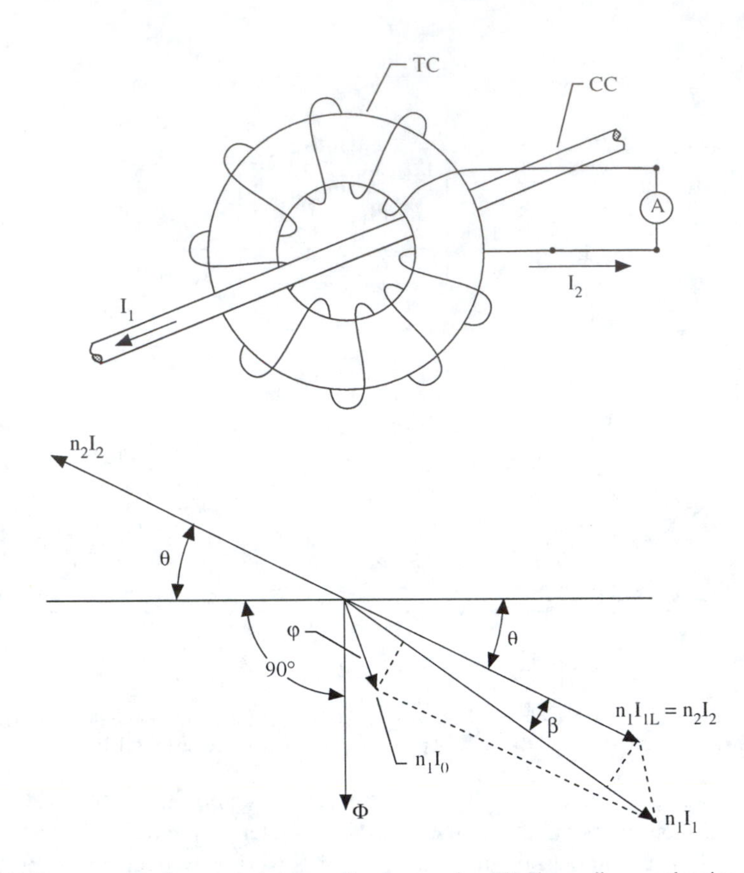

FIGURE 8.41 (A) A current transformer used at powerline frequencies. (B) Phasor diagram showing the current and flux vectors in a typical current transformer. See text for explanation of symbols. (Figure 8.41B reproduced from Stout, M.B., *Basic Electrical Measurements*, 2nd ed., Prentice-Hall, Englewood Cliffs, NJ, 1960. With permission.)

The *clamp-on ammeter* is a portable version of the current transformer in which the secondary load impedance is high, so sufficient secondary EMF is developed to operate an AC millivoltmeter, or a D'Arsonval/rectifier-type AC voltmeter. Another meter movement used with a clamp-on ammeter is the rotating iron vane type used in the Columbia Electric Manufacturing Co. "Tong Test Ammeter". In the Columbia meter movement, magnetic flux induced in the iron jaws by enclosing a current-carrying conductor acts to draw an iron vane into an airgap at the base of the jaws. This torque acts against a torsion spring, producing a pointer deflection which is a nonlinear function of primary current. Mechanical damping is produced by eddy currents induced in an aluminum damping pan. Ranges are changed in the Columbia clamp-on AC ammeter by plugging in different meter assemblies into the clamp-on jaws' airgap. Full-scale currents from 20 to 1000 A rms are available. The bottom 20% of the scales of the Columbia meters is not useful, because of deflection nonlinearity.

Yet another means of measuring AC current makes use of a *compensated thermoelement*, illustrated in Figure 8.42. Two large copper blocks serve as attachment points for the input current line and as thermal reference points. The heater element is attached between the blocks, and the hot junction of the thermocouple is spot welded to the middle of the heater. The ends of the thermocouple wires are attached to copper strips which are in intimate thermal contact with the copper blocks, but electrically insulated from them by mica sheets. The leads to the D'Arsonval microammeter are attached to the copper strips. The AC (or DC) input current causes a temperature rise in the heating element which is sensed by the thermocouple. The voltage drop across the thermoelement heater is designed to be 0.15 V at full-scale current. Hence this type of ammeter

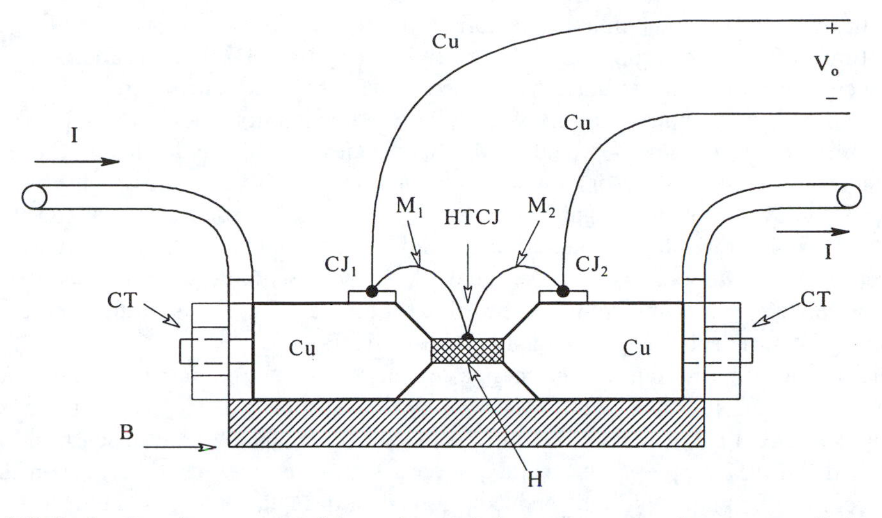

FIGURE 8.42 Section through a compensated thermoelement device used to measure the mean-squared values of DC or AC currents. B, bakelite base; Cu, copper blocks; CT, current input and output terminals; H, heater element; HTCJ, hot thermocouple junction; CJ1 and CJ2, compensating (cold) thermocouple junctions.

absorbs 150 mW for each ampere of full-scale range. This type of ammeter operates at DC, and at AC frequencies up to 50 MHz, and for full-scale currents of 5 to 50 A rms (Stout, 1950).

8.5.2 Electronic and Magneto-Optic AC Ammeters

To measure AC currents smaller than the full-scale currents required to operate the inefficient electrodynamometer and iron vane meter movements (about 50 mA rms) or the full-scale current for vacuum thermocouple AC ammeters (about 3 mA rms), it is necessary to amplify the current, or a voltage proportional to the current. The operational transresistor op amp circuit offers a means of measuring very small AC currents flowing into ground, ranging from picoamperes to about 10 mA. In this case, the op amp's summing junction appears as a virtual ground, as shown in Figure 8.43. The AC voltage at the output of the operational transresistor is simply $V_O = -R_F I_S$, and can now be displayed by any of the conventional AC, analog indicating meters we have discussed above, or it can be displayed digitally, as described in Chapter 9.

Another means of measuring low-level AC currents is to use a current transformer followed by an AC amplifier, followed by an AC, analog indicating meter, or digital readout. Pearson Electronics, Inc. makes a variety of wide-bandwidth current sensors which have a voltage output. For example, the Pearson model 2877 current sensor has a 50 ohm output impedance, an open-circuit

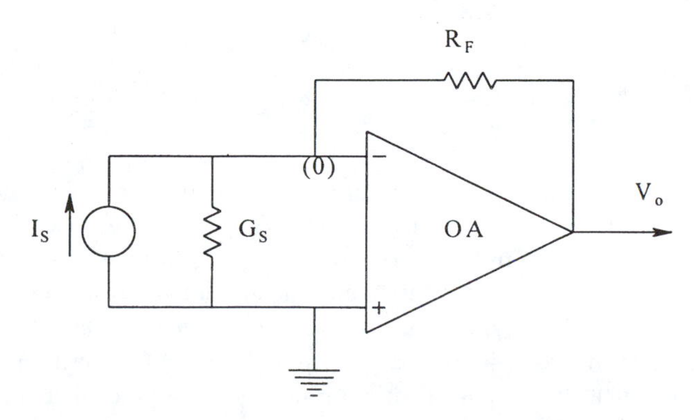

FIGURE 8.43 The operational transresistor circuit is used to convert small currents to voltages.

transresistance of 1 V/A, a maximum peak current of 100 A, maximum rms current of 2.5 A, a 2 nsec rise time, a 0.2%/μsec droop, and –3 dB frequencies of 300 Hz and 200 MHz. At the other end of the current measurement range, the Pearson model 1423 has an open circuit transresistance of 0.001 V/A, a maximum peak current of 500,000 A, a maximum rms current of 2,500 A, a 300 nsec rise time, a 0.7%/msec droop, and –3 dB frequencies of 1 Hz to 1.2 MHz. Current transformers, such as the Pearson sensors, find application in measuring current transients in insulation breakdown studies, and in investigations of high-power pulse circuits using silicon controlled rectifiers (SCRs), hydrogen thyratrons, strobe tubes, klystrons, and magnetrons. Of course, steady-state sinusoidal currents can also be measured. Since the maximum current sensitivity of the Pearson model 2877 sensor is 1 V/A, it is possible to resolve AC currents in the microamp range using appropriate low-noise amplifiers and band-pass filters.

Another means of measuring AC current in a conductor is to *measure the magnetic field around the conductor.* A sensitive *Hall effect sensor probe* (see Section 6.3.5.2) can be put next to the conductor to intercept the solenoidal B field. The AC output voltage of the Hall sensor is amplified and then used to drive an appropriate meter. For more sensitive current sensing, a flux concentrating, iron "C" core can be put around the conductor, as shown in Figure 8.44, and the Hall sensor put in its airgap. The ferromagnetic material and the length of the airgap must be chosen so that the magnetic material does not go into saturation at the desired, full-scale, AC current to be measured.

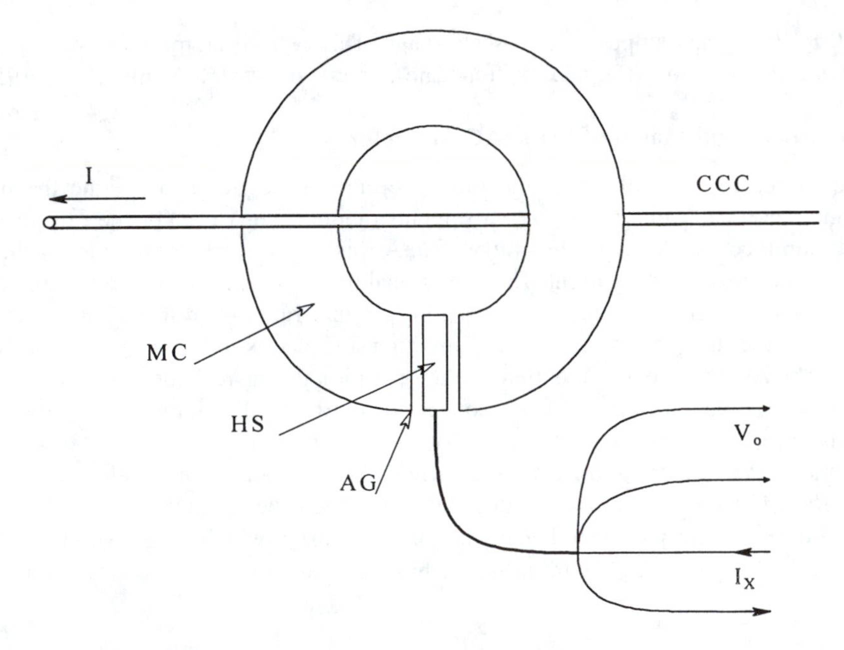

FIGURE 8.44 A Hall sensor used to measure the AC current in a conductor. The Hall output voltage is proportional to the product of B_Y I_H, hence I. MC, magnetic core; AG, airgap; HS, Hall sensor; CCC, current-carrying conductor.

The final AC current measuring system we will consider is the *Faraday magneto-optic sensor,* discussed in some detail in Section 6.6.1 of this text. A good physical discussion of the Faraday magneto-optic effect can be found in Chapter 7 of the optics text by K. D. Möller (1988). Some of the earliest work on fiber optic Faraday effect current sensors was done by A. J. Rogers (1973). As shown in Figure 6.29, the Rogers system uses a ratio detector to make the system insensitive to amplitude variations in the laser source. Equations 6.78 to 6.82 describe the operation of the ratio detector, and show that the output of the ratio detector is a sinusoidal voltage at powerline frequency with peak amplitude proportional to the peak amplitude of the conductor by the Verdet constant, as used in Equation 6.55. After amplification, this line frequency voltage can be displayed by one of the many AC responding meters we have described above. Other Faraday magneto-optic current sensor designs have been described by Cease and Johnston (1990) and Nicati and Robert (1988).

8.6 MAGNETIC FIELD MEASUREMENTS

There are many applications in the design of rotating electric machines (alternators, DC generators and motors, brushless DC motors, stepping motors, induction motors, etc.) where it is useful to be able to measure the magnetic field strength in air gaps in order to verify designs. There are also many applications in physics and in chemistry where it is necessary to measure magnetic fields; for example, in mass spectrometers, magnetic resonance imaging systems, and in particle accelerators. Recently, powerline frequency magnetic fields have been implicated as possible contributing causes in health problems including birth defects, leukemia, and cancer (*EPRI Journal*, Jan/Feb 1990). Thus, measurement of the spatial distribution of 60 Hz B fields in areas containing people, experimental animals, or cell cultures is important for the clarification of the significance of 60 Hz B fields in health problems.

The magnetic field vector, B, is also called the magnetic induction or the magnetic flux density. Its units are newtons per coulomb divided by meters per second, or equivalently, volt-seconds per square meter, or webers per square meter. The more common units for B are gauss (1 Wb/m^2 = 10^4 G), or teslas (10^4 G = 1 T). Magnetic flux is defined as the integral of B over an area perpendicular to B, or

$$\Phi = \int_S \mathbf{B} \cdot \mathbf{dA} \tag{8.109}$$

The units of magnetic flux are webers (MKS), or Maxwells, found by integrating 1 G over 1 cm^2. The vector **H** is called the magnetic field. It it related to **B** by the well-known relation, **H** = **B**/μ. The MKS units of **H** are newtons/weber or amps/meter, and the CGS units of **H** are oersteds (1 A/m = $4\pi \times 10^{-3}$ Oe).

There are a variety of magnetic sensors which can be used to measure DC and AC **B** fields in space. The first and most basic means is the solenoidal search coil. If the **B** field is homogeneous over the area of the coil, a time-varying **B** field will induce an open-circuit EMF given by the well-known relation:

$$E_M = NA\dot{B}\cos(\theta) \tag{8.110}$$

where N is the number of turns in the coil, A is the area of the coil, θ is the angle **B** makes with respect to a normal to the coil's area, and $\dot{B}$ = dB/dt. In most powerline B fields, $B(t) = B_o \sin(\omega t)$, so:

$$e_M(t) = NA\cos(\theta)\omega B_o \cos(\omega t) \tag{8.111}$$

If the **B** field is constant, e.g., the Earth's magnetic field, an EMF may be induced in the coil proportional to B by rotating the coil at an angular velocity of ω_c r/s. In this case, the cos(θ) term becomes $\cos(\omega_c t)$, and the induced EMF is given by

$$e_M(t) = NAB\omega_c \sin(\omega_c t) \tag{8.112}$$

Several variations on the rotating coil magnetic field meter are described by Lion (1959). For example, a coil can be mechanically sinusoidally oscillated around a center position in which a

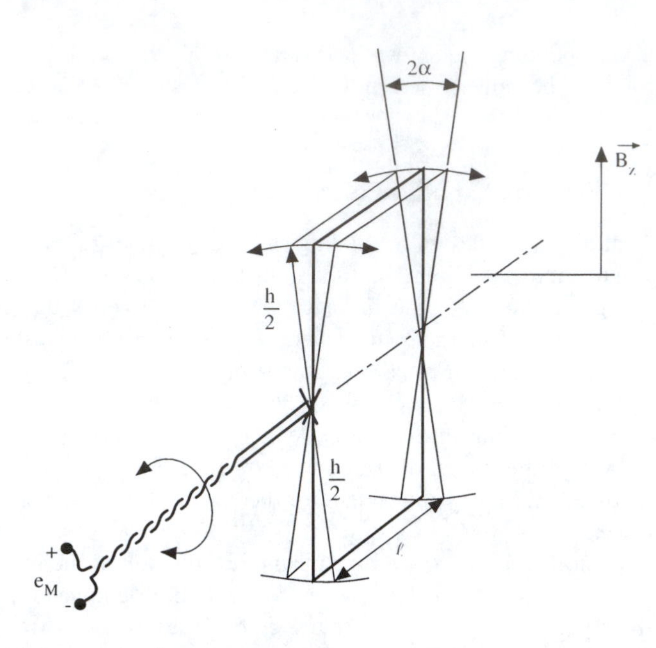

FIGURE 8.45 An oscillating search coil used to measure a DC magnetic field (after Lion, 1959).

constant B vector is parallel wth the plane of the coil, as shown in Figure 8.45. The EMF induced in the coil can be shown to be given by

$$e_M(t) = NAB\alpha_o\omega_r\cos(\omega_r t) \tag{8.113}$$

Here, α_o is the angle over which the coil is oscillated at mechanical radian frequency, ω_r, so $\alpha(t) = \alpha_o\sin(\omega_r t)$. We assume that $\alpha_o < 15°$.

Three orthogonal search coils can be used to determine the magnitude and direction of a powerline frequency **B** field. The **B** vector can be resolved into three orthogonal vector components, $\mathbf{B_X}$, $\mathbf{B_Y}$, and $\mathbf{B_Z}$. Each component may be considered to produce an EMF in its respective search coil. Referring to Figure 8.46, we see that in spherical coordinates, the three components of **B** can be written as scalars:

$$B_Z = B\cos(\phi) \tag{8.114A}$$

$$B_X = B\sin(\phi)\cos(\theta) \tag{8.114B}$$

$$B_Y = B\sin(\phi)\sin(\theta) \tag{8.114C}$$

where ϕ is the co-latitude angle measured between the positive y-axis and B, and θ is the longitude angle, measured between the positive x-axis and the projection of B on the x-y plane. This projection is B_r, and is equal to:

$$B_r = B\sin(\phi) \tag{8.115}$$

From Equation 8.113 above, we can write expressions for the instantaneous EMFs induced in the coils:

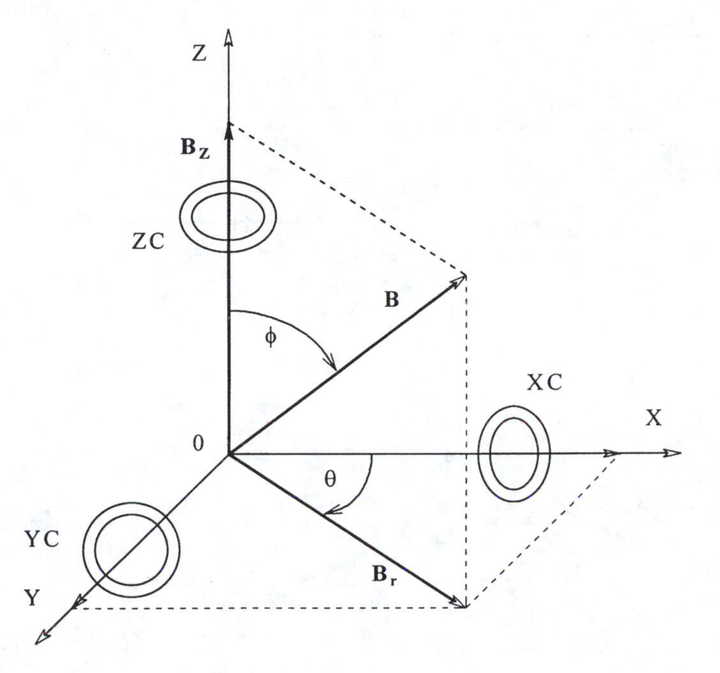

FIGURE 8.46 Three orthogonal search coils used to measure the X, Y, and Z components of a time-varying magnetic field.

$$e_{MX}(t) = NA\omega B \sin(\phi)\cos(\theta)\cos(\omega t) \qquad 8.116A$$

$$e_{MY}(t) = NA\omega B \sin(\phi)\sin(\theta)\cos(\omega t) \qquad 8.116B$$

$$e_{MZ}(t) = NA\omega B \cos(\phi)\cos(\omega t) \qquad 8.116C$$

Now it is easy to show that if we square each of the coil EMFs with four-quadrant analog multipliers, then average each multiplier output by passing it through a low-pass filter, add together the mean-square voltages, then take the square root, we will obtain a DC voltage proportional to |**B**|, regardless of the orientation of the three-coil probe in the B field. This process is illutrated in Figure 8.47. Numerically, B can be shown to be

$$B = \frac{\sqrt{2}V_O}{NA\omega} \qquad 8.117$$

(We assume all coils to have identical Ns and As.) The angles ϕ and θ can also be found relative to the probe's x-, y-, and z-axes from trigonometric identities; however, this is tedious. A better way to find the direction of **B** is to orient the probe so that $e_{MX}(t)$ is maximum and $e_{MY}(t)$ and $e_{MZ}(t)$ are minimum. Now the angle of **B** is parallel with the probe's x-axis.

Analog Hall effect sensors provide another major means of measuring either constant or time-varying magnetic fields. The physics governing the behavior of Hall sensors was described in detail in Section 6.3.5.2. It was shown that the Hall sensor EMF is given by:

$$E_H(t) = \frac{B(t)\cos(\theta)I_x(t)R_H}{h} \qquad 8.118$$

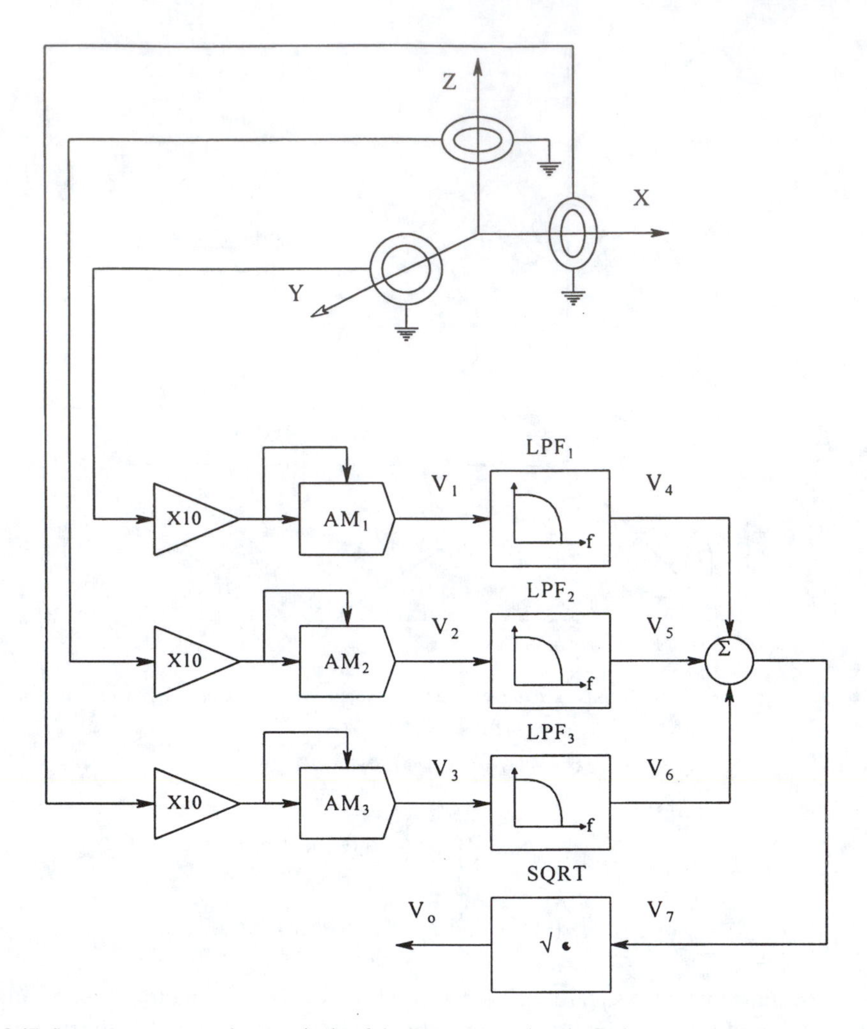

FIGURE 8.47 System to compute the magnitude of the **B**(t) vector using the Pythagorean theorem.

where θ is the angle B makes with the x-z plane, $I_x(t)$ is the current flowing in the + x-direction (electrons in n-semiconductor drift in the negative x-direction), and R_H is the Hall constant: $R_H = -1/qd$ for n semiconductor, and $R_H = 1/qd$ for p semiconductor. q is the magnitude of the electron charge, d is the density of the carrier doping in the semiconductor, and h is the thickness of the Hall slab in the y direction. In the operation of most analog Hall sensors, I_x is generally DC. However, in the measurement of a low-level, constant B, a sinusoidal or square wave $I_x(t)$ can be used to permit noise reduction by the amplification of $e_H(t)$ out of the 1/f noise region, and also the use of synchronous rectification to further improve SNR.

Modern analog Hall sensors can be purchased with built-in, IC DC amplifiers, or as the basic sensor chip. Unamplified Hall sensors have conversion gains ranging from 10 to 55 mV/kG. Through the use of a 9", flux concentrating iron bar, the F. W. Bell model BH-850 Hall sensor has a gain of 18 mV/gauss. Normal operating current for the Bell BH-850 sensor is 200 mA. The resistance in its current path, and its Thevenin output resistance for E_H are both 3.5 ohms, giving good, low-noise performance.

Various companies make gaussmeters for various applications having either analog or digital meter readouts. For example, the Bell model 615 digital gaussmeter, a benchtop instrument, reads 10 to 10^6 G in five ranges, has a DC to 2 kHz frequency range, is accurate to ±0.5% of FS, and has a readout resolution of 1 part in 10^3. The Bell model 4048 is a hand-held, portable, digital

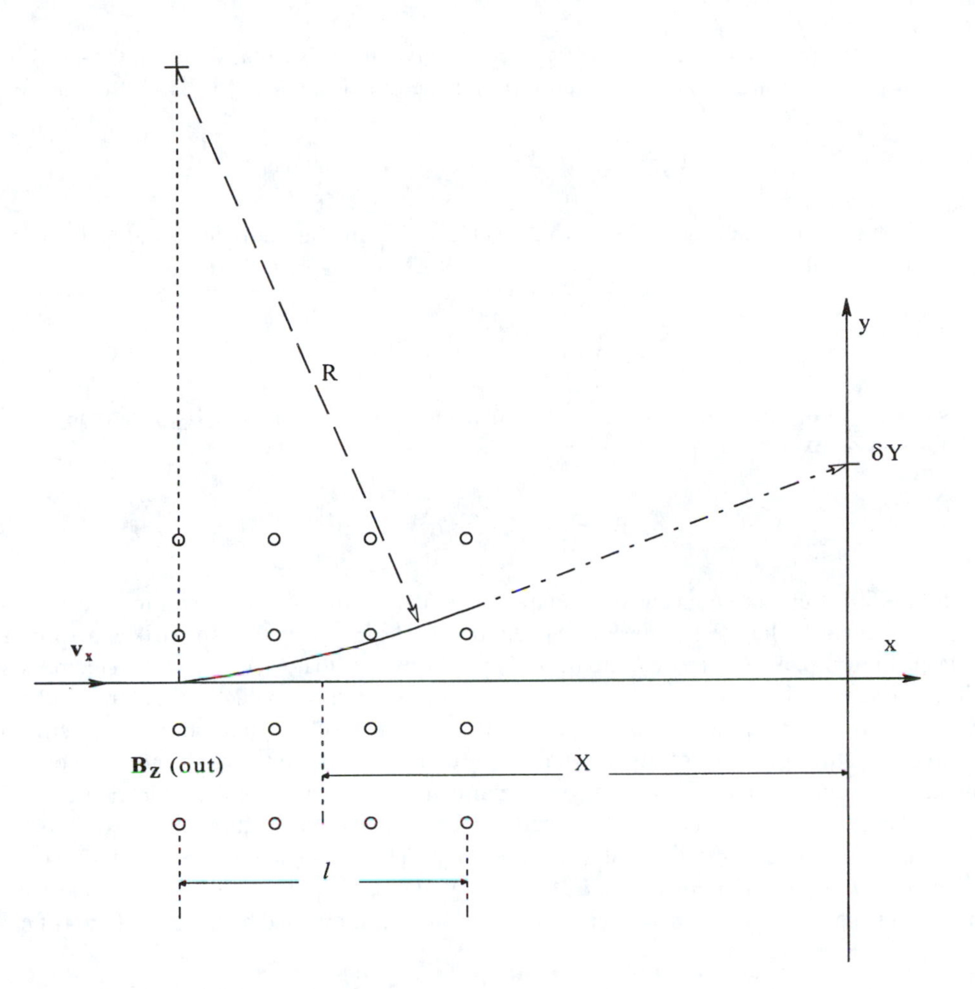

FIGURE 8.48 Use of a cathode ray tube to measure B_Z. The Lorentz force on the electrons passing through the orthogonal B field causes a deflection, ΔY. See text for details.

gauss/tesla meter with three full-scale ranges; 200, 2000, and 2×10^4 G. Accuracy is ± 2% of reading for DC fields, ±2.5% for 45 to 100 Hz AC fields, ±2% for 100 Hz to 3 kHz, and ±6% for 3 to 5 kHz fields. This instrument is available with transverse and axial Hall probes.

The practical lower limit of Hall sensor sensitivity is set by noise, both from the sensor's thermal and shot noise, and from the signal conditioning amplifier's e_{na} and i_{na}. Milligauss resolution is practical with most modern Hall sensors. To achieve higher sensitivity in magnetic field measurements, it is necessary to use sensors which utilize different physical principles.

For example, at microwave frequencies, many materials exhibit a very strong Faraday magneto-optic effect (see Section 6.6.1). The Verdet constant of certain ferromagnetic ferrites at 9 GHz is on the order of 0.1°/(oersted cm). By the use of such a ferrite, magnetic flux density variations on the order of 10^{-5} G have been detected (Lion, 1950).

The deflection of a collimated, low-energy electron beam by a B field is also an effective means of measuring low values of B, ranging from 2×10^{-6} to 2×10^{-4} G (Lion, 1950). Figure 8.48 illustrates the geometry of the CRT used for such a measurement. Note that B must be perpendicular to the electron beam axis and the deflection axis. The electron beam is accelerated through a potential of V volts giving the electrons in the beam an average velocity of

$$\mathbf{v}_x = \sqrt{2V(q/m)}\,\mathbf{i}_x \quad \text{m/s} \qquad 8.119$$

where q/m is the ratio of electron charge to electron mass, and $\mathbf{i}_x$ is a unit vector in the +x direction. The beam passes through l meters of a uniform magnetic field of strength B_Z. A Lorentz force of

$$\mathbf{F}_Y = -q(\mathbf{v} \times \mathbf{B}) = -(qv_X B_Z)\mathbf{i}_Y \text{ newtons} \quad 8.120$$

acts on each electron, causing the beam to deflect along a circular path with a radius R. R may be shown to be equal to

$$R = (mv_X)/(qB_Z) \text{ meters} \quad 8.121$$

Kraus (1953) shows that the beam deflection, δy, can be used to measure B_Z. Referring to Figure 8.48, B_Z can be expressed as

$$B_Z = \frac{\delta y}{IX}\sqrt{2V(m/q)} \text{ Wb/m}^2 \quad 8.122$$

Lion (1950) illustrates a means of detecting beam deflections, δy, too small to visualize on the phosphor screen of the electron beam magnetometer. The collimated beam strikes a split anode, shown in Figure 8.49. At zero B_Z, the beam is positioned so that exactly half its electrons strike each anode plate. The anode currents are converted to two voltages which are subtracted by a DC differential amplifier. In the presence of a $B_Z > 0$, the beam deflects, and more current strikes one plate than the other, causing an output signal. If this output signal is integrated, and then conditioned to make a current to drive a deflection coil, the effective flux density of the deflection coil will be equal to $-B_Z$ in the steady state, giving a null output. The integrator output is proportional to B_Z.

The most sensitive magnetic field sensors are the *SQUID magnetometers*. SQUID stands for *Superconducting QUantum Interference Device*. SQUIDS find application in biophysical measurements in neurophysiolgy. The nerve action potentials are accompanied by transient flows of electric

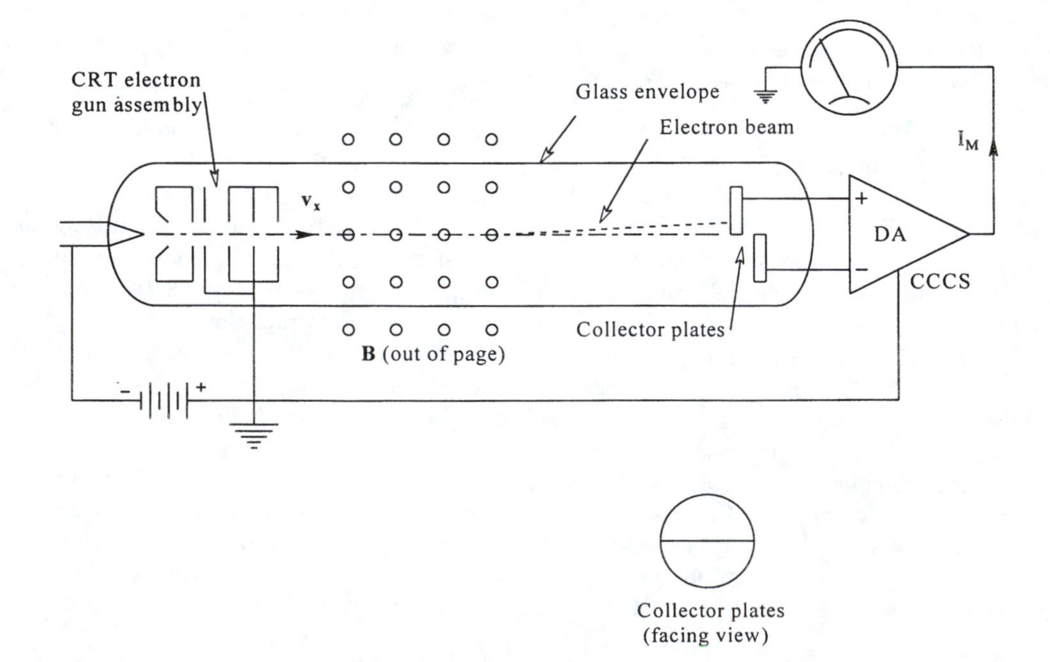

FIGURE 8.49 A sensitive cathode ray magnetometer described by Lion, 1959.

currents, generally carried by ions. These transient currents generate minute, time-varying, magnetic fields which can be measured and located spatially in the brain by arrays of SQUID sensors (Ribeiro et al., 1988). Besides studies in neuromagnetism, SQUIDS can be used in magnetocardiography, detection of changes in the Earth's magnetic field in geological studies, etc.

SQUIDS are made from various superconductor materials, such as niobium or lead, and are operated at the boiling point of liquid He at atmospheric pressure (4.2 K). There is the possibility that SQUIDS can be made from special, high-temperature semiconductors which can operate at temperatures up to 125 K (Clarke and Koch, 1988).

The conventional DC SQUIDs is the most sensitive magnetometer available. Its superconducting ring has an inductance in the picoHenry range, and is joined with two Josephson junctions, as shown in Figure 8.50A. The SQUID is really a four-terminal device; two terminals are used to input a DC bias current, I_B, and the same two terminals are used to monitor the output voltage, V_O. V_O remains zero until the bias current reaches a critical value, I_O. Then the output voltage increases with current, and is also a function of the magnetic flux linking the SQUID ring. The DC bias current, I_B, is made greater than I_O. The superconducting SQUID ring circuit undergoes the phenomenon of *fluxoid quantization* in which the magnetic flux linking the SQUID is given by $n\Phi_o$, where n is an integer, and Φ_o is the flux quantum, equal to $h/2q = 2 \times 10^{-15}$ Wb. If we apply an additional flux, Φ_i, through the SQUID ring, a supercurrent, $I_S = -\Phi_i/L$ is set up in the ring to create a flux which cancels Φ_i. In other words, $LI_S = -\Phi_i$. From Figure 8.50B, we see that at a constant bias current, the SQUID output voltage varies periodically as a function of Φ/Φ_o. This phenomenon is a lot like the generation of intensity variations due to interference rings in an optical system, such as a Sagnac gyro. The SQUID sensor is often operated as a feedback device, as shown in Figure 8.50C. A bias flux is adjusted so that V_O is linearly proportional to small changes in input flux, $\delta\Phi$. V_O is then conditioned and integrated and used to control the current in a feedback coil such that a flux $-\delta\Phi$ is generated to cancel the input flux. This system operates the SQUID as a null-flux detector, and is called a flux-locked SQUID. This closed-loop operation permits a large dynamic range of input flux to be measured. As shown in Figure 8.50A, a superconducting input coil must be used to couple the flux to be measured into the SQUID toroid. It appears that optimum SQUID sensitivity occurs when the inductance of the pickup loop, L_P, is equal to that of the coupling loop, L_C. Assuming noiseless flux coupling, Clarke and Koch (1988) show that the equivalent flux density noise of the SQUID can be given by

$$B_N(f) = \frac{2\sqrt{2L_P}}{\pi r_p^2}\sqrt{\frac{\varepsilon(f)}{\alpha^2}} \ \text{Wb/m}^2/\sqrt{\text{Hz}} \qquad 8.123$$

where α is the coupling coefficient between the SQUID and the input coil, r_p is the radius of the input coil, L_P is the inductance of the pickup coil, and $\varepsilon(f)$ is the white noise energy/Hz of the SQUID:

$$\varepsilon(f) = 9kTL/R \ \text{J/Hz} \qquad 8.124$$

k is Boltzmann's constant, T is in degrees Kelvin, L is the SQUID's inductance, and R is the resistance shunting each Josephson junction in the SQUID ring. For a typical Nb SQUID with $100 < L < 500$ pH, the noise energy is about 1 to 5×10^{-32} J/Hz (Clarke and Koch, 1988). With such low noise, it is possible to resolve magnetic flux densities on the order of $10^{-14}\ \text{T}/\sqrt{\text{Hz}}$ ($1\ \text{T} = 10^4$ G). A very low white noise of $5\ \text{fT}/\sqrt{\text{Hz}}$ has been achieved using thin-film technology. Of course, external magnetic noise pickup can confound sensitive measurements, so SQUID sensors are normally well shielded with layers of Mu metal, aluminum, and even superconducting materials.

To further reduce the pickup of unwanted magnetic field noise, use is made of a gradiometer coil, shown in Figure 8.51A. This coil responds to spatial nonuniformities in the field under

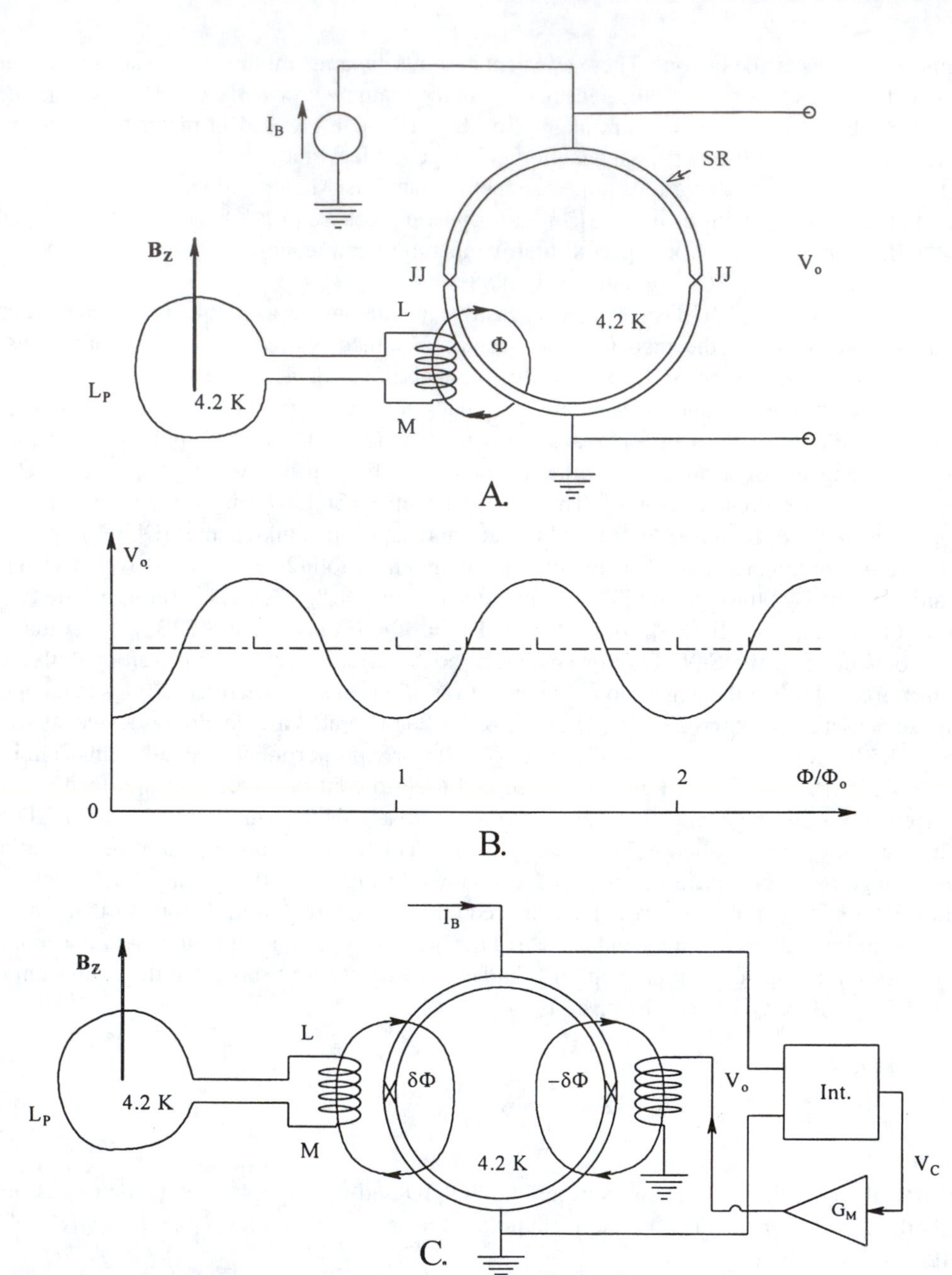

FIGURE 8.50 (A) Schematic of a DC SQUID device. SR, superconducting ring of lead or niobium; JJ, Josephson junctions; IB, DC bias current; L, inductance of the coupling coil; M, mutual inductance between coupling coil and the SR. (B) Variation of SQUID output voltage at constant bias current as a function of the ratio of the applied flux to the flux quantum. (C) A feedback or flux-locked SQUID.

measurement, i.e., $\partial B/\partial z$. First-order gradients from a magnetic dipole fall off as $1/r^4$, so the gradiometer coil discriminates against distant sources of interference in favor of local dipoles. Some workers have even used second derivative gradiometer coils which respond to $\partial^2 B/\partial z^2$ to obtain even better resolution of local sources (Ribeiro et al., 1988). A second derivative coil is shown in Figure 8.51B.

A comparison of the magnetic flux density resolution of common magnetometers is shown in Figure 8.52, adapted from Clarke and Koch (1988). Note that the nonsuperconducting coil gives

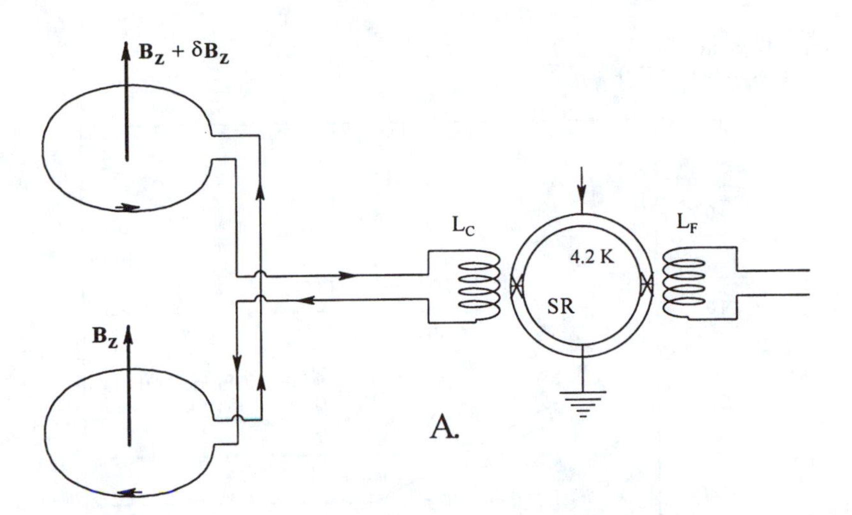

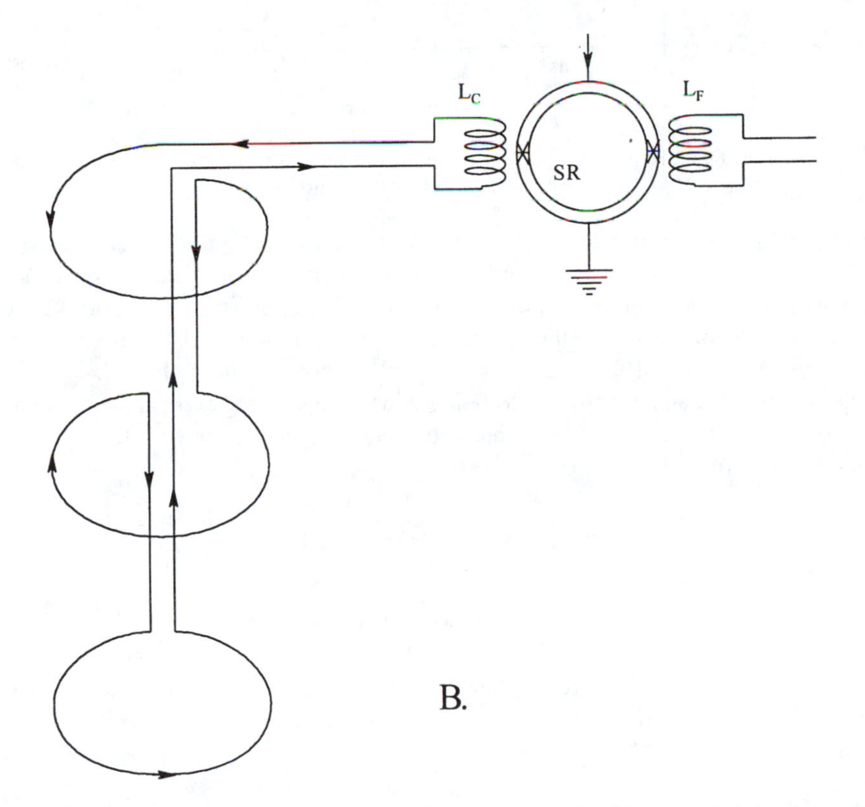

FIGURE 8.51 (A) A SQUID with a gradiometer flux pickup. (B) A SQUID with a second derivative gradiometer flux pickup.

very high resolution at frequencies above 1 Hz, and is generally robust and simple. The extra sensitivity of SQUID sensors is needed, however, for such biomedical applications as neuromagnetism and magnetocardiography.

8.7 PHASE MEASUREMENTS

It is often necessary to measure the phase difference between two periodic signals of the same frequency. In power distribution systems, the power factor angle between the AC supply voltage

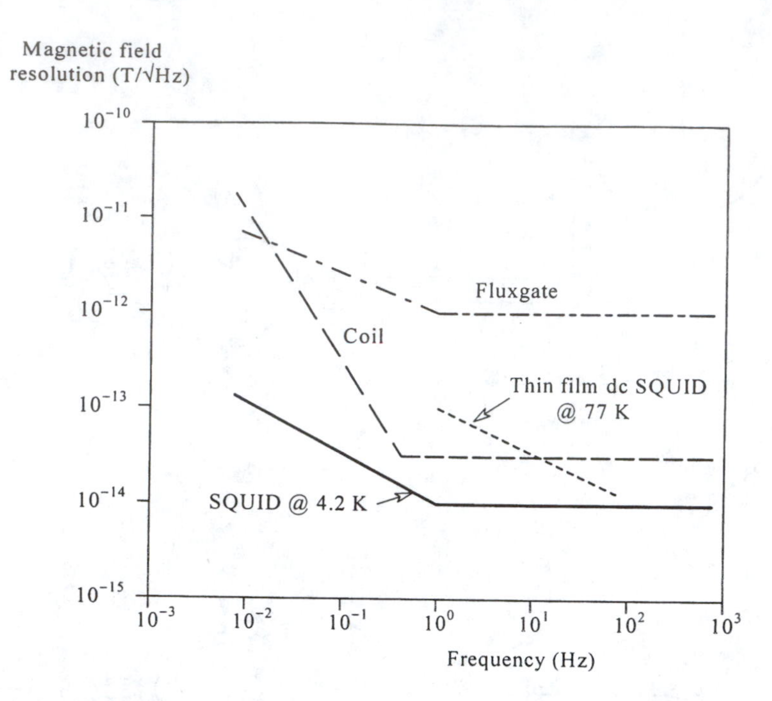

FIGURE 8.52 A comparison of the resolution and bandwidths of various sensitive magnetic field sensors. (From Clark, J. and Koch, R.H., *Science*, 88, 217–223, 1988. With permission.)

and the load current is of great economic and practical importance to the power generator and the end user. We have seen that several magneto-optical electric current sensing systems have AC outputs in which the phase difference is proportional to an optical polarization angle change induced by the magnetic field produced by the current under measurement. Also, phase information is important in the description of feedback system performance and filter behavior.

Phase measurements can be classified as being done by analog or digital systems. To illustrate the concept of phase difference, let us define two periodic functions having the same frequency, but not coincident in time origin:

$$v_1(t) = v_1(t+T) = v_1(t+kT) \tag{8.125A}$$

$$v_2(t) = v_2(t+T+\Delta T) = v_2(t+kT+\Delta T) \tag{8.125B}$$

where ΔT is the time difference between v_1 and v_2, T is the period of v_1 and v_2, and k is an integer. Figure 8.53A illustrates two sinewaves having a phase difference. v_2 is said to lead v_1 by a phase difference of φ, where $\varphi = 2\pi(\Delta T/T)$ radians,

$$\varphi = 2\pi(\Delta T/T) \text{ radians} \tag{8.126A}$$

or

$$\varphi = 360(\Delta T/T) \text{ degrees} \tag{8.126B}$$

if

$$v_1(t) = V_1 \sin(\omega t) \tag{8.127A}$$

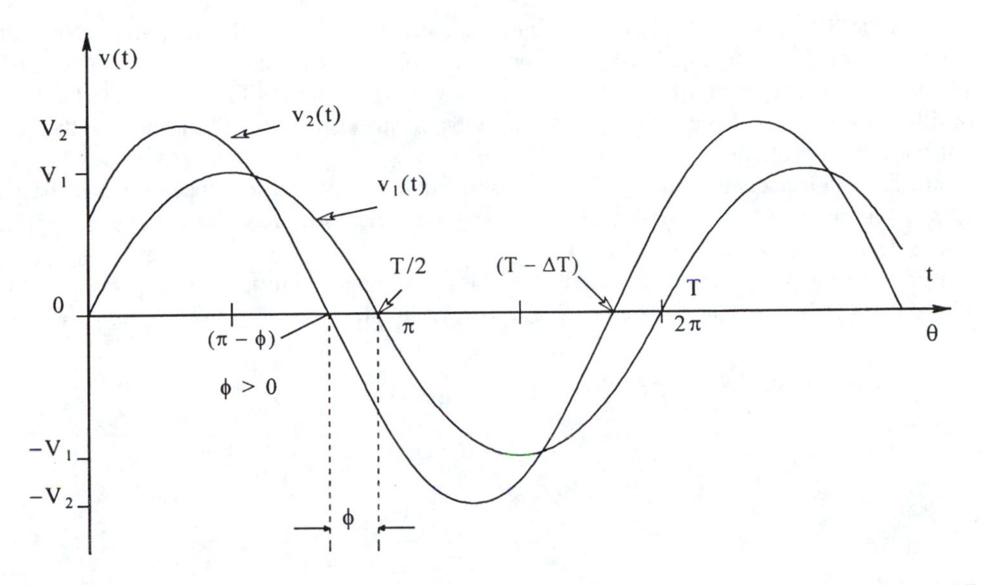

FIGURE 8.53 (A) Figure illustrating the phase difference between two sinewaves having the same frequency and different amplitudes.

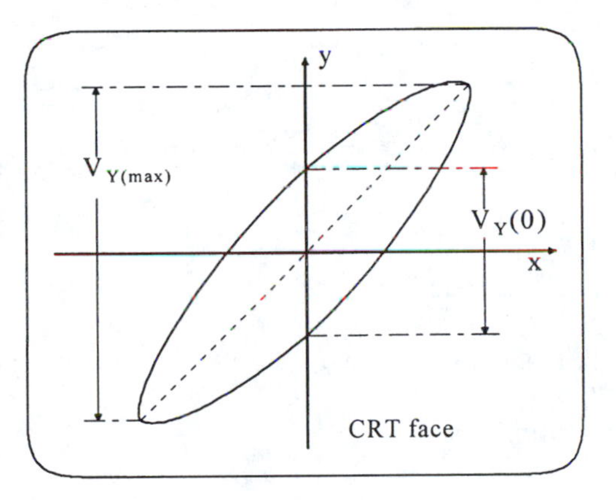

FIGURE 8.53 (B) Lissajous ellipse on CRT face, showing intercepts used in calculating the phase shift.

and

$$v_2(t) = V_2 \sin(\omega t + \varphi) \quad 8.127B$$

8.7.1 Analog Phase Measurements

Probably the simplest and easiest direct measurement of phase angle can be made on an oscilloscope using the *Lissajous figure* formed when two sinewaves, given by Equation 8.127 are input to the x (horizontal deflection)-axis and the y (vertical deflection)-axis, respectively. In general, we see an ellipse on the CRT screen, as shown in Figure 8.53B. The ellipse must be centered at the origin (center) of the CRT screen. At t = 0, $v_1 = 0$, and $v_2 = V_2\sin(\varphi)$. Thus, it is easy to show that the phase angle is given by:

$$\varphi = \sin^{-1}\left[V_Y(0)/V_Y(\max)\right] \quad 8.128$$

From Equation 8.128, when the Lissajous figure is a straight line, $\varphi = 0°$, and when it is a circle, $\varphi = 90°$. Accuracy of the Lissajous figure method of phase measurement is poor near ±90°. At phase angles near integer multiples of 180°, accuracy is better, and is limited by the thickness of the oscilloscope trace, and one's ability to estimate the distances to its intersections with the vertical axis of the CRT's graticule.

Somewhat more accurate analog phase measurement techniques have been devised based on waveform averaging. Most of these techniques have been used in phase-lock loops to implement the phase detector element (Northrop, 1990).

The first analog phase detector we will describe is the analog multiplier/low-pass filter. In this system, shown in Figure 8.54, the input signals must be in quadrature for proper operation. That is:

$$v_1(t) = V_1 \sin(\omega_1 t) \qquad 8.129A$$

and

$$v_2(t) = V_2 \sin(\omega t + \varphi) \qquad 8.129B$$

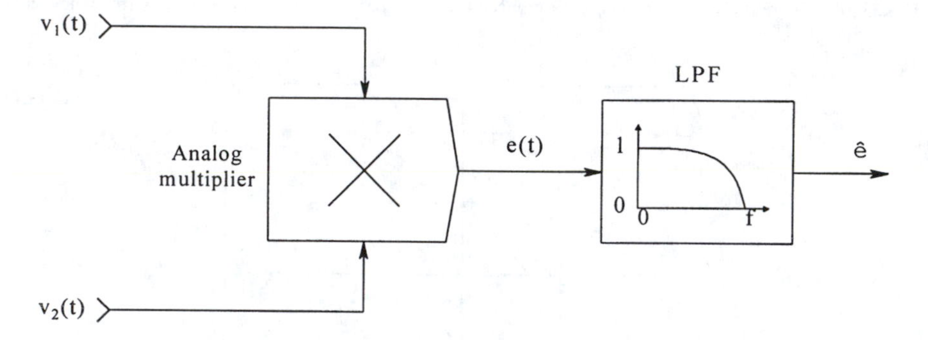

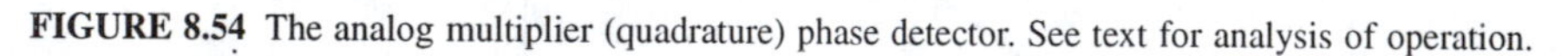
FIGURE 8.54 The analog multiplier (quadrature) phase detector. See text for analysis of operation.

The output of the analog multiplier is

$$e(t) = (V_1 V_2/20)[\sin(2\omega_1 t + \varphi) - \sin(\varphi)] \qquad 8.130$$

After passing through the low-pass filter, we have the DC component of e(t):

$$\bar{e} = -(V_1 V_2/20)\sin(\varphi) \cong -(V_1 V_2/20)\varphi \quad (\varphi \text{ in radians for } \varphi < .25\text{R}) \qquad 8.131$$

The sign change is the result of the trigonometrical algebra. Linearity in this phase detector is preserved for $\varphi < 12°$, otherwise an inverse sine nonlinearity must be used to recover a voltage proportional to φ.

A *NAND gate flip-flop (FF) phase detector* which works with transistor-transistor logic (TTL) signals is shown in Figure 8.55A. The input analog sinewaves are converted to TTL square waves by amplitude comparators connected as zero-crossing detectors with hysteresis for noise rejection. The TTL square wave outputs from the comparators are further conditioned by one-shot multivibrators such as 74LS123s, which have narrow complimentary outputs with dwell times generally much less than T/100. A simple NAND gate R-S flip-flop produces an output whose duty cycle depends on the phase difference between the input signals. As seen from the waveforms of Figure 8.55B, the FF output has a 50% duty cycle when φ is 180°. The FF output, Z(t), is inverted and

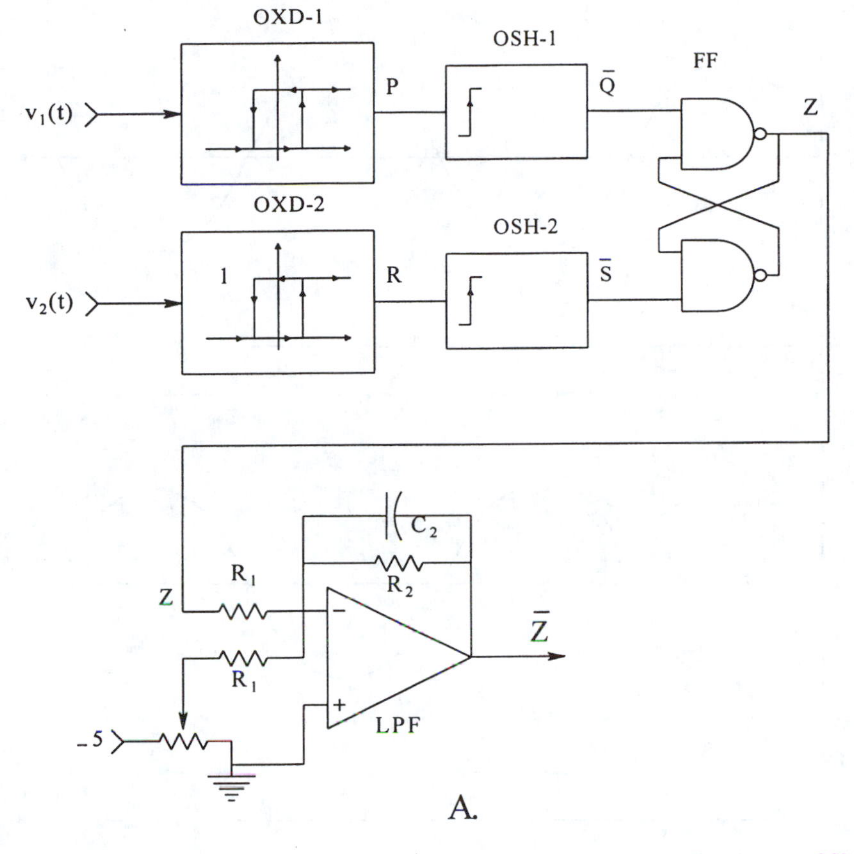

FIGURE 8.55 (A) Schematic of a NAND gate, RS flip-flop phase detector. 0XD, zero crossing detector; OSMV, one-shot multivibrator (e.g., 1/2-74123); FF, RS flip-flop; LPF, low-pass filter.

averaged by the op amp low-pass filter, and the DC level is adjusted so that when $\varphi = 180°$, the op amp output voltage $\bar{Z}$ is zero. A plot of $\bar{Z}$ vs. φ is shown in Figure 8.55C; note that it is periodic.

An *exclusive NOR gate* can also be used for an analog phase detector. The input signals are converted to TTL square waves for input to the ENOR gate; the gate's output is a double-frequency TTL pulse train whose duty cycle varies linearly with φ. As in the case of the RSFF phase detector, the ENOR output, W(t), is averaged by low-pass filtering, and the DC level is adjusted. The DC op amp output, W, is also a periodic function of φ.

A third, logic-based phase detector utilizes a special logic IC, the Motorola MC4044 (Motorola Semiconductor Products, Inc., Phoenix, AZ). Figure 8.56A illustrates the innards of the 4044, and shows how its two outputs can be connected to an op amp circuit which does low-pass filtering and DC adjustment. The I/O characteristics of the 4044 phase detector are shown in Figure 8.56B. Note that unlike the analog phase detectors described above, the output of the 4044 phase detector system is zero when $\varphi = 0$, and the linear range extends to a full ±360°.

Other analog phase detectors are of the *switched-amplifier design.* Such switched detectors are also of the quadrature type. In effect, the signal $v_2(t) = V_2\cos(\omega_1 t + \varphi)$ is multiplied by the function, $SGN[v_1(t)]$. SGN[x] is the well-known signum function; $SGN[x < 0] = -1$, $SGN[x \geq 0] = +1$. Thus:

$$Z(t) = K_A SGN[v_1(t)] V_2 \cos(\omega_1 t + \varphi) \tag{8.132}$$

From a consideration of the Fourier series for the $SGN[v_1]$ square wave, it is easy to show that for $\varphi < 12°$, the average of Z(t), $\bar{Z} = K_D\varphi$. An example of a switched-amplifier phase detector is illustrated in Figure 8.57.

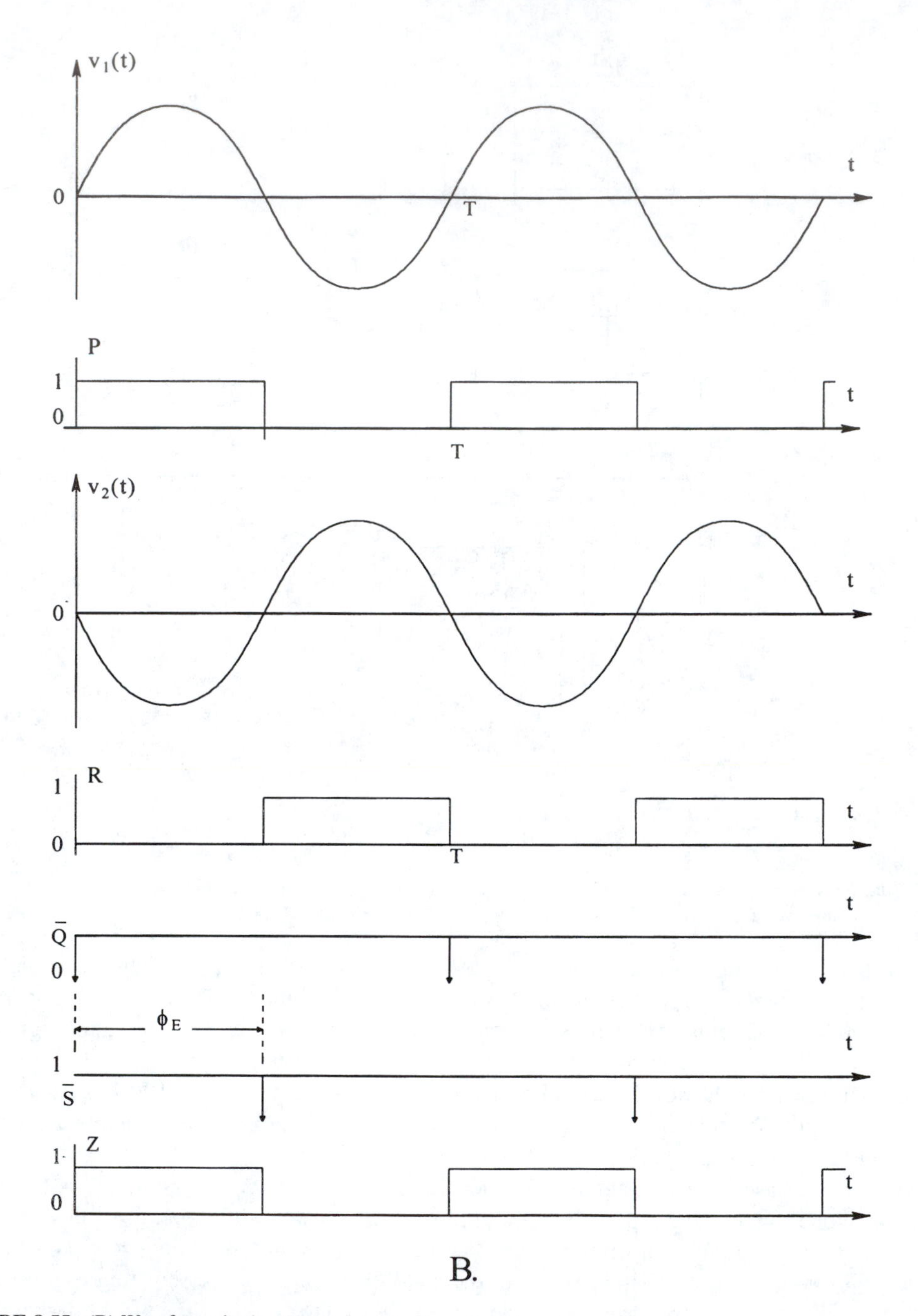

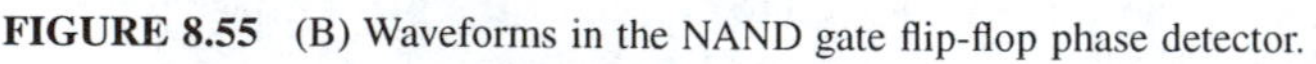
FIGURE 8.55 (B) Waveforms in the NAND gate flip-flop phase detector.

8.7.2 Digital Phase Detectors

We have seen above that analog phase detectors all generate waveforms whose average values are proportional to the phase shift, φ. Digital phase meters, on the other hand, measure the time interval, ΔT, as set forth in Equation 8.125B. The time interval is measured by counting clock pulses generated from a precision quartz crystal oscillator between successive, positive-going, zero crossings of v_1 and v_2. For example, the Stanford Research Systems (SRS) Model SR620 Universal Time Interval Counter uses a precision 90 MHz clock to count the intervals between the successive positive zero-crossings of v_1 and v_2. It also measures the period, T_1, of v_1. An internal microprocessor system then computes and displays:

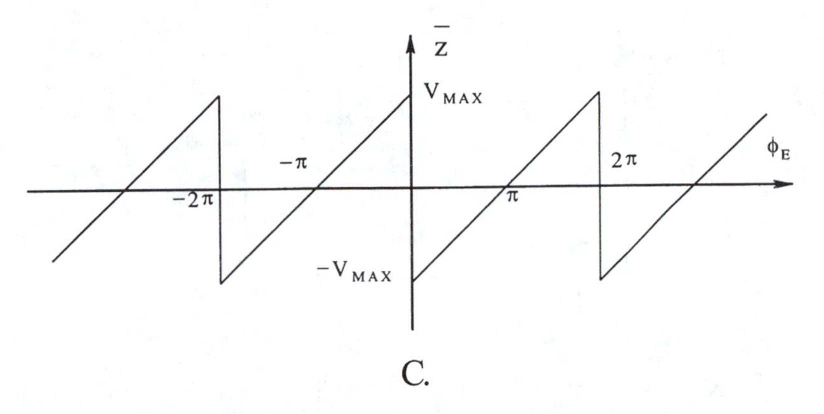

FIGURE 8.55 (C) Note the periodic form of this phase detector's transfer curve which has zeros at odd multiples of 180° phase difference.

$$\varphi = 360° \, \Delta T/T \text{ degrees} \tag{8.133}$$

The SRS SR260 phase meter works from 0 to 100 MHz, has a total range of ±180°, a resolution of $[25 \times 10^{-12} \times (1/T_1) \times 360 + 0.001]°$, and an error of less than $\pm[10^{-9} \times (1/T_1) \times 360 + 0.001]°$. The SR620 permits averaging a preset number of φs to compute a mean φ for greater accuracy in the presence of phase noise. The SRS SR620 is an example of a state-of-the-art instrument having an IEEE488 instrument/computer interface, and internally generated display voltages so that in the absence of a host computer, the instrument output and status can be displayed on any XY lab oscilloscope.

We next consider a millidegree phase meter designed by Northrop and Du. This system was described in detail in Section 7.3.2, and is illustrated in Figure 7.41. A phase-lock loop is run from the reference input signal as a phase-locked frequency multiplier. Originally designed to be used in an optical system in which the chopping frequency was 100 Hz, the Northrop and Du system generates a clock frequency which is 360,000 times the input frequency, or 36 MHz. Because of the 360,000:1 frequency multiplication of the PLL, there is 1 clock pulse per millidegree of input phase, regardless of any small drift in input frequency. The high-frequency PLL clock pulses are gated to a cascade of three, high-speed, TTL counters (74F579) during the interval, ΔT, where $v_1 > 0$ and $v_2 \leq 0$. The total count is obviously equal to φ in millidegrees. To improve accuracy in the presence of phase noise, the cumulative counts for $N = 2^k$ intervals, ΔT_n, can be averaged ($k = 0, 1, \ldots 8$, and $n = 1, 2, \ldots N$). The upper bound on the input frequency of the Northrop and Du phase measuring system is set by the PLL VCO (50 MHz for an NE564) or the counters (115 MHz for 74F579s). Also, the range of frequency drift tolerated by the system is set by the lock range of the PLL. Another system limitation is set by counter overflow; a φ over 46.6° will cause the counters to exceed their maximum count of 2^{24}.

8.8 MEASUREMENTS OF FREQUENCY AND PERIOD (TIME)

There are two basic definitions of frequency: *average frequency*, measured as events (e.g., zero crossings) per unit time, and *instantaneous frequency*, measured as the reciprocals of the time intervals between successive events. Most modern frequency meters are digital systems which measure average frequency by counting events (cycles) over an accurately determined gate time. However, there are several other less accurate means of measuring frequency which may be used. Some of these methods compare the unknown frequency with a frequency standard by some means. These include the Lissajous figure viewed on an oscilloscope, the zero-beat heterodyne method, the fast-fourier transform (FFT) spectrum analyzer, and the tunable filter.

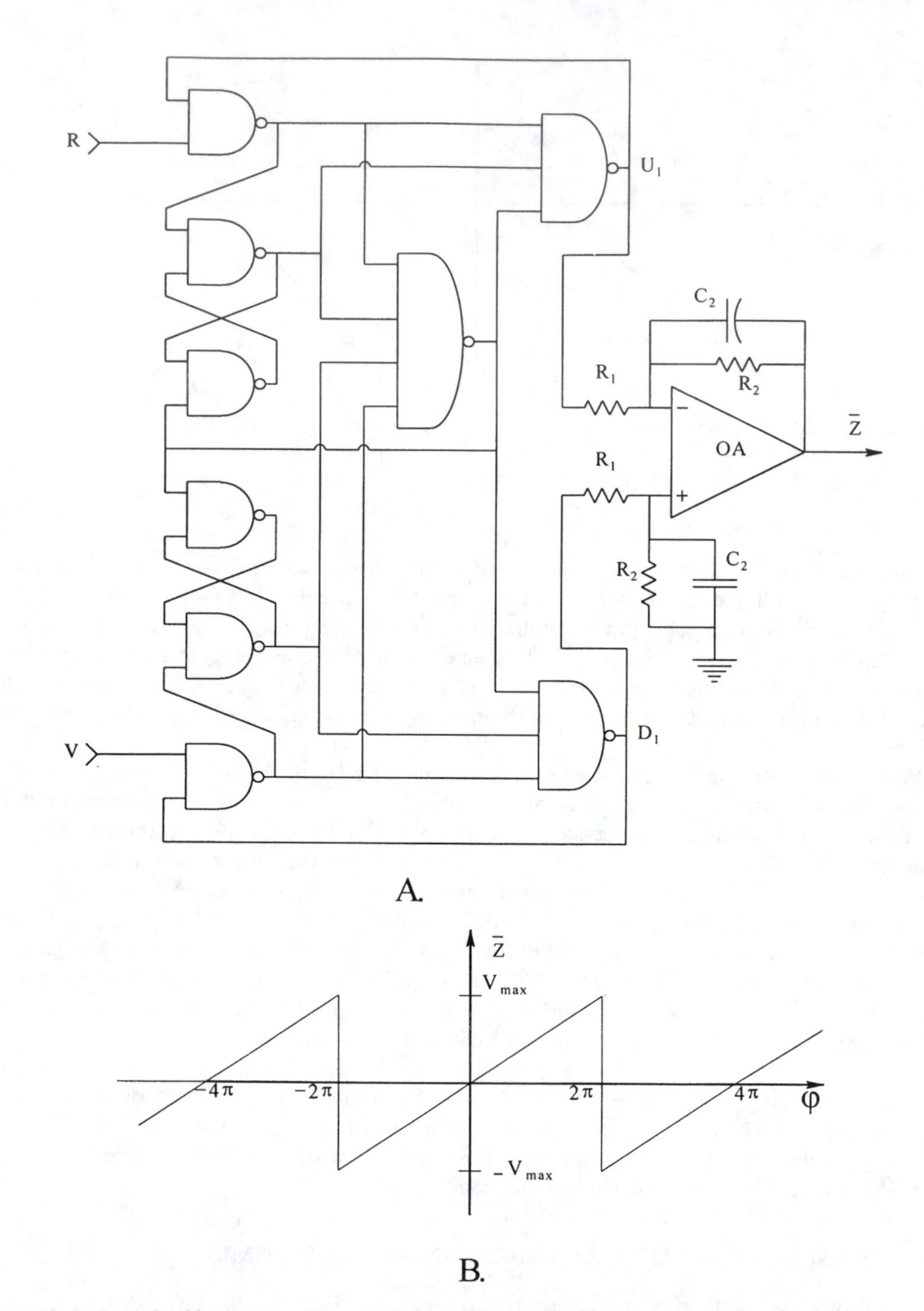

FIGURE 8.56 (A) Combinational logic used in the Motorola MC4044 digital phase detector. (Motorola Semiconductor Products, Inc., Phoenix, AZ). (B) I/O characteristic of the MC4044 phase detector.

The oscilloscope can be used to measure frequency by comparing the unknown frequency to an accurate, variable frequency standard by a direct observation of the Lissajous figure on its CRT. The Lissajous pattern is made by putting the waveform with the unknown frequency on the y (vertical)-axis, and the known frequency waveform on the x (horizontal)-axis. When the known frequency is adjusted so that the observed pattern is either a stationary straight line, ellipse, or

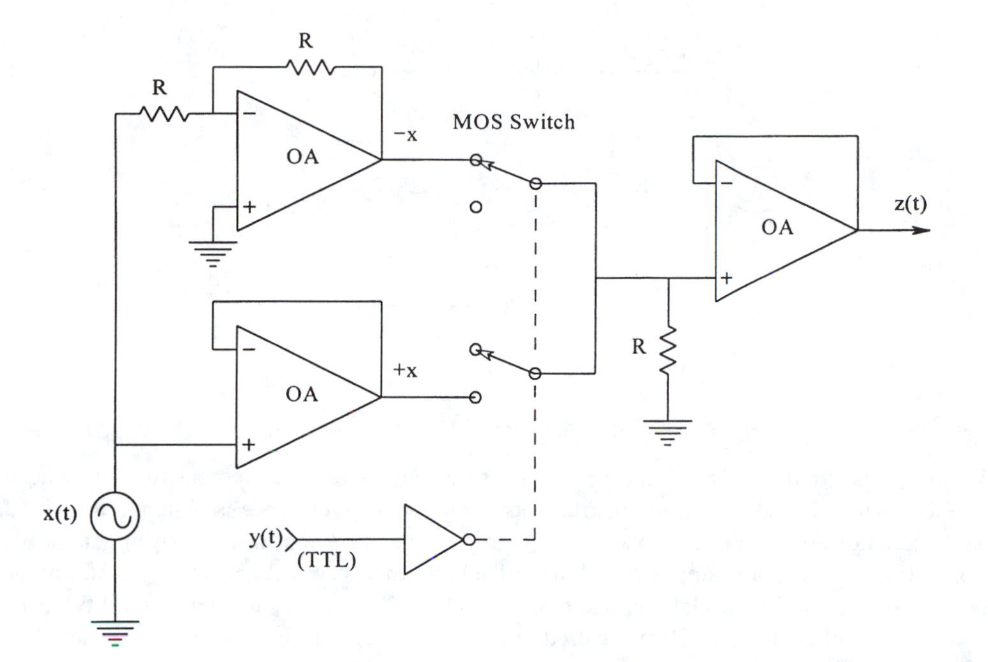

FIGURE 8.57 A switched-amplifier phase detector. This is a quadrature detector with zeros at odd multiples of 90° phase difference between the sinewave input at x(t), and the TTL squarewave at y(t). The output z(t) must be low-pass filtered to recover V_o proportional to $\sin(\theta_E)$, where θ_E is the phase difference between a cosine wave applied at x, and a "sinusoidal" square wave at y.

circle, the known frequency equals the unknown frequency, exactly. The Lissajous method works best at audio and low radio frequencies.

The zero-beat method also makes use of a variable-frequency, standard source which is mixed with the unknown frequency signal. Recall that mixing, in its purest form is multiplication. Mixing can also be accomplished by adding the two signals together, and then passing them through a nonlinear amplitude transfer function, such as $y = a + bx + cx^2 + dx^3$, etc. The output from the mixing process can be shown to contain sinewaves having frequencies which are the sum and difference of the two input frequencies. Thus as the variable, known frequency is adjusted close to the unknown frequency, the frequency of the output difference frequency term approaches zero. This zero-beat phenomenon can be detected by actually listening to the audio beat frequency resulting from mixing the standard frequency source with the unknown. The beat frequency method has an apparent dead zone where the mixed difference frequency lies below the human range of hearing. Visual observation of the beat frequency signal on an oscilloscope or DC voltmeter can partially overcome the dead zone problem.

Another means of estimating the frequency of a coherent signal is to observe its rms spectrum on the display of an FFT spectrum analyzer. In such instruments, the frequency resolution is determined by the sampling frequency and the number of samples. Resolution can be as high as one part in 1024 in a 1024 point transform, and is obviously lower for fewer points.

The Wien bridge was discussed in Section 5.4.1.5 as a means of measuring capacitance. The Wien bridge also is unique in that its output also exhibits a null at its tuned frequency, behaving like a passive notch filter. When the bridge is configured as shown in Figure 8.58, it is easy to show that the output null occurs when $f = 1/(2\pi RC)$ Hz. Again, the null can be detected at audio frequencies by headphones, or by an AC voltmeter. Typically, capacitors $C_1 = C_2 = C$ are varied together to obtain null. Alternately, the resistors $R_1 = R_2 = R$ can be varied together with the Cs constant. The accuracy of the Wien bridge method depends on the sensitivity of the null detector, and the calibration of bridge components. It is best used at audio frequencies.

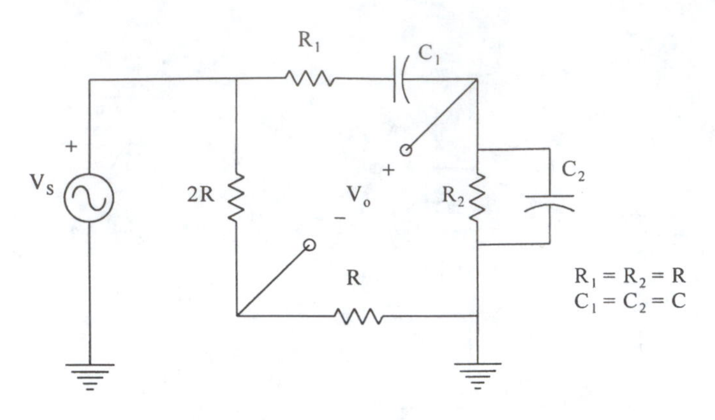

FIGURE 8.58 A Wien bridge used as a frequency-sensitive circuit.

Fundamental to all frequency and period measurements is the requirement for a very accurate frequency source. In most benchtop instruments, the frequency reference is obtained from a quartz crystal oscillator operated under conditions of strict temperature control in an oven generally kept to ±0.01°C or better. For example, the Hewlett-Packard (San Jose, CA) HP105B quartz frequency standard has buffered sinusoidal outputs at 5, 1, and 0.1 MHz, and a short-term stability measured over at 5 MHz of 5 parts in 10^{12} (measured over a 1 s averaging time). The "aging rate" is $<5 \times 10^{-10}$ per 24 h for this oscillator.

Accurate variable reference frequencies can be derived from accurate fixed-frequency oscillators by several means. Direct frequency multiplication uses the generation of harmonics of the reference frequency waveform. The reference waveform is passed through an amplitude comparator to generate a TTL or emitter coupled logic (ECL) waveform of the same frequency. This logic waveform is then passed through a one-shot multivibrator to create a train of narrow pulses having a Fourier spectrum containing many harmonics. The pulse train is then made the input to a high-Q tuned circuit which has its center frequency equal to that of the desired harmonic. It is practical to recover harmonics up to order 9 with this technique. One can also use synchronous binary logic "rate multipliers," such as the 7497, to reduce the oscillation frequency to f_R, given by the relation,

$$f_R = f_o(M/64) \tag{8.134}$$

where f_o is the oscillator frequency, M is a six-bit binary number, e.g.,

$$M = D_0(2^0) + D_1(2^1) + D_2(2^2) + D_3(2^3) + D_4(2^4) + D_5(2^5) \tag{8.135}$$

and D_k = 1 or 0. More complex frequency synthesis techniques make use of two or more interconnected phase-lock loops, mixers, rate multipliers, and crystal oscillator sources. Figure 8.59 illustrates an example of a vernier loop synthesizer. A detailed treatment of PLL frequency synthesizers may be found in the texts by Egan (1981) and by Kinley (1980). The Hewlett-Packard Model 8360 series of microwave frequency synthesized sweep generators make use of a single, high accuracy 10 MHz quartz reference source, and uses three PLLs and a feedback scheme to obtain 1 Hz resolution over a range of 2 MHz to 40 GHz (HP83642A). The HP 8360 synthesizer architecture is shown in Figure 8.60. The four subsystems in this synthesizer are a 2 to 7.8 GHz YIG-tuned oscillator (YTO), a reference PLL, a fractional-N PLL, and a sampler loop. A portion of the RF output from the YTO is sent to a sampler where it is mixed with a high-order harmonic of the 200 to 220 MHz PLL forming a 20 to 40 MHz IF signal. A phase detector compares this IF signal to the output of the fractional-N PLL. This phase detector output is summed into the tuning control for the YIG oscillator. The sampler loop can be incremented in 500 kHz steps, and

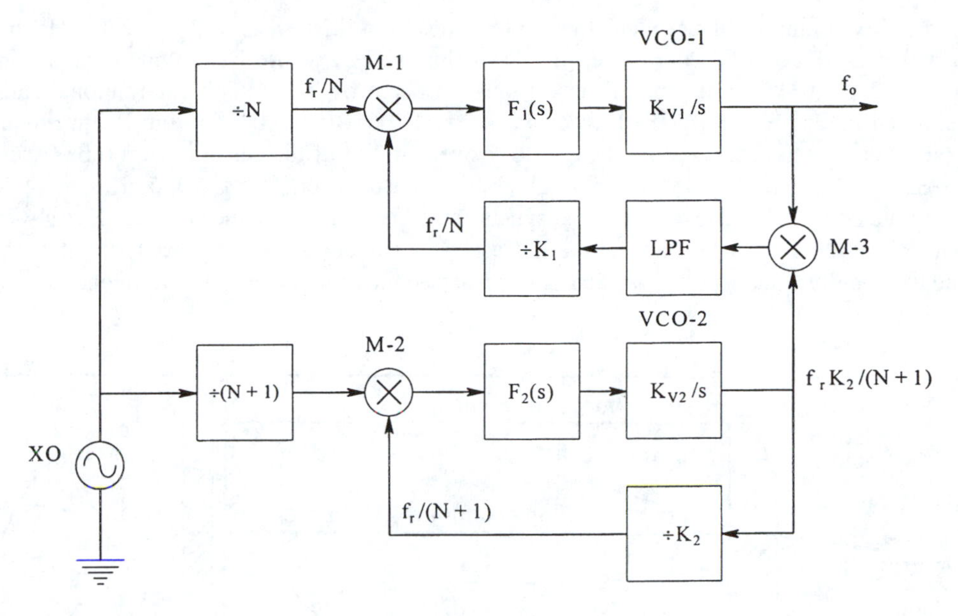

FIGURE 8.59 A phaselock, vernier-loop, frequency synthesizer.

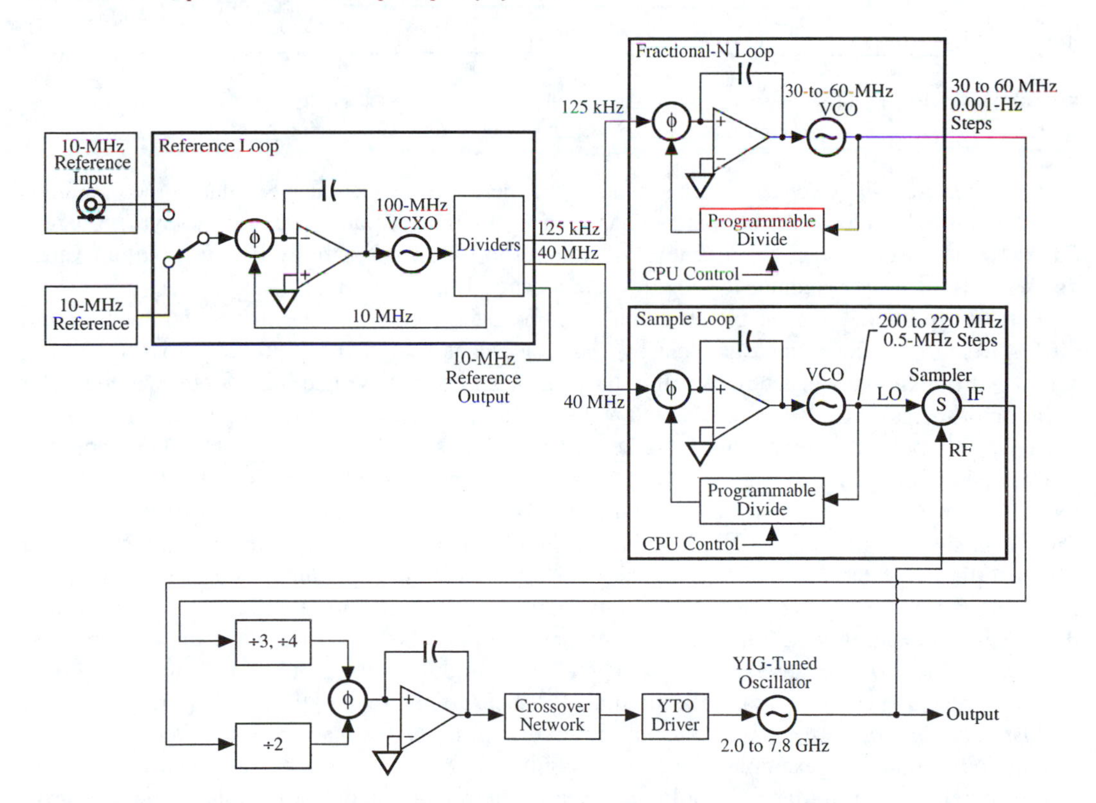

FIGURE 8.60 System diagram of a Hewlett-Packard HP8360 frequency synthesizer. See text for explanation. (Used with permission from Hewlett-Packard Co., San Jose, CA.)

the fractional-N loop can be stepped in 0.001 Hz increments. This mode of operation permits this family of microwave frequency synthesizers to maintain 1 Hz resolution over their entire output range.

The primary time/frequency standard in current use is the cesium-133 atomic beam clock. A diagram of the cesium clock is shown in Figure 8.61. This standard was discussed in detail in

Chapter 2. To summarize the cesium clock's properties, it oscillates at 9.192 631 770 GHz, having an effective Q of 2×10^8. Coordinated universal time (UTC) whose basic unit is the second, is defined as 9,192,631,770 periods of the cesium-133 beam oscillator. This international standard was adopted in October 1967. The Hewlett-Packard HP5061B(Opt 004) Cesium Beam Frequency Standard has a long-term stability of $\pm 2 \times 10^{-12}$ over the life of the cesium beam tube. Accuracy is also $\pm 2 \times 10^{-12}$. The HP5051B cesium clock has sinusoidal outputs at 10, 5, 1, and 0.1 MHz. HP cesium clocks are used to calibrate and synchronize the SATNAV, Omega, and LORAN-C radio navigation systems for boats and aircraft. Cesium beam clocks are ordinarily used to adjust secondary rubidium and quartz oscillators used as secondary standards for frequency or period determination.

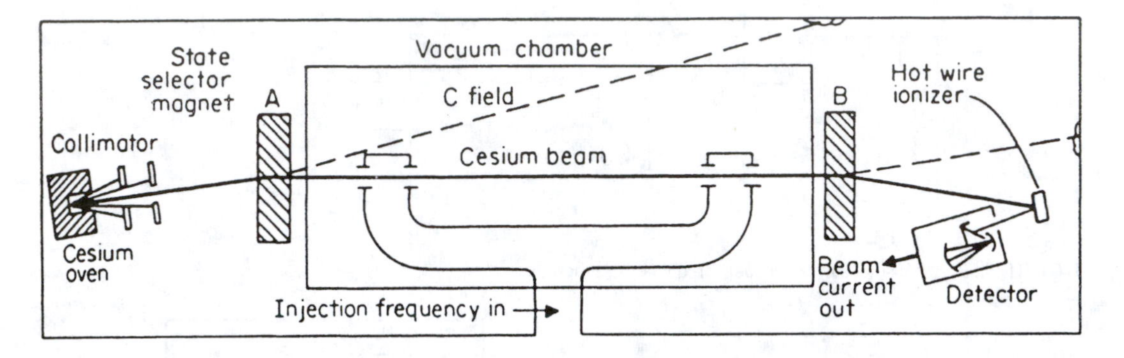

FIGURE 8.61 Diagram of a cesium beam frequency standard clock. (From Oliver, B.M. and Cage, J.M., *Electronic Measurements and Instrumentation*, McGraw-Hill, New York, 1971. With permission.)

Rubidium frequency standards are second in the hierarchy of accuracy. Similar to the operation of a cesium beam clock, the atomic resonance of a rubidium vapor cell is used to synchronize a quartz crystal oscillator in a frequency-lock loop. The long-term stability of the rubidium vapor oscillator is $\pm 1 \times 10^{-11}$/month. It, too, has outputs at 5, 1, and 0.1 Mhz.

The two principal modes of operation of universal counter-timer instruments, such as the HP5345A, are frequency measurement by counting events such as positive zero crossings of the signal with unknown frequency over the duration of an accurately determined gate window. The duration of the gate window is determined by logic circuits which count a preset number of internal master clock cycles. The gate dwell can be set from 1000 s to 100 nsec in the HP5345A counter. The frequency range on channel A is 0.05 to 500 MHz, and periods can be measured from 2 nsec to 2×10^4 s. The master quartz clock in the HP5345A counter-timer runs at 10 MHz, and has a long-term stability of $<\pm\ 5 \times 10^{-10}$/d, and a 1 s stability of $<\pm\ 1 \times 10^{-11}$. Figure 8.62 illustrates the Hz uncertainty of the HP5345A as a function of applied frequency, input noise, and gate time. Note that the measured frequency uncertainty increases with input noise and input frequency, and decreases with gate time. Counter-timer instruments can also be operated to measure the time, ΔT, between positive (or negative) zero crossings of two periodic inputs, thus providing phase information. They also can be run as accurate timers to measure the time between input trigger pulses.

Instantaneous frequency measurements are used to describe the unevenness or frequency noise in a periodic signal. An example of such a noisy frequency source is the human electrocardiogram (Northrop et al., 1967). Many physiological factors control and modulate the rate of the heartbeat; these include blood pCO_2, emotions, exercise, etc. Other applications of instantaneous frequency measurements have been to characterize nerve impulse sequences (Northrop and Horowitz, 1966) and the variations in the rotational speed of a turbine.

The instantaneous pulse frequency demodulator (IPFD) was described in Section 7.1.3, and two circuits for IPFDs are shown in Figures 7.11 and 7.12. The jth element of instantaneous frequency, r_j, is defined as the reciprocal of the jth period or inter-event interval. That is,

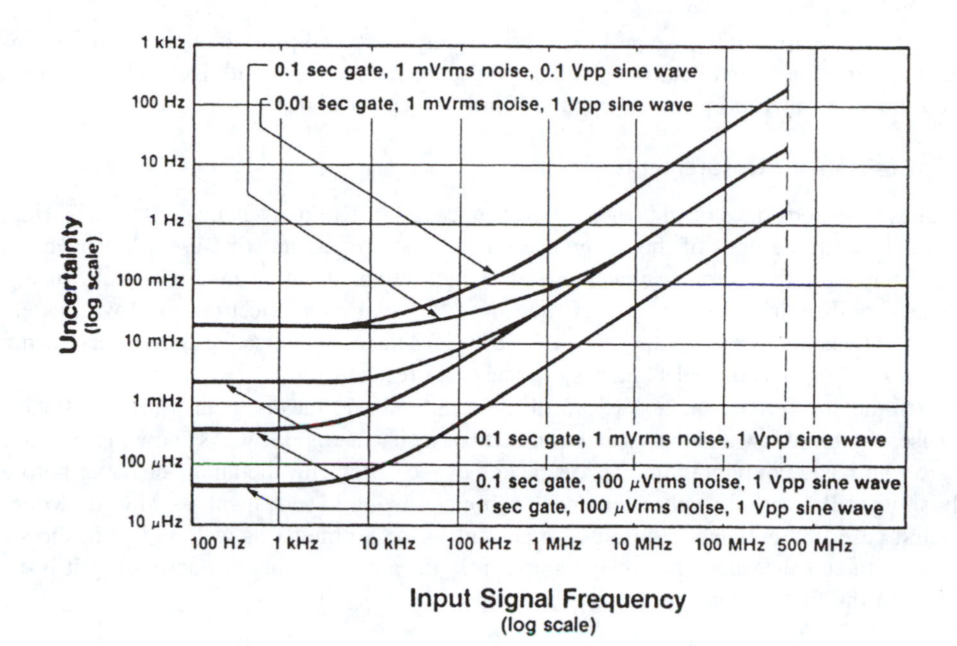

FIGURE 8.62 Frequency resolution error of the HP5345A counter-timer instrument. Noise accompanying the input signal and internal uncertainties affect the accuracy of frequency and period measurements. (Used with permission of Hewlett-Packard Co., San Jose, CA.)

$$r_j \equiv 1/(t_j - t_{j-1}) \qquad 8.136$$

where t_j is the time of occurrence of the jth event and t_{j-1} is the time of occurrence of the next previous event. Any periodic waveform can thus be characterized in the time domain by a number sequence, $\{r_i\}$, which describes the variation of frequency of the waveform on an interval by interval basis. As seen in Equation 7.48, the instantaneous frequency number sequence can be converted to a stepwise analog voltage output, given by:

$$q(t) = k\sum_{i=1}^{i=\infty}\left[1/(t_i - t_{i-1})\right]\left[U(t - t_i) - U(t - t_{i+1})\right] \qquad 8.137$$

where U(t − a) is the unit step function, defined as 0 for $t < a$, and 1 for $t \geq a$.

IPFDs can be of either analog or digital design, and have generally found applications in characterizing the irregularities of frequency in audio frequency waveforms.

In summary, we see that there are a variety of methods which can be used to measure or to estimate a signal's frequency, which can range from 10^{-3} Hz (mHz) to the hundreds of GHz ($>10^{11}$ Hz). Frequency measurements can be made by direct counting of events per unit time, or by comparison with an accurate, calibrated oscillator. At very low frequencies, it is more effective to measure the period of the unknown signal, rather than to count its cycles over some fixed interval. Averaging of either period or frequency measurements improves accuracy.

8.9 MEASUREMENT OF RESISTANCE, CAPACITANCE, AND INDUCTANCE

In this section, we will examine and review the various means of measuring resistance, capacitance, and inductance. We have already examined null methods of measuring R, C, and L at audio frequencies using various bridge circuits in Chapters 4 and 5. These methods will be referred to, but not repeated here. Rather, we will examine various active and passive ohmeter circuits, means of characterizing linear and nonlinear (voltage-variable) capacitances with DC and high-frequency AC, and means of measuring the properties of inductances at high frequencies.

8.9.1 Resistance Measurements

Techniques have been developed to measure resistances from 10^{-7} ohms to over 10^{14} ohms. Needless to say, at the extreme ends of this range specialized instruments must be used. We have already seen in Chapter 4 that very accurate resistance measurements are commonly made using DC Wheatstone or Kelvin bridges and a DC null detector such as an electronic nanovoltmeter. The values of the resistances used in the arms of these bridges must, of course, be known very accurately. Below we shall discuss other DC means of measuring resistance.

The voltmeter-ammeter method is probably the most basic means of measuring resistance. It makes use of Ohm's law and the assumption that the resistance is linear. As shown in Figure 8.63, there are two basic configurations for this means of measurement; the ammeter being before R_X, which is in parallel with the voltmeter, and the ammeter being in series with R_X after the voltmeter. In the first case, the ammeter measures the current in the voltmeter as well as R_X; in the second case, the voltmeter measures the voltage drop across the ammeter plus that across R_X. It is easy to show that in the first case, R_X is given by:

$$R_X = \frac{V_X}{I - V_X/R_{VM}} \tag{8.138}$$

where V_X is the voltmeter reading, I is the ammeter reading, and R_{VM} is the resistance of the voltmeter (ideally, infinite). In the second case, we find

$$R_X = V/I_X - R_{AM} \tag{8.139}$$

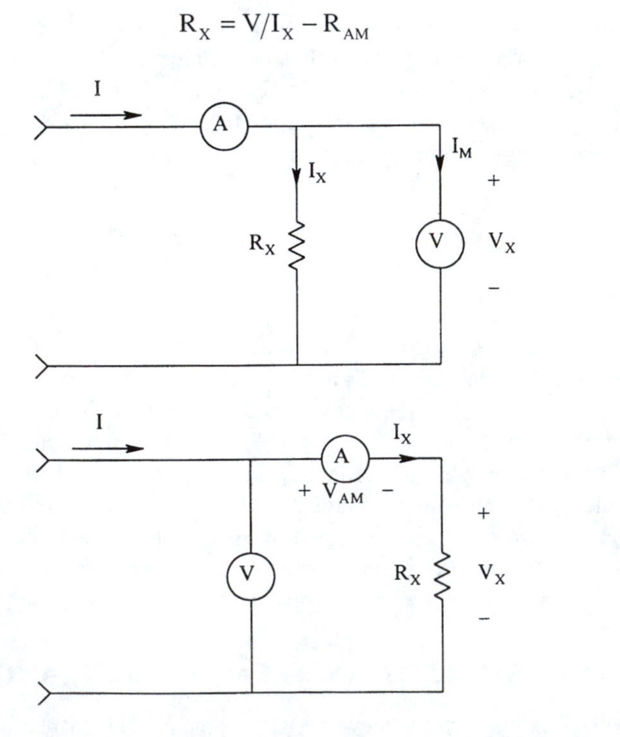

FIGURE 8.63 Two circuits that can be used to measure resistance using a voltmeter and ammeter.

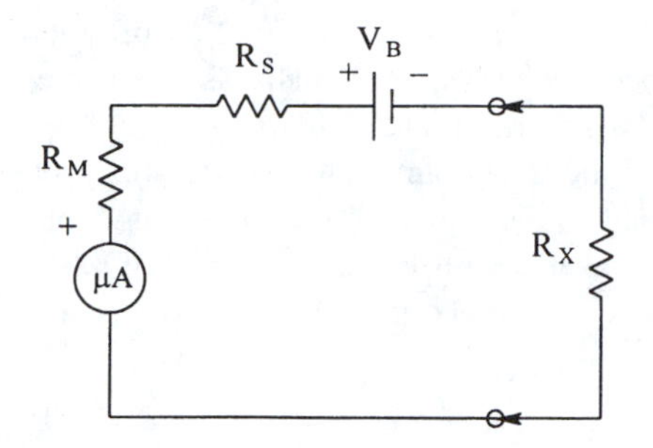

FIGURE 8.64 A simple series ohmmeter circuit.

where V is the voltmeter reading, I_X is the ammeter reading (current through R_X) and R_{AM} is the resistance of the ammeter. The first method is best for measuring low values of R_X where $R_{VM} >> R_X$, and the second method has less error for $R_{AM} << R_X$. Obviously, R_X must be found by calculation, and the accuracy of the result depends on the accuracy of the meters.

The voltmeter-ammeter method of measuring resistance can be used to measure very low resistances, on the order of microohms. An electronic current source is used to pass a known amount of DC current through the unknown, low resistance. An electronic DC nanovoltmeter is connected across R_X, which is directly proportional to V_M.

Common passive benchtop volt-ohm-milliammeters, or VOM, milliammeters, or VOMs (multimeters) use a simple series ohmeter circuit, shown in simplest form in Figure 8.64, have a nonlinear (hyperbolic) scale on the D'Arsonval microammeter with zero ohms full-scale and infinite ohms at zero deflection. A typical benchtop VOM series ohmmeter may have five R_H ranges, ranging typically from 12 ohms to 120 kiloohms. Inspection of the simple circuit shows that the DC meter current is given by:

$$I_M = \frac{V_B}{(R_M + R_S) + R_X} \tag{8.140}$$

and the full-scale meter current is:

$$I_{M(FS)} = V_B/(R_M + R_S) \tag{8.141}$$

We note that $(R_M + R_S)$ is the Thevenin resistance the resistor R_X "sees." It is convenient to define $(R_M + R_S)$ as R_H, the half-deflection resistance. That is, when $R_X = R_H$, $I_M = I_{M(FS)}/2$. Further analysis of the simple series ohmmeter is made simpler by defining the ohmmeter's fractional meter deflection, F, as:

$$F = \frac{\theta}{\theta_{FS}} = \frac{I_M}{I_{M(FS)}} = \frac{R_M + R_S}{R_M + R_S + R_X} = \frac{1}{1 + R_X/R_H} \tag{8.142}$$

Here θ is the meter deflection angle, θ_{FS} is the full-scale deflection angle, and the Rs are as shown in Figure 8.64.

To find the most sensitive part of the series ohmmeter scale, we define the ohmmeter sensitivity S as dF/dR_X. This derivative is easily found from Equation 8.142 above. It is:

$$S = \frac{dF}{dR_X} = \frac{R_H}{(R_H + R_X)^2} \tag{8.143}$$

S clearly has a maximum with respect to R_H. If we find dS/dR_H and set it equal to zero, the peak in the sensitivity is seen to occur when $R_H = R_X$. That is, the center of the series ohmmeter scale is the most sensitive (and most accurate) part.

Because the voltage, V_B, of the series ohmmeter's battery drops as the battery is used and ages, a variable resistor, R_V, is added in parallel with the microammeter to compensate for the drop in V_B. This practical series ohmmeter circuit is shown in Figure 8.65. R_V is adjusted with the meter leads shorted ($R_X = 0$) so $I_M = I_{MFS}$. In this circuit, R_H is still the Thevenin resistance R_X sees; this is:

$$R_H = R_S + \frac{R_M R_V}{R_M + R_V} \quad 8.144$$

An increase of R_V to compensate for a drop in V_B will thus change R_H from its design value and cause a small meter calibration error.

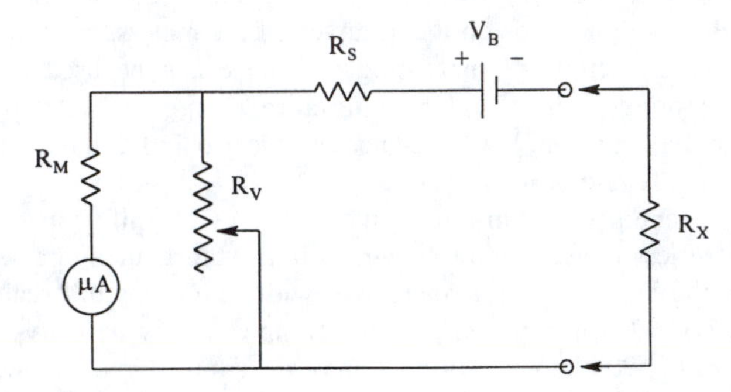

FIGURE 8.65 A practical series ohmmeter circuit; R_V is used to set the ohmmeter to zero in compensation for changes in the value of V_B.

Shunt ohmmeters are less frequently encountered than are series ohmmeters. Shunt ohmmeters are used to measure low resistances, and have R_Hs ranging from 0.5 to 50 ohms. They are primarily used to measure resistances associated with coils such as DC motor armature windings. The circuit of a shunt ohmmeter is shown in Figure 8.66. In this circuit, the battery must supply a substantial current, often in the ampere range. Consequently, a robust battery, such as a lead-acid motorcycle-type battery, is used. As in the case of the series ohmmeter, we can define the half-deflection resistance, R_H, as the Thevenin resistance seen by R_X:

$$R_H = \frac{R_M R_S}{R_M + R_S} \quad 8.145$$

Full-scale meter current occurs when $R_X = \infty$, and is simply

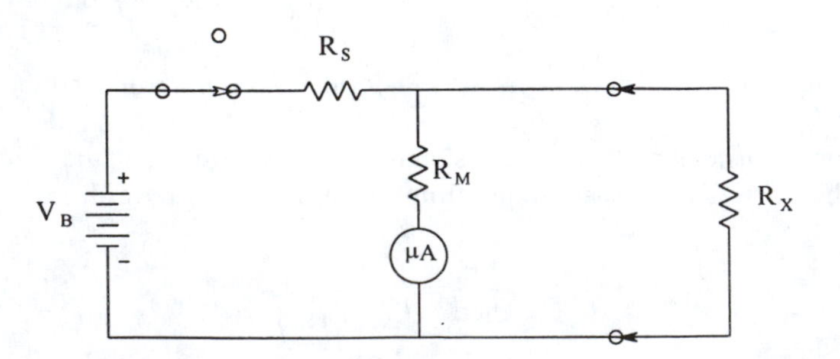

FIGURE 8.66 Circuit of a simple shunt ohmmeter.

$$I_{M(FS)} = V_B/(R_M + R_S) \quad 8.146$$

In general, the meter current is:

$$I_M = \frac{V_B R_X/(R_X + R_S)}{R_M + R_X R_S/(R_X + R_S)} = \frac{V_B R_X/(R_M + R_S)}{R_X + R_M R_S/(R_M + R_S)} \quad 8.147$$

Zero meter deflection occurs for $R_X = 0$ (short-circuited input). The fractional deflection factor of the shunt ohmmeter is found to be:

$$F = \frac{\theta}{\theta_{FS}} = \frac{I_M}{I_{M(FS)}} = \frac{R_X}{R_X + R_H} \quad 8.148$$

It can also be shown that the center of the shunt ohmmeter's scale is also the point of greatest sensitivity.

Both series and shunt ohmmeters have relatively low accuracy, generally 2% to 5% at center scale, which is adequate for most noncritical applications, such as measuring a motor coil's resistance, checking circuit continuity, or verifying resistor values (for those persons who do not know the resistor color code).

We next examine several electronic ohmmeter circuits. The readout of such circuits can be either analog or digital. In general, their accuracy is an order of magnitude better than that obtained with a series or shunt ohmmeter. Electronic ohmmeters can measure resistances ranging from 0.1 microohm up to 10^{18} ohms. Obviously, special techniques and instruments must be used to measure resistances at the extreme ends of the range cited above. The first electronic circuit we shall consider is the *"normal-mode" ohmmeter*, shown in Figure 8.67. The op amp can be an IC, electrometer type, having a bias current on the order of 40 fA. A calibrated current source, easily made from a battery and a large resistor, injects a DC current, I_S, into the unknown resistance, R_X. The voltage at the op amp's noninverting input is given by KCL:

$$V_1(G_X + sC_i) + I_B = I_S \quad 8.149$$

or

$$V_1 = \frac{(I_S - I_B)R_X}{(1 + sC_i R_X)} \quad 8.150$$

It is easy to see that the op amp's output, $V_O = V_1(1 + R_1/R_2)$, so

$$V_O = (1 + R_1/R_2)\frac{(I_S - I_B)R_X}{(1 + sC_i R_X)} \quad 8.151$$

If the switch is opened at t = 0, the input may be considered to be $(I_S - I_B)/s$, and $v_o(t)$ will rise exponentially with time constant $C_i R_X$ to a steady-state value given by:

$$v_{o(SS)} = (1 + R_1/R_2)(I_S - I_B)R_X \text{ volts} \quad 8.152$$

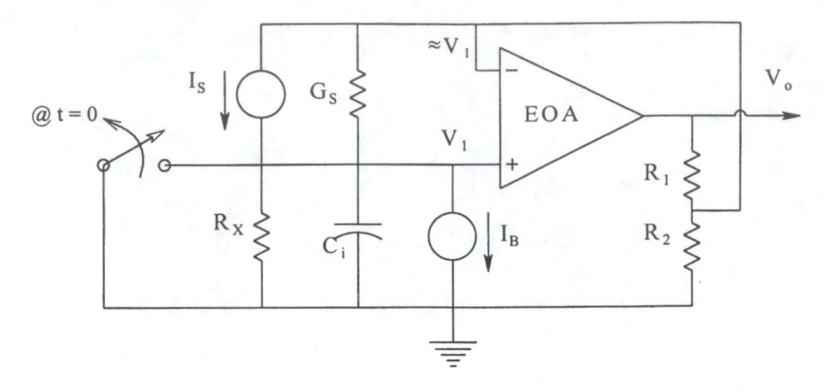

FIGURE 8.67 Schematic of a "normal-mode" electronic ohmmeter. I_S and G_S represent the Norton equivalent of a floating current source. See text for analysis.

Thus, the op amp output voltage is directly proportional to R_X, and a linear ohms scale can be used on an analog output meter. Two problems arise with this circuit when measuring R_Xs over 10^{10} ohms: the bias current must be kept much less than I_S, and if C_i is on the order of several hundred picofarads, the ohmmeter's response time constant can become significantly long (tens or hundreds of seconds), requiring excessive time to obtain a steady-state reading. If we assume that full-scale deflection of the normal-mode electronic ohmmeter is $v_{o(SS)} = 1$ V, then full-scale resistances from 1 to 10^{10} ohms can be measured with I_S values ranging from 10 mA to 1 pA, respectively, with $(1 + R_1/R_2) = 100$.

To measure resistors in excess of 10^{10} ohms, the *fast-mode electronic ohmmeter* configuration, shown in Figure 8.68, is generally used. Now the unknown resistor is placed in the feedback loop of the op amp. One end of R_X is at virtual ground at the summing junction, and the other end is connected to the V_2 node. The voltage V_2 is clearly equal to $-(I_S + I_B)R_X$. By KCL, we may write:

$$V_2(G_2 + G_1) - V_O G_1 - I_S = 0 \tag{8.153A}$$

$$-(I_S - I_B)R_X(G_2 + G_1) - V_O G_1 - I_S = 0 \tag{8.153B}$$

$$-I_S[R_X(G_2 + G_1) + 1] - I_B R_X(G_2 + G_1) = V_O G_1 \tag{8.153C}$$

and assuming $R_X >> R_1, R_2$, we can solve for the output voltage:

$$V_O = -R_X I_S(1 + R_1/R_2) - R_X I_B(1 + R_1/R_2) \tag{8.154}$$

The second term is negligible as long as $I_S >> I_B$. If the output current is to be read by the microammeter, M, we note that

$$I_O = (V_O - V_2)/R_1 = -I_O R_X(G_2 + G_1) + I_S R_X G_1 = -I_S R_X G_2 \tag{8.155}$$

For example, if $R_X = 10^{14}$ ohms, $I_S = 1$ pA, and $R_2 = 10^4$ ohms, then $I_O = -10$ mA.

High megohm resistors (1 G Ω or higher) are central to the operation of many electrometer instruments, including feedback picoammeters, as described in Section 8.3.2 of this text. Such resistors are typically made from metal oxides, and are enclosed in sealed glass tubes. The voltage coefficient of this type of resistor, defined as $(\Delta R/R)/V$, is typically less than 5 ppm up to 100 V

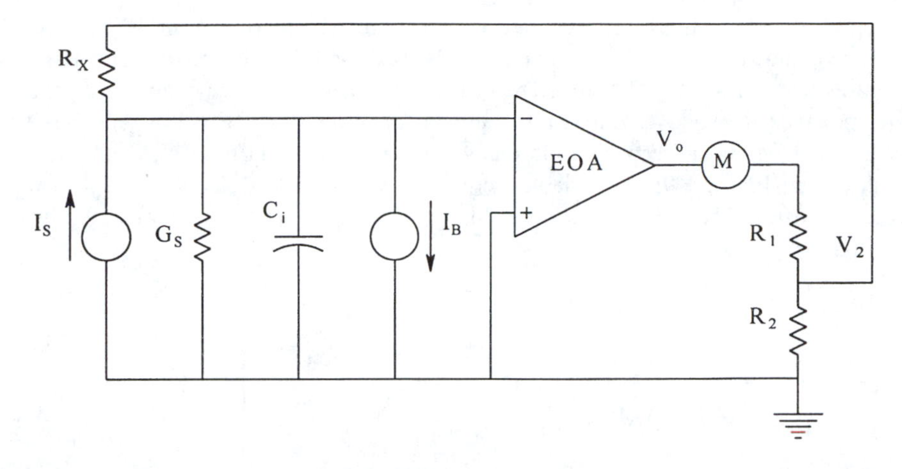

FIGURE 8.68 Schematic of a "fast-mode" electronic ohmmeter. R_X is placed in the op amp's feedback loop. The summing junction is at virtual ground.

applied voltage, and the tempcos of high-meg resistors are on the order of 100 ppm/°C for 10^8 ohms, and 1000 ppm/°C for 10^{11} ohms. High meg resistors are delicate, and must be protected from mechanical shock, and their glass envelopes must be kept free of dirt and fingerprint oils and salt by cleaning with pure alcohol after handling.

Measurement of high meg resistors with fast-mode electronic ohmmeters is not as simple as just connecting R_X to the meter and taking the reading. The reading can be influenced by static and slowly changing electrostatic fields. Even on humid days, most persons have a DC electric field associated with their bodies, even when cotton clothing is worn. Such fields can influence the measurement of R_X, destroying accuracy. To uncouple R_X from DC electrostatic influences, and to prevent pickup of powerline E and B fields by the measurement apparatus, R_X should be thoroughly shielded both electrostatically and magnetically. (The 60 Hz fields can induce voltages that may saturate the amplifier, giving incorrect DC readings of resistance.)

8.9.2 Capacitance Measurements

We have seen that accurate capacitance measurements can be made at audio frequencies (generally 1 kHz) using a variety of bridge circuits, described in Section 5.4.1 of this text. It is often important to characterize capacitors at radio frequencies, and at ultra-low frequencies; the latter range being important in the characterization of capacitor dielectrics, and dielectrics used as electrical insulation. There are also voltage-variable capacitances associated with semiconductor pn junctions which are important to measure. These include the capacitance of varactor diodes (reverse-biased pn junction diodes), the gate-source capacitance, gate-used and ages, a variable resistor, R_V, is added in parallel with the microammeter to drain capacitance, drain-source capacitance, C_{iss}, and C_{rss} of FETs, and the C_π and C_μ of BJTs. We will describe below some of the non-bridge means to measure capacitors and to characterize their equivalent circuits.

8.9.2.1 The Q-Meter Used for Capacitance Measurement

At radio frequencies, the equivalent circuit for a capacitor will not only include a resistance to account for dielectric power losses, but also losses due to skin effect in the capacitor's leads. Equivalent inductances may also appear at VHF, UHF, and SHFs due to lead geometry and physical layout of the plates and dielectric. Figure 8.69 illustrates one such equivalent circuit. A very practical instrument for measuring the net capacitance of a capacitor at a radio frequency is the Q-meter. (The Q-meter can also be used to measure inductor Q and inductance, as described in the next section.) There are several commercial Q-meters: the Boonton-type 260A covers the frequency range from 50 kHz to 50 MHz. The Boonton-type 190A Q-meter covers from 20 to 260 MHz, the

Marconi-type TF-1245 is used with separate voltage sources covering from 40 kHZ to 50 MHz (type TF-1246), and from 20 to 300 MHz (type TF-1247), and the Hewlett-Packard HP 4342A Q-meter operates from 22 kHZ to 70 MHz. The Q-meter is basically a very simple circuit, consisting of a low impedance RF voltage source, a variable capacitor, and an RF voltmeter. The Q-meter is used with standard coils and capacitors in many of its measurement modes. A schematic of a Q-meter is shown in Figure 8.70. A variable-frequency RF current is passed through a 0.02 ohm, noninductive resistor. The RF voltage across this resistor appears to be from a nearly ideal RF voltage source, V_1.

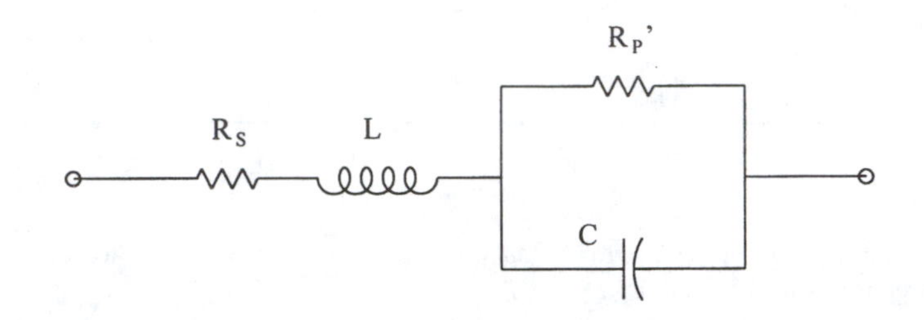

FIGURE 8.69 High-frequency, lumped-parameter equivalent circuit of a practical capacitor.

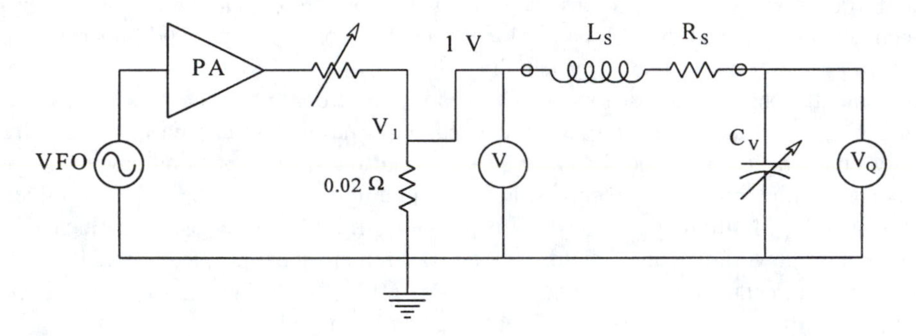

FIGURE 8.70 Circuit of basic Q-meter. VFO, variable radio frequency (RF) oscillator; PA, power amplifier. See text for description.

Measurement of an unknown capacitance, C_X, can be done in several ways. In the first method, C_X is placed across the internal capacitance, C_V, and a standard series inductor of appropriate value is placed in series. The circuit is resonated at f_o and the capacitance of the variable capacitor is recorded as C_1. Next, C_X is removed, and the variable capacitor again tuned to C_2 to resonate the circuit. It is easy to see that $C_X = C_2 - C_1$. This technique works as long as $C_X < \max C_V$. If C_X is larger than $\max C_V$, then a known standard capacitor, C_S, can be placed in series with C_X to make the series combination less than $\max C_V$. Once the capacitance of the series combination of C_X and C_S is known, C_X can be calculated.

Cooper (1978) gives a more general method for measuring C_X and its equivalent series resistance, R_{XS}, on a Q-meter. Refer to Figure 8.71, in which the unknown capacitor is placed in series with the standard inductor having inductance L_S and series resistance R_S. As in the parallel technique described above, two measurements must be made: first, the unknown capacitor is shorted out and the Q-meter is resonated with C_V at f_o. The Q and C_V for this case are recorded as Q_1 and C_1. Next, the short is removed and resonance is again observed, giving Q_2 and C_2. Now in the first case, $X_C = X_L$, or

$$\frac{1}{2\pi f_o C_1} = 2\pi f_o L_S \tag{8.156}$$

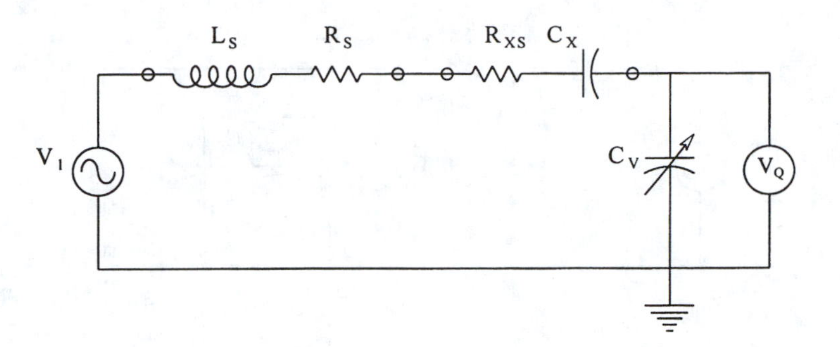

FIGURE 8.71 Series Q-meter circuit used to measure an unknown capacitance at RFs. See text for analysis.

and

$$Q_1 = \frac{2\pi f_o L_S}{R_S} = \frac{1}{2\pi f_o C_1 R_S} \tag{8.157}$$

For the second measurement, it can be shown that the reactance of the unknown capacitor is given by

$$X_X = \frac{1}{2\pi f_o C_X} = \frac{C_2 - C_1}{2\pi f_o C_2 C_1} \tag{8.158}$$

Hence

$$C_X = \frac{C_1 C_2}{C_2 - C_1} \tag{8.159}$$

The series resistive component of the unknown capacitance can be shown to be given by (Cooper, 1978):

$$R_{XS} = \frac{C_1 Q_1 - C_2 Q_2}{2\pi f_o C_1 C_1 C_2 Q_1 Q_2} \tag{8.160}$$

The series dissipation factor of the unknown capacitor, $D_{SX} = \omega C_X R_{XS}$, can be written from the above relations:

$$D_{SX} = \frac{C_1 Q_1 - C_2 Q_2}{Q_1 Q_2 (C_2 - C_1)} \tag{8.161}$$

It is also possible to place the unknown capacitor (impedance) in parallel with C_V, as shown in Figure 8.72. This configuration is used when $X_X > X_{C1}$. As in the series case treated above, we first remove the unknown capacitor and resonate the Q-meter by adjusting C_V. At resonance, $X_L = X_{C1}$, thus:

$$C_1 = 1/(\omega^2 L_S) \tag{8.162}$$

and

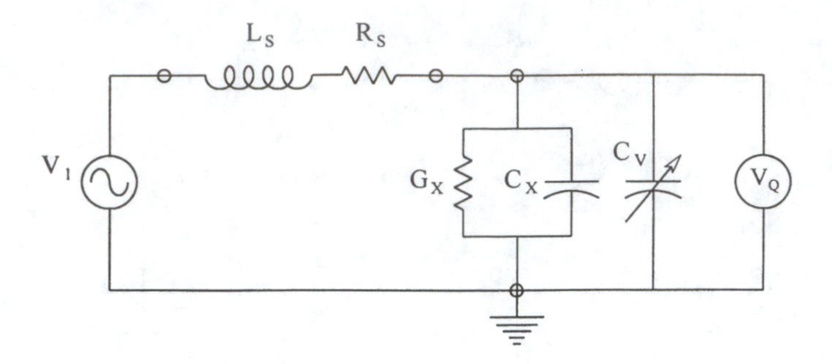

FIGURE 8.72 Parallel connection of the Q-meter used to measure an unknown capacitance.

$$Q_1 = \omega L_S / R_S = 1/(\omega C_1 R_S) \tag{8.163}$$

Next, the unknown capacitor is placed in parallel with C_V, and C_V is adjusted to again obtain resonance. At resonance,

$$X_L = \omega L_S = \frac{X_{C2} X_X}{X_{C2} - X_X} = \frac{1}{\omega C_1} \tag{8.164}$$

Equation 8.164 can be manipulated to yield:

$$X_X = \frac{1}{\omega(C_1 - C_2)} \tag{8.165}$$

From which it is easy to write

$$C_X = C_1 - C_2 \tag{8.166}$$

In addition, Cooper (1978) shows that the equivalent shunt conductance of the unknown capacitance can be written as:

$$G_X = \omega C_1 \frac{Q_1 - Q_2}{Q_1 Q_2} \tag{8.167}$$

and the dissipation factor of the unknown capacitor can be calculated from

$$D_{XP} = \frac{1}{\omega C_X R_X} = \frac{C_1(Q_1 - Q_2)}{Q_1 Q_2 (C_1 - C_2)} = \frac{C_1}{Q_1 Q_2} \frac{\Delta Q}{\Delta C} \tag{8.168}$$

The Q-meter is seen to be a versatile instrument for measuring the properties of unknown capacitors at radio frequencies. The accuracy is modest, however (1 to 2%), and calculations are required to obtain the equivalent circuit parameters.

8.9.2.2 Capacitance Measurement by Q/V

Capacitance is defined by the basic relation, C = Q/V. Hence, if one applies a DC voltage, V_C, to an initially uncharged capacitor, and then measures the total charge, Q, accumulated in the steady state, the ratio of Q to V_C is, by definition, the capacitance in farads. This basic definition is used in the design of several commercial instruments used to measure capacitors. The Keithley Model 595 Quasistatic CV Digital Capacitance Meter uses a novel, stepwise programming of the voltage applied to the capacitor. A sequence of N steps of voltage, each step of height ΔV, where ΔV is selectable to be 0.01, 0.02, 0.05, or 0.10 V, can be applied to C_X. The maximum range of voltage is ±20 V. As each step of voltage is applied to the unknown capacitor, an electrometer charge amplifier circuit integrates the current flowing into the capacitor as a result of each voltage step. The Model 595 meter samples the voltage proportional to charge at the integrator output just before each voltage step is applied, and it samples the charge amplifier's output twice at an interval ΔT after its output reaches steady state from charging C_X. This sampling is shown in Figure 8.73. The op amp's DC bias current may thus be estimated from:

$$I_B = (Q_3 - Q_2)/\Delta T \tag{8.169}$$

This bias current must be subtracted from the current charging the capacitor under measurement. The charge amplifier's output can be shown to be:

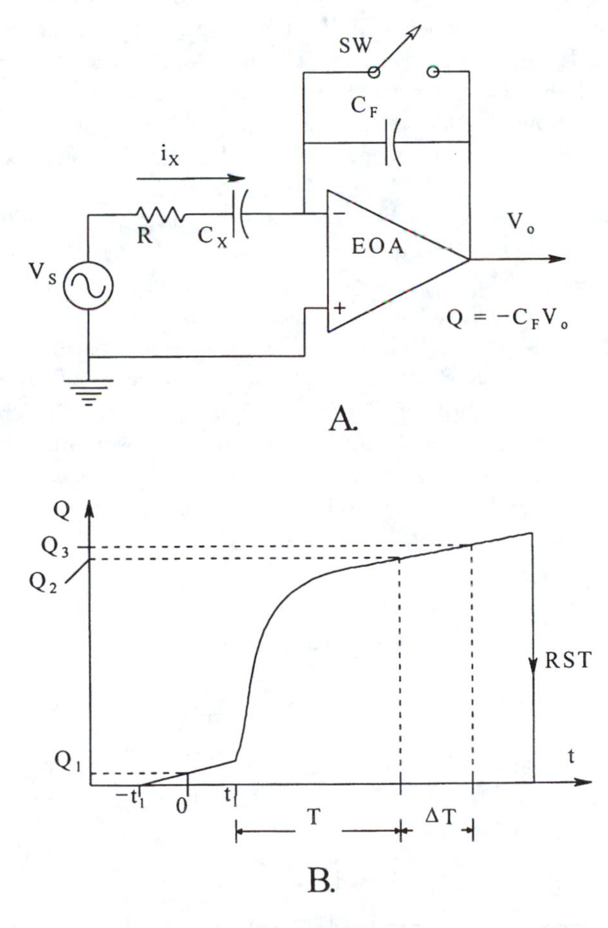

FIGURE 8.73 (A) Electrometer op amp charge amplifier circuit used to measure voltage-variable capacitances by the ΔQ/ΔV method. (B) Charge amplifier output waveform in response to one input voltage step, showing sampling points used in the Keithley model 595 CV capacitance meter. See text for description.

$$v_o(t) = -\frac{I_B}{C_F} - \frac{\Delta V C_X}{C_F}\{1 - \exp[-(t - t_1)/RC_X]\} \quad 8.170$$

Referring again to Figure 8.73 and to Equation 8.170, we see that in the steady state, $C_X(V)$ is given by:

$$C_X(V) = \frac{(Q_3 - Q_1) - I_B(T + t_1)}{\Delta V} \quad 8.171$$

C_X can also be expressed in terms of the charge amplifier's steady-state output voltage:

$$C_X(V) = -\frac{C_F}{\Delta V}[v_{o(SS)} @ T + t_1] - \frac{I_B(T + t_1)}{\Delta V} \quad 8.172$$

The V reported by the Keithley 595 CV meter is (V_{LAST} + 0.5 ΔV).

If the capacitor is linear, then $C_X(V)$ is constant over the range of V. If the capacitance is a function of voltage, such as the capacitance of a reverse-biased pn junction, then $C_X(V)$ can be plotted vs. V with the Keithley 595 instrument in order to characterize the capacitor's voltage dependency. The model 595 CV meter can also be used to measure the junction capacitances of BJTs and FETs, as well as metal oxide semiconductor (MOS) chip capacitors. In spite of its digital display, sampling, microprocessor control and IEEE-488 bus, the 595 is basically an analog instrument, and has 1% accuracy.

Another application of the C = Q/V method to characterize capacitor (and other) dielectrics has been described by Mopsik (1984). The procedure has been called "*time domain spectroscopy*," or TDS. The basic circuit for the Mopsik TDS system is shown in Figure 8.74. An electrometer charge amplifier (integrator) is connected to a node between the capacitor and dielectric under investigation and a reference, low-loss capacitor, C_R. C_R is variable and has a dry air dielectric. The input to C_X, the capacitor under test (CUT) is a positive step voltage, V_C, from an op amp. $-V_C$ is applied simultaneously from another op amp to C_R. The purpose of C_R is to permit a charging current equal and opposite to that of C_X to be summed at the charge amplifier's summing junction. In analyzing the Mopsik TDS system, we will consider two cases, the first where $C_X = C_R$, i.e., there is perfect compensation, and the second where $C_X \neq C_R$. The second case is more realistic, as perfect matching of C_X and C_R will not ordinarily be obtained in practice.

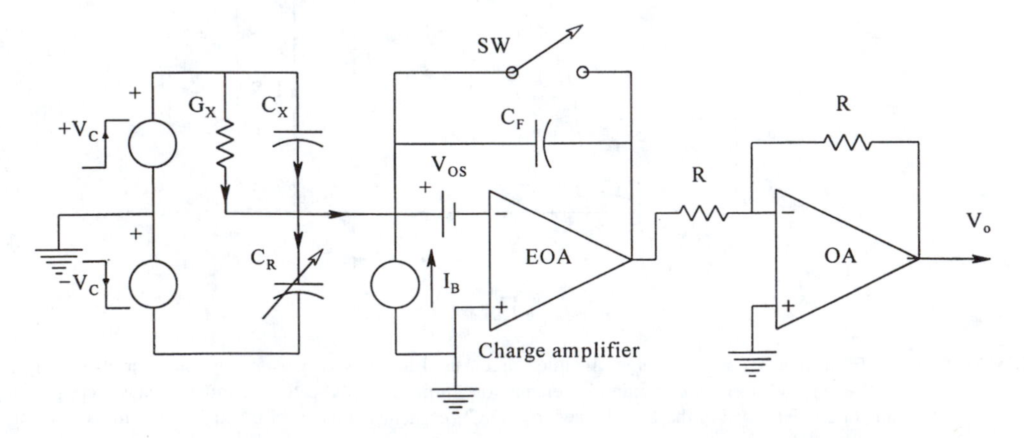

FIGURE 8.74 The Mopsik (1984) circuit for characterization of capacitances in the frequency domain.

In the first case, there will be three components to the system output: one due to the DC "leakage current" through an equivalent conductance, G_X, shunting C_X. G_X describes the behavior of the C_X dielectric; G_X may have a purely ohmic component, as well as time- and voltage-dependent nonlinear components. Such behavior may arise from the presence of "free" electrons and holes in the dielectric, as well as mobile positive and negative ions, and impurity centers capable of "trapping" charged particles and ions until sufficient energy releases them. The other two components at the output are due to the op amp's bias current, I_B, and offset voltage, V_{OS}. If we assume that C_F has zero charge at t = 0 when the steps of $\pm V_C$ are applied to the C_X and C_R, then we may write in the time domain:

$$v_o(t) = \int_0^t \frac{V_C}{C_F} G_X(V_C, t)\, dt + \frac{I_B}{C_F} t + V_{OS} U(t) \qquad 8.173$$

A plot of these components is shown in Figure 8.75. Note that in the case of a good low-conductivity dielectric, I_B may exceed $V_C G_X$ in magnitude, making the bias current error a significant component at the output. V_{OS} can generally be nulled out at a particular operating temperature, however.

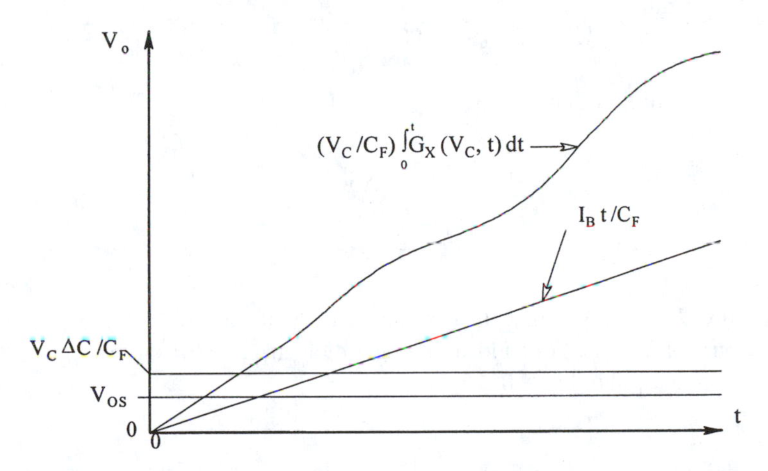

FIGURE 8.75 Plot of the electrometer charge amplifier's output in the Mopsik system. The four components of Equation 8.174 are shown.

In the second case, $\Delta C = C_X - C_R \neq 0$, so there will also be a capacitive charging component to the output of the charge amplifier. If we assume that the capacitors charge instantaneously, then the charge amplifier output will be

$$v_o(t) = \int_0^t \frac{V_C}{C_F} G_X(V_C, t)\, dt + \frac{I_B}{C_F} t + \left(V_{OS} + \frac{V_C \Delta C}{C_F} \right) U(t) \qquad 8.174$$

Note that the ΔC term can be positive or negative.

In the steady state, for a linear capacitor, we have the well-known relation, $C = Q/V_C$. Here Q is the total charge accumulated in an ideal capacitor in the steady state. Mopsik (1984) noted that the charge Q does not really reach a steady state, but continues to increase after the pure capacitance component, C_X, has charged up to the applied DC voltage, V_C. The continuing increase in Q is due to dielectric phenomena. Mopsik defines a "time-variable capacitance" as:

$$C(t) \equiv Q(t)/V_C = v_o(t) C_F / V_C \qquad 8.175$$

In order to characterize the time-variable capacitor given by Equation 8.175 in the frequency domain, Mopsik takes the Fourier transform of the time derivative of the $v_o(t)$ transient, given by Equation 8.174, times C_F/V_C. He defines this Fourier transform as:

$$\mathbf{C}^*(j\omega) \equiv \mathbf{F}\{dC(t)/dt\} = j\omega\mathbf{C}(j\omega) = \mathbf{G}_X(j\omega, V_C) + \frac{I_B}{j\omega V_C} + \left[\frac{V_{OS}C_F}{V_C} + \Delta C\right] \qquad 8.176$$

Note that $\mathbf{C}^*(j\omega)$ as defined by Mopsik *is not* a capacitance; it has the dimensions of a complex admittance. Mopsik defines the real and imaginary parts of $\mathbf{C}^*(j\omega)$ by Equation 8.177:

$$\mathbf{C}^*(j\omega) \equiv C(\omega)' - jC(\omega)'' \qquad 8.177$$

At a given frequency, we can show that the admittance $\mathbf{C}^*(j\omega)$ can also be described by a standard, parallel R-C model. Assume an AC voltage, V_C, is impressed across $\mathbf{C}^*$, and the parallel model. Thus, the current is:

$$\mathbf{I}^* = V_C\mathbf{C}^*(j\omega) = V_C[C'(\omega) - jC''(\omega)] = \mathbf{I_P} = V_C[G_P + j\omega C_P] \qquad 8.178$$

By comparing terms, we find that

$$C' = G_P \qquad 8.179A$$

$$C'' = -\omega C_P \qquad 8.179B$$

Referring to Figure 8.76, we see that the loss angle tangent, $\tan(\delta)$, at frequency ω must be the same for both capacitor models. For an ideal, lossless dielectric, $\tan(\delta)$ should approach zero, which implies that $C'(\omega)$ and G_P should $\to 0$.

$$\tan(\delta^*) = \frac{C'(\omega)}{-C''(\omega)} \tan(\delta) = \frac{G_P}{\omega C_P} \qquad 8.180$$

The CUT can also be expressed in terms of complex permittivity, ε^*. Here we assume that $\varepsilon^*(j\omega) \equiv \varepsilon'(\omega) - j\varepsilon''(\omega)$. Hence:

$$\mathbf{C}(j\omega) = \mathbf{C}^*(j\omega)/j\omega = \frac{A\varepsilon^*(j\omega)}{d} = \frac{1}{j\omega}[C'(\omega) - C''(\omega)] \qquad 8.181$$

Thus

$$\varepsilon^*(j\omega) = \frac{d}{j\omega A}[C'(\omega) - C''(\omega)] \qquad 8.182$$

Here we assume the CUT is an ideal parallel plate capacitor, where A is the equivalent area of its plates, and d is its plate separation.

The Mopsik system calculates $\tan[\delta(\omega)]$ for the unknown capacitor and dielectric from the ratio of $C'(\omega)$ to $-C''(\omega)$. $C'(\omega)$ is the real part of the admittance, $\mathbf{C}^*(j\omega)$, which is found by taking the Fourier transform of $\dot{v}_o(t)C_F/V_C$, as written in Equation 8.176 above. $C'(\omega)$ can be expressed as:

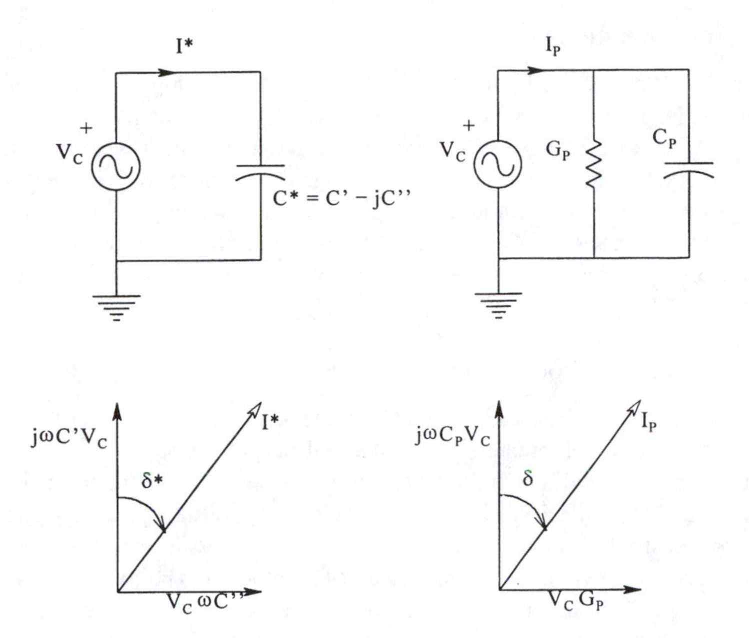

FIGURE 8.76 Equivalence of the loss tangent for the familiar parallel R-C capacitor model, and the complex capacitor model of Mopsik ($I^* = I_P$).

$$C'(\omega) = RE\{C^*(j\omega)\} = RE\{G_X(j\omega, V_C)\} + \frac{V_{OS}C_F}{V_C} + \Delta C \quad 8.183$$

The imaginary part of $\mathbf{C}^*(j\omega)$ is:

$$C''(\omega) = IM\{C^*(j\omega)\} = -IM\{G_X(j\omega, V_C)\} + \frac{I_B}{\omega V_C} \quad 8.184$$

Equations 8.183 and 8.184 above illustrate the source of artifacts in the Mopsik system. Clearly, I_B can have a major effect on the $\tan[\delta(\omega)]$ result, as can offset voltage and a non-zero difference between C_R and C_X. When we can assume that $C_R = C_X$, and V_{OS} and $I_B = 0$, then the loss tangent is:

$$\tan[\delta(\omega)] = \frac{RE\{G_X(j\omega, V_C)\}}{IM\{G_X(j\omega, V_C)\}} \quad 8.185$$

The Mopsik approach to characterizing a capacitor and its dielectric is unique. Because the CUT is assumed to be proportional to the charge, q(t), measured under constant V_C, it is a function of time and can be Fourier transformed. Hence, we have a complex capacitance which is a function of frequency, and which will have a real and an imaginary part. Mopsik Fourier transforms the *derivative* of C(t), hence his C* is really an admittance. It is doubtful in practice that the CUT's $\tan[\delta(\omega)]$ calculated by the Mopsik method is independent of the electrometer's I_B. Such independence would be possible only if the current $V_CG_X >> I_B$. I_B for a typical solid-state electrometer op amp is about 30 fA. Also, the initial step transient artifacts due to V_{OS} and ΔC will contribute errors to the high-frequency portion of $\tan[\delta(\omega)]$.

8.9.3 Inductance Measurements

Measurement of inductances can be made from subsonic to ultra-high frequencies with a variety of means. Meaningful measurements can be made over the range of nanohenries to kilohenries, but not at the same frequency. Circuits used to characterize inductances were introduced in Section 5.1, and audio frequency inductance-measuring bridges were described in Section 5.4.2. Audio frequency bridges are the most accurate means of measuring inductance at audio frequencies. In this section, we will describe two other means of measuring inductance: the AC voltmeter method, and the Q-meter. Inductances can also be measured with vector impedance meters, described in Section 8.10 below.

8.9.3.1 AC Voltmeter Method of Estimating Inductance

At powerline frequencies, it is possible to find the inductance of a coil by first measuring the coil's DC resistance, R'_L, with an ohmmeter. The actual real part of the series impedance of the coil at the measurement frequency will generally be slightly higher than R'_L due to magnetic core losses and skin effect at radio frequencies. The effective AC, series coil resistance is defined to be R_L. A simple series circuit with the coil and a known external resistor, R, is used (see Figure 8.77A). The only instrument needed is a high input impedance AC voltmeter. The excitation voltage (V_S), the voltage across the coil (V_L), and the voltage across the resistor (V_R) are measured. By Kirchoff's voltage law, we know the vector sum of V_L and V_R must equal V_S. Also, we know that the phase angle of the current in a series inductive circuit *lags* the phase of the voltage across the circuit. We can thus draw the voltage vector (phasor) diagram as shown in Figure 8.77B. The three known voltages form the three sides of a triangle. The angle of V_R, $-\theta$, is also the angle of the current in the circuit, I. The angle of V_S is assumed to be zero. By Ohm's law, the magnitude of the current is:

$$I = V_R/R \tag{8.186}$$

The magnitude of the voltage across the coil can be written:

$$V_L = |\mathbf{IZ}_L| = \frac{V_R}{R}\sqrt{\omega^2 L^2 + R_L^2} \tag{8.187}$$

Equation 8.187 can be solved for L, since we know ω, R_L, R, V_R and V_L:

$$L = (1/\omega)\sqrt{R^2 V_L^2/V_R^2 - R_L^2} \tag{8.188}$$

L can also be solved for in terms of the current angle, $-\theta$, ω, R, and R_L, and also in terms of V_S, V_R , ω, R, and R_L. (The derivation of these expressions are left as exercises for the reader.)

8.9.3.2 Use of the Q-Meter to Measure Inductance and Q

We have already seen in Section 8.9.2.1 how a Q-meter can be used to measure an unknown capacitor at radio frequencies. Measurement of an unknown series inductance and its Q are quite simple. Referring to the basic Q-meter schematic shown in Figure 8.70, we replace the standard inductor with a short-circuit; the unknown inductor goes in the Z_S position, and Z_P is made infinite. The frequency of V_1 is set to the desired value, and V_1 is adjusted to be 1.0 V. Next, C_V is varied to obtain series resonance in the circuit. At resonance

$$\frac{V_Q}{V_1} = \frac{V_C}{V_1} = \frac{IX_C}{IR_L} = \frac{IX_L}{IR_L} = \frac{\omega L}{R_L} = Q_L \tag{8.189}$$

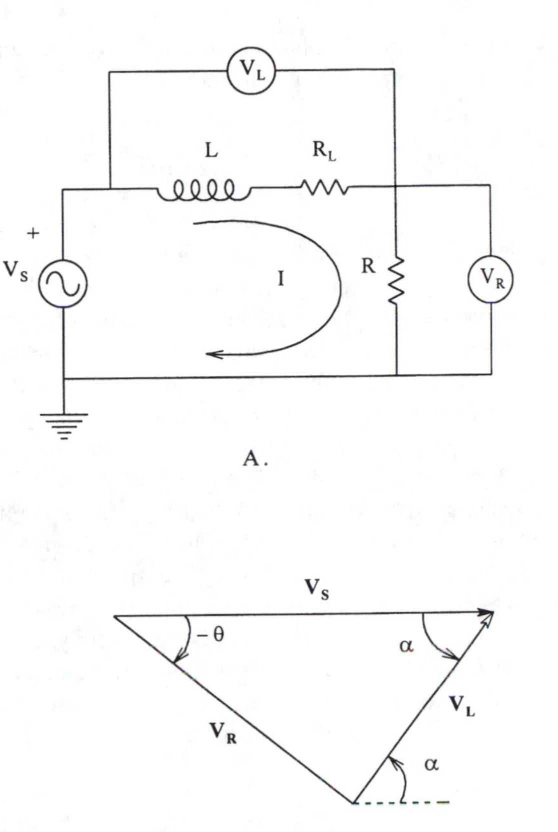

FIGURE 8.77 (A) Circuit of voltmeter method of estimating inductance at audio and powerline frequencies. (B) Vector (phasor) voltages in the circuit of (A).

Hence, V_Q in volts equals the inductor's Q. This simple analysis assumes that the range of C_V is adequate to resonate with L, i.e., $f = 1/(2\pi\sqrt{LC_V})$. If resonance is not possible within the working range of C_V, then, depending on whether we are dealing with an inductor at the lower end of the range (90 nH) or one at the upper end of the range (130 mH), we can add a known standard series inductance in series with the unknown inductor, or a known standard capacitor in parallel with C_V, respectively. The Boonton Model 260A Q-meter allows measurement of inductances >5 μH to ±3%, and Qs to ±5%.

It should be stressed that the Q-meter measures the effective inductance, and the Q is determined as the ratio of the effective inductive reactance to the effective series real part of the loop impedance. At HF and VHF, the effective parallel capacitance of the coil acts to reduce the effective Q (see Equation 5.3).

8.10 VECTOR IMPEDANCE METERS

A linear, two-terminal, passive, electrical component may be characterized at a given frequency by its vector impedance, or admittance. Impedance, **Z**, and admittance, **Y**, are defined in rectangular vector form by:

$$\frac{\mathbf{V}}{\mathbf{I}} = \mathbf{Z} = R + jX = \frac{1}{\mathbf{Y}} = \frac{1}{G + jB} \tag{8.190}$$

It is understood that **V** is the sinusoidal voltage across the two-port, and **I** is the vector current through it. Impedance can be measured at a given frequency by computing the vector ratio:

$$\mathbf{Z} = Z\angle\theta = \frac{V\angle\theta_1}{I\angle\theta_2} = Z\angle(\theta_1 - \theta_2) \qquad 8.191$$

If the magnitude of **I** is held constant, then measurement of **V** and the phase difference between **V** and **I** will characterize **Z** at a given frequency. Figure 8.78 shows the block diagram of a basic audio and video frequency vector impedance meter (VIM). The VIM is operated in the constant current magnitude mode, although some VIMs can operate in the constant voltage mode. A wide-band current transformer is used to sense the current flowing through the unknown impedance to ground. The voltage output of the current transformer is conditioned and used to parametrically adjust the applied voltage to maintain constant current. The microcomputer in the VIM calculates the impedance magnitude from the magnitudes of **V** and **I**, and also samples and presents the output of the phase meter. In the phase meter, the analog **V** and **I** signals are converted to TTL signals by high-speed comparators, and these signals are inputs to a NAND gate digital phase comparator, such as shown in Figure 8.56. The admittance, Y, can be easily found by changing the sign of the phase angle of **Z**, and taking the reciprocal of the magnitude of Z. VIMs can also give the user impedance in rectangular vector form, i.e., R and X, by multiplying |**Z**| by cos(θ) and sin(θ), respectively. Once X is known, and the impedance is known to be inductive, the equivalent L can be found by dividing X by the known 2πf. A similar means can be used to obtain a numerical value of C when the impedance is known to be from a capacitor.

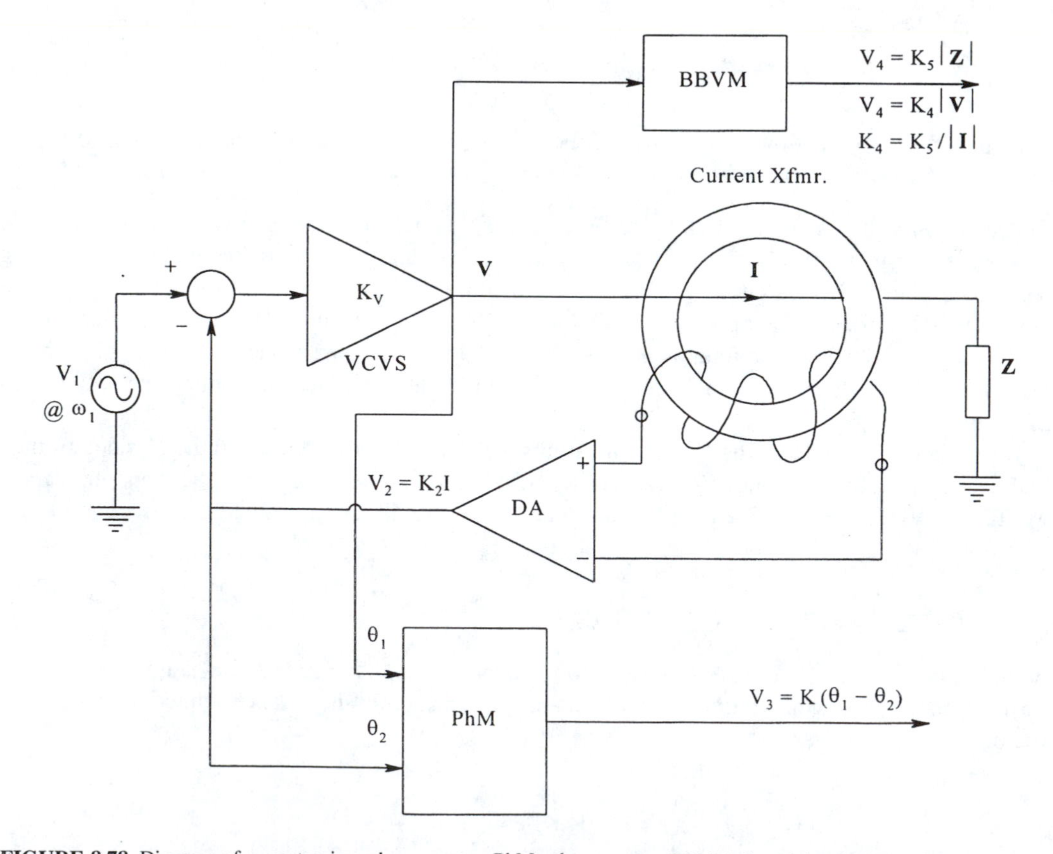

FIGURE 8.78 Diagram of a vector impedance meter. PhM, phase meter; BBVM, broad-band (RF) voltmeter (e.g., a thermocouple type); OSC, oscillator; VCVS, voltage-controlled voltage source.

Hewlett-Packard (San Jose, CA) offers many VIMs covering a wide range of frequencies, useful for a variety of applications, including the screening of critical components in manufacturing processes. We will discuss the specifications of one representative HP VIM to give the reader a feeling for the state-of-the-art in this type of instrument. The HP model 4192A LF Impedance Analyzer (VIM) operates from 5 Hz to 13 MHz. This instrument can measure 11 impedance-related parameters over its frequency range: |**Z**|, |**Y**|, θ, R, X, G, B, L, C, Q, and D. Outputs are read on two-, four-, and half-digit numerical displays, and can be accessed through the HP-IB (IEEE-488) bus. The range of Z, X, and R is from 0.1 mΩ to 1.3 MΩ, and the range of Y, G, and B is from 1 nS to 13 S. Frequency accuracy is ±50 ppm, and frequency synthesizer steps are 0.001 Hz from 5 Hz to 10 kHz, 0.01 Hz from 10 kHz to 100 kHz, 0.1 Hz from 100 kHz to 1 MHz, and 1 Hz from 1 MHz to 13 MHz. Series or parallel equivalent circuit models may be user-selected for impedance and admittance measurements. The complete range of inductance which can be measured is from 10 μH (at high frequencies) to 1 kH (at low frequencies) with a basic accuracy of ±0.27%. Capacitance can be measured from 0.1 fF to 199 mF, with a basic accuracy of ±0.15%. D = 1/Q can be measured from 10^{-4} to 20 with an accuracy of 10^{-3} for C measurements, and 0.003 for L measurements. Maximum percent accuracy in measuring the 11 impedance parameters is a function of frequency and the value of the parameter being measured. The best accuracy generally occurs between 100 Hz and 1 MHz, and it varies inversely with parameter magnitude (e.g., best phase accuracy is obtained when measuring low values of Y, G, or B, and when measuring low values of |Z|, R, or X; likewise, best accuracy in measuring Z, R, or X occurs for low ohm values of these parameters, etc.).

The HP 4191A RF impedance analyzer is another VIM intended to work in the 1 to 1000 MHz range. In addition to measuring the 11 impedance parameters mentioned above, this meter also measures the reflection coefficient, Γ(ω) of a standard transmission line with characteristic impedance, Z_O, terminated with an unknown impedance, Z_L. Γ(ω) is defined as:

$$\Gamma(\omega) \equiv \frac{\mathbf{V}_r}{\mathbf{V}_i} = \frac{\mathbf{Z}_L - \mathbf{Z}_O}{\mathbf{Z}_L + \mathbf{Z}_O} = \frac{\text{VSWR} - 1}{\text{VSWR} + 1} = \Gamma\angle\theta \qquad 8.192$$

The HP 4191A VIM uses two microwave directional couplers to isolate the input wave, V_i, and the reflected wave, V_r. Hence Γ(ω) can be calculated from the vector quotient, and knowing $\mathbf{Z}_O$, we can calculate $\mathbf{Z}_L$. In addition to presenting Γ in polar form, the HP 4191A VIM also can give Γ_X and Γ_Y over a range of 10^{-4} to 1 with a resolution of 10^{-4}. Typical accuracy in measuring Γ(ω) is ±0.2% from 1 to 100 MHz, rising to about ±0.5% at 1 GHz.

In summary, vector impedance meters are seen to be versatile, modern, microprocessor-based instruments, capable of measuring a wide variety of impedance-related parameters, including inductance, capacitance, Q, and D over a wide range of frequencies and values. Modern VIMs have IEEE 488 bus I/O, which allows them to be directly controlled by, and report their measurements to, a computer system. They also are relatively expensive.

8.11 CHAPTER SUMMARY

Chapter 8 has dealt with descriptions of the common means of measuring the electrical parameters of DC voltage, electric field strength, DC current, AC voltage, AC current, phase, frequency, resistance, capacitance, inductance, impedance, and admittance. The most accurate means of measurement are seen to be null methods, in which standards are used for comparison. Null methods were treated in detail in Chapters 4 and 5.

The tremendous range of component values used in electrical engineering systems design generally means that no one instrument type, design, or frequency is suitable to make accurate measurements over the entire practical range of a parameter's values. Because of the presence of parasitic capacitance in inductors, and parasitic inductance in capacitors, high-frequency measurements of inductance and capacitance were seen to be more realistically treated as complex impedance or admittance measurements. Finally, we described the organization of modern, microprocessor-controlled, vector impedance meters which are ubiquitous instruments for the characterization of two-terminal device impedances over wide ranges of frequencies.

9

Digital Interfaces in Measurement Systems

9.0 INTRODUCTION

As we have seen in Figure 1.1, a modern instrumentation system generally includes a digital computer which is used to supervise, coordinate, and control the measurements, and which often is used to store data (data logging), condition data, and to display it in a meaningful, summary form on a monitor. In this chapter we will describe the hardware associated with the conversion of analog information to digital formats, and noise and resolution problems associated with the analog-to-digital conversion (ADC) process.

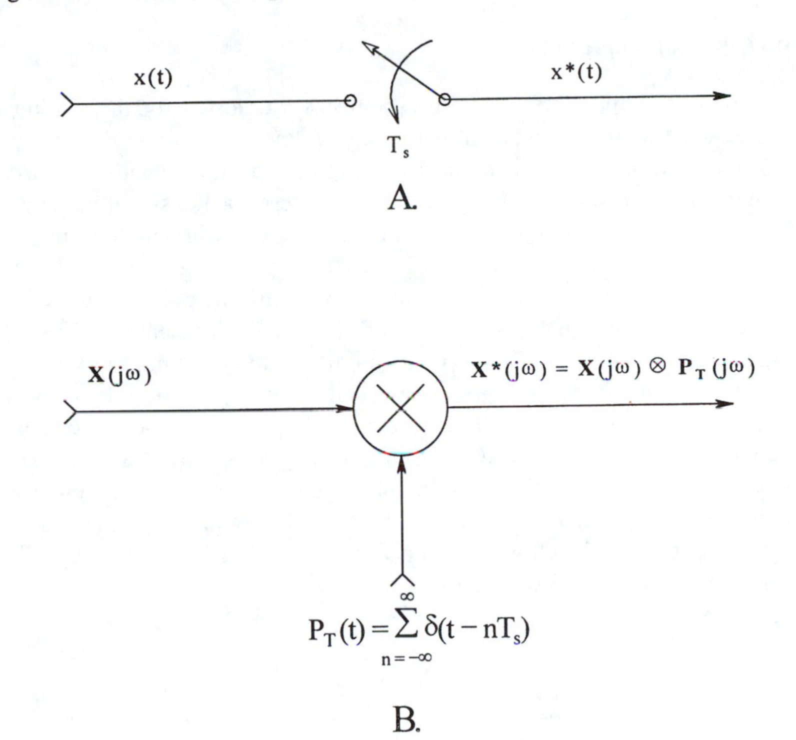

FIGURE 9.1 (A) Symbol for an ideal sampling process (analog-to-digital conversion). T_s is the period between successive samples. (B) Impulse modulation equivalent to ideal sampling. $\mathbf{X}^*(j\omega)$ = Fourier transform of the sampled signal. $\mathbf{X}^*(j\omega)$ can be expressed as the complex convolution of $\mathbf{X}(j\omega)$ and $\mathbf{P_T}(j\omega)$.

We also examine various means of digital communication with instruments, including the IEEE-488.2 bus, serial data protocols, and other parallel bus architectures used with PCs. Examples of commercially available interface cards are described, and problems associated with sending both analog and digital data long distances at high data rates on cables are analyzed.

In describing data conversion interfaces, it is expedient to first consider digital-to-analog converters (DACs) because these systems are used in the designs of several kinds of analog-to-digital converters (ADCs). Data conversion from analog to digital form or digital to analog form is generally done continuously and periodically. Periodic data conversion has a great effect on the information content of the converted data, as we will show in the next section. A noise-free analog signal sample theoretically has infinite resolution. Once the analog signal sample has been converted to digital form, it is represented by a digital (binary) number of a finite number of bits (e.g., 12), which limits the resolution of the sample (one part in 4096 for 12 bits). This obligatory rounding-off of the digital sample is called *quantization*, and the resulting errors can be thought of as being caused by a quantization noise.

Integrated circuit ADCs and DACs are available in a wide spectrum of specifications. They are available with quantization levels ranging from 6 to 20 bits, and conversion rates ranging from low audio frequencies to less than 1 nsec per sample. Most ADCs and DACs have parallel digital outputs or inputs, respectively, although a few are designed to use serial data I/O protocols.

Before discussing data conversion hardware, we will first discuss the dynamics of data conversion, and quantization noise.

9.1 THE SAMPLING THEOREM

The sampling theorem is important because it establishes a criterion for the minimum sampling rate which must be used to digitize a signal with a given, low-pass, power density spectrum. The relation between the highest significant frequency component in the signal's power density spectrum, and the sampling frequency neccessary to accurately describe the signal in the digital domain is called the *Nyquist criterion*. The implications of the Nyquist criterion and aliasing are discussed below.

Because analog-to-digital data conversion is generally a periodic process, we will first analyze what happens when an analog signal, x(t), is periodically and ideally sampled. The ideal sampling process generates a data sequence from x(t) defined only at the sampling instants, when $t = nT_s$, where n is an integer ranging from $-\infty$ to $+\infty$, and T_s is the sampling period. It is easy to show that an ideal sampling process is mathematically equivalent to impulse modulation, as shown in Figure 9.1. Here the continuous analog signal x(t) is multiplied by an infinite train of unit impulses or delta functions which occur only at the sampling instants. This multiplication process produces a periodic number sequence, x*(t), at the sampler output. In the frequency domain, $\mathbf{X}^*(j\omega)$ is given by the complex convolution of $\mathbf{X}(j\omega)$ with the Fourier transform of the pulse train, $\mathbf{P}_T(j\omega)$. In the time domain, the pulse train can be written as:

$$P_T(t) = \sum_{n=-\infty}^{\infty} \delta(t - nT_s) \qquad 9.1$$

The periodic function, P_T, can also be represented in the time domain by a Fourier series in complex form:

$$P_T(t) = \sum_{n=-\infty}^{\infty} C_n \exp(-j\omega_s t) \qquad 9.2$$

where

$$\omega_s = \frac{2\pi}{T_s} \text{ r/s} \qquad 9.3$$

T_s and the complex-form Fourier series coefficients are given by:

$$C_n = \frac{1}{T_s} \int_{-T_s/2}^{T_s/2} P_T(t) \exp(+j\omega_s t)\, dt = \frac{1}{T_s} \tag{9.4}$$

Thus, the complex Fourier series for the pulse train is found to be:

$$P_T(t) = \frac{1}{T_s} \sum_{n=-\infty}^{\infty} \exp(-jn\omega_s t) \tag{9.5}$$

The sampler output is the time-domain product of Equation 9.5 and x(t):

$$\begin{aligned} x^*(t) &= x(t) \frac{1}{T_s} \sum_{n=-\infty}^{\infty} \exp(-jn\omega_s t) \\ &= \frac{1}{T_s} \sum_{n=-\infty}^{\infty} x(t) \exp(-jn\omega_s t) \end{aligned} \tag{9.6}$$

The Fourier theorem for complex exponentiation is:

$$\mathbf{F}\{y(t)\ \exp(-jat)\} \equiv \mathbf{Y}(j\omega - ja) \tag{9.7}$$

Using this theorem, we can write the Fourier transform for the sampler output:

$$x^*(j\omega) = \frac{1}{T_s} \sum_{n=-\infty}^{\infty} \mathbf{x}(j\omega - jn\omega_s) \tag{9.8}$$

Equation 9.8 for $\mathbf{X}^*(j\omega)$ is the result of the complex convolution of $\mathbf{X}(j\omega)$ and $\mathbf{P}_T(j\omega)$; it is in the *Poisson sum form*, which helps us to visualize the effects of (ideal) sampling in the frequency domain, and to understand the phenomenon of aliasing. In Figure 9.2A, we plot the magnitude of $\mathbf{X}(j\omega)$; note that $\mathbf{X}(j\omega)$ is assumed to have negligible power above the Nyquist frequency, $\omega_s/2$. When x(t) is sampled, we see a periodic spectrum for $\mathbf{X}^*(j\omega)$.

Note that x(t) can be recovered from the sampler output by passing x*(t) through an ideal low-pass filter (LPF), as shown in Figure 9.2A. In Figure 9.2B, we have assumed that the base-band spectrum of x(t) extends beyond $\omega_s/2$. When such an x(t) is sampled, the resultant $\mathbf{X}^*(j\omega)$ is also periodic, but the high-frequency corners of the component spectra overlap. An ideal LPF thus cannot uniquely recover the base-band spectrum, and thus x(t). This condition of overlapping spectral components in $\mathbf{X}^*(j\omega)$ is called *aliasing,* and it can lead to serious errors in digital signal processing.

Because of the problem of aliasing, all properly designed ADC systems used in fast-Fourier transform (FFT) spectrum analyzers and related equipment must operate on input signals that obey the Nyquist criterion, i.e., the power density spectrum of x(t), $S_{XX}(f)$, must have no significant power at frequencies above one half the sampling frequency. One way to ensure that this criterion is met is to use properly designed, analog low-pass, *anti-aliasing filters* immediately preceding the sampler. Anti-aliasing filters are generally high-order, sharp cut-off, linear phase, LPFs that attenuate the input signal at least by 40 dB at the Nyquist frequency. According to Northrop (1990):

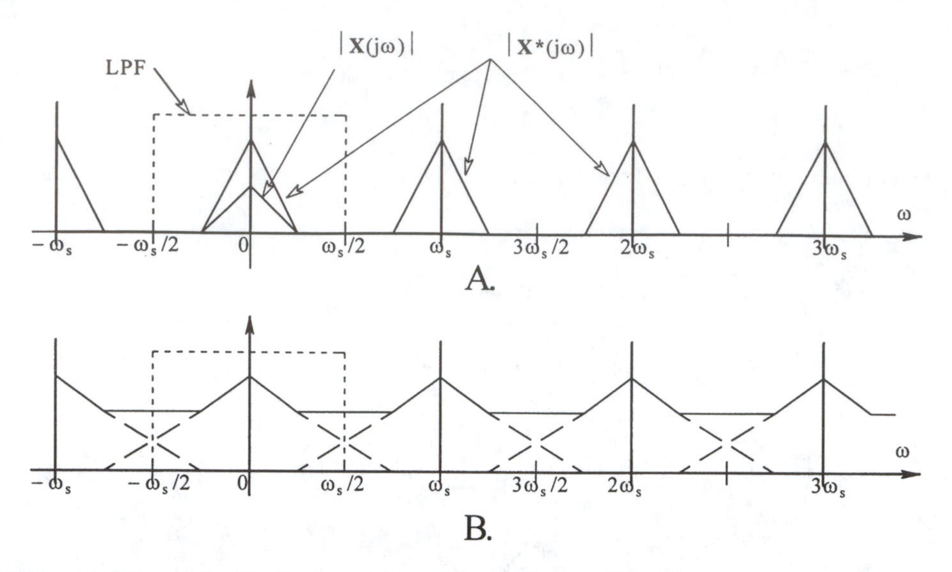

FIGURE 9.2 (A) Magnitude of the Fourier transform of the sampler output. Note the spectrum is periodic at the sampling frequency. The input spectrum contains no power above the Nyquist frequency, $\omega_s/2$. (B) Magnitude of the Fourier transform of the sampler output when the input signal's spectrum extends beyond the Nyquist frequency. This condition produces aliasing.

Many designs are possible for high-order anti-aliasing filters. For example, Chebychev filters maximize the attenuation cutoff rate at the cost of some passband ripple. Chebychev filters can achieve a given attenuation cutoff slope with a lower order (fewer poles) than other filter designs. It should be noted that in the limit as passband ripple approaches zero, the Chebychev design approaches the Butterworth form of the same order, which has no ripple in the passband. Chebychev filters are designed in terms of their order, n, their cutoff frequency, and the maximum allowable peak-to-peak ripple (in decibels) of their passband.

If ripples in the frequency response stopband of an anti-aliasing filter are permissible, then elliptical or Cauer filter designs may be considered. With stopband ripple allowed, even sharper attenuation in the transition band than obtainable with Chebychev filters of a given order can be obtained. Elliptic LPFs are specified in terms of their order, their cutoff frequency, the maximum peak-to-peak passband ripple, and their minimum stopband attenuation.

Bessel or Thomson filters are designed to have linear phase in the passband. They generally have a "cleaner" transient response, that is, less ringing and overshoot at their outputs, given transient inputs.

As an example of anti-aliasing filter design, Franco (1988) shows a sixth-order Chebychev anti-aliasing filter made from three Sallen and Key (quadratic) low-pass modules. This filter was designed to have an attenuation of 40 dB at the system's 20 kHz Nyquist frequency, a corner frequency of 12.8 kHz, a –3 dB frequency of 13.2 kHz, and a ± 1 dB passband ripple.

One problem in the design of analog anti-aliasing filters to be used with instruments having several different sampling rates is adjusting the filters to several different Nyquist frequencies. One way of handling this problem is to have a fixed filter for each separate sampling frequency. Another way is to make the filters easily tunable, either by digital or analog voltage means. Some approaches to the tunable filter problem are discussed in Chapter 10 of the text by Northrop (1990).

Anti-aliasing LPFs are generally not used at the inputs of digital oscilloscopes, because one can see directly if the sampled waveform is sampled too slowly for resolution. An optional analog LPF is often used to cut high-frequency interference on digital oscilloscope inputs; this, however, is not an anti-aliasing filter.

9.2 QUANTIZATION NOISE

When an a band-limited analog signal is sampled and then converted by an N bit ADC to digital form, an uncertainty in the signal level exists which can be considered to be equivalent to a broad-band, quantization noise added to the analog signal which is digitized by an infinite-bit ADC. To illustrate the properties of quantization noise, refer to the quantization error generating model of Figure 9.3. A Nyquist-limited, analog signal, x(t), is sampled and digitized by an N-bit ADC. The ADC's digital output is the input to an N-bit DAC. The quantization error is defined at sampling instants as the difference between DAC output and the analog input signal.

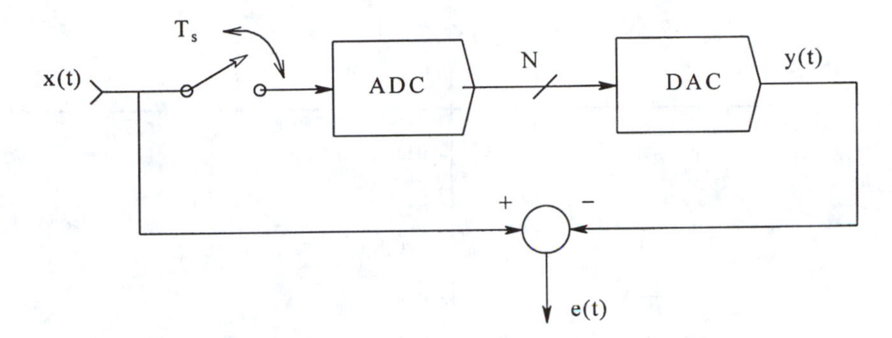

FIGURE 9.3 Block diagram of the system used to observe quantization error in the A/D process.

Figure 9.4 illustrates the nonlinear function relating sampler input, x*(t), to the DAC output, y(t). In this example, N = 4. The quantizer has (2^N) levels and ($2^N - 1$) steps. Compared to the direct path, the error $e(nT_s)$ can range over ±q/2, where q is the voltage step size of the ADC. It is easy to see that

$$q = \frac{V_{MAX}}{(2^N - 1)} \qquad 9.9$$

where V_{MAX} is the maximum (peak-to-peak) range of the ADC/DAC system. For example, if a 10 bit ADC is used to convert a signal ranging from −5 to +5 V, then by Equation 9.9, q = 9.775 mV. If x(t) has zero mean, and its probability density function (PDF) has a standard deviation, σ_x, which is large compared to q, then it can be shown that the PDF of $e(nT_s)$ is well modeled by a uniform (rectangular) density over e = ±q/2. This rectangular PDF is shown in Figure 9.5. The mean-squared error voltage is found from:

$$E\{e^2\} = \overline{e^2} = \int_{-q/2}^{q/2} e^2(1/q)\,de = \left.\frac{(1/q)e^3}{3}\right|_{-q/2}^{q/2} = \frac{q^2}{12} \text{ MSV} \qquad 9.10$$

Thus it is possible to treat quantization error noise as a zero-mean, broad-band noise with standard deviation of $q/\sqrt{12}$ volts, added to x(t).

In order to minimize the effects of quantization noise for an N-bit ADC, it is important that the analog input signal, x(t), use nearly the full dynamic range of the ADC. In the case of a zero-mean, time-varying signal which is Nyquist band-limited, gains and sensitivities should be chosen so that the peak expected x(t) does not exceed the maximum voltage limits of the ADC. Also, if x(t) has a Gaussian PDF, the dynamic range of the ADC should be about ±3 standard deviations of the signal. Under this condition, it is possible to derive an expression for the mean-squared signal-to-

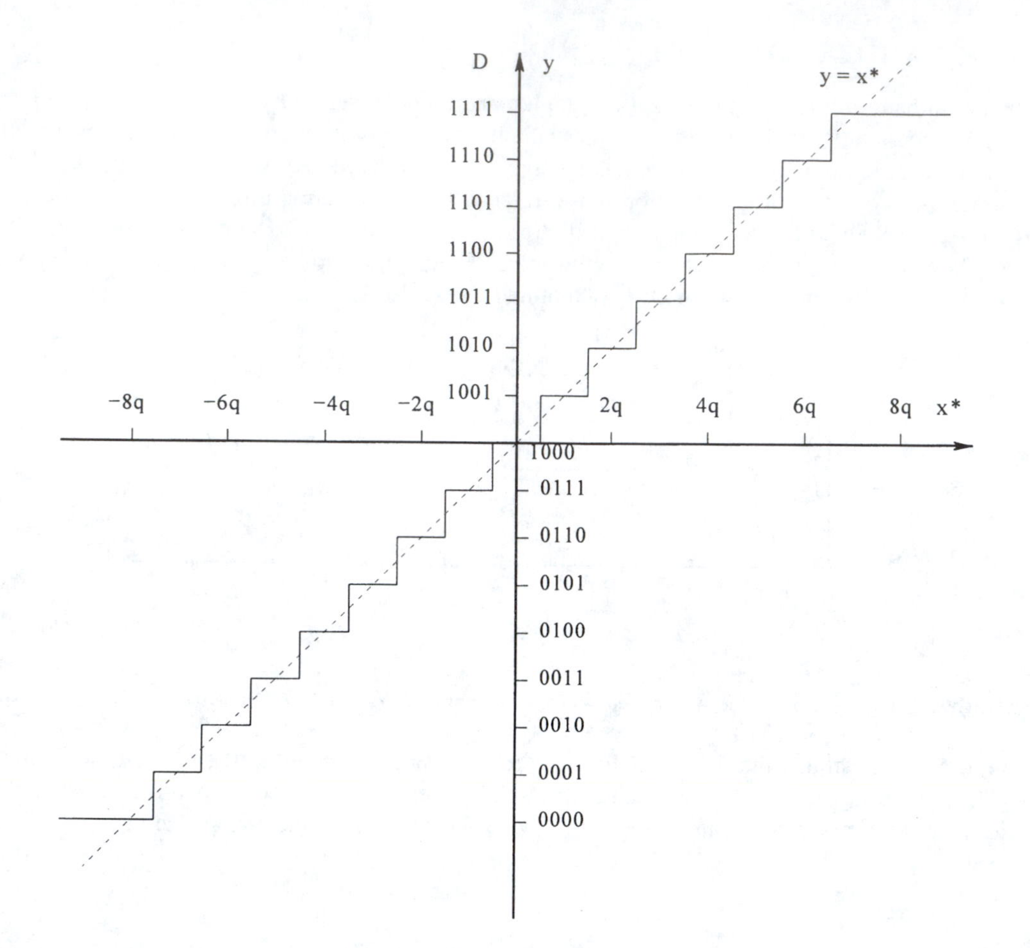

FIGURE 9.4 I/O diagram of a 4-bit quantizer (ADC/DAC in Figure 9.3). Note that there are $2^4 = 16$ levels, and $(2^4 - 1) = 15$ steps.

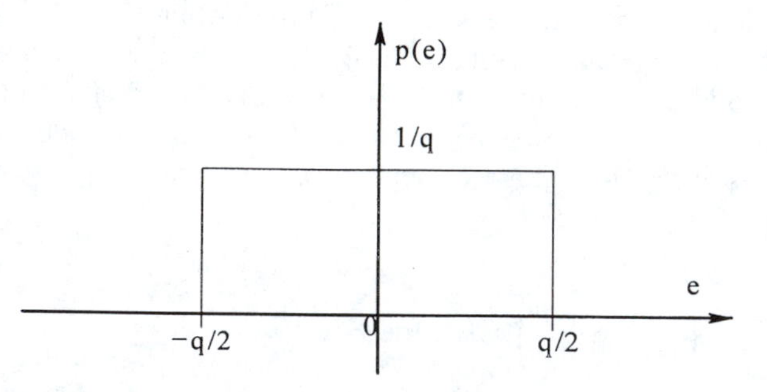

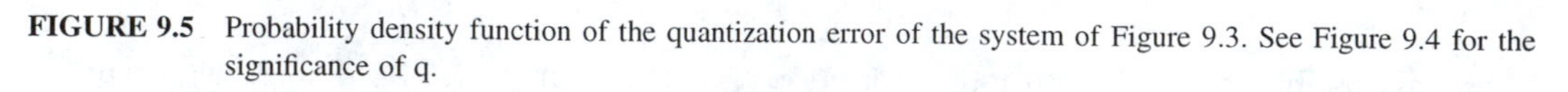

FIGURE 9.5 Probability density function of the quantization error of the system of Figure 9.3. See Figure 9.4 for the significance of q.

noise ratio of the ADC and its quantization noise. Let the signal have an rms value σ_x volts. From Equation 9.9, we see that the quantization step size can be written:

$$q \approx \frac{6\sigma_x}{(2^N - 1)} \text{ volts} \qquad 9.11$$

This relation for q can be substituted into Equation 9.10 for the variance of the quantization noise. Thus the output noise is:

$$N_o = \frac{q^2}{12} = \frac{36\sigma_x^2}{12(2^N - 1)} \text{ MSV} \quad 9.12$$

Thus the mean-squared signal-to-noise ratio of the quantizer is:

$$\text{SNRq} = (2^N - 1)/3 \text{ MSV/MSV} \quad 9.13$$

Note that the quantizer SNR is independent of σ_x as long as σ_x is held constant under the dynamic range constraint described above. In dB, SNRq = 10 log[(2^N – 1)/3]. Table 9.1 summarizes the SNRq of the quantizer for different bit values.

TABLE 9.1 SNR Values for an N-bit ADC Treated as a Quantizer

N	dB SNRq
6	31.2
8	43.4
10	55.4
12	67.5
14	79.5
16	91.6

Note: Total input range is assumed to be 6 σ_x volts.

16-bit ADCs are routinely used in modern digital audio systems because of their low quantization noise.

9.3 DIGITAL-TO-ANALOG CONVERTERS

A digital-to-analog converter (DAC) is an IC device which converts an N bit digital word to an equivalent analog voltage or current. DACs can operate at very high sampling rates, or can act as static DC voltage sources. A DAC allows digital information which has been processed and/or stored by a digital computer to be realized in analog form. DACs are essential components in the design of CRT terminals, modern control systems, and digital audio systems, to mention a few applications.

As we mentioned in the Introduction to this chapter, it is necessary to understand DACs before we examine the designs of certain analog-to-digital converters (ADCs) which use DACs as components. Most modern DACs are designed to accept straight binary inputs, although some units have been built that require binary-coded decimal (BCD) or gray code inputs. Most DACs use metal oxide semiconductor (MOS) or bipolar junction transistor (BJT) switches to pass current through selected resistors in an R-2R resistance ladder circuit. Either voltage or current reference sources are used to power the ladder. Either internal or external op amps are used to condition the DAC's output currents. DACs can be configured to have unipolar or bipolar outputs, depending on their application.

All DACs have a certain time required for the analog output to reach steady-state value following the digital input. This conversion time can be as long as 10 msec or as short as 1 nsec. Immediately following a change in digital input, there is a DAC output transient, or glitch. The glitch can arise from the transient response of the op amp(s) used to condition the output voltage, and also from internal switching transients in the DAC.

There are many designs for DACs which have evolved over the past 20 years or so. At present, most DACs use the current-mode, R-2R ladder design, as shown in Figure 9.6. This design has low switching transients, and is relatively fast. The switches are typically MOS transistors with low ON resistance. V_R is the DAC's DC reference voltage, generally +5 or +10 V, although in some designs, V_R can be a bipolar, time-varying signal, allowing two-quadrant multiplication of V_R times the digital input. A DAC is called a *multiplying DAC* (MDAC) when it is operated in this mode. The DAC of Figure 9.6 has two output lines in which current must flow to ground, or an op amp's summing junction virtual ground at all times. Because of the grounded output lines, the current in the switched 2R resistors remains constant, hence the node voltages, $V_1 \ldots V_N$, remain constant, and there are minimum glitches caused by charging distributed capacitances from IC circuit elements to the IC substrate. Note that V_R sees a resistance of R ohms looking into the DAC, and thus the MSB current through S1 is simply $V_R/2R$. Each successive switch current is 1/2 of that through the preceding switch. Thus the maximum output current, $I_{o(1)MAX}$ is:

$$\begin{aligned} I_{o(1)MAX} &= (V_R/R)(1/2^1 + 1/2^2 + 1/2^3 + \cdots 1/2^N) \\ &= (V_R/R)\frac{(2^N - 1)}{2^N} \end{aligned} \qquad 9.14$$

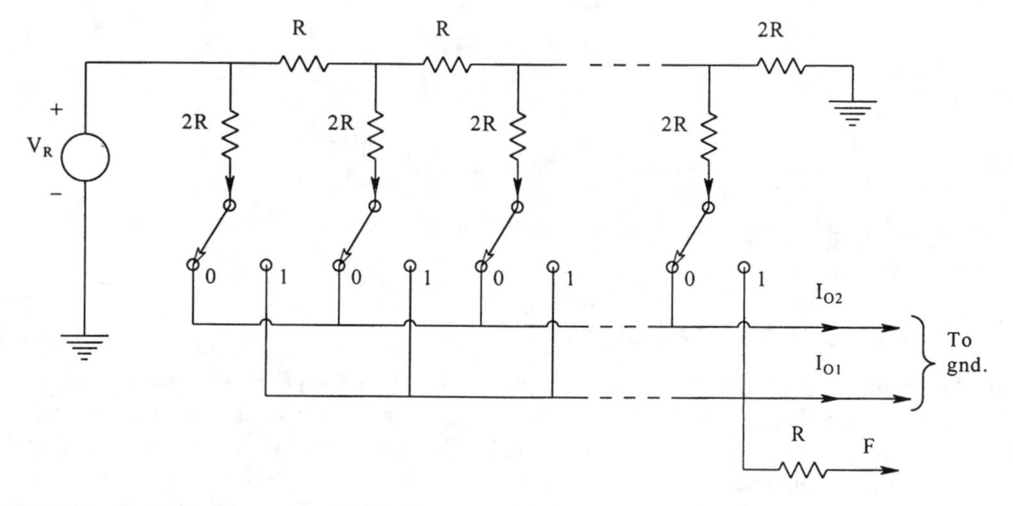

FIGURE 9.6 An R-2R current-scaling DAC ladder. S1...SN are SPDT MOS switches actuated by the binary input signal.

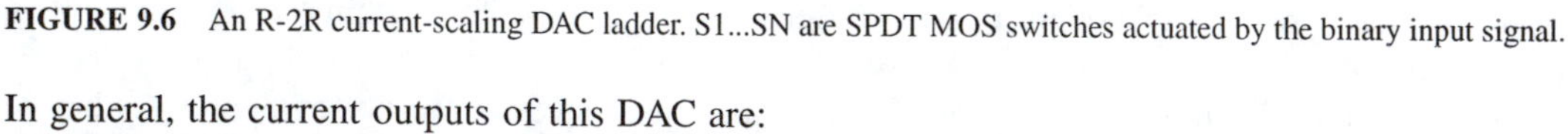

In general, the current outputs of this DAC are:

$$I_{o(1)} = (V_R/R)\sum_{k=1}^{k=N}(D_k/2^k) \qquad 9.15A$$

$$I_{o(2)} = \overline{I_{o(1)}} = (V_R/R)\left[\frac{(2^N - 1)}{2^N} - \sum_{k=1}^{k=N}(D_k/2^k)\right] \qquad 9.15B$$

Here D_k is the logic level controlling the kth switch, either 0 or 1. k = 1 is the most significant bit (MSB), and k = N is the least significant bit (LSB) of the input word.

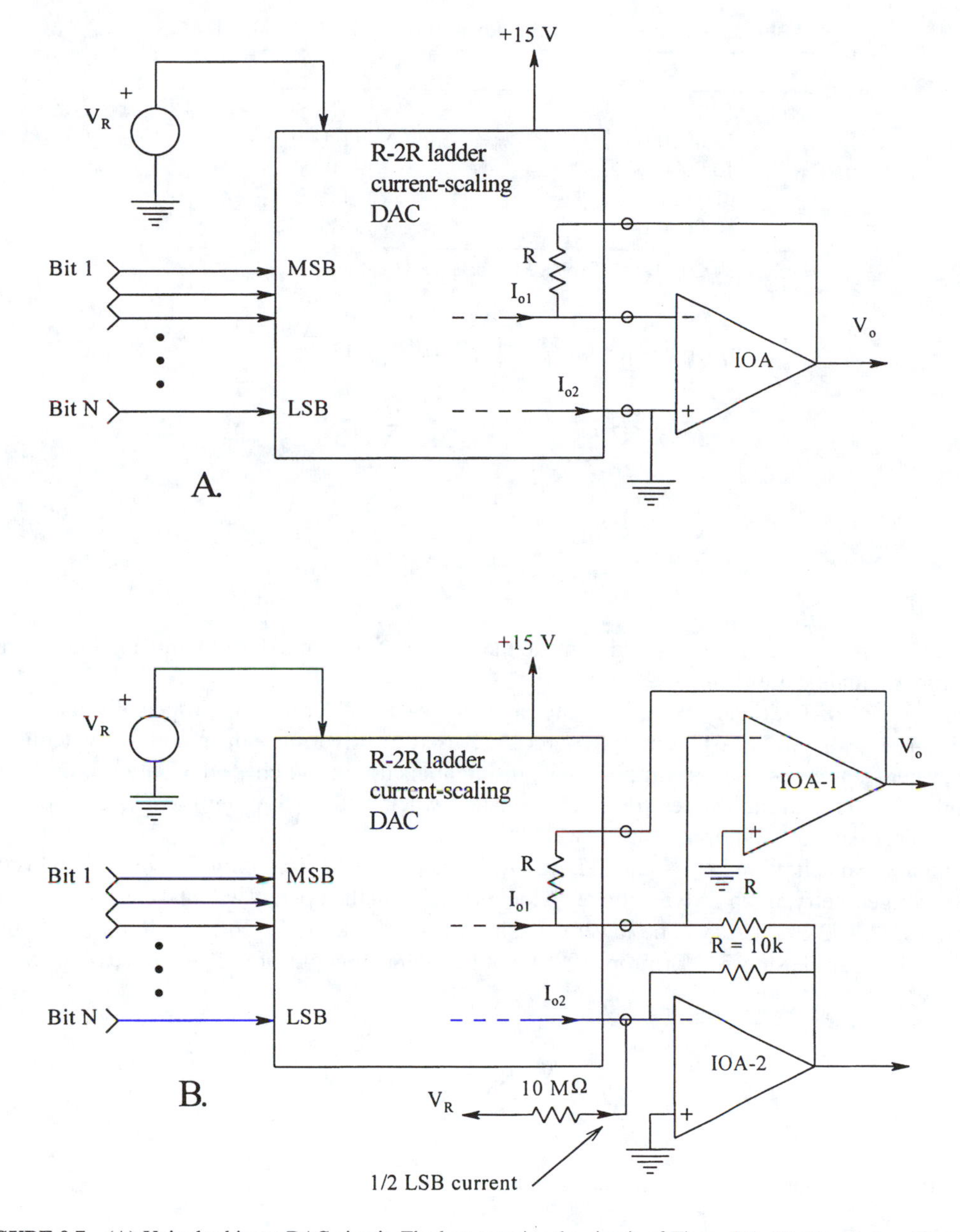

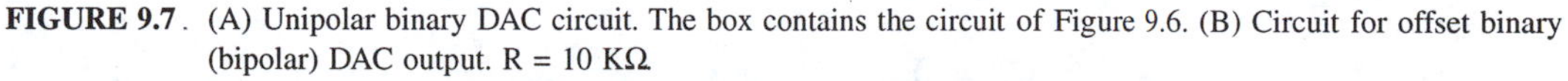

FIGURE 9.7. (A) Unipolar binary DAC circuit. The box contains the circuit of Figure 9.6. (B) Circuit for offset binary (bipolar) DAC output. R = 10 KΩ.

Figure 9.7 illustrate how op amps are used to convert the output currents given by Equations 9.15A and 9.15B to voltages. In the DAC circuit of Figure 9.6, note that there is an extra resistor of resistance R connected to the $I_{o(1)}$ node. This resistor is used as the feedback resistor for the op amp transresistor of Figure 9.7A. Negative unipolar voltage output is obtained because $V_o = -RI_{o(1)}$, where $I_{o(1)}$ is given by Equation 9.15A above. (A second inverter can be used to obtain a positive V_o.) Figure 9.7B illustrates the same DAC given a bipolar output so V_o ranges over approximately $\pm V_R$. In this case, the input word must be in offset binary format, as illustrated in Table 9.2. In the general case, the output of the offset binary DAC can be written:

TABLE 9.2 Coding and output for a 10-bit, Offset Binary DAC

D	Vo
11 1111 1111	$-V_R$ (511/512)
10 0000 0001	$-V_R$ (1/512)
10 0000 0000	0
01 1111 1111	$+V_R$ (1/512)
00 0000 0000	$+V_R$ (511/512)

Note: The 1/2 LSB offset voltage has been neglected for simplicity.

$$V_o = -RI_{o(1)} - V_2 = -RI_{o(1)} - \left(-RI_{o(2)}\right) + \frac{\Delta V_o}{2}$$

$$= -V_R \sum_{k=1}^{k=N} \frac{D_k}{2^k} + V_R \left(\frac{2^N - 1}{2^N} - \sum_{k=1}^{k=N} \frac{D_k}{2k} \right) + \frac{\Delta V_o}{2} \tag{9.16}$$

$$= -2V_R \sum_{k=1}^{k=N} \frac{D_k}{2^k} + V_R \left(\frac{2^N - 1}{2^N} \right) + \frac{\Delta V_o}{2}$$

The $\Delta V_o/2$ term is a 1/2 LSB step used to make the bipolar DAC's transfer function an odd function to minimize quantization error.

Many other DAC designs exist. In many practical R-2R ladder circuits, especially those used for high-speed operation, BJT current sources are used to drive the ladder nodes. Switching of these current sources is often accomplished by differential current switches which allow the currents through the BJT current sources to remain constant whether they flow into the ladder nodes or ground. This design is illustrated in Figure 9.8.

Switched-capacitor or charge-scaling DACs which use MOS IC technology are available. Instead of a ladder geometry, a capacitive voltage divider is used with the upper, switched capacitors having the values, C, $C/2^1$, … $C/2^{N-1}$. The lower capacitor of the divider is C pF. This circuit is shown in Figure 9.9. It can be shown (Northrop, 1990) that the voltage output of the charge-scaling DAC is given by:

$$V_o = \frac{V_R}{C_T} \sum_{k=1}^{k=N} D_k \left(C/2^{k-1} \right) \tag{9.17}$$

where C_T is the total capacitance of the switched array:

$$C_T = \sum_{k=1}^{k=N} \left(C/2^{k-1} \right) + C/2^{N-1} = 2C \tag{9.18}$$

It is not practical to build switched-capacitor DACs for Ns much above 8 because of the large range in sizes required between the MSB and LSB capacitors.

The reader who is interested in the details of DAC designs and how DACs are specified should consult Chapter 14 in Northrop (1990), Chapter 11 in Franco (1988), Chapter 4 in Tompkins and Webster (1988), or Chapter 2 in Zuch (1981).

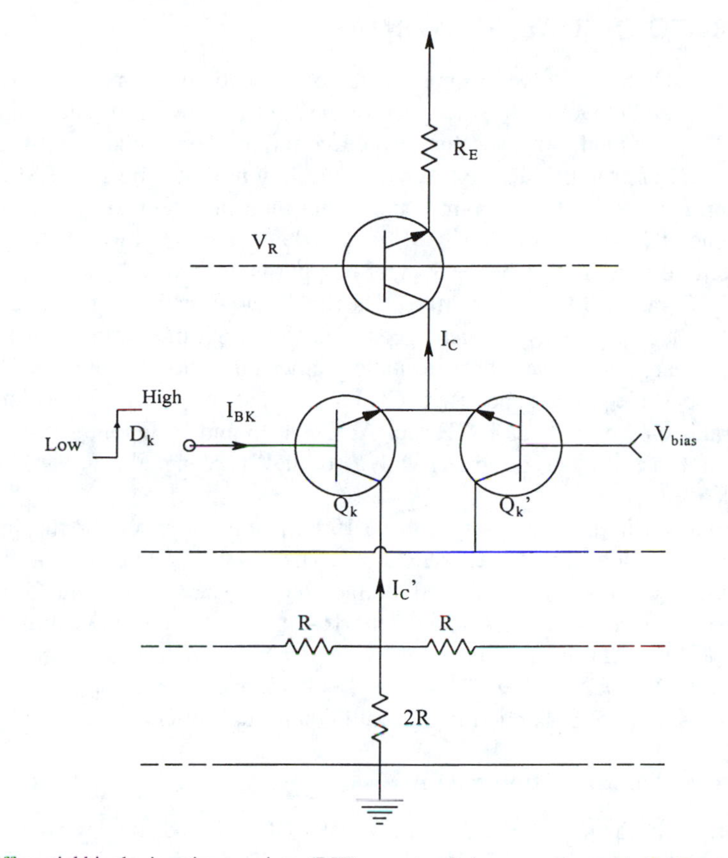

FIGURE 9.8 Differential bipolar junction transistor (BJT) current switch commonly used in R-2R, current-scaling DACs. D_k = kth digital input. The actual kth ladder current is $I'_C = I_C - I_{Bk}$.

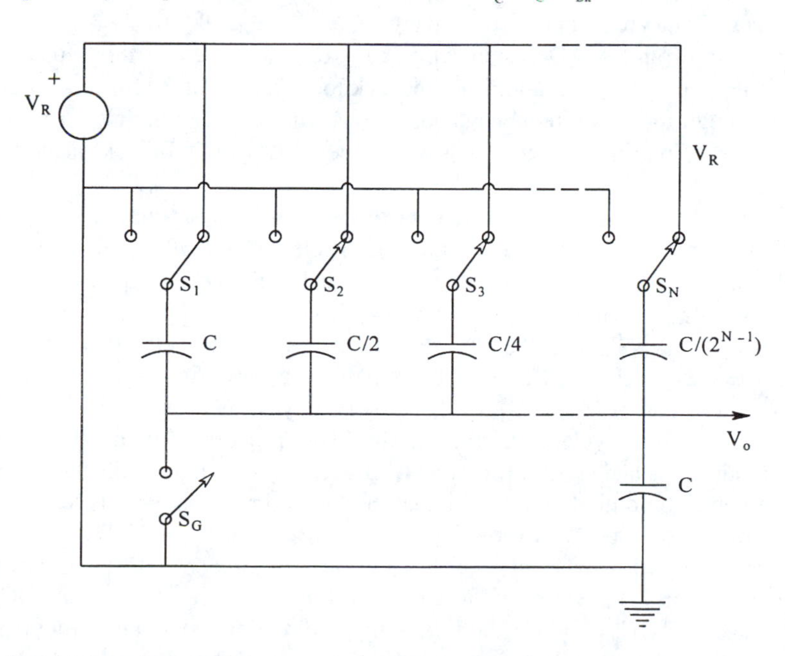

FIGURE 9.9 Simplified circuit of a switched-capacitor or charge-scaling DAC. The entire circuit is realized with MOS IC technology.

9.4 ANALOG-TO-DIGITAL CONVERTERS

Various types of ADCs have evolved to meet specific applications in instrumentation, communications, control, and audio. Obviously, the speed of the A-to-D conversion rate is not critical when digitizing a "DC" measurand, however, the resolution may be very important, so a large number of bits may be used. For example, a 16-bit ADC has resolution limited by ±1/2 LSB, or ±7.63 ppm, a 20-bit ADC can resolve to ±0.477 ppm. Commercial plug-in data acquisition cards for PCs of various types generally do not use ADCs of greater than 16 bits. Thus custom ADC interface systems must be used if resolution better than ±7.63 ppm is desired.

The fastest ADCs are the flash converters, described below, which can convert 8 bits with a sampling period of less than 1 nsec. Such speed is useful when measuring transient phenomena such as partial discharges in power cable insulation, other insulation breakdown phenomena, and transient events in particle physics and lasers. Commercial data acquisition cards for PCs generally have sampling rates on the order of 1 MHz for ADCs up to and including 12 bits. Signatec, Inc. offers a combined, 8-bit, 100 MHz digitizer with a 16-bit, digital signal processor on a PC/AT bus compatible card.

One might wonder why it is necessary to have 16-bit resolution when the display resolution of terminals and graphics plotters rarely exceeds 9 bits. The answer to this question lies in the lower quantization noise of the 16-bit systems, and the fact that the statistics computed on the measurements have lower variances when calculated using 16-bit data. We will describe the organization, features, and applications in measurements of five major types of ADCs: (1) successive-approximation converters; (2) tracking (servo) types; (3) dual-slope, integrating converters; (4) flash (parallel) converters; and (5) dynamic range, floating point converters.

9.4.1 Successive-Approximation (SA) ADCs

SA ADCs are probably the most widely used class of ADC, with accuracies ranging from 8 to 16 bits. They are available as single, LSI, digital/analog ICs. Their advantages include low cost, and moderate conversion speed. However, they can have missing output codes, require a sample-and-hold input interface, and are difficult to auto-zero.

A block diagram of an SA ADC is shown in Figure 9.10. Note that this type of ADC uses a DAC and an analog comparator in a logical feedback loop to control its conversion algorithm. SA ADCs are fast enough for audio frequency applications, with conversion times ranging from about 2 to 20 μsec. Conversion time depends, as we will see, on the digital clock of the SA ADC, and its number of bits.

The conversion cycle of an 8-bit SA ADC begins with the analog input signal being sampled and held at $t = 0$. Simultaneously, the output register is cleared and all D_k are set to 0. At the next clock cycle, D1 is set to 1 with all other $D_k = 0$. This makes the DAC output $V_o = V_R$ (128/256) = $V_R/2$. The comparator performs the operation $\text{SGN}(V_X - V_R/2)$. If $\text{SGN}(V_X - V_R/2) = 1$, then D_1 is kept 1; if $\text{SGN}(V_X - V_R/2) = -1$, then D_1 is set to 0. This completes the first (MSB) cycle in the conversion process. Next, D_2 is set to 1 (D_1 remains the value found in the first cycle). The comparator output signals if $V_X > [D_1(V_R/2) + V_R/4]$. If yes, then D_2 is set to 1, if no, $D_2 = 0$, completing the conversion cycle of the second bit. This process continues until all N bits are converted, and then stops and signals Data Ready. It is easily seen that N clock cycles are required to convert V_X to an N-bit digital word. Note that the final conversion error is less than ±1/2 LSB. The SA ADC conversion process is shown in flow-chart form in Figure 9.11.

SA ADCs are designed to work at audio frequencies and higher, with their conversion cycles under computer control. Often their output registers use tristate logic, so that an ADC can be used with other tristate SA ADCs and DACs on the same, bidirectional data bus. Note that if N, N-bit SA ADCs are on the same data bus, and their starting times are staggered one after another, an ADC output can be read into the computer every clock cycle.

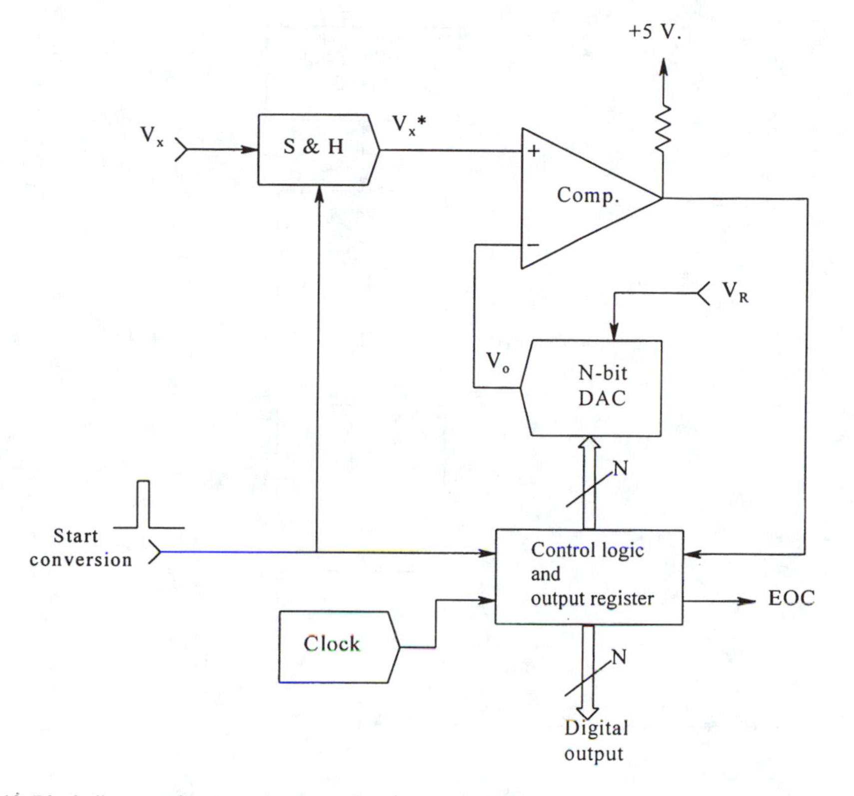

FIGURE 9.10 Block diagram of a successive approximation ADC. S&H is an analog sample-and-hold (or track-and-hold) circuit used to "freeze" V_X during the A/D conversion process.

9.4.2 Tracking or Servo ADCs

The tracking ADC is a relatively slow ADC, best suited for the digitization of DC and low audio frequency measurands. Figure 9.12A illustrates the block diagram of a tracking ADC. Figure 9.12B illustrates the input voltage, $V_X(t)$, and the DAC output, $V_o(t)$ in a T-ADC when it is first turned on. Note that the DAC output slews up to V_X at a rate set by the DAC's clock. When V_o exceeds V_X , the comparator's output goes low which causes the counter to count down one clock cycle to bring $V_o < V_X$. Then the comparator goes high, signaling an up count, etc. In following a very slow AC or DC V_X waveform, V_o is seen to have a ±1 LSB limit cycle around the true value of V_X. This means that in realizing a digital output, the true LSB digit must be dropped because of the limit cycle oscillation. Thus, an 11 bit DAC is required to realize a 10-bit output T-ADC.

A T-ADC has a maximum slew rate, η, in volts per second, at which it can follow a rapidly changing $V_X(t)$ without gross error (Northrop, 1990). This is:

$$\eta = \frac{V_{o(MAX)}}{2^N T_C} \text{ V/s} \qquad 9.19$$

As an example, let N = 10 bits, $V_{o(MAX)}$ = 10 V, and T_C = 1 μsec. η is found to be 9.766×10^3 V/s. Thus, $dV_X(t)/dt$ must be less than this η for no gross error. The T-ADC, in addition to slew-rate limitations, performs poorly in the presence of high-frequency noise on V_X. Thus, low-pass filtering of V_X before it is seen by the comparator improves T-ADC performance. This type of ADC is suitable for the design of low-cost, digital, DC voltmeters and ammeters, etc.

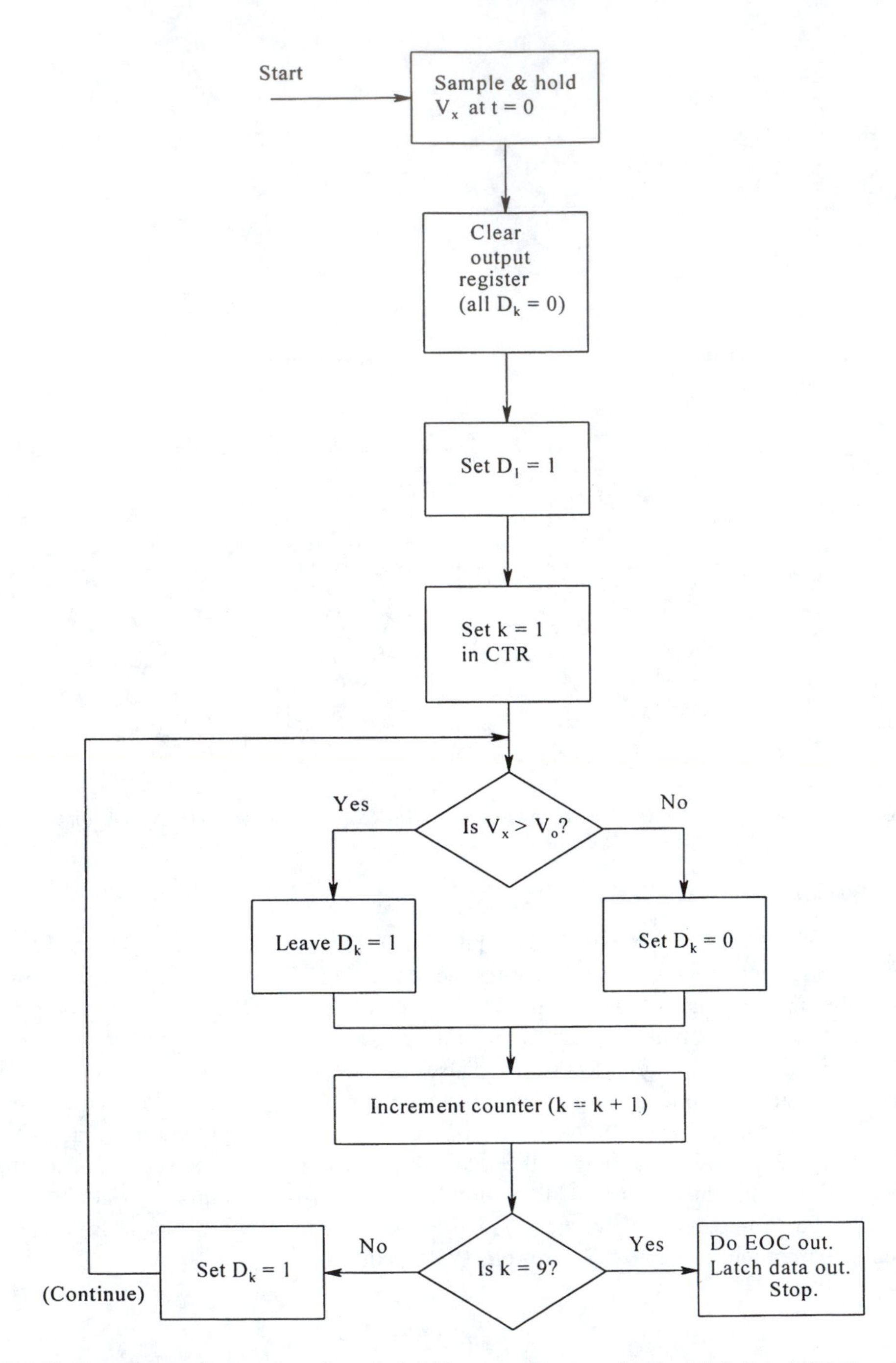

FIGURE 9.11 Flowchart for the operation of a typical, 8-bit, successive approximation ADC. D_1 = MSB, D_8 = LSB.

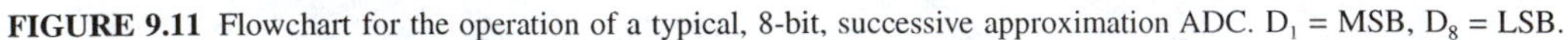

9.4.3 Dual-Slope Integrating ADCs

DSI-ADCs have the advantages of high inherent accuracy (up to 22 bits output), excellent high-frequency noise rejection, they do not require a sample and hold at their input, and there is no possibility of missing output codes. They are widely used in inexpensive DC digital instruments. Their conversion times are long, ranging from about 10 msec to 0.333 s. They generally have an autozeroing mode, discussed below. Figure 9.13 illustrates a DSI-ADC which converts a unipolar DC input. Note that a DAC is not used in this ADC design, but a high-performance analog integrator

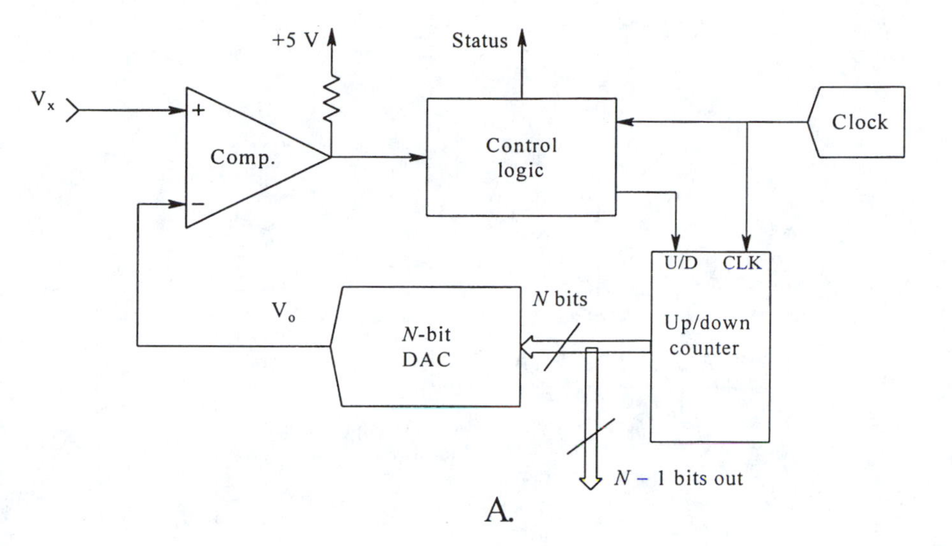

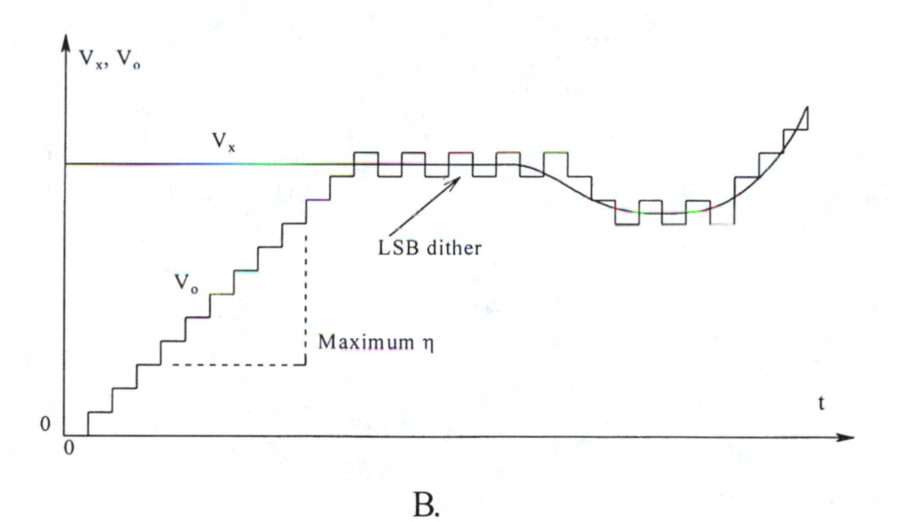

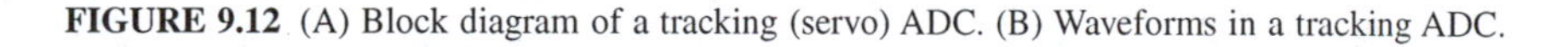

FIGURE 9.12 (A) Block diagram of a tracking (servo) ADC. (B) Waveforms in a tracking ADC.

is required. The control logic operates MOS analog switches, S1 and S2. The conversion cycle of a DSI-ADC begins with the control logic first clearing the counter and setting the charge on the integrator capacitor equal to zero with switch S2. Next, $V_X \geq 0$ is integrated for 2^N clock cycles. At the end of this integration time, T_1, the integrator output, V_2, can be expressed as:

$$V_2 = -\frac{1}{RC}\int_0^{T_1} \frac{V_x(t)}{T_1} dt\ T_1 = -\frac{T_1}{RC}\langle V_x \rangle_{T_1} \qquad 9.20$$

where $T_1 = 2^N\ T_C$. Switch S1 now connects the integrator input to $-V_R$. V_2 now ramps linearly toward 0 volts with positive slope. The counter is made to count clock cycles from T_1 to T_F, when V_2 reaches 0 volts. At T_F, the counter count, M, is latched into the output register. At T_F we can write:

$$V_2 = 0 = -(T_1/RC)\langle V_x \rangle_{T_1} + T_2 V_R/RC \qquad 9.21$$

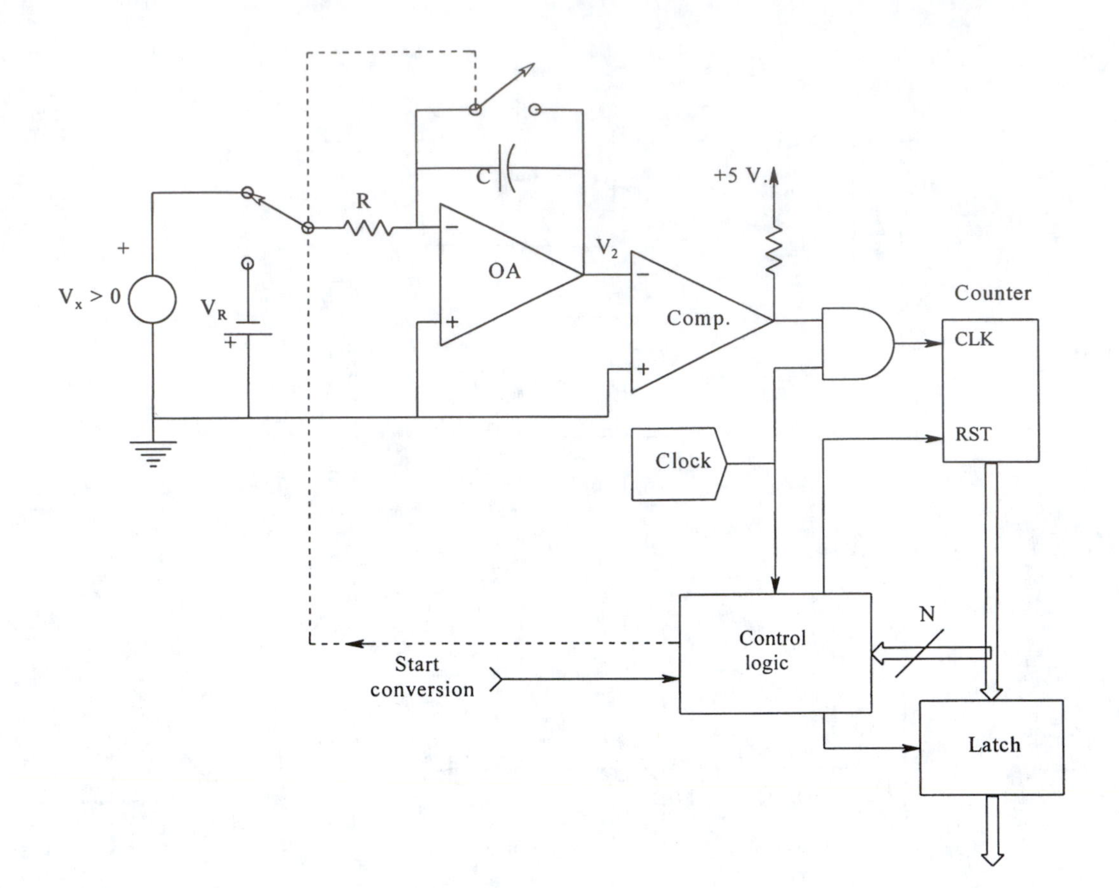

FIGURE 9.13 Diagram of a unipolar, dual-slope, integrating ADC.

where $T_2 = T_F - T_1$. However,

$$T_2 = M T_C \tag{9.22}$$

so we can write:

$$M T_C V_R / RC = 2^N T_C \langle V_x \rangle_{T_1} / RC \tag{9.23}$$

which leads to:

$$M = \langle V_x \rangle_{T_1} 2^N / V_R \tag{9.24}$$

Thus we see that the numerical count, M, is proportional to V_X averaged over T_1 seconds.

Calibration of the DSI-ADC is seen to be independent of values of the clock period, R, and C. It does require a temperature-stabilized V_R, and an integrator op amp and comparator with low offset voltage drift. In addition, the integrator op amp must have negligible DC bias current. DSI-ADCs are widely used in digital multimeters. They are available as LSI ICs, and generally have an autozeroing mode. In one autozeroing strategy, the output count, M_o, measured with the input shorted to ground ($V_X = 0$), is read every other conversion cycle, and then subtracted from M obtained with V_X connected. Another autozeroing design is analog-based in which a special auto-zero capacitor is charged to the net offset voltages of the buffer amplifier, integrator op amp and the comparator during a special autozero cycle with inputs shorted to ground. The autozero capacitor, charged to the net offset voltage, is then switched in series with the input during the measurement cycles to effectively reduce the net ADC offset voltage to less than 10 μV (this autozero architecture is used in the Intersil ICL7106/7107 DSI-ADC).

It can be shown that if T_1 is made an integer number, K, of powerline frequency periods, i.e., $T_1 = 2^N T_C = K/60$ seconds, then powerline hum contaminating V_X can be rejected by as much as 70 dB. Other modifications of the DSI-ADC include a connection for bipolar input signals which gives an offset binary output code, and charge-balancing ADCs in which a reference current source switched into the integrator's summing junction replaces V_R in the DSI-ADC previously described (Northrop, 1990).

Some commercially available, dual-slope ADCs include the Analog Devices' 22 bit AD1175K; the Intersil ICL7106/7107, 3 1/2 digit, autozeroing ADC for LCD display digital multimeters; the Intersil ICL7129, 4 1/2 digit DMM DSI-ADC; the Teledyne TSC7135, 4 1/2 digit DMM DSI-ADC; the Teledyne TSC808 auto-ranging, AC/DC, 3 1/2 digit DMM IC.

9.4.4 Flash (Parallel) ADCs

Flash ADCs (FADCs) are the cornerstone of the ultra-high speed, digitizing front ends of certain digital storage oscilloscopes (DSOs), such as manufactured by LeCroy, Nicolet, Hewlett-Packard, and Tektronix. Eight-bit digitizing rates in excess of 1 gigasamples/s (10^9 Sa/s) are currently available for specialized DSOs. The digitized outputs of such flash ADCs must be buffered by emitter-coupled logic (ECL) or proprietary logic registers, which store and download the digital data rates acceptable by computer bus structures, logic and RAM memories. Downloading rates around 1 megabyte/s are typical for DMA transfers to PCs with 25 MHz clocks.

An FADC can be realized as a hybrid or LSI IC. From Figure 9.14 we see that an FADC is composed of four subsystems: (1) an analog track-and-hold or sample-and-hold circuit which is used to sample and "freeze" the input voltage, and present that sample to the inputs of the high-speed, analog, amplitude comparators; (2) a stable, DC reference voltage source and a voltage divider ladder that supplies the switching reference voltages to the analog comparators; (3) 2^N, high-speed, analog voltage comparators having negligible DC offset voltages; (4) a combinational logic circuit composed of OR, NOR, and inverter gates that accepts the logic level outputs of the 2^N comparators, and that generates a parallel digital word at the FADC output.

Practical considerations in the design of integrated circuit FADCs having the architecture of Figure 9.14 limit the number of bits to eight. Problems exist with the input capacitances of the 2^8 comparators loading the track-and-hold circuit, signal propagation delays, and maintaining low DC offset voltages on the comparators.

A design for a FADC having a 12-bit output is shown in Figure 9.15. This system uses two 6-bit FADCs, a very fast 6-bit DAC, and two sample-and-hold circuits. Conversion speed of this system will be slower than either 6-bit FADC alone because of the conversion lag and settling time of the DAC and $K_V = 2^6$ amplifier. A conversion cycle of the 12-bit two-stage FADC operates as follows: S&H-1 samples the analog input signal, giving an analog output V_X. V_X is flash-converted by the first 6-bit FADC and simultaneously reconverted to analog form by the 6-bit DAC. The DAC output, V_1, is subtracted from V_X, giving the analog quantization error, V_E, of the first ADC. V_E is amplified by a factor of $2^6 = 64$, again sampled and held, then converted by the second FADC to give the 6 least significant bit output. In mathematical terms:

$$V_1 = V_{x(MAX)} \sum_{n=1}^{6} \frac{D_n}{2^n} \qquad 9.25$$

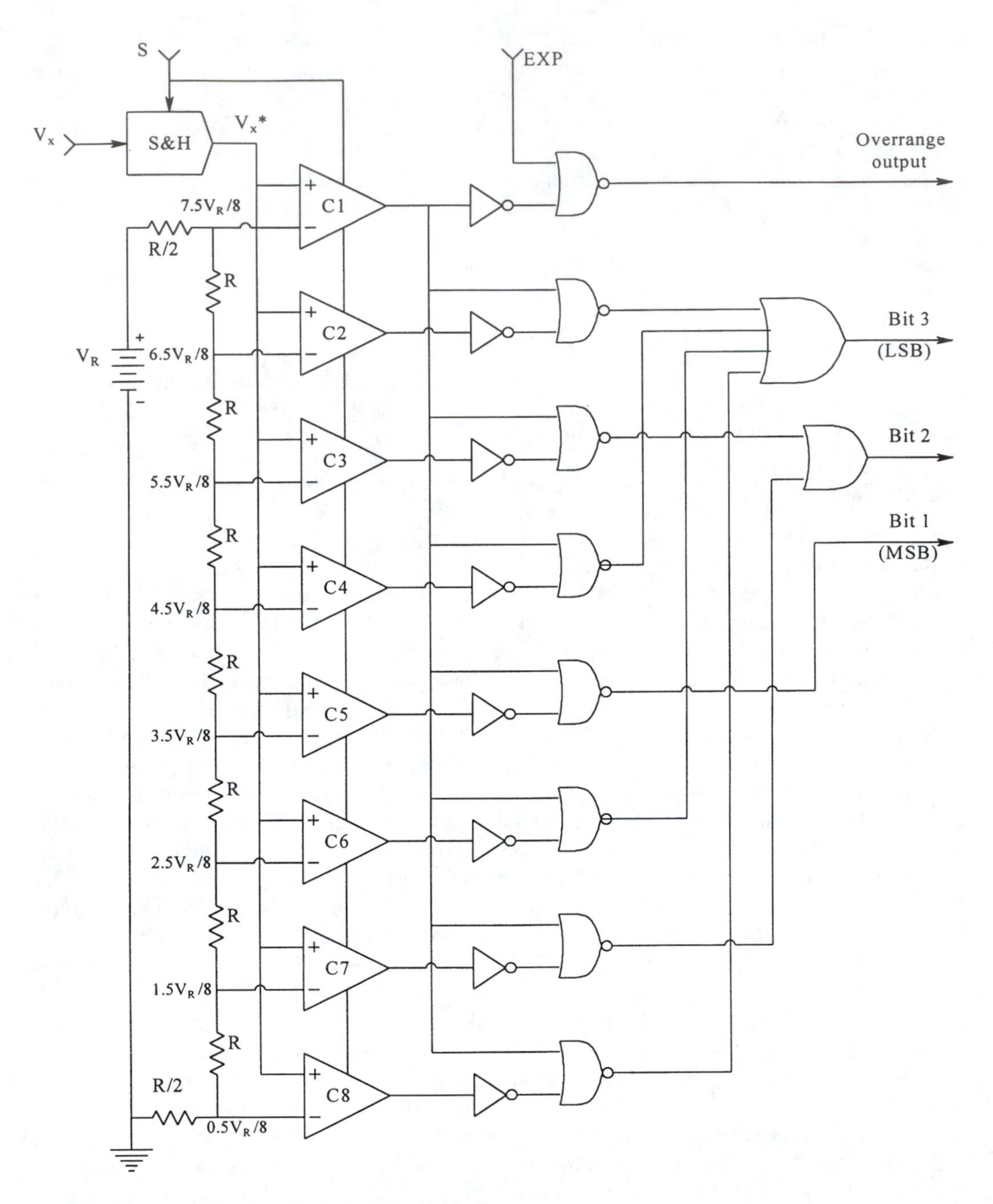

FIGURE 9.14 Schematic of a 3-bit, flash ADC with binary output.

$$V_2 = 2^6 V_E = 2^6 (V_x - V_1) = 2^6 \left(V_x - V_{x(MAX)} \sum_{n=1}^{6} \frac{D_n}{2^n} \right) \tag{9.26}$$

In general, V_E can be as large as $V_{X(MAX)}/2^6$, hence the second 6-bit converter generates a binary code on the quantized remnant, V_E. Internally, the two 6-bit FADCs each require $(2^6 - 1)$ comparators; thus, this 12-bit FADC design is less complex than even an 8-bit FADC. It can be shown that the total coded output of this system can be written:

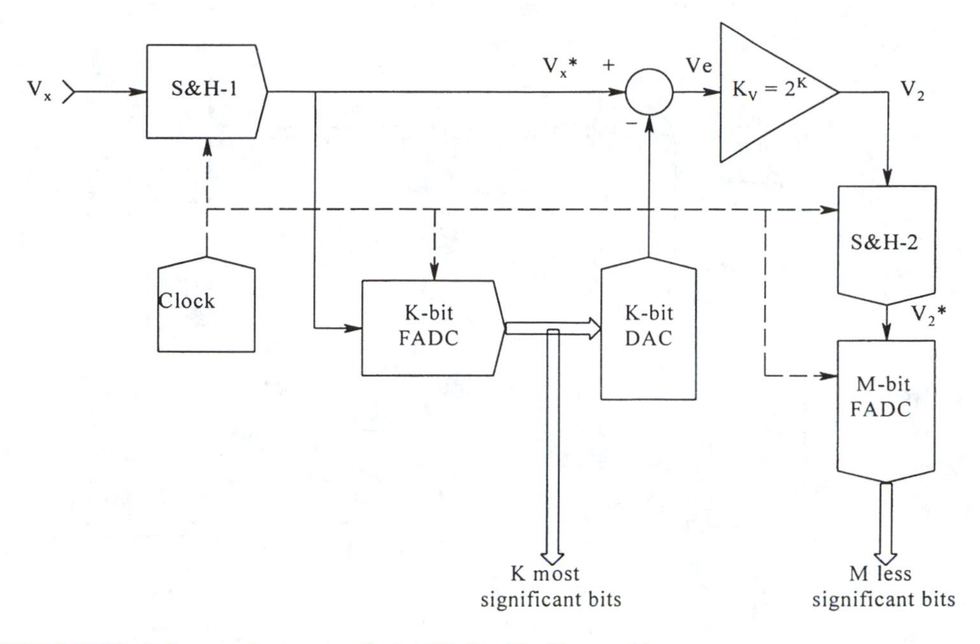

FIGURE 9.15 Block diagram of a two-step, flash ADC. K + M = N output bits.

$$V_x = V_{x(MAX)} \sum_{k=1}^{12} \frac{D_k}{2^k} \tag{9.27}$$

In general, FADCs are expensive, specialized ADCs whose principal application is the digitization of analog signals with frequencies from 10 MHz to over 1 GHz. They can, of course, operate at slower sampling rates and are widely used in the front-ends of digital storage oscilloscopes which operate on signals having frequencies from milliHz to 500 MHz. Eight-bit resolution is typical.

Some examples of FADC ICs include the Burr-Brown ADC600, a 12-bit, 10 megasamples/s (MSPS) circuit which uses the two-stage architecture described above, the Analog Devices AD770 8-bit, 200 MSPS IC, and the Sony CXA1096P 8-bit, 20 MSPS converter. The state of the art in flash converters probably lies close to the proprietary digitizing system offered by LeCroy. The LeCroy Model 6880B waveform digitizer system samples at a fixed 1.35 GSPS, giving either an 8-bit or 11-bit output, stored in a 10,016-word buffer memory holding 7.42 μsec of sampled data.

9.4.5 Dynamic Range, Floating Point ADCs

Figure 9.16 illustrates a block diagram of the Micro Networks model MN-5420, DRFP-ADC. This ADC has an accuracy better than 2 ppm, and can do 3.2×10^5 conversions/second. It has a 16-bit output consisting of a 12-bit mantissa and a 4-bit exponent. The 4-bit FADC codes the exponent, and its output sets the gain of a programmable gain amplifier (PGA) having gains $K_V = 2^k$, k = 1, 2, 4, … 128, 256. The settled amplifier output goes through a track-and-hold circuit, and is converted by a fast, 1 μsec, 12-bit, SA ADC. When the analog input signal is small, the gain of the PGA is high. For example, for $V_X < 19.5$ mV, the gain is 256. The value of a mantissa LSB at this gain is 9.5 μV. As designed, the maximum input voltage of the MN-5420 DRFP-ADC is ±5 V. Table 9.3 gives this ADC's gain switching points and the mantissa's LSB values. When V_X goes negative, the exponent coding is the same as for positive V_X; however, the MSB of the 12-bit mantissa goes from 0 to 1.

The MN-5420 DRFP-ADC design is an ingenious approach to high-resolution, high-speed analog-to-digital conversion. The heart of the design is the precision, auto-zeroing, PGA. The

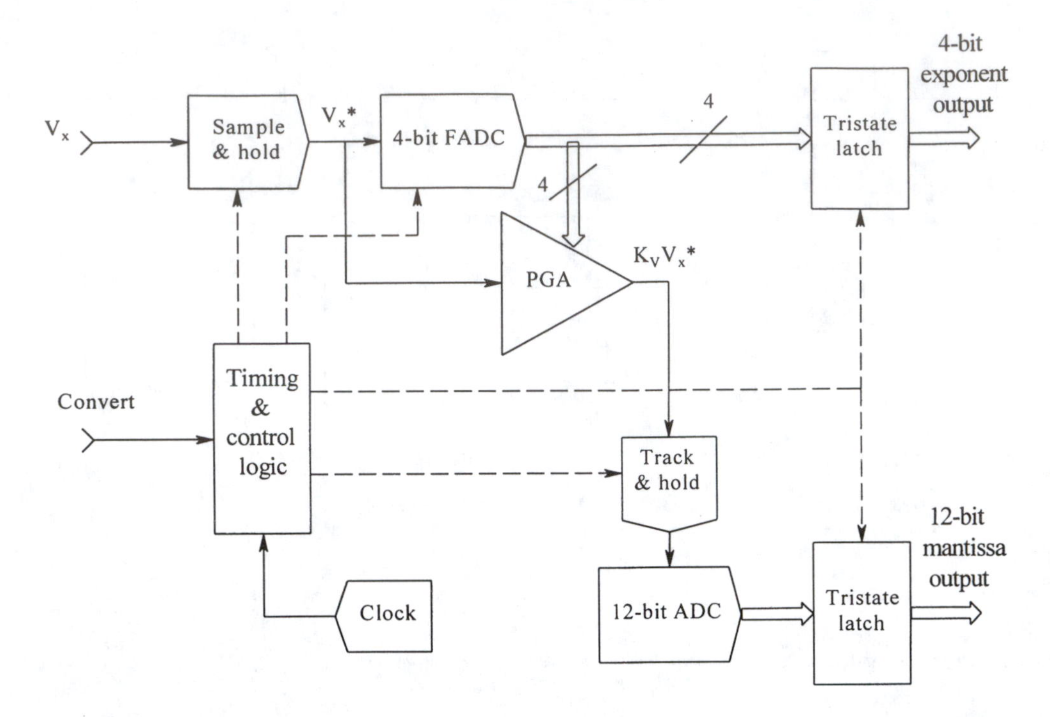

FIGURE 9.16 Block diagram of the MN-5420, dynamic range, floating point ADC. The programmed gain amplifier (PGA) has an autozeroing feature (not shown).

overall ADC speed is limited by the speed at which gains can be switched, and the speed of the 12-bit SA ADC. Note from Table 9.3, that the resolution of the DRFP ADC is 1.9 ppm on its most sensitive autorange.

TABLE 9.3 MN-5420, Dynamic-Range, Floating-Point ADC PGA Switching Points and Exponent Coding for $V_X > 0$ (V_X Range Is ±5 V)

PGA Switching Voltage, V_X	PGA Gain	Exponent	LSB Voltage
$+5 \leq V_X \leq 2.5$ V	1	1 0 0 0	2.44 mV
$2.5 < V_X \leq 1.25$ V	2	0 1 1 1	1.22 mV
$1.25 < V_X \leq 0.625$ V	4	0 1 1 0	610 μV
$0.625 < V_X \leq 0.3125$ V	8	0 1 0 1	305 μV
$0.3125 < V_X \leq 0.15625$ V	16	0 1 0 0	153 μV
$0.15625 < V_X \leq 0.078125$ V	32	0 0 1 1	76 μV
$0.078125 < V_X \leq 0.039063$ V	64	0 0 1 0	38 μV
$0.039063 < V_X \leq 0.019531$ V	128	0 0 0 1	19 μV
$0.019531 < V_X \leq 0$ V	256	0 0 0 0	9.5 μV

9.5 THE IEEE-488 INSTRUMENTATION BUS (GPIB)

In computer nomenclature, a "bus" is "a collection of unbroken signal lines that interconnect computer modules (the connections are made by taps on the lines)" (Stone, 1982). In general, a bus contains data lines, control and signaling lines, and in some cases, address lines. There are many parallel bus architectures and protocols which have been developed since the advent of the microcomputer. These include, but are not limited to: the Intel Multibus (IEEE-796 bus), the VME bus, the Centronics printer bus, the DEC Unibus, the LSI-11 bus, the IBM PC bus (IBM PC/XT/AT), the MacIntosh II NuBus, the MicroVAX Q-bus, the IBM PS2 Micro-Channel Architecture (MCA) bus, the Intel iSBX bus, the ISA and EISA PC buses, and of course, the IEEE-488 Instrumentation Bus (GPIB) which we describe below.

The IEEE-488 bus was developed by Hewlett-Packard in the early 1970s as a standard, 8-bit, bidirectional, asynchronous bus (HP-IB) to enable a number of HP-IB-compatible instruments to communicate with a controlling computer, and with each other. Not only can measured data be sent to the host computer for storage and processing, but in certain GPIB-compatible instruments, the computer can be used to set the instrument's front panel controls and to control the measurements. The original IEEE-488 standard was defined by the IEEE Standards Committee in 1975, and it was revised in 1978, and again in 1987. The GPIB is unique in that it uses unique, stackable (hermaphrodite) connectors on its cables, as shown in Figure 9.17. Many semiconductor manufacturers offer LSI IC interfaces for the GPIB. Some of these include the three Intel i829X series ICs (the i8291 talker/listener, the i8292 GPIB controller, and the i2893 bus transceiver-driver), the Texas Instruments TMS9914, the Motorola 68488, and the Signetics HEF4738.

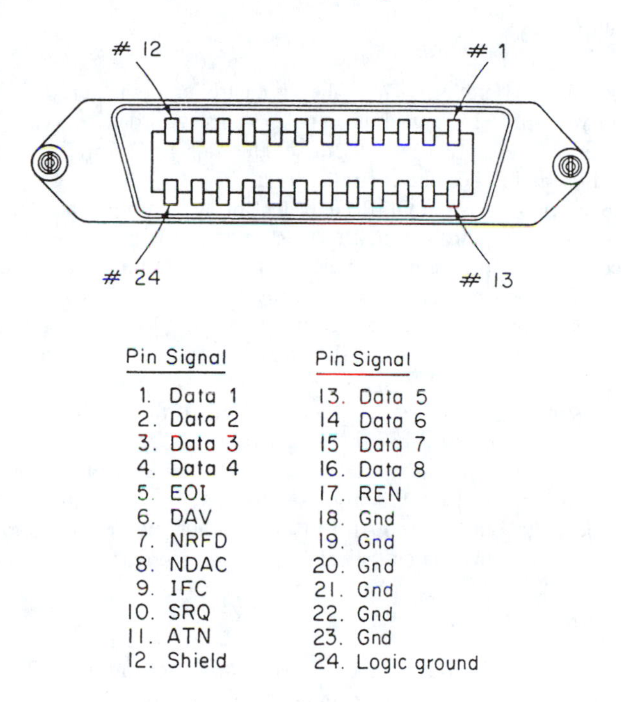

FIGURE 9.17 Standard, IEEE-488 instrumentation bus stackable connector and its pin assignments. (From Helfrick, A.D. and Cooper, W.D., *Modern Electronic Instrumentation and Measurement Techniques*, Prentice-Hall, Englewood Cliffs, NJ, 1990. with permission.)

9.5.1 The GPIB Bus Structure

The GPIB consists of eight, tristate, bidirectional data lines, and eight control lines that select, deselect, and otherwise coordinate the asynchronous communications between the host computer and satellite instruments. The maximum practical data transmission rate in most GPIB systems is about 250 kB/s, equivalent to about 2 Mbits/s. Practical considerations on loading a typical IEEE-488 bus limit the number of devices on the bus to about 15, all which should be located within 3 m or so of the host computer. All lines on the GPIB use a complimentary TTL logic protocol. An active LOW is a voltage less than 0.8 V, and may be treated as a logical TRUE or "1". An active HI is a voltage greater than 2.5 v, and is a logic FALSE or "0". In the tristate, Hi-Z condition, pullup resistors tied to +5 V make the data lines assume a high impedance, HI voltage. All commands and most data on the 8-bit data I/O lines are generally sent using the 7-bit ASCII code set, in which case the eighth bit is used for parity, or is unused.

The eight control lines, their acronyms and functions are described below: they may be subdivided into Handshake Lines and Interface Management Lines.

Handshake Lines

1. **DAV** (data valid): When a selected device on the GPIB supplies an 8-bit word to the data lines (i.e., is a talker), DAV is set LOW (or TRUE) to indicate to all on the bus that the data byte on the bus is ready to be read to a listener device. DAV is one of the three "handshaking" lines that control data transmission on the GPIB.
2. **NRFD** (not ready for data): This second handshaking line is pulled LOW by a selected listener to indicate it is ready to accept data. (NRFD would be better called "RFD").
3. **NDAC** (not data accepted): This third handshaking line is set LOW by the selected listener device when the data byte transmitted has been accepted and new data may now be supplied.

Interface Management Lines

4. **ATN** (attention): This line is pulled LOW by the bus controller (computer) to signal that it is sending a command. It drives ATN HI to signal that a talker can send it data messages.
5. **IFC** (interface clear): The GPIB controller drives this line LOW to reset the status of all other devices on the GPIB, and to become controller-in-charge (CIC).
6. **SRQ** (service request): This is the GPIB equivalent of an interrupt line. Any device on the bus (except of course, the controller) can pull the SRQ line LOW, signaling to the controller that it requires "service". Devices can be programmed to signal SRQ for various reasons, such as being over-loaded, out of paper, having completed an internal task such as computing a fast-Fourier transform (FFT) spectrum, etc. The controller, upon sensing SRQ LOW, must poll the devices on the GPIB to determine which one sent the SRQ, and then take appropriate control action (service).
7. **REN** (remote enable): This controller output line is set LOW to allow the controller to take over front-panel controls of a selected instrument on the GPIB (if that instrument has the capability of having its front-panel settings taken over by the controller).
8. **EOI** (end or identify): This line has two functions — a talker uses EOI LOW to mark the end of a multi-byte data message. EOI LOW is also used by the bus controller to initiate a parallel poll. When the controller sets both EOI and ATN LOW, a parallel poll causing devices configured for a parallel poll to present status bits on the data bus, which are read by the controller.

9.5.2 GPIB Operation

As we have already seen, the 8-bit data bus can carry data or commands. The controller pulls the ATN line LOW, to signal it is sending commands. There are four types of command: *addressed, listen, talk,* and *universal.* Commands are generally represented by 7-bit ASCII characters on the bus. The type of command being sent is signaled by bits 5, 6, and 7 on the data bus. If bits 5, 6, and 7 are HI voltages (logic 0), then the command is an addressed command. If bit 5 is LOW, and 6 and 7 are HI, then the command is universal. If bit 6 is LOW and 7 is HI, then the command is a listen command, and if bit 6 is HI and 7 is LOW, then the command is a talk command. If both bits 6 and 7 are LOW, the command byte is a secondary command. Secondary commands are used for sending secondary addresses or setting up devices for a parallel poll.

As suggested by their name, universal commands affect all devices on the GPIB. There are five universal commands.

1. **LLO** (local lockout): LLO is represented by 11h on the bus; it is used to disable the front panel controls of all devices on the bus. Under this command, the controller will have complete control of instrument settings, etc.
2. **DCL** (device clear): DCL resets all devices on the GPIB. It is 14h (h for hex) on the data bus.
3. **PPU** (parallel poll unconfigure): A PPU command (15h on the data bus) resets the parallel poll responses of all devices on the GPIB, allowing new parallel poll responses to be specified.

4. **SPE** (serial poll enable): the controller responds to an SRQ by sending out an SPE (18h) on the data bus. The SPE causes each device on the GPIB to prepare a single status word to be sent back to the controller when it is sequentially interrogated by the controller. When the controller finds the device that sent the SRQ, it sends out an SPD command.
5. **SPD** (serial poll disable) is sent out by the controller (19h) on the data bus following the SRQ identification. The SPD command resets all the devices' buses to the normal mode following the SPE so that the SRQ problem can be serviced.

Addressed commands are vectored to those certain devices that have been sent listen commands. There are five addressed commands in the GPIB protocol; they, too, are represented by hex words on the data bus. **GTL** (go to local) (01h) cancels the universal command, LLO, and restores local, front panel control to addressed devices. **SDC** (selected device clear): (04h) SDC clears those devices that have received listen commands. **PPC** (parallel port configure): (05h) the PPC command is sent to a specific device to tell it which data bit it will use to give its status. A secondary word is sent following the addressed PPC command to specify which data line will have the status signal on it, and whether a 1 or a 0 will be used to signal that the device needs service. **GET** (group trigger): (08h) GET synchronizes the operation of several instruments on the GPIB, such as starting a group of related measurements. **TCT** (take control): (09h) this enables the main controller to pass control of the GPIB to a secondary controller. The device which is to become the new controller is first sent a listen command, followed by TCT. Figure 9.18 summarizes the IEEE-488 bus command codes. The interested reader can explore the GPIB in greater depth in the texts by Stone (1988), Tompkins and Webster (1988), Helfrick and Cooper (1990), and the National Instruments 1992 Databook on IEEE-488 and VXIbus Control, Data Acquisition, and Analysis.

Many manufacturers of interface cards for PCs and MACs offer IEEE-488.2 compatible cards and controlling software. Some of the newer GPIB systems have hardware and software that permit one to make a very fast and flexible measurement control system. For example, the National Instruments' NB-GPIB interface board for the MacIntosh II has the latest IEEE-488.2 functions, and can read and write data at a sustained 800 kBytes/s. The National Instruments' LabVIEW 2™ software for the Mac II and PCs is a powerful, graphical programming language that permits the creation of virtual instruments on the monitor whose settings can be altered by mouse, and sent to the actual instrument by GPIB. The NB-GPIB card is compatible with NuBus DMA operations and the RTSI bus. The National Instruments' GPIB-SE card for the Mac SE is another advanced GPIB card that claims data bus transfer rates up to 1 Mbytes/s. This card, too, is compatible with the LabVIEW software. Metrabyte also offers IBM PC-compatible IEEE-488 cards and handling software, the MBC-488 card and the IE-488 card. These cards support up to 15 devices on the bus, and do DMA data transfers at 450 kBytes/s.

9.6 SERIAL DATA COMMUNICATIONS LINKS

While the GPIB is eminently suited for managing a large measurement system, there is often need to couple the output of a single instrument to a computer interface over very long distances using few electrical conductors. Serial data transmission has well-established protocols, and historically predates the GPIB by at least 10 years. The RS-232C serial interface was originally developed in the early 1960s to send data to a CRT terminal or teletypewriter over telephone wires. Its low maximum bit transfer rates now make the RS-232C serial interface generally obsolete for the control of the modern equivalents of terminals, printers, and modern instruments. The RS-232C interface currently finds use in coupling mice, trackballs, joysticks, and other slow input devices to computers, and computers to certain plotters and non-graphics printers. Few measurement instruments are designed to use a serial interface; it lacks the speed and flexibility of the IEEE-488.2 bus. However, for specific applications, such as reading in the data from remote DC sensors, and sending data to loggers, it can be useful.

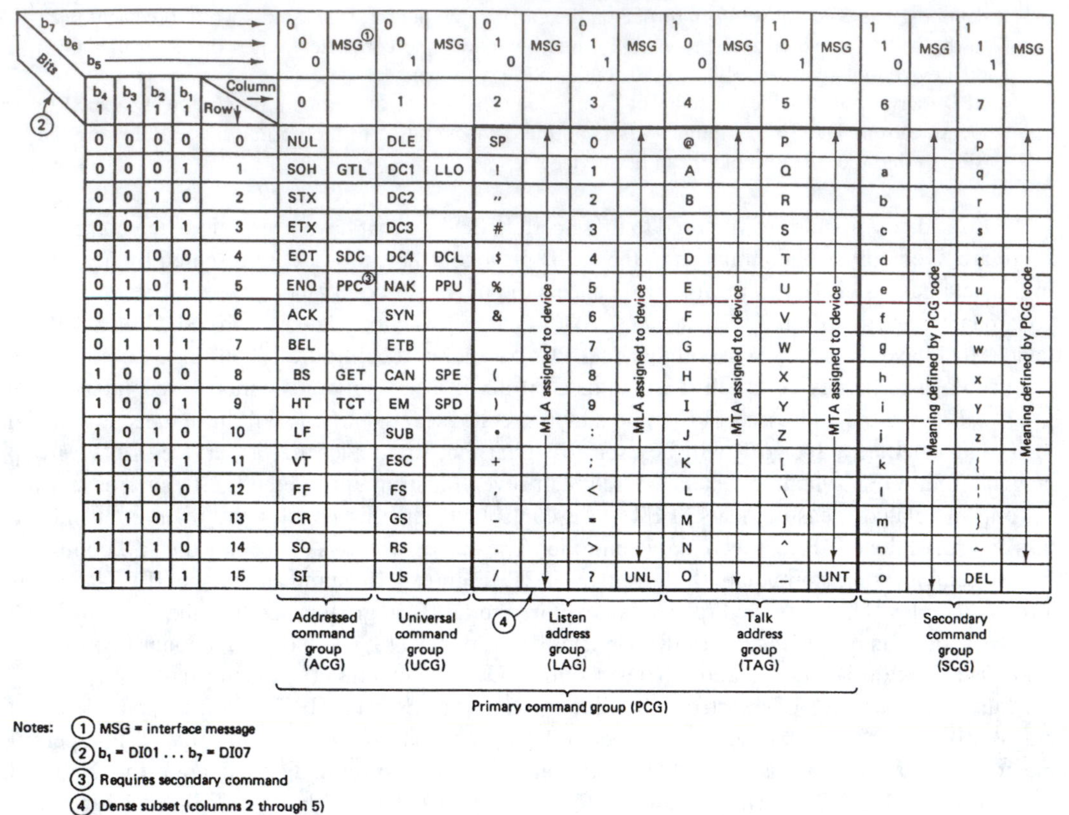

(Sent and received with ATN = 1)

b7 b6 b5 →						0 0 0	MSG①	0 0 1	MSG	0 1 0	MSG	0 1 1	MSG	1 0 0	MSG	1 0 1	MSG	1 1 0	MSG	1 1 1	MSG
b4	b3	b2	b1		Row ↓ / Column →	0		1		2		3		4		5		6		7	
0	0	0	0		0	NUL		DLE		SP		0		@		P				p	
0	0	0	1		1	SOH	GTL	DC1	LLO	!		1		A		Q		a		q	
0	0	1	0		2	STX		DC2		"		2		B		R		b		r	
0	0	1	1		3	ETX		DC3		#		3		C		S		c		s	
0	1	0	0		4	EOT	SDC	DC4	DCL	$		4		D		T		d		t	
0	1	0	1		5	ENQ	PPC③	NAK	PPU	%		5		E		U		e		u	
0	1	1	0		6	ACK		SYN		&		6		F		V		f		v	
0	1	1	1		7	BEL		ETB				7		G		W		g		w	
1	0	0	0		8	BS	GET	CAN	SPE	(		8		H		X		h		x	
1	0	0	1		9	HT	TCT	EM	SPD	)		9		I		Y		i		y	
1	0	1	0		10	LF		SUB		*		:		J		Z		j		z	
1	0	1	1		11	VT		ESC		+		;		K		[		k		{	
1	1	0	0		12	FF		FS		,		<		L		\		l		|	
1	1	0	1		13	CR		GS		-		=		M		]		m		}	
1	1	1	0		14	SO		RS		.		>		N		^		n		~	
1	1	1	1		15	SI		US		/		?	UNL	O		_	UNT	o		DEL	

FIGURE 9.18 Command codes for the IEEE-488 (1978) instrumentation bus. (From Tompkins, W.J. and Webster, J.G., Eds., *Interfacing Sensors to the IBM*™ PC, Prentice-Hall, Englewood Cliffs, NJ, 1988. with permission.)

Other, more modern, serial, asynchronous data transmission protocols include the RS-422, RS-423, RS-449, and RS-485. We will first describe the RS-232C serial interface, as it is the oldest serial bus, it is widely used, and its protocol is used with little variation in the later serial bus designs. RS stands for *Recommended Standard* (of the Electronic Industries Association, Washington, D.C.).

9.6.1 The RS-232C Interface

In describing the RS-232C interface, we encounter some new acronyms. RS-232C serial communications are sent by *data terminal equipment* (DTE) and *data communications equipment* (DCE). DTE include such things as computer terminals, teleprinters, computers, and digital instruments. DCE are devices (modems) that encode the serial digital signals into low-bandwidth (sinusoidal) formats compatible with transmission on voice band telephone lines. The rate at which modems transmit data is given in baud, which stands for *bit rate audio*. (Modern modems used for serial telephone line communications between PCs and instruments typically run between 2,400 baud and 14,400 baud, depending on the limiting bandwidth of the transmission system. FAX modems typically run at 14,400 baud.) Just as the GPIB has a unique, standard connector, the RS-232C interface uses a standard male or female 25-pin, DB25P (male) or DB25S (female) connector, shown in Figure 9.19.

The RS-232C interface operates either in the simplex, half-duplex, or full-duplex modes. In the simplex mode, data transmission is unidirectional, e.g., from the computer to a printer. In the half-

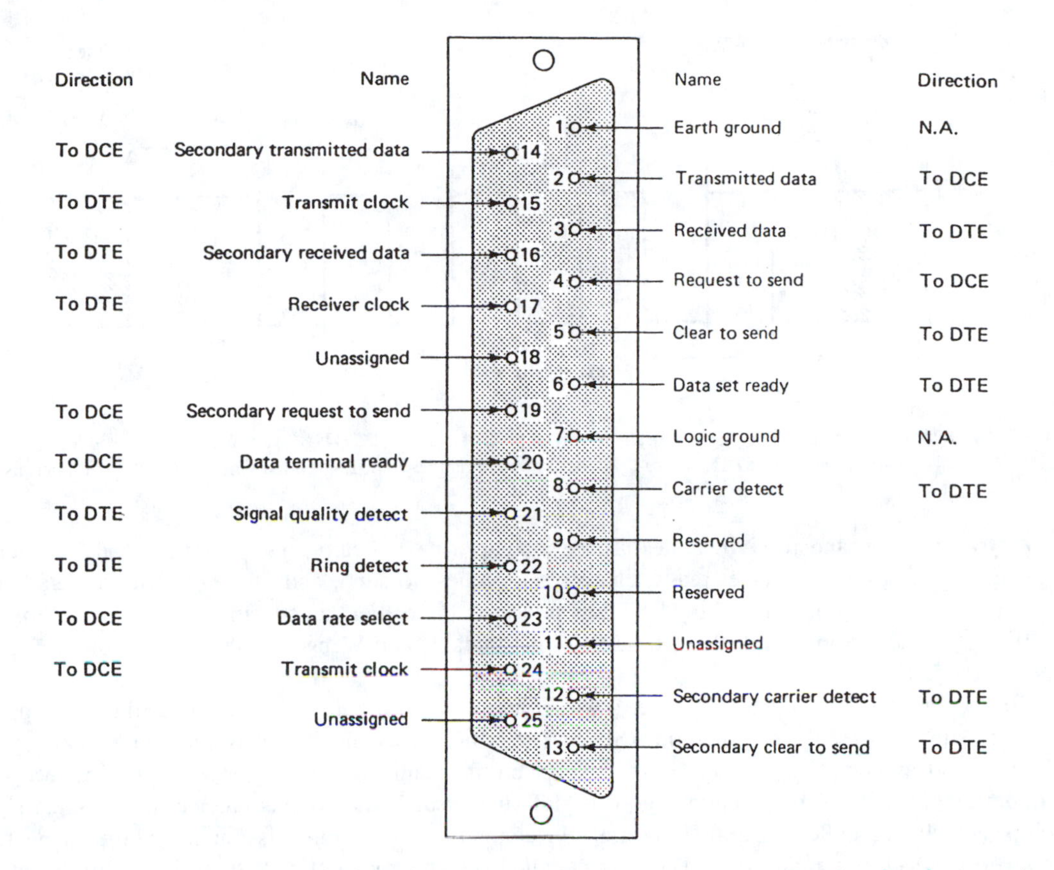

FIGURE 9.19 Standard pin connections for the RS-232C serial data interface. Pin nomenclature based on the connections in DTE equipment. (From Wolf, S., *Guide to Electronic Measurements and Laboratory Practice*, 2nd ed., Prentice-Hall, Englewood Cliffs, NJ, 1983. with permission.)

duplex mode, serial data can be sent in both directions, but in only one direction at a time. Full-duplex operation permits simultaneous, bidirectional, serial data transmission.

Data is transmitted as 8-bit, ASCII words signaled by high or low logic voltages on the transmitted data (TD) line. An example of this process is shown in Figure 9.20. In the idle state, the TD line is held high. At the beginning of data word transmission, the TD line goes low, and stays low for one clock period. This is the start bit, which is always low. The receiving equipment senses the high-low transition of the start bit, and to verify start, samples the received data (RD) one-half clock period later. If low, the start bit is verified. The state of the received data line is then sampled eight times at intervals of one clock period. The last (eighth) sample is the MSB and is called the parity bit. In setting up an RS-232C interface, the user can specify the use of odd, even, or no parity in the data transmission process. If even parity is used, the parity (8th) bit is set so that the total number of logical 1s in the transmitted word, including the parity bit, is even. If the parity check algorithm in the receiver counts an odd number of highs in a received word, it declares a parity error. The parity error can terminate transmission, or require that the same character be sent again. Similarly, if an even number of 1s is counted in a received word when using odd parity, an error is declared. Parity is generally used when serial data is transmitted under noisy conditions over long distances. It is normally not used in situations where the receiver is connected by a short cable to the transmitter in a low-noise environment.

At the end of data transmission, after the parity bit is sent, one or two high stop bits are sent before the TD line is declared idle, and is ready to transmit the next word (ASCII character). Note that there are a number of handshaking functions on the RS-232C lines which accompany the serial

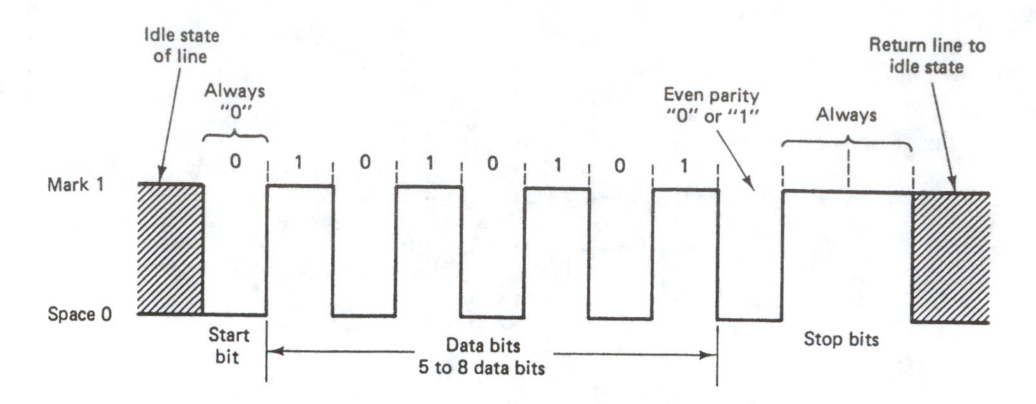

FIGURE 9.20 Example of an 8-bit, serial data signal sent on pin 2 of an RS-232C DTE. (From Wolf, S., *Guide to Electronic Measurements and Laboratory Practice*, 2nd ed., Prentice-Hall, Englewood Cliffs, NJ, 1983. with permission.)

data transmission and receiving lines. These include: ring indicator, data terminal ready, carrier detect, signal ground, data set ready, clr to send, request to send, and others. For more detailed descriptions of the operation of the RS-232C interface, the reader can consult Section 5.2 in Stone (1988), Chapter 18 in Wolf (1983), Chapter 6 in Tompkins and Webster (1988), or Motorola App. Note AN-781A.

In the earliest days of RS-232C use, it was common for computers to be connected to teletypewriters such as the Western Electric ASR33. The ASR33 used no IC electronic components; all logical and switching operations were done electromechanically at very low baud rates. More modern serial data communications used UART (universal asynchronous receiver/transmitter) IC chip sets which handle the parallel to serial conversion of data for transmission, and at the receiver, convert the RD to parallel form. The present state of the art for the RS-232C protocol limits data transmission rates to less than 20 kbaud, and wire cables are seldom effective over 15 m in length. Also, RS-232C protocol restricts the dV/dt of the TD line data to a maximum of 30 V/μsec. To circumvent these problems, other serial data transmission standards were developed, making use of improvements in hardware to achieve higher baud rates and longer transmission lines. These interfaces are described in the next section.

9.6.2 The RS-422, RS-423, and RS-485 Interfaces

These interfaces make use of our knowledge about the transmission properties of transmission lines for transient signals and improved IC designs to realize improved data transmission rates over longer distances than possible with the 30+-year-old technology of the RS-232 interface. Their properties are summarized in the following paragraphs, and are illustrated in Figure 9.21.

The *RS-422A Interface* uses a balanced, twisted-pair transmission line, terminated in the characteristic impedance of the line. A balanced or differential amplifier line driver as well as a differential line receiver are used. The RS-422A interface can transmit data at up to 10 Mbaud, and can have lengths of up to 1200 m. There is a trade-off between baud rate and cable length: 10 Mbaud is possible on a 12 m cable, 1 Mbaud on a 100 m cable, 100 kbaud on a 1 km cable, and up to 80 kbaud on a 1.2 km cable. The TI 9636 and 9637 ICs, as well as the Motorola MC3487 differential receiver and the MC3486 differential driver can be used to realize an RS-422A link.

The mechanical connections for the RS-422A interface are specified by the RS-449 standard, which specifies a 37-pin connector supporting the mandatory twisted pair lines for receive ready, test mode, data mode, request to send, clear to send, receive data, and send data (Stone, 1982).

The *RS-423A Interface* uses an unbalanced single line, similar to the RS-232C link. Even so, the RS-423A link is faster than the RS-232C; it was designed, according to Stone (1982), to provide

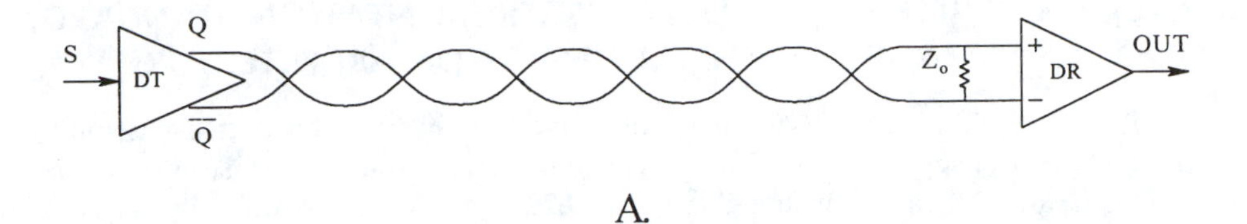

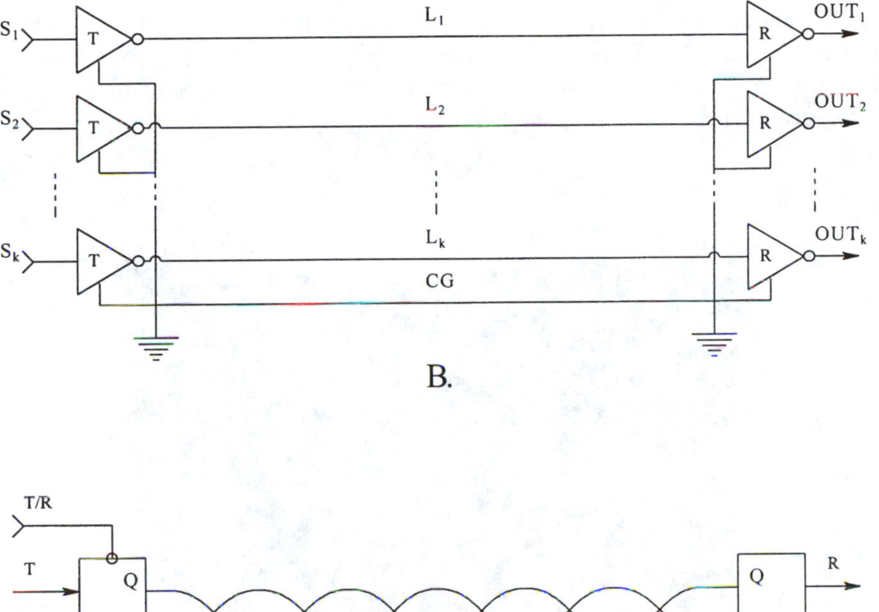

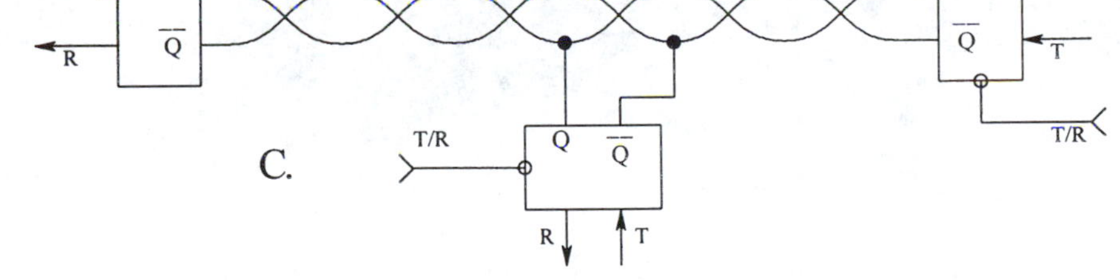

FIGURE 9.21 (A) RS-422A, balanced, twisted-pair transmission line. DT, differential transmitter; DR, differential receiver; Z_0, characteristic impedance of the line. (B) RS-423A, unbalanced serial data transmission lines. CG, common ground. (C) RS-485, balanced, bidirectional, twisted pair serial data transmission system. Differential tri-state transceivers are used.

a linkage between the old RS-232C interface, and the RS-422A interface. Its operational protocol is very similar to RS-232C, although improvements in the driver and receiver electronics allow it to operate at rates up to 100 kbaud over short cables (<30 m) and at significantly slower rates over cables up to 1.2 km.

The *RS-485 Interface* is a balanced (twisted pair), "party line" on which a number of secondary receivers and transmitters can operate. In this respect, it is effectively a data bus. Although the RS-485 interface can transmit data up to 10 Mbaud, practical considerations of line resistance, loading, and terminations set practical limits on line lengths. SN75172 and SN75174 differential drivers and SN74173 and SN75175 differential receivers are used to implement the RS-485 protocol.

9.7 THE CAMAC (IEEE-583) MODULAR INSTRUMENTATION STANDARD, AND THE VXI MODULAR INSTRUMENTATION ARCHITECTURE

The IEEE-583, -595, and -596 CAMAC standards define a hardware and data transmission system which is used to house, support, and communicate with various compatible instrumentation modules. A typical CAMAC "crate", illustrated in Figure 9.22, is a physical package which has 25 slots or powered stations for compatible, plug-in instruments. Generally, the two right-hand slots are used to house the CAMAC controller module. The controller module allows the individual modules to be coordinated and controlled by an external computer using the IEEE-488.2 bus, or in some cases, an RS232C interface. Commands and data transfers to and from the modules and the controller are made over the CAMAC DATAWAY, which uses 24-bit data words. Figure 9.23 illustrates the (internal) DATAWAY connections to three modules in a CAMAC system. The read and write lines are each 24 bits wide. Two lines are used for the timing strobes; there are three common control lines, three status lines, four address bus lines, five function lines, and one station number line per module, for a total of 66 lines per module!

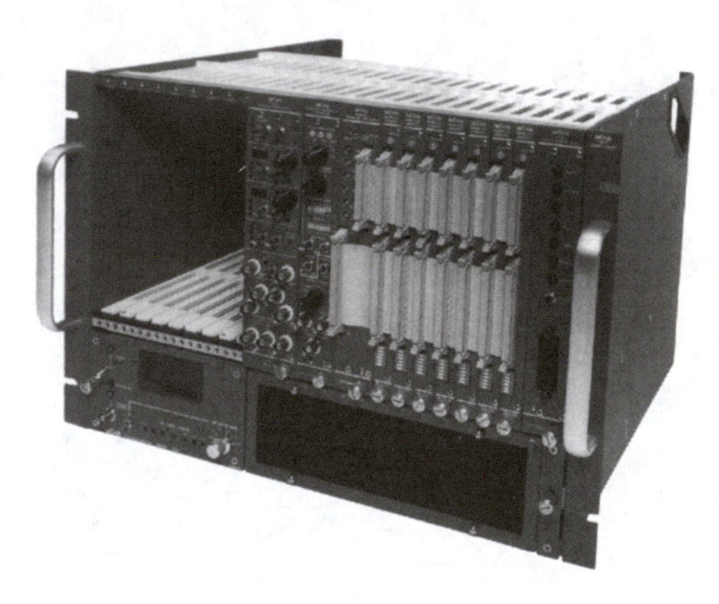

FIGURE 9.22 A partially filled CAMAC "crate". (Figure courtesy of LeCroy Corp., Chestnut Ridge, NY.)

Lack of space prevents going into the details of the CAMAC DATAWAY control and command operation protocols. The interested reader can find their descriptions in the LeCroy instruments catalog (1990). The major advantage of CAMAC-housed instrumentation is its ease of use through the GPIB, and its speed in local data manipulation. Individual instrument modules are managed by the CAMAC controller over the DATAWAY, and only one GPIB bus is required. CAMAC systems represent top-of-the-line instrumentation, and as such, are expensive. They are used extensively in large industrial applications and for measurements in nuclear and plasma physics.

The recently developed, VXIbus measurement systems architecture is similar in many respects to the CAMAC system, but improves on certain aspects of CAMAC system design. VXI stands for **V**MEbus **EX**tensions for **I**nstrumentation. The VXIbus is a nonproprietary, open architecture bus for modular instrumentation systems that fully incorporates the VMEbus standard. It includes not only mechanical and electrical interfacing, but also communications protocols between compatible devices and an external controller. The VXIbus architecture was developed by a consortium of instrumentation companies, including Textronix, Hewlett-Packard, and National Instruments. As

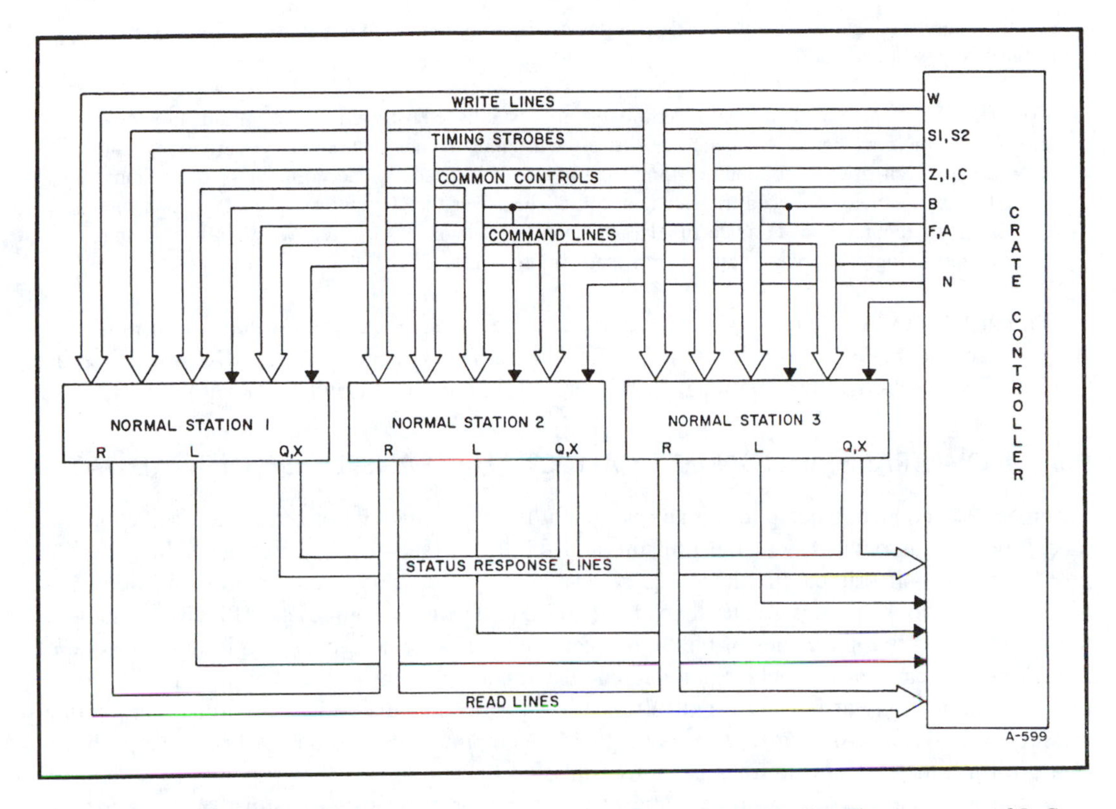

FIGURE 9.23 DATAWAY signal paths to three CAMAC modules from a crate controller (Figure courtesy of LeCroy Corp., Chestnut Ridge, NY.)

in CAMAC systems, special VXI instrument cabinets are used to support a wide variety of modular instruments. All members of the HP 75000 VXIbus instrument family use the Standard Commands for Programmable Instruments (SCPI), a new industry standard, for command, control, and communication. HP SCPI conforms to the new, IEEE-488.2 bus standard. In addition, all HP 75000 products support the HP Interactive Test Generator (HP ITG) software. There are many VXIbus instrumentation modules available from Hewlett-Packard and other companies. All of these modules are characterized by having input and output connectors, but no knobs, dials, pushbuttons, or switches. All settings and control operations are accomplished by software over the VXIbus. Some of the VXIbus module functions available from Hewlett-Packard include command modules and controllers, digital multimeters, counters, function generators, amplifiers, D/A converter, digital I/O, digitizing oscilloscopes, power meter, various switches and multiplexers, a serial port adapter for RS-232C/422 protocols, and a VXIbus to MXIbus extender to tie VXI mainframes together and to an MXIbus computer interface. (MXI stands for **M**ultisystem **EX**tension **I**nterface. It was developed, and announced as an open industry standard, in April 1989 by National Instruments Corp.)

Tektronix (1990) gives a good summary of VXI communications protocols:

> Two communications protocols are defined in VXI for Message Based Devices: Word Serial Protocol (WSP) and Shared Memory Protocol (SMP). The other type of communication possible in VXI, is low-level, binary data transfers which is defined for Register Based Device communication. Communication in a VXIbus system has a hierarchical commander/servant structure which is established by the system Resource Manager when the system powers up. A Commander is any device with one or more lower level devices (servants). The Commander initiates communication with its servants. A servant is a device which operates under the control of a Commander. All message based devices must respond to a WSP, but can also respond to the more advanced communication Shared Memory Protocol. With WSP, the required communication is 16 bit words transferred serially by reading or writing to one of the

communication registers resident on a servant device. There is a WSP command set defined extensively in the VXIbus Specification.

SMP defines the operation of a shared memory channel established between two communicating devices with shared memory space. A channel can be asynchronous or synchronous. Asynchronous channels are used for high priority communications such as warnings, events, and commands which must be immediately executed. The data transfer is accomplished over the ComputerBus in large data blocks. SMP is an efficient method of transferring large amounts of data in a VXI system whether the information is command strings or results obtained from performing a test.

Similar to CAMAC systems, VXIbus equipment is expensive, and best suited for an industrial, ATE environment. For further details of VXIbus instrument systems and the MXIbus, the reader should consult catalogs from National Instruments, Tektronix, and Hewlett-Packard.

9.8 HOW TRANSMISSION LINES AFFECT THE TRANSFER OF DIGITAL DATA

Whenever we connect a computer to an instrument through a parallel or serial interface cable, we face problems associated with the transmission of digital data on a transmission line. Similar problems exist on printed circuit boards when we couple data and address lines between ICs, whenever the clock speed is greater than about 10 MHz. The main sources of difficulty come when the transmission line properties of the high-speed data paths are not given weight in the design of interfaces. We can observe delays, attenuation and rounding of pulses, as well as glitches caused by signal reflections at branches and terminations of the lines caused by improper terminating resistances and impedance mismatches at signal branch points on the lines. All of these problems serve to limit the rate of data transfer on transmission lines.

In this section, we will examine some of the major properties of transmission lines used in computer interfaces. Such lines include coaxial cables, shielded twinax cables, twisted pairs of wires, and ribbon cables. It will be seen that all transmission lines can be described by a common model.

9.8.1 The Transmission Line Model

In our treatment of transient signal propagation on short transmission lines, we will assume that the lines are lossless (i.e., $R_S = G_P = 0$ in Figure 9.24). It can be shown (Lathi, 1965) that the characteristic impedance of the lossless transmission line is given by:

$$Z_O = \sqrt{\frac{L}{C}} \text{ ohms} \tag{9.28}$$

and the velocity of wave propagation on the line is:

$$v = \frac{1}{\sqrt{LC}} \text{ m/s} \tag{9.29}$$

where L is the inductance per unit length of the uniform line, and C is the capacitance per unit length between conductors. Most coaxial cables are made to have either 50 or 75 ohm characteristic impedances, and they come in a variety of physical sizes and voltage ratings. For example, the coaxial cable most commonly encountered in instrumentation systems is the ubiquitous RG58A/U, which has $Z_O = 50$ ohms, C = 29.5 pF/ft, and v = 0.649 c (c is the speed of light *in vacuo*). In the case of ribbon cables having alternate conductors grounded, wires running parallel to a ground plane, and parallel wire, twin-lead antenna wire, Z_Os generally range from 100 to 300 ohms or

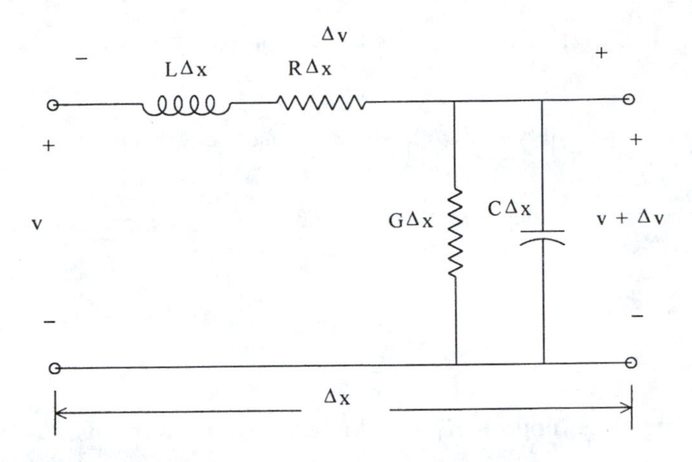

FIGURE 9.24 Lumped-parameter, equivalent circuit of a length, Δx, of a transmission line.

more. Figure 9.25 illustrates the end views of common transmission line geometries. From electrostatics, the capacitances per unit length of the lines can be shown to be (Taub and Schilling, 1977):

$$C = \frac{2\pi\varepsilon}{\ln(R/r)} \text{ F/m, for coaxial cables} \qquad 9.30$$

$$C = \frac{2\pi\varepsilon_0}{\ln(2h/r)} \text{ F/m, for round wire over a ground plane} \qquad 9.31$$

$$C = \frac{\pi\varepsilon_0}{\ln(D/r)} \text{ F/m, for parallel wires in space} \qquad 9.32$$

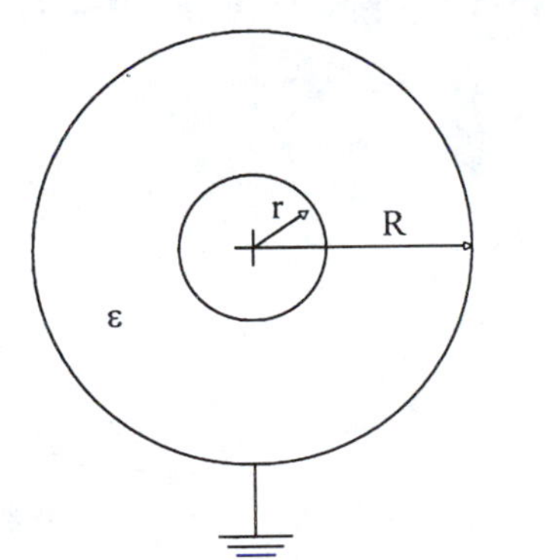

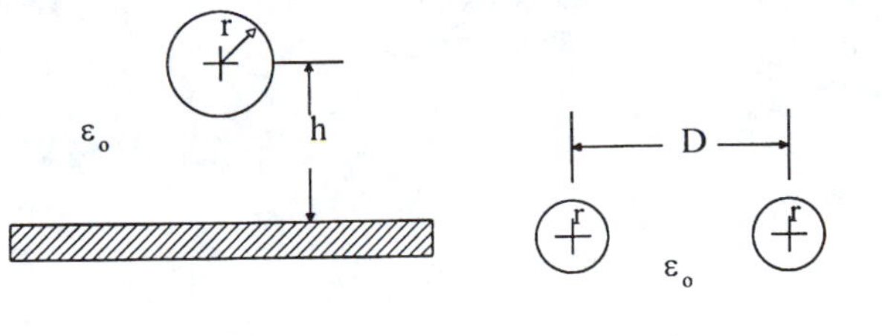

FIGURE 9.25 End views of three common transmission line geometries: (left) a coaxial cable; (center) a wire over a ground plane; (right) a parallel balanced line.

The inductance of these models can also be calculated (Kraus, 1953; Motorola AN-270). (Results are good approximations when h/r and D/r >> 1.)

$$L = \frac{\mu}{2\pi}\ln(R/r) \text{ Hy/m, for coaxial cables} \qquad 9.33$$

$$L = 11.7\ \log 10(2h/r)\ \text{nH/in., for a wire in space over a ground plane} \quad 9.34$$

$$L = \frac{\mu_O}{\pi}\left[1/4 + \ln(D/r)\right]\ \text{Hy/m, for parallel wires in space} \quad 9.35$$

As an example, we use Equations 9.28, 9.30, and 9.33 to find an expression for the characteristic impedance of (lossless) coaxial cables:

$$Z_O = \frac{\ln(R/r)}{2\pi}\sqrt{\frac{\mu}{\varepsilon}}\ \text{ohms} \quad 9.36$$

9.8.2 Reflections on an Improperly Terminated, Lossless Transmission Line

The result of sending digital (pulsatile) data waveforms over improperly terminated transmission lines is the creation of transient glitches at the receiving end, which, along with other noise on the line, can cause errors in data transfer. A transient waveform will propagate without distortion or attenuation on a lossless transmission line with a velocity $v = 1/\sqrt{LC}$ m/s. On most coaxial cables used for digital data transmission, $0.65 \leq v/c \leq 0.83$, depending on cable design. However, when the transient voltage propagating on the line reaches a cable end, a reflection, or backwards-propagating wave can arise at this end to satisfy certain boundary conditions. This first reflection propagates back to the source end of the line with velocity v, where a second reflection can arise and propagate at velocity v to the terminating end, etc. This process can be described mathematically in some detail; the interested reader is referred to the Text by Lathi (1965) for an excellent derivation of the relations we use below.

Refer to Figure 9.26, which shows a lossless transmission line of characteristic impedance, Z_O. To understand the dynamics of reflections on the line, we consider a voltage wave (step) which has propagated to the end of the line. V_d is the instantaneous voltage of the propagating wave just before it "sees" Z_L, and V_r is the instantaneous voltage of the back-propagating, reflected wave originating at Z_L to satisfy Ohm's law boundary conditions. Thus, at the Z_L termination, at $t = \tau = L/v$, we have:

$$V_L = V_d + V_r \quad 9.37$$

$$I_L = (V_d - V_r)/Z_O \quad 9.38$$

The boundary condition at Z_L will be satisfied if:

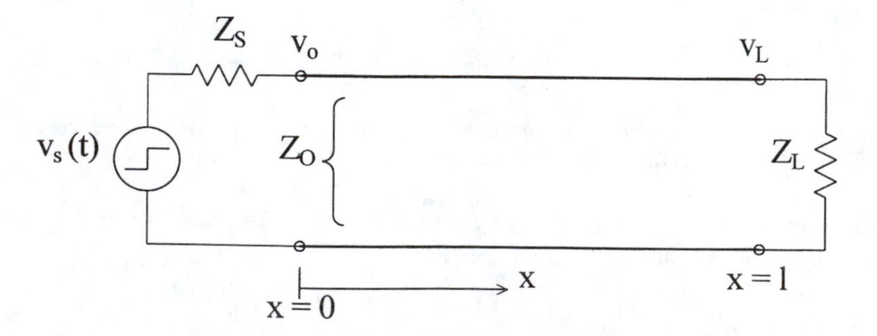

FIGURE 9.26 Schematic of a lossless transmission line of characterictic impedance Z_o and length l.

$$\frac{V_L}{I_L} = Z_L = Z_O \frac{V_d + V_r}{V_d - V_r} \qquad 9.39$$

Solving Equation 9.39 for the ratio of V_r to V_d, we obtain the reflection coefficient, $\rho_L(s)$, defined for the line's termination:

$$\rho_L(s) = \frac{Z_L(s) - Z_O(s)}{Z_L(s) + Z_O(s)} = \frac{V_r}{V_d} \qquad 9.40$$

A similar reflection coefficient can be found for the source end of the line:

$$\rho_S(s) = \frac{Z_S(s) - Z_O(s)}{Z_S(s) + Z_O(s)} \qquad 9.41$$

where Z_S is the Thevenin source impedance of the line.

In **example 1**, we examine what happens on a mismatched line when a positive, 5.0 volt step is generated by $v_s(t)$ at t = 0. We assume zero initial conditions on the line, and let Z_O = 75 ohms, $Z_S = R_S$ = 75 ohms, and $Z_L = R_L = \infty$ (open circuit). The step voltage is acted on by a voltage divider at x = t = 0. The source-end voltage divider attenuates by a factor of 75/(75 + 75) = 0.5. Hence the initial propagating voltage, V_O, is 2.5 V. The voltage at R_L is 0 until $t = \tau = L/v$. At t = L/v, the wave reaches R_L, and boundary conditions require that a reflected wave be generated. The termination reflection coefficient is found to be:

$$\rho_L = \frac{R_L - Z_O}{R_L + Z_O} = \frac{-75}{+75} = +1 \qquad 9.42$$

and the source reflection coefficient is:

$$\rho_S = \frac{R_S - Z_O}{R_S + Z_O} = \frac{75 - 75}{75 + 75} = 0 \qquad 9.43$$

Hence the first reflected wave has a value of

$$V_r(1) = 1.000 \times 2.5 = 2.5 \text{ V} \qquad 9.44$$

The voltage at the load is thus V_L = 2.5 + 2.5 = 5.0 V over $l/v \leq t \leq 3\ l/v$. At $t = 2L/v = 2\tau$, the reflected wave of 2.5 V reaches the source end of the line. There, there will be no reflection, because $\rho_S = 0$. The steady state has been reached. In the steady state, we expect the voltage everywhere on the line to be 5.0 V, as no current flows on the line. The waveforms for V_O and V_L are shown in Figure 9.27A.

In **example 2**, there are reflections at both ends of the line. We let $Z_O = 50\ \Omega$, $R_S = 25\ \Omega$, and $R_L = 100\ \Omega$, giving a moderately mismatched and underdamped line. Now the reflection coefficients are: ρ_L = (100 – 50)/(100 + 50) = 1/3, and ρ_S = (25 – 50)/(25 + 50) = –1/3. The amplitude of the initial wave at t = 0 is $V_O(0+) = 5 \times 50/(50 + 25)$ = 3.333 V. The 3.33 V step travels at v m/s to the load end of the line, arriving at $t = \tau = L/v$ seconds. There, to satisfy boundary conditions, a reflected wave, $V_r(1)$, is generated, propagating toward the source. Its amplitude is given by

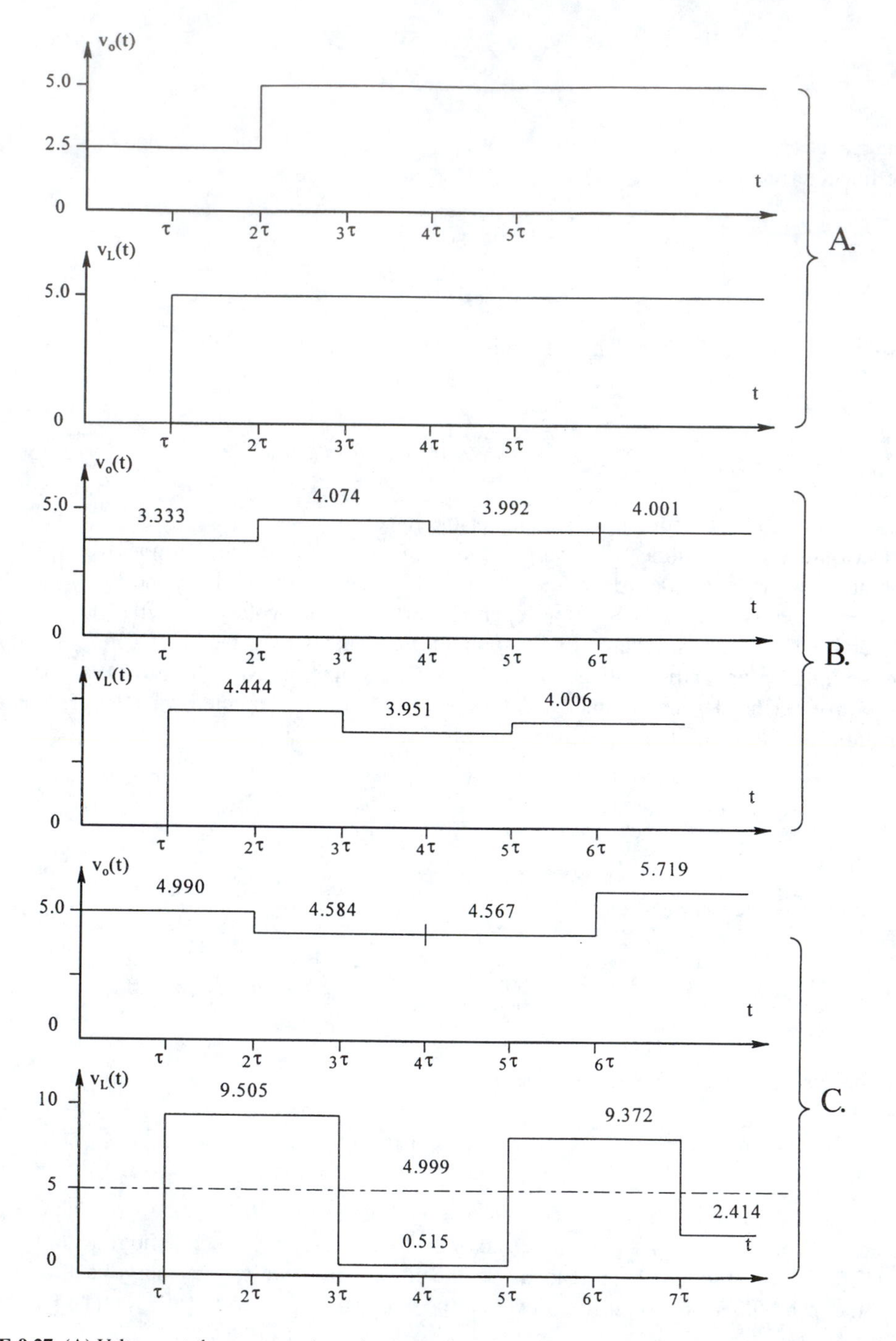

FIGURE 9.27 (A) Voltages at the source and termination ends of a transmission line. See text example 1 for conditions. (B) Transients on the transmission line of example 2. (C) Transients on the transmission line of example 3.

$$V_r(1) = 3.333 \times \rho_L = 3.333/3 = 1.111 \text{ V} \quad 9.45$$

Now the voltage across R_L is $V_L = 3.333 + 1.111 = 4.444$ V for $\tau \le t \le 3\tau$. V_r (1) arrives at the source end of the line at $t = 2L/v$. A second reflection originates at $t = 2L/v$ at the source end of the line, propagating toward R_L. The process of waves of diminishing amplitude being reflected from end to end of the transmission line continues, until practically, a DC steady state is reached.

Table 9.4 illustrates a systematic way to find the amplitudes of the reflections at the ends of a mismatched, lossless line. It is seen that a type of power series in the reflection coefficients describes the line's behavior. In Table 9.4, $v_S(t)$ is the Thevenin (open-circuit) source voltage, V_0 is the voltage at the source end (x = 0) of the line, and V_L is the voltage at the load end (x = L) of the line. $v_S(t)$ is assumed to be a 5 V step for simplicity. Other numerical parameters are from example 2. Data from Table 9.4 are plotted in Figure 9.27B. Note that both $v_0(t)$ and $v_L(t)$ closely reach steady state in three reflections.

TABLE 9.4 Systematic Means to Calculate the Reflections on a Mismatched, Lossless Transmission Line (Parameters are from Example 2)

Time $\tau = L/v$	V_0	V_L
0	$V_O(1) = V_S(0+)\frac{Z_O}{Z_O + Z_S} = 5\frac{50}{50+25}$ $V_O(1) = 3.33333$	0
t	3.33333	$V_L(1) = V_0(1)[1 + \rho_L] = 3.33333[1.33333]$ $= 4.44444$
2τ	$V_O(2) = V_L(1) + V_O(1)[\rho_L^1 \rho_S^1]$ $= 4.44444 - 3.33333[0.11111]$ $= 4.07407$	4.44444
3τ	4.074074	$V_L(3) = V_O(2) + V_O(1)[\rho_L^2 \rho_S^1]$ $= 4.07407 - 3.33333[0.037037]$ $= 3.9506173$
4τ	$V_O(4) = V_L(3) + V_O(1)[\rho_L^2 \rho_S^2]$ $= 3.95062 + 3.33333[0.0123457]$ $= 3.9917696$	3.95062
5τ	3.99177	$V_L(5) = V_O(4) + V_O(1)[\rho_L^3 \rho_S^2]$ $= 3.99177 + 3.33333[0.0041152]$ $= 4.0054870$
6τ	$V_O(6) = V_L(5) + V_O(1)[\rho_L^3 \rho_S^3]$ $= 4.00549 - 3.33333[0.0013717]$ $= 4.00091$	4.00549
∞	$V_{O(SS)} = V_{L(SS)} = V_S(0+)\frac{Z_L}{Z_L + Z_S} = 5.0\frac{100}{100+25} = 4.00000$ V	

A situation that can be troublesome in data transmission is the case where logic signals are sent to a receiver along a long, severely mismatched transmission line. In this **third example**, we assume a coaxial line is driven from an op amp buffer having a very low output resistance; $R_S = 0.1$ ohm. The signal is received by a linear amplifier with an input resistance of R_L = 1000 ohms. The characteristic impedance of the coaxial cable is 50 ohms, its length is 3 m, and the propagation delay at 1 MHz is known to be 5.136 nsec/m. The reflection coefficients are found to be: ρ_L = 0.90476190, $\rho_S = -0.9960798$. The amplitude of the initial wave is:

$$V_O = 5.0\frac{50}{50+0.1} = 4.99002 \text{ V} \qquad 9.49$$

Following the procedure used in Table 9.4, we can calculate and plot the voltage transient at the load end of the cable, $V_L(t)$. The time for a wave to travel down the cable is $\tau = L/v = 15.408$

nsec. For $0 \le t \le \tau$, $v_L(t) = 0$, as the wave has not yet reached R_L. At $t = \tau$, the V_0 wave reaches R_L, and the first reflection is generated, etc. Table 9.5 summarizes the V_0 and V_L values for this cable:

TABLE 9.5 Reflections on the cable of Example 3

t	V_o	V_L
0	4.99002	0
t	4.99002	9.50480
2τ	4.58415	9.50480
3τ	4.58415	0.51537
4τ	4.56762	0.51537
5τ	4.56762	9.37157
6τ	5.71910	9.37157
7τ	5.71910	2.41447
∞	4.99950	4.99950

Note: See Figure 9.27C for plots.

Of particular importance in this severe case of transmission line mismatch is the fact that the received voltage goes low 46.2 nsec after going high. Such behavior will render data transmission virtually impossible, or at least severely limit its rate to well below the reciprocal of the time it takes $v_L(t)$ to reach values above the logic HI level at the receiver.

By the simple expedient of making $R_S = R_L = Z_O$, all reflections on the line are eliminated, and the only problem is the 15.4 nsec initial delay in the signal reaching R_L. Of course, V_S now should be made 10 V because the steady-state voltage at R_L will be $V_S/2$ by the R_S, R_L voltage divider.

In the examples above, we have assumed a short (<10 m), lossless transmission line. If longer lines are used, in addition to possible reflection problems due to mismatched terminating and source impedances, there is noticeable attenuation of high frequencies due to cable losses, and a resulting rounding and attenuation of pulses as they travel along the line. The longer the line, the more severe the attenuation and loss of high frequencies, and the slower pulsed data must be sent in order to resolve individual 1,0 pulses. Voltage and current transfer functions for lossy transmission lines have been derived by Lathi (1965). They can be expanded into power series, and then inverse Laplace-transformed to obtain the voltage and current at any distance along a finite line, i.e., v(x,t) and i(x,t), given an input transient at t = 0.

To model the behavior of transmission line in the time domain, two approaches are commonly used. The first approach uses a series of lumped-parameter, per unit length RL/CG circuits as shown in Figure 9.24. Their behavior is solved with an electronic circuit analysis program such as MicroCap or SPICE. The second approach is to use FFT techniques with the voltage transfer function, $\mathbf{H}_V(x,jw)$, to directly find the voltage on the line, v(x,t).

9.9 DATA TRANSMISSION ON FIBER OPTIC CABLES

A relatively new technology, fiber optic cables (FOCs) are the emerging means for high-speed, serial data transmission. They are replacing conventional wire transmission lines in point-to-point applications. FOC systems are used in increasing numbers in instrumentation systems because of their low cost, high reliability, low loss, and wide bandwidth.

Fiber optic cables are dielectric waveguides in which information is generally transmitted as 0,1 amplitude-modulated electromagnetic waves of lightwave frequencies. A major application of FOC systems is in robust telecommunications. The channel capacity of fiber optic communications links, however calculated, far exceeds that of conventional wire communications (e.g., coaxial cables, twisted pairs) and present satellite microwave transponders. The present bandwidth of FOC data transmission systems is limited by the power, wavelength, and speed of the LED or laser diode (LAD) sources, and the sensitivity and response time of the PIN or avalanche photodiode detectors coupled to the FOC. Theoretically, FOCs can support data bandwidths approaching a terahertz (THz); the present state of the art, limited by terminal photonic devices, is in the low GHz range. Such high bandwidths are seldom needed in instrumentation systems, where 5 Hz to 125 MHz signal bandwidths are more commonly used.

As in the case of transmission lines, an FOC behaves in the large like a low-pass filter with a transport lag. The longer a given FOC, the greater the transmission lag and the attenuation at the receiver, and the lower the system bandwidth for data transmission. A practical criterion for FOC bit rate can be established by considering how close in time two input current pulses to the LED or LAD source can be in order to just resolve the two pulses as a "101", instead of a 111 or 110, etc., at the photodiode output.

FOCs are made from glasses or plastic. Fiber diameters range widely with the cable application. In telecommunications, FOCs are rapidly replacing copper twisted pairs for all but local subscriber lines. FOCs are far less expensive than wire cables, they are immune to electrical and electromagnetic interference, and they have bandwidths greater than conductive transmission lines. A basic FOC, serial communications channel is shown schematically in Figure 9.28.

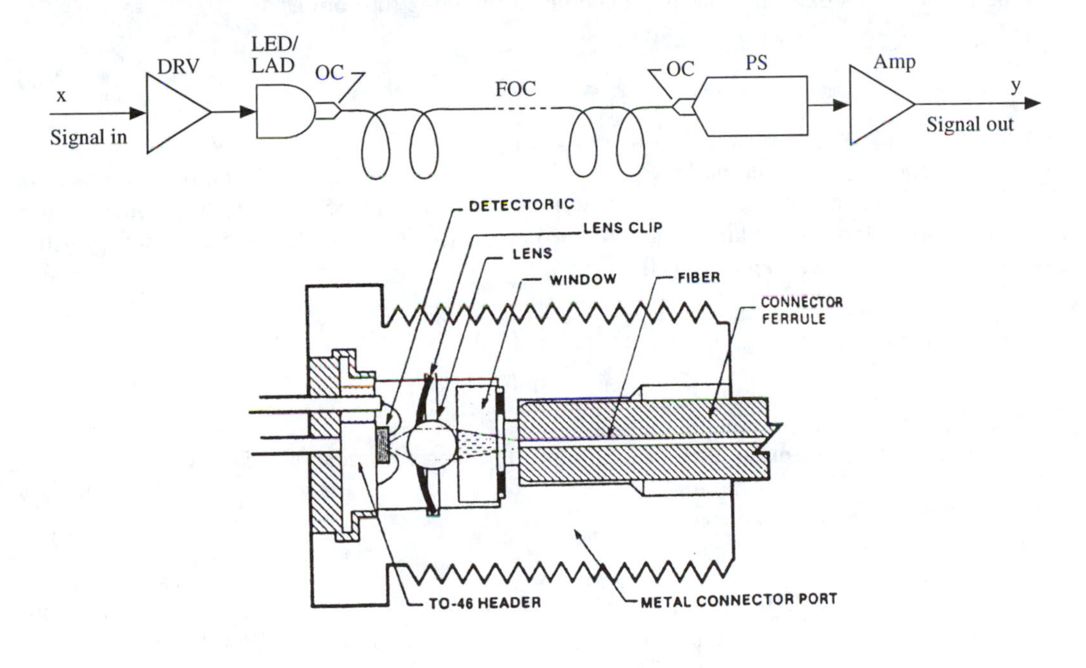

FIGURE 9.28 (A) A basic, unidirectional, fiber optic serial communications channel. DRV, electronic driver amplifier; LED/LAD, light-emitting diode or laser diode; OC, optical coupler; FOC, fiber optic cable; PS, photosensor; AMP, amplifier. (B) Section through a Hewlett-Packard, model HFBR-2204 fiber optic receiver. A spherical lens concentrates the light on the sensor. (Figure courtesy of Hewlett-Packard Co., San Jose, CA.)

9.9.1 Fiber Optic Cable Basics

Figure 9.29 illustrates a typical step-index FOC. The core of a step-index FOC has a uniform (in x, y, and z coordinates) optical index of refraction n_1. The FOC core carries the light energy used to signal data. The light energy propagates in the core in a variety of modes, as we will see. To understand how light is entrained in the core, we need to review some elementary optical theory governing refraction and reflection at plane interfaces of two dielectric media having different indices of refraction. The index of refraction of an optical medium is the ratio of the speed of light *in vacuo* (free space) to the speed of light in the medium. The refractive index of the FOC core is:

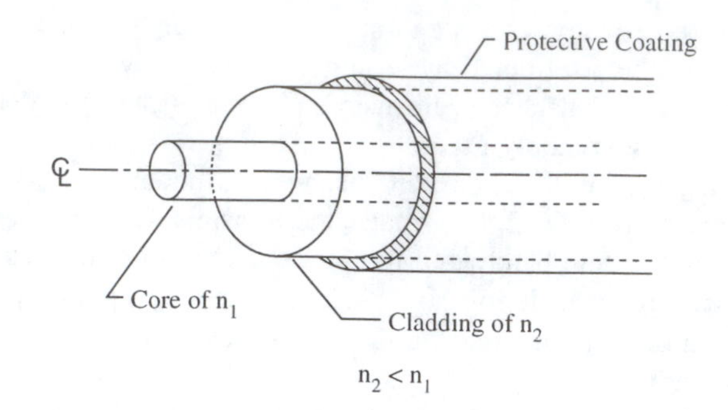

FIGURE 9.29 Diagram of the cut-back end of a step-index fiber optic cable.

$$n_1 = c/v_1 \tag{9.50}$$

and the refractive index of the cladding material surrounding the core is:

$$n_2 = c/v_2 \tag{9.51}$$

Obviously, n_0 for air $\cong 1$. In an FOC, $n_1 > n_2 > 1$ by design. Figure 9.30 shows what happens to light rays in the core incident on the core/cladding interface, here modeled as a plane surface for simplicity. A critical angle, θ_C, exists such that if the angle of incidence, $\theta_1 < \theta_C$, then some fraction of light energy enters the cladding and is trapped and dissipated in it. The angle relations for this case are given by the well-known Snell's law:

$$\frac{n_1}{n_2} = \frac{\sin(\theta_2)}{\sin(\theta_1)} \tag{9.52}$$

A critical angle, $\theta_1 = \theta_C$, exists such that $\theta_2 = 90°$. In this case, Equation 9.52 gives us:

$$\theta_C = \sin^{-1}(n_2/n_1) \tag{9.53}$$

Rays in the core with θ_1s $> \theta_C$ experience "total reflection" at the core/cladding boundary, and thus most of their energy remains within the core, bouncing back and forth off the boundary as they propagate to the receiver end of the FOC. Such entrained rays are called meridional rays because they traverse the axis of the fiber.

It can be shown from detailed consideration of the Maxwell equations for electromagnetic waves incident on plane dielectric boundaries, that some portion of an incident ray's energy will be reflected, even if $0 \le \theta_1 \le 90°$, and some portion will be transmitted. The proportions reflected and refracted can be shown to depend on the polarization state of the incident light, on the indices of refraction of the two media, and on the angle of incidence. For example, the fraction of unpolarized light normally incident on the end of a step-index FOC is given by (Sears, 1949):

$$I_r = I_i \frac{(n_1 - 1)^2}{(n_1 + 1)^2} \tag{9.54}$$

For $n_1 = 1.5$, $I_r/I_i = 0.040$, or 4% of the normally incident light energy does not enter the cable core. In passing from a higher to a lower refractive index medium, there is a very steep increase in I_r/I_i as the angle of incidence approaches θ_C. This behavior is why light travelling in the FOC

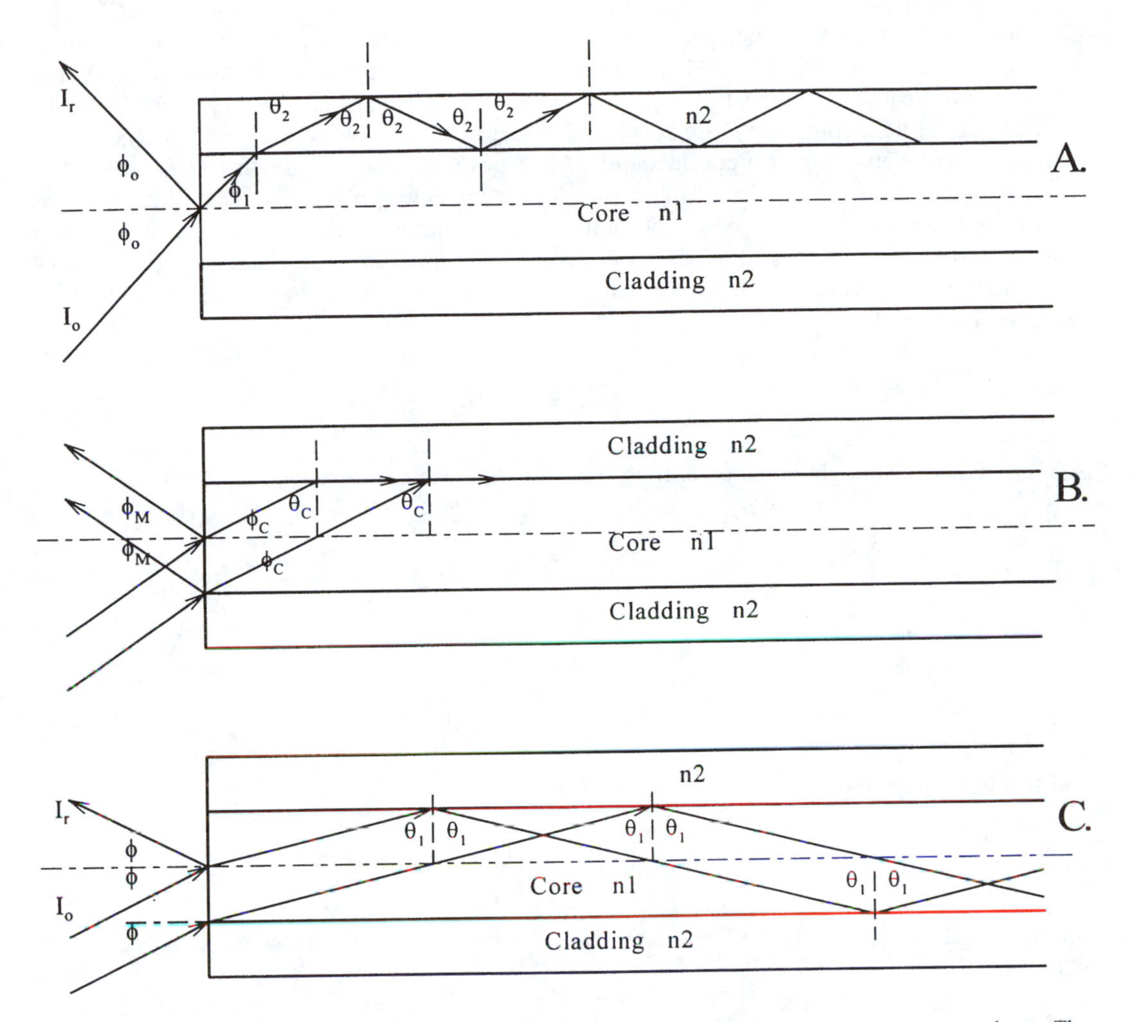

FIGURE 9.30 (A) Lateral view of a step-index FOC. Here an incident ray enters the cable at a very steep angle, φ_o. The angle of incidence of the internal ray on the core/cladding boundary is less than the critical angle, φ_c, and most of its energy enters the cladding and is trapped there. (B) Critical angle condition for the core/cladding boundary. (C) Now the ray is sufficiently normal to the end of the core so that most of its energy propagates down the core by alternately reflecting off the core/cladding boundary. (The situation is made more complex by the cylindrical geometry.)

core tends to remain in it, and not enter the cladding. On the other hand, there is a slow increase in I_r/I_i as the angle of incidence approaches and passes through θ_C as unpolarized light travelling in a low refractive index medium is incident on a plane boundary with a high refractive index medium. Thus some fraction of the light trapped in the cladding can reenter the core. The reflection vs. refraction relations (Fresnel's formulae) are given in detail in Sears (1949).

For the special case of light entering on the z-axis of the FOC ($\phi = 0°$, $\theta_1 = 90°$), it takes the axial ray $\tau_a = Ln_1/c$ seconds to reach the end of the FOC at z = L meters. From simple geometry, photons following a bouncing path can be seen to take:

$$\tau_{\theta 1} = L\,n_1/\left[c\sin(\theta_1)\right] \qquad 9.55$$

to reach the end of the cable. The difference between $\tau_{\theta 1}$ and τ_a is called the *modal dispersion.* Modal dispersion causes broadening of transmitted pulses, and thus is a factor reducing the transmission bandwidth of multimode, step-index FOCs. The reader who is interested in a rigorous treatment of modal propagation in plane (rectangular) waveguides using Maxwell's equations should consult texts on FOCs by Cheo (1990), Gowar (1984), Barnoski (1976), or Keiser (1983). The mathematical treatment of modal propagation in cylindrical waveguides (FOCs) is complicated by the fact that all field components are coupled. The text by Cheo (1990) gives such analyses.

Another geometrical optical consideration of FOCs is the numerical aperture (NA) of a cable. The NA of a FOC is related to the largest value of the ray entrance angle to the core from space, ϕ_M, that will permit a ray to propagate down the core. In other words, a ray entrance angle greater than ϕ_M results in $\theta_1 < \theta_C$, and nearly all its energy is coupled into the cladding and lost. Total internal reflections occur for $\phi < \phi_M$, and modal propagation occurs. At the critical entrance angle, we have, by Snell's law:

$$n_0 \sin(\phi_M) = n_1 \sin(\phi_C) \qquad 9.56$$

Also, at the critical angle in the core, from Equation 9.53, we can write:

$$\sin^2(\theta_C) = (n_2/n_1)^2 \qquad 9.57$$

By trigonometric identity:

$$\cos(\theta_C) = [1 - \sin^2(\theta_C)]^{1/2} \qquad 9.58$$

and from the right triangle:

$$\sin(\phi_C) = \cos(\theta_C) \qquad 9.59$$

Substituting Equations 9.56, 9.57, and 9.59 into 9.58, we obtain an expression for the step index FOC's NA (note $n_0 = 1$):

$$NA = \sin(\phi_M) = (n_1^2 - n_2^2)^{1/2} \qquad 9.60$$

Figure 9.31 shows a geometrical interpretation of NA in terms of θ_C, n_1, and n_2. Note that the FOC NA is not a function of the core diameter.

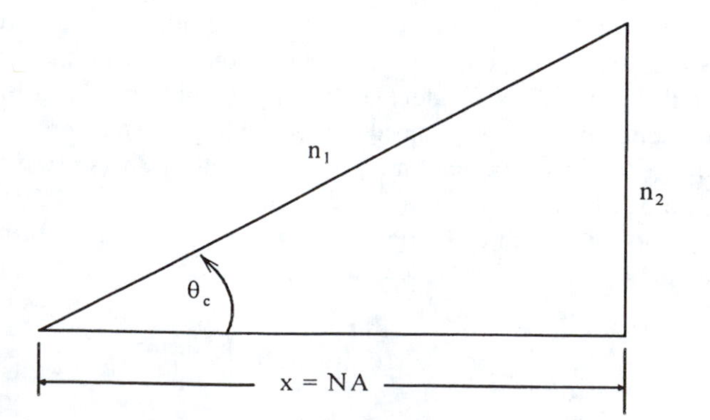

FIGURE 9.31 A geometrical interpretation of a fiber's numerical aperture (NA).

FOCs can be subdivided into multimode and single-mode fibers. Multimode fibers generally have core diameters ranging from 50 μm to as great as 1000 μm. It is physically easier to couple light into and out of multimode fibers because of their greater core diameters. Single-mode FOCs have core diameters from 3 to 12 μm, and consequently, I/O coupling requires great precision. The maximum number of modes in a planar, dielectric waveguide (as a first approximation to a cylindrical FOC) can be shown to be approximated by (Stark et al., 1988):

$$M_{MAX} = 2d(NA)/(\lambda n_1) - 1 \qquad 9.61$$

where d is the core diameter, λ is the wavelength of light on the core, and NA is the fiber's numerical aperture, given by Equation 9.60. From Equation 9.55, the maximum modal dispersion can be shown to be:

$$\Delta\tau = \tau_c - \tau_a = \frac{L\,n_1}{c\,n_2}(n_1 - n_2) \text{ seconds} \qquad 9.62$$

While single-mode fibers do not suffer from modal dispersion, they, as well as multimode fibers exhibit material dispersion, which is caused by the fact that n_1 is not a constant, but is, in fact, a function of the wavelength of the light in the core. Cheo (1990) shows that the per-unit length difference in group delay, $\Delta\tau_G$, for transmitted components separated by $\Delta\lambda$ in wavelength is given by:

$$\Delta\tau_G = \frac{\lambda\sigma_S}{c}\frac{d^2n}{d\lambda^2} \text{ seconds/m} \qquad 9.63$$

where σ_S is the rms spectral width of the source, given by:

$$\sigma_S = \int_0^{\infty} (\lambda - \lambda_O)^2 S(\lambda)\,d\lambda \qquad 9.64$$

$$\lambda_O = \int_0^{\infty} \lambda S(\lambda)\,d\lambda \qquad 9.65$$

and $S(\lambda)$ is the spectral distribution of the source. Thus, the material dispersion is proportional to the source's spectral width and the second derivative of $n_1(\lambda)$ around the mean wavelength, λ_o. Interestingly, $\Delta\tau_G \to 0$ at λ around 1300 nm for silica fiber cores. Hence sources operating around this wavelength offer improved bandwidth for single-mode FOCs.

A detailed, theoretical derivation of FOC bandwidth, based on a two pulse (101), intersymbol interference under noise-free conditions is given by Stark et al. (1988). These authors assumed a source λ_o of 1000 nm, the second derivative of the phase constant, $\beta''(\lambda_o) = 3 \times 10^{-26}$ s²/m, and an interpulse interference ratio, $\delta = 0.2$, at a point midway between the two received, unit pulses. (The tails of the unit height pulses sum to 0.2 halfway between them.) Their cable bandwidth, B, defined as the reciprocal of the interpulse interval required to make $\delta = 0.2$, increases with the spectral purity of the source, and decreases with cable length. Figure 9.32, adapted from Stark et al., illustrates FOC bandwidth B, as a function of source bandwidth, $\Delta\lambda$ around 1000 nm, and cable length.

The graded index fiber (GRIN) represents a design which mitigates the effect of modal dispersion found in step-index fibers and yet keeps the large core diameter permitting simpler I/O connections.

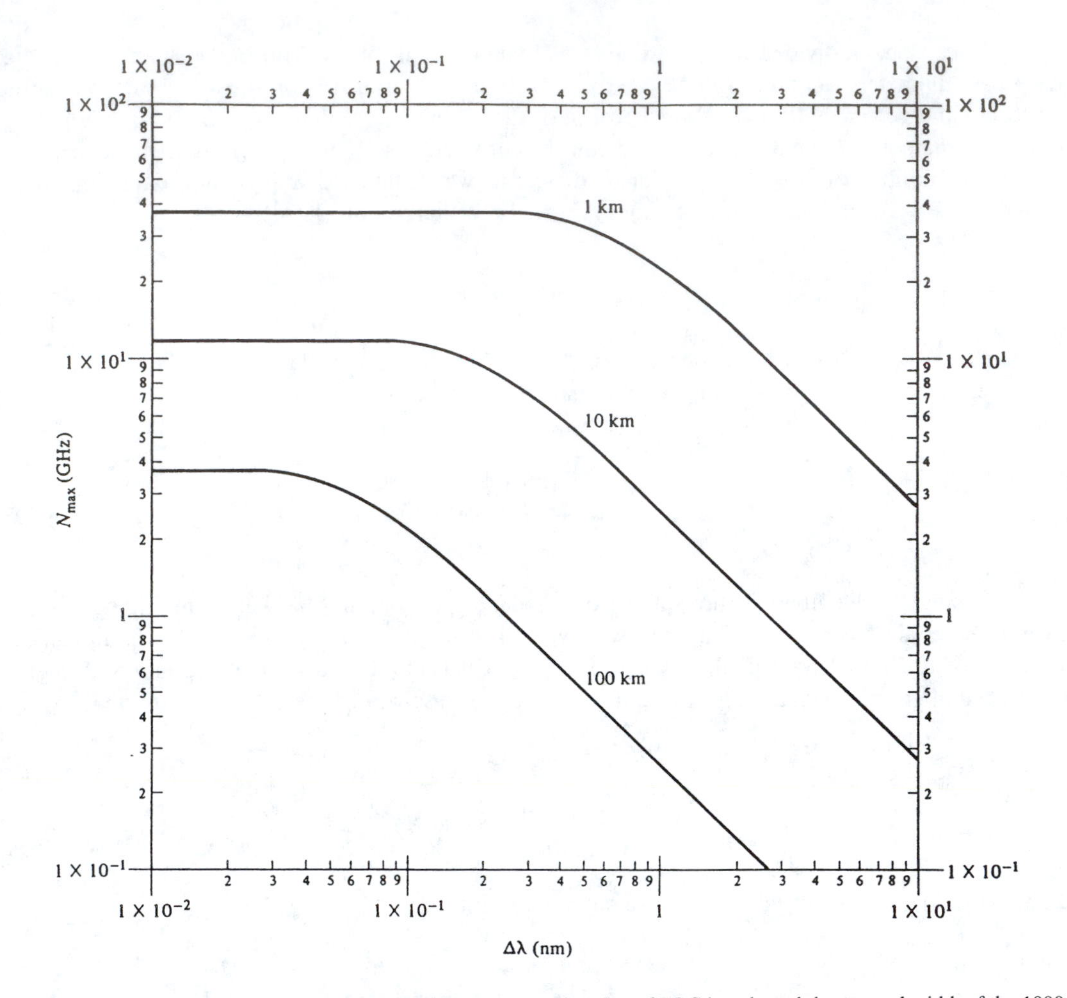

FIGURE 9.32 Plots of FOC signal bandwidth, B, in GHz as a function of FOC length, and the spectral width of the 1000 nm source, Δλ. (Figure adapted from Stark et al., 1988, with permission.) Note that as the optical source's bandwidth increases, the permissible transmission rate decreases, underscoring the value of IR laser diodes as FOC system sources.

Figure 9.33 illustrates a median section of a graded index FOC, and its refractive index profile. There are three principal GRIN core index profiles: parabolic profile (Equation 9.66), α profile (Equation 9.67), and hyperbolic or "selfoc" profile (Equation 9.68).

$$n(r) = n_{1(MAX)}\left(1 - \varepsilon r^2\right)^{1/2}, \quad 0 \le r \le r_1 \tag{9.66}$$

$$n(r) = n_{1(MAX)}\left[1 - 2\Delta\left(r/r_1\right)^{\alpha}\right]^{1/2} \tag{9.67}$$

$$n(r) = n_{1(MAX)} \operatorname{sec} h\left(r\Delta/r_1\right) \tag{9.68}$$

$$\Delta = \left(n_{1(MAX)} - n_2\right)/n_{1(MAX)} \tag{9.69}$$

Modal dispersion is greatly reduced in GRIN FOCs because instead of reflecting at the core/cladding boundary of a step index fiber, oblique rays are bent so that they follow a smooth, periodic

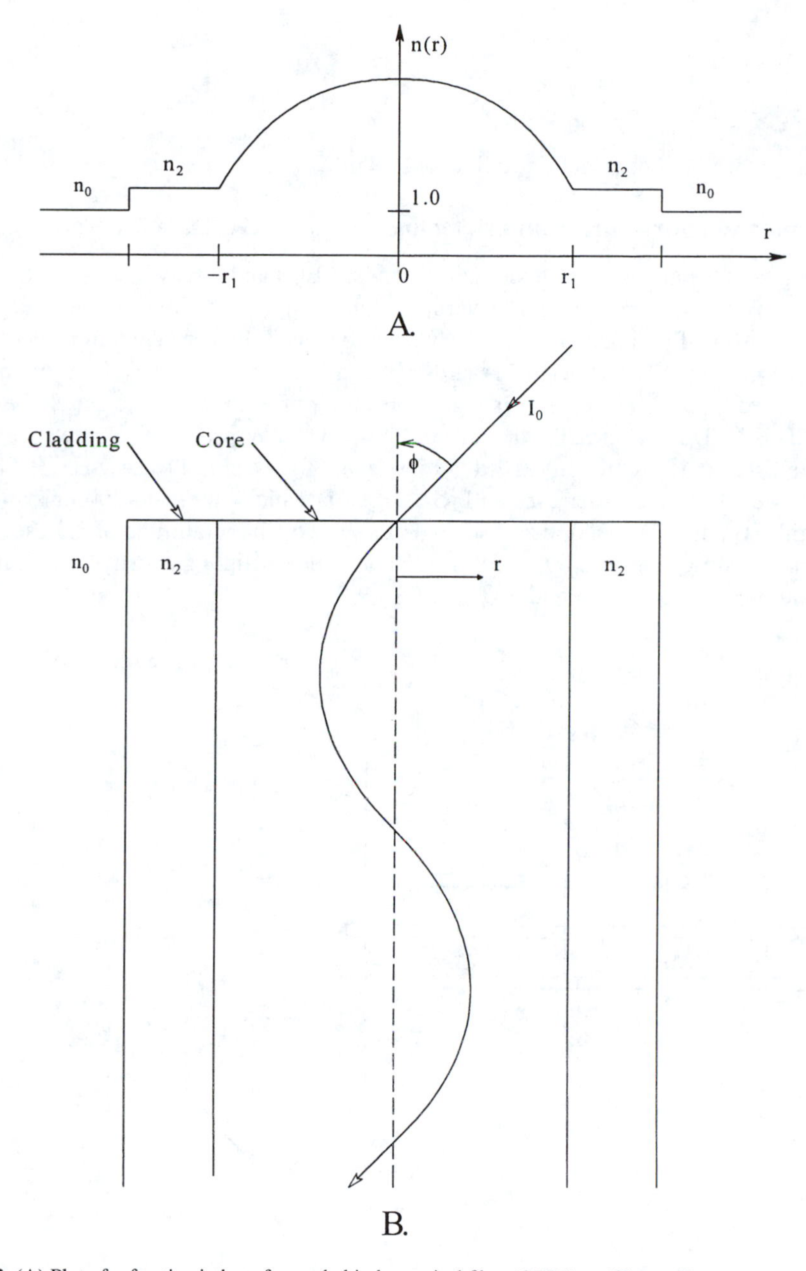

FIGURE 9.33 (A) Plot of refractive index of a graded index optical fiber (GRIN) vs. fiber radius. (B) Path of an obliquely entering ray as it propagates down the GRIN FOC. See text for description.

pathway down the core, as shown in Figure 9.33B. An axial ray travels down the shortest path in the center of the core at the slowest velocity, $v = c/n_{1(MAX)}$. An oblique ray travels a longer path, most of which lies in the higher velocity periphery of the core. Thus, both oblique and direct rays tend to arrive at the receiving end at the same time, minimizing modal dispersion.

The differential delay between the fastest and slowest modes in an α-profile, GRIN fiber, for a monochromatic (LAD) source can be shown to be:

$$\Delta\tau_G = \left(n_{1(MAX)}\Delta^2\right)/8c \text{ seconds/m} \tag{9.70}$$

where:

$$\alpha = 2(1 - 1.2\Delta) \tag{9.71}$$

for minimum intermodal dispersion (Stark et al., 1988).

9.9.2 Semiconductor Sources and Detectors Used in FOC Data Transmission

Two principal sources are used for fiber optic systems, LEDs and laser diodes (LADs). LEDs used in FOC systems are generally double heterojunction designs, as shown in Figure 9.34. A thin recombination layer of p-doped GaAs is sandwiched between thicker layers of p-type $Al_XGa_{1-X}As$ and n-type $Al_YGa_{1-Y}As$. The Hewlett-Packard HFBR-1204 high efficiency fiber optic transmitter (Hewlett-Packard Co., San Jose, CA) is a good example of a commercial, double heterojunction FOC LED. This LED emits at 820 nm, and can supply 316 μW peak optical power into a 100 μm core diameter, step index, FOC. Its 10% to 90% rise and fall time is 11 nsec. The HP HFBR-1100 general purpose LED transmitter uses an InGaAsP LED which emits at 1300 nm with a spectral bandwidth of 130 nm. It has an optical rise time of 1.2 nsec, and a fall time of 2.3 nsec. It supports a maximum signaling rate of 200 MBd. Modulation of the light intensity is accomplished by switching the diode's current on and off.

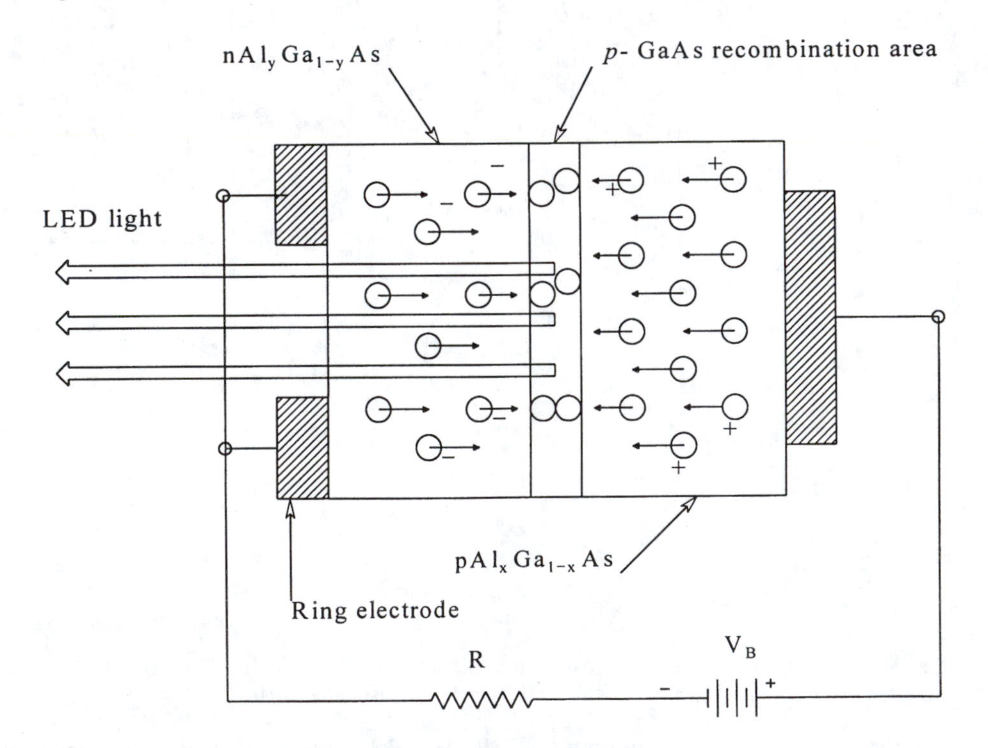

FIGURE 9.34 Layer cake diagram of a double heterojunction AlGaAs LED source.

Laser diodes are used in single-mode FOC systems designed to operate at extremely high data rates. An example of a double heterostructure LAD is shown schematically in Figure 9.35. As long as the LAD's supply current is held constant along with its temperature, its light output remains constant and of constant spectral purity. The envelope of a LAD's output line spectrum narrows with increased output power, and its peak shifts to a slightly longer wavelength, e.g., a spectrum peak at 830 nm at 4.6 mW output shifts to a peak at 833 nm at 12 mW output. This particular LAD had about 2.5 spectral lines/nm (Cheo, 1990). Cheo describes a constricted double heterostructure

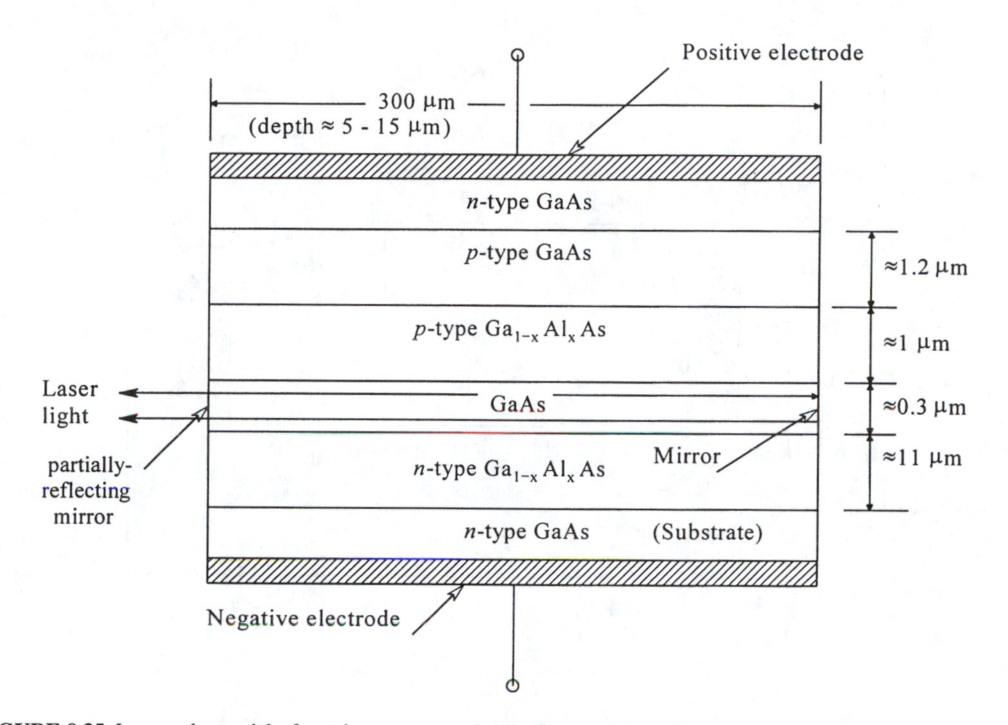

FIGURE 9.35 Layer cake model of a stripe geometry, double heterojunction, GaAs laser diode. (From Stark, H. et al., *Modern Electrical Communications: Analog, Digital and Optical Systems*, Prentice-Hall, Englewood Cliffs, NJ, 1988. with permission.)

LAD which produces a single, narrow, spectral output peak at about 837.5 nm at an input current of 100 mA.

The two major types of semiconductor photodetectors used in FOC systems are the PIN photodiode, and the avalanche photodiode. All photodetectors must have high quantum efficiency at the operating wavelength, high responsivity, rapid response time, and small size. The quantum efficiency, η, is defined as the number of carrier pairs (hole + electron) generated per incident photon. Ideally, $\eta = 1$. In practice, we see $0.3 \leq \eta \leq 0.95$. The responsivity, $R(\lambda)$, is defined as the ratio of primary photocurrent to incident optical power.

$$R(\lambda) = \eta e/h\nu \qquad 9.72$$

For an ideal photodiode, $\eta = 1$ and $R = \lambda/1.24$ amperes/watt. Typical values for R are 0.65 μA/μW for Si at 800 nm, and 0.45 μA/μW for Ge at 1300 nm (Stark et al., 1988).

The I in PIN stands for *intrinsic* semiconductor material, sandwiched between a p- and n-layer. The PIN photodiode is generally operated in reverse bias (see Figure 9.36). The photocurrent, I_P, is added to the dark (leakage) current, I_L. The objective is to resolve a small change in the total diode reverse current (due to a pulse of absorbed photons) above the shot noise and thermal noise in the diode circuit. PIN photodiode responsivity is a function of the materials from which it is made. Silicon PINs operate effectively from 200 to 1000 nm, with response rising with shorter wavelengths. Germanium PINs are useful in the range from 1000 to 1800 nm. PINs made from $Al_XGa_{1-X}AsSb$ have ηs as high as 80% in the 1.8 to 2.5 μm range (Cheo, 1990). The HP HFBR-2100, 1300 nm, PIN, FOC detector has 1 nsec output rise and fall time, enabling it to be used in a 1300 nm, 200 MBd, FOC system.

The avalanche photodiode (APD) is a p+ipn, four-layered structure, shown schematically in Figure 9.37. Its increased sensitivity results from an internal, intrinsic gain mechanism that greatly

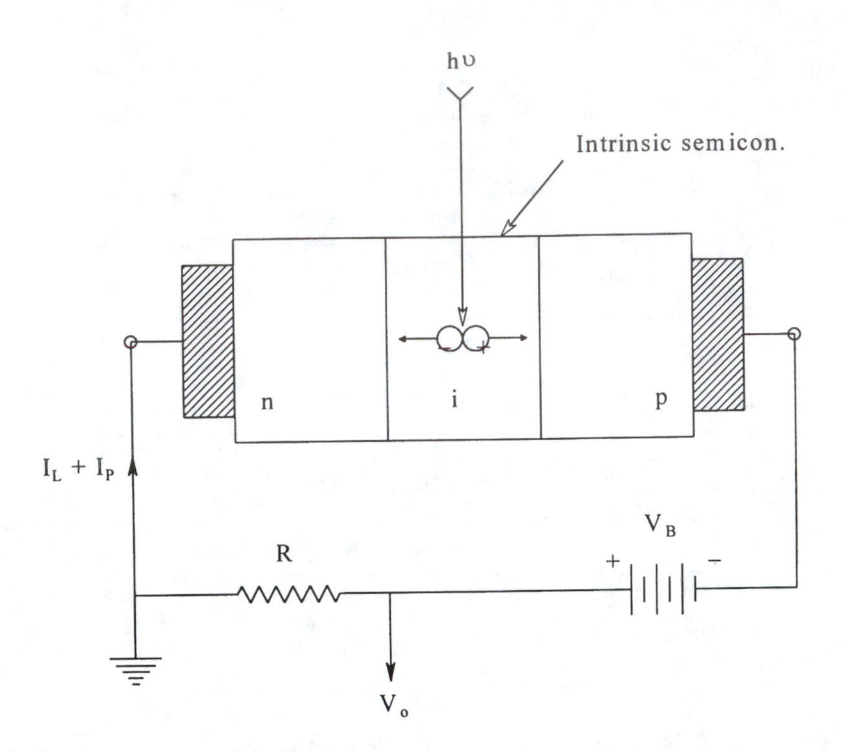

FIGURE 9.36 Diagram of a PIN photodiode.

increases the number of mobile carriers above the original photoinduced holes and electrons. The DC electric field conditions in the reverse-biased APD are such that the depletion region reaches into the i-region. The electric field at the pn junction is only 5% to 10% lower than that necessary for avalanche breakdown of the junction. Mobile photoelectrons and holes gain enough kinetic energy in the high E field region of the p layer to produce additional mobile electrons and holes through inelastic collisions which impart enough energy to promote new carriers to the conduction band. These secondary carriers, in turn, can gain enough kinetic energy to cause further ionization, etc. Thus n primary photo-electrons can give rise to αn electrons at the diode output. n primary photo-holes results in αn holes contributing to I_P. The net APD current multiplication factor, M, is a function of the internal electric field distribution, diode geometry, doping, and the wavelength of the incident photons. Cheo (1990) gives a good analytical treatment of APD behavior and noise. APDs are made fron Si, Ge, InGaAsP, etc., depending on the desired operating wavelength. At wavelengths below 1 μm, Si APDs have been built with net current gains of about 100, quantum efficiencies approaching 100%, and response times around 1 ns. In general, APDs exhibit faster rise times than PIN diodes; however, each requires careful setting of its operating voltage and constant operating temperature.

9.9.3 FOC Systems

The instrumentation system designer who wishes to interconnect groups of instruments, such as a VXI cluster, over a long distance to a controlling computer by fiber optic link no longer has the burden of assembling custom components to build the link. Hewlett-Packard markets a wide variety of plastic and glass FOC components, including single and double (duplex) FOCs, LED transmitters, and PIN receivers and signal conditioners. Hewlett-Packard also offers FOC system evaluation kits for designers to test feasibility with 1000 μm core plastic cable and 5 MBd, TTL-compatible transmitters and receivers. HP FOC link equipment ranges from 125 m, 40 kBd, TTL systems to 2 km, 200 MBd, ECL systems. This FOC equipment is described in the HP Optoelectronics Designer's Catalog.

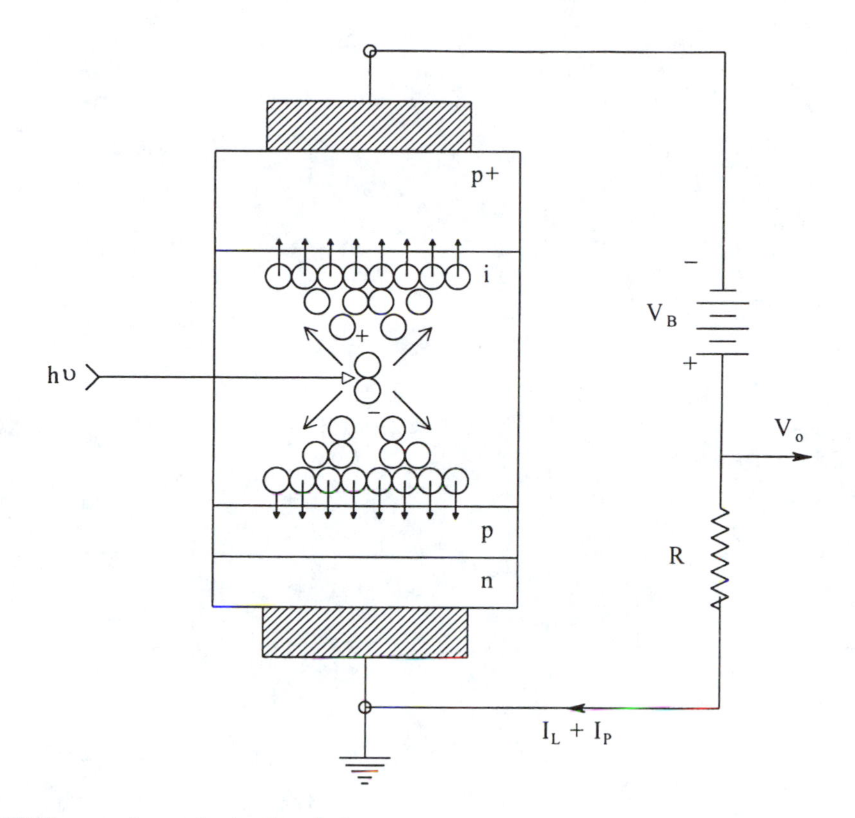

FIGURE 9.37 Diagram of an avalanche photodiode.

9.10 CHAPTER SUMMARY

In this chapter we have introduced the primary limitations of converting signals from analog to digital format, namely the introduction of an equivalent broad-band noise due to quantization, and the loss of high-frequency information due to aliasing. Quantization was seen to be caused by the numerical roundoff in representing analog data samples with digital numbers of finite bit length. Aliasing was shown to be the result of sampling a bandwidth-limited signal too slowly, i.e., at a rate below twice the highest frequency in the signal's power density spectrum.

The R-2R digital-to-analog converter (DAC) architecture was introduced as representative of most modern DACs. Five types of analog-to-digital converters were described, and their applications were discussed relative to instrumentation systems. Some mention was made of PC-compatible data conversion cards.

Digital data buses were reviewed, and the IEEE-488, GPIB, parallel instrumentation bus was described in some detail. Serial data communication links, including the venerable RS-232C bus were described, and their limitations discussed. The CAMAC instumentation packaging protocol and its related internal bus were covered.

Finally, we discussed transmission lines used for digital data transfer, and showed how improper impedance termination can lead to voltage reflections on the line that cause errors and significantly slow data transmission rates. Long transmission lines pose an extra problem because line losses cease to be negligible, and there is not only delay in transmission, but transmitted pulses become rounded and attenuated as they traverse the line. The same problems of delay, attenuation, and distortion are seen on fiber optic cables, but they were seen to have signal bandwidths far exceeding conventional, low-loss, coaxial cables. Fiber optic cables also are immune to various forms of electromagnetic interference because of their dielectric waveguide properties.

10

Introduction to Digital Signal Conditioning

10.0 INTRODUCTION

As we have seen in the preceding chapter, it is common practice to convey the analog-conditioned output of a sensor to an analog-to-digital converter, and then pass that sampled data to a digital computer for further processing, storage, and display. In this chapter, we introduce some of the basic digital signal processing (DSP) operations done on sampled data, and describe how they are accomplished. They include, but are not limited to, *smoothing or low-pass filtering, integration, differentiation, computation of the rms value of a data sequence, notch or band-reject filtering* (to eliminate coherent interference in the signal bandwidth), *computation of the discrete Fourier transform of a data sequence* to characterize it in the (discrete) frequency domain, and *interpolation and extrapolation of data sequences.* Some of the DSP algorithms we have just mentioned exist as options in specialized data acquisition software packages, and unless an instrumentation engineer is developing a system from scratch, there is often no need to write custom software for a specific, dedicated DSP task; one simply uses a packaged DSP routine on an input sequence to obtain the conditioned output sequence.

10.1 DIGITAL FILTERS AND THE Z-TRANSFORM

Synchronous operations on sampled data can be described in the time domain as difference equations, or in the frequency domain using the complex variable, z. To illustrate how discrete DSP operations are described using the z-transform, we first consider a well-behaved, continuous analog signal, x(t), defined for $0 \leq t \leq \infty$. For example, let $x(t) = Ae^{-bt}$. As we have seen in Chapter 9, sampling x(t) is equivalent to impulse modulation. We can write the sampled x(t), x*(t), in a more compact form, where T is the sample period and n is the sample number:

$$x^*(t) = \sum_{n=0}^{\infty} x(nT)\delta(t - nT) \quad 10.1$$

If we take the Laplace transform of x*(t), we get:

$$x^*(s) = \sum_{n=0}^{\infty} x(nT)e^{-snT} \quad 10.2$$

x(nT) is the value of x(t) at t = nT. Here x(nT) is treated as the area of the nth delta function, and the Laplace transform of the nth delta funtion is simply e^{-snT}. Equation 10.2 is easily changed to the open-sequence form of the z-transformed x(t) by letting the complex variable, z, be defined as:

$$z \equiv e^{sT} \quad 10.3$$

Hence, we can write:

$$X(z) = \sum_{n=0}^{\infty} x(nT)z^{-n} \tag{10.4}$$

For the exponential example given above, we can write:

$$X(z) = A\left[1 + e^{-bT}z^{-1} + e^{-2bT}z^{-2} + e^{-3bT}z^{-3} + \cdots\right] \tag{10.5}$$

A power series of this form can expressed in closed form as:

$$X(z) = \frac{A}{1 - e^{-bT}z^{-1}} = \frac{Az}{z - e^{-bT}} \tag{10.6}$$

A general, closed-form definition of the z-transform of any Laplace transformable f(t) can be derived from a consideration of the impulse modulation process, which is multiplication of f(t) in the time domain with a periodic, unit delta function train, $P_T(t)$, beginning at t = 0. This multiplication gives f*(t). In the frequency domain, F*(s) is found by the complex convolution of F(s) with the Laplace transform of the pulse train, $P_T(s)$. $P_T(s)$ can be written:

$$P_T(s) = 1 + e^{-sT} + e^{-2sT} + e^{-3sT} + \cdots = \sum_{n=0}^{\infty} e^{-nsT} \tag{10.7}$$

In closed form, $P_T(s)$ is:

$$P_T(s) = \frac{1}{1 - e^{-sT}} \rightarrow \frac{1}{1 - z^{-1}} = \frac{z}{z - 1} \tag{10.8}$$

Now by complex convolution:

$$F^*\left(e^{-sT}\right) = \frac{1}{2\pi j}\oint \frac{F(p)}{1 - e^{-sT}e^{pT}}\,dp \tag{10.9}$$

Using the definition of z above and treating z as a constant in the convolution integral:

$$F_z(z) = \frac{1}{2\pi j}\oint \frac{zF(p)}{z - e^{pT}}\,dp \tag{10.10}$$

F(p) is the Laplace transform of f(t) with s = p. Thus the contour integral, Equation 10.10, can be used to find the z transform of any f(t) having an F(s). Table 10.1 shows the Laplace and z-transforms of some typical deterministic time functions.

A digital filter is a linear, discrete operator which alters a periodic input number sequence, x(k), producing an output number sequence, y(k). Digital filters are generally described as rational polynomials in the complex variable, z. They may be subdivided into two major categories: (1) *recursive filters*, in which the present filter output, y(t), depends not only on the present and past input samples [x(0), x(T), x(2T),..., x(kT),...] but also on filter output samples, [y(0), y(T), y(2T),..., y(kT),...]. (2) *Moving average, transversal,* or *FIR* (finite duration impulse response)

TABLE 10.1 Some Common Laplace and z-Transforms

Time Function ($t \ge 0$)	Laplace Transform	z-Transform
$\delta(t)$	1	1
$\delta(t - nT)$	e^{-snT}	z^{-n}
$U(t)$	$1/s$	$\frac{z}{z-1}$
$t\,U(t)$	$1/s^2$	$\frac{Tz}{(z-1)^2}$
e^{-bt}	$\frac{1}{s+b}$	$\frac{z}{z-e^{-bT}}$
$t\,e^{-bt}$	$\frac{1}{(s+b)^2}$	$\frac{Tze^{-bT}}{(z-e^{-bT})^2}$
$\sin(\omega t)$	$\frac{\omega}{s^2+\omega^2}$	$\frac{z\sin(\omega T)}{z^2-2z\cos(\omega T)+1}$
$\cos(\omega t)$	$\frac{s}{s^2+\omega^2}$	$\frac{z^2-z\cos(\omega T)}{z^2-2z\cos(\omega T)+1}$
$e^{-bt}\sin(\omega t)$	$\frac{\omega}{(s+b)^2+\omega^2}$	$\frac{e^{-bT}z\sin(\omega T)}{z^2-2ze^{-bT}\cos(\omega T)+e^{-2bT}}$
$e^{-bT}\cos(\omega t)$	$\frac{s+b}{(s+b)^2+\omega^2}$	$\frac{z^2-ze^{-bT}\cos(\omega T)}{z^2-2ze^{-bT}\cos(\omega T)+e^{-2bT}}$

filters. FIR filters use only the present and past values of $x^*(t)$; their impulse responses are truly finite, i.e., after M sample periods following a "1" input at $t = 0$, their output is identically zero. The impulse response of a recursive filter generally approaches zero asymptotically as k approaches infinity. In the frequency (z) domain, the z-transformed output of a linear digital filter can be written as the product of the filter's transfer function, $G_Z(z)$, and the z transform of the filter's input, that is:

$$Y(z) = X(z)G_z(z) \qquad 10.11$$

In the time domain, the kth output sequence member from a digital filter can be found by the process of real, discrete convolution. This can be written as

$$y(kT) = y_k = \sum_{i=-\infty}^{\infty} g_i x_{k-i} \qquad 10.12$$

Here $\{g_i\}$ are the elements of the discrete filter's unit impulse response, and $\{x_{k-i}\}$ are the elements of $\{x_i\}$ reversed in time and shifted k sample periods. The real convolution process is illustrated for a simple example in Figure 10.1.

Recursive filters used in linear DSP operations can be expressed as rational polynomials in the complex variable, z. For example, $G_Z(z)$ can be written in a general, unfactored, polynomial form as:

$$\frac{Y(z)}{X(z)} = G_z(z) = \frac{a_m z^m + a_{m-1} z^{m-1} + \cdots + a_1 z^1 + a_0 z^0}{b_n z^n + b_{n-1} z^{n-1} + \cdots + b_1 z^1 + b_0 z^0},\ n \ge m \qquad 10.13$$

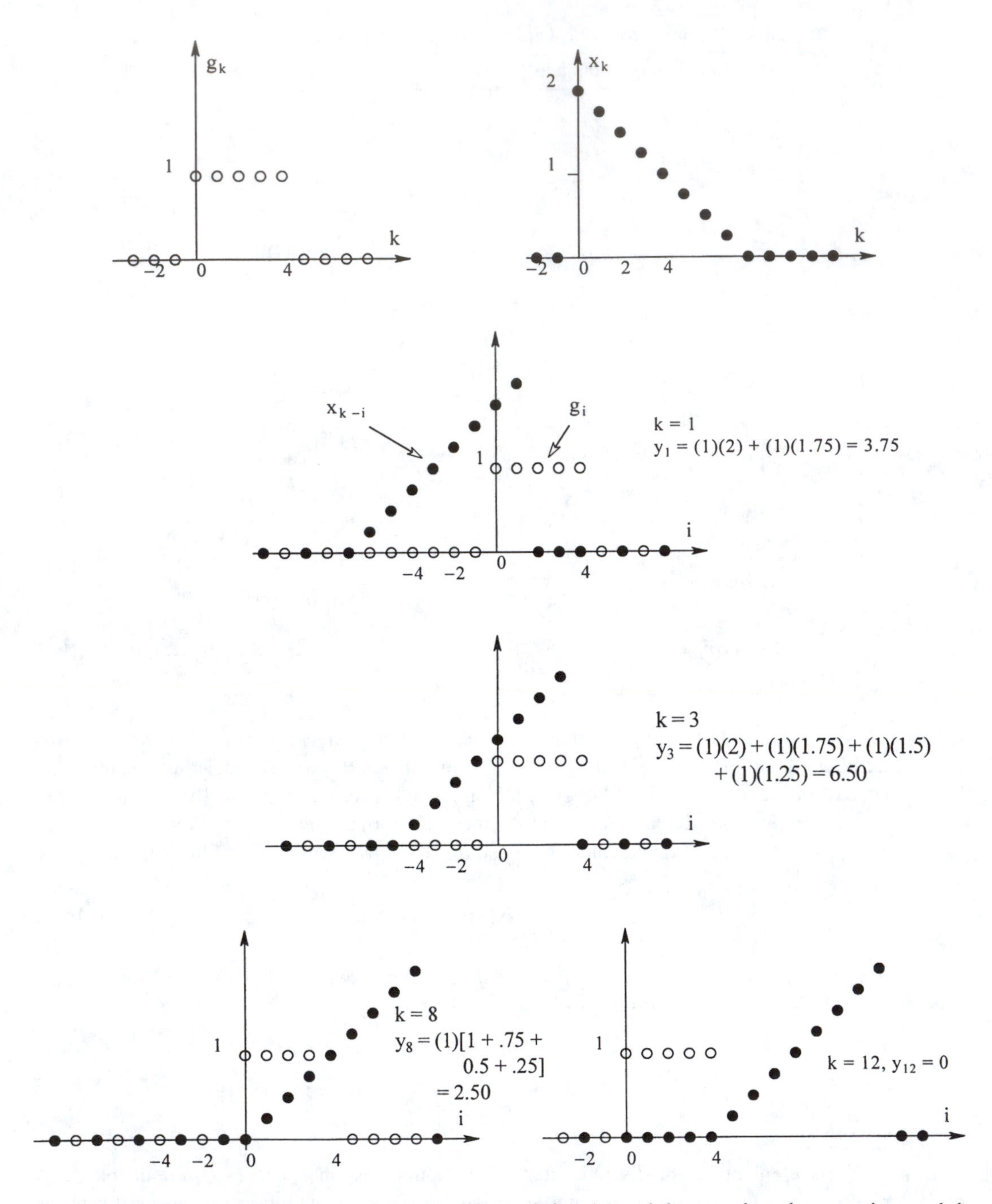

FIGURE 10.1 Illustration of the process of discrete, real convolution. A sampled rectangular pulse, g_k, and a sampled finite ramp waveform, x_k, are shown on the top line. In the next four lines, steps in the the real convolution operation given by Equation 10.12 are shown. g_k becomes g_i, and x_k becomes x_{k-i}. That is, x–i is x_k reversed in time i, and shifted by constant k.

Both the numerator and denominator polynomial may have roots at the origin of the z-plane, their roots may occur in complex-conjugate pairs, or lie on the negative or positive real axis in the z-plane. The coefficients, $\{a_k\}$ and $\{b_k\}$, are real numbers. The poles (denominator roots) of a stable G(z) will lie inside the unit circle in the z-plane.

To appreciate how a recursive $G_Z(z)$ can be implemented on a computer in real time, we change $G_Z(z)$ to be in powers of 1/z or e^{-sT}, the delay operator. Simple algebra gives:

$$\frac{Y(z)}{X(z)} = G_z(z) = \frac{\left(a_m + a_{m-1}z^{-1} + \cdots + a_1 z^{-m+1} + a_0 z^{-m}\right)z^{-(n-m)}}{b_n + b_{n-1}z^{-1} + \cdots + b_1 z^{-n+1} + b_0 z^{-n}}, \quad n \geq m \qquad 10.14$$

From Equation 10.14, it is easy to write:

$$Y(z)\left(b_n + b_{n-1}z^{-1} + \cdots + b_1 z^{-n+1} + b_0 z^{-n}\right) = X(z)\left(a_m z^{-(n-m)} + a_{m-1}z^{-(n-m+1)} + \cdots + a_1 z^{-(n+1)} + a_0 z^{-n}\right) \qquad 10.15$$

To simplify notation, let the kth past sample in $x^*(t)$, $x(t - kT)$ be written as x_{-k}, and the pth past sample in $y^*(t)$ be written as y_{-p}. Now we can write Equation 10.15 in the time domain:

$$b_n y + b_{n-1}y_{-1} + \cdots + b_1 y_{-n+1} + b_0 y_{-n} = a_m x_{-n+m} + a_{m-1}x_{-n+m-1} + \cdots + a_1 x_{-n+1} + a_0 x_{-n} \qquad 10.16$$

Equation 10.16 can easily be solved for the present filter output, y, in terms of present and past weighted output and input samples. To illustrate this process, we use a simple example:

$$G_z(z) = \frac{a_2 z^2 + a_1 z^1 + a_0}{b_3 z^3 + b_2 z^2 + b_1 z^1 + b_0}, \qquad n - m = 1 \qquad 10.17$$

Following the procedure described above, the present output of $G_Z(z)$ can be written:

$$y = (1/b_3)\left(a_2 x_{-1} + a_1 x_{-2} + a_0 x_{-3} - b_2 y_{-1} + b_1 y_{-2} + b_0 y_{-3}\right) \qquad 10.18$$

Implementation of this recursive filter example is shown in Figure 10.2. Note that in this particular example, no *present* value of x is used in computing the present value of y.

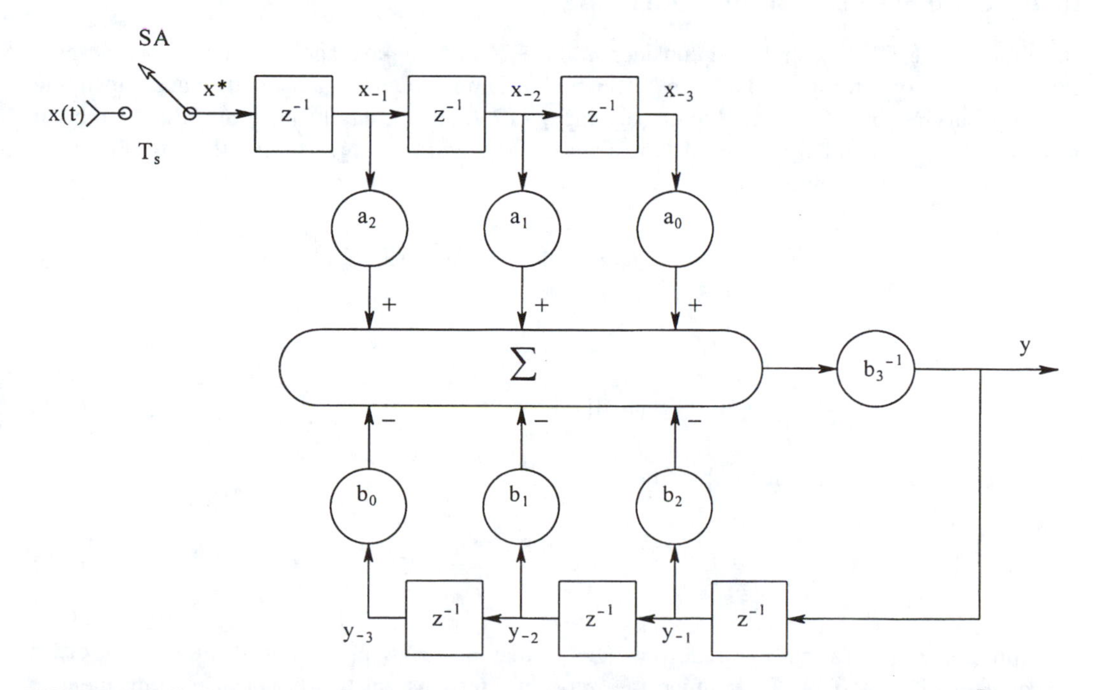

FIGURE 10.2 Block diagram showing the implementation of a third-order, recursive filter. z^{-1} represents e^{sT_s}, the unit delay operator. T_s is the system sampling period.

The impulse response of a digital filter, $\{g_k\}$, is simply its output number sequence, given an input x = 1 at t = 0. $\{g_k\}$ can also be found by long division of the numerator of the filter's G(z) by its denominator. This division generally gives a sequence of terms with coefficients of the form, z^{-k}. In the time domain, this is

$$\{g_k\} = \sum_{k=0}^{\infty} g_k \delta(t - kT) \tag{10.19}$$

For stable transfer functions, $g_k \to 0$ as $k \to \infty$.

An nth order FIR digital filter transfer function can be written in the form:

$$\frac{Y(z)}{X(z)} = \frac{a_n z^n + a_{n-1} z^{n-1} + \cdots + a_1 z^1 + a_0}{z^n}, \qquad m = n,\ b_n = 1 \tag{10.20}$$

By taking the inverse z-transform of Equation 10.20, we obtain the difference equation:

$$y = a_n x + a_{n-1} x_{-1} + \cdots + a_1 x_{-n+1} + a_0 x_{-n} \tag{10.21}$$

It is easy to see that if a "1" is input at t = 0, the FIR filter's impulse response will have, including the present sample, M = (n + 1) terms.

Before going on to consider some typical applications of digital filters, we remind the reader that in using DSP routines, we generally assume that the analog input signal, x(t), is bandwidth limited by an anti-aliasing low-pass filter so that it contains no significant spectral energy above the Nyquist frequency (one-half the sampling frequency).

10.2 SOME SIMPLE DSP ALGORITHMS

The first simple DSP example we consider is the FIR, moving-average, smoothing or low-pass filter. This type of filter is often used to "clean up" or remove high-frequency noise accompanying a slowly varying or DC signal. Moving average, FIR filters generally have an odd number of polynomial terms, and have a unity DC response. For example, a three-term, Hanning, FIR low-pass filter is described in the time domain by

$$y(nT) = (1/4)\left[x(nT) + 2x(nT - T) + x(nT - 2T)\right]$$

or

$$y = (1/4)\left(x + 2x_{-1} + x_{-2}\right) \tag{10.22}$$

We can write the z-transform for this filter as:

$$G_z(z) = \frac{Y(z)}{X(z)} = \frac{0.25z^2 + 0.5z^1 + 0.25}{z^2} = 0.25\left(1 + 2z^{-1} + z^{-2}\right) \tag{10.23}$$

The numerator of $G_Z(z)$ has two real roots (zeros) at z = −1. A z-plane plot of the roots of $G_Z(z)$ is shown in Figure 10.3A. To examine the frequency response of this $G_Z(z)$, it is convenient to substitute $z = e^{sT} = e^{j\omega T}$ into Equation 10.23, and vary ω.

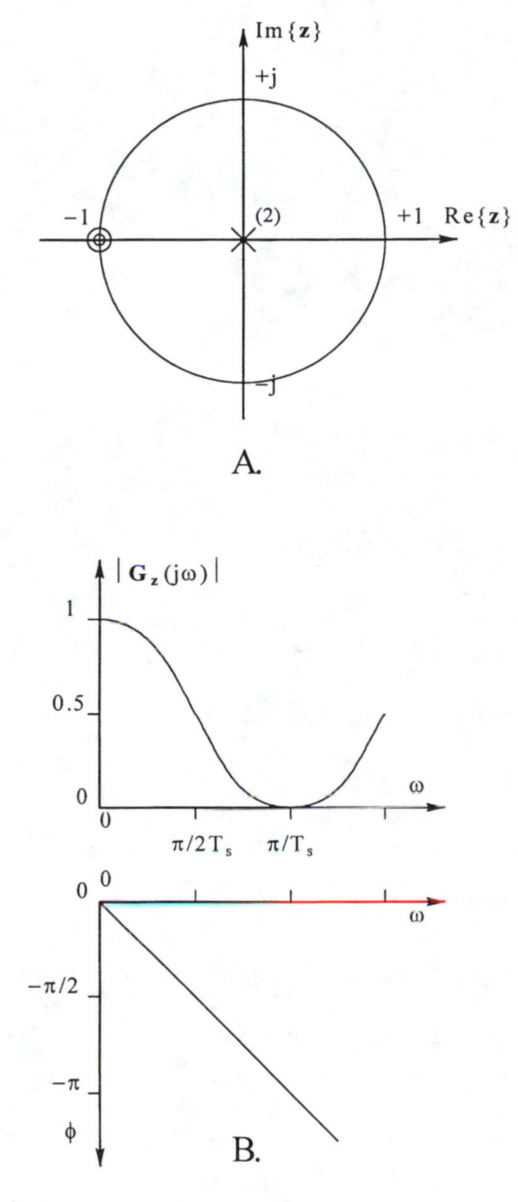

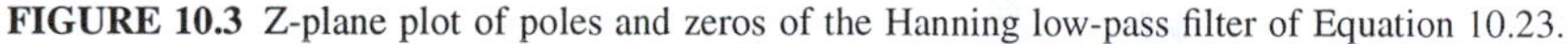
FIGURE 10.3 Z-plane plot of poles and zeros of the Hanning low-pass filter of Equation 10.23.

$$\mathbf{G}_z(j\omega T) = 0.25\left(1 + 2e^{-j\omega T} + e^{-j2\omega T}\right) = 0.5\left[1 + \cos(\omega T)\right]e^{-j\omega T} \qquad 10.24$$

The radian sampling frequency is $\omega_s = 2\pi/T$, and the Nyquist frequency is $\omega_N = \omega_s/2 = \pi/T$. Since we assume x(t) is anti-alias filtered before sampling, we need only consider $0 \le \omega \le \omega_N = \pi/T$. The magnitude and phase of $\mathbf{G}_Z(j\omega T)$ is plotted in Figure 10.3B. Note that the phase is linear in ω, and the DC magnitude response of the Hanning filter is unity, and decreases to zero at $\omega = \omega_N$, and then increases again.

Many other higher-order, FIR smoothing filters can be designed. In general, the higher the order of the FIR filter, the sharper the attenuation can be made. Many criteria exist for the design of FIR smoothing filters. For example, Tompkins and Webster (1981) derive a five-term, FIR, low-pass filter based on a least squared error criterion to fit the five data samples used in the filter with the best parabola. This gives a time domain filter function of:

$$y = (1/35)(-3x + 12x_{-1} + 17x_{-2} + 12x_{-3} - 3x_{-4}) \quad 10.25$$

In the z-domain, this algorithm becomes:

$$\frac{Y(z)}{X(z)} = G_z(z) = (1/35)(-3 + 12z^{-1} + 17z^{-2} + 12z^{-3} - 3z^{-4}) \quad 10.26$$

The frequency response magnitude of this $\mathbf{G_Z}(j\omega T)$ and several other FIR low-pass filters are illustrated in Tompkins and Webster (1981). These authors also derive a five-point FIR low-pass filter that satisfies the criteria:

$$\mathbf{G_z}(j0) = 1 \quad 10.27A$$

$$\frac{d\mathbf{G_z}(j0)}{d\omega T} = 0 \quad 10.27B$$

$$\frac{d^2\mathbf{G_z}(j0)}{d\omega T^2} = 0 \quad 10.27C$$

That is, the filter's frequency response is maximally flat at DC. This filter is shown to have the difference equation:

$$y = (1/16)(-1x + 4x_{-1} + 10x_{-2} + 4x_{-3} - 1x_{-4}) \quad 10.28$$

Its frequency reponse magnitude is:

$$\mathbf{G_z}(j\omega T) = 5/8 + 4/8\cos(\omega T) - 1/8\cos(2\omega T) \quad 10.29$$

Clearly, $\mathbf{G_Z}(j0) = 1$, and $\mathbf{G_Z}(j\pi) = 0$ for this filter.

Often in instrumentation applications we wish to eliminate a specific, coherent frequency, ω_i, from a low-frequency signal. We assume that $\omega_I < \omega_N$. To create a digital notch filter, we need to place a pair of complex-conjugate zeros of G(z) on the unit circle in the z-plane. This can be done most simply with a quadratic FIR filter of the form:

$$G_z(z) = 0.5(1 + z^{-2}) \quad 10.30$$

which is implemented as

$$y = 0.5(x + x_{-2}) \quad 10.31$$

Substituting $e^{j\omega T}$ for z in Equation 10.30 gives us the frequency response of this simple notch filter:

$$\mathbf{G_z}(j\omega T) = \cos(\omega T)e^{-j\omega T} \quad 10.32$$

It is easy to see that $\mathbf{G_Z}(j0) = \mathbf{G_Z}(j\omega_N T) = 1$, and $\mathbf{G_Z}(j(\omega_N/2)T) = 0$, i.e., the notch frequency, ω_i, is 1/2 the Nyquist frequency in this example. Attenuation around the notch frequency is certainly not sharp for this filter. Tuning the notch is accomplished by adjusting the sampling period, T.

To obtain a sharp, high-Q notch, we can use a recursive notch filter design, adapted from Widrow and Stearns (1985).

$$F_Z(z) = \frac{z^2 - z2\cos(\phi) + 1}{z^2 - z2\eta\cos(\phi) + \eta^2} \qquad 10.33$$

where $\eta = 1 - \varepsilon$, and $0 < \varepsilon << 1$, and the angle $\phi = (\omega_i/\omega_N)\pi$ is shown on the unit circle in Figure 10.4. ω_i, as in the example above, is the frequency of the interfering sinusoid that we wish to eliminate. Consideration of the numerator of F(z) shows that the filter has complex-conjugate zeros on the unit circle at:

$$z_{01} = \cos(\phi) + j\sin(\phi) \qquad 10.34A$$

$$z_{02} = \cos(\phi) - j\sin(\phi) \qquad 10.34B$$

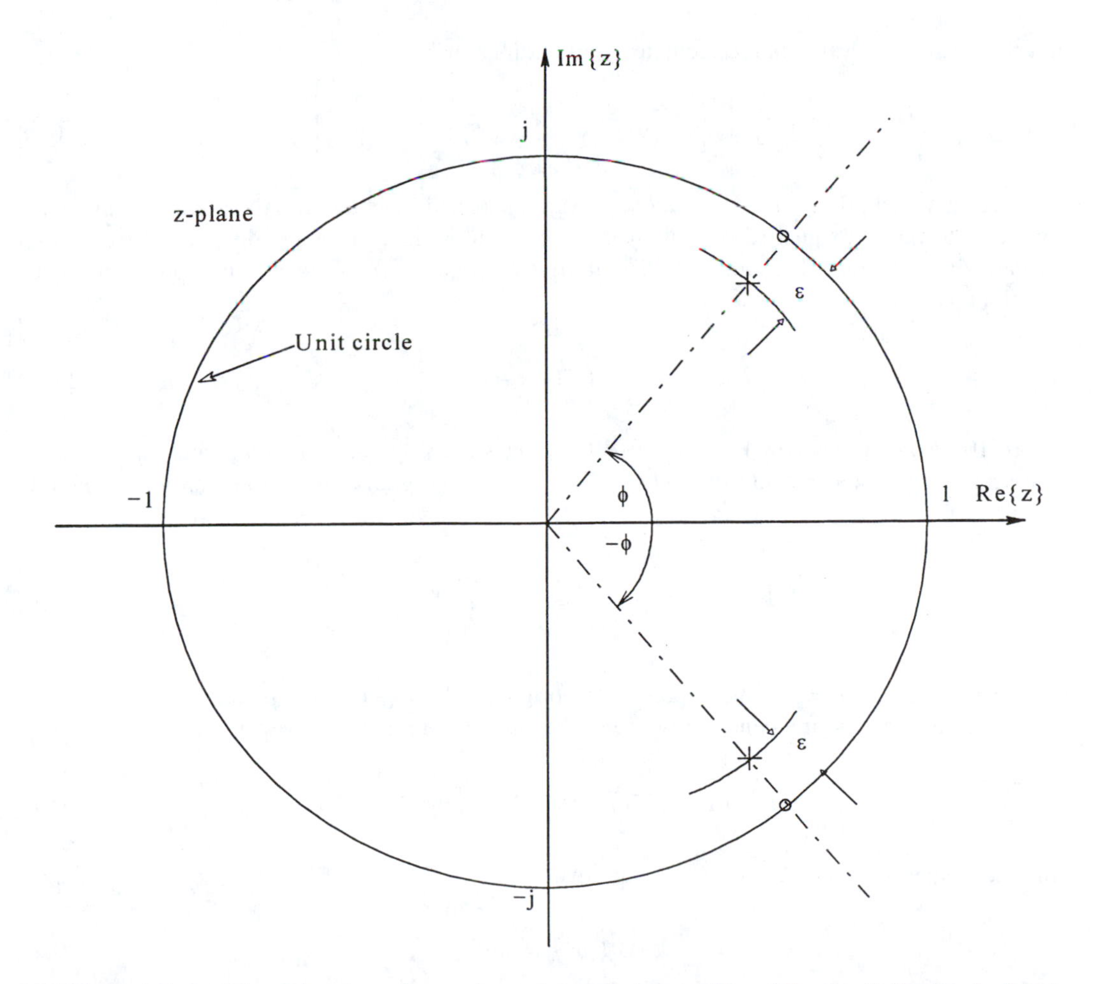

FIGURE 10.4 Z-plane plot of the poles and zeros of recursive, quadratic notch filter of Equation 10.33, given by Widrow and Stearns (1985).

The poles of $F_Z(z)$ are easily shown to be just inside the unit circle, lying at $\pm\phi$ with respect to the real axis. Factoring the denominator yields:

$$z_{P1} = \eta(z_{01}) \qquad 10.35A$$

$$z_{P2} = \eta(z_{02}) \qquad 10.35B$$

Following Widrow and Stearns (1985), we can show that the recursive notch filter's Q is given by:

$$Q = \frac{\text{Center frequency}}{-3 \text{ dB bandwidth}} = \frac{\omega_i T}{2(1-\eta)} \qquad 10.36$$

(The higher the Q, the sharper the notch.) The Widrow and Stearns adaptive filter can be tuned to any ω_i by calculating:

$$\cos(\phi) = \cos[(\omega_i/\omega_N)\pi] \qquad 10.37$$

and then using this value in the recursion formula for y(nT):

$$y = \eta[x + (y_{-1} - x_{-1})2\cos(\phi) + x_{-2} - \eta y_{-2}] \qquad 10.38$$

Often in signal processing applications, we are interested in estimating the points in time when a bandwidth-limited signal reaches its maxima or minima, i.e., the points in time where its first derivative goes to zero. The simplest form of *digital differentiator* is the FIR, two-point difference equation:

$$y = (1/T)(x - x_{-1}) \qquad 10.39$$

Figure 10.5A illustrates how Equation 10.39 estimates the slope of x(t) by a straight-line approximation using the present and first past sample of x(t). The digital transfer function for this simple routine is:

$$D_z(z) = \frac{Y(z)}{X(z)} = \frac{z-1}{Tz} \qquad 10.40$$

$D_Z(z)$ has a zero at z = +1, and a pole at the origin of the z-plane. By substituting $e^{j\omega T}$ for z in $D_Z(z)$, we can find the frequency response of the simple differentiator over $0 \le \omega \le \omega_N$.

$$\mathbf{D_z}(j\omega T) = (2/T)\sin(\omega T/2)\exp[j(\pi/2 - \omega T/2)] \qquad 10.41$$

For low ω such that $\sin(\omega T/2) \approx \omega T/2$, in radians:

$$\mathbf{D_z}(j\omega T) \cong \omega\, e^{j\pi/2} \qquad 10.42$$

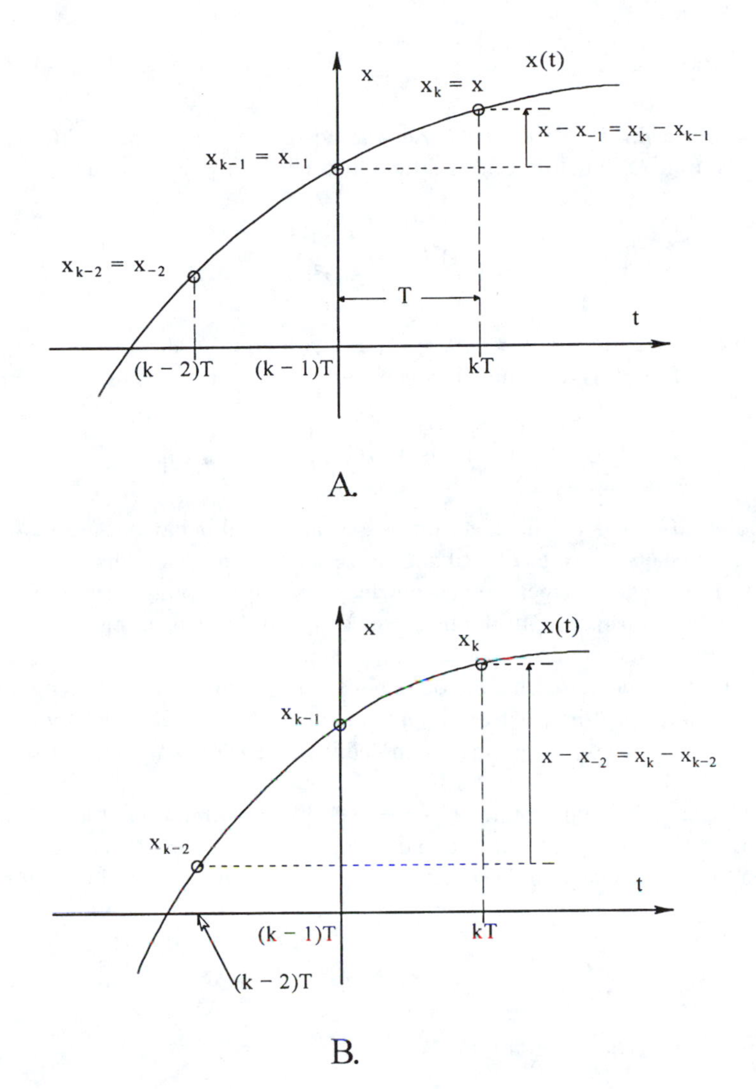

FIGURE 10.5 (A) Ilustration of the simple, two-point, digital differentiator operating on x(t). (B) Illustration of the three-point, central difference, digital differentiator, given by Equation 10.47.

which is the frequency response function of an ideal (analog) differentiator. As ω increases, $|\mathbf{D_Z}(j\omega T)|$ reaches a peak at the Nyquist frequency, and then decreases.

Other digital differentiation routines are sometimes used. All digital differentiators depart from ideality as ω approaches the nyquist frequency. In some cases, it is desirable to have the differentiator frequency response reach a peak and roll off at frequencies higher than $\omega_N/2$ to minimize the effects of noise on x*(t). These routines include (Tompkins and Webster, 1981):

$$y = (1/2T)\left(x - x_{-2}\right) \quad \text{(three-point central difference)} \tag{10.43}$$

$$y = 0.1\left(2x + x_{-1} - x_{-3} - 2x_{-4}\right) \quad \text{(five-point, least squares, parabolic fit)} \tag{10.44}$$

$$y = (1/28)\left(3x + 2x_{-1} + x_{-2} - x_{-4} - 2x_{-5} - 3x_{-6}\right) \quad \text{(seven-point, LSPF)} \tag{10.45}$$

$$y = (1/60)(4x + 3x_{-1} + 2x_{-2} + x_{-3} - x_{-5} - 2x_{-6} - 3x_{-7} - 4x_{-8}) \quad \text{(nine-point)} \qquad 10.46$$

How the *three-point central difference differentiator* estimates the slope of x(t) is illustrated in Figure 10.5B. The transfer function of this differentiator is easily seen to be:

$$D_z(z) = \frac{z^2 - 1}{2T z^2} \qquad 10.47$$

This transfer function has zeros at z = 1, z = −1, and two poles at the origin of the z plane. The frequency response of $D_Z(z)$ is determined by substituting $e^{j\omega T}$ for z, and using the Euler relations:

$$\mathbf{D}_z(j\omega T) = (1/T)\sin(\omega T)\exp[j(\pi/2 - \omega T)] \qquad 10.48$$

Like the two-point differentiator, this algorithm has nearly ideal behaviour at low frequencies. Its magnitude response peaks at $\omega = \omega_N/2$, and goes to zero at $\omega = \omega_N$. Note that the gains of the two- and three-point differentiators depend on a knowledge of the sampling period, but the five-point and higher least squares, parabolic fit differentiator gains do not depend on T.

Digital integration routines allow us to estimate the area under a curve, using sampled data. In chemometric instrumentation systems, the outputs of various types of chromatographs and spectrographs are often integrated in order to estimate the quantity of a chemical species in a sample. Another obvious application of integrating a sampled variable is to compute its mean value, or rms value.

The simplest form of digital integrator is the *rectangular integrator*, the action of which is illustrated in Figure 10.6A. All digital integration routines are recursive, i.e., they add a present estimate of area to the cumulative old estimate of area. That is, the present integrator output, i(nT) = i is given by:

$$i = i_{-1} + \Delta i \qquad 10.49$$

In the case of the rectangular integrator:

$$i = i_{-1} + Tx_{-1} \qquad 10.50$$

In terms of a z-transform transfer function, Equation 10.50 can be written:

$$\frac{I(z)}{X(z)} = H_{ZR}(z) = \frac{T}{z - 1} \qquad 10.51$$

Substituting $e^{j\omega T}$ for z gives us the frequency response of the rectangular integrator:

$$\mathbf{H}_{ZR}(j\omega T) = \frac{T/2}{\sin(\omega T/2)}\exp[-j(\pi/2 + \omega T/2)] \qquad 10.52$$

At low frequencies, $\mathbf{H}_{ZR}(j\omega T)$ behaves like:

$$\mathbf{H}_{ZR}(j\omega T) \cong (1/\omega)\exp[-j\pi/2] \qquad 10.53$$

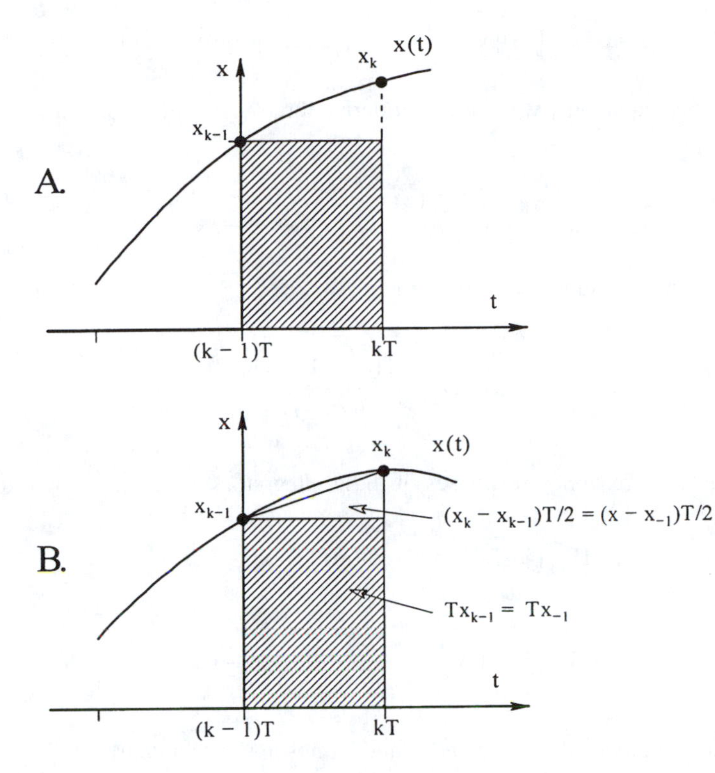

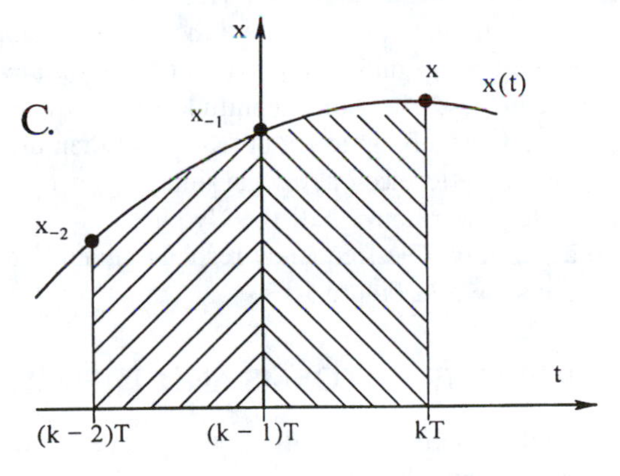

FIGURE 10.6 (A) Illustration of three common, digital integration routines. Simple rectangular integration is shown here. Its transfer function is given by Equation 12.51. (B) Trapezoidal integration, See Equation 10.57. (C) Simpson's rule integration. See Equation 10.58.

which is the behavior of an (ideal) analog integrator. At the Nyquist frequency:

$$\mathbf{H}_{\mathbf{ZR}}\left(j\omega_N T\right) = \left(T/2\right)\exp\left[-j\pi\right] \qquad 10.54$$

Many other integration routines exist, some of them quite complex. Here we will review two other simple integration algorithms commonly used in DSP:

$$i = i_{-1} + \left[Tx_{-1} + \left(T/2\right)\left(x - x_{-1}\right)\right] \quad \text{(Trapezoidal integration)} \qquad 10.55$$

$$i = i_{-2} + (T/3)\left[x + 4x_{-1} + x_{-2}\right] \quad \text{(Simpson's } ^1/_3 \text{ rule)} \qquad 10.56$$

From these difference equations, we can easily derive the transfer functions in z:

$$H_{ZT}(z) = \frac{T(z-1)}{2(z-2)} \qquad 10.57$$

for *trapezoidal integration*, and

$$H_{ZS}(z) = \frac{T\left(z^2 + 4z + 1\right)}{3\left(z^2 - 1\right)} \qquad 10.58$$

for *Simpson's rule*. The frequency responses of these integrators are easily shown to be:

$$\mathbf{H}_{ZT}(j\omega T) = (T/2)\cot(\omega T/2)\exp[-j\pi/2] \qquad 10.59$$

$$\mathbf{H}_{ZS}(j\omega T) = \frac{T[2 + \cos(\omega T)]}{3\sin(\omega T)}\exp[-j\pi/2] \qquad 10.60$$

Note that the phase for both the trapezoidal and Simpson's rule integrators is that of an ideal integrator. However, their frequency response magnitudes depart from an ideal integrator at higher frequencies. Simpson's rule follows $1/\omega$ well out to $\omega = \omega_N/2$, and then reverses slope and increases sharply as ω increases above $2\omega_N/3$. Thus, Simpson's rule behaves poorly on any x(t) containing noise power above 2/3 the Nyquist frequency. The gain magnitude of the trapezoidal integrator follows the ideal frequency response curve closely up to about $\omega_N/3$, and then drops off to zero at ω_N. Thus, the trapezoidal integrator gives better accuracy when integrating noisy signals. If the power density spectrum of x(t) has little power above half the Nyquist frequency, the Simpson's integration rule may perform more accurately. The frequency response magnitude curves for the three integrators we have discussed are shown in Figure 10.7.

10.3 DISCRETE AND FAST FOURIER TRANSFORMS AND THEIR APPLICATIONS

Just as we can represent continuous, time-domain, signals in the frequency domain using the continuous Fourier transform (CFT) pair, shown below:

$$\mathbf{F}(j\omega) = \int_{-\infty}^{\infty} f(t)e^{-j\omega t}\,dt \qquad 10.61$$

$$f(t) = \frac{1}{2\pi}\int_{-\infty}^{\infty} \mathbf{F}(j\omega)e^{j\omega t}\,d\omega \qquad 10.62$$

we can also characterize discrete time signals of finite length in the frequency domain using the discrete Fourier transform (DFT) and the inverse discrete Fourier transform (IDFT). In the CFT, f(t) can exist over all time. In the DFT, we assume a finite number of samples, N, is taken of x(t),

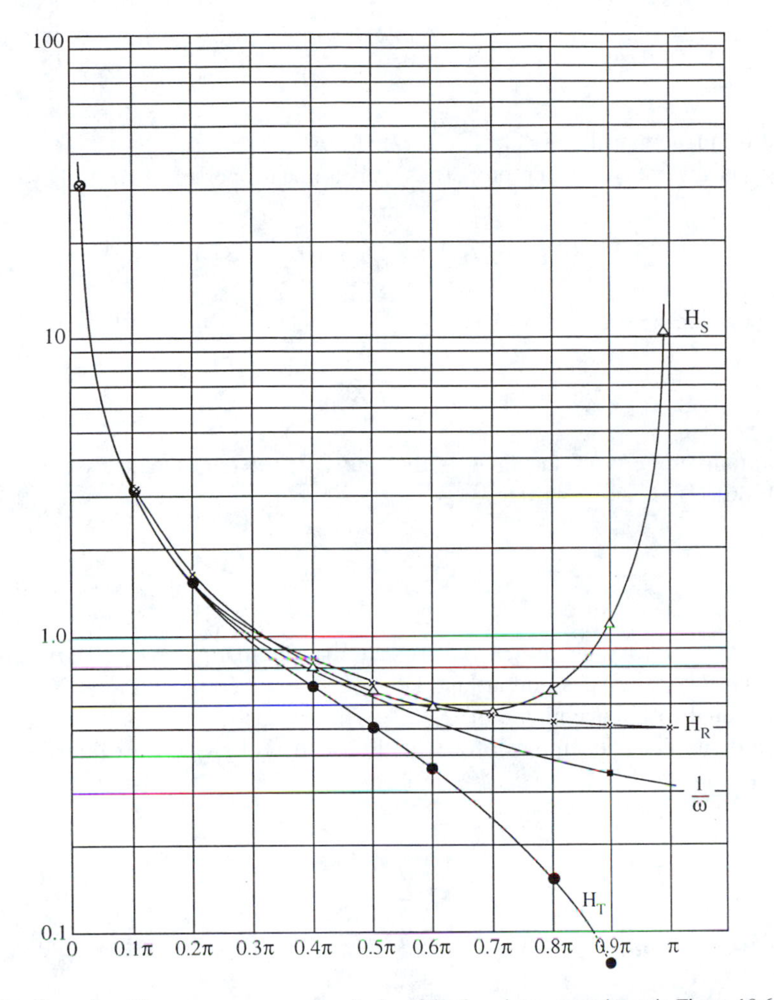

FIGURE 10.7 Nondimensional frequency response magnitudes of the three integrators shown in Figure 10.6. The frequency response magnitude of an ideal (analog) integrator, 1/ωT, is shown for comparison. HR, rectangular integrator; HT, trapezoidal integrator; HS, Simpson's rule integrator.

each at an interval of T_S seconds. The total length of time window or epoch over which x(t) is sampled is $T_E = (N - 1)T_S$. In computing the DFT, we assume $x(t) = x(t + NT_S)$, i.e., x(t) is periodic with a period of NT_S seconds.

The N elements of $\{x_n\}_N$ give rise to N elements of the DFT according to the relations:

$$\mathbf{X}_k = \sum_{m=0}^{N-1} x_m \exp\left(-j\frac{2\pi}{N}km\right), \quad k = 0, 1, \cdots, N-1 \qquad 10.63A$$

$$\mathbf{X}_k = \sum_{m=0}^{N-1} x_m \cos(2\pi k\, m/N) - j\sum_{m=0}^{N-1} x_m \sin(2\pi k\, m/N) \qquad 10.63B$$

The N terms of the DFT are complex numbers; they have real and imaginary parts or equivalently, magnitudes and angles. The frequency spacing between adjacent terms of the spectrum, $\{\mathbf{X}_n\}_N$, is given by Equation 10.64:

$$\Delta\omega = \frac{2\pi}{T_E} = \frac{2\pi}{(N-1)T_S} \text{ r/s} \quad 10.64$$

and the total frequency span of $\{X_k\}_N$ is 0 to $2\pi/T_S$ r/s.

The DFT has several interesting properties: it is a linear operator, i.e., it obeys superposition:

$$DFT_N\{Ap_k + Bq_k\} = A\,DFTN\{p_k\} + B\,DFTN\{q_k\} \quad 10.65$$

The DFT is periodic with period N. That is:

$$\mathbf{X}_k = \mathbf{X}_{k+N} \quad 10.66$$

A very important property of the DFT is that the DFT of real signals is conjugate-symmetric (Williams, 1986). This can be written:

$$\mathbf{X}_k = \mathbf{X}^*_{N-1-k}, \qquad 0 \le k \le N/2 \quad 10.67$$

Another way of describing this property is to state that the real part of $\mathbf{X}_k$ is even about k = N/2, and the imaginary part of $\mathbf{X}_k$ is odd around k = N/2. Thus the magnitude of $\mathbf{X}_k$ is even around k = N/2, and is equal to the magnitude of $\mathbf{X}^*_{N-1-k}$. This means that one half of a DFT spectrum of $\{x_n\}_N$ is redundant. Thus N samples of x(t) yield N/2 useful spectrum values ranging from 0 to $f_N = 1/2T_S$ Hz.

The inverse DFT can be written:

$$x_k = \frac{1}{N}\sum_{m=0}^{N-1} \mathbf{X}_m \exp\left(+j\frac{2\pi}{N}mk\right) \quad 10.68$$

The IDFT, also, is linear and periodic in N.

As an example of setting up a DFT routine, let us assume we need to resolve and examine in the frequency domain two coherent (sinusoidal) signals to which broadband noise has been added. The signals are at 400 and 420 Hz. We wish to examine the magnitudes of $\{\mathbf{X}_k\}_N$ to determine the relative levels of the signals and the noise. First, let us set the Nyquist frequency of the system to be 1000 Hz. The anti-aliasing filters must have attenuated the signals plus noise to a negligible output power level at and above 1 kHz. The sampling rate required is thus 2000 Sa/s. The signal spacing is 20 Hz. This implies that we should be able to resolve a Δf = 2 Hz. Thus $T_E = 1/\Delta f$ = 0.5 s, and $N = T_E/T_S + 1$ = 1001 samples. (Typically, 1024 samples would be used in calculating $\{\mathbf{X}_k\}_N$.)

10.3.1 Use of Data Windows to Improve Spectral Resolution

Direct computation of $\{\mathbf{X}_k\}_N$ using Equation 10.63 is always done with a *windowing function* operating on the sampled data, $\{x_k\}_N$. That is, the product

$$x'_k = w_k x_k \quad 10.69$$

is computed in the time domain, where w_k is the windowing function. The purpose of a windowing function is to reduce a phenomenon known as "spectral leakage." If a pure, sinusoidal x(t) is sampled and DFTd, we generally observe symmetrical, non-zero sidelobes in $\{\mathbf{X}_k\}_N$ around the main peak at the sinusoid's frequency. This phenomenon was called spectral leakage because spectral energy at the sinusoid's frequency effectively "leaks" into the adjacent frequency terms. What windows do in general is to reduce the magnitude of the leakage sidelobes at the expense of the sharpness of the main spectral peak at the input sinusoid's frequency.

In general, in the frequency domain:

$$^{w}\mathbf{X}(j\omega) = \mathbf{W}(j\omega) \otimes \mathbf{X}(j\omega) \tag{10.70}$$

That is, the FT of the windowed, sampled data sequence can be expressed as the complex convolution of the FT of the windowing function with the FT of x(t).

To illustrate how windowing works, we first consider the case of the simplest, rectangular window consisting of N unit impulses, spaced T_S seconds apart. The rectangular window function is a truncated unit impulse train which multiplies (modulates) x(t) in the time domain to generate the sampler output. The rectangular window function can be written as:

$$w(t) = \{w_k\}_N = \sum_{k=0}^{N-1} \delta(t - kT_S) \tag{10.71}$$

A fundamental property of Fourier transforms is that the CFT of an even function in t is even in ω, and has zero phase. It is therefore algebraically convenient to consider both the window and x(t) to be even functions. This process is shown in Figure 10.8. We choose N to be an odd number so M will be an integer. Note that the epoch length T_E is:

$$T_E = (N-1)T_S = 2M\,T_S \tag{10.72}$$

Hence:

$$M = (N-1)/2 \tag{10.73}$$

The CFT of the rectangular window can be shown to be (Papoulis, 1977):

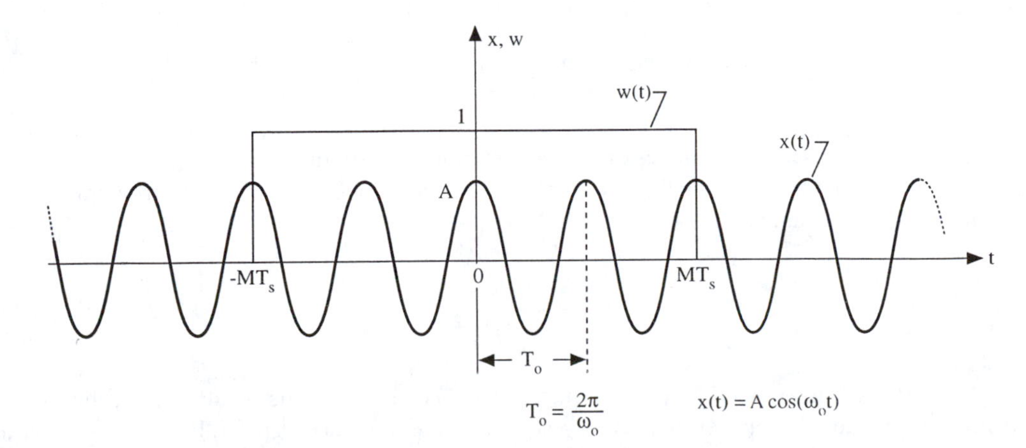

FIGURE 10.8 A rectangular (even) window function and a cosine wave. A finite number of samples of the cosine wave effectively multiplies it by the sampled, rectangular window function.

$$W(w) = \frac{\sin[\omega(2M+1)T_S/2]}{\sin(\omega T_S/2)} = \frac{\sin(\omega N T_S/2)}{\sin(\omega T_S/2)} \quad 10.74$$

For purposes of this example, let us assume that x(t) is a cosine wave (even function) and x(t) = A $\cos(\omega_o t)$, where $\omega_o < \omega_N$. The CFT of the cosine x(t) is well known:

$$\mathbf{X}(\omega) = A[\pi\delta(\omega + \omega_0) + \pi\delta(\omega - \omega_0)] \quad 10.75$$

Complex convolution now becomes real convolution, where the kernel only exists at $\omega = \pm\omega_o$. Thus the CFT of the finite, sampled, rectangular-windowed cosine wave is:

$$^{w}\mathbf{X}(\omega) = \frac{A}{2}\left\{\frac{\sin[(\omega+\omega_0)NT_S/2]}{\sin[(\omega+\omega_0)T_S/2]} + \frac{\sin[(\omega-\omega_0)NT_S/2]}{\sin[(\omega-\omega_0)T_S/2]}\right\} \quad 10.76$$

The main spectral peaks in $^{W}\mathbf{X}(\omega)$ are seen to occur at $\omega = \pm\omega_o$. The leakage is due to the side lobes of Equation 10.74. Figure 10.9 illustrates normalized plots of the DFTs of several window functions (see caption for details). Clearly, to reduce leakage and to get best resolution, we must find a window function whose CFT has a narrow main lobe and low sidelobes compared to the main lobe peak. Many such window functions have been devised and used. These include, but are not limited to (Papoulis, 1977): Bartlett, Tukey, Hamming, parabolic, maximum energy concentration, minimum energy moment, minimum amplitude moment, von Hann (Hanning), and Kaiser. (The latter two windows are described in Williams, 1986.) The Hamming window is one of the most widely used windows in DFT and FFT computations. Its FT has low major sidelobes. It consists of a cosine on a pedestal. For N ≥ 64 samples, it can be shown that the pedestal height approaches 0.536 and the cosine term is multiplied by 0.464. Each sample of x_k is multiplied by a corresponding w_k:

$$w_k = 0.536 + 0.464\cos\{\pi[2k - N + 1]/N\}, \quad k = 0, 1, \cdots, (N-1) \quad 10.77$$

In the symmetrical, even function case, we center the window function at m = 0, and we can write:

$$w_m = 0.536 + 0.464\cos[\pi 2m/(N-1)], \quad -N/2 \le m \le (N/2 - 1) \text{ integer m} \quad 10.78$$

Figure 10.10 illustrates the major lobes of several common window functions. The DFTs have been normalized. Note that the base width, B, of the major spectral lobe of the Bartlett (triangular) and Hamming windows is about:

$$B \approx \frac{8\pi}{(N-1)T_S} \text{ r/s} \quad 10.79$$

while the B of the Parzen window is somewhat wider. The choice of the window algorithm to be used involves a tradeoff between main lobe width, and the amount of spectral leakage seen at a given distance from the frequency of the main peak. Generally, the narrower the main lobe, the greater the leakage. Selection of a window algorithm can be somewhat subjective, and the window selected can influence one's ability to interpret the DFT spectrograms. Most commercial DFT

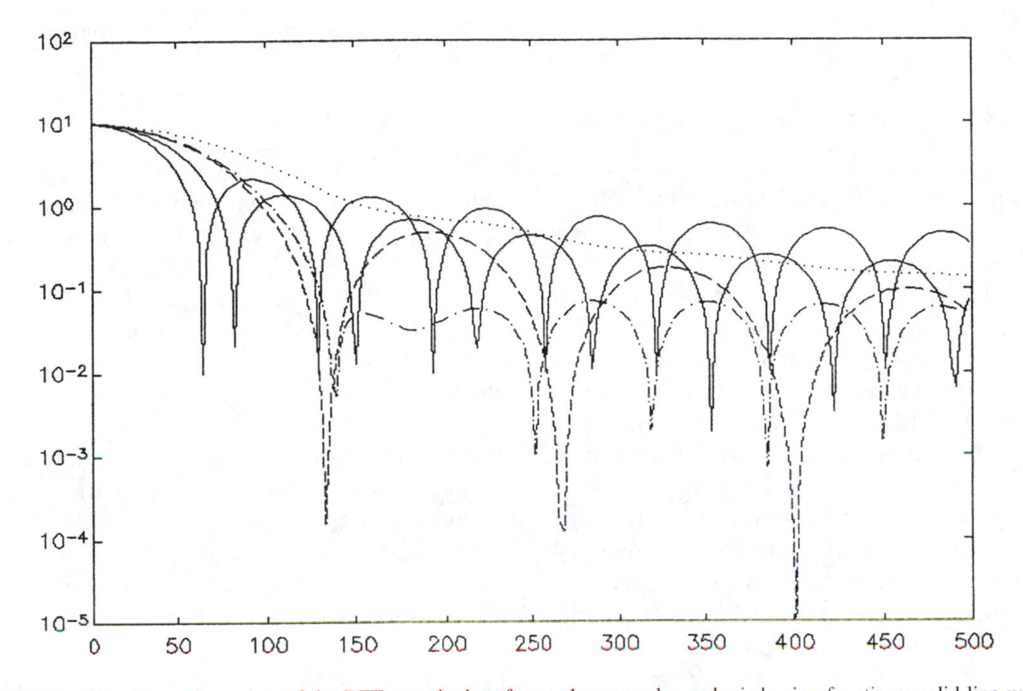

FIGURE 10.9 Normalized plots of the DFT magnitudes of several commonly used windowing functions: solid line with narrowest main lobe and highest side lobes are the rectangular window. Key: solid line with wider main lobe and lower side lobes, Kaiser window; dashes, triangular (Bartlett) window; dash-dot, Hamming window; dots, triangle squared (Parzen). A 1024-point FFT was computed with Matlab™. Log magnitude vs. linear frequency scale.

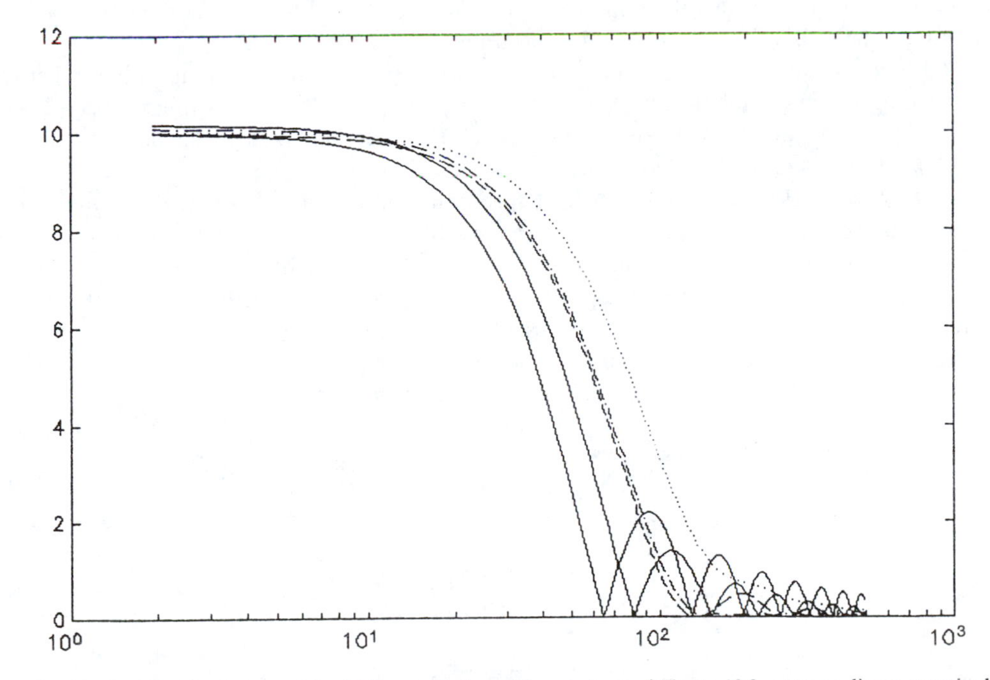

FIGURE 10.10 Normalized plots of the DFT magnitudes of the windows of Figure 10.9, except a linear magnitude vs. log frequency scale was used. Note the tradeoff between main lobe narrowness and side-lobe height. The same key used in Figure 10.9 is used.

spectrum analyzers offer the user the choice of a plain, rectangular window, a cosine window such as the Tukey, Hanning, or Hamming, or the Kaiser window.

10.3.2 Use of the DFT to Characterize Random Signals and Noise

A major use of the modern spectrum analyzer is to allow us to visualize the distribution of power with frequency in an analog waveform. This feature can be useful in vibration analysis of mechanical structures, measurement of noise, determination of linear system transfer functions by the white noise method, etc.

In Chapter 3, we saw that the power density spectrum (PDS) is used to describe the power content at a particular frequency of an infinite-length, analog waveform. The units of the PDS are mean-squared volts (units) per Hz. In modern engineering practice, we use the DFT with an appropriate window to estimate the square root of the PDS (RPDS) of an analog waveform. In this case, the RPDS units are rms volts (units)/$\sqrt{\text{Hz}}$.

The discrete RPDS can be calculated in the following manner: first, the analog waveform, x(t), is passed through an anti-aliasing low-pass filter to remove any significant spectral power at and above the Nyquist frequency, $\omega_N = \omega_S/2 = \pi/T_S$ r/s. Next, a finite length of the anti-aliasing filter's analog output, x (t), is sampled at a rate of $1/T_S$ samples/second. N values of x(nT) are stored in an array, $\{x_k\}_N$, in the computer. Each x_k is multiplied by the corresponding window weighting constant, w_k, giving a windowed data array, $\{^W x_k\}_N$. N elements of the DFT of $\{^W x_k\}$ are now calculated using symmetrical coefficients, giving $\{^W \mathbf{X}_m\}_N$. Because of the conjugate symmetry property of DFT output, only N/2 points in $\{^W \mathbf{X}_m\}_N$ are unique. We store a buffer array consisting of $\{^W \mathbf{X}_m\}_{N/2}$ for m = 0 to m = (N/2 – 1). m = 0 corresponds to f = 0, and k = (N/2 – 1) corresponds to $f = f_{MAX} = f_S/2 = f_N = 1/2T_S$ Hz in the spectrum. The frequency spacing between Fourier coefficients is $\Delta f = 1/(N-1)T_S$ Hz. The coefficients of a hypothetical N = 8 point DFT are shown in Figure 10.11. We note that the DFT coefficient magnitudes can be found from the Pythagorean theorem. Also we observe that the important parameter, T_S, does not appear in the DFT operation, Equation 10.63. T_S is used to calculate the spectrum scale factors, Δf and f_{MAX}.

If x(t) is a broadband (noisy) waveform, there will a great deal of natural variation between corresponding spectrum coefficient magnitudes in the computation of M "similar" spectra. Averaging the corresponding elements of M spectra permits a reduction of the standard deviation of a given element by a factor of about $1/\sqrt{M}$.

In the Hewlett-Packard Model 3582A DFT spectrum analyzer, noisy spectra are averaged in the following manner: M spectra are computed sequentially. Two arrays of length N are set up. One holds the average of the squares of the corresponding real parts of the DFT coefficients, the other the average of the squares of the corresponding imaginary parts of the DFT coefficients. After M spectra are calculated and the mean squares of the corresponding coefficients are stored, the corresponding mean squares are added, and the square root is taken of each corresponding sum, giving a true, rms, RPDS. These steps can be shown mathematically:

$$\overline{\left[\text{Re}\{\mathbf{X}_k\}\right]^2} = \frac{1}{M}\sum_{i=1}^{M}\left[\text{Re}\{{}^i\mathbf{X}_k\}\right]^2 \qquad 10.80$$

$$\overline{\left[\text{Im}\{\mathbf{X}_k\}\right]^2} = \frac{1}{M}\sum_{i=1}^{M}\left[\text{Im}\{{}^i\mathbf{X}_k\}\right]^2 \qquad 10.81$$

$$\text{rms}\,\mathbf{X}_k = \sqrt{\overline{\left[\text{Re}\{\mathbf{X}_k\}\right]^2} + \overline{\left[\text{Im}\{\mathbf{X}_k\}\right]^2}}, \quad N/2 \le k \le (N-1) \text{ integer } k \qquad 10.82$$

The HP 3582A spectrum analyzer also offers the option of an exponential, moving averager to reduce the uncertainty in calculating the spectra of non-stationary, noisy signals. Non-stationarity

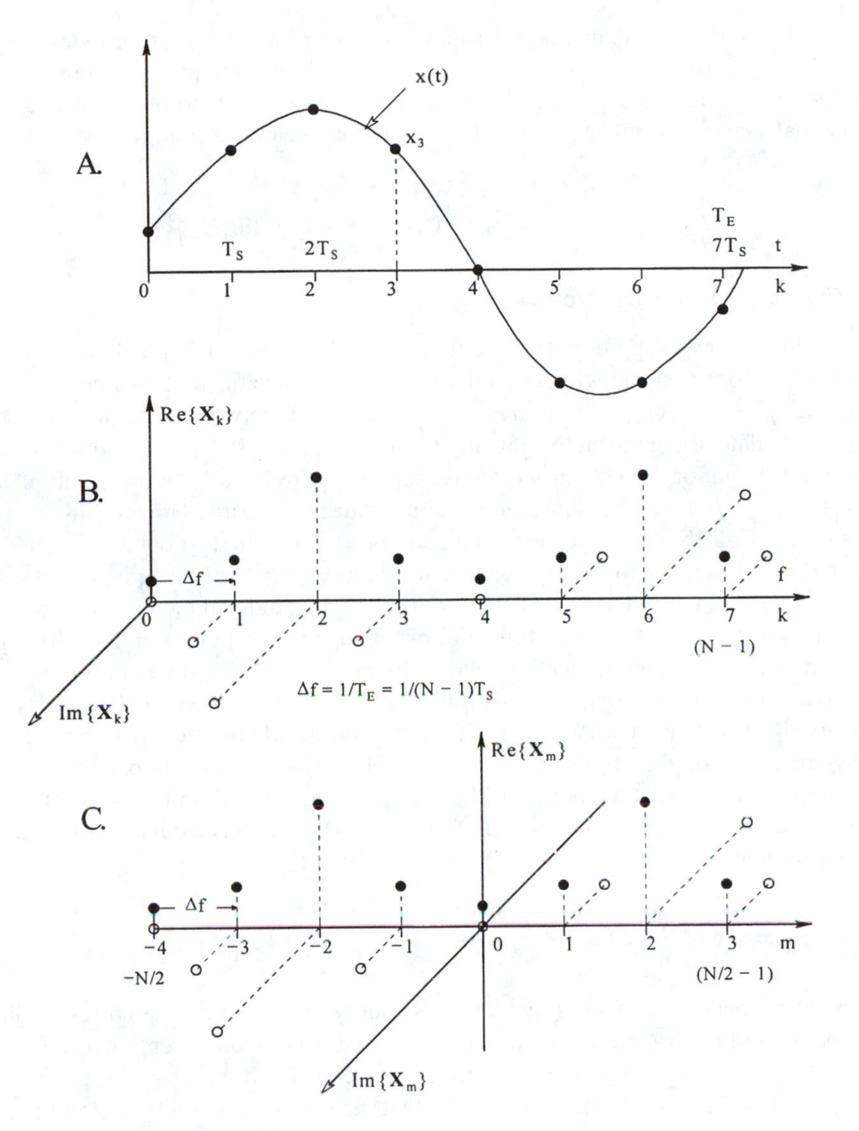

FIGURE 10.11 (A) Eight samples of a continuous time function, x(t), are taken. A rectangular window is used. (B) The hypothetical DFT of the eight-point, x_k is shown. Real and imaginary components of the DFT are shown for $0 \le k \le 7$ (eight samples). Note that the Pythagorean magnitude of the DFT is repeated for $0 \le k \le (N/2 - 1)$, and for $N/2 \le k \le (N - 1)$. This is a result of the property of conjugate symmetry. (C) By translating DFT coordinates, we see that we only need 1/2 of the N-point DFT to characterize x* in the frequency domain.

means the underlying physical processes producing the noise in x(t) are changing in time, e.g., a resistor may be heating up, and its increasing temperature produces an increasing resistance and increasing thermal noise. The exponential averaging process weights the most recent samples more heavily than old samples. We illustrate it below for the real parts of the kth spectral element:

$$\overline{\text{Exp}\left[\text{Re}\{\mathbf{X}_k\}\right]^2} = (1/4)\left\{\left[\text{Re}\{{}^1\mathbf{X}_k\}\right]^2 + (3/4)^1\left[\text{Re}\{{}^2\mathbf{X}_k\}\right]^2 + (3/4)^2\left[\text{Re}\{{}^3\mathbf{X}_k\}\right]^2 + \cdots + (3/4)^{M-2}\left[\text{Re}\{{}^{M-1}\mathbf{X}_k\}\right]^2 + (3/4)^{M-1}\left[\text{Re}\{{}^{M}\mathbf{X}_k\}\right]^2\right\} \quad 10.83$$

Here the most recent DFT calculation is given the index "1" and the first, or oldest, the index M. Note that $(3/4)^8 = 0.100$ and $(3/4)^{16} = 0.0100$, so the "tail" of the moving average has negligible weight after computing 16 consecutive DFTs from x(t). As in the case of calculating the rms $\mathbf{X}_k$, the exponentially averaged rms $\mathbf{X}_k$ is found by the Pythagorean theorem:

$$\text{Exp}(\text{rms}\,\mathbf{X}_k) = \sqrt{\overline{\text{Exp}\left[\text{Re}\{\mathbf{X}_k\}\right]^2} + \overline{\text{Exp}\left[\text{Im}\{\mathbf{X}_k\}\right]^2}} \quad 11.84$$

10.3.3 The Fast Fourier Transform

In order to obtain a good approximation to the CFT of a bandwidth-limited waveform, we must calculate the DFT of the signal with a sampling rate above twice the highest frequency in the PDS of x(t), and use enough samples to insure a close spacing, Δf, between adjacent values of $\mathbf{X}_k$. It is common to calculate spectra using Ns ranging from $256 = 2^8$ to $4096 = 2^{12}$, and higher. Use of the DFT algorithm, Equation 10.63A, as written is seen to require 2^N calculations (multiplications and additions) to calculate each $\mathbf{X}_k$ coefficient's real and imaginary parts. However, there are N coefficients, so a total of $2N^2$ multiplications and additions are required to obtain the DFT, $\{\mathbf{X}_k\}_N$. If the magnitudes of $\{\mathbf{X}_k\}_N$ are to be displayed, then we must apply the Pythagorean theorem N times as well. It is clear that a large DFT calculated by the direct method will involve many high-precision computer operations, and take a long time. For example, computation of an N = 4096 point DFT involves over 33.5 million multiplications and additions. This can be time-consuming.

To overcome the computational burden of direct computation of large DFTs, Cooley and Tukey (1965) devised a fast Fourier transform (FFT) means of calculating the DFT. Actually, there are now many variations on the original Cooley-Tukey FFT. It is not our purpose here to describe in detail the steps involved in calculating an FFT/DFT. Rather, we will outline the strategy used and explain why it is efficient. Using the complex form of the DFT, direct calculation of the kth complex term can be written out as:

$$\mathbf{X}_k = x_0 e^{-j(2\pi k0/N)} + x_1 e^{-j(2\pi k1/N)} + \cdots + x_i e^{-j(2\pi ki/N)} + \cdots + x_{N-1} e^{-j(2\pi k[N-1]/N)} \quad 10.85$$

Each of the N exponential terms in Equation 10.85 defines an angle for the corresponding real x_k. Thus, N vector summations are required in order to find $\mathbf{X}_k$ in the conventional DFT, and a total of N^2 vector summations must be done to find the entire DFT, $\{\mathbf{X}_k\}_N$.

In computing an FFT, use is made of the fact that the exponential angle operator is periodic in k_i/N. This periodicity leads to redundancy in the DFT calculations which is taken advantage of in an FFT algorithm by rearranging the order of the calculations so that there are $\log_2 N$ columns of vector summers, each of which contains N summers, for a total of N $\log_2(N)$ complex summers in an FFT array. Each summer sums two vectors. In FFT operations, N is generally made a power of 2, e.g., $N = 2^b$, where b is a positive integer, such as b = 10 for N = 1024. The angles of the vectors must still be calculated when the FFT array is set up, and stored as constants. Their values depend only on k, i, and N. It is not practical to try to illustrate the signal flow graph of an FFT operation for Ns over 8. Figure 10.12 shows an implementation of an 8-point FFT. Note the symmetrical rearrangement of the x_ks at the FFT input. In some versions of the FFT, the output coefficients are rearranged.

The efficiency, η, of the FFT vs. the DFT calculated in the direct method can be calculated by assuming that computation speed is inversely proportional to the number of vector additions required in each case. So we can write:

$$\eta = \frac{S_{FFT}}{S_{DFT}} = \frac{N}{\log_2(N)} \quad 10.86$$

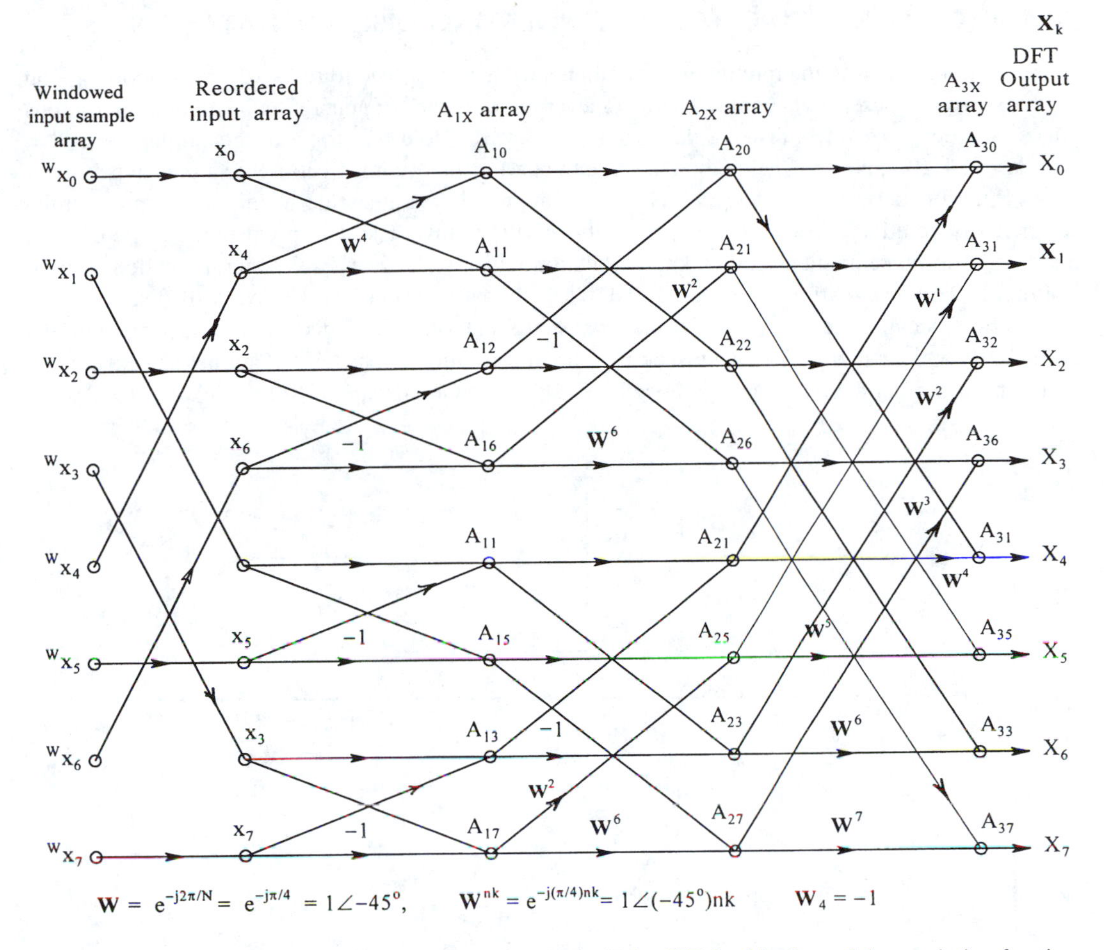

FIGURE 10.12 Flow-graph showing implementation of the Cooley/Tukey FFT algorithm for an eight-sample time function. Note that the FFT output values in the A_{3x} array are complex, i.e., they have real and imaginary parts.

For a 4096-point FFT, $\eta = 341.3$; for N = 1024, $\eta = 102.4$, etc.

There are many commercial FFT software routines available in various computer languages and programs, including FORTRAN, C, Matlab, etc. Also, companies offering data acquisition interface cards and DSP cards for PCs generally offer their software versions of FFT routines. Thus, there is little probability that a person setting up a DSP capability in a measurement system will have to write their own FFT routine from scratch.

There are a number of commercially available, plug-in, DSP coprocessor boards for PCs and workstations. Designs for these boards are constantly evolving, improving their speeds and capabilities. Some of the commercially available FFT cards are quite fast. For example, the Array Microsystems model a66540 Frequency Domain Array Processor (FDAP) card for VME bus computers can do a 1024-point real FFT in 13.3 μsec. Their a66550 card for PC/AT computers will do a 1024-point FFT in 125.9 μsec. The Ixthos IXD7232 VME bus DSP card can run at a sustained 50 MFLOPS and do a 1024-point, complex FFT (radix 4) in 770 μsec. National Instruments' AFFA system uses two plug-in cards to make a powerful, audio frequency FFT system for the Macintosh II PC with NuBus. The AFFA system does real-time audio spectrum analysis up to 20 kHz. It can do 64, 126, 256, 512, or 1024-point FFTs at sampling rates of 22.05, 24, 32, 44.1 or 48 k Sa/s. It has rather elegant, menu-controlled operating software, which includes choice of rectangular, Hanning, Hamming, ExactBlackman, Blackman, and Blackman-Harris data windows. Many other companies, too numerous to list here, make DSP, coprocessor, plug-in boards for PCs, and they offer operating software that enables flexible calculation of FFTs.

10.4 DIGITAL ROUTINES FOR INTERPOLATING DISCRETE DATA

Often in making and in interpreting measurements we have the need to reconstruct a smooth analog curve between discrete data points. There are a variety of means of generating a smooth, continuous curve between a set of N discrete data points, $\{x_k, y_k\}_N$. Note that for a time-sampled y(t), $x_k = kT_S$, $k = 0, 1, 2\ldots, N-1$. One method of smooth curve generation is to compute a piecewise-linear approximation between the data points. In this approach, we generate a number of polynomial segments of relatively low order (0,1,2,3) between the points. The data point pairs at the ends of each polynomial segment are called knots. Polynomials of order 2 or 3 are generally called splines. Polynomial approximations of order 0 and 1 are shown in Figure 10.13. We will discuss cubic splines here because they are generally the most useful in terms of a tradeoff between computational complexity and accuracy. The reader seeking more information on the field of splines should consult texts on numerical analysis such as those by Casulli and Greenspan (1988) or Scheid (1990).

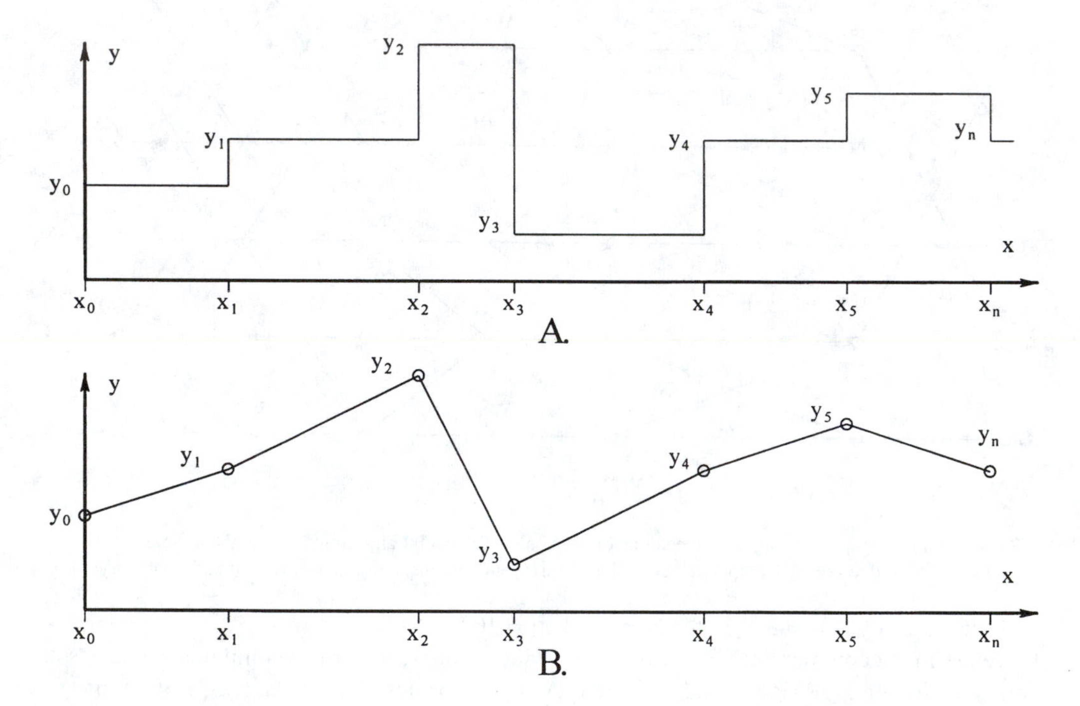

FIGURE 10.13 Illustration of zero-order (A), and first-order (B), polynomial approximations for interpolating between a set of discrete data points in order to regenerate the continuous time function from which the samples were taken.

A cubic spline is a cubic polynomial which is made to fit three consecutive data points: (x_k, y_k), (x_{k+1}, y_{k+1}), and (x_{k+2}, y_{k+2}), or if y is sampled data, (kT_S, y_k), $([k+1]T_S, y_{k+1})$, and $([k+2]T_S, y_{k+2})$. The polynomial may be written:

$$y = a + bx + cx^2 + dx^3 \qquad 10.87$$

and

$$\dot{y} = b + 2cx + 3dx^2 \qquad 10.88$$

We assume that $x_0 = t = 0$ (local time origin), and that the intervals $x_i - x_{i-1} = T_S$ over the entire data set. Thus $x_0 = 0$, $x_1 = T_S$, $x_k = kT_S$, etc. We also assume that the first derivative, $\dot{y}_0$, is known or well estimated. To calculate the cubic spline for the first three data points $k = 0,1,2$, we can write:

$$y_0 = a \qquad \text{10.89A}$$

$$y_1 = a + bT_S + cT_S^2 + dT_S^3 \qquad \text{10.89B}$$

$$y_2 = a + b2T_S + c4T_S^2 + d8T_S^3 \qquad \text{10.89C}$$

also, the first derivative at $x_0 = t = 0$ is, from Equation 10.87 above:

$$\dot{y}_0 = b \qquad \text{10.89D}$$

We wish to find the coefficients a, b, c, and d for the $k = 0,1,2$ spline. This may be done by simultaneously solving Equations 10.89A, 10.89B. Doing so gives:

$$a = y_0 \qquad \text{10.90A}$$

$$b = \dot{y}_0 \qquad \text{10.90B}$$

$$c = \left[-7y_0 + 8y_1 - \dot{y}_2 - 6T_S y_0\right] / \left(4T_S^2\right) \qquad \text{10.90C}$$

$$d = \left[3y_0 - 4y_1 + \dot{y}_2 + 2T_S y_0\right] / \left(4T_S^3\right) \qquad \text{10.90D}$$

Substituting these values into the cubic spline formula yields:

$$y(t) = y_0 + \dot{y}_0 t + \frac{-7y_0 + 8y_1 - \dot{y}_2 - 6T_S\dot{y}_0}{4T_S^2} t^2 + \frac{3y_0 - 4y_1 + y_2 - 2T_S\dot{y}_0}{4T_S^3} t^3 \qquad \text{10.91}$$

over the interval $0 \le t \le 2T_S$. By recursion, we can write a general cubic spline formula valid for any triplet of data points T_S seconds apart (y_k, y_{k+1}, y_{k+2}):

$$\begin{aligned} S_3(t) = y_k + \dot{y}_k\left(t - kT_S\right) + \frac{-7y_k + 8y_{k+1} - y_{k+2} - 6T_S\dot{y}_k}{4T_S^2}\left(t - kT_S\right)^2 \\ + \frac{3y_k - 4y_{k+1} + y_{k+2} - 2T_S\dot{y}_k}{4T_S^3}\left(t - kT_S\right)^3 \end{aligned} \qquad \text{10.92}$$

The first derivative estimate needed for the cubic spline of the next, adjacent triad of data points is given by:

$$\dot{y}_{k+2} = \dot{y}_k + \frac{2\left(y_{k+2} - 2y_{k+1} + y_k\right)}{T_S} \qquad \text{10.93}$$

Note that the first "knot" is at k = 0, the second knot is at k = 2, the third at k = 4, etc.

As an example of calculating the cubic splines of a small data set using Equations 10.92 and 10.93, let $T_S = 1$ s, $\dot{y}_0 = 1$, $y_0 = 0$, $y_1 = 1$, $y_2 = 2$, $y_3 = 1$, $y_4 = 0$. Over the first interval, $0 \le t \le 2T_S = 2$ s, we have:

$$S_{0,2}(t) = 0 + (1)t + (0)t^2 + (0)t^3 = t \qquad 10.94$$

The derivative estimate at k = 2 is:

$$\dot{y}_2 = 1 + \frac{2[2 - 2(1) + 0]}{1} = 1 \qquad 10.95$$

and over the interval, $2T_S \le t \le 4T_S$, the spline is:

$$\begin{aligned} S_{2,4}(t) &= 2 + (1)(t-2) + \frac{[-7(2) + 8(1) - (6)(1)(1)]}{4(1)}(t-2)^2 + \frac{[3(2) - 4(1) + 0 + 2(1)(1)]}{4(1)}(t-2)^3 \\ &= 2 + (t-2) - 3(t-2)^2 + (t-2)^3 \end{aligned} \qquad 10.96$$

Many other types of spline functions exist, as do variations on the cubic spline. For example, if both the first and second derivative of the sampled variable are known at t = 0, then the cubic spline, S(t), can be found between consecutive pairs of data points instead of between triplets of points. Using a development similar to that above for three-point, cubic splines, it is possible to show that for $kT_S \le t \le (k + 1)T_S$:

$$Sk(t) = \dot{y}_k + y_k(t - kT_S) + (\ddot{y}_k/2)(t - kT_S)^2 + \frac{y_{k+1} - y_k - T_S\dot{y}_k - (T_S^2/2)\ddot{y}_k}{T_S^3}(t - kT_S)^3 \qquad 10.97$$

and

$$\dot{y}_k = \dot{y}_{k-1} + T_S\ddot{y}_{k-1} + 3\frac{y_k - y_{k-1} - T_S\dot{y}_{k-1} - (T_S^2/2)\ddot{y}_{k-1}}{T_S} \qquad 10.98$$

$$\ddot{y}_k = \ddot{y}_{k-1} + 6\frac{y_k - y_{k-1} - T_S\dot{y}_{k-1} - (T_S^2/2)\ddot{y}_{k-1}}{T_S^2} \qquad 10.99$$

Although the scope of this discussion has been an introduction to cubic splines, it should be noted that the technique of splines is easily applied to data having nonuniform sampling periods, the resulting spline approximations can then be "sampled" at a uniform rate, and these M samples can be used to calculate a conventional DFT of $\{x_k y_k\}_N$, with M > N. Also, spline recursion equations can be put in tridiagonal matrix form for simultaneous solution of Sk(t) from the entire data set at once (Scheid, 1990). Some of the other interpolation methods that exist for analog data interpolation include the methods of LaGrange, Hermite, Gauss, Stirling, Newton, and the least squares method of fitting a continuous linear equation to a data set, $\{x_k, y_k\}_N$ (described in Chapter 1 of this text).

10.4.1 Estimating Missing Data at the Sampling Instants

We occasionally have a data sequence where one or more data points are missing, perhaps because the experimenter neglected to record them, or because a burst of noise completely masked the true data value. If data point y_k is missing in a large data set, and the data set does not vary abruptly in time, then the missing data point can be found by linear interpolation:

$$y_k = (y_{k+1} + y_{k-1})/2 \qquad 10.100$$

That is, the missing y_k can be approximated by the average of its nearest neighbors. In the event that the missing data point is part of a curve with nonzero second derivative, a more accurate estimate can be derived from the solution of four simultaneous, cubic algebraic equations (Williams, 1986). Two adjacent data points on either side of the missing data point are used:

$$y_k = (-y_{k-2} + 4y_{k-1} 4y_{k+1} - y_{k+2})/6 \qquad 10.101$$

Because cubic splines work with nonuniform data x-values, they also can be used to find a missing y_k. We assume y_{k-2}, y_{k-1}, and y_{k+1} are known, and can be used in the spline calculation. $\dot{y}_{k-2}$ also must be known or estimated.

A data point missing at the beginning or end of a finite data sequence, $\{x_k,y_k\}_N$, requires numerical extrapolation, which can also be carried out by a polynomial estimation routine using existing data (Williams, 1986). The accuracy of extrapolation decreases as the interval (x_N,x_{N-1}) increases, and as the bandwidth of $\mathbf{Y}(j\omega)$ increases. If data are changing monotonically and are not noisy, then a reasonable value for y_N can be calculated.

We give an example of a simple extrapolator that uses the three-point central difference differentiator to estimate the slope of y_k at the end of the sequence, where $k = N - 1$. This slope is found from Equation 10.43.

$$m_{k-1} = (y_{N-1} - y_{N-3})/2T_S \qquad 10.102$$

Simple algebra tells us that the extrapolated value of y at $t = NT_S$ is

$$y_N = y_{N-1} + m_{k-1}T_S = 1.5y_{N-1} - 0.5y_{N-3} \qquad 10.103$$

This same, basic procedure can be turned around to find an estimate of y_{-1} at $t = -T_S$.

The three-point, cubic spline approach can also be used to find an estimate for y_N at the end of the sequence $\{x_k,y_k\}_N$. The trick here is to redefine the local time origin so that the new $y_1 = y_N$ (value to be extrapolated), $y_0 = y_{N-1}$ (the last data sample of the sequence), $y_{-1} = y_{N-2}$, and $y_{-2} = y_{N-3}$ in the sequence. The first derivative at y_0 is estimated by the three-point central difference method:

$$\dot{y}_0 = (y_0 - y_{-2})/2T_S = (y_{N-1} - y_{N-3})/2T_S \qquad 10.104$$

Equations 10.89 can now be used with appropriate changes of indices and the derivative estimate of Equation 10.104 to show that in terms of the original data set,

$$y_N = 3y_{N-2} - 2y_{N-3} \qquad 10.105$$

Proof of Equation 10.105 is left as an exercise for the interested reader. Note that y_{N-1} curiously drops out of this cubic spline extrapolation formula.

10.5 CHAPTER SUMMARY

Digital signal processing and its related mathematical field of numerical analysis are of considerble importance to all branches of engineering, and particularly to electrical and systems engineers, and to measurement and instrumentation engineers. Many textbooks have been written in these areas; we cite a representative group in the bibliography and references for this chapter. Our purpose in this chapter was to introduce some key concepts in DSP of particular use in instrumentation and measurement system design.

We first discussed the z-transform and digital filters as a linear means of operating on a sequence of sampled data in order to do a discrete filtering operation analogous to an analog filter, such as low-pass filtering. We showed that the relevant range of frequencies in digital filtering is from 0 to the Nyquist frequency, $f_S/2$. Examples of finite impulse response (FIR) low-pass and band-pass filters were given. We also examined the time-domain structure of the filtering process, showing how a computer program running in real time can implement a filtering algorithm by a series of synchronous delays, multiplications by constants, and summations. Recursive, infinite (duration) impulse response (IIR) filters were introduced, and a second-order, high-Q, notch filter implementation was given as an example.

In measurement systems, the operations of digital differentiation and integration are also important. Several basic differentiation and integration routines were discussed and their algorithms described in the frequency domain. Integration routines were shown to be recursive, while differentiators generally use FIR algorithms.

A large part of the chapter was devoted to introducing the discrete Fourier transform (DFT). We examined how it can approximate the continuous Fourier transform (CFT) and how window functions prevent sidelobe leakage. The basics of the fast Fourier transform (FFT) algorithm to compute the DFT were described, an eight-point FFT example was illustrated and an expression for FFT/DFT efficiency was derived. The availability of commercially manufactured DSP and FFT plug-in boards for PCs was cited as the state-of-the-art way to accelerate FFT and digital filter calculations in real time.

Finally, we considered the problem of reconstructing continuous analog data between data samples. We introduced cubic splines as an efficient way to reconstruct analog data, as well as to interpolate missing data points and to extrapolate a data point at the end of a data sequence.

11

Examples of the Design of Measurement Systems

11.0 INTRODUCTION

In this final chapter, we describe the needs for, the systems design philosophies, and the means of realizing, systems used to measure: (1) the location and intensity of partial discharges in the insulation of coaxial, high-voltage, power cables; (2) an electronic system used to count fish passing through a downstream bypass facility at a hydroelectric power plant; (3) a system used to measure the ocular pulse, the minute pulsations of the cornea of the eye due to blood flow in the eye; (4) a closed-loop, pulsed laser ranging system and velocimeter that operates on a feedback principle that maintains constant a phase difference between the transmitted pulses and the received pulses. These systems have been chosen to illustrate certain design principles, as well as to describe four diverse measurement systems.

The author and his graduate students have designed these unique measurement systems; therefore, their designs reflect their personal bias toward certain circuit architectures and manufacturers' components. It is well-appreciated that systems design in electrical engineering is an art as well as a science. The designer must balance criteria on system performance with cost and ease of manufacturing, safety, system reliability, ergonomics, esthetics, and environmental concerns. There are often several designs that can lead to the same end product, and no clear set of factors appears to favor one design over the others. It is in these fuzzy cases that a designer is more free to show personal preferences, i.e., exhibit his or her "design style."

In the first design example described below, we describe the need for, the design approach, the specifications, and a critique of the performance of a partial discharge measurement system.

11.1 DESIGN OF A SYSTEM TO DETECT, MEASURE, AND LOCATE PARTIAL DISCHARGES IN HIGH-VOLTAGE, COAXIAL POWER CABLES

In the process of aging, the insulation of high-voltage, underground, coaxial power cables may develop physical defects such as water trees and other cavities. Under certain conditions, these defects can serve as the foci for massive insulation failures (arc-overs). Long before massive insulation failure occurs, the water trees and other defect sites, subject to periodic electric field stress in the dielectric, begin to arc over internally, producing *partial discharges* (PDs). PDs are not fatal events for the cable; they can lead to one in time, however. An equivalent circuit for a power cable with a PD site is shown in Figure 11.1. The partial discharge occurs when the gas or vapor dielectric in the defect cavity breaks down under electrostatic stress, permitting the charge stored in the defect walls to dissipate, temporarily reducing the electric field in the defect. At the next AC cycle, the PD is repeated, etc. The instantaneous applied voltage on the cable at which a given PD occurs is called the *inception voltage.* At 60 Hz, a coaxial high-voltage power cable behaves like a large capacitor. A typical cable has a capacitance of about 60 pF/ft. In the frequency band from 20 kHz to about 20 MHz, the cable behaves like a lossy, RLC transmission line (see Section 11.8), with a characteristic impedance, Z_0, of around 45 ohms.

When the water tree or other insulation defect arcs over at the AC inception voltage, the resulting PD injects what is effectively an impulse of charge into the transmission line equivalent circuit of the cable. The PD discharge transient is over in picoseconds; however, a voltage pulse propagates away from the site in both directions on the cable. The direct PD voltage transient recorded at the

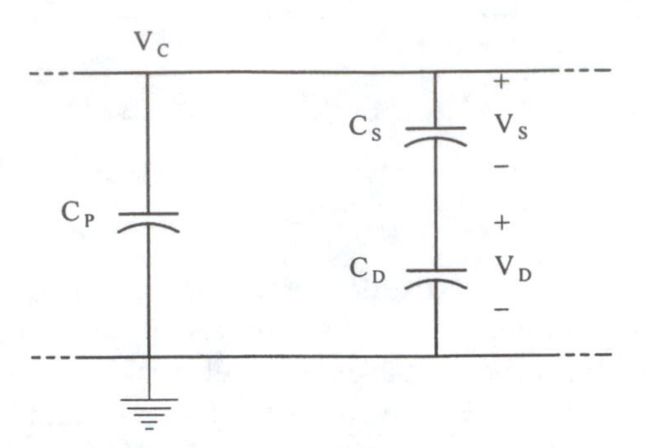

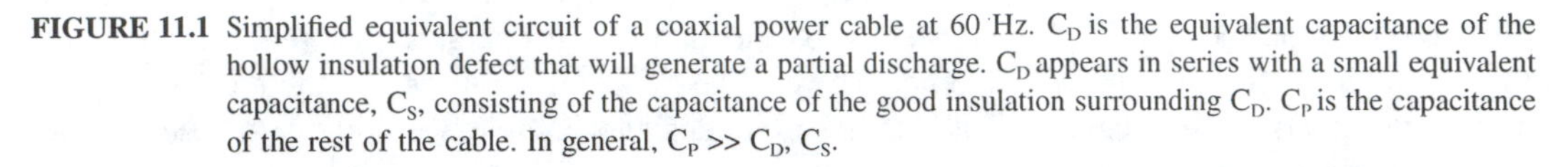

FIGURE 11.1 Simplified equivalent circuit of a coaxial power cable at 60 Hz. C_D is the equivalent capacitance of the hollow insulation defect that will generate a partial discharge. C_D appears in series with a small equivalent capacitance, C_S, consisting of the capacitance of the good insulation surrounding C_D. C_P is the capacitance of the rest of the cable. In general, $C_P >> C_D, C_S$.

excitation/measurement (E/M) end of the cable is typically about 1 mV peak amplitude and several hundred nanoseconds in duration. It is superimposed on the kilovolt range, 60 Hz cable excitation voltage, which presents a problem in its separation and measurement.

The direct PD transient first arrives at the excitation/measurement (E/M) end of the power cable at time T_0. The transient which travels to the far end of the cable is reflected at its open-circuit end, and then travels back past the PD site to the E/M end. The Thevenin impedance of the high-voltage source driving the E/M end of the cable is higher than the cable's Z_o, so that all PD transients arriving at the E/M end are reflected to the open-circuited, far end of the cable, where they are again reflected, etc. By examining the *reflection series* of a PD pulse at the E/M end of the cable, and knowing the physical length of the cable, it is possible to estimate the location of the PD site. A typical reflection series from a single PD, recorded from a 15 kV URD, XLPE dielectic, coaxial power cable in the author's laboratory is shown in Figure 11.2. Note how successive pulses are attenuated and broadened by the cable's low-pass filtering characteristics. In the field, if a PD is judged to be severe, the section of cable containing the PD site is excised, and a new cable section is spliced in, or the entire cable may be replaced with a new one.

The initial voltage transient at the PD site can be modeled, following the lumped-parameter development of Mason (1965). Three capacitors are used in Mason's model: C_D is the equivalent capacitance of the insulation defect giving rise to the PD. C_S is the equivalent series capacitance between C_D and the cable's outer shield and the cable's center conductor. C_P is the total cable capacitance, less C_S and C_D in series. Generally, $C_P >> C_S, C_D$. A PD occurs when the 60 Hz voltage on the cable reaches the inception voltage, V_{inc}. At this time, the dielectric of the insulation defect breaks down, and the voltage across C_D quickly goes to zero as its charge is neutralized. This situation is illustrated in Figure 11.3. At t = 0–, just before the PD occurs, the charge stored in C_P is $Q_P(0-) = C_P$ Vinc. It is easy to see that the charges stored on C_S and C_D are equal, and equal to $Q_{SD} = V_{inc}C_SC_D/(C_S + C_D)$. Thus the voltage across the defect cavity at breakdown is:

$$V_D = Q_{SD}/C_D = V_{inc}\frac{C_S}{C_S + C_D} \qquad 11.1$$

The voltage across C_S is:

$$V_S = Q_{SD}/C_S = V_{inc}\frac{C_D}{C_S + C_D} \qquad 11.2$$

Obviously, $V_{inc} = V_D + V_S$.

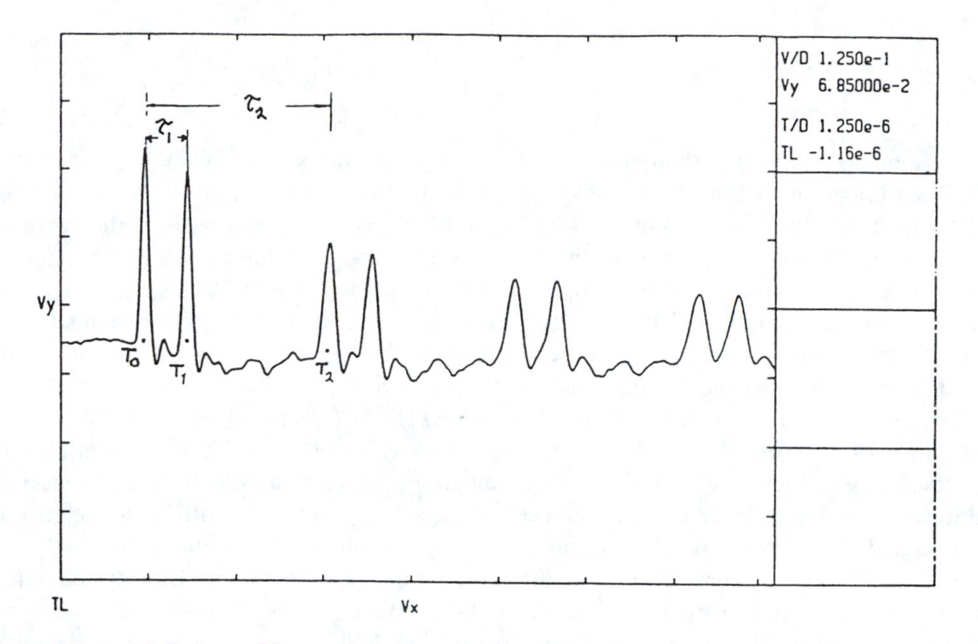

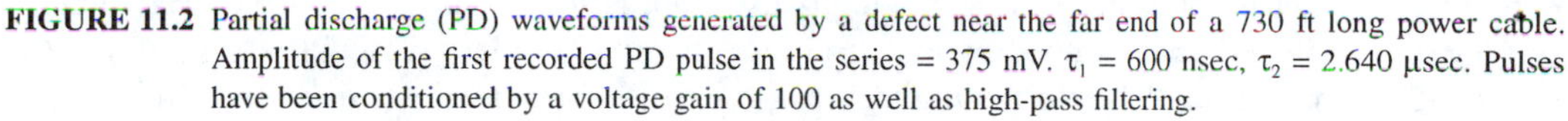

FIGURE 11.2 Partial discharge (PD) waveforms generated by a defect near the far end of a 730 ft long power cable. Amplitude of the first recorded PD pulse in the series = 375 mV. τ_1 = 600 nsec, τ_2 = 2.640 μsec. Pulses have been conditioned by a voltage gain of 100 as well as high-pass filtering.

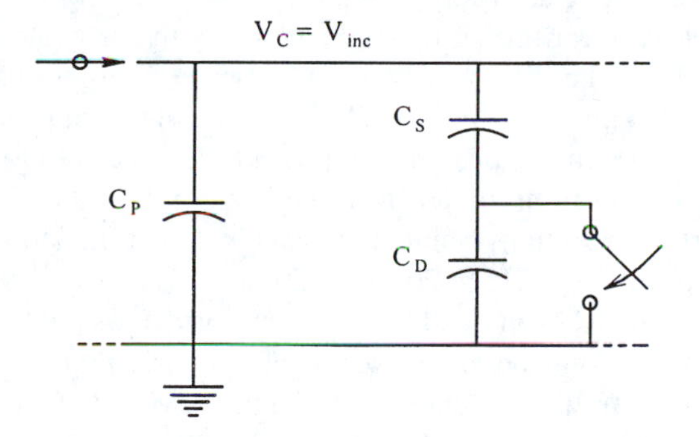

FIGURE 11.3 Equivalent circuit of a power cable when its applied voltage equals the inception voltage. The PD effectively instantaneously discharges (shorts out) C_D.

The PD occurs at t = 0. At t = 0+, $V_D \to 0$, $Q_D \to 0$, and $V_S \to V_{inc}$. Now the total charge on C_P and C_S is:

$$Q_{PS} = C_P V_{inc} + Q_{SD}\ (\text{on } C_S) \qquad 11.3$$

Assuming the transmission line series inductance prevents instantaneous longitudinal charge flow, the voltage across the PD site at t = 0+ is now:

$$V' = \frac{Q_{PS}}{C_P + C_S} = \frac{V_{inc}\left[C_P + C_S C_D/(C_S + C_D)\right]}{C_P + C_S} \qquad 11.4$$

The initial voltage change is nearly instantaneous, with a rise time estimated to be in picoseconds. The voltage change at the site is approximately:

$$\Delta V = V' - V_{inc} = V_{inc}\left[\frac{C_P + C_S C_D/(C_S + C_D)}{C_P + C_S} - 1\right] \cong \frac{-V_{inc} C_S^2}{C_P(C_S + C_D)} \qquad 11.5$$

Let us calculate a typical PD initial amplitude: Let C_P = 10 nF, V_{inc} = 2000 V, C_S = 1 pF, C_D = 3 pF. From Equation 11.5 above, ΔV = –50 mV. The 60 Hz excitation on the power cable is over 92 dB larger than the initial PD amplitude. As the PD "spike" propagates along the cable, it is attenuated and rounded due to the attenuation of high frequencies by the cable's transfer function. This attenuation and rounding are illustrated in the PDs shown in Figure 11.2.

Measurements on the PD reflection series requires the separation of the PD transients from the 60 Hz excitation voltage. Figure 11.4 illustrates a block diagram of the signal acquisition portion of the PD detection, location, and characterization system developed by Mashikian et al. (1989b). The cable is excited from a clean, variable-voltage, 60 HZ HV source. There must be no corona or PDs in the HV power source which can be confused with or mask PD signals from the cable. A Tettex AG type 3380/100/70, 10 nF, 70 kV, compressed gas standard high-voltage capacitor is used to couple voltages on the cable under test to the first high-pass filter (HPF1) used to condition the PD signal. A low-noise, op amp, signal conditioning amplifier (A1) with a gain of 10 is used to boost the PD signal at the output of HPF1. In the initial (1986) design of the system, an AD 380 op amp was used; at present, we use an AD 840 op amp because it has lower voltage noise and lower cost. A1 is followed by a second high-pass filter, HPF2, to further reduce 60 Hz voltage. The output of HPF2 is conditioned by a second broad-band amplifier (A2) with a gain of 10. A2 drives a 50 ohm coaxial cable to the digital storage oscilloscope (DSO) which serves to digitize and record the PD transients. Digital records from the DSO's memory or floppy disks are passed on to the Macintosh II PC used for further DSP operations by the general-purpose instrumentation bus (GPIB). The schematics of the HPFs are shown in Figure 11.5. Figure 11.6 shows the frequency response of the two cascaded high-pass filters and amplifiers. The gain of the combined filters and amplifiers is –221 dB at 60 Hz. In practice, more 60 Hz interference voltage is seen from common-mode, ground loop effects than from the 60 Hz voltage that passes through the filters. To reduce the common-mode, 60 Hz hum, a unity-gain, broadband isolation transformer (RFT) (North Hills Electronics, Inc. model 0016PA, 20 Hz to 20 MHz, 50 ohm Z_{in} and Z_{out}) is used.

The signal-to-noise ratio (SNR) of weak PDs is improved by signal averaging on a Nicolet Model 4094 DSO. Averaging is only possible given reliable, consistent triggering of the DSO sweep by the peak of the first (largest) PD transient recorded at the E/M end of the energized power cable. A special, narrow-bandpass filter is used on the PD waveforms to more reliably extract the triggering pulses for signal averaging on the DSO. When the PD SNR is too low for reliable averaging on the DSO, we use a cross-correlation technique on the PC to establish a reference time T = 0 at the start of each noisy PD pulse group so averaging can take place reliably. The DSO sampling period is normally set at 20 nsec/sample; 16,000, 8-bit samples are taken on a delayed input signal, so the full extent of the first (trigger) PD pulse can be seen.

After a suitable PD reflection series or averaged reflection series is acquired, it can be stored on the Nicolet DSO's floppy disks and then transferred by the IEEE 488 bus to the MacIntosh II PC for further DSP operations which include the PD site location algorithms. One location algorithm deconvolves the broad, recorded pulse series with the cable transfer function in order to better estimate the true times of occurence of the pulses in the reflection series. Another location algorithm uses a maximum likelihood (ML) method with cross-correlation (Knapp et al., 1990). The ML location algorithm has been found to be robust in the presence of noise. PD site location is based on the *a priori* knowledge of the cable's length. From the PD reflection series shown in Figure 11.2, we can find the PD propagation velocity on the cable from the fact that that it takes τ_2 seconds for a pulse to travel from the E/M end of the cable to the far end and then return, a distance of 2L meters. Thus:

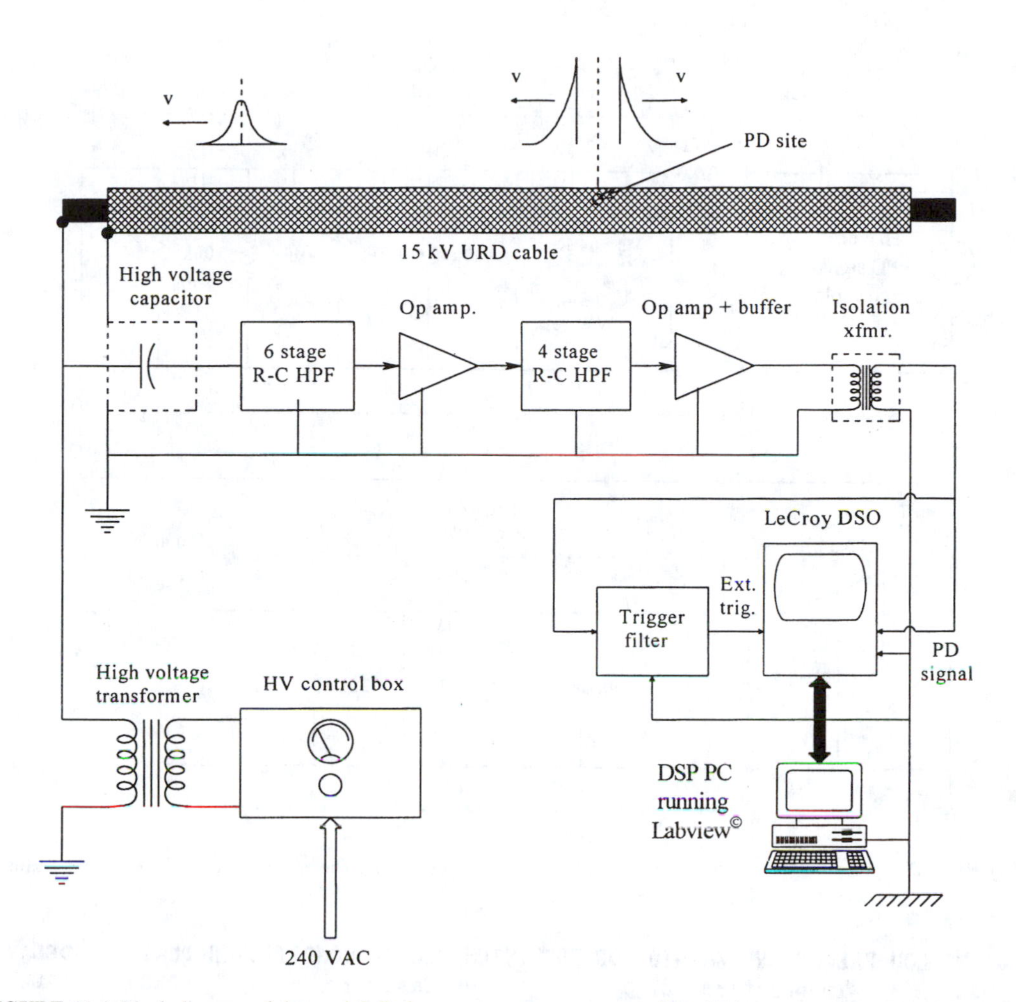

FIGURE 11.4 Block diagram of the partial discharge measurement system of Mashikian et al. (1990). The high voltage capacitor is 10 nF. The op amp gain stages each have a gain of 10, and –3 dB bandwidths of 5 MHz. The high-pass filters (HPFs) are shown in detail in Figure 11.5.

$$c_c = 2L/\tau_2 \ \text{m/s} \tag{11.6}$$

c_c on most XLPE high-voltage cables is about 200 m/μsec, or 0.667 times the speed of light.

The true time of origin of the PD is initially unknown, so we work with the time differences between the first three pulses in the reflection series. Thus:

$$\tau_1 = T_1 - T_0 \tag{11.7}$$

and

$$\tau_2 = T_2 - T_0 \tag{11.8}$$

The time differences are typically measured using the maxima of the pulses corrected for broadening and attenuation by the cable's transfer function. Equation 11.7 can be used to write:

$$\tau_1 = (2L - x_d)/c_c - x_d/c_c \tag{11.9}$$

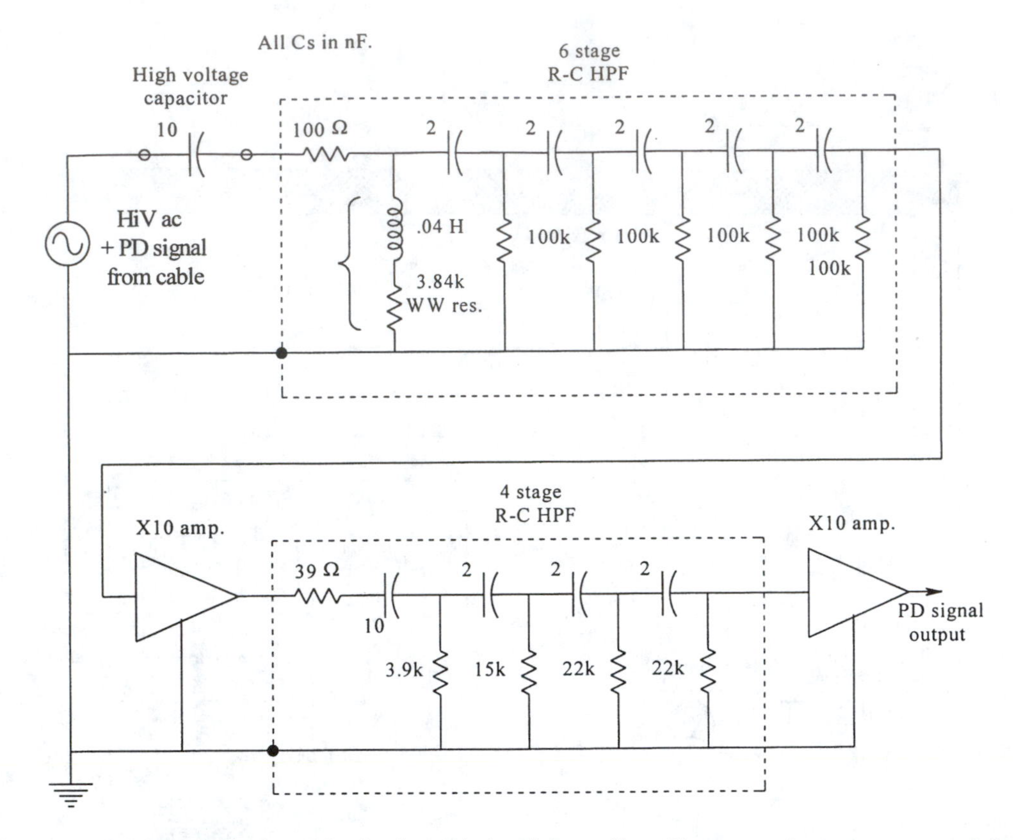

FIGURE 11.5 Schematic diagram of the signal conditioning high-pass filters. The op amp circuits are represented simply by gain of 10 VCVSs.

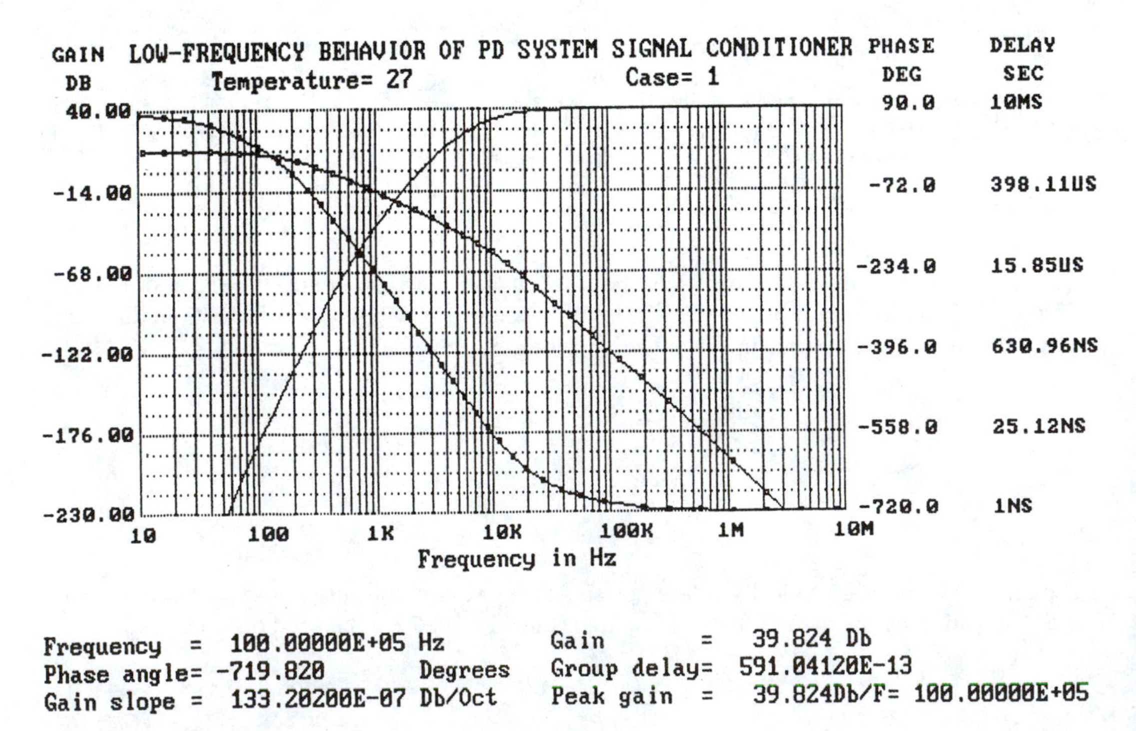

FIGURE 11.6 Frequency response of the circuit of Figure 11.5.

Equation 11.9 is solved for x_d, the distance from the E/M end of the cable to the site of PD origin. c_c is eliminated by substituting Equation 11.6 into Equation 11.9. The final result is:

$$x_d = L(1 - \tau_1/\tau_2) \quad 11.10$$

Thus, we have only to know the cable length, L, and estimate the time differences τ_1 and τ_2 in order to find an estimate for x_d.

In its present embodiment, the partial discharge measurement and locating system is able to locate artificial PD sites at known locations on a cable with an accuracy of 99.5% (Knapp et al. 1989). When the PD site is very near either end of the cable (<30 m from the ends), the PD pulse peaks merge, and estimation of the pulse origin times is not ordinarily possible. By splicing an extra 30 m of defect-free, identical cable to the far end of the cable under test, we obtain distinct separation of the PD pulses and improved location resolution for PD sites closer than about 30 m from the cable ends.

Another problem that affects the SNR of the system and hence the resolution of x_d is the pickup of coherent interference in the system's 20 kHz to 10 MHz signal bandwidth. Such interference is from RF sources such as local AM broadcast stations. In certain cases, when the cable is above ground, such RFI can be of the same order of magnitude as the PD reflection series, confounding measurements. Our approach to this problem is to do a spectral analysis of the system noise to locate the frequencies of the coherent interference; we then implement one or more digital, FIR notch (band-reject) filters on the system computer to remove the interference, leaving the PD waveforms largely intact. Of course, we follow good grounding, shielding, and guarding practices. As mentioned above, we also use a broadband isolation transformer to reduce common-mode noise. However, RFI can be picked up by the cable under test in spite of these practices. System broadband random noise arises in the headstage amplifier and in the high-pass filters. We now use AD840 op amps for broadband amplifiers. These op amps have f_T = 400 MHz, η = 400 V/μsec, and e_{na} = 4 nV/$\sqrt{Hz}$. Their broadband noise is 10 μV rms in a 10 Hz to 10 MHz bandwidth.

Our present instrumentation allows the detection and reliable site estimation (x_d) of PD pulses as low as 80 μV peak at the E/M end of the cable using a 5 MHz signal bandpass. Although the estimated PD site location is found in terms of percent of the total, known cable length, the actual location of the site is complicated by the fact that the cable plunges into its trench and then snakes a certain amount along its path. In practice, we dig a hole to the cable at the expected site, and then inject a transient test pulse into the unenergized cable by magnetic induction. A fast current pulse is passed through a coil with about 25 turns around the outside of the cable. This induces a PD-like transient, and a corresponding reflection series which can be compared to the stored PD series. A coincidence of pulse peaks in the two series means the test coil is very near the PD inception site. Accuracy with the current pulse "vernier" method is limited by the DSO sampling rate. For repetitive signals, such as the current test pulses, the effective sampling period of the DSO is 2 nsec/S. This interval corresponds to a distance on the cable of 0.4 m, giving a comparison location error of 0.2% on a 200 m cable.

The results of ongoing development of DSP algorithms for pulse time difference estimation suggest that a routine location accuracy of 99.8% will be possible. This accuracy translates into a ±0.5 m error in finding a PD site on a 250 m cable. The use of sophisticated DSP filtering techniques on noisy data is what has made the PD measurement instrument a successful, field-portable system.

11.2 DESIGN OF A CONDUCTIVITY-BASED FISH COUNTER

In this example, we describe the design of a system that detects the transient conductivity change when a fish swims in water flowing between two electrodes. The fish counter system was developed as part of a research project supported by the Northeast Utilities Service Corp., in which it was desired to develop a simple yet reliable, automatic means of counting the number of adult shad

(*Alosa sapidissima*) passing through a downstream fish bypass. The downstream bypass at Holyoke, MA, returned downstream-migrating, spawned shad to the Connecticut River from an industrial canal system in which they became trapped. It is not our purpose here to discuss the details of how the fish entered the canals, or the system by which they were forced to enter the downstream bypass entrance.

After entering the bypass entrance, the shad were flushed by a heavy flow of water down a two-foot diameter steel pipe, eventually reaching the Connecticut River. A horizontal section of the steel pipe was replaced with an insulated, three foot long, fiberglass pipe inside of which were two, 1" wide, stainless steel hoop electrodes placed 20 inches apart. Under normal operation of the fish bypass, the pipe is full of water flowing at about 6 ft/s.

The principle of operation of the fish counter is that in the steady state, with no fish passing between the electrodes, the self-balancing AC conductance bridge nulls itself regardless of the conductance of the water/electrode circuit. When one (or more) fish passes between the electrodes, it causes a transient change in the conductivity, ΔG_x, of the water between the electrodes (generally an increase in Connecticut River water). This ΔG_x transiently unbalances the bridge, and the resultant ouput transient is discriminated and counted. The self-balancing dynamics are relatively slow (steady state is reached in about 2 sec.). The self-balancing feature is intended to maintain maximum bridge sensitivity at null as the conductivity between the electrodes changes slowly due to water temperature, salinity, and level changes (Northrop et al., 1982).

Figure 11.7 shows a simplified circuit diagram of the self-balancing bridge. The bridge excitation signal is a 2 Vppk, 180 Hz sinewave. OA-4 is a buffer/driver for the "hot" electrode. The summing junction of OA-1 is at virtual ground; OA-1 is a current-to-voltage converter whose AC output voltage is proportional to the conductivity between the electrodes. DA is an IC, differential, instrumentation amplifier that subtracts the AC output of OA-1 from the AC output of the variable-gain element, AM (an analog multiplier). The AC output of DA is a double-sideband, supressed carrier signal. It is detected by a phase-sensitive rectifier IC (PSR) (such as the AD630) followed by a low-pass filter. The DC output of the low-pass filter is proportional to the difference between the outputs of OA-1 and the AM; it is the system error signal. OA-2 is an integrator, and OA-3 shifts the DC level of the integrator output in order to obtain a DC bias at the analog multiplier's DC input.

Under steady-state conditions, the DC error signal is integrated by OA-2. The DC voltage, V_3, changes the AC output of AM so the error signal (αV_1) goes to zero. This self-balancing action occurs over a wide range of conductance between the electrodes. The capacitor in the feedback path of OA-1 is adjusted to remove any phase difference between V_x and V_r. Such a phase difference can arise from the capacitive action of the water/electrode interface.

Under normal conditions, a fish passing between the electrodes causes a transient increase in the conductivity for about 100 msec. The self-balancing bridge cannot respond fast enough to null this transient, and a voltage pulse appears at V_o. Because of water turbulence, there is also a constant, broadband noise voltage at V_o. In order that a passing fish be counted, the bridge's output pulse must be above a preset DC threshold set above the noise peaks. When a pulse exceeds this threshold, an analog comparator output goes high, triggering a one-shot multivibrator to produce a standard 1 msec TTL pulse which is counted by a dedicated personal computer.

Figure 11.8 shows the equivalent system's block diagram of the fish counter bridge. The dynamic response of the bridge to small conductance changes can be derived from this diagram. The bridge output voltage is:

$$\begin{aligned} V_o &= K_a K_d (V_x - V_r) \\ &= K_a K_d (-V_1 G_x / G_r - \alpha V_1 V_3 / 10) \end{aligned} \quad 11.11$$

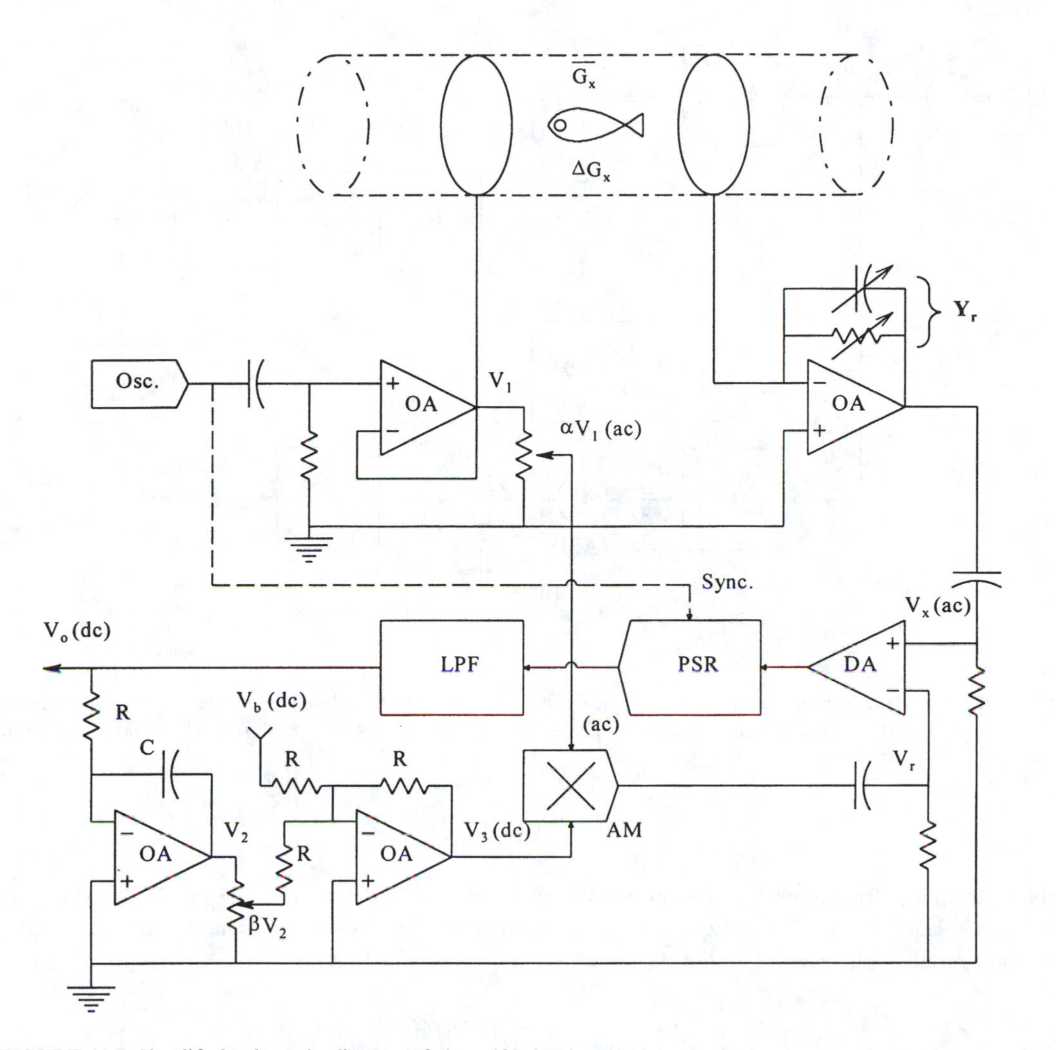

FIGURE 11.7 Simplified schematic diagram of the self-balancing conductance bridge used in the fish counter. The oscillator is 180 Hz, phase locked to the 60 Hz powerline (see text). LPF, Sallen and Key low-pass filter with 20 Hz fn; PSR, phase sensitive rectifier; AM, analog multiplier. (From Northrop, R.B., *Analog Electronic Circuits: Analysis and Applications*, Addison-Wesley, Reading, MA, 1990. With permission.)

For simplicity, we assume here that Y_x and Y_r are pure conductances, G_x and G_r, and we neglect the PSR's low-pass filter's dynamics. The DC control voltage, V_3, can be written:

$$V_3 = -\left(V_b - \alpha V_o / sRC\right) \tag{11.12}$$

Substituting Equation 11.12 into Equation 11.11, we find:

$$V_o = \frac{sK_aK_dV_1\left[\alpha V_b/10 - \left(\overline{G}_x + \Delta G_x\right)/G_r\right]}{s + K_aK_d\alpha\beta/(10RC)} \tag{11.13}$$

In this transfer function, both V_b and $\overline{G}_x$ are constants. Application of the Laplace final value theorem shows that the steady-state value of V_o is zero, hence self-balancing. If we make V_b equal to:

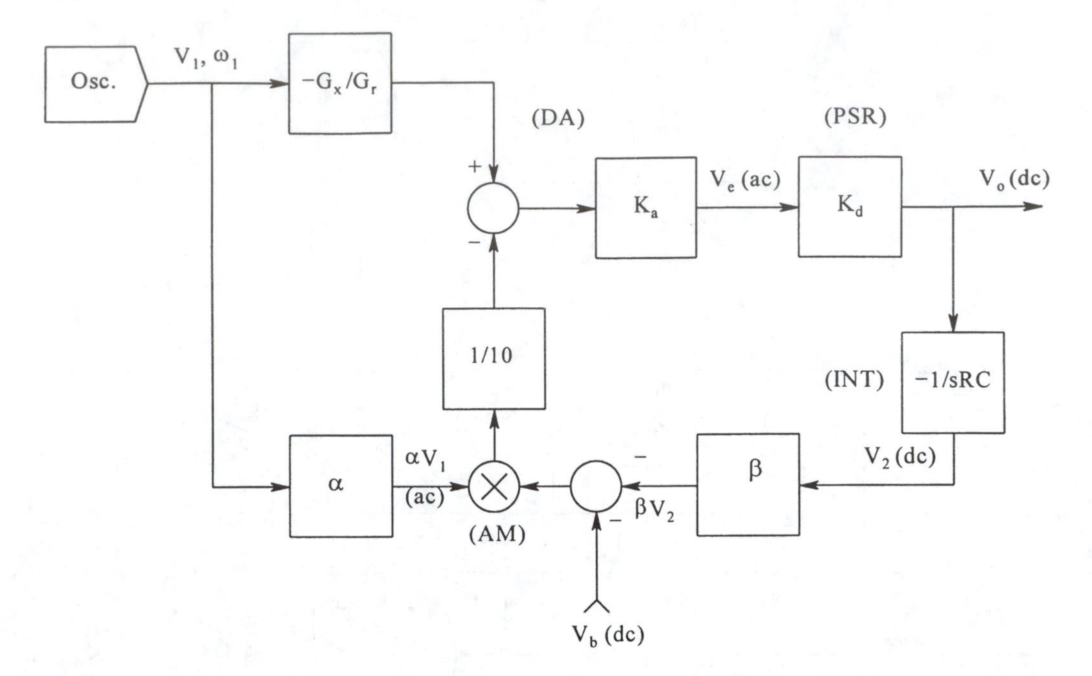

FIGURE 11.8 Systems block diagram describing the dynamics of the conductance bridge of Figure 11.7. (From Northrop, R.B., *Analog Electronic Circuits: Analysis and Applications*, Addison-Wesley, Reading, MA, 1990. With permission.)

$$V_b = 10G_x/(\alpha G_r) \qquad 11.14$$

then the integrator output has a symmetrical swing around 0 V for Gx = $\overline{G}_x$ ± ΔGx. When we substitute Equation 11.14 into Equation 11.13, and treat the parametric conductance change, $\Delta G_x(t)$, as the system input, we obtain the transient unbalance transfer function:

$$\frac{V_o}{\Delta Gx}(s) = \frac{sK_aK_dV_1/G_r}{s + K_aK_d\alpha\beta/(10RC)} \qquad 11.15$$

Note that this transfer function is that of a simple high-pass filter; it has a flat frequency response above its break frequency given by 20 $\log(K_aK_dV_1/G_r)$ dB. A step input, $\Delta G_x(t) = \Delta G_xU(t)$, gives an exponential voltage output pulse given by:

$$V_o(t) = -\Delta G_x(K_aK_dV/G_r)\exp[-(K_aK_dV_1\alpha\beta/10RC)t] \qquad 11.16$$

The autobalancing conductance bridge can be made very sensitive; inspection of Equation 11.16 shows that if $G_r = 10^{-4}$ S, $K_a = 100$, $K_d = 10$, and $V_1 = 2$ V, then a conductance step of $\Delta G_x = 5.0 \times 10^{-8}$ S will produce $V_o = 1$ V peak. We found experimentally that $\overline{G}_x$ was typically 0.004 S, and the peak $\Delta G_x/\overline{G}_x$ for shad was about 2%. $\Delta Gx/\overline{G}_x$ for the smaller blueback herrings was about 0.4%.

There are several sources of error in operating the conductivity-based fish counter. These include the baseline noise, which can cause infrequent false counts and false omissions. A second source of error is the attitude of the fish in the water flowing past the electrodes. Nearly all the fish are unconscious, being temporarily stunned by the anodal electrotaxis system that attracts them to the bypass (Northrop, 1967). Therefore some fish pass through the electrodes at pitch and yaw angles ranging from 0° to 90° or more. A fish perpendicular to the water flow will produce a smaller pulse than one parallel to it. Occasionally, we have seen a fish recover from electronarcosis and begin to swim as it enters the fiberglass pipe. Swimming produces a nearly sinusoidal voltage which reaches the pulse counting threshold, giving several false positive counts. A third source of error is from two or more fish passing the electrodes at once, giving one large, irregular pulse. A fourth source of error comes from blueback herring, fish about half the size of shad that sometimes migrate with shad. A large herring can be confused with a small shad. Generally, the counter threshold was set high enough to ignore about 98% of the herring pulses. A fifth source of error was found to be excess V_0 baseline noise caused by water turbulence and bubbles passing through the electrodes under conditions of a not completely filled bypass pipe. Such excess noise occurred infrequently when the water level in the canal fell below the normal, regulated level.

In order to "calibrate" the statistics of the counter, the following distasteful procedure was followed in June 1982. Freshly dead shad, taken from the upstream fish elevator system at Holyoke, were dumped into the bypass entrance in groups of 5, 10, 25, 50, and 100 fish. The counter count was recorded for each group, and a regression analysis was done. The coefficient of determination, r^2, was found to be 0.987, the Y-intercept of the regression line was 0.446, and its slope was 0.910. The percent standard deviation was 2.11%. $N = 31$ groups of fish were used. The counter was tested with groups of fish because fish attracted to the bypass were stunned and entered it in small schools. A vortex or whirlpool at the bypass entrance acted to separate the fish so that they generally entered the pipe one at a time; an example of serendipity in design.

In the early field trials of the fish counter bridge, we observed that at certain times, a slow, periodic variation of the V_0 baseline occurred, in addition to the noise from water turbulence, etc. The slow baseline oscillations effectively modulated the pulse discrimination threshold, giving rise to gross counting errors. Because the counter was physically housed in the alternator room of a hydroelectric generator station, we suspected that third harmonics from powerline frequency magnetic fields were somehow mixing with the nominal 180 Hz local oscillator used for the bridge. The solution to this problem was to use a phaselock loop to synchronize the 180 Hz bridge oscillator to *exactly* three times the local 60 Hz powerline frequency. Once this change was made, there was no more trouble from low-frequency beat voltages in the bridge output.

In summary, the conductivity-based fish counter was found to be accurate and easy to use. The major disadvantage to its design was that nonstationarity in its noise caused the counter's statistical accuracy to change. Thus, its effective use required constant flow conditions in the bypass pipe.

11.3 DESIGN OF A NO-TOUCH, OCULAR PULSE MEASUREMENT SYSTEM

This section describes the design of an experimental, noninvasive medical instrument which can measure the minute pulsations of the clear, elastic cornea of the eye as the heart forces blood into the internal circulation of the eye, including the retina. The waveshape and amplitude of the ocular (corneal) pulse (OP) has been shown to be an important indicator of intraocular blood pulse pressure and flow, and as such, can be used as a diagnostic tool for the detection of glaucoma, or of arteriosclerosis of the internal carotid arteries which supply blood to the eyes and the brain. The impairment of blood flow in the internal carotid arteries by atherosclerotic plaques can lead to strokes, and can be symptomatic of cerebrovascular disease. The ocular pulse waveform follows the period of the heartbeat which in general, is not constant. The normal, peak-to-peak deflection of the cornea is approximately 15 μm in both rabbits and humans. Glaucoma (elevated intraocular pressure) reduces the amplitude of the OP, and alters its waveform.

Many techniques for the measurement of ocular pulse have been developed; all require physical contact with the cornea, which in turn requires sterile technique and anesthetization of the cornea. A number of these techniques have been reviewed by Northrop and Nilakhe (1977). The no-touch ocular pulse system described in this section measures the relative displacement of the cornea with repect to a pair of ultrasound transducers suspended in front of the eye. 900 kHz, CW ultrasound is directed at the cornea through the air. A small fraction of the ultrasonic energy is reflected from

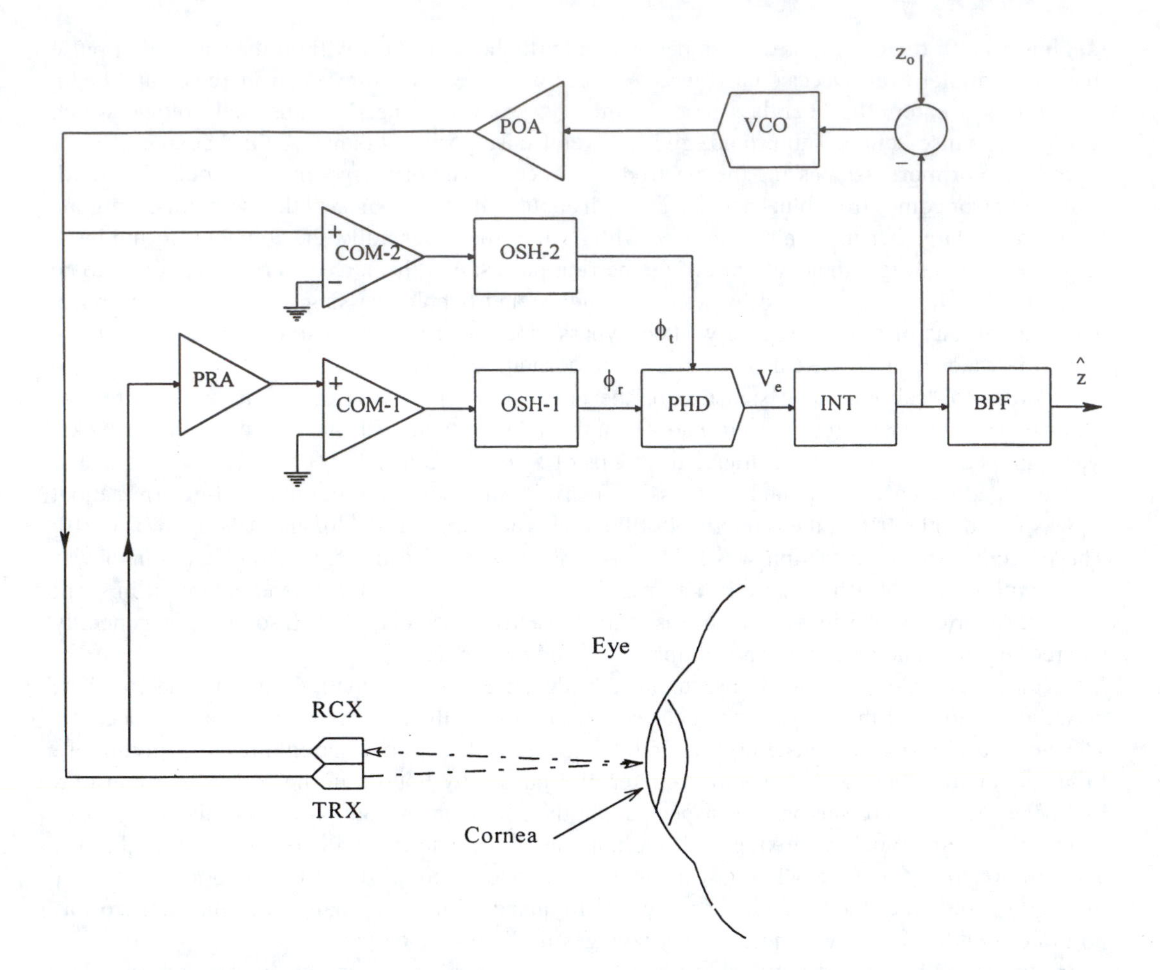

FIGURE 11.9 Block diagram of the no-touch, ocular pulse measurement system of Northrop and Nilakhe (1977). See text for description.

the cornea back to a receiving transducer. Both transducers are LZT disks, 1/4" in diameter, with 3 MHz resonant frequencies. The sensing of the ocular pulse is accomplished by a feedback system, shown in Figure 11.9.

A conventional voltage-controlled oscillator (VCO) with sinusoidal output is the signal source for a power amplifier (POA) which drives TRX, the transmitting transducer which is suspended about 1 cm from the cornea. About 3 mW of ultrasonic energy is radiated from TRX. Passing through the air, the ultrasound waves are largely reflected from the cornea because of the gross mismatch in the acoustic impedance of air vs. the cornea. The reflected ultrasound signal is converted to a voltage by the receiving transducer (RCX) and amplified by a low-noise preamplifier, PRA. PRA's sinusoidal output is converted to a TTL squarewave by a high-speed, amplitude comparator with hystereses, COM1. COM1 acts as a zero-crossing detector. Its output is conditioned by a one-shot multivibrator (74LS123) (OSH-1) giving narrow, 20 nsec, complimentary output pulses. Similarly, the reference sinusoidal output of the POA is converted to a TTL squarewave by COM2 and thence to narrow complimentary pulses by OSH-2. The narrow, complimentary pulses, ϕ_1 and ϕ_2, are the inputs to a simple R-S flip-flop phase detector, PHD. (See Section 8.7 and Figure 8.55 in this text for a detailed description of this type of phase detector.) The PHD output is a heart-rate signal, V_e. V_e is integrated by an analog (op amp) integrator, and the integrator output is fed back to the VCO. The time-varying part of the integrator output can be shown to be proportional to the small changes in the total ultrasound airpath length, hence the ocular pulse. The integrator output, z, is conditioned by a band-pass filter (BPF) which blocks DC and frequencies above 30 Hz. The output of the BPF, z_o, is the system output.

The action of the feedback system is to preserve the same phase relation between the ϕ_1 and ϕ_2 signals. Hence, if the cornea bulges outward, the total ultrasound airpath is incrementally shortened. The phase tracking loop raises the VCO output frequency slightly so that the same number of wavelengths of sound are present in the airpath. When the cornea flattens, the airpath lengthens, and the phase tracking loop adjusts the VCO frequency to a slightly lower value, to again preserve the same number of sound wavelengths in the airpath. The dynamics of the system are nonlinear. However, because $\Delta X/\overline{X}$ is on the order of 0.15%, a linear relation is observed between z_o and ΔX. Northrop and Nilakhe (1977) show that the steady-state integrator output can be written as:

$$z = \bar{z} + \Delta z = \left[z_{dc} - \frac{2\pi(N + 1/2)c}{2\overline{X}K_3} \right] + \Delta X \left[\frac{2\pi(N + 1/2)c}{2\overline{X}^2 K_3} \right] \tag{11.17}$$

Here N is the integer number of ultrasound wavelengths in the average airpath, $\overline{X}$, K_3 is the VCO gain in r/s/V, c is the velocity of sound in air, and z_{dc} is the DC input to the VCO that sets the desired output frequency and N for $\Delta X = 0$. N can be found from:

$$N = \text{INT}\left[\frac{2\overline{X}K_3 z_{dc}/2\pi}{c} \right] \tag{11.18}$$

The zero-deflection, closed-loop VCO frequency was shown to be given by:

$$\omega_{CL} = \frac{2\pi(N + 1/2)c}{2\overline{X}} = K_3\bar{z} \ \text{r/s} \tag{11.19}$$

Figure 11.10 illustrates a simplified systems block diagram of the ocular pulse measurement system. The blocks to the left of the dotted line describe the propagation delays due to the finite velocity of sound, and the Doppler shift of the moving cornea. The 1/s blocks represent the conversion of frequency to phase, which is detected by the PHD. Subtraction of the $2\pi(N + 1/2)$ phase term is required to account for the transport lag in the airpath and the phase detector's periodic characteristic (see Figure 8.55C in this text).

In evaluating the prototype no-touch ocular pulse measurement system, we found that the output signal was very noisy. Sources of noise in the output included electronic noise from the preamplifier, micronystagmus of the eye, and fluttering of the eyelids. Micronystagmus is a constant, naturally occurring, small motion of the eyeball; eyelid flutter occurs when we try to conciously hold our eyes open for half a minute, while fixating our gaze on a spot. To reduce this noise to a level that allowed clinically effective recording of the ocular pulse, we averaged the OP over 32 heartbeats (pulse cycles), and triggered the signal averager with the QRS spikes from the subject's electrocardiogram. Signal averaging reduced the noise on the OP waveform to about an equivalent of ±1/4 μm on a 15 μm ocular pulse waveform. Everyone's heartbeat is irregular, the period being modulated even under resting conditions by anxiety, breathing, or breath holding, and unknown causes. The exceptions to this rule are persons with pacemakers, or persons taking cardioregulatory drugs, such as digoxin. Thus, in cases where a subject's pulse was very irregular, there was little improvement in OP SNR by averaging.

The no-touch ocular pulse system was first evaluated on the eyes of anesthetized, normal, adult rabbits. Little variation of the OP waveshape was seen between individuals. When informed, consenting, normal, conscious human subjects were used, we observed a wide range of OP waveforms between individuals. An individual's OP waveshape remained relatively the same from day

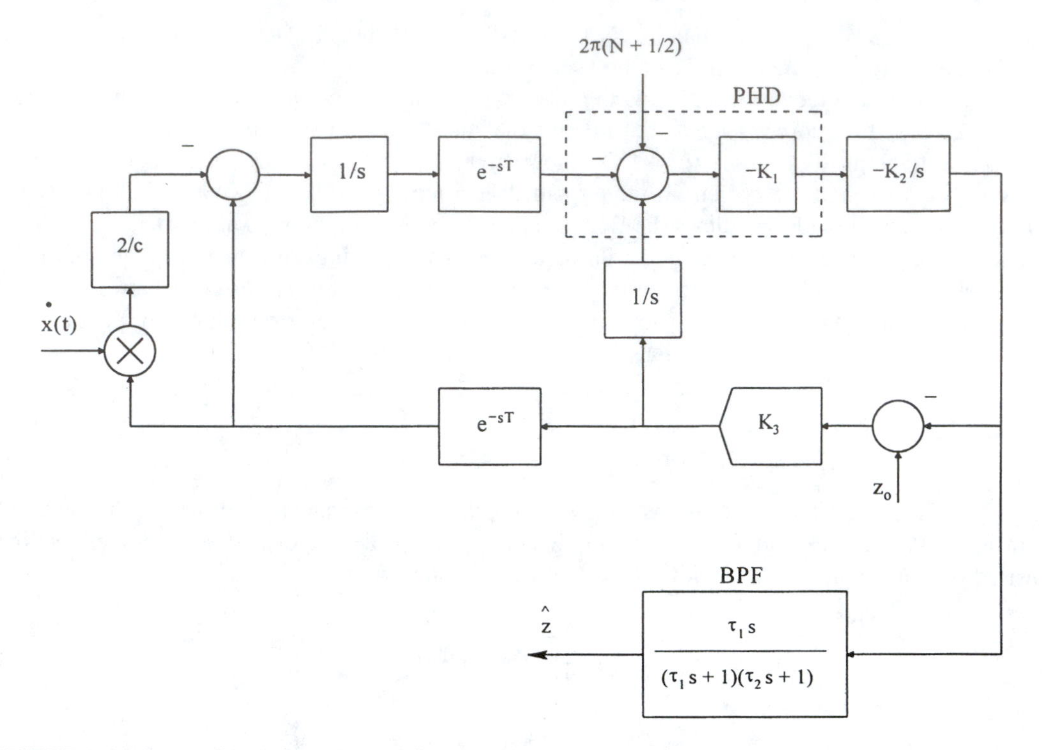

FIGURE 11.10 Systems block diagram describing the dynamics of the system shown in Figure 11.9. $\dot{x}$ is the corneal velocity input. (From Northrop, R.B. and Nilakhe, S.S., *IEEE TRANS. BIOMED. ENG.*, 24(2):139–148, 1977. With permission.)

to day. We found that the amplitude of the OP wave is very labile, depending on the CO_2 level in the subject's blood. By holding the breath for as little as 30 s, blood CO_2 would rise, and a subject's central nervous system (CNS) compensated for the rise by increasing the circulation in the CNS by dilating cerebral arteries. This dilation caused a reduced pressure in the internal carotid system, hence a lower pulse pressure to the eye, and a lower amplitude OP. This effect was verified by having OP subjects breath known air/CO_2 gas mixtures (Northrop and Decker, 1977). Figure 11.11 shows averaged OP waves from two normal individuals; see caption for details.

In closing, it should be noted that the system architecture of the no-touch OP measurement system shown in Figure 11.9 can be adapted to other wave modalities and frequencies to measure small displacements. Using 40 kHz ultrasound, the system was adapted to make an infant apnea and convulsion monitor (Northrop, 1980). We note that microwaves can be used in a similar, closed-loop, phase tracking system to penetrate soil and rubble in order to detect living persons buried by earthquakes or avalanches. The ΔX signal could come from respiration movements and heartbeats.

11.4 DESIGN OF A CLOSED-LOOP, CONSTANT PHASE, PULSED LASER RANGING SYSTEM AND VELOCIMETER

Laser velocimeters are used by police to measure vehicle velocities on highways. They are also used to measure the relative velocity and range between two space vehicles during docking maneuvers. In the police application, the pulsed laser is generally directed from a stationary transmitter/receiver in line with vehicle travel, as from a bridge overpass or on a curve, to a nonspecular surface on the moving vehicle. A telescope collimated with the laser is used to collect the reflected signal and direct it to a fast, sensitive photodetector, such as an avalanche photodiode. There is a Doppler shift on the frequency of the return signal, but this is not what is measured. The time delay between transmission and the detection of the return pulse is what is quantified. In

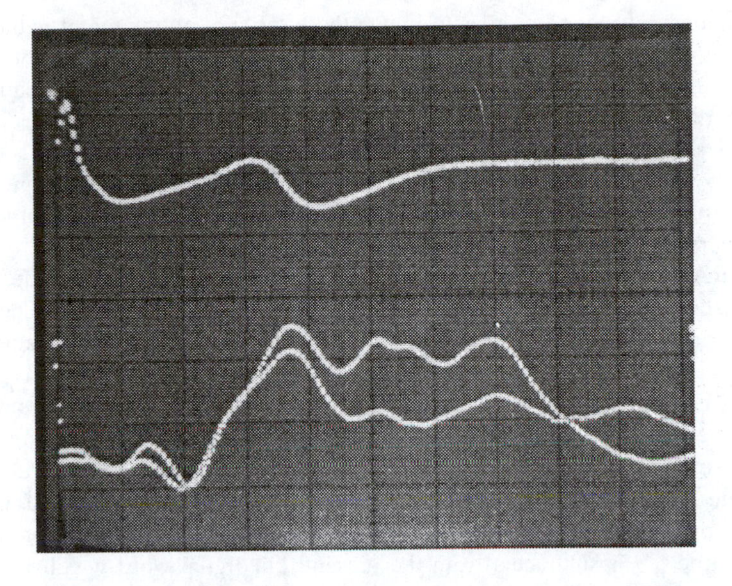

A.

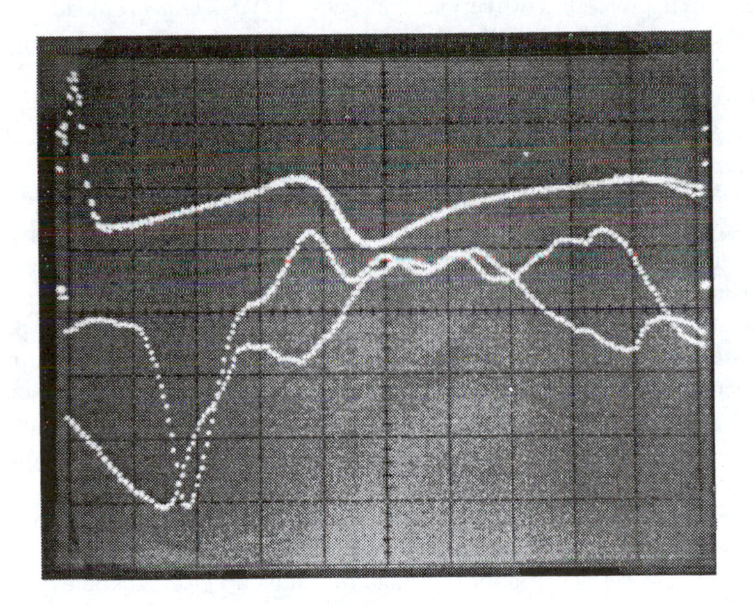

B.

FIGURE 11.11 Ocular pulse waveforms obtained by averaging the ocular pulses from each eye for 32 heartbeats. Vertical scale: 10 μm/cm, upward = corneal expansion. Horizontal scale: 80 ms/cm. (A) OPs from a 20-year-old male. Top trace, subject's ECG; second trace, OP of right eye; lowest trace, OP of left eye. Note that the OP waveshapes are similar, but the right eye's OP is larger. The reason for this inequality is unknown. (B) OPs from a 39-year-old female. Same scale. Top trace, ECG; second trace, OP of left eye; third trace, OP of right eye. Baseline separation is arbitrary. Note the unique, fast, inward deflection of the left cornea following the QRS complex of the ECG. The cause of this feature, too, is unknown.

3.33 nsec, light travels 1 m. Unlike Doppler radar velocimeters, LAser VElocimeter and RAnging (LAVERA) systems are highly directional, subject only to the transmitted laser beam's divergence. It is this specificity that reduces target ambiguity when one or more other vehicles are in close proximity to the desired target. In order to image the laser spot on the target vehicle, the vehicle surface should be a nonspecular (unlike a mirror), diffuse reflector with a high albedo (brightness). A mirror-like finish at an angle to the transmitted beam will cause the reflected beam to be directed away from the receiver telescope, and not be sensed, generating a "stealth" condition. Thus, windshields and chrome bumpers in modern automobiles make poor target surfaces. Black vinyl bumpers are also poor targets because of their low albedos. Such poorly responding targets are called "uncooperative." A retroreflecting surface makes an ideal, "cooperative" target, as nearly all of the laser's energy is directed back parallel to its beam.

Early LAVERA systems used by police generally operate in an open-loop mode, the laser being pulsed at a constant rate (e.g., 381 pps in a Laser Atlanta Optics, Inc. unit). The time between transmitting any one pulse and the detection of its return image from the moving vehicle is measured and stored in a digital memory. One way of measuring the return time, which is proportional to the target range, is to have the kth transmitted pulse start a fast Miller integrator circuit (similar to the horizontal sweep circuit in an analog oscilloscope) generating a ramp of voltage. A fast, track-and-hold (T&H) circuit follows the ramp until the kth return pulse triggers the T&H circuit to hold the ramp voltage at that instant. The held ramp voltage is then converted to digital form by an ADC, whose output is stored in digital memory. The previous, $(k-1)$th digital word, proportional to the range at time t_{k-1} is subtracted from the present, kth digital word which is proportional to the range at time t_k. The difference, appropriately scaled, is proportional to vehicle velocity. In practice, a finite impulse response (FIR) smoothing routine operates on the sequence of differences to reduce noise. This approach has two serious drawbacks: (1) it does not provide range information, and (2) its bandwidth, hence its ability to respond to sudden changes in vehicle velocity, is limited by the low, constant pulse rate of the laser source. Another open-loop laser LAVERA system, described by Koskinen et al. (1992), used a higher, fixed, pulse rate (between 1 and 50 Mhz) and an emitter coupled logic (ECL) time to amplitude converter (TAC), which is in effect, a precise phase detector.

The components of a typical LAVERA system consist of a pulsed, laser diode source, usually about 1.5 μm wavelength, collimating optics, a target (moving or stationary, cooperative or uncooperative), a receiving telescope, a high-speed photosensor, an electronic signal conditioning amplifier, a time to voltage converter to estimate target range, signal processing circuitry (digital and/or analog), and a readout display. In the description of the closed-loop, constant-phase LAVERA system below, we describe the systems architecture of the range and velocity estimating circuits, and do not consider the details of the pulse circuits, the laser, optics, the sensor, or preamplifier.

Figure 11.12 shows a block diagram of our system. There are many possible digital phase detector circuits that can be used. We use a digital phase detector based on the Motorola 4044 IC (Motorola Semiconductor Products, Inc., Phoenix, AZ); it compares the phase between the kth transmitted pulse and its "echo" received $\tau_k = 2L_k/c$ second later, where L_k is the distance to the target (assumed constant over τ_k), and c is the velocity of light in air. If the frequency of the pulses, f_c, is constant, the phase lag between received and transmitted pulses can be expressed as:

$$\varphi = 360\frac{2L}{cT_c} = 2.40\times10^{-6}\ L\ f_c \text{ degrees (L in m, } f_c \text{ in Hz)} \qquad 11.20$$

The *average* voltage at the output of the low-pass filter conditioning the output of the digital phase detector is $V_\varphi = K_d\varphi$. V_φ is next conditioned by a simple linear integrator, K_i/s. The output of the integrator, V_L, has a DC level, V_r, subtracted from it to form V_c, the input to the *voltage-to-period converter* (VPC). Do not confuse the VPC with the well-known VCO or voltage controlled oscillator IC. The frequency output of a VCO behaves according to the rule: $f_c = K_V V_C + c$. In the VPC, it is the *period* of the oscillator output which is directly proportional to V_c. Thus, for the VPC we can write:

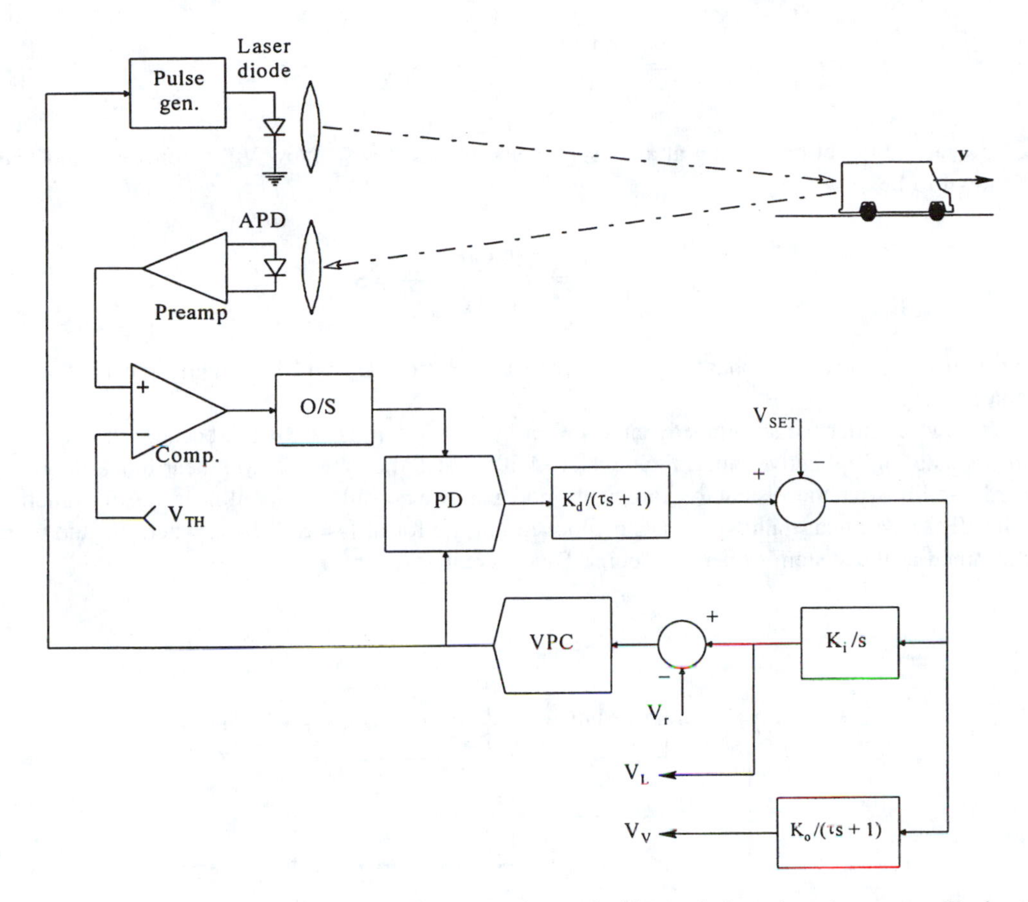

FIGURE 11.12 Block diagram of a closed-loop, constant-phase, pulsed laser velocity and ranging system. The heart of the system is the voltage-to-period converter-controlled oscillator (VPC). PD is the digital phase detector. V_V is the output voltage proportional to target velocity, V_L is the output voltage proportional to target range, L.

$$T_c = 1/f_c = K_V V_C + d \qquad 11.21$$

The VPC output serves to pulse the laser diode and act as a reference input to the phase detector IC. The entire system operates as a closed-loop, constant-phase control system. The period of the VPC output is automatically adjusted as L changes to maintain a constant φ and V_φ. Because the closed-loop system has a pole at the origin of the s-plane, it is a so-called type I control system that has zero steady-state phase error for constant L (stationary targets), and a small steady-state phase error for targets moving at a constant velocity ($dL/dt = v_o$). We can use this general property of a type I control system to examine the steady-state relationships governing the output voltage proportional to velocity, V_V, and the output voltage proportional to range, V_L. If the target is at rest at range L_o, then V_V will be zero, and the steady-state V_L is found to be:

$$V_L = \frac{K_d 2.40 \times 10^{-6} L_o}{K_V V_{SET}} + (V_r - d/K_V) = S_L L_o + \lambda \qquad 11.22$$

where λ is a constant. Let the desired, steady-state phase difference in the system be φ_o. Because $V_i = 0$ in the steady state, $V_\varphi = K_d \varphi_o = V_{SET}$. If we substitute this relation for V_{SET} into Equation 11.21 above, we obtain:

$$V_L = \frac{2.40 \times 10^{-6} L_o}{K_V \varphi_o} + (V_r - d/K_V) \tag{11.23}$$

Now let the target be moving at a constant velocity, $dL/dt = v_o$. Now V_V is non-zero, and may be shown to be:

$$V_V \cong \frac{dV_L/dt}{K_i} = \frac{2.40 \times 10^{-6} v_o}{K_i K_V \varphi_o'} = S_V v_o \tag{11.24}$$

Note that the steady-state phase lag, φ_o', when $L(t) = v_o t + L_o$, will be slightly greater than φ_o when $L = L_o$.

We now consider the system's dynamics. Clearly, the system is nonlinear, because the input, L, appears as a multiplicative input. Figure 11.13A illustrates the system's nonlinear block diagram. In order to linearize the system to examine its loop gain, we consider the nonlinear transfer function of the VPC as a linear gain with a DC component, of the form, $f_c = S_V V_c + e$, where S_V and e are determined at the system's operating point. By differentiation:

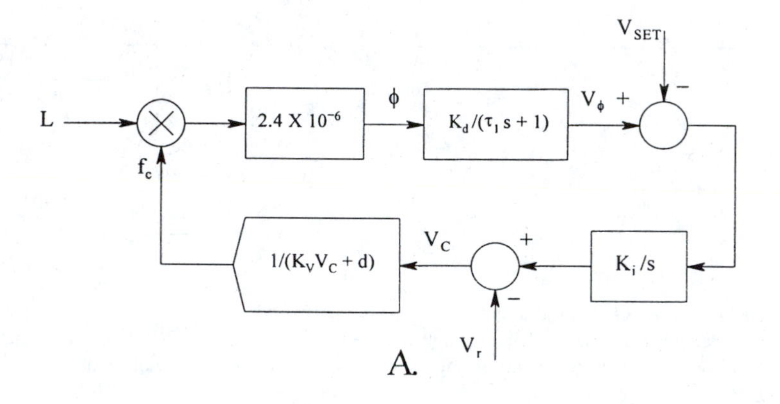

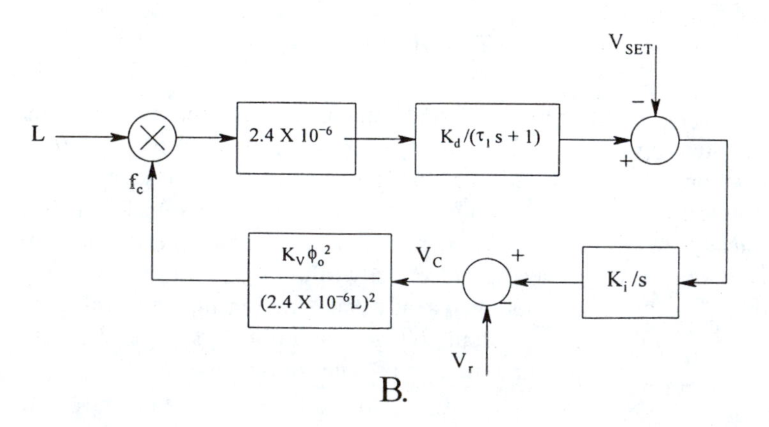

FIGURE 11.13 (A) Systems block diagram of our LAVERA system. Note that the system is highly nonlinear. (B) Linearized LAVERA system block diagram, used to calculate the closed-loop system's natural frequency and damping factor.

$$S_V = df_c/dV_C = -K_V/(K_V V_C + d)^2 = -K_V(f_c)^2 = -K_V(\varphi_o/[2.40 \times 10^{-6} L])^2 \qquad 11.25$$

The linearized system is shown in Figure 11.13B. Its loop gain is second-order, and is given by:

$$A_L(s) = -\frac{\varphi_o^2 K_V K_d K_i}{2.40 \times 10^{-6} L s(\tau s + 1)} \qquad 11.26$$

It can easily be shown that the closed-loop, nonlinear system's natural frequency[2] is:

$$\omega_n^2 = K_V \varphi_o^2 K_d K_i/(\tau 2.40 \times 10^{-6} L) \ (r/s)^2 \qquad 11.27$$

The system's damping factor is also nonlinear, and is given by:

$$\zeta = (1/2\tau)\sqrt{\frac{2.40 \times 10^{-6} L \tau}{\varphi_o^2 K_V K_d K_i}} \qquad 11.28$$

As an example, let the following parameters be used: L_o = 100 m, φ_o = 25.0°, K_i = 500, τ = 0.001 s, K_d = 0.01, K_V = 4.75E – 6. We calculate S_V = 2.4E – 6/($K_i K_V \varphi_o$) = 4.042E – 5 V/m/s. The range sensitivity is: S_L = 2.4E – 6/($K_V \varphi_o$) = 2.0211E – 2 V/m. The undamped natural frequency is $\omega_n = \sqrt{K_V \varphi_o^2 K_d K_i/(2.4E-6\tau L_o)}$ = 248.7 r/s, or f_n = 39.6 Hz. The closed-loop system damping factor at a range of 100 m is $\zeta = 1/(2\ \tau\omega_n)$ = 2.011 (overdamped). If L varies from 10 to 300 m, the SS VPC frequency varies from 1.0417 MHz to 34.72 kHz.

An experimental investigation of the LAVERA system described above is currently (12/95) underway in the author's laboratory. We have also done simulations of the nonlinear, Type I LAVERA system using *Simnon*™. (Simnon is a nonlinear ODE solver which is well suited to study the dynamics of nonlinear systems.) The Simnon program used to simulate our LAVERA system is shown in Figure 11.14. Note that the program treats the VPC's output frequency as a continuous variable rather than as discrete pulses. This approach is possible because the VPC frequency is orders of magnitude larger than the system's natural frequency. The program also simulates op amp saturation at ±14 V. To illustrate system performance, we plot in Figure 11.15 V_V, f_c (in hundreds of kHz), L (in hundreds of meters), and V_L (in volts) for a target position that starts at L = 20 m, and at t = 2 s starts to recede at v = 25 m/s. Note that the frequency of the VPC decreases hyperbolically as target distance increases. Figure 11.16 shows a family of V_V curves for various target velocities. Note that V_V is constant for constant target velocites. The sharp upward breaks on the V_V curves occur when one or more op amps in the simulated system saturate at ±14 V.

The key to the system's linearity (V_L being proportional to L and V_V being proportional to dL/dt) lies in the use of a *voltage-to-period converter*-controlled oscillator rather than a conventional VCO. If a conventional VCO is used, V_V will no longer be a simple linear function of dL/dt; V_L will also depend on L in a nonlinear manner. A design for a simple VPC is shown in Figure 11.17. An RS flip-flop made from two 3-input NOR gates generates a squarewave which is an input to an op amp fast integrator which generates a sawtooth (± ramp) output with slope ±V_{in}/RC. V_{in} is the effective ± input voltage to the integrator; in this case, about 1.8 V. When the integrator input is HI, a negative-going ramp is generated at the integrator output, V_5. When V_5 reaches –10 V, comparator Q2 goes HI, triggering one-shot multivibrator Q3 to make a fast positive pulse which resets the Q output of the RS flip-flop to LOW. The net input to the integrator is now negative, so V_5 begins to

```
CONTINUOUS SYSTEM ladar6   "Type 1 Feedback laser velocimeter.
STATE VL Vv Vphi           " 8/08/95 Simulation for Text.
DER  dVL dVv dVphi         " Params chosen for Vc=+-10V, 1E4<fc<1E6, PHIss = " 30 degrees.
TIME t
"
dVL = Ki*(-Vset + Vphisat)  " Vo = integrator output
Vc = VLsat - Vr             " Vc = input to VPC.
fc = 1/(Kv*Vcsat + d)       " Voltage-to-period conversion.
Vi = Vphisat - Vset
phi = 2.4e-6*L*fc           "Phase lag due to deltaT = 2L/c, degrees
dVphi = phi*Kd/tau1 - Vphisat/tau1   " Phase detector LPF.
dVv = -Vv/tau2 + Ko*Visat/tau2  " Low-pass filter at output: Vv prop. to v.
" SCALED VARIABLES
fo = Kf*fc
Ls = L*KL
" SYSTEM INPUT
L = if t < to then Lo else (V*(t - to) + Lo)
" L = Lo + L1 + L2 + L3 + L4 + L5 + L6 + L7 + L8 + L9 + L10 "Stepwise L(t).
"L1 = if t > to then Lo else 0
"L2 = if t > 2*to then Lo else 0
"L3 = if t > 3*to then Lo else 0
"L4 = if t > 4*to then Lo else 0
"L5 = if t > 5*to then Lo else 0
"L6 = if t > 6*to then Lo else 0
"L7 = if t > 7*to then Lo else 0
"L8 = if t > 8*to then Lo else 0
"L9 = if t > 9*to then Lo else 0
"L10 = if t > 10*to then Lo else 0
" SATURATION OF OP AMP OUTPUT VOLTAGES:
VLsat = IF VL > 14 then 14 else if VL < -14 then -14 else VL
Vphisat = IF Vphi > 14 then 14 else if Vphi < -14 then -14 else Vphi
Visat = IF Vi > 14 then 14 else if Vi < -14 then -14 else Vi
Vcsat = IF Vc > 14 then 14 else if Vc < -14 then -14 else Vc
" CONSTANTS:
zero:0
Vset:0.25
to:2
delL:100
Lo:20
Vr:1.10526
tau1:0.001
tau2:0.01
Kv:4.75E-6
Kl:.01
V:25
Kf:1E-5
Ko:2985
Kd:1.E-2
d:5.25E-6
Ki:500
"
END
```

FIGURE 11.14 Simnon™ program listing used to model the dynamic behavior of the nonlinear LAVERA system. Note that the program models frequency as a continuous variable. Op amp saturation is also simulated.

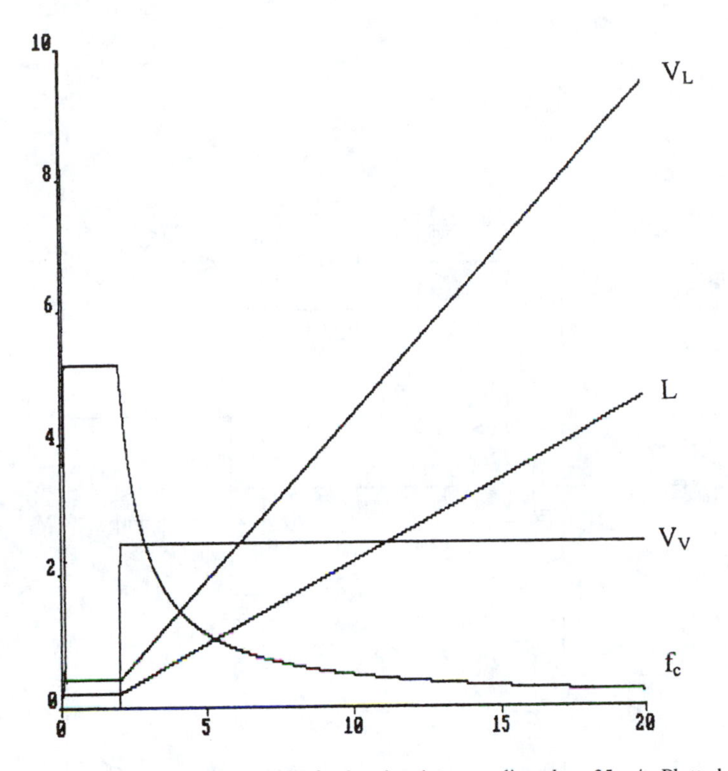

FIGURE 11.15 In this simulation, L remains at 20 m for 2 s, then increases linearly at 25 m/s. Plotted are L(t) in hundeds of meters, V_V in volts, V_L in volts, and f_c in hundreds of kilohertz. Note that f_c decreases hyperbolically as L increases, tending to keep the set phase difference, φ_o, constant.

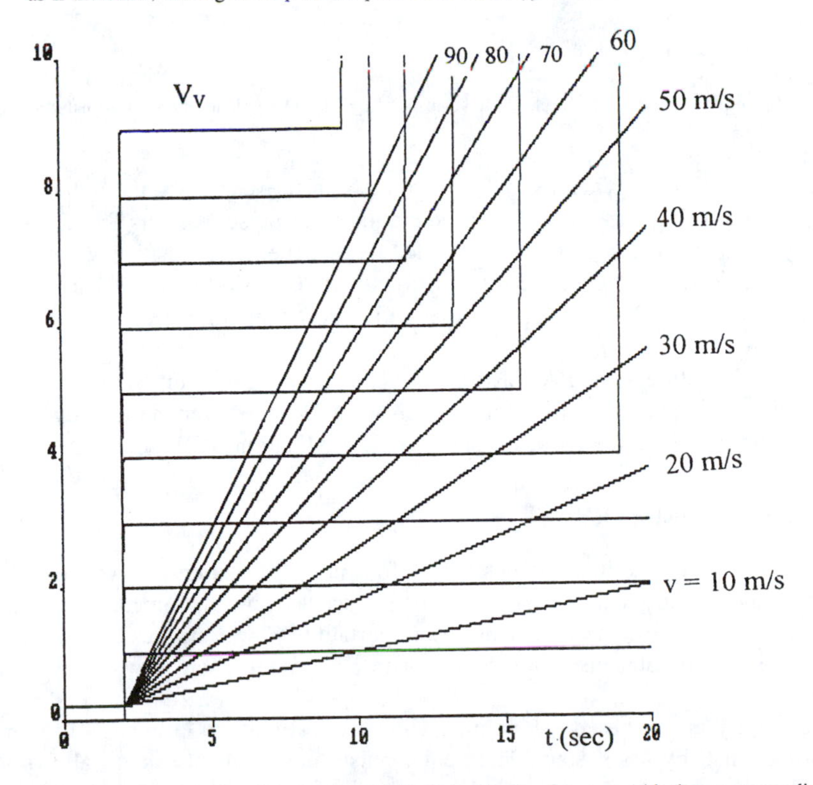

FIGURE 11.16 A family of curves showing simulated L(t) in hundreds of meters with the corresponding $V_V(t)$ plots. Abrupt breaks in the V_V plots occur when the simulated system's op amps saturate.

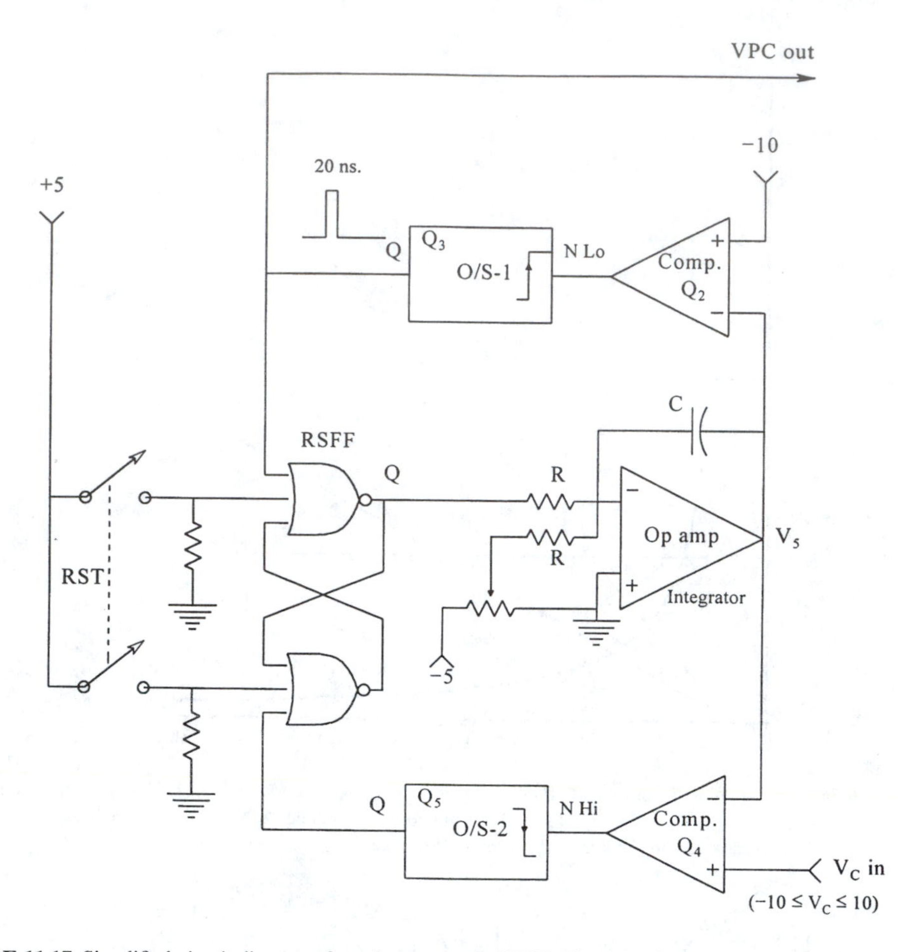

FIGURE 11.17 Simplified circuit diagram of a voltage-to-period VCO. The potentiometer is adjusted to get symmetrical slopes on V_5.

ramp up until it reaches V_C. When the positive-going V_5 just exceeds V_C, the output of comparator Q4 goes LOW, triggering one-shot Q5 to produce the reset pulse that sets the flip-flop output HI again. The cycle then repeats itself. It is easy to see that the more positive V_5 is, the longer the period of oscillation will be, giving a relation for the VPC period of the form: $T_c = K_V V_C + d$. From consideration of the V_5 waveform, it is easy to show that $K_V = 2RC/V_{in}$ and $d = 20RC/V_{in}$, $-10 < V_C < 10$.

The system architecture of the LAVERA system can be extended to ultrasound to simultaneously measure velocity and distance of objects moving in air or water, or even in the body. At long ranges, pulsed ultrasound bursts can be used; for short-range applications, CW sound is indicated.

11.5 CHAPTER SUMMARY

In this chapter we have described the designs of four novel measurement systems. One has been designed to measure and locate partial discharges caused by insulation faults on coaxial, high-voltage, power cables. These measurements are important to power utilities companies in that they indicate the sites where fatal insulation failures can occur, and where new cable sections should be spliced in.

The conductivity bridge-based fish counter allows an estimate to be made of the effectiveness of a downstream fish bypass system. The ocular pulse measurement system allows a diagnostic physiological measurement to be made without touching the eye. The closed-loop, constant-phase, pulsed laser velocity and range measuring system has novel design in its use of a voltage-to-period converter-controlled oscillator. The same system architecture can be extended to ultrasonic ranging systems, using either pulsed or CW ultrasound.

The accuracy of all four systems was seen to be limited by noise. In all cases, low-noise amplification must be used. Further improvement of system performance in noise was achieved in the partial discharge measurement system by implementing digital notch filters to remove coherent interference, using cross-correlation to align noisy partial discharge waveforms for ensemble averaging, and other DSP algorithms for PD site location. The effects of noise in the ocular pulse circuit could be adequately reduced in most cases by signal averaging triggered by the subject's ECG QRS wave. The effects of noise in the LAVERA system are minimized by using a high power laser, low noise preamplifiers and op amps, and fast logic.

REFERENCES and BIBLIOGRAPHY

Allocca, J.A. and A. Stuart (1983) *Electronic Instrumentation,* Reston Publishers, Reston, VA.

Angelo, E.J. (1969) *Electronics: BJTs, FETs, and Microcircuits*, McGraw-Hill, New York.

Balanis, C.A. (1989) *Advanced Engineering Electromagnetics*, John Wiley & Sons, New York.

Barnes Engineering (1983) *Handbook of Infrared Radiation Measurements,* Barnes Engineering Co., Stamford, CT.

Barnes, J.R. (1987) *Electronic System Design: Interference and Noise Control Techniques,* Prentice-Hall, Englewood Cliffs, NJ.

Barney, G.C. (1985) *Intelligent Instrumentation: Microprocessor Applications in Measurement and Control,* Prentice-Hall, Englewood Cliffs, NJ.

Barnoski, M.K., Ed. (1976) *Fundamentals of Optical Fiber Communications*, Academic Press, Orlando, FL.

Barwicz, A. and W.J. Bock (1990) An electronic high-pressure measuring system using a polarimetric fiber optic sensor, *IEEE Trans. Instrum. Meas.,* 39(6):976-981.

Beckwith, T.G. and N.L. Buck (1961) *Mechanical Measurements*, Addison-Wesley, Reading, MA.

Bock, W.J., T.R. Wolinsli and A. Barwicz (1990) Development of a polarimetric optical fiber sensor for electronic measurement of high pressure, *IEEE Trans. Instrum. Meas.,* 39(5):715-721.

Bradford, M. (1975) *Helicopter Electrostatic Discharge Control System*, MS Dissertation, University of Connecticut, Storrs, CT.

Cage, M.E., R.F. Dziuba and B.F. Field (1985) A test of the quantum Hall effect as a resistance standard, *IEEE Trans. Instrum. Meas.*, 34(2):301-303.

Cease, T.W. and P. Johnston (1990) A magneto-optic current transducer, *IEEE Trans. Power Delivery,* 5(2):548-555.

Cheo, P.K. (1990) *Fiber Optics and Optoelectronics,* Prentice-Hall, Englewood Cliffs, NJ.

Clarke, J. and R.H. Koch (1988) The impact of high temperature superconductivity on SQUID magnetometers, *Science,* 14 Oct. 88:217-223.

Cooley, J.W. and J.W. Tukey (1965) An algorithm for the machine computation of complex Fourier series, *Math Comput.,* 19:297-301.

Cooper, W.D. (1978) *Electronic Instrumentation and Measurement Techniques,* 2nd ed., Prentice-Hall, Englewood Cliffs, NJ.

Coté, G.L., M.D. Fox and R.B. Northrop (1992) Noninvasive optical polarimetric glucose sensing using a true phase measurement technique, *IEEE Trans. Biomed. Eng.*, 39(7):752-756.

Coté, G.L., M.D. Fox and R.B. Northrop (1990) Optical polarimetric sensor for blood glucose measurement, *Proc. 16th Annu. Northeast Bioengineering Conf.*, R. Gaumond, Ed., IEEE Press, New York, 102-103.

Culshaw, B. (1984) *Optical Fiber Sensing and Signal Processing,* Peter Peregrinus, London.

Culshaw, B., E.L. Moore and Z. Zhipeng, Eds. (1992) *Advances in Optical Fiber Sensors,* SPIE-Optical Engineering Press, Bellingham, WA.

Cutting, C.L., A.C. Jason and J.L. Wood (1955) *J. Sci. Instr.*, 32:425.

Cutosky, R.D. (1974) New NBS measurements of the absolute farad and ohm, *IEEE Trans. Instrum. Meas.,* 23:305-309.

Dahake, S.L. et al. (1983) Progress in the realization of the units of capacitance, resistance and inductance at the National Physical Laboratory, India, *IEEE Trans. Instrum. Meas.,* 32(1):5-8.

Dandridge, A. (1991) Fiber optic sensors based on the Mach-Zender and Michelson interferometers, *Fiber Optic Sensors: An Introduction for Engineers and Scientists,* Eric Udd, Ed., John Wiley & Sons, New York.

Davis, C.M. (1986) *Fiberoptic Sensor Technology Handbook,* Optical Technologies, Inc., Herndon, VA.

Delahaye, F., A. Fau, D. Dominguez and M. Bellon (1987) Absolute determination of the farad and the ohm, and measurement of the quantized Hall resistance R_H (2) at LCIE, *IEEE Trans. Instrum. Meas.,* 36(2):205-207.

Drake, A.D. and D.C. Leiner (1984) A fiber Fizeau interferometer for measuring minute bioligical displacements, *IEEE Trans. Biomed. Eng.,* 31(7):507-511.

Drysdale, C.V., A.C. Jolley and G.F. Tagg (1952) *Electrical Measuring Instruments,* Part 1, 2nd ed. John Wiley & Sons, New York.

Du, Z. (1993) *A Frequency Independent High Resolution Phase Meter,* MS Dissertation, University of Connecticut, Storrs, CT.

Durst, F., et al. (1974) *Lectures for International Short Course on Laser Velocimetry*, 25-29 March, Purdue University, West Lafayette, IN.

Egan, W.F. (1981) *Frequency Synthesis by Phaselock,* John Wiley & Sons, New York.

Endo, T., M. Koyangi and A. Nakamura (1983) High-accuracy Josephson potentiometer, *IEEE Trans. Instrum. Meas.,* 32(1):267-271.

Fisher, P.D. (1992) Improving on police radar, *IEEE Spectrum,* July, 38-43.

Fox, M.D. and L.G. Puffer (1976) Analysis of transient plant movements by holographic interferometry, *Nature,* 261:488-490.

Fox, M.D. (1978) Multiple crossed-beam ultrasound Doppler velocimetry, *IEEE Trans. Sonics Ultrasonics,* 25(5):281-286.

Fox, M.D. and J.F. Donnelly (1978) Simplified method for determining piezoelectric constants for thickness mode transducers, *J. Acoust. Soc. Amer.,* 64(5):1261-1265.

Fox, M.D. and L.G. Puffer (1978) Model for short term movements in *Stapelia variegata* L., *Plant Physiol.,* 61:209-212.

Fox, M.D. and W.M. Gardiner (1988) Three-dimensional Doppler velocimetry of flow jets, *IEEE Trans. Biomed. Eng.,* 35(10):834-841.

Franco, S. (1988) *Design with Operational Amplifiers and Analog Integrated Circuits,* McGraw-Hill, New York.

Gähler, C., S. Friedrich, R.O. Miles and H. Melchior (1991) Fiber optic temperature sensor using sampled homodyne detection, *Appl. Optics,* 30(21):2938-2940.

Geddes, L.A. and L.E. Baker (1968) *Principles of Applied Biomedical Instrumentation,* John Wiley & Sons, New York.

George, W.K. and J.L. Lumley (1973) The laser Doppler velocimeter and its application to the measurement of turbulence, *J. Fluid Mechanics,* 60(2):321-362.

Gilham, E.J. (1957) A high-precision photoelectric polarimeter, *J. Sci. Instrum.,* 34:435-439.

Gowar, J. (1984) *Optical Fiber Communications,* Prentice-Hall, Englewood Cliffs, NJ.

Graeme, J.G. (1974) *Applications of Operational Amplifiers*, McGraw-Hill, New York.

Greenspan, D. and V. Casulli (1988) *Numerical Analysis for Applied Mathematics, Science and Engineering*, Addison-Wesley, Reading, MA.

Grubb, W.T. and L.H. King (1980) Palladium-palladium oxide pH electrodes, *Anal. Chem.,* 52:270-273.

Hall, H.T. (1958) *Rev. Sci. Instrum.,* 29:267.

Hamilton, C.A. et al. (1991) A 24 GHz Josephson array voltage standard, *IEEE Trans. Instrum. Meas.,* 40(2):301-304.

Hansen, A.T. (1983) Fiber-optic pressure transducers for medical application, *Sensors Actuators,* 4:545-554.

Helfrick, A.D. and W.D. Cooper (1990) *Modern Electronic Instrumentation and Measurement Techniques,* Prentice-Hall, Englewood Cliffs, NJ.

Hewlett-Packard (1991-1992) *Optoelectronics Designer's Catalog.*

Hey, J.C. and W.P. Kram, Eds. (1978) *Transient Voltage Suppression Manual, 2nd ed.*, General Electric Co., Auburn, NY.

Hochberg, R.C. (1986) Fiber-optic sensors, *IEEE Trans. Instrum. Meas.,* 35(4):447-450.

Ibuka, M., S. Naito and T. Furuya (1983) Point-Contact Josephson Voltage Standard, *IEEE Trans. Instrum. Meas.,* 32(1):276-279.

Igarishi, T., Y. Koizumi and M. Kanno (1968) Determination of an absolute capacitance by a horizontal cross-capacitor, *IEEE Trans. Instrum. Meas.,* 17(4):226-231.

Jaeger, K.B., P.D. Levine and C.A. Zack (1991) Industrial experience with a quantized Hall effect system, *IEEE Trans. Instrum. Meas.,* 40(2):256-261.

James, H.M., N.B. Nichols and R.S. Phillips (1947) *Theory of Servomechanisms*, McGraw-Hill, New York.

Janata, J. (1989) *Principles of Chemical Sensors,* Plenum Press, New York.

Jones, L.D. and A.F. Chin (1983) *Electronic Instruments and Measurements,* John Wiley & Sons, New York.

Keiser, G. (1983) *Optical Fiber Communications,* McGraw-Hill, New York.

Keithley Instruments staff (1992) Low level Measurements, 4th ed., Keithley Instruments, Inc., Cleveland, OH.

Kersey, A.D. and A. Dandridge (1990) Application of fiber-optic sensors, *IEEE Trans. Components, Hybrids Manu. Technol.,* 13(1):137-143.

Kersey, A.D. and W.K. Burns (1993) Fiber optic gyroscopes put a new spin on navigation, *Photonics Spectra,* Dec. 72-76.

Kibble, B.P. (1983) Realizing the ampere by levitating a superconducting mass — a suggested procedure, *IEEE Trans. Instrum. Meas.,* 32(1):144.

Kibble, B.P., R.C. Smith and I.A. Robinson (1983) The NPL moving-coil ampere determination, *IEEE Trans. Instrum. Meas.* 32(1):141-143.

Kinley, H. (1980) *The PLL Synthesizer Cookbook,* Tab Books, Blue Ridge Summit, PA.

Kitchin, C. and L. Counts (1983) *RMS to DC Conversion Application Guide,* Analog Devices, Inc., Norwood, MA.

Kim, Y.B. and H.J. Shaw (1986) Fiber-optic gyroscopes, *IEEE Spectrum,* March, 54-60.

Knapp, C.H., R. Bansal, M.S. Mashikian and R.B. Northrop (1990) Signal processing techniques for partial discharge site location in shielded cables, *IEEE Trans. on Power Delivery,* 5(2):859-865.

Ko, W.H. et al., Eds. (1985) *Implantable Sensors for Closed-Loop Prosthetic Systems,* Futura Publishing, Mt. Kisco, NY.

Koskinen, M., J. Typpo and J. Kostamovaara (1992) A fast time-to-amplitude converter for pulsed time-of-flight laser rangefinding, *SPIE Vol. 1633 Laser Radar VII,* 128-136.

Kostamovaara, J., K. Määttä, M. Kiskinen and R. Myllylä (1992) Pulsed laser radars with high modulation frequency in industrial applications, *SPIE Vol. 1633 Laser Radar VII,* 114-127.

Kraus, J.D. (1953) *Electromagnetics,* McGraw-Hill, New York.

Lathi, B.P. (1965) *Signals, Systems and Communication,* John Wiley & Sons, New York.

LeCroy 1994 Research Instrumentation Catalog, Lecroy Corp., Chestnut Ridge, NY.

Lion, K.S. (1959) *Instrumentation in Scientific Research,* McGraw-Hill, New York.

Liu, C.C. and M.R. Neuman (1982) Fabrication of miniature pO_2 and pH sensors using microelectronic techniques, *Diabetes Care,* 5:275.

Macovski, A. (1983) *Medical Imaging Systems,* Prentice-Hall, Englewood Cliffs, NJ.

Malchow, D., Ed. (1985) *Lock-In Applications Anthology,* EG&G Princeton Applied Research, Princeton, NJ.

Marcuse, D. (1981) *Principles of Optical Fiber Measurements,* Academic Press, New York.

Maron, S.H. and C.F. Prutton (1958) *Principles of Physical Chemistry,* Macmillan, New York.

Mashikian, M.S., R. Bansal and R.B. Northrop (1989a) Location and characterization of partial discharge sites on shielded power cables, *Proc. 1989 IEEE Conf. on Power Transmission and Distribution,* New Orleans, April 3-7.

Mashikian, M.S., R.B. Northrop, R. Bansal and C.L. Nikias (1989b) *Method and Instrumentation for the Detection, Location and Characterization of Partial Discharges and Faults in Electric Power Cables*, U.S. Patent 4,887,041, 12 Dec.

Mashikian, M.S., R.B. Northrop and D. Sui (1995) *Cable Fault Detection Using a High Voltage Alternating Polarity DC Signal Superposed with a System Frequency AC Signal,* U.S. Patent 5,448,176, 5 Sept.

Mason, J.H. (1965) Discharge detection and measurements, *Proc. IEEE,* 112(7):1407-1422.

McDonald, B.M. and R.B. Northrop (1993) Two-phase lock-in amplifier with phase-locked loop vector tracking. Paper 32.1. 11th European Conference on Circuit Theory and Design, Davos, Switzerland, Aug. 30-Sept. 3.

Medlock, R.S. (1990) Review of modulating techniques for fiber optic sensors, *The Distributed Fiber Optic Sensing Handbook,* J.P. Dakin et al., Eds., IFS Publications, Springer-Verlag, Berlin; Millman, J. (1979) *Microelectronics*, McGraw-Hill, New York.

Millman, J. (1979) *Microelectronics*, McGraw-Hill, New York.

Möller, K.D. (1988) *Optics,* University Science Books, Mill Valley, CA.

Mopsik, F.I. (1984) Precision time-domain dielectric spectrometer, *Rev. Sci. Instrum.,* 55(1):79-87.

Moseley, P.T., J. Norris and D. Williams (1991) *Techniques and Mechanisms in Gas Sensing,* Adam Hilger, New York.

Motorola Application Note AN-781-A, *RS-232-C Interfacing.*

Muller, R.S. et al., Eds. (1991) *Microsensors,* IEEE Press, New York.

Nanavati, R.P. (1975) *Semiconductor Devices,* Intext Educational Publishers, NY.

Navon, D.H. (1975) *Electronic Materials and Devices,* Houghton-Mifflin, Boston.

Nemat, A. (1990) A digital frequency independent phase meter, *IEEE Trans. Instrum. Meas.,* 39(4):665-666.

Neuman, M.R. (1988) Neonatal monitoring, *Encyclopedia of Medical Devices and Instrumentation,* J.G. Webster, Ed., John Wiley & Sons, New York, 2015-2034.

Nicati, P.A. and P. Robert (1988) Stabilised current sensor using Sagnac interferometer, *J. Phys. E Sci. Instrum.,* 21:791-796.

Northrop, R.B. (1966) An instantaneous pulse frequency demodulator for neurophysiological applications, *Proc. Symp. Biomed. Eng.*, Milwaukee, 1:5-8.

Northrop, R.B. (1967) Electrofishing, *IEEE Trans. Biomed. Eng.* 14(3):191-200.

Northrop, R.B., J.-M. Wu and H.M. Horowitz (1967) An instantaneous frequency cardiotachometer, *Digest 7th Intl. Conf. on Eng. Med. Biol.*, Stockholm, 417.

Northrop, R.B. and B.M. Decker (1977) Design of a no-touch infant apnea monitor, *Proc. 5th Annu. Northeast Bioengineering Conf.*, Pergamon Press, New York, 245-248.

Northrop, R.B. and S.S. Nilakhe (1977) A no-touch ocular pulse measurement system for the diagnosis of carotid occlusions, *IEEE Trans. Biomed. Eng.*, 24(2):139-148.

Northrop, R.B. and B.M. Decker (1978) Assessment of cerebral hemodynamics by no-touch ocular pulse, *Proc. 6th New England Bioengineering Conf.*, D. Jaron, Ed., 105-107. Pergamon Press, Elmsford, NY.

Northrop, R.B. (1980) *Ultrasonic Respiration/Convulsion Monitoring Apparatus and Method for Use,* U.S. Patent 4,197,856, 15 April.

Northrop, R.B. (1982) D.L. Camarata and S.S. Nilakhe (1982) A conductivity-based fish counter for use in fish bypass systems at hydroelectric facilities, *Proc. 10th Ann. Northeast Bioengineering Conf.*, 245-248. IEEE Press.

Northrop, R.B. (1990) *Analog Electronic Circuits: Analysis and Applications,* Addison-Wesley, Reading, MA.

Oliver, B.M. and J.M. Cage (1971) *Electronic Measurements and Instrumentation,* McGraw-Hill, New York.

Olsen, H.F. (1943) *Elements of Acoustical Engineering,* D. Van Nostrand Co., New York.

Ott, H.W. (1976) *Noise Reduction Techniques in Electronic Systems,* John Wiley & Sons, New York.

Pallas-Areny, R. and J.G. Webster (1991) *Sensors and Signal Conditioning,* John Wiley & Sons, New York.

Papoulis, A. (1977) *Signal Analysis,* McGraw-Hill, New York..

Patiño, N.M. and M.E. Valentinuzzi (1992) Lion's twin-T circuit revisited, *IEEE Eng. Med. Biol. Mag.*, 11(3):61-66.

Patranabis, D. (1983) *Principles of Industrial Instrumentation*, Tata McGraw-Hill, Delhi, India.

Peslin. R. et al. (1975) Frequency response of the chest: modelling and parameter estimation, *J. Appl. Phsiol.*, 39(4):523-534.

Peterson, J.I. and S.R. Goldstein (1982) A miniature fiberoptic pH sensor potentially suitable for glucose measurements, *Diabetes Care,* 5(3):272-274.

Peura, R.A. and Y. Mendelson (1984) Blood glucose sensors: an overview, *Proc. IEEE/NSF Symp. Biosensors,* 63-68.

Pimmel, R.L. et al. (1977) Instrumentation for measuring respiratory impedance by forced oscillations, *IEEE Trans. Biomed. Eng.*, 24(2):89-93.

Piquemal, F. et al. (1991) Direct comparison of quantized Hall resistances, *IEEE Trans. Instrum. Meas.*, 40(2):234-236.

Pöpel, R. et al. (1991) Nb/Al_2O_3/Nb Josephson voltage standards at 1 V and 10 V, *IEEE Trans. Instrum. Meas.*, 40(2):298-300.

Post, E.J. (1967) Sagnac effect, *Rev. Mod. Phys.*, 39(2):475-493.

Quinn, T.J. (1991) The kilogram: the present state of our knowledge, *IEEE Trans. Instrum. Meas.*, 40(2):81-85.

Reymann, D. (1991) A practical device for 1 nV accuracy measurements with Josephson arrays. *IEEE Trans. Instrum. Meas.*, 40(2):309-311.

Ribeiro, P.C., S.J. Williamson and L. Kaufman (1988) SQUID arrays for simultaneous magnetic measurements: calibration and source localization performance, *IEEE Trans. Biomed. Eng.*, 35(7):551-559.

Rogers, A.J. (1973) Optical technique for measurement of current at high voltage, *Proc. IEEE,* 120(2):261-267.

Rogers, A.J. (1976) Method for simultaneous measurement of voltage and current on high voltage lines using optical techniques, *Proc. IEEE,* 123(10):957-960.

Rogers, A.J. (1977) The electrogyration effect in crystalline quartz, *Proc. R. Soc. London Ser. A,* 353:177-192.

Rogers, A.J. (1979) Optical measurement of current and voltage on power systems, *Electrical Power Applications,* 2(4):22-25.

Sagnac, G. (1914) Effet tourbillonnaire optique la circulation de l'éther lumineux dans un interférographe tournante, *J. Phys. Radium,* 4:177-195.

Scheid, F.S. (1990) *2000 Solved Problems in Numerical Analysis*, McGraw-Hill, New York.

Sears, F.W. (1949) *Optics,* Addison-Wesley, Cambridge, MA.

Shida, K. et al. (1987) Determination of the quantized Hall resistance value by using a calculable capacitor at ETL, *IEEE Trans. Instrum. Meas.*, 36(2):214-217.

Shida, K. et al. (1989) SI value of quantized Hall resistance based on ETL's calculable capacitor, *IEEE Trans. Instrum. Meas.,* 38(2):252-255.

Shields, J.Q., R.F. Dziuba and H.P. Layer (1989) New realization of the ohm and farad using the NBS calculable capacitor, *IEEE Trans. Instrum. Meas.,* 38(2):249-251.

Sirohi, R.S. and H.P. Kothiyal (1991) *Optical Components, Systems and Measurement Techniques,* Marcel Dekker, New York.

Stark, H., F.B. Tuteur and J.B. Anderson (1988) *Modern Electrical Communications: Analog, Digital and Optical Systems,* Prentice-Hall, Englewood Cliffs, NJ.

Stone, H.S. (1982) *Microcomputer Interfacing,* Addison-Wesley, Reading, MA.

Stout, M.B. (1950) *Basic Electrical Measurements,* Prentice-Hall, New York.

Stout, M.B. (1960) *Basic Electrical Measurements,* 2nd ed. Prentice-Hall, Englewood Cliffs, NJ.

Streetman, B.G. (1972) *Solid State Electronic Devices,* Prentice-Hall, Englewood Cliffs, NJ.

Takahashi, H., C. Masuda, A. Ibaraki and K. Miyaji (1986) An application for optical measurements using a rotating linearly polarized light source, *IEEE Trans. Instrum. Meas.,* 35(3):349-353.

Taub, H. and D.L. Schilling (1977) *Digital Integrated Electronics,* McGraw-Hill, New York.

Taylor, B.N. (1990) New international representations of the volt and ohm effective January 1, 1990, *IEEE Trans. Instrum. Meas.,* 39(1):2-5.

Taylor, B.N. (1991) The possible role of the fundamental constants in replacing the kilogram, *IEEE Trans. Instrum. Meas.,* 40(2):86-91.

Tektronix Product Catalog (1990) Tektronix, Inc., Wilsonville, OR.

Thompson, A.M. and D.G. Lampard (1956) A new theorem in electrostatics and its application to calculable standards of capacitance, *Nature,* 177:888.

Tompkins, W.J., Ed. (1993) *Biomedical Digital Signal Processing,* Prentice-Hall, Englewood Cliffs, NJ.

Tompkins, W.J. and J.G. Webster, Eds. (1981) *Design of Microcomputer-Based Medical Instrumentation,* Prentice-Hall, Englewood Cliffs, NJ.

Tompkins, W.J. and J.G. Webster, Eds. (1988) *Interfacing Sensors to the IBM™ PC,* Prentice- Hall, Englewood Cliffs, NJ.

Turner, J.D. (1988. *Instrumentation for Engineers,* Springer-Verlag, NY.

Udd, Eric, Ed. (1991) *Fiber Optic Sensors: An Introduction for Engineers and Scientists.* John Wiley & Sons, NY.

van der Ziel, A. (1974) *Introductory Electronics,* Prentice-Hall, Englewood Cliffs, NJ.

von Klitzing, K., G. Dorda and M. Pepper (1980) New method for high accuracy determination of the fine-structure constant based on quantized Hall resistance, *Phys. Rev. Lett.,* 45:494-496.

von Klitzing, K. (1986) The quantum Hall effect, *The Physics of the Two-Dimensional Electron Gas,* J.T. Devreese and F.M. Peeters, Eds., Plenum Press, New York.

Webster, J.G., Ed. (1978) *Medical Instrumentation: Application and Design.* Houghton-Mifflin, Boston.

Webster, J.G., Ed. (1992) *Medical Instrumentation: Application and Design,* 2nd ed., Houghton-Mifflin, Boston.

Widrow, B. and S.D. Stearns (1985) *Adaptive Signal Processing,* Prentice-Hall, Englewood Cliffs, NY.

Williams, C.S. (1986) *Designing Digital Filters,* Prentice-Hall, Englewood Cliffs, NJ.

Wingard, L.B. et al. (1982) Potentiometric measurement of glucose concentration with an immobilized glucose oxidase/catalase electrode, *Diabetes Care,* 5(3):199-202.

Wise, D.L., Ed. (1989) *Applied Biosensors.* Butterworths, Boston.

Wolf, S. (1983) *Guide to Electronic Measurements and Laboratory Practice,* 2nd ed. Prentice-Hall, Englewood Cliffs, NJ, chap. 16.

Wolfbeis, O.S. (1991) *Fiber Optic Chemical Sensors and Biosensors,* Vol. 1, CRC Press, Boca Raton, FL.

Yang, E.S. (1988) *Microelectronic Devices,* McGraw-Hill, New York.

Yeh, Y. and H.Z. Cummins (1964) Localized fluid flow measurements with a HeNe laser spectrometer, *Appl. Phys. Lett.,* 4(10):176.

Zuch, E.L. (1981) *Data Acquisition and Conversion Handbook,* 4th ed., Datel-Intersil, Inc., Mansfield, MA.

INDEX

T

V

W

Z